Statistical Mechanics: Entropy, Order Parameters, and Complexity

OXFORD MASTER SERIES IN PHYSICS

The Oxford Master Series is designed for final year undergraduate and beginning graduate students in physics and related disciplines. It has been driven by a perceived gap in the literature today. While basic undergraduate physics texts often show little or no connection with the huge explosion of research over the last two decades, more advanced and specialized texts tend to be rather daunting for students. In this series, all topics and their consequences are treated at a simple level, while pointers to recent developments are provided at various stages. The emphasis is on clear physical principles like symmetry, quantum mechanics, and electromagnetism which underlie the whole of physics. At the same time, the subjects are related to real measurements and to the experimental techniques and devices currently used by physicists in academe and industry. Books in this series are written as course books, and include ample tutorial material, examples, illustrations, revision points, and problem sets. They can likewise be used as preparation for students starting a doctorate in physics and related fields, or for recent graduates starting research in one of these fields in industry.

CONDENSED MATTER PHYSICS

1. M.T. Dove: *Structure and dynamics: an atomic view of materials*
2. J. Singleton: *Band theory and electronic properties of solids*
3. A.M. Fox: *Optical properties of solids, second edition*
4. S.J. Blundell: *Magnetism in condensed matter*
5. J.F. Annett: *Superconductivity, superfluids, and condensates*
6. R.A.L. Jones: *Soft condensed matter*
17. S. Tautz: *Surfaces of condensed matter*
18. H. Bruus: *Theoretical microfluidics*
19. C.L. Dennis, J.F. Gregg: *The art of spintronics: an introduction*
21. T.T. Heikkilä: *The physics of nanoelectronics: transport and fluctuation phenomena at low temperatures*
22. M. Geoghegan, G. Hadziioannou: *Polymer electronics*

ATOMIC, OPTICAL, AND LASER PHYSICS

7. C.J. Foot: *Atomic physics*
8. G.A. Brooker: *Modern classical optics*
9. S.M. Hooker, C.E. Webb: *Laser physics*
15. A.M. Fox: Quantum optics: *an introduction*
16. S.M. Barnett: *Quantum information*
23. P. Blood: *Quantum confined laser devices*

PARTICLE PHYSICS, ASTROPHYSICS, AND COSMOLOGY

10. D.H. Perkins: *Particle astrophysics, second edition*
11. Ta-Pei Cheng: *Relativity, gravitation and cosmology, second edition*
24. G. Barr, R. Devenish, R. Walczak, T. Weidberg: *Particle physics in the LHC era*
26. M. E. Peskin: *Concepts of elementary particle physics*

STATISTICAL, COMPUTATIONAL, AND THEORETICAL PHYSICS

12. M. Maggiore: *A modern introduction to quantum field theory*
13. W. Krauth: *Statistical mechanics: algorithms and computations*
14. J.P. Sethna: *Statistical mechanics: entropy, order parameters, and complexity, second edition*
20. S.N. Dorogovtsev: *Lectures on complex networks*
25. R. Soto: *Kinetic theory and transport phenomena*

Statistical Mechanics

Entropy, Order Parameters, and Complexity

SECOND EDITION

James P. Sethna

Laboratory of Atomic and Solid State Physics
Cornell University, Ithaca, NY

OXFORD
UNIVERSITY PRESS

OXFORD
UNIVERSITY PRESS

Great Clarendon Street, Oxford, OX2 6DP,
United Kingdom

Oxford University Press is a department of the University of Oxford.
It furthers the University's objective of excellence in research, scholarship,
and education by publishing worldwide. Oxford is a registered trade mark of
Oxford University Press in the UK and in certain other countries

First Edition published in 2006
Second Edition published in 2021

Published in the United States of America by Oxford University Press
198 Madison Avenue, New York, NY 10016, United States of America

British Library Cataloguing in Publication Data

Data available

Library of Congress Control Number: 2020947122

ISBN 978–0–19–886524–7 (hbk.)
ISBN 978–0–19–886525–4 (pbk.)

DOI: 10.1093/oso/9780198865247.001.0001

Printed and bound by
CPI Group (UK) Ltd, Croydon, CR0 4YY

Preface to the second edition

The second edition of *Statistical Mechanics: Entropy, Order Parameters, and Complexity* features over a hundred new exercises, plus refinement and revision of many exercises from the first edition. The main chapters are largely unchanged, except for a refactoring of my discussion of the renormalization group in Chapter 12. Indeed, the chapters are designed to be the stable kernel of their topics, while the exercises cover the growing range of fascinating applications and implications of statistical mechanics.

This book reflects "flipped classroom" innovations, which I have found to be remarkably effective. I have identified a hundred pre-class questions and in-class activities, the former designed to elucidate and reinforce sections of the text, and the latter designed for group collaboration. These are denoted with the symbols ⓟ and @, in an extension of the difficulty rating ①–⑤ used in the first edition. *Human correlations*, *Fingerprints*, and *Crackling noises* are some of my favorite activities. These exercises, plus a selection of less-specialized longer exercises, form the core of the undergraduate version of my course.

Extensive online material [181] is now available for the exercises. Mathematica and python notebooks provide hints for almost fifty computational exercises, allowing students to tackle serious new research topics like *Conformal invariance*, *Subway bench Monte Carlo*, and *2D turbulence and Jupiter's great red spot*, while getting exposed to good programming practices. Handouts and instructions facilitate activities such as *Pentagonal frustration* and *Hearing chaos*. The answer key for the exercises now is polished enough that I regret not being able to share it with any but those teaching the course.

Finally, the strength of the first edition was in advanced exercises, which explored in depth the subtleties of statistical mechanics and the broad range of its application to various fields of science. Many substantive exercises continue this trend, such as *Nucleosynthesis and the arrow of time*, *Word frequencies and Zipf's law*, *Pandemic*, and *Kinetic proofreading in cells*.

I again thank the National Science Foundation and Cornell's physics department for making possible the lively academic atmosphere at Cornell and my amazing graduate students; both were crucial for the success of this endeavor. Thanks to the students and readers who stamped out errors and obscurities. Thanks to my group members and colleagues who contributed some of the most creative and insightful exercises presented here—they are acknowledged in the masterpieces that they crafted. Thanks to Jaron Kent-Dobias for several years of enthusiasm, insight, and suggestions. A debt is gratefully due to Matt Bierbaum; many of the best exercises in this text make use of his wonderfully interactive Ising [28] and mosh pit [32] simulations.

Enormous thanks are due to my lifelong partner, spouse, and love, Carol Devine, who tolerates my fascination with solving physics puzzles and turning them into exercises, because she sees it makes me happy.

James P. Sethna
Ithaca, NY
September 2020

Preface

The purview of science grows rapidly with time. It is the responsibility of each generation to join new insights to old wisdom, and to distill the key ideas for the next generation. This is my distillation of the last fifty years of statistical mechanics—a period of grand synthesis and great expansion.

This text is careful to address the interests and background not only of physicists, but of sophisticated students and researchers in mathematics, biology, engineering, computer science, and the social sciences. It therefore does not presume an extensive background in physics, and (except for Chapter 7) explicitly does not assume that the reader knows or cares about quantum mechanics. The text treats the intersection of the interests of all of these groups, while the exercises encompass the union of interests. Statistical mechanics will be taught in all of these fields of science in the next generation, whether wholesale or piecemeal by field. By making statistical mechanics useful and comprehensible to a variety of fields, we enrich the subject for those with backgrounds in physics. Indeed, many physicists in their later careers are now taking excursions into these other disciplines.

To make room for these new concepts and applications, much has been pruned. Thermodynamics no longer holds its traditional key role in physics. Like fluid mechanics in the last generation, it remains incredibly useful in certain areas, but researchers in those areas quickly learn it for themselves. Thermodynamics also has not had significant impact in subjects far removed from physics and chemistry: nobody finds Maxwell relations for the stock market, or Clausius–Clapeyron equations applicable to compression algorithms. These and other important topics in thermodynamics have been incorporated into a few key exercises. Similarly, most statistical mechanics texts rest upon examples drawn from condensed matter physics and physical chemistry—examples which are then treated more completely in other courses. Even I, a condensed matter physicist, find the collapse of white dwarfs more fun than the low-temperature specific heat of metals, and the entropy of card shuffling still more entertaining.

The first half of the text includes standard topics, treated with an interdisciplinary slant. Extensive exercises develop new applications of statistical mechanics: random matrix theory, stock-market volatility, the KAM theorem, Shannon entropy in communications theory, and Dyson's speculations about life at the end of the Universe. The second half of the text incorporates Monte Carlo methods, order parameters, linear response and correlations (including a classical derivation of the fluctuation-dissipation theorem), and the theory of abrupt and continuous phase transitions (critical droplet theory and the renormalization group).

This text is aimed for use by upper-level undergraduates and graduate students. A scientifically sophisticated reader with a familiarity with partial derivatives and introductory classical mechanics should find this text accessible, except for Chapter 4 (which demands Hamiltonian mechanics), Chapter 7 (quantum mechanics), Section 8.2 (linear algebra), and Chapter 10 (Fourier methods, introduced in the Appendix). An undergraduate one-semester course might cover Chapters 1–3, 5–7, and 9. Cornell's hard-working first-year graduate students covered the entire text and worked through perhaps half of the exercises in a semester. I have tried to satisfy all of these audiences through the extensive use of footnotes: think of them as optional hyperlinks to material that is more basic, more advanced, or a sidelight to the main presentation. The exercises are rated by difficulty, from ①(doable by inspection) to ⑤(advanced); exercises rated ④ (many of them computational laboratories) should be assigned sparingly. Much of Chapters 1–3, 5, and 6 was developed in an sophomore honors "waves and thermodynamics" course; these chapters and the exercises marked ① and ② should be accessible to ambitious students early in their college education. A course designed to appeal to an interdisciplinary audience might focus on entropy, order parameters, and critical behavior by covering Chapters 1–3, 5, 6, 8, 9, and 12. The computational exercises in the text grew out of three different semester-long computational laboratory courses. We hope that the computer exercise hints and instructions on the book website [181] will facilitate their incorporation into similar courses elsewhere.

The current plan is to make individual chapters available as PDF files on the Internet. I also plan to make the figures in this text accessible in a convenient form to those wishing to use them in course or lecture presentations.

I have spent an entire career learning statistical mechanics from friends and colleagues. Since this is a textbook and not a manuscript, the presumption should be that any ideas or concepts expressed are not mine, but rather have become so central to the field that continued attribution would be distracting. I have tried to include references to the literature primarily when it serves my imagined student. In the age of search engines, an interested reader (or writer of textbooks) can quickly find the key ideas and articles on any topic, once they know what it is called. The textbook is now more than ever only a base from which to launch further learning. My thanks to those who have patiently explained their ideas and methods over the years—either in person, in print, or through the Internet.

I must thank explicitly many people who were of tangible assistance in the writing of this book. I thank the National Science Foundation and Cornell's Laboratory of Atomic and Solid State Physics for their support during the writing of this text. I thank Pamela Davis Kivelson for the magnificent cover art. I thank Eanna Flanagan, Eric Siggia, Saul Teukolsky, David Nelson, Paul Ginsparg, Vinay Ambegaokar, Neil Ashcroft, David Mermin, Mark Newman, Kurt Gottfried, Chris Henley, Barbara Mink, Tom Rockwell, Csaba Csaki, Peter Lepage, and Bert Halperin for helpful and insightful conversations. Eric Grannan, Piet Brouwer, Michelle Wang, Rick James, Eanna Flanagan, Ira Wasserman, Dale Fixsen, Rachel Bean, Austin Hedeman, Nick Trefethen, Sarah Shandera, Al Sievers, Alex Gaeta, Paul Ginsparg, John Guckenheimer, Dan Stein, and Robert Weiss were of important assistance in developing various exercises. My approach to explaining the renormalization group (Chapter 12) was developed in collaboration with Karin Dahmen, Chris Myers, and Olga Perković. The students in my class have been instrumental in sharpening the text and debugging the exercises; Jonathan McCoy, Austin Hedeman, Bret Hanlon, and Kaden Hazzard in particular deserve thanks. Adam Becker, Surachate (Yor) Limkumnerd, Sarah Shandera, Nick Taylor, Quentin Mason, and Stephen Hicks, in their roles of proofreading, grading, and writing answer keys, were powerful filters for weeding out infelicities. I thank Joel Shore, Mohit Randeria, Mark Newman, Stephen Langer, Chris Myers, Dan Rokhsar, Ben Widom, and Alan Bray for reading portions of the text, providing invaluable insights, and tightening the presentation. I thank Julie Harris at Oxford University Press for her close scrutiny and technical assistance in the final preparation stages of this book. Finally, Chris Myers and I spent hundreds of hours together developing the many computer exercises distributed through this text; his broad knowledge of science and computation, his profound taste in computational tools and methods, and his good humor made this a productive and exciting collaboration. The errors and awkwardness that persist, and the exciting topics I have missed, are in spite of the wonderful input from these friends and colleagues.

I especially thank Carol Devine, for consultation, insightful comments and questions, and for tolerating the back of her spouse's head for perhaps a thousand hours over the past two years.

James P. Sethna
Ithaca, NY
February, 2006

Contents

List of figures

What is statistical mechanics?

Many systems in nature are far too complex to analyze directly. Solving for the behavior of all the atoms in a block of ice, or the boulders in an earthquake fault, or the nodes on the Internet, is simply infeasible. Despite this, such systems often show simple, striking behavior. Statistical mechanics explains the simple behavior of complex systems.

The concepts and methods of statistical mechanics have infiltrated into many fields of science, engineering, and mathematics: ensembles, entropy, Monte Carlo, phases, fluctuations and correlations, nucleation, and critical phenomena are central to physics and chemistry, but also play key roles in the study of dynamical systems, communications, bioinformatics, and complexity. Quantum statistical mechanics, although not a source of applications elsewhere, is the foundation of much of physics. Let us briefly introduce these pervasive concepts and methods.

Ensembles. The trick of statistical mechanics is not to study a single system, but a large collection or *ensemble* of systems. Where understanding a single system is often impossible, one can often calculate the behavior of a large collection of similarly prepared systems.

For example, consider a random walk (Fig. 1.1). (Imagine it as the trajectory of a particle in a gas, or the configuration of a polymer in solution.) While the motion of any given walk is irregular and typically impossible to predict, Chapter 2 derives the elegant laws which describe the set of all possible random walks.

Chapter 3 uses an ensemble of all system states of constant energy to derive equilibrium statistical mechanics; the collective properties of temperature, entropy, and pressure emerge from this ensemble. In Chapter 4 we provide the best existing mathematical justification for using this constant-energy ensemble. In Chapter 6 we develop *free energies* which describe parts of systems; by focusing on the important bits, we find new laws that emerge from the microscopic complexity.

Entropy. Entropy is the most influential concept arising from statistical mechanics (Chapter 5). It was originally understood as a thermodynamic property of heat engines that inexorably increases with time. Entropy has become science's fundamental measure of disorder and information—quantifying everything from compressing pictures on the Internet to the heat death of the Universe.

Quantum statistical mechanics, confined to Chapter 7, provides the microscopic underpinning to much of astrophysics and condensed

Statistical Mechanics: Entropy, Order Parameters, and Complexity. James P. Sethna, Oxford University Press (2021). © Oxford University Press. DOI:10.1093/oso/9780198865247.003.0001

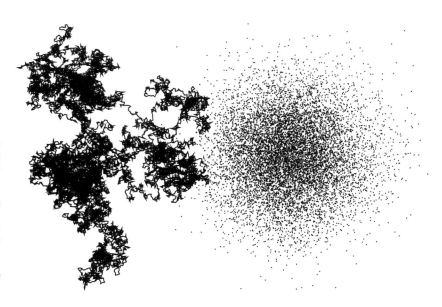

Fig. 1.1 Random walks. The motion of molecules in a gas, and bacteria in a liquid, and photons in the Sun, are described by *random walks*. Describing the specific trajectory of any given random walk (left) is not feasible. Describing the statistical properties of a large number of random walks is straightforward (right, showing endpoints of many walks starting at the origin). The deep principle underlying statistical mechanics is that it is often easier to understand the behavior of these *ensembles* of systems.

matter physics. There we use it to explain metals, insulators, lasers, stellar collapse, and the microwave background radiation patterns from the early Universe.

Monte Carlo methods allow the computer to find ensemble averages in systems far too complicated to allow analytical evaluation. These tools, invented and sharpened in statistical mechanics, are used everywhere in science and technology—from simulating the innards of particle accelerators, to studies of traffic flow, to designing computer circuits. In Chapter 8, we introduce Monte Carlo methods, the Ising model, and the mathematics of Markov chains.

Phases. Statistical mechanics explains the existence and properties of phases. The three common phases of matter (solids, liquids, and gases) have multiplied into hundreds: from superfluids and liquid crystals, to vacuum states of the Universe just after the Big Bang, to the pinned and sliding "phases" of earthquake faults. We explain the deep connection between phases and perturbation theory in Section 8.3. In Chapter 9 we introduce the *order parameter field*, which describes the properties, excitations, and topological defects that emerge in a given phase.

Fluctuations and correlations. Statistical mechanics not only describes the average behavior of an ensemble of systems, it describes the entire distribution of behaviors. We describe how systems fluctuate and evolve in space and time using *correlation functions* in Chapter 10. There we also derive powerful and subtle relations between correlations, response, and dissipation in equilibrium systems.

Abrupt phase transitions. Beautiful spatial patterns arise in statistical mechanics at the transitions between phases. Most such transitions are abrupt; ice is crystalline and solid until (at the edge of the ice cube) it becomes unambiguously liquid. We study the nucleation

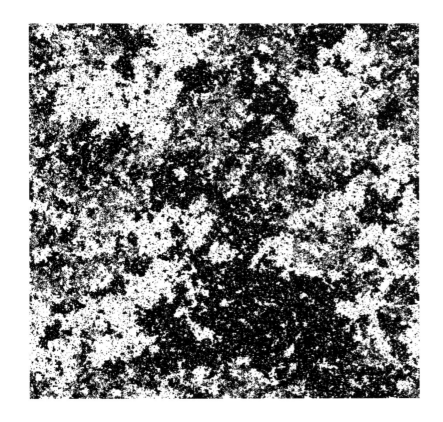

Fig. 1.2 Ising model at the critical point. The two-dimensional Ising model of magnetism at its transition temperature T_c. At higher temperatures, the system is nonmagnetic; the magnetization is on average zero. At the temperature shown, the system is just deciding whether to magnetize upward (white) or downward (black).

of new phases and the exotic structures that can form at abrupt phase transitions in Chapter 11.

Criticality. Other phase transitions are continuous. Figure 1.2 shows a snapshot of a particular model at its phase transition temperature T_c. Notice the self-similar, fractal structures; the system cannot decide whether to stay gray or to separate into black and white, so it fluctuates on all scales, exhibiting *critical phenomena*. A random walk also forms a self-similar, fractal object; a blow-up of a small segment of the walk looks statistically similar to the original (Figs. 1.1 and 2.2). Chapter 12 develops the scaling and renormalization-group techniques that explain these self-similar, fractal properties. These techniques also explain *universality*; many properties at a continuous transition are surprisingly system independent.

Science grows through accretion, but becomes potent through distillation. Statistical mechanics has grown tentacles into much of science and mathematics (see, e.g., Fig. 1.3). The body of each chapter will provide the distilled version: those topics of fundamental importance to all fields. The accretion is addressed in the exercises: in-depth introductions to applications in mesoscopic physics, astrophysics, dynamical systems, information theory, low-temperature physics, statistics, biology, lasers, and complexity theory.

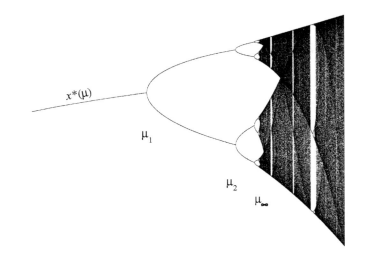

Fig. 1.3 The onset of chaos. Mechanical systems can go from simple, predictable behavior (left) to a chaotic state (right) as some external parameter μ is tuned. Many disparate systems are described by a common, *universal* scaling behavior near the onset of chaos (note the replicated structures near μ_∞). We understand this scaling and universality using tools developed to study continuous transitions in liquids and gases. Conversely, the study of chaotic motion provides statistical mechanics with our best explanation for the increase of entropy.

Exercises

Two exercises, *Emergence* and *Emergent vs. fundamental*, illustrate and provoke discussion about the role of statistical mechanics in formulating new laws of physics.

Four exercises review probability distributions. *Quantum dice and coins* explores discrete distributions and also acts as a preview to Bose and Fermi statistics. *Probability distributions* introduces the key distributions for continuous variables, convolutions, and multidimensional distributions. *Waiting time paradox* uses public transportation to concoct paradoxes by confusing different ensemble averages. And *The birthday problem* calculates the likelihood of a school classroom having two children who share the same birthday.

Stirling's approximation derives the useful approximation $n! \sim \sqrt{2\pi n}(n/e)^n$; more advanced students can continue with *Stirling and asymptotic series* to explore the zero radius of convergence for this series, often found in statistical mechanics calculations.

Five exercises demand no background in statistical mechanics, yet illustrate both general themes of the subject and the broad range of its applications. *Random matrix theory* introduces an entire active field of research, with applications in nuclear physics, mesoscopic physics, and number theory, beginning with histograms and ensembles, and continuing with level repulsion, the Wigner surmise, universality, and emergent symmetry. *Six degrees of separation* introduces the ensemble of *small world networks*, popular in the social sciences and epidemiology for modeling interconnectedness in groups; it introduces network data structures, breadth-first search algorithms, a continuum limit, and our first glimpse of *scaling*. *Satisfactory map colorings* introduces the challenging computer science problems of *graph colorability* and *logical satisfiability*: these search through an ensemble of different choices just as statistical mechanics averages over an ensemble of states. *Self-propelled particles* discusses emergent properties of *active matter*. *First to fail: Weibull* introduces the statistical study of *extreme value statistics*, focusing not on the typical fluctuations about the average behavior, but the rare events at the extremes.

Finally, statistical mechanics is to physics as statistics is to biology and the social sciences. Four exercises here, and several in later chapters, introduce ideas and methods from statistics that have particular resonance with statistical mechanics. *Width of the height distribution* discusses maximum likelihood methods and bias in the context of Gaussian fits. *Fisher information and Cramér Rao* introduces the Fisher information metric, and its relation to the rigorous bound on parameter estimation. And *Distances in probability space* then uses the local difference between model predictions (the metric tensor) to generate total distance estimates between different models.

(1.1) **Quantum dice and coins.**[1] (Quantum) @
You are given two unusual three-sided dice which, when rolled, show either one, two, or three spots. There are three games played with these dice: *Distinguishable*, *Bosons*, and *Fermions*. In each turn in these games, the player rolls the two dice, starting over if required by the rules, until a legal combination occurs. In *Distinguishable*, all rolls are legal. In *Bosons*, a roll is legal only if the second of the two dice shows a number that is is larger or equal to that of the first of the two dice. In *Fermions*, a roll is legal only if the second number is strictly larger than the preceding number. See Fig. 1.4 for a table of possibilities after rolling two dice.

Our dice rules are the same ones that govern the quantum statistics of noninteracting identical particles.

Fig. 1.4 Quantum dice. Rolling two dice. In *Bosons*, one accepts only the rolls in the shaded squares, with equal probability 1/6. In *Fermions*, one accepts only the rolls in the darkly shaded squares (not including the diagonal from lower left to upper right), with probability 1/3.

(a) *Presume the dice are fair: each of the three numbers of dots shows up 1/3 of the time. For a legal turn rolling a die twice in the three games (Distinguishable, Bosons, and Fermions), what is the probability $\rho(5)$ of rolling a 5?*

(b) *For a legal turn in the three games, what is the probability of rolling a double?* (Hint: There is a Pauli exclusion principle: when playing *Fermions*, no two dice can have the same number of dots showing.) Electrons are fermions; no two noninteracting electrons can be in the same quantum state. Bosons are gregarious (Exercise 7.9); noninteracting bosons have a larger likelihood of being in the same state.

Let us decrease the number of sides on our dice to $N = 2$, making them quantum coins, with a head H and a tail T. Let us increase the total number of coins to a large number M; we flip a line of M coins all at the same time, repeating until a legal sequence occurs. In the rules for legal flips of quantum coins, let us make $T < H$. A legal Boson sequence, for example, is then a pattern $TTTT \cdots HHHH \cdots$ of length M; all legal sequences have the same probability.

(c) *What is the probability in each of the games, of getting all the M flips of our quantum coin the same (all heads $HHHH \cdots$ or all tails $TTTT \cdots$)?* (Hint: How many legal sequences are there for the three games? How many of these are all the same value?)

The probability of finding a particular legal sequence in Bosons is larger by a constant factor due to discarding the illegal sequences. This factor is just one over the probability of a given toss of the coins being legal, $Z = \sum_\alpha p_\alpha$ summed over legal sequences α. For part (c), all sequences have equal probabilities $p_\alpha = 2^{-M}$, so $Z_{\text{Dist}} = (2^M)(2^{-M}) = 1$, and Z_{Boson} is 2^{-M} times the number of legal sequences. So for part (c), the probability to get all heads or all tails is $(p_{TTT\ldots} + p_{HHH\ldots})/Z$. The normalization constant Z in statistical mechanics is called the *partition function*, and will be amazingly useful (see Chapter 6).

Let us now consider a biased coin, with probability $p = 1/3$ of landing H and thus $1 - p = 2/3$ of landing T. Note that if two sequences are legal in both Bosons and Distinguishable, their relative probability is the same in both games.

(d) *What is the probability $p_{TTT\ldots}$ that a given toss of M coins has all tails (before we throw out the illegal ones for our game)? What is Z_{Dist}? What is the probability that a toss in Distinguishable is all tails? If Z_{Bosons} is the probability that a toss is legal in Bosons, write the probability that a legal toss is all tails in terms of Z_{Bosons}. Write the probability $p_{TTT\ldots HHH}$ that a toss has $M - m$ tails followed by m heads (before throwing out the illegal ones). Sum these to find Z_{Bosons}. As M gets large, what is the probability in Bosons that all coins flip tails?*

We can view our quantum dice and coins as noninteracting particles, with the biased coin having a lower energy for T than for H (Section 7.4). Having a nonzero probability of having all the

[1] This exercise was developed in collaboration with Sarah Shandera.

bosons in the single-particle ground state T is Bose condensation (Section 7.6), closely related to superfluidity and lasers (Exercise 7.9).

(1.2) Probability distributions. ②

Most people are more familiar with probabilities for discrete events (like coin flips and card games), than with probability distributions for continuous variables (like human heights and atomic velocities). The three continuous probability distributions most commonly encountered in physics are: (i) *uniform*: $\rho_{\text{uniform}}(x) = 1$ for $0 \leq x < 1$, $\rho(x) = 0$ otherwise (produced by random number generators on computers); (ii) *exponential*: $\rho_{\text{exponential}}(t) = e^{-t/\tau}/\tau$ for $t \geq 0$ (familiar from radioactive decay and used in the collision theory of gases); and (iii) *Gaussian*: $\rho_{\text{gaussian}}(v) = e^{-v^2/2\sigma^2}/(\sqrt{2\pi}\sigma)$, (describing the probability distribution of velocities in a gas, the distribution of positions at long times in random walks, the sums of random variables, and the solution to the diffusion equation).

(a) Likelihoods. *What is the probability that a random number uniform on $[0, 1)$ will happen to lie between $x = 0.7$ and $x = 0.75$? That the waiting time for a radioactive decay of a nucleus will be more than twice the exponential decay time τ? That your score on an exam with a Gaussian distribution of scores will be greater than 2σ above the mean?* (Note: $\int_2^\infty (1/\sqrt{2\pi}) \exp(-v^2/2)\,dv = (1 - \text{erf}(\sqrt{2}))/2 \sim 0.023$.)

(b) Normalization, mean, and standard deviation. *Show that these probability distributions are normalized: $\int \rho(x)\,dx = 1$. What is the mean x_0 of each distribution? The standard deviation $\sqrt{\int (x - x_0)^2 \rho(x)\,dx}$?* (You may use the formulæ $\int_{-\infty}^{\infty}(1/\sqrt{2\pi}) \exp(-v^2/2)\,dv = 1$ and $\int_{-\infty}^{\infty} v^2(1/\sqrt{2\pi}) \exp(-v^2/2)\,dv = 1$.)

(c) Sums of variables. *Draw a graph of the probability distribution of the sum $x + y$ of two random variables drawn from a uniform distribution on $[0, 1)$. Argue in general that the sum $z = x + y$ of random variables with distributions $\rho_1(x)$ and $\rho_2(y)$ will have a distribution given by $\rho(z) = \int \rho_1(x)\rho_2(z - x)\,dx$ (the convolution of ρ with itself).*

Multidimensional probability distributions. In statistical mechanics, we often discuss probability distributions for many variables at once (for

example, all the components of all the velocities of all the atoms in a box). Let us consider just the probability distribution of one molecule's velocities. If v_x, v_y, and v_z of a molecule are independent and each distributed with a Gaussian distribution with $\sigma = \sqrt{kT/M}$ (Section 3.2.2) then we describe the combined probability distribution as a function of three variables as the product of the three Gaussians:

$$
\begin{aligned}
\rho(v_x, v_y, v_z) &= \frac{1}{(2\pi(kT/M))^{3/2}} \exp(-M\mathbf{v}^2/2kT) \\
&= \sqrt{\frac{M}{2\pi kT}} \exp\left(\frac{-Mv_x^2}{2kT}\right) \\
&\quad \times \sqrt{\frac{M}{2\pi kT}} \exp\left(\frac{-Mv_y^2}{2kT}\right) \\
&\quad \times \sqrt{\frac{M}{2\pi kT}} \exp\left(\frac{-Mv_z^2}{2kT}\right). \quad (1.1)
\end{aligned}
$$

(d) *Show, using your answer for the standard deviation of the Gaussian in part (b), that the mean kinetic energy is $kT/2$ per dimension. Show that the probability that the speed is $v = |\mathbf{v}|$ is given by a Maxwellian distribution*

$$
\rho_{\text{Maxwell}}(v) = \sqrt{2/\pi}(v^2/\sigma^3) \exp(-v^2/2\sigma^2). \quad (1.2)
$$

(Hint: What is the shape of the region in 3D velocity space where $|\mathbf{v}|$ is between v and $v + \delta v$? The surface area of a sphere of radius R is $4\pi R^2$.)

(1.3) Waiting time paradox.[2] @

Here we examine the *waiting time paradox*: for events happening at random times, the average time until the next event equals the average time between events. If the average waiting time until the next event is τ, then the average time since the last event is also τ. Is the mean total gap between two events then 2τ? Or is it τ, the average time to wait starting from the previous event? Working this exercise introduces the importance of different *ensembles*.

On a highway, the average numbers of cars and buses going east are equal: each hour, on average, there are 12 buses and 12 cars passing by. The buses are scheduled: each bus appears exactly 5 minutes after the previous one. On the other hand, the cars appear at random. In a short interval dt, the probability that a car comes by is dt/τ, with $\tau = 5$ minutes. This

---

[2]The original form of this exercise was developed in collaboration with Piet Brouwer.

leads to a distribution $P(\delta^{\text{Car}})$ for the arrival of the *first* car that decays exponentially, $P(\delta^{\text{Car}}) = 1/\tau \exp(-\delta^{\text{Car}}/\tau)$.

A pedestrian repeatedly approaches a bus stop at random times t, and notes how long it takes before the first bus passes, and before the first car passes.

(a) *Draw the probability density for the ensemble of waiting times δ^{Bus} to the next bus observed by the pedestrian. Draw the density for the corresponding ensemble of times δ^{Car}. What is the mean waiting time for a bus $\langle \delta^{\text{Bus}} \rangle_t$? The mean time $\langle \delta^{\text{Car}} \rangle_t$ for a car?*

In statistical mechanics, we shall describe specific physical systems (a bottle of N atoms with energy E) by considering *ensembles* of systems. Sometimes we shall use two different ensembles to describe the same system (all bottles of N atoms with energy E, or all bottles of N atoms at that temperature T where the mean energy is E). We have been looking at the time-averaged ensemble (the ensemble $\langle \cdots \rangle_t$ over random times t). There is also in this problem an ensemble average over the gaps between vehicles ($\langle \cdots \rangle_{\text{gap}}$ over random time intervals); these two give different averages for the same quantity.

A traffic engineer sits at the bus stop, and measures an ensemble of time gaps Δ^{Bus} between neighboring buses, and an ensemble of gaps Δ^{Car} between neighboring cars.

(b) *Draw the probability density of gaps she observes between buses. Draw the probability density of gaps between cars. (Hint: Is it different from the ensemble of car waiting times you found in part (a)? Why not?) What is the mean gap time $\langle \Delta^{\text{Bus}} \rangle_{\text{gap}}$ for the buses? What is the mean gap time $\langle \Delta^{\text{Car}} \rangle_{\text{gap}}$ for the cars?* (One of these probability distributions involves the Dirac δ-function[3] if one ignores measurement error and imperfectly punctual public transportation.)

You should find that the mean waiting time for a bus in part (a) is half the mean bus gap time in (b), which seems sensible—the gap seen by the pedestrian is the sum of the $\delta_+^{\text{Bus}} + \delta_-^{\text{Bus}}$ of the waiting time and the time since the last bus. However, you should also find the mean waiting time for a car *equals* the mean car gap time. The equation $\Delta^{\text{Car}} = \delta_+^{\text{Car}} + \delta_-^{\text{Car}}$ would seem to imply that the average gap seen by the pedestrian

is twice the mean waiting time.

(c) *How can the average gap between cars measured by the pedestrian be different from that measured by the traffic engineer? Discuss.*

(d) *Consider a short experiment, with three cars passing at times $t = 0$, 2, and 8 (so there are two gaps, of length 2 and 6). What is $\langle \Delta^{\text{Car}} \rangle_{\text{gap}}$? What is $\langle \Delta^{\text{Car}} \rangle_t$? Explain why they are different.*

One of the key results in statistical mechanics is that predictions are *independent of the ensemble* for large numbers of particles. For example, the velocity distribution found in a simulation run at constant energy (using Newton's laws) or at constant temperature will have corrections that scale as one over the number of particles.

(1.4) **Stirling's formula.** (Mathematics) @
Stirling's approximation, $n! \sim \sqrt{2\pi n}(n/\text{e})^n$, is remarkably useful in statistical mechanics; it gives an excellent approximation for large n. In statistical mechanics the number of particles is so large that we usually care not about $n!$, but about its logarithm, so $\log(n!) \sim n \log n - n + \frac{1}{2}\log(2\pi n)$. Finally, n is often so large that the final term is a tiny fractional correction to the others, giving the simple formula $\log(n!) \sim n \log n - n$.

(a) *Calculate $\log(n!)$ and these two approximations to it for $n = 2$, 4, and 50. Estimate the error of the simpler formula for $n = 6.03 \times 10^{23}$. Discuss the fractional accuracy of these two approximations for small and large n.*

Note that $\log(n!) = \log(1 \times 2 \times 3 \times \cdots \times n) = \log(1) + \log(2) + \cdots + \log(n) = \sum_{m=1}^{n} \log(m)$.

(b) *Convert the sum to an integral, $\sum_{m=1}^{n} \approx \int_0^n \mathrm{d}m$. Derive the simple form of Stirling's formula.*

(c) *Draw a plot of $\log(m)$ and a bar chart showing $\log(\text{ceiling}(n))$. (Here ceiling(x) represents the smallest integer larger than x.) Argue that the integral under the bar chart is $\log(n!)$. (Hint: Check your plot: between $x = 4$ and 5, ceiling$(x) = 5$.)*

The difference between the sum and the integral in part (c) should look approximately like a collection of triangles, except for the region between zero and one. The sum of the areas equals the error in the simple form for Stirling's formula.

---

[3]The δ-function $\delta(x-x_0)$ is a probability density which has 100% probability of being in any interval containing x_0; thus $\delta(x-x_0)$ is zero unless $x = x_0$, and $\int f(x)\delta(x - x_0)\,\mathrm{d}x = f(x_0)$ so long as the domain of integration includes x_0. Mathematically, this is not a function, but rather a distribution or a measure.

(d) *Imagine doubling these triangles into rectangles on your drawing from part (c), and sliding them sideways (ignoring the error for m between zero and one). Explain how this relates to the term $\frac{1}{2} \log n$ in Stirling's formula* $\log(n!) - (n \log n - n) \approx \frac{1}{2} \log(2\pi n) = \frac{1}{2} \log(2) + \frac{1}{2} \log(\pi) + \frac{1}{2} \log(n)$.

(1.5) **Stirling and asymptotic series.**[4]   (Mathematics, Computation) ③

Stirling's formula (which is actually originally due to de Moivre) can be improved upon by extending it into an entire series. It is not a traditional Taylor expansion; rather, it is an *asymptotic series*. Asymptotic series are important in many fields of applied mathematics, statistical mechanics [171], and field theory [172].

We want to expand $n!$ for large n; to do this, we need to turn it into a continuous function, interpolating between the integers. This continuous function, with its argument perversely shifted by one, is $\Gamma(z) = (z-1)!$. There are many equivalent formulæ for $\Gamma(z)$; indeed, any formula giving an analytic function satisfying the recursion relation $\Gamma(z+1) = z\Gamma(z)$ and the normalization $\Gamma(1) = 1$ is equivalent (by theorems of complex analysis). We will not use it here, but a typical definition is $\Gamma(z) = \int_0^\infty e^{-t} t^{z-1} \, dt$; one can integrate by parts to show that $\Gamma(z+1) = z\Gamma(z)$.

(a) *Show, using the recursion relation $\Gamma(z+1) = z\Gamma(z)$, that $\Gamma(z)$ has a singularity (goes to $\pm\infty$) at zero and all the negative integers.*

Stirling's formula is extensible [18, p. 218] into a nice expansion of $\Gamma(z)$ in powers of $1/z = z^{-1}$:

$$\Gamma[z] = (z-1)!$$
$$\sim (2\pi/z)^{\frac{1}{2}} e^{-z} z^z (1 + (1/12)z^{-1}$$
$$+ (1/288)z^{-2} - (139/51840)z^{-3}$$
$$- (571/2488320)z^{-4}$$
$$+ (163879/209018880)z^{-5}$$
$$+ (5246819/75246796800)z^{-6}$$
$$- (534703531/902961561600)z^{-7}$$
$$- (4483131259/86684309913600)z^{-8}$$
$$+ \ldots). \tag{1.3}$$

This looks like a Taylor series in $1/z$, but is subtly different. For example, we might ask what the radius of convergence [174] of this series is.

The radius of convergence is the distance to the nearest singularity in the complex plane (see note 26 on p. 227 and Fig. 8.7(a)).

(b) *Let $g(\zeta) = \Gamma(1/\zeta)$; then Stirling's formula is something times a power series in ζ. Plot the poles (singularities) of $g(\zeta)$ in the complex ζ plane that you found in part (a). Show that the radius of convergence of Stirling's formula applied to g must be zero, and hence no matter how large z is Stirling's formula eventually diverges.*

Indeed, the coefficient of z^{-j} eventually grows rapidly; Bender and Orszag [18, p. 218] state that the odd coefficients ($A_1 = 1/12$, $A_3 = -139/51840, \ldots$) asymptotically grow as

$$A_{2j+1} \sim (-1)^j 2(2j)!/(2\pi)^{2(j+1)}. \tag{1.4}$$

(c) *Show explicitly, using the ratio test applied to formula 1.4, that the radius of convergence of Stirling's formula is indeed zero.*[5]

This in no way implies that Stirling's formula is not valuable! An asymptotic series of length n approaches $f(z)$ as z gets big, but for fixed z it can diverge as n gets larger and larger. In fact, asymptotic series are very common, and often are useful for much larger regions than are Taylor series.

(d) *What is 0!? Compute 0! using successive terms in Stirling's formula (summing to A_N for the first few N).* Considering that this formula is expanding about infinity, it does pretty well! Quantum electrodynamics these days produces the most precise predictions in science. Physicists sum enormous numbers of Feynman diagrams to produce predictions of fundamental quantum phenomena. Dyson argued that quantum electrodynamics calculations give an asymptotic series [172]; the most precise calculation in science takes the form of a series which cannot converge. Many other fundamental expansions are also asymptotic series; for example, Hooke's law and elastic theory have zero radius of convergence [35, 36] (Exercise 11.15).

---

[4]Hints for the computations can be found at the book website [181].

[5]If you do not remember about radius of convergence, see [174]. Here you will be using every other term in the series, so the radius of convergence is $\lim_{j\to\infty} \sqrt{|A_{2j-1}/A_{2j+1}|}$.

(1.6) **Random matrix theory.**[6] (Mathematics, Quantum, Computation) ③

One of the most active and unusual applications of ensembles is *random matrix theory*, used to describe phenomena in nuclear physics, mesoscopic quantum mechanics, and wave phenomena. Random matrix theory was invented in a bold attempt to describe the statistics of energy level spectra in nuclei. In many cases, the statistical behavior of systems exhibiting complex wave phenomena—almost any correlations involving eigenvalues and eigenstates—can be quantitatively modeled using ensembles of matrices with completely random, uncorrelated entries!

The most commonly explored ensemble of matrices is the Gaussian orthogonal ensemble (GOE). Generating a member H of this ensemble of size $N \times N$ takes two steps.

- Generate an $N \times N$ matrix whose elements are independent random numbers with Gaussian distributions of mean zero and standard deviation $\sigma = 1$.

- Add each matrix to its transpose to symmetrize it.

As a reminder, the Gaussian or normal probability distribution of mean zero gives a random number x with probability

$$\rho(x) = \frac{1}{\sqrt{2\pi}\sigma} e^{-x^2/2\sigma^2}. \tag{1.5}$$

One of the most striking properties that large random matrices share is the distribution of level splittings.

(a) *Generate an ensemble with $M = 1{,}000$ or so GOE matrices of size $N = 2$, 4, and 10. (More is nice.) Find the eigenvalues λ_n of each matrix, sorted in increasing order. Find the difference between neighboring eigenvalues $\lambda_{n+1} - \lambda_n$, for n, say, equal to[7] $N/2$. Plot a histogram of these eigenvalue splittings divided by the mean splitting, with bin size small enough to see some of the fluctuations.* (Hint: Debug your work with $M = 10$, and then change to $M = 1{,}000$.)

What is this dip in the eigenvalue probability near zero? It is called *level repulsion*.

For $N = 2$ the probability distribution for the eigenvalue splitting can be calculated pretty simply. Let our matrix be $M = \begin{pmatrix} a & b \\ b & c \end{pmatrix}$.

(b) *Show that the eigenvalue difference for M is $\lambda = \sqrt{(c-a)^2 + 4b^2} = 2\sqrt{d^2 + b^2}$ where $d = (c-a)/2$, and the trace $c+a$ is irrelevant. Ignoring the trace, the probability distribution of matrices can be written $\rho_M(d, b)$. What is the region in the (b, d) plane corresponding to the range of eigenvalue splittings $(\lambda, \lambda + \Delta)$? If ρ_M is continuous and finite at $d = b = 0$, argue that the probability density $\rho(\lambda)$ of finding an eigenvalue splitting near $\lambda = 0$ vanishes (level repulsion).* (Hint: Both d and b must vanish to make $\lambda = 0$. Go to polar coordinates, with λ the radius.)

(c) *Calculate analytically the standard deviation of a diagonal and an off-diagonal element of the GOE ensemble (made by symmetrizing Gaussian random matrices with $\sigma = 1$). You may want to check your answer by plotting your predicted Gaussians over the histogram of H_{11} and H_{12} from your ensemble in part (a). Calculate analytically the standard deviation of $d = (c-a)/2$ of the $N = 2$ GOE ensemble of part (b), and show that it equals the standard deviation of b.*

(d) *Calculate a formula for the probability distribution of eigenvalue spacings for the $N = 2$ GOE, by integrating over the probability density $\rho_M(d, b)$.* (Hint: Polar coordinates again.)

If you rescale the eigenvalue splitting distribution you found in part (d) to make the mean splitting equal to one, you should find the distribution

$$\rho_{\text{Wigner}}(s) = \frac{\pi s}{2} e^{-\pi s^2/4}. \tag{1.6}$$

This is called the *Wigner surmise*; it is within 2% of the correct answer for larger matrices as well.[8]

(e) *Plot eqn 1.6 along with your $N = 2$ results from part (a). Plot the Wigner surmise formula against the plots for $N = 4$ and $N = 10$ as well.* Does the distribution of eigenvalues depend in detail on our GOE ensemble? Or could it be *universal*, describing other ensembles of real symmetric matrices as well? Let us define a ± 1 ensemble of real symmetric matrices, by generating an $N \times N$ matrix whose elements are independent random variables, ± 1 with equal probability.

---

[6]This exercise was developed with the help of Piet Brouwer. Hints for the computations can be found at the book website [181].
[7]Why not use all the eigenvalue splittings? The mean splitting can change slowly through the spectrum, smearing the distribution a bit.
[8]The distribution for large matrices is known and universal, but is much more complicated to calculate.

(f) *Generate an ensemble of $M = 1,000$ symmetric matrices filled with ± 1 with size $N = 2$, 4, and 10. Plot the eigenvalue distributions as in part (a). Are they universal (independent of the ensemble up to the mean spacing) for $N = 2$ and 4? Do they appear to be nearly universal[9] (the same as for the GOE in part (a)) for $N = 10$? Plot the Wigner surmise along with your histogram for $N = 10$.*

The GOE ensemble has some nice statistical properties. The ensemble is invariant under orthogonal transformations:

$$H \to R^\top H R \quad \text{with} \quad R^\top = R^{-1}. \quad (1.7)$$

(g) *Show that $\mathrm{Tr}[H^\top H]$ is the sum of the squares of all elements of H. Show that this trace is invariant under orthogonal coordinate transformations (that is, $H \to R^\top H R$ with $R^\top = R^{-1}$).* (Hint: Remember, or derive, the cyclic invariance of the trace: $\mathrm{Tr}[ABC] = \mathrm{Tr}[CAB]$.)

Note that this trace, for a symmetric matrix, is the sum of the squares of the diagonal elements plus *twice* the squares of the upper triangle of off-diagonal elements. That is convenient, because in our GOE ensemble the variance (squared standard deviation) of the off-diagonal elements is half that of the diagonal elements (part (c)).

(h) *Write the probability density $\rho(H)$ for finding GOE ensemble member H in terms of the trace formula in part (g). Argue, using your formula and the invariance from part (g), that the GOE ensemble is invariant under orthogonal transformations: $\rho(R^\top H R) = \rho(H)$.*

This is our first example of an *emergent symmetry*. Many different ensembles of symmetric matrices, as the size N goes to infinity, have eigenvalue and eigenvector distributions that are invariant under orthogonal transformations *even though the original matrix ensemble did not have this symmetry*. Similarly, rotational symmetry emerges in random walks on the square lattice as the number of steps N goes to infinity, and also emerges on long length scales for Ising models at their critical temperatures.

(1.7) **Six degrees of separation.**[10] (Complexity, Computation) ④

One of the more popular topics in random network theory is the study of how connected they

are. *Six degrees of separation* is the phrase commonly used to describe the interconnected nature of human acquaintances: various somewhat uncontrolled studies have shown that any random pair of people in the world can be connected to one another by a short chain of people (typically around six), each of whom knows the next fairly well. If we represent people as nodes and acquaintanceships as neighbors, we reduce the problem to the study of the relationship network. Many interesting problems arise from studying properties of randomly generated networks. A network is a collection of *nodes* and *edges*, with each edge connected to two nodes, but with each node potentially connected to any number of edges (Fig. 1.5). A random network is constructed probabilistically according to some definite rules; studying such a random network usually is done by studying the entire ensemble of networks, each weighted by the probability that it was constructed. Thus these problems naturally fall within the broad purview of statistical mechanics.

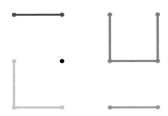

Fig. 1.5 Network. A network is a collection of nodes (circles) and edges (lines between the circles).

In this exercise, we will generate some random networks, and calculate the distribution of distances between pairs of points. We will study *small world networks* [140, 206], a theoretical model that suggests how a small number of shortcuts (unusual international and intercultural friendships) can dramatically shorten the typical chain lengths. Finally, we will study how a simple, universal scaling behavior emerges for large networks with few shortcuts.

---

[9]Note the spike at zero. There is a small probability that two rows or columns of our matrix of ± 1 will be the same, but this probability vanishes rapidly for large N.

[10]This exercise and the associated software were developed in collaboration with Christopher Myers. Hints for the computations can be found at the book website [181].

Constructing a small world network. The L nodes in a small world network are arranged around a circle. There are two kinds of edges. Each node has Z short edges connecting it to its nearest neighbors around the circle (up to a distance $Z/2$). In addition, there are $p \times L \times Z/2$ shortcuts added to the network, which connect nodes at random (see Fig. 1.6). (This is a more tractable version [140] of the original model [206], which rewired a fraction p of the $LZ/2$ edges.)

(a) *Define a network object on the computer. For this exercise, the nodes will be represented by integers. Implement a network class, with five functions:*

(1) `HasNode(node)`, *which checks to see if a node is already in the network;*

(2) `AddNode(node)`, *which adds a new node to the system (if it is not already there);*

(3) `AddEdge(node1, node2)`, *which adds a new edge to the system;*

(4) `GetNodes()`, *which returns a list of existing nodes; and*

(5) `GetNeighbors(node)`, *which returns the neighbors of an existing node.*

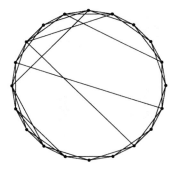

Fig. 1.6 Small world network *with $L = 20$, $Z = 4$, and $p = 0.2$.*[11]

Write a routine to construct a small world network, which (given L, Z, and p) adds the nodes and the short edges, and then randomly adds the shortcuts. Use the software provided to draw this small world graph, and check that you have im-

plemented the periodic boundary conditions correctly (each node i should be connected to nodes $(i - Z/2) \bmod L, \dots, (i + Z/2) \bmod L$).[12]

Measuring the minimum distances between nodes. The most studied property of small world graphs is the distribution of shortest paths between nodes. Without the long edges, the shortest path between i and j will be given by hopping in steps of length $Z/2$ along the shorter of the two arcs around the circle; there will be no paths of length longer than L/Z (halfway around the circle), and the distribution $\rho(\ell)$ of path lengths ℓ will be constant for $0 < \ell < L/Z$. When we add shortcuts, we expect that the distribution will be shifted to shorter path lengths.

(b) *Write the following three functions to find and analyze the path length distribution.*

(1) `FindPathLengthsFromNode(graph, node)`, *which returns for each `node2` in the graph the shortest distance from `node` to `node2`. An efficient algorithm is a breadth-first traversal of the graph, working outward from `node` in shells. There will be a `currentShell` of nodes whose distance will be set to ℓ unless they have already been visited, and a `nextShell` which will be considered after the current one is finished (looking sideways before forward, breadth first), as follows.*

   – *Initialize $\ell = 0$, the distance from `node` to itself to zero, and `currentShell = [node]`.*

   – *While there are nodes in the new `currentShell`:*

      * *start a new empty `nextShell`;*

      * *for each neighbor of each node in the current shell, if the distance to `neighbor` has not been set, add the node to `nextShell` and set the distance to $\ell + 1$;*

      * *add one to ℓ, and set the current shell to `nextShell`.*

   – *Return the distances.*

This will sweep outward from `node`, measuring the shortest distance to every other node in the network. (Hint: Check your code with a network with small N and small p, compar-

---

[11]There are seven new shortcuts, where $pLZ/2 = 8$; one of the added edges overlapped an existing edge or connected a node to itself.

[12]Here $(i - Z/2) \bmod L$ is the integer $0 \leq n \leq L - 1$, which differs from $i - Z/2$ by a multiple of L.

ing a few paths to calculations by hand from the graph image generated as in part (a).)

(2) `FindAllPathLengths(graph)`, *which generates a list of all lengths (one per pair of nodes in the graph) by repeatedly using* `FindPathLengthsFromNode`. *Check your function by testing that the histogram of path lengths at $p = 0$ is constant for $0 < \ell < L/Z$, as advertised. Generate graphs at $L = 1,000$ and $Z = 2$ for $p = 0.02$ and $p = 0.2$; display the circle graphs and plot the histogram of path lengths. Zoom in on the histogram; how much does it change with p? What value of p would you need to get "six degrees of separation"?*

(3) `FindAveragePathLength(graph)`, *which computes the mean $\langle \ell \rangle$ over all pairs of nodes. Compute ℓ for $Z = 2$, $L = 100$, and $p = 0.1$ a few times; your answer should be around $\ell = 10$. Notice that there are substantial statistical fluctuations in the value from sample to sample. Roughly how many long bonds are there in this system? Would you expect fluctuations in the distances?*

(c) *Plot the average path length between nodes $\ell(p)$ divided by $\ell(p = 0)$ for $Z = 2$, $L = 50$, with p on a semi-log plot from $p = 0.001$ to $p = 1$.* (Hint: Your curve should be similar to that of with Watts and Strogatz [206, Fig. 2], with the values of p shifted by a factor of 100; see the discussion of the continuum limit below.) *Why is the graph fixed at one for small p?*

Large N and the emergence of a continuum limit. We can understand the shift in p of part (c) as a continuum limit of the problem. In the limit where the number of nodes N becomes large and the number of shortcuts $pLZ/2$ stays fixed, this network problem has a nice limit where distance is measured in radians $\Delta\theta$ around the circle. Dividing ℓ by $\ell(p = 0) \approx L/(2Z)$ essentially does this, since $\Delta\theta = \pi Z \ell / L$.

(d) *Create and display a circle graph of your geometry from part (c) ($Z = 2$, $L = 50$) at $p = 0.1$; create and display circle graphs of Watts and Strogatz's geometry ($Z = 10$, $L = 1,000$) at $p = 0.1$ and $p = 0.001$. Which of their systems looks statistically more similar to yours? Plot (perhaps using the scaling collapse routine provided) the rescaled average path length $\pi Z \ell / L$*

versus the total number of shortcuts $pLZ/2$, for a range $0.001 < p < 1$, for $L = 100$ and 200, and for $Z = 2$ and 4.

In this limit, the average bond length $\langle \Delta\theta \rangle$ should be a function only of M. Since Watts and Strogatz [206] ran at a value of ZL a factor of 100 larger than ours, our values of p are a factor of 100 larger to get the same value of $M = pLZ/2$. Newman and Watts [144] derive this continuum limit with a renormalization-group analysis (Chapter 12).

(e) *Real networks. From the book website [181], or through your own research, find a real network[13] and find the mean distance and histogram of distances between nodes.*

Fig. 1.7 Betweenness Small world network with $L = 500$, $K = 2$, and $p = 0.1$, with node and edge sizes scaled by the square root of their betweenness.

In the small world network, a few long edges are crucial for efficient transfer through the system (transfer of information in a computer network, transfer of disease in a population model, ...). It is often useful to measure how crucial a given node or edge is to these shortest paths. We say a node or edge is *between* two other nodes if it is along a shortest path between them. We measure the *betweenness* of a node or edge as the total number of such shortest paths passing through it, with (by convention) the initial and final nodes included in the count of between nodes; see Fig. 1.7. (If there are K multiple shortest paths of equal length between two nodes, each path adds $1/K$ to its intermediates.) The efficient algorithm to measure betweenness is a depth-first traversal quite analogous to the shortest-path-length algorithm discussed above.

---

[13]Examples include movie-actor costars, *Six degrees of Kevin Bacon*, or baseball players who played on the same team.

(f) *Betweenness (advanced). Read [68, 141], which discuss the algorithms for finding the betweenness. Implement them on the small world network, and perhaps the real world network you analyzed in part (e). Visualize your answers by using the graphics software provided on the book website [181].*

(1.8) **Satisfactory map colorings.**[14] (Computer science, Computation, Mathematics) ③

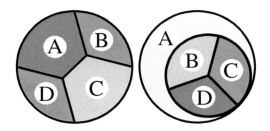

Fig. 1.8 Graph coloring. Two simple examples of graphs with $N = 4$ nodes that can and cannot be colored with three colors.

Many problems in computer science involve finding a good answer among a large number of possibilities. One example is *3-colorability* (Fig. 1.8). Can the N nodes of a graph be colored in three colors (say red, green, and blue) so that no two nodes joined by an edge have the same color?[15] For an N-node graph one can of course explore the entire ensemble of 3^N colorings, but that takes a time exponential in N. Sadly, there are no known shortcuts that fundamentally change this; there is no known algorithm for determining whether a given N-node graph is three-colorable that guarantees an answer in a time that grows only as a power of N.[16]

Another good example is *logical satisfiability* (**SAT**). Suppose one has a long logical expression involving N boolean variables. The logical expression can use the operations NOT ($\neg$), AND ($\wedge$), and OR ($\vee$). It is *satisfiable* if there is some assignment of *True* and *False* to its variables that makes the expression *True*. Can we solve a general satisfiability problem with N variables in a worst-case time that grows less quickly than exponentially in N? In this exercise, you will show that logical satisfiability is in a sense computationally at least as hard as 3-colorability. That is, you will show that a 3-colorability problem with N nodes can be mapped onto a logical satisfiability problem with $3N$ variables, so a polynomial-time (nonexponential) algorithm for the **SAT** would imply a (hitherto unknown) polynomial-time solution algorithm for 3-colorability.

If we use the notation A_R to denote a variable which is true when node A is colored red, then $\neg(A_R \wedge A_G)$ is the statement that node A is not colored both red and green, while $A_R \vee A_G \vee A_B$ is true if node A is colored one of the three colors.[17]

There are three types of expressions needed to write the colorability of a graph as a logical satisfiability problem: A has some color (above), A has only one color, and A and a neighbor B have different colors.

(a) *Write out the logical expression that states that A does not have two colors at the same time. Write out the logical expression that states that A and B are not colored with the same color. Hint: Both should be a conjunction (AND, $\wedge$) of three clauses each involving two variables.*

Any logical expression can be rewritten into a standard format, the *conjunctive normal form*. A *literal* is either one of our boolean variables or its negation; a logical expression is in conjunctive normal form if it is a conjunction of a series of clauses, each of which is a disjunction (OR, $\vee$) of literals.

(b) *Show that, for two boolean variables X and Y, that $\neg(X \wedge Y)$ is equivalent to a disjunction of literals $(\neg X) \vee (\neg Y)$. (Hint: Test each of the four cases). Write your answers to part (a) in conjunctive normal form. What is the maximum*

---

[14]This exercise and the associated software were developed in collaboration with Christopher Myers, with help from Bart Selman and Carla Gomes. Computational hints can be found at the book website [181].

[15]The famous four-color theorem, that any map of countries on the world can be colored in four colors, shows that all planar graphs are 4-colorable.

[16]Because 3-colorability is **NP**–complete (see Exercise 8.15), finding such a polynomial-time algorithm would allow one to solve traveling salesman problems and find spin-glass ground states in polynomial time too.

[17]The operations AND ($\wedge$) and NOT $\neg$ correspond to common English usage ($\wedge$ is true only if both are true, $\neg$ is true only if the expression following is false). However, OR ($\vee$) is an *inclusive or*—false only if both clauses are false. In common English usage *or* is usually *exclusive*, false also if both are true. ("Choose door number one or door number two" normally does not imply that one may select both.)

number of literals in each clause you used? Is it the maximum needed for a general 3-colorability problem?

In part (b), you showed that any 3-colorability problem can be mapped onto a logical satisfiability problem in conjunctive normal form with at most three literals in each clause, and with three times the number of boolean variables as there were nodes in the original graph. (Consider this a hint for part (b).) Logical satisfiability problems with at most k literals per clause in conjunctive normal form are called **kSAT** problems.

(c) *Argue that the time needed to translate the 3-colorability problem into a* **3SAT** *problem grows at most quadratically in the number of nodes M in the graph (less than αM^2 for some α for large M). (Hint: the number of edges of a graph is at most M^2.) Given an algorithm that guarantees a solution to any N-variable* **3SAT** *problem in a time $T(N)$, use it to give a bound on the time needed to solve an M-node 3-colorability problem. If $T(N)$ were a polynomial-time algorithm (running in time less than N^x for some integer x), show that 3-colorability would be solvable in a time bounded by a polynomial in M.*

We will return to logical satisfiability, **kSAT**, and **NP**–completeness in Exercise 8.15. There we will study a statistical ensemble of **kSAT** problems, and explore a phase transition in the fraction of satisfiable clauses, and the divergence of the typical computational difficulty near that transition.

(1.9) **First to fail: Weibull.**[18]   (Mathematics, Statistics, Engineering) ③

Suppose you have a brand-new supercomputer with $N = 1,000$ processors. Your parallelized code, which uses all the processors, cannot be restarted in mid-stream. How long a time t can you expect to run your code before the first processor fails?

This is example of *extreme value statistics* (see also exercises 12.23 and 12.24), where here we are looking for the smallest value of N random variables that are all bounded below by zero. For large N the probability distribution $\rho(t)$ and survival probability $S(t) = \int_t^\infty \rho(t')\,dt'$ are often

given by the *Weibull distribution*

$$S(t) = e^{-(t/\alpha)^\gamma},$$
$$\rho(t) = -\frac{dS}{dt} = \frac{\gamma}{\alpha}\left(\frac{t}{\alpha}\right)^{\gamma-1} e^{-(t/\alpha)^\gamma}. \qquad (1.8)$$

Let us begin by assuming that the processors have a constant rate Γ of failure, so the probability density of a single processor failing at time t is $\rho_1(t) = \Gamma\exp(-\Gamma t)$ as $t \to 0$, and the survival probability for a single processor $S_1(t) = 1 - \int_0^t \rho_1(t')\,dt' \approx 1 - \Gamma t$ for short times. (a) *Using $(1 - \epsilon) \approx \exp(-\epsilon)$ for small ϵ, show that the the probability $S_N(t)$ at time t that all N processors are still running is of the Weibull form (eqn 1.8). What are α and γ?*

Often the probability of failure per unit time goes to zero or infinity at short times, rather than to a constant. Suppose the probability of failure for one of our processors

$$\rho_1(t) \sim Bt^k \qquad (1.9)$$

with $k > -1$. (So, $k < 0$ might reflect a breaking-in period, where survival for the first few minutes increases the probability for later survival, and $k > 0$ would presume a dominant failure mechanism that gets worse as the processors wear out.)

(b) *Show the survival probability for N identical processors each with a power-law failure rate (eqn 1.9) is of the Weibull form for large N, and give α and γ as a function of B and k.*

The parameter α in the Weibull distribution just sets the scale or units for the variable t; only the exponent γ really changes the shape of the distribution. Thus the form of the failure distribution at large N only depends upon the power law k for the failure of the individual components at short times, not on the behavior of $\rho_1(t)$ at longer times. This is a type of *universality*,[19] which here has a physical interpretation; at large N the system will break down soon, so only early times matter.

The Weibull distribution, we must mention, is often used in contexts not involving extremal statistics. Wind speeds, for example, are naturally always positive, and are conveniently fit by Weibull distributions.

[18]Developed with the assistance of Paul (Wash) Wawrzynek
[19]The Weibull distribution is part of a family of extreme value distributions, all of whom are universal. See Chapter 12 and Exercise 12.24.

(1.10) Emergence. ⓟ

We begin with the broad statement "Statistical mechanics explains the simple behavior of complex systems." New laws emerge from bewildering interactions of constituents.

Discuss which of these emergent behaviors is probably not *studied using statistical mechanics.*

(a) The emergence of the wave equation from the collisions of atmospheric molecules,

(b) The emergence of Newtonian gravity from Einstein's general theory,

(c) The emergence of random stock price fluctuations from the behavior of traders,

(d) The emergence of a power-law distribution of earthquake sizes from the response of rubble in earthquake faults to external stresses.

(1.11) Emergent vs. fundamental. ⓟ

Statistical mechanics is central to condensed matter physics. It is our window into the behavior of materials—how complicated interactions between large numbers of atoms lead to physical laws (Fig. 1.9). For example, the theory of sound emerges from the complex interaction between many air molecules governed by Schrödinger's equation. More is different [10].

For example, if you inhale helium, your voice gets squeaky like Mickey Mouse. The dynamics of air molecules change when helium is introduced—the same law of motion, but with different constants.

(a) Look up the wave equation for sound in gases. How many constants are needed? Do the details of the interactions between air molecules matter for sound waves in air?

Statistical mechanics is tied also to particle physics and astrophysics. It is directly important in, e.g., the entropy of black holes (Exercise 7.16), the microwave background radiation (Exercises 7.15 and 10.1), and broken symmetry and phase transitions in the early Universe (Chapters 9, 11, and 12). Where statistical mechanics focuses on the *emergence* of comprehensible behavior at low energies, particle physics searches for the *fundamental* underpinnings at high energies (Fig. 1.10). Our different approaches reflect the complicated science at the atomic scale of chemistry and nuclear physics. At higher energies, atoms are described by elegant field theories (the *standard model* combining electroweak theory for electrons, photons, and neutrinos with QCD for quarks and gluons); at lower energies effective laws emerge for gases, solids, liquids, superconductors, ...

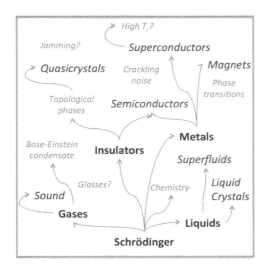

Fig. 1.9 Emergent. New laws describing macroscopic materials emerge from complicated microscopic behavior [177].

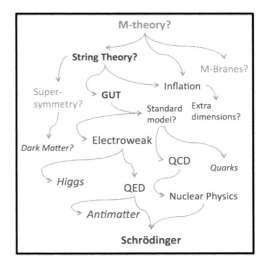

Fig. 1.10 Fundamental. Laws describing physics at lower energy emerge from more fundamental laws at higher energy [177].

The laws of physics involve parameters—real numbers that one must calculate or measure, like the speed of sound for a each gas at a given density and pressure. Together with the initial conditions (e.g., the density and its rate of change for a gas), the laws of physics allow us to predict how our system behaves.

Schrödinger's equation describes the Coulomb interactions between electrons and nuclei, and their interactions with electromagnetic field. It can in principle be solved to describe almost all of materials physics, biology, and engineering, apart from radioactive decay and gravity, using a Hamiltonian involving only the parameters $\hbar$, e, c, m_e, and the the masses of the nuclei.[20] Nuclear physics and QCD in principle determine the nuclear masses; the values of the electron mass and the fine structure constant $\alpha = e^2/\hbar c$ could eventually be explained by even more fundamental theories.

(b) *About how many parameters would one need as input to Schrödinger's equation to describe materials and biology and such?* Hint: There are 253 stable nuclear isotopes.

(c) *Look up the Standard Model—our theory of electrons and light, quarks and gluons, that also in principle can be solved to describe our Universe (apart from gravity). About how many parameters are required for the Standard Model?*

In high-energy physics, fewer constants are usually needed to describe the fundamental theory than the low-energy, effective emergent theory—the fundamental theory is more elegant and beautiful. In condensed matter theory, the fundamental theory is usually less elegant and messier; the emergent theory has a kind of parameter compression, with only a few combinations of microscopic parameters giving the governing parameters (temperature, elastic constant, diffusion constant) for the emergent theory.

Note that this is partly because in condensed matter theory we confine our attention to one particular material at a time (crystals, liquids, superfluids). To describe all materials in our world, and their interactions, would demand many parameters.

My high-energy friends sometimes view this from a different perspective. They note that the methods we use to understand a new superfluid, or a topological insulator, are quite similar to the ones they use to study the Universe. They admit a bit of envy—that we get a new universe to study every time an experimentalist discovers another material.

(1.12) **Self-propelled particles.**[21] (Active matter) ③ Exercise 2.20 investigates the statistical mechanical study of flocking—where animals, bacteria, or other active agents go into a collective state where they migrate in a common direction (like moshers in circle pits at heavy metal concerts [31, 188–190]). Here we explore the transition to a migrating state, but in an even more basic class of active matter: particles that are self-propelled but only interact via collisions. Our goal here is to both study the nature of the collective behavior, and the nature of the transition between disorganized motion and migration.

We start with an otherwise equilibrium system (damped, noisy particles with soft interatomic potentials, Exercise 6.19), and add a propulsion term

$$F_i^{\text{speed}} = \mu(v_0 - v_i)\hat{v}_i, \qquad (1.10)$$

which accelerates or decelerates each particle toward a target speed v_0 without changing the direction. The damping constant μ now controls how strongly the target speed is favored; for $v_0 = 0$ we recover the damping needed to counteract the noise to produce a thermal ensemble.

This simulation can be a rough model for crowded bacteria propelling themselves around, or for artificially created *Janus* particles that have one side covered with a platinum catalyst that burns hydrogen peroxide, pushing it forward.

Launch the mosh pit simulator [32]. If necessary, reload the page to the default setting. Set all particles active (Fraction Red to 1), set the Particle count to $N = 200$, Flock strength = 0, Speed $v_0 = 0.25$, Damping = 0.5 and Noise Strength = 0, Show Graphs, and click Change. After some time, you should see most of the particles moving along a common direction. (Increase Frameskip to speed the process.) You can increase the Box size and number maintaining the density if you have a powerful computer, or

---

[20]The gyromagnetic ratio for each nucleus is also needed in a few situations where its coupling to magnetic fields are important.
[21]This exercise was developed in collaboration with David Hathcock. It makes use of the mosh pit simulator [32] developed by Matt Bierbaum for [190].

decrease it (but not below 30) if your computer is struggling.

(a) *Watch the speed distribution as you restart the simulation.* Turn off Frameskip to see the behavior at early times. *Does it get sharply peaked at the same time as the particles begin moving collectively? Now turn up frameskip to look at the long-term motion. Give a qualitative explanation of what happens. Is more happening than just selection of a common direction?* (Hint: Understanding why the collective behavior maintains itself is easier than explaining why it arises in the first place.)

We can study this emergent, collective flow by putting our system in a box—turning off the periodic boundary conditions along x and y. Reload parameters to default, then all active, $N = 300$, flocking = 0, speed $v_0 = 0.25$, raise the damping up to 2 and set noise = 0. Turn off the periodic boundary conditions along both x and y, set the frame skip to 20, and Change. Again, box sizes as low as 30 will likely work.

After some time, you should observe a collective flow of a different sort. You can monitor the average flow using the angular momentum (middle graph below the simulation).

(b) *Increase the noise strength. Can you disrupt this collective behavior? Very roughly, a what noise strength does the transition occur? (You can use the angular momentum as a diagnostic.)*

A key question in equilibrium statistical mechanics is whether a qualitative transition like this is continuous (Chapter 12) or discontinuous (Chapter 11). Discontinuous transitions usually exhibit both bistability and hysteresis: the observed transition raising the temperature or other control parameter is higher than when one lowers the parameter. Here, if the transition is abrupt, we should have a region with three states—a melted state of zero angular momentum, and a collective clockwise and counterclockwise state.

Return to the settings for part (b) to explore more carefully the behavior near the transition.

(c) *Use the angular momentum to measure the strength of the collective motion (taken from the center graph, treating the upper and lower bounds as ± 1). Graph it against noise as you raise the noise slowly and carefully from zero, until it vanishes.* (You may need to wait longer when you get close to the transition.) *Graph it again as*

you lower the noise. Do you find the same transition point on heating and cooling (raising and lowering the noise)? Is the transition abrupt, or continuous? Did you ever observe switches between the clockwise and anti-clockwise states?

(1.13) **The birthday problem.** ②
Remember birthday parties in your elementary school? Remember those years when two kids had the same birthday? How unlikely!

How many kids would you need in class to get, more than half of the time, at least two with the same birthday?

(a) Numerical. *Write* `BirthdayCoincidences(K, C)`, *a routine that returns the fraction among C classes for which at least two kids (among K kids per class) have the same birthday.* (Hint: By sorting a random list of integers, common birthdays will be adjacent.) *Plot this probability versus K for a reasonably large value of C. Is it a surprise that your classes had overlapping birthdays when you were young?*

One can intuitively understand this, by remembering that to avoid a coincidence there are $K(K-1)/2$ pairs of kids, all of whom must have different birthdays (probability $364/365 = 1 - 1/D$, with D days per year).

$$P(K, D) \approx (1 - 1/D)^{K(K-1)/2} \qquad (1.11)$$

This is clearly a crude approximation—it doesn't vanish if $K > D$! Ignoring subtle correlations, though, it gives us a net probability

$$P(K, D) \approx \exp(-1/D)^{K(K-1)/2}$$
$$\approx \exp(-K^2/(2D)) \qquad (1.12)$$

Here we've used the fact that $1 - \epsilon \approx \exp(-\epsilon)$, and assumed that K/D is small.

(b) Analytical. *Write the exact formula giving the probability, for K random integers among D choices, that no two kids have the same birthday.* (Hint: What is the probability that the second kid has a different birthday from the first? The third kid has a different birthday from the first two?) *Show that your formula does give zero if $K > D$. Converting the terms in your product to exponentials as we did above, show that your answer is consistent with the simple formula above, if $K \ll D$. Inverting eqn 1.12, give a formula for the number of kids needed to have a 50% chance of a shared birthday.*

Some years ago, we were doing a large simulation, involving sorting a lattice of $1,000^3$ random fields (roughly, to figure out which site on the lattice would trigger first). If we want to make sure that our code is unbiased, we want different random fields on each lattice site—a giant birthday problem.

Old-style random number generators generated a random integer (2^{32} "days in the year") and then divided by the maximum possible integer to get a random number between zero and one. Modern random number generators generate all 2^{52} possible double precision numbers between zero and one.

(c) *If there are 2^{32} distinct four-byte unsigned integers, how many random numbers would one have to generate before one would expect coincidences half the time? Generate lists of that length, and check your assertion.* (Hints: It is faster to use array operations, especially in interpreted languages. I generated a random array with N entries, sorted it, subtracted the first $N-1$ entries from the last $N-1$, and then called *min* on the array.) *Will we have to worry about coincidences with an old-style random number generator? How large a lattice $L \times L \times L$ of random double precision numbers can one generate with modern generators before having a 50% chance of a coincidence?*

(1.14) **Width of the height distribution.**[22] (Statistics) ③

In this exercise we shall explore statistical methods of fitting models to data, in the context of fitting a Gaussian to a distribution of measurements. We shall find that *maximum likelihood* methods can be *biased*. We shall find that all sensible methods converge as the number of measurements N gets large (just as thermodynamics can ignore fluctuations for large numbers of particles), but a careful treatment of fluctuations and probability distributions becomes important for small N (just as different ensembles become distinguishable for small numbers of particles). The Gaussian distribution, known in statistics as the *normal* distribution

$$\mathcal{N}(x|\mu,\sigma^2) = \frac{1}{\sqrt{2\pi\sigma^2}} e^{-(x-\mu)^2/(2\sigma^2)} \qquad (1.13)$$

is a remarkably good approximation for many properties. The heights of men or women in a given country, or the grades on an exam in a large class, will often have a histogram that is well described by a normal distribution.[23] If we know the heights x_n of a sample with N people, we can write the likelihood that they were drawn from a normal distribution with mean μ and variance σ^2 as the product

$$P(\{x_n\}|\mu,\sigma) = \prod_{n=1}^{N} \mathcal{N}(x_n|\mu,\sigma^2). \qquad (1.14)$$

We first introduce the concept of *sufficient statistics*. Our likelihood (eqn 1.14) does not depend independently on each of the N heights x_n. What do we need to know about the sample to predict the likelihood?

(a) *Write $P(\{x_n\}|\mu,\sigma)$ in eqn 1.14 as a formula depending on the data $\{x_n\}$ only through N, $\overline{x} = (1/N)\sum_n x_n$ and $S = \sum_n (x_n - \overline{x})^2$.*

Given the model of independent normal distributions, its likelihood is a formula depending only on[24] $\overline{x}$ and S, the sufficient statistics for our Gaussian model.

Now, suppose we have a small sample and wish to estimate the mean and the standard deviation of the normal distribution.[25] *Maximum likelihood* is a common method for estimating model parameters; the estimates $(\mu_{\mathrm{ML}}, \sigma_{\mathrm{ML}})$ are given by the peak of the probability distribution P.

(b) *Show that $P(\{x_n\}|\mu_{\mathrm{ML}},\sigma_{\mathrm{ML}})$ takes its maximum value at*

$$\mu_{\mathrm{ML}} = \frac{\sum_n x_n}{N} = \overline{x}$$
$$\sigma_{\mathrm{ML}} = \sqrt{\sum_n (x_n - \overline{x})^2/N} = \sqrt{S/N}. \qquad (1.15)$$

(Hint: It is easier to maximize the log likelihood; $P(\boldsymbol{\theta})$ and $\log(P(\boldsymbol{\theta}))$ are maximized at the same point $\boldsymbol{\theta}_{\mathrm{ML}}$.)

If we draw samples of size N from a distribution of known mean μ_0 and standard deviation σ_0,

---

[22]This exercise was developed in collaboration with Colin Clement.

[23]This is likely because one's height is determined by the additive effects of many roughly uncorrelated genes and life experiences; the central limit theorem would then imply a Gaussian distribution (Chapter 2 and Exercise 12.11).

[24]In this exercise we shall use $\overline{X}$ denote a quantity averaged over a single sample of N people, and $\langle X \rangle_{\mathrm{samp}}$ denote a quantity also averaged over many samples.

[25]In physics, we usually estimate measurement errors separately from fitting our observations to theoretical models, so each experimental data point d_i comes with its error σ_i. In statistics, the estimation of the measurement error is often part of the modeling process, as in this exercise.

how do the maximum likelihood estimates differ from the actual values? For the limiting case $N = 1$, the various maximum likelihood estimates for the heights vary from sample to sample (with probability distribution $\mathcal{N}(x|\mu, \sigma^2)$), since the best estimate of the height is the sampled one). Because the average value $\langle \mu_{\mathrm{ML}} \rangle_{\mathrm{samp}}$ over many samples gives the correct mean, we say that μ_{ML} is *unbiased*. The maximum likelihood estimate for σ^2_{ML}, however, is biased. Again, for the extreme example $N = 1$, $\sigma^2_{\mathrm{ML}} = 0$ for every sample!

(c) *Assume the entire population is drawn from some (perhaps non-Gaussian) distribution of variance* $\langle x^2 \rangle_{\mathrm{samp}} = \sigma^2_0$. *For simplicity, let the mean of the population be zero. Show that*

$$\langle \sigma^2_{\mathrm{ML}} \rangle_{\mathrm{samp}} = (1/N) \left\langle \sum_{n=1}^{N} (x_n - \overline{x})^2 \right\rangle_{\mathrm{samp}}$$
$$= \frac{N-1}{N} \sigma^2_0. \qquad (1.16)$$

that the variance for a group of N people is on average smaller than the variance of the population distribution by a factor $(N-1)/N$. (Hint: $\overline{x} = (1/N) \sum_n x_n$ is not necessarily zero. Expand it out and use the fact that x_m and x_n are uncorrelated.)

The maximum likelihood estimate for the variance is biased on average toward smaller values. Thus we are taught, when estimating the standard deviation of a distribution[26] from N measurements, to divide by $\sqrt{N-1}$:

$$\sigma^2_{\mathrm{N}-1} \approx \frac{\sum_n (x_n - \overline{x})^2}{N-1}. \qquad (1.17)$$

This correction $N \to N - 1$ is generalized to more complicated problems by considering the number of independent degrees of freedom (here $N - 1$ degrees of freedom in the vector $x_n - \overline{x}$ of deviations from the mean). Alternatively, it is interesting that the bias disappears if one does not estimate both σ^2 and μ by maximizing the joint likelihood, but integrating (or *marginalizing*) over μ and then finding the maximum likelihood for σ^2.

(1.15) **Fisher information and Cramér–Rao.**[27] (Statistics, Mathematics, Information geometry) ④

Here we explore the geometry of the space of probability distributions. When one changes the external conditions of a system a small amount, how much does the ensemble of predicted states change? What is the *metric* in probability space? Can we predict how easy it is to detect a change in external parameters by doing experiments on the resulting distribution of states? The metric we find will be the *Fisher information matrix* (FIM). The *Cramér–Rao bound* will use the FIM to provide a rigorous limit on the precision of any (unbiased) measurement of parameter values.

In both statistical mechanics and statistics, our models generate probability distributions $P(\mathbf{x}|\boldsymbol{\theta})$ for behaviors $\mathbf{x}$ given parameters $\boldsymbol{\theta}$.

- A crooked gambler's loaded die, where the state space is comprised of discrete rolls $\mathbf{x} \in \{1, 2, \ldots, 6\}$ with probabilities $\boldsymbol{\theta} = \{p_1, \ldots, p_5\}$, with $p_6 = 1 - \sum_{j=1}^{5} \theta_j$.
- The probability density that a system with a Hamiltonian $\mathcal{H}(\boldsymbol{\theta})$ with $\boldsymbol{\theta} = (T, P, N)$ giving the temperature, pressure, and number of particles, will have a probability density $P(\mathbf{x}|\boldsymbol{\theta}) = \exp(-\mathcal{H}/k_B T)/Z$ in phase space (Chapter 3, Exercise 6.22).
- The height of women in the US, $\mathbf{x} = \{h\}$ has a probability distribution well described by a normal (or Gaussian) distribution $P(\mathbf{x}|\boldsymbol{\theta}) = 1/\sqrt{2\pi\sigma^2} \exp(-(x - \mu)^2/2\sigma^2)$ with mean and standard deviation $\boldsymbol{\theta} = (\mu, \sigma)$ (Exercise 1.14).
- A least squares model $y_i(\boldsymbol{\theta})$ for N data points $d_i \pm \sigma$ with independent, normally distributed measurement errors predicts a likelihood for finding a value $\mathbf{x} = \{x_i\}$ of the data $\{d_i\}$ given by

$$P(\mathbf{x}|\boldsymbol{\theta}) = \frac{e^{-\sum_i (y_i(\theta) - x_i)^2/2\sigma^2}}{(2\pi\sigma^2)^{N/2}}. \qquad (1.18)$$

(Think of the theory curves you fit to data in many experimental labs courses.)

How "distant" is a loaded die is from a fair one? How "far apart" are the probability distributions of particles in phase space for two small system at different temperatures and pressures? How

---

[26] Do not confuse this with the estimate of the error in the *mean* $\overline{x}$.
[27] This exercise was developed in collaboration with Colin Clement and Katherine Quinn.

hard would it be to distinguish a group of US women from a group of Pakistani women, if you only knew their heights?

We start with the least-squares model.

(a) *How big is the probability density that a least-squares model with true parameters $\boldsymbol{\theta}$ would give experimental results implying a different set of parameters $\boldsymbol{\phi}$? Show that it depends only on the distance between the vectors $|\mathbf{y}(\boldsymbol{\theta}) - \mathbf{y}(\boldsymbol{\phi})|$ in the space of predictions.* Thus the predictions of least-squares models form a natural manifold in a behavior space, with a coordinate system given by the parameters. The point on the manifold corresponding to parameters $\boldsymbol{\theta}$ is $\mathbf{y}(\boldsymbol{\theta})/\sigma$ given by model predictions rescaled by their error bars, $\mathbf{y}(\boldsymbol{\theta})/\sigma$.

Remember that the metric tensor $g_{\alpha\beta}$ gives the distance on the manifold between two nearby points. The squared distance between points with coordinates $\boldsymbol{\theta}$ and $\boldsymbol{\theta} + \epsilon\boldsymbol{\Delta}$ is $\epsilon^2 \sum_{\alpha\beta} g_{\alpha\beta} \Delta_\alpha \Delta_\beta$.

(b) *Show that the least-squares metric is $g_{\alpha\beta} = (J^T J)_{\alpha\beta}/\sigma^2$, where the Jacobian $J_{i\alpha} = \partial y_i / \partial\theta_\alpha$.*

For general probability distributions, the natural metric describing the distance between two nearby distributions $P(\mathbf{x}|\boldsymbol{\theta})$ and $Q = P(\mathbf{x}|\boldsymbol{\theta} + \epsilon\boldsymbol{\Delta})$ is given by the FIM:

$$g_{\alpha\beta}(\boldsymbol{\theta}) = -\left\langle \frac{\partial^2 \log P(\mathbf{x}|\boldsymbol{\theta})}{\partial\theta_\alpha \partial\theta_\beta} \right\rangle_{\mathbf{x}} \tag{1.19}$$

Are the distances between least-squares models we intuited in parts (a) and (b) compatible with the the FIM?

(c) *Show for a least-squares model that eqn 1.19 is the same as the metric we derived in part (b).* (Hint: For a Gaussian distribution $\exp((x - \mu)^2/(2\sigma^2))/\sqrt{2\pi\sigma^2}$, $\langle x \rangle = \mu$.)

If we have experimental data with errors, how well can we estimate the parameters in our theoretical model, given a fit? As in part (a), now for general probabilistic models, how big is the probability density that an experiment with true parameters $\boldsymbol{\theta}$ would give results perfectly corresponding to a nearby set of parameters $\boldsymbol{\theta} + \epsilon\boldsymbol{\Delta}$?

(d) *Take the Taylor series of $\log P(\boldsymbol{\theta} + \epsilon\boldsymbol{\Delta})$ to second order in ϵ. Exponentiate this to estimate how much the probability of measuring values corresponding to the predictions at $\boldsymbol{\theta} + \epsilon\boldsymbol{\Delta}$ fall off compared to $P(\boldsymbol{\theta})$.* Thus to linear order the FIM $g_{\alpha\beta}$ estimates the range of likely measured

parameters around the true parameters of the model.

The *Cramér–Rao bound* shows that this estimate is related to a rigorous bound. In particular, errors in a multiparameter fit are usually described by a *covariance matrix* Σ, where the variance of the likely values of parameter θ_α is given by $\Sigma_{\alpha\alpha}$, and where $\Sigma_{\alpha\beta}$ gives the correlations between two parameters θ_α and θ_β. One can show within our quadratic approximation of part (d) that the covariance matrix is the inverse of the FIM $\Sigma_{\alpha\beta} = (g^{-1})_{\alpha\beta}$. The *Cramér–Rao bound* roughly tells us that no experiment can do better than this at estimating parameters. In particular, it tells us that the error range of the individual parameters from a sampling of a probability distribution is bounded below by the corresponding element of the inverse of the FIM

$$\Sigma_{\alpha\alpha} \geq (g^{-1})_{\alpha\alpha}. \tag{1.20}$$

(if the estimator is *unbiased*, see Exercise 1.14). This is another justification for using the FIM as our natural distance metric in probability space. In Exercise 1.16, we shall examine *global* measures of distance or distinguishability between potentially quite different probability distributions. There we shall show that these measures all reduce to the FIM to lowest order in the change in parameters. In Exercises 6.23, 6.21, and 6.22, we shall show that the FIM for a Gibbs ensemble as a function of temperature and pressure can be written in terms of thermodynamic quantities like compressibility and specific heat. There we use the FIM to estimate the *path length in probability space*, in order to estimate the entropy cost of controlling systems like the Carnot cycle.

(1.16) **Distances in probability space.**[28] (Statistics, Mathematics, Information geometry) ③

In statistical mechanics we usually study the behavior expected given the experimental parameters. Statistics is often concerned with estimating how well one can deduce the parameters (like temperature and pressure, or the increased risk of death from smoking) given a sample of the ensemble. Here we shall explore ways of measuring distance or distinguishability between distant probability distributions.

Exercise 1.15 introduces four problems (loaded dice, statistical mechanics, the height distribu-

---

[28]This exercise was developed in collaboration with Katherine Quinn.

tion of women, and least-squares fits to data), each of which have parameters $\boldsymbol{\theta}$ which predict an ensemble probability distribution $P(\mathbf{x}|\boldsymbol{\theta})$ for data $\mathbf{x}$ (die rolls, particle positions and momenta, heights, ...). In the case of least-squares models (eqn 1.18) where the probability is given by a vector $x_i = y_i(\boldsymbol{\theta}) \pm \sigma$, we found that the distance between the predictions of two parameter sets $\boldsymbol{\theta}$ and $\boldsymbol{\phi}$ was naturally given by $|\mathbf{y}(\boldsymbol{\theta})/\sigma - \mathbf{y}(\boldsymbol{\phi})/\sigma|$. We want to generalize this formula—to find ways of measuring distances between probability distributions given by arbitrary kinds of models.

Exercise 1.15 also introduced the Fisher information metric (FIM) in eqn 1.19:

$$g_{\mu\nu}(\boldsymbol{\theta}) = -\left\langle \frac{\partial^2 \log(P(\mathbf{x}))}{\partial\theta_\alpha \partial\theta_\beta} \right\rangle_{\mathbf{x}} \qquad (1.21)$$

which gives the distance between probability distributions for nearby sets of parameters

$$d^2(P(\boldsymbol{\theta}), P(\boldsymbol{\theta} + \epsilon\boldsymbol{\Delta})) = \epsilon^2 \sum_{\mu\nu} \Delta_\mu g_{\mu\nu} \Delta_\nu. \quad (1.22)$$

Finally, it argued that the distance defined by the FIM is related to how distinguishable the two nearby ensembles are—how well we can deduce the parameters. Indeed, we found that to linear order the FIM is the inverse of the covariance matrix describing the fluctuations in estimated parameters, and that the Cramér–Rao bound shows that this relationship between the FIM and distinguishability works even beyond the linear regime.

There are several measures in common use, of which we will describe three—the Hellinger distance, the Bhattacharyya "distance", and the Kullback–Liebler divergence. Each has its uses. The Hellinger distance becomes less and less useful as the amount of information about the parameters becomes large. The Kullback–Liebler divergence is not symmetric, but one can symmetrize it by averaging. It and the Bhattacharyya distance nicely generalize the least-squares metric to arbitrary models, but they violate the triangle inequality and embed the manifold of predictions into a space with Minkowski-style time-like directions [155].

Let us review the properties that we ordinarily demand from a distance between points P and Q.

- We expect it to be positive, $d(P,Q) \geq 0$, with $d(P,Q) = 0$ only if $P = Q$.
- We expect it to be symmetric, so $d(P,Q) = d(Q,P)$.
- We expect it to satisfy the *triangle inequality*, $d(P,Q) \leq d(P,R) + d(R,Q)$—the two short sides of a triangle must extend at total distance enough to reach the third side.
- We want it to become large when the points P and Q are extremely different.

All of these properties are satisfied by the least-squares distance of Exercise 1.15, because the distances between points on the surface of model predictions is the Euclidean distance between the predictions in data space.

Our first measure, the Hellinger distance at first seems ideal. It defines a *dot product* between probability distributions P and Q. Consider the discrete gambler's distribution, giving the probabilities $\mathbf{P} = \{P_j\}$ for die roll j. The normalization $\sum P_j = 1$ makes $\{\sqrt{P_j}\}$ a unit vector in six dimensions, so we define a dot product $P \cdot Q = \sum_{j=1}^6 \sqrt{P_j}\sqrt{Q_j} = \int \mathrm{d}\mathbf{x} \sqrt{P(\mathbf{x})}\sqrt{Q(\mathbf{x})}$. The Hellinger distance is then given by the squared distance between points on the unit sphere:[29]

$$\begin{aligned} d_{\mathrm{Hel}}^2(P,Q) &= (P-Q)^2 = 2 - 2P \cdot Q \\ &= \int \mathrm{d}\mathbf{x} \left(\sqrt{P(\mathbf{x})} - \sqrt{Q(\mathbf{x})}\right)^2. \end{aligned} \qquad (1.23)$$

(a) *Argue, from the last geometrical characterization, that the Hellinger distance must be a valid distance function. Show that the Hellinger distance does reduce to the FIM for nearby distributions, up to a constant factor. Show that the Hellinger distance never gets larger than $\sqrt{2}$. What is the Hellinger distance between a fair die $P_j \equiv 1/6$ and a loaded die $Q_j = \{1/10, 1/10, \ldots, 1/2\}$ that favors rolling 6?*

The Hellinger distance is peculiar in that, as the statistical mechanics system gets large, or as one adds more experimental data to the statistics model, all pairs approach the maximum distance $\sqrt{2}$.

(b) *Our gambler keeps using the loaded die. Can the casino catch him? Let $P_N(\mathbf{j})$ be the probability that rolling the die N times gives the sequence $\mathbf{j} = \{j_1, \ldots, j_N\}$. Show that*

$$P_N \cdot Q_N = (P \cdot Q)^N, \qquad (1.24)$$

---

[29]Sometimes it is given by *half* the distance between points on the unit sphere, presumably so that the maximum distance between two probability distributions becomes one, rather than $\sqrt{2}$.

and hence

$$d^2_{\text{Hel}}(P_N, Q_N) = 1 - (P \cdot Q)^N. \qquad (1.25)$$

After $N = 100$ rolls, how close is the Hellinger distance from its maximum value?

From the casino's point of view, the certainty that the gambler is cheating is becoming squeezed into a tiny range of distances. (P_N and Q_N becoming increasingly orthogonal does not lead to larger and larger Hellinger distances.) In an Ising model, or a system with N particles, or a cosmic microwave background experiment with N measured areas of the sky, even tiny changes in parameters lead to orthogonal probability distributions, and hence Hellinger distances near its maximum value of one.[30]

The Hellinger overlap $(P \cdot Q)^N = \exp(N \log(P \cdot Q))$ keeps getting smaller as we take N to infinity; it is like the exponential of an extensive quantity.

Our second measure, the Bhattacharyya distance, can be derived from a limit of the Hellinger distance as the number of data points N goes to zero:

$$
\begin{aligned}
d^2_{\text{Bhatt}}(P, Q) &= \lim_{N \to 0} \tfrac{1}{2} d^2_{\text{Hel}}(P_N, Q_N)/N \\
&= -\log(P \cdot Q) \qquad (1.26) \\
&= -\log \left(\sum_{\mathbf{x}} \sqrt{P(\mathbf{x})} \sqrt{Q(\mathbf{x})} \right).
\end{aligned}
$$

We sometimes say that we calculate the behavior of N *replicas* of the system, and then take $N \to 0$. Replica theory is useful, for example, in disordered systems, where we can average $F = -k_B T \log(Z)$ over disorder (difficult)

by finding the average of Z^N over disorder (not so hard) and then taking $N \to 0$.

(d) *Derive eqn 1.26.* (Hint: $Z^N \approx \exp(N \log Z) \approx 1 + N \log Z$ for small N.)

The third distance-like measure we introduce is the *Kullback–Leibler divergence* from Q to P.

$$d_{\mathbf{KL}}(Q|P) = - \int d\mathbf{x} P(\mathbf{x}) \log(Q(\mathbf{x})/P(\mathbf{x})). \qquad (1.27)$$

(c) *Show that the Kullback–Liebler divergence is positive, zero only if $P = Q$, but is not symmetric. Show that, to quadratic order in ϵ in eqn 1.22, that the Kullback–Liebler divergence does lead to the FIM.*

The Kullback–Liebler divergence is sometimes symmetrized:

$$
\begin{aligned}
d_{\mathbf{sKL}}&(Q, P) \qquad\qquad\qquad (1.28) \\
&= \tfrac{1}{2}(d_{\mathbf{KL}}(Q|P) + d_{\mathbf{KL}}(P|Q)) \\
&= \int d\mathbf{x}(P(\mathbf{x}) - Q(\mathbf{x})) \log(P(\mathbf{x})/Q(\mathbf{x})).
\end{aligned}
$$

The Bhattacharyya distance and the symmetrized Kullback–Liebler divergence share several features, both good and bad.

(d) *Show that they are* intensive *[155]—that the distance grows linearly with repeated measurements[31] (as for repeated rolls in part (b)). Show that they do not satisfy the triangle inequality. Show that they does satisfy the other conditions for a distance. Show, for the nonlinear least-squares model of eqn 1.18, that they equal the distance in data space between the two predictions.*

---

[30]The problem is that the manifold of predictions is being curled up onto a sphere, where the short-cut distance between two models becomes quite different from the geodesic distance within the model manifold.

[31]This also makes these measures behave nicely for large systems as in statistical mechanics, where small parameter changes lead to nearly orthogonal probability distributions.

Random walks and emergent properties

2

What makes physics possible? Why are the mathematical laws that describe our macroscopic world so simple? Our physical laws are not direct statements about the underlying reality of the Universe. Rather, our laws emerge out of far more complex microscopic behavior.[1] Statistical mechanics provides a set of powerful tools for understanding simple behavior that emerges from underlying complexity.

In this chapter we will explore the emergent behavior for *random walks*. Random walks are paths that take successive steps in random directions. They arise often in statistical mechanics: as partial sums of fluctuating quantities, as trajectories of particles undergoing repeated collisions, and as the shapes for long, linked systems like polymers. They introduce two kinds of emergent behavior. First, an individual random walk, after a large number of steps, becomes *fractal* or *scale invariant* (explained in Section 2.1). Secondly, the endpoint of the random walk has a probability distribution that obeys a simple continuum law, the *diffusion equation* (introduced in Section 2.2). Both of these behaviors are largely independent of the microscopic details of the walk; they are *universal*. Random walks in an external field provide our first examples of *conserved currents*, *linear response*, and *Boltzmann distributions* (Section 2.3). Finally, we use the diffusion equation to introduce *Fourier* and *Green's function* techniques (Section 2.4). Random walks neatly illustrate many of the themes and methods of statistical mechanics.

[1]You may think that Newton's law of gravitation, or Einstein's refinement to it, is more fundamental than the diffusion equation. You would be correct; gravitation applies to everything. But the simple macroscopic law of gravitation emerges, presumably, from a quantum exchange of immense numbers of virtual gravitons just as the diffusion equation emerges from large numbers of long random walks. The diffusion equation and other continuum statistical mechanics laws are special to particular systems, but they emerge from the microscopic theory in much the same way as gravitation and the other fundamental laws of nature do. This is the source of many of the surprisingly simple mathematical laws describing nature [211].

2.1 Random walk examples: universality and scale invariance

Statistical mechanics often demands sums or averages of a series of fluctuating quantities: $s_N = \sum_{i=1}^{N} \ell_i$. The energy of a material is a sum over the energies of the molecules composing the material; your grade on a statistical mechanics exam is the sum of the scores on many individual questions. Imagine adding up this sum one term at a time. The path $s_1, s_2, \ldots$ forms an example of a one-dimensional random walk. We illustrate random walks with three examples: coin flips, the drunkard's walk, and polymers.

Coin flips. For example, consider flipping a coin and recording the difference s_N between the number of heads and tails found. Each coin

Statistical Mechanics: Entropy, Order Parameters, and Complexity. James P. Sethna, Oxford University Press (2021). © Oxford University Press. DOI:10.1093/oso/9780198865247.003.0002

flip contributes $\ell_i = \pm 1$ to the total. How big a sum $s_N = \sum_{i=1}^{N} \ell_i =$ (heads $-$ tails) do you expect after N flips?

The average of s_N is not a good measure for the sum, because it is zero (positive and negative steps are equally likely). We could measure the average[2] absolute value $\langle |s_N| \rangle$, but it turns out that a nicer characteristic distance is the root-mean-square (RMS) of the sum, $\sqrt{\langle s_N^2 \rangle}$. After one coin flip, the mean square

$$\langle s_1^2 \rangle = 1 = \tfrac{1}{2}(-1)^2 + \tfrac{1}{2}(1)^2; \tag{2.1}$$

and after two and three coin flips

$$\begin{aligned} \langle s_2^2 \rangle &= 2 = \tfrac{1}{4}(-2)^2 + \tfrac{1}{2}(0)^2 + \tfrac{1}{4}(2)^2, \\ \langle s_3^2 \rangle &= 3 = \tfrac{1}{8}(-3)^2 + \tfrac{3}{8}(-1)^2 + \tfrac{3}{8}(1)^2 + \tfrac{1}{8}(3)^2 \end{aligned} \tag{2.2}$$

(for example, the probability of having two heads in three coin flips is three out of eight, HHT, THT, and TTT). Can you guess what $\langle s_7^2 \rangle$ will be, without computing it?

Does this pattern continue? We can try writing the RMS after N steps in terms of the RMS after $N-1$ steps, plus the last step. Because the average of the sum is the sum of the average, we find

$$\langle s_N^2 \rangle = \langle (s_{N-1} + \ell_N)^2 \rangle = \langle s_{N-1}^2 \rangle + 2\langle s_{N-1} \ell_N \rangle + \langle \ell_N^2 \rangle. \tag{2.3}$$

Now, ℓ_N is ± 1 with equal probability, independent of what happened earlier (and thus independent of s_{N-1}). Thus $\langle s_{N-1} \ell_N \rangle = \tfrac{1}{2} s_{N-1}(+1) + \tfrac{1}{2} s_{N-1}(-1) = 0$. We also know that $\ell_N^2 = 1$, so

$$\langle s_N^2 \rangle = \langle s_{N-1}^2 \rangle + 2\underbrace{\langle s_{N-1} \ell_N \rangle}_{} + \langle \ell_N^2 \rangle = \langle s_{N-1}^2 \rangle + 1. \tag{2.4}$$

If we assume $\langle s_{N-1}^2 \rangle = N - 1$ we have proved by induction on N that $\langle s_N^2 \rangle = N$.[3]

Hence the RMS of (heads $-$ tails) is equal to the square root of the number of coin flips:

$$\sigma_s = \sqrt{\langle s_N^2 \rangle} = \sqrt{N}. \tag{2.5}$$

Drunkard's walk. Random walks also arise as trajectories that undergo successive random collisions or turns; for example, the trajectory of a perfume molecule in a sample of air[4] (Exercise 2.4). Because the air is dilute and the interactions are short ranged, the molecule will basically travel in straight lines, with sharp changes in velocity during infrequent collisions. After a few substantial collisions, the molecule's velocity will be uncorrelated with its original velocity. The path taken by the molecule will be a jagged, random walk through three dimensions.

The random walk of a perfume molecule involves random directions, random velocities, and random step sizes (Exercise 2.22). It is more convenient to study steps at regular time intervals, so we will instead consider the classic problem of a drunkard's walk (Fig. 2.1). The drunkard is presumed to start at a lamp-post at $x = y = 0$. He takes steps

---

[2] We use angle brackets $\langle \cdot \rangle$ to denote averages over ensembles. Here our ensemble contains all 2^N possible sequences of N coin flips.

[3] The mean of the absolute value $\langle |s_N| \rangle$ is not nearly as simple to calculate. This is one example of why the root-mean-square is nicer to work with than the average absolute value.

[4] Real perfume in a real room will primarily be transported by convection; in liquids and gases, diffusion dominates usually only on short length scales. Solids do not convect, so thermal or electrical conductivity would be a more accurate—but less vivid—illustration of random walks.

$\boldsymbol{\ell}_N$ each of length L, at regular time intervals. Because he is drunk, the steps are in completely random directions, each uncorrelated with the previous steps. This lack of correlation says that the average dot product between any two steps $\boldsymbol{\ell}_m$ and $\boldsymbol{\ell}_n$ is zero, since all relative angles θ between the two directions are equally likely: $\langle \boldsymbol{\ell}_m \cdot \boldsymbol{\ell}_n \rangle = L^2 \langle \cos(\theta) \rangle = 0.$[5] This implies that the dot product of $\boldsymbol{\ell}_N$ with $\mathbf{s}_{N-1} = \sum_{m=1}^{N-1} \boldsymbol{\ell}_m$ is zero. Again, we can use this to work by induction:

$$\langle \mathbf{s}_N^2 \rangle = \langle (\mathbf{s}_{N-1} + \boldsymbol{\ell}_N)^2 \rangle = \langle \mathbf{s}_{N-1}^2 \rangle + \langle 2\mathbf{s}_{N-1} \cdot \boldsymbol{\ell}_N \rangle + \langle \boldsymbol{\ell}_N^2 \rangle$$
$$= \langle \mathbf{s}_{N-1}^2 \rangle + L^2 = \cdots = NL^2, \tag{2.6}$$

so the RMS distance moved is $\sqrt{N}L$.

Random walks introduce us to the concepts of *scale invariance* and *universality*.

Scale invariance. What kind of path only goes $\sqrt{N}$ total distance in N steps? Random walks form paths which look jagged and scrambled. Indeed, they are so jagged that if you blow up a small corner of one, the blown-up version looks just as jagged (Fig. 2.2). Each of the blown-up random walks is different, just as any two random walks of the same length are different, but the ensemble of random walks of length N looks much like that of length $N/4$, until N becomes small enough that the individual steps can be distinguished. Random walks are *scale invariant*: they look the same on all scales.[6]

Universality. On scales where the individual steps are not distinguishable (and any correlations between steps is likewise too small to see) we find that all random walks look the same. Figure 2.2 depicts a drunkard's walk, but any two-dimensional random walk would give the same behavior (statistically). Coin tosses of two coins (penny sums along x, dime sums along y) would produce, statistically, the same random walk ensemble on lengths large compared to the step sizes. In three dimensions, photons[7] in the Sun (Exercise 2.2) or in a glass of milk undergo a random walk with fixed speed c between collisions.[8] After a few steps all of these random walks are statistically indistinguishable from that of our variable-speed perfume molecule. This independence of the behavior on the microscopic details is called *universality*.

Random walks are simple enough that we could show directly that each individual case behaves like the others. In Section 2.2 we will generalize our argument that the RMS distance scales as $\sqrt{N}$ to simultaneously cover both coin flips and drunkards; with more work we could include variable times between collisions and local correlations to cover the cases of photons and molecules in a gas. We could also calculate properties about the jaggedness of paths in these systems, and show that they too agree with one another after many steps. Instead, we will wait for Chapter 12 (and specifically Exercise 12.11), where we will give a deep but intuitive explanation of why each of these problems is scale invariant, and why all of these problems share the same behavior on long length scales. Universality and scale invariance will be explained there using renormalization-group methods, originally developed to study continuous phase transitions.

[5]More generally, if two variables are uncorrelated then the average of their product is the product of their averages; in this case this would imply $\langle \boldsymbol{\ell}_m \cdot \boldsymbol{\ell}_n \rangle = \langle \boldsymbol{\ell}_m \rangle \cdot \langle \boldsymbol{\ell}_n \rangle = \mathbf{0} \cdot \mathbf{0} = 0$.

Fig. 2.1 Drunkard's walk. The drunkard takes a series of steps of length L away from the lamp-post, but each with a random angle.

[6]They are also fractal with dimension two, in all spatial dimensions larger than two. This just reflects the fact that a random walk of "volume" $V = N$ steps roughly fits into a radius $R \sim s_N \sim N^{1/2}$ (see Fig. 2.2). The fractal dimension D of the set, defined by $R^D = V$, is thus two.

[7]In case you have not heard, a photon is a quantum of light or other electromagnetic radiation.

[8]Even stock prices (Fig. 2.3) are roughly described as random walks.

Fig. 2.2 Random walk: scale invariance.   Random walks form a jagged, fractal pattern which looks the same when rescaled. Here each succeeding walk is the first quarter of the previous walk, magnified by a factor of two; the shortest random walk is of length 31, the longest of length 32,000 steps. The left side of Fig. 1.1 is the further evolution of this walk to 128,000 steps.

Polymers. Finally, random walks arise as the shapes for *polymers*. Polymers are long molecules (like DNA, RNA, proteins, and many plastics) made up of many small units (called monomers) attached to one another in a long chain. Temperature can introduce fluctuations in the angle between two adjacent monomers; if these fluctuations dominate over the energy,[9] the polymer shape can form a random walk. Here each step does not increase the time, but rather advances the monomer number along the chain by one.

The random walks formed by isolated polymers are *not* the same as those in our first two examples; they are in a different *universality class*. This is because the polymer cannot intersect itself; a walk that would cause two monomers to overlap is not allowed. Polymers undergo *self-avoiding* random walks. In two and three dimensions, it turns out that the effects of these self-intersections is not a small, microscopic detail, but changes the properties of the random walk in an essential way.[10] One can show that these forbidden intersections would often arise on far-separated regions of the polymer, and that they change the dependence of squared radius $\langle s_N^2 \rangle$ on the number of segments N (Exercise 2.10). In particular, the power law $\sqrt{\langle s_N^2 \rangle} \sim N^\nu$ changes from the ordinary random walk value $\nu = 1/2$ to a higher value ($\nu = 3/4$ in two dimensions and $\nu \approx 0.588$ in three dimensions [129]). Power laws are central to the study of scale-invariant systems; ν is our first example of a *universal critical exponent* (Chapter 12).

2.2   The diffusion equation

In the *continuum limit* of long length and time scales, simple behavior emerges from the ensemble of irregular, jagged random walks; their evolution is described by the *diffusion equation*:[11]

$$\frac{\partial \rho}{\partial t} = D \nabla^2 \rho = D \frac{\partial^2 \rho}{\partial x^2}. \tag{2.7}$$

The diffusion equation can describe the evolving density $\rho(x, t)$ of a local cloud of perfume as the molecules random walk through collisions with the air molecules. Alternatively, it can describe the probability density of an individual particle as it random walks through space; if the particles are noninteracting, the probability distribution of one particle describes the density of all particles.

In this section, we derive the diffusion equation by taking a continuum limit of the ensemble of random walks. Consider a general, uncorrelated random walk where at each time step Δt the particle's position x changes by a step ℓ:

$$x(t + \Delta t) = x(t) + \ell(t). \tag{2.8}$$

Let the probability distribution for each step be $\chi(\ell)$.[12] We will assume that χ has mean zero and standard deviation a, so the first few

Fig. 2.3 S&P 500, normalized. Standard and Poor's 500 stock index daily closing price since its inception, corrected for inflation, divided by the average 6.4% return over this time period. Stock prices are often modeled as a biased random walk (Exercise 2.11). Notice that the fluctuations (risk) in individual stock prices will typically be much higher. By averaging over 500 stocks, the random fluctuations in this index are reduced, while the average return remains the same; see [124, 125]. For comparison, a one-dimensional multiplicative random walk is also shown.

[9]Polymers do not always form random walks. Polymeric plastics at low temperature can form crystals; functional proteins and RNA often pack tightly into well-defined shapes. Molten plastics and denatured proteins, though, do form self-avoiding random walks. Double-stranded DNA is rather stiff; the step size for the random walk of DNA in solution is many nucleic acids long (Exercise 2.10).

[10]Self-avoidance is said to be a *relevant* perturbation that changes the universality class. In (unphysical) spatial dimensions higher than four, self-avoidance is irrelevant; hypothetical hyper-polymers in five dimensions would look like regular random walks on long length scales.

[11]In the remainder of this chapter we specialize for simplicity to one dimension. We also change variables from the sum s to position x.

[12]In our two examples the distribution $\chi(\ell)$ is discrete; we can write it using the Dirac δ-function (see note 3 on p. 7). For coin flips, $\chi(\ell) = \frac{1}{2}\delta(\ell + 1) + \frac{1}{2}\delta(\ell - 1)$; for the drunkard, $\chi(\boldsymbol{\ell}) = \delta(|\boldsymbol{\ell}| - L)/(2\pi L)$, evenly spaced around the circle.

moments[13] of χ are

$$\int \chi(z)\, dz = 1,$$
$$\int z\chi(z)\, dz = 0, \tag{2.9}$$
$$\int z^2\chi(z)\, dz = a^2.$$

What is the probability distribution $\rho(x, t + \Delta t)$ at the next time step, given the probability distribution $\rho(x', t)$?

For the particle to go from x' at time t to x at time $t + \Delta t$, the step $\ell(t)$ must be $x - x'$. This happens with probability $\chi(x - x')$ times the probability density $\rho(x', t)$ that it started at x'. Integrating over original positions x', we have

$$\rho(x, t + \Delta t) = \int_{-\infty}^{\infty} \rho(x', t)\chi(x - x')\, dx'$$
$$= \int_{-\infty}^{\infty} \rho(x - z, t)\chi(z)\, dz, \tag{2.10}$$

Fig. 2.4 Continuum limit for random walks. We suppose the typical step sizes a are small compared to the broad ranges on which $\rho(x)$ varies, so we may do a Taylor expansion in gradients of ρ.

where we change variables to $z = x - x'$.[14]

Now, suppose ρ is broad; the step size is very small compared to the scales on which ρ varies (Fig. 2.4). We may then do a Taylor expansion of eqn 2.10 in z:

$$\rho(x, t + \Delta t) \approx \int \left[\rho(x, t) - z\frac{\partial \rho}{\partial x} + \frac{z^2}{2}\frac{\partial^2 \rho}{\partial x^2} \right] \chi(z)\, dz$$

$$= \rho(x, t) \int \chi(z)\, dz - \frac{\partial \rho}{\partial x} \int z\chi(z)\, dz + \frac{1}{2}\frac{\partial^2 \rho}{\partial x^2} \int z^2\chi(z)\, dz$$

$$= \rho(x, t) + \frac{1}{2}\frac{\partial^2 \rho}{\partial x^2}a^2, \tag{2.11}$$

using the moments of χ in eqn 2.9. Now, if we also assume that ρ is slow, so that it changes only slightly during this time step, we can approximate $\rho(x, t + \Delta t) - \rho(x, t) \approx (\partial\rho/\partial t)\,\Delta t$, and we find

$$\frac{\partial \rho}{\partial t} = \frac{a^2}{2\Delta t}\frac{\partial^2 \rho}{\partial x^2}. \tag{2.12}$$

This is the diffusion equation[15] (eqn 2.7), with

$$D = a^2/2\Delta t. \tag{2.13}$$

The diffusion equation applies to all random walks, so long as the probability distribution is broad and slowly varying compared to the size and time of the individual steps.

[15]D must be greater than zero. Random walks and diffusion tend to even out the hills and valleys in the density. Hills have negative second derivatives ($\partial^2\rho/\partial x^2 < 0$) and should flatten ($\partial\rho/\partial t < 0$), valleys have positive second derivatives and fill up.

[16] Something is locally conserved if its integral (zeroth moment) is independent of time, and if the substance only moves continuously from one place to another (no "teleportation" allowed). For example, the probability density $\rho(x)$ of a single particle undergoing a random walk is also conserved; like particle density, probability cannot be created or destroyed, it can only slosh around.

2.3   Currents and external forces

As the particles in our random walks move around, they are never created or destroyed; they are *locally conserved*.[16] If $\rho(x)$ is the density of a conserved quantity, we may write its evolution law (see Fig. 2.5) in terms of the current $J(x)$ passing a given point x:

$$\frac{\partial \rho}{\partial t} = -\frac{\partial J}{\partial x}. \tag{2.14}$$

Here the current J is the rate at which stuff flows to the right through the point x; since the stuff is conserved, the only way the density can change is by flowing from one place to another. The diffusion eqn 2.7 results from current conservation (eqn 2.14) and a current[17] that is proportional to the local gradient in the density:

$$J_{\text{diffusion}} = -D\frac{\partial \rho}{\partial x}, \tag{2.15}$$

as we would expect in general from linear response.[18] Particles diffuse (random walk) on average from regions of high density toward regions of low density.

In many applications one has an average drift term along with a random walk. In some cases (like the total grade in a multiple-choice test, Exercise 2.1) there is naturally a nonzero mean for each step in the random walk. In other cases, there is an external force F that is biasing the steps to one side; the mean net drift is $F\,\Delta t$ times a mobility[19] γ:

$$x(t + \Delta t) = x(t) + F\gamma\Delta t + \ell(t). \tag{2.16}$$

We can derive formulæ for this mobility given a microscopic model. On the one hand, if our air is dense and the diffusing molecule is large, we might treat the air as a viscous fluid of dynamical viscosity η; if we also simply model the molecule as a sphere of radius r, a fluid mechanics calculation tells us that the mobility is $\gamma = 1/(6\pi\eta r)$. On the other hand, if our air is dilute and the diffusing molecule is small, we can model the trajectory as free acceleration between collisions separated by equal times Δt, and we can assume that the collisions completely scramble the sign of the initial velocity v_0.[20] In this case, the net motion due to the external force is half the acceleration F/m times the time squared: $\frac{1}{2}(F/m)(\Delta t)^2 = F\,\Delta t(\Delta t/2m)$ so $\gamma = (\Delta t/2m)$. Using eqn 2.13, we find

$$\gamma = \frac{\Delta t}{2m}\left(D\frac{2\Delta t}{a^2}\right) = \frac{D}{m(a/\Delta t)^2} = \frac{D}{mv_0^2}, \tag{2.17}$$

where $a = v_0\Delta t$ is the size of the unbiased random-walk step.

Starting from eqn 2.16, we can repeat our analysis of the continuum limit (eqns 2.10–2.12) to derive the diffusion equation in an external force:[21]

$$\frac{\partial \rho}{\partial t} = -\gamma F\frac{\partial \rho}{\partial x} + D\frac{\partial^2 \rho}{\partial x^2}, \tag{2.18}$$

which follows from the current

$$J = \gamma F\rho - D\frac{\partial \rho}{\partial x}. \tag{2.19}$$

The sign of the new term can be explained intuitively: if ρ is increasing in space (positive slope $\partial \rho/\partial x$) and the force is dragging the particles

$J(x) \rightarrow \boxed{\rho(x)\,\Delta x} \rightarrow J(x + \Delta x)$

Fig. 2.5 Conserved current. Let $\rho(x, t)$ be the density of some conserved quantity (number of molecules, mass, energy, probability, ...), varying in one spatial dimension x, and $J(x)$ be the net rate at which the quantity is passing to the right through a point x. The amount of stuff in a small region $(x, x+\Delta x)$ is $n = \rho(x)\,\Delta x$. The flow of particles into this region from the left is $J(x)$ and the flow out is $J(x + \Delta x)$, so

$$\frac{\partial n}{\partial t} = J(x) - J(x + \Delta x) \approx \frac{\partial \rho}{\partial t}\Delta x,$$

and we derive the conserved current relation:

$$\frac{\partial \rho}{\partial t} = -\frac{J(x + \Delta x) - J(x)}{\Delta x} = -\frac{\partial J}{\partial x}.$$

[17]The diffusion equation implies a current $J = -D\,\partial\rho/\partial x + C$, but a constant background current C independent of ρ is not physical for random walks.

[18]See note 43 on p. 156.

[19]The ratio of velocity over force is called the mobility.

[20]More realistic models have a distribution of velocities and step sizes. This changes some of the formulas by factors of two. See Exercise 2.22.

[21]Warning: If the force is not constant in space, the evolution also depends on the gradient of the force:

$$\frac{\partial \rho}{\partial t} = -\frac{\partial J}{\partial x} = -\gamma\frac{\partial F(x)\rho(x)}{\partial x} + D\frac{\partial^2 \rho}{\partial x^2}$$
$$= -\gamma\rho\frac{\partial F}{\partial x} - \gamma F\frac{\partial \rho}{\partial x} + D\frac{\partial^2 \rho}{\partial x^2}.$$

Similar problems can arise if the diffusion constant depends on space or density. *When working with a conserved property, write your equations first in terms of the current, to guarantee that it is conserved:* $J = -D(\rho, \mathbf{x})\nabla\rho + \gamma(\mathbf{x})F(\mathbf{x})\rho(\mathbf{x})$. The author has observed himself and several graduate students wasting up to a week at a time when this rule is forgotten.

forward ($F > 0$), then ρ will decrease with time because the high-density regions ahead of x are receding and the low-density regions behind x are moving in.

The diffusion equation describes how systems of random-walking particles approach equilibrium (see Chapter 3). The diffusion equation in the absence of an external force describes the evolution of perfume density in a room. A time-independent equilibrium state ρ^* obeying the diffusion eqn 2.7 must have $\partial^2 \rho^*/\partial x^2 = 0$, so $\rho^*(x) = \rho_0 + Bx$. If the perfume cannot penetrate the walls, $\partial \rho^*/\partial x \propto J = 0$ at the boundaries, so $B = 0$. Thus, as one might expect, the perfume evolves to a rather featureless equilibrium state $\rho^*(x) = \rho_0$, evenly distributed throughout the room.

In the presence of a constant external force (like gravitation) the equilibrium state is more interesting. Let x be the height above the ground, and $F = -mg$ be the force due to gravity. By eqn 2.18, the equilibrium state ρ^* satisfies

$$0 = \frac{\partial \rho^*}{\partial t} = \gamma mg \frac{\partial \rho^*}{\partial x} + D \frac{\partial^2 \rho^*}{\partial x^2}, \tag{2.20}$$

[22] Nonzero B would correspond to a constant-density rain of perfume.

which has general solution $\rho^*(x) = A \exp[-(\gamma/D)mgx] + B$. We assume that the density of perfume B in outer space is zero,[22] so the density of perfume decreases exponentially with height:

$$\rho^*(x) = A \exp\left(-\frac{\gamma}{D}mgx\right). \tag{2.21}$$

[23] In Chapter 6 we shall derive the Boltzmann distribution, implying that the probability of having energy $mgh = E$ in an equilibrium system is proportional to $\exp(-E/k_B T)$, where T is the temperature and k_B is Boltzmann's constant. This has just the same form as our solution (eqn 2.21), if

$$D/\gamma = k_B T. \tag{2.22}$$

This is called the *Einstein relation*. The constants D and γ in the (nonequilibrium) diffusion equation are related to one another, because the density must evolve toward the equilibrium distribution dictated by statistical mechanics. Our rough derivation (eqn 2.17) also suggested that $D/\gamma = mv^2$, which with eqn 2.22 suggests that $k_B T$ must equal twice the mean kinetic energy along x; this is also true, and is called the *equipartition theorem* (Section 3.2.2).

The perfume molecules are pulled downward by the gravitational force, and remain aloft only because of the random walk. If we generalize from perfume to oxygen molecules (and ignore temperature gradients and weather) this gives the basic explanation for why it becomes harder to breathe as one climbs mountains.[23]

2.4   Solving the diffusion equation

We take a brief mathematical interlude to review two important methods for solving the diffusion equation: Fourier transforms and Green's functions. Both rely upon the fact that the diffusion equation is linear; if a family of solutions $\rho_n(x, t)$ are known, then any linear combination of these solutions $\sum_n a_n \rho_n(x, t)$ is also a solution. If we can then expand the initial density $\rho(x, 0) = \sum_n a_n \rho_n(x, 0)$, we have formally found the solution.

[24] One should note that much of quantum field theory and many-body quantum mechanics is framed in terms of things also called Green's functions. These are distant, fancier cousins of the simple methods used in linear differential equations (see Exercise 10.9).

Fourier methods are wonderfully effective computationally, because of fast Fourier transform (FFT) algorithms for shifting from the real-space density to the solution space. Green's function methods are more important for analytical calculations and as a source of approximate solutions.[24]

2.4.1 Fourier

The Fourier transform method decomposes ρ into a family of *plane wave* solutions $\widetilde{\rho}_k(t)e^{ikx}$.

The diffusion equation is homogeneous in space; our system is *translationally invariant*. That is, if we have a solution $\rho(x, t)$, another equally valid solution is given by $\rho(x - \Delta, t)$, which describes the evolution of an initial condition translated by Δ in the positive x direction.[25] Under very general circumstances, a linear differential equation describing a translation-invariant system will have solutions given by plane waves $\rho(x, t) = \widetilde{\rho}_k(t)e^{ikx}$.

We argue this important truth in detail in Section A.4. Here we just try it. Plugging a plane wave[26] into the diffusion eqn 2.7, we find

$$\frac{\partial \rho}{\partial t} = \frac{d\widetilde{\rho}_k}{dt}e^{ikx} = D\frac{\partial^2 \rho}{\partial x^2} = -Dk^2\widetilde{\rho}_k e^{ikx}, \tag{2.23}$$

$$\frac{d\widetilde{\rho}_k}{dt} = -Dk^2\widetilde{\rho}_k, \tag{2.24}$$

$$\widetilde{\rho}_k(t) = \widetilde{\rho}_k(0)e^{-Dk^2 t}. \tag{2.25}$$

Now, these plane wave solutions by themselves are unphysical; we must combine them to get a sensible density. First, they are complex; we must add plane waves at k and $-k$ to form cosine waves, or subtract them and divide by 2i to get sine waves. Cosines and sines are also not by themselves densities (because they go negative), but they in turn can be added to one another (for example, added to a $k = 0$ constant background ρ_0) to make for sensible densities. Indeed, we can superimpose all different wavevectors to get the general solution

$$\rho(x, t) = \frac{1}{2\pi}\int_{-\infty}^{\infty} \widetilde{\rho}_k(0)e^{ikx}e^{-Dk^2 t}\, dk. \tag{2.26}$$

Here the coefficients $\rho_k(0)$ we use are just the Fourier transform of the initial density profile (eqn A.9):

$$\widetilde{\rho}_k(0) = \int_{-\infty}^{\infty} \rho(x, 0)e^{-ikx}\, dx, \tag{2.27}$$

and we recognize eqn 2.26 as the inverse Fourier transform (eqn A.10) of the solution time evolved in Fourier space (eqn 2.25). Thus, by writing ρ as a superposition of plane waves, we find a simple law: the short-wavelength parts of ρ are "squelched" as time t evolves, with wavevector k being suppressed by a factor $e^{-Dk^2 t}$.

2.4.2 Green

The Green's function method decomposes ρ into a family of solutions $G(x - y, t)$ where G describes the evolution of an initial state concentrated at one point, here representing the diffusing particles all starting at a particular point y.

[25]Make sure you know that $g(x) = f(x - \Delta)$ shifts the function in the positive direction; for example, the new function $g(\Delta)$ is at Δ what the old one was at the origin, $g(\Delta) = f(0)$.

[26]Many readers will recognize this method of calculation from wave equations or Schrödinger's equation. Indeed, Schrödinger's equation in free space is the diffusion equation with an imaginary diffusion constant.

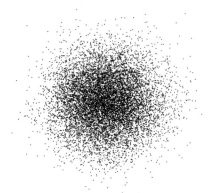

Fig. 2.6 Many random walks. 10,000 endpoints of random walks, each 1,000 steps long. Notice that after 1,000 steps, the distribution of endpoints looks quite Gaussian. Indeed after about five steps the distribution is extraordinarily close to Gaussian, except far in the tails.

[27]Physicists call $\exp(-x^2)$ a Gaussian. In statistics, the corresponding probability distribution $(1/\sqrt{2\pi})\exp(-x^2/2)$ is called a *normal distribution*. It is useful to remember that the Fourier transform of a normalized Gaussian $(1/\sqrt{2\pi}\sigma)\exp(-x^2/2\sigma^2)$ is another Gaussian, $\exp(-\sigma^2 k^2/2)$ of standard deviation $1/\sigma$ and with no prefactor (eqn 2.32).

[28]Take the exponent $ikx - Dk^2 t$ in eqn 2.30 and complete the square to $-Dt(k - ix/(2Dt))^2 - x^2/(4Dt)$, and then change variables to $\kappa = k - ix/(2Dt)$:

$$G(x,t) = \frac{1}{2\pi} e^{-\frac{x^2}{4Dt}}$$

$$\int_{-\infty-ix/(2Dt)}^{\infty-ix/(2Dt)} e^{-Dt\kappa^2}\, d\kappa. \qquad (2.31)$$

If we could shift the limits of integration downward to the real axis, the integral would give $\sqrt{\pi/Dt}$, yielding a derivation of eqn 2.32. This last step (shifting the limits of integration), is not obvious; we must rely on Cauchy's theorem, which allows one to deform the integration contour in the complex plane (Fig. 10.11 in Section 10.9). This is done backward (real to Fourier space) in note 21, Exercise A.4.

Let us first consider the case where all particles start at the origin. Suppose we have one unit of perfume, released at the origin at time $t = 0$. What is the initial condition $\rho(x, t = 0)$? It is zero except at $x = 0$, but the integral $\int \rho(x,0)\,dx = 1$, so $\rho(0,0)$ must be really, really infinite. This is the Dirac delta-function $\delta(x)$ (see note 3 on p. 7) which mathematically (when integrated) is a linear operator on functions returning the value of the function at zero:

$$\int f(y)\delta(y)\, dy = f(0). \qquad (2.28)$$

Let us define the Green's function $G(x,t)$ to be the time evolution of the density $G(x,0) = \delta(x)$ with all the perfume at the origin. Naturally, $G(x,t)$ obeys the diffusion equation $\partial G/\partial t = D\partial^2 G/\partial x^2$. We can use the Fourier transform methods of the previous section to solve for $G(x,t)$. The Fourier transform at $t = 0$ is

$$\widetilde{G}_k(0) = \int G(x,0)e^{-ikx}\, dx = \int \delta(x)e^{-ikx}\, dx = 1 \qquad (2.29)$$

(independent of k). Hence the time-evolved Fourier transform is $\widetilde{G}_k(t) = e^{-Dk^2 t}$, and the time evolution in real space is

$$G(x,t) = \frac{1}{2\pi}\int e^{ikx}\widetilde{G}_k(0)e^{-Dk^2 t}\, dk = \frac{1}{2\pi}\int e^{ikx}e^{-Dk^2 t}\, dk. \qquad (2.30)$$

This last integral is the inverse Fourier transform of a Gaussian,[27] which can be performed[28] giving another Gaussian

$$G(x,t) = \frac{1}{\sqrt{4\pi Dt}}\, e^{-x^2/4Dt}. \qquad (2.32)$$

This is the Green's function for the diffusion equation. The Green's function directly tells us the distribution of the endpoints of random walks centered at the origin (Fig. 2.6).

- The Green's function gives us the whole probability distribution of distances. For an N-step random walk of step size a, we saw in Section 2.1 that $\sqrt{\langle x^2 \rangle} = \sqrt{N}a$; does this also follow from our Green's function? At time t, the Green's function (eqn 2.32) is a Gaussian with standard deviation $\sigma(t) = \sqrt{2Dt}$; substituting in our diffusion constant $D = a^2/2\Delta t$ (eqn 2.13), we find an RMS distance of $\sigma(t) = a\sqrt{t/\Delta t} = a\sqrt{N}$, where $N = t/\Delta t$ is the number of steps taken in the random walk; our two methods do agree.

- Finally, since the diffusion equation has translational symmetry, we can solve for the evolution of random walks centered at any point y; the time evolution of an initial condition $\delta(x - y)$ is $G(x - y, t)$. Since we can write any initial condition $\rho(x, 0)$ as a superposition of δ-functions:

$$\rho(x,0) = \int \rho(y,0)\delta(x - y)\, dy = \int \rho(y,0)G(x - y, 0)\, dy, \qquad (2.33)$$

we can write a general solution $\rho(x, t)$ to the diffusion equation:

$$\rho(x, t) = \int \rho(y, 0) G(x - y, t) \, dy$$
$$= \int \rho(y, 0) \exp(-(x-y)^2/4Dt)/\sqrt{4\pi Dt} \, dy. \qquad (2.34)$$

This equation states that the current value of the density is given by the original values of the density in the neighborhood, smeared sideways (convolved) with the function G. Thus by writing ρ as a superposition of point sources, we find that the diffusion equation smears out all the sharp features in the initial condition. The distribution after time t is the initial distribution averaged over a range given by the typical random walk distance $\sqrt{2Dt}$.

Equation 2.32 is the *central limit theorem*: the sum of many independent random variables has a probability distribution that converges to a Gaussian.[29]

[29] Perhaps this is why statisticians call the Gaussian a normal distribution; under normal circumstances a sum or average of many measurements will have fluctuations described by a Gaussian.

Exercises

Random walks in grade space, *Photon diffusion in the Sun*, and *Molecular motors* describe random walks in diverse contexts. *Perfume walk* explores the atomic trajectories in molecular dynamics. *Generating random walks* numerically explores emergent symmetries and the central limit theorem for random walks. *Continuous time walks: Ballistic to diffusive* explores random walks where the steps happen with a distribution of step sizes and time intervals. *Random walks and generating functions* introduces a common calculational technique. *Polymers and random walks* explores self-avoiding random walks; in two dimensions, we find that the constraint that the walk must avoid itself gives new critical exponents and a new universality class (see also Chapter 12).

Five exercises, *Drifting random walk*, *Diffusion of non-conserved particles*, *Density dependent diffusion*, *Local conservation*, and *Absorbing boundary conditions*, briefly review and extend our theory of random walks and diffusion.

Fourier and Green and *Periodic diffusion* illustrate the qualitative behavior of the Fourier and Green's function approaches to solving the diffusion equation. *Thermal diffusion* and *Frying pan* derive the diffusion equation for thermal conductivity, and apply it to culinary physics.

Stocks, volatility, and diversification quantifies the fluctuations in the stock-market, and explains why diversification lowers your risk without changing your mean asset

growth. *Computational finance: pricing derivatives* focuses on a single step of a random walk in stock prices, to estimate the value of stock option, introducing the Black–Scholes model. *Building a percolation network* introduces another ensemble (percolating networks) that, like random walks, exhibits self-similarity and power-law scaling; we will study the (much more subtle) continuum limit for percolation in Chapter 12 and Exercise 12.12. *Lévy flight* studies random walks where the steps have long tails in their lengths. *Flocking* explores the emergent collective motions of active agents (like birds and wildebeests). Finally, *Run and tumble* studies the random-walk strategy taken by bacteria in their search for food.

(2.1) **Random walks in grade space.** ⓟ
Many students complain about multiple-choice exams, saying that it is easy to get a bad grade just by being unlucky in what questions get asked. While easy and unambiguous to grade, are they good measures of knowledge and skill? A course is graded using multiple choice exams, with ten points for each problem. A particular student has a probability 0.7 of getting each question correct.
(a) *Generalize the coin-flip random walk discussion (near eqn 2.3), to calculate the student's mean score* $\langle s_N \rangle$*, the mean square*

score $\langle s_N^2 \rangle$, and the standard deviation $\sigma_N = \sqrt{\langle (s_N - \langle s_N \rangle)^2 \rangle}$ for a test with N questions. (Note that $\langle s_{N-1}\ell_n \rangle \neq 0$ in this case.)

Measuring instruments in physics are often characterized by the signal-to-noise ratio. One can view the random choice of questions as a source of noise, and your calculation in part (a) as an estimate of that noise.

(b) *An introductory engineering physics exam with ten ten-point multiple choice questions has a class mean of 70 and a standard deviation of 15. How much of the standard deviation is attributable to random chance (noise), as you calculate in part (a)? If the remainder of the variation between people is considered the signal, is it a good test, with a high signal to noise ratio?*

(2.2) **Photon diffusion in the Sun.** (Astrophysics) @

If fusion in the Sun turned off today, how long would it take for us to notice? This question became urgent some time back when the search for solar neutrinos failed.[30] Neutrinos, created in the same fusion reaction that creates heat in the Solar core, pass through the Sun at near the speed of light without scattering—giving us a current status report. The rest of the energy takes longer to get out.

Most of the fusion energy generated by the Sun is produced near its center. The Sun is 7×10^5 km in radius. Convection probably dominates heat transport in approximately the outer third of the Sun, but it is believed that energy is transported through the inner portions (say from $R = 0$ to a radius $R = 5 \times 10^8$ m) through a random walk of X-ray photons. (A photon is a quantized package of energy; you may view it as a particle which always moves at the speed of light c. Ignore for this exercise the index of refraction of the Sun.) There are a range of estimates for the mean free path for a photon in the Sun. For our purposes, assume photons travel at the speed of light, but bounce in random directions (without pausing) with a step size of $\ell = 0.1$ cm $= 10^{-3}$ m.

About how many random steps N will the photon take of length ℓ to get to the radius R where convection becomes important? About how many years Δt will it take for the photon to get there? Related formulæ: $c = 3 \times 10^8$ m/s; $\langle x^2 \rangle \approx 2Dt$; $\langle s_n^2 \rangle = n\sigma^2 = n\langle s_1^2 \rangle$. There are

$31{,}556{,}925.9747 \approx \pi \times 10^7 \approx 3 \times 10^7$ s in a year.

(2.3) **Molecular motors and random walks.**[31] (Biology) ②

Fig. 2.7 Motor protein. As it carries some cargo along the way (or builds an RNA or protein, ...) the molecular motor moves against an external force f_{ext} and consumes ATP molecules, which are hydrolyzed to ADP and phosphate (P).

Inside your cells, there are several different molecular motors, which move and pull and copy (Fig. 2.7). There are molecular motors which contract your muscles, there are motors which copy (*transcribe*) your DNA into RNA and copy (*translate*) your RNA into protein, there are motors which transport biomolecules around in the cell. All of these motors share some common features: (1) they move along some linear track (microtubule, DNA, ...), hopping forward in discrete jumps between low-energy positions; (2) they consume energy (burning ATP or NTP) as they move, generating an effective force pushing them forward; and (3) their mechanical properties can be studied by seeing how their motion changes as the external force on them is changed. For transcription of DNA into RNA, the motor moves on average one base pair (A, T, G, or C) per step; Δx is about 0.34 nm. The motor must cross an asymmetric energy barrier as it attaches another nucleotide to the RNA (Fig. 2.8). Wang and co-authors (Fig. 2.9) showed that the motor stalls at an external force of about 27 pN (pico-Newton).

(a) *At that force, what is the energy difference between neighboring wells due to the external force from the bead? Let us assume that this stall force is what is needed to balance the natural force downhill that the motor develops to propel the transcription process. What does this imply about the ratio of the forward rate to the backward rate, in the absence of the external*

---

[30]The missing electron neutrinos, as it happened, "oscillated" into other types of neutrinos.
[31]This exercise was developed with the assistance of Michelle Wang.

force from the laser tweezers, at a temperature of $300\,\mathrm{K}$? (The population in a well of free energy G is proportional to $\exp(-G/k_BT)$, where $k_B = 1.381 \times 10^{-23}\,\mathrm{J/K}$; we shall derive this in Chapter 6. Hints: If the population was in thermal equilibrium the net flux would be equal going forward and backward; the net flux out of a well is the population in that well times the rate; a given motor does not know whether it is part of an equilibrium ensemble.)

Fig. 2.8 Effective potential for moving along DNA. The energy (or rather the Gibbs free energy) for the molecular motor as a function of distance along the DNA. The motor is in a low-energy state just after it transcribes one nucleotide into RNA. The energy barrier V needs to be crossed in order to transcribe the next nucleotide. The energy asymmetry δ is a sum of contributions from the bonding of the RNA nucleotide, the burning of ATP, and the detachment of the apparatus at the completed end. The experiment changes this asymmetry by adding an external force tilting the potential to the left, retarding the transcription.

The natural force downhill is coming from the chemical reactions which accompany the motor moving one base pair; the motor burns up an NTP molecule into a PP_i molecule, and attaches a nucleotide onto the RNA. The net energy from this reaction depends on details, but varies between about 2 and 5 times 10^{-20} J. This is actually a Gibbs free energy difference, but for this exercise treat it as just an energy difference.

(b) *The motor is not perfectly efficient; not all the chemical energy is available as motor force. From your answer to part (a), give the efficiency of the motor as the ratio of force-times-distance produced to energy consumed, for the range of consumed energies given.*

Many of the models for these motors are based on Feynman's *Ratchet and pawl* discussion [61, I.46], where he (presciently) speculates about how gears and ratchets would work on a molecular level.

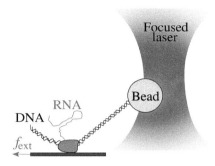

Fig. 2.9 Laser tweezer experiment. The laser beam is focused at a point (the *laser trap*); the polystyrene bead is pulled (from dielectric effects) into the intense part of the light beam. The *track* is a DNA molecule attached to the bead, the motor is an RNA polymerase molecule, and the force is applied by a glass cover slip to which the motor is attached. As the motor copies DNA onto RNA, it pulls the DNA track toward itself, dragging the bead out of the trap, generating a force resisting the motion.

(2.4) **Perfume walk.**[32] (Computation) ②
The trajectory of a perfume molecule in still air, or more generally any molecule in a dilute gas, is a chaotic path of nearly straight segments followed by collisions—a random walk. Download the molecular dynamics software.

Run a simulation of an interacting dilute gas, setting the average velocity of the atoms to zero.[33] Watch the motion of a single perfume atom. Notice that as it passes the edge of the container, it reappears at the opposite face; this simulation uses *periodic boundary conditions*.[34] Your software should have options to

---

[32]This exercise and the associated software were developed in collaboration with Christopher Myers. Computational hints can be found on the book website [181].

[33]The atoms interact via a Lennard–Jones pair potential, which is a good approximation for the forces between noble gas molecules like argon.

[34]Periodic boundary conditions are an artificial method which allows a small simulation to mimic infinite space, by mathematically identifying the opposite faces of a square region; $(x, y, z) \equiv (x \pm L, y, z) \equiv (x, y \pm L, z) \equiv (x, y, z \pm L)$.

plot and analyze the trajectory $\mathbf{r}_u = (x_u, y_u, z_u)$ of a given atom unfolded into a continuous path which ignores the periodic boundary conditions.

(a) *Does the trajectory of the perfume atom appear qualitatively like a random walk? Plot $x_u(t)$ versus t, and $x_u(t)$ versus $y_u(t)$.* The time it takes the atom to completely change direction (lose memory of its original velocity) is the collision time, and the distance it takes is the collision length. *Crudely estimate these.*

(b) *Plot $\mathbf{r}_u^2(t)$ versus t, for several individual particles (making sure the average velocity is zero). Do they individually grow with time in a regular fashion? Plot $\langle \mathbf{r}_u^2 \rangle$ versus t, averaged over all particles in your simulation. Does it grow linearly with time? Estimate the diffusion constant D.*

(2.5) **Generating random walks.**[35]   (Computation) ③

One can efficiently generate and analyze random walks on the computer.

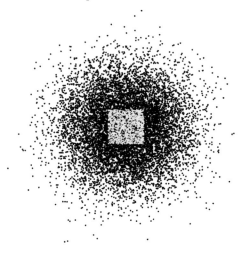

Fig. 2.10 Emergent rotational symmetry. Endpoints of many random walks, with one step (central square of bright dots) and ten steps (surrounding pattern). Even though the individual steps in a random walk break rotational symmetry (the steps are longer along the diagonals), multistep random walks are spherically symmetric. The rotational symmetry emerges as the number of steps grows.

(a) *Write a routine to generate an N-step random walk in d dimensions, with each step uniformly distributed in the range $(-1/2, 1/2)$ in each dimension. (Generate the steps first as an $N \times d$ array, then do a cumulative sum.) Plot x_t versus t for a few 10,000-step random walks. Plot x versus y for a few two-dimensional random walks, with $N = 10$, 1,000, and 100,000. (Try to keep the aspect ratio of the XY plot equal to one.) Does multiplying the number of steps by one hundred roughly increase the net distance by ten?*

Each random walk is different and unpredictable, but the *ensemble* of random walks has elegant, predictable properties.

(b) *Write a routine to calculate the endpoints of W random walks with N steps each in d dimensions. Do a scatter plot of the endpoints of 10,000 random walks with $N = 1$ and 10, superimposed on the same plot. Notice that the longer random walks are distributed in a circularly symmetric pattern, even though the single step random walk $N = 1$ has a square probability distribution (Fig. 2.10).*

This is an *emergent symmetry*; even though the walker steps longer distances along the diagonals of a square, a random walk several steps long has nearly perfect rotational symmetry.[36]

The most useful property of random walks is the *central limit theorem*. The endpoints of an ensemble of N step one-dimensional random walks with RMS step-size a has a Gaussian or normal probability distribution as $N \to \infty$,

$$\rho(x) = \frac{1}{\sqrt{2\pi}\sigma} \exp(-x^2/2\sigma^2), \qquad (2.35)$$

with $\sigma = \sqrt{N}a$.

(c) *Calculate the RMS step-size a for one-dimensional steps uniformly distributed in $(-1/2, 1/2)$. Write a routine that plots a histogram of the endpoints of W one-dimensional random walks with N steps and 50 bins, along with the prediction of eqn 2.35, for x in $(-3\sigma, 3\sigma)$. Do a histogram with $W = 10,000$ and $N = 1, 2, 3,$ and 5. How quickly does the Gaussian distribution become a good approximation to the random walk?*

---

[35]This exercise and the associated software were developed in collaboration with Christopher Myers. Hints for the computations can be found at the book website [181].

[36]The square asymmetry is an *irrelevant perturbation* on long length and time scales (Chapter 12). Had we kept terms up to fourth order in gradients in the diffusion equation $\partial\rho/\partial t = D\nabla^2\rho + E\nabla^2\left(\nabla^2\rho\right) + F\left(\partial^4\rho/\partial x^4 + \partial^4\rho/\partial y^4\right)$, then F is square symmetric but not isotropic. It will have a typical size $\Delta t/a^4$, so is tiny on scales large compared to a.

(2.6) **Fourier and Green.** @

An initial density profile $\rho(x, t = 0)$ is perturbed slightly away from a uniform density ρ_0, as shown in Fig. 2.11. The density obeys the diffusion equation $\partial\rho/\partial t = D\partial^2\rho/\partial x^2$, where $D = 0.001\,\mathrm{m}^2/\mathrm{s}$. The lump centered at $x_0 = 5$ is a Gaussian $\exp(-(x - x_0)^2/2)/\sqrt{2\pi}$, and the wiggle centered at $x = 15$ is a smooth envelope function multiplying $\cos(10x)$.

As a first step in guessing how the pictured density will evolve, let us consider just a cosine wave.

(a) Fourier. *If the initial wave were $\rho_{\cos}(x, 0) = \cos(10x)$, what would it be at $t = 10\,s$?* Related formulæ: $\tilde{\rho}(k, t) = \tilde{\rho}(k, t')\tilde{G}(k, t - t')$; $\tilde{G}(k, t) = \exp(-Dk^2 t)$.

As a second step, let us check how long an initial δ-function would take to spread out as far as the Gaussian on the left.

(b) Green. *If the wave at some earlier time $-t_0$ were a δ-function at $x = 0$, $\rho(x, -t_0) = \delta(x)$, what choice of the time elapsed t_0 would yield a Gaussian $\rho(x, 0) = \exp(-x^2/2)/\sqrt{2\pi}$ for the given diffusion constant $D = 0.001\,\mathrm{m}^2/\mathrm{s}$?* Related formulæ: $\rho(x, t) = \int \rho(y, t')G(y - x, t - t')\,\mathrm{d}y$; $G(x, t) = (1/\sqrt{4\pi Dt})\exp(-x^2/(4Dt))$.

Fig. 2.11 Initial profile of density deviation from average.

(c) Pictures. *Now consider time evolution for the next ten seconds. The initial density profile $\rho(x, t = 0)$ is as shown in Fig. 2.11. Which of the choices (A)–(E) represents the density at $t = 10\,s$?* (Hint: Compare $t = 10\,\mathrm{s}$ to the time t_0 from part (b).) Related formulæ: $\langle x^2\rangle \sim 2Dt$.

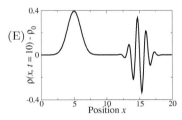

(d) *How many of these solutions can you rule out by inspection (without doing a calculation)? In particular, diffusion:*

- *Conserves density: ρ is neither created or destroyed*
- *Does not mix wavelengths: the different wavevectors $\tilde{\rho}(k)$ evolve separately*
- *Squelches short wavelengths faster than long wavelengths*
- *Is symmetric under inversion $x \leftrightarrow -x$*

Which of these basic properties of diffusion rule out which of the answers?

(2.7) **Periodic diffusion.** ②
Consider a one-dimensional diffusion equation $\partial\rho/\partial t = D\partial^2\rho/\partial x^2$, with initial condition periodic in space with period L, consisting of a δ-function at every $x_n = nL$: $\rho(x,0) = \sum_{n=-\infty}^{\infty}\delta(x-nL)$.
(a) *Using the Green's function method, give an approximate expression for the the density, valid at short times and for $-L/2 < x < L/2$, involving only one term (not an infinite sum).* (Hint: How many of the Gaussians are important in this region at early times?)
(b) *Using a Fourier series,*[37] *give an approximate expression for the density, valid at long times, involving only two terms (not an infinite sum).* (Hint: How many of the wavelengths are important at late times?)
(c) *Give a characteristic time τ in terms of L and D, such that your answer in (a) is valid for $t \ll \tau$ and your answer in (b) is valid for $t \gg \tau$.*

(2.8) **Thermal diffusion.** ②
The rate of energy flow in a material with thermal conductivity k_t and a temperature field $T(x,y,z,t) = T(\mathbf{r},t)$ is $\mathbf{J} = -k_t\nabla T$.[38] Energy is locally conserved, so the energy density E satisfies $\partial E/\partial t = -\nabla\cdot\mathbf{J}$.
(a) *If the material has constant specific heat c_p and density ρ, so $E = c_p\rho T$, show that the temperature T satisfies the diffusion equation $\partial T/\partial t = k_t/(c_p\rho)\nabla^2 T$.*
(b) *By putting our material in a cavity with microwave standing waves, we heat it with a periodic modulation $T = \sin(kx)$ at $t = 0$, at which time the microwaves are turned off. Show that the amplitude of the temperature modulation decays exponentially in time. How does the amplitude decay rate depend on wavelength $\lambda = 2\pi/k$?*

(2.9) **Frying pan.** ②
An iron frying pan is quickly heated on a stove top to 400 degrees Celsius. Roughly how long it will be before the handle is too hot to touch (within, say, a factor of two)? (Adapted from [163, p. 40].)
Do this three ways.

(a) *Guess the answer from your own experience. If you have always used aluminum pans, consult a friend or parent.*
(b) *Get a rough answer by a dimensional argument. You need to transport heat $c_p\rho V\Delta T$ across an area $A = V/\Delta x$. How much heat will flow across that area per unit time, if the temperature gradient is roughly assumed to be $\Delta T/\Delta x$? How long δt will it take to transport the amount needed to heat up the whole handle?*
(c) *Model the problem as the time needed for a pulse of heat at $x = 0$ on an infinite rod to spread out a RMS distance $\sigma(t)$ equal to the length of the handle, and use the Green's function for the heat diffusion equation (Exercise 2.8).*
Note: For iron, the specific heat $c_p = 450\,\mathrm{J/(kg\ C)}$, the density $\rho = 7{,}900\,\mathrm{kg/m^3}$, and the thermal conductivity $k_t = 80\,\mathrm{W/(m\ C)}$.

(2.10) **Polymers and random walks.**[39] (Computation, Condensed matter) ③
Polymers are long molecules, typically made of identical small molecules called *monomers* that are bonded together in a long, one-dimensional chain. When dissolved in a solvent, the polymer chain configuration often forms a good approximation to a random walk. Typically, neighboring monomers will align at relatively small angles; several monomers are needed to lose memory of the original angle. Instead of modeling all these small angles, we can produce an equivalent problem focusing all the bending in a few hinges; we approximate the polymer by an uncorrelated random walk of straight segments several monomers in length. The equivalent segment size is called the *persistence length*.[40]
(a) *If the persistence length to bending of DNA is $50\,\mathrm{nm}$, with $3.4\,\text{Å}$ per nucleotide base pair, what will the RMS distance $\sqrt{\langle R^2\rangle}$ be between the ends of a gene in solution with 100,000 base pairs, if the DNA is accurately represented as a random walk?*
Polymers are not accurately represented as pure random walks, however. Random walks, particularly in low dimensions, often intersect themselves. Polymers are best represented as *self-avoiding* random walks: a polymer samples only those configurations that do not cross them-

---

[37]You can use a Fourier transform, but you will find $\tilde{\rho}(k,0)$ is zero except at the values $k = 2\pi m/L$, where it is a δ-function.
[38]We could have derived this law of thermal conductivity from random walks of phonons, but we have not done so.
[39]A link to the software can be found at the book website [181].
[40]Some seem to define the persistence length with a different constant factor.

selves.

Let us investigate whether self-avoidance will change the basic nature of the polymer configuration in two dimensions. In particular, does the end-to-end typical distance continue to scale with the square root of the length L of the polymer, $R \sim \sqrt{L}$?

(b) *Two-dimensional self-avoiding random walk. Give a convincing, short argument explaining whether or not a typical, non self-avoiding random walk in two dimensions will come back after large numbers of monomers and cross itself.* (Hint: How big a radius does it extend to? How many times does it traverse this radius?)

Find or write a simulation of self-avoiding random walks, that allows one to generate multiple walks of a given length (see [181]). The simulation should model a two-dimensional random walk as a connected line between neighboring lattice points on the square lattice of integers, starting random walks at the origin and discarding them when they hit the same lattice point twice. Average the squared length of the surviving walks as a function of the number of steps.

(c) *Measure for a reasonable length of time, print out the current state, and enclose it. Did the simulation give $R \sim \sqrt{L}$? If not, what is the estimate that your simulation gives for the exponent ν relating R to L? How does it compare with the two-dimensional theoretical exponent $\nu = \frac{3}{4}$?*

(2.11) **Stocks, volatility, and diversification.**[41] (Finance, Computation) ②

Stock prices are fairly good approximations to random walks. The Standard and Poor's 500 index is a weighted average of the prices of five hundred large companies in the United States stock-market.

Download SandPConstantDollars.dat and the hints files. Each line in the data file represents a weekday (no prices are listed on Saturday or Sunday). The first column is time t (in days, since mid-October 1982), and the second column is the Standard and Poor's index $SP(t)$ for that day, corrected for inflation (using the consumer price index for that month).

Are the random fluctuations in the stock-market due to external events?

(a) *Plot the price index versus time. Notice the large peak near year 2000. On September 11, 2001 the World Trade Center was attacked (day number 6,903 in the list). Does it seem that the drop in the stock-market after 2000 is due mostly to this external event?*

Sometimes large fluctuations are due to external events; the fluctuations in ecological populations and species are also quite random, but the dinosaur extinction was surely caused by a meteor.

What do the steps look like in the random walk of Standard and Poor's index? This depends on how we define a step; do we ask how much it has changed after a year, a month, a week, or a day? A technical question arises: do we measure time in days, or in trading days? We shall follow the finance community, and consider only trading days. So, we will define the lag variable ℓ to be one trading day for a daily percentage change (even if there is a weekend or holiday in between), five for a weekly percentage change, and 252 for a yearly percentage change (the number of trading days in a typical year).

(b) *Write a function P_ℓ that finds all pairs of time points from our data file separated by a time interval $\Delta t = \ell$ and returns a list of per cent changes*

$$P_\ell(t) = 100 \frac{SP(t + \ell) - SP(t)}{SP(t)}$$

over that time interval. Plot a histogram of the daily changes, the weekly changes, and the yearly changes. Which of the three represents a reasonable time for you to stay invested in the Standard and Poor's index (during which you have mean percentage growth larger than a tiny fraction of the fluctuations)? Also, why do you think the yearly changes look so much more complicated than the other distributions? (Hint for the latter question: How many years are there in the data sample? Are the steps $SP(n) - SP(n - \ell)$ independent from $SP(m) - SP(m - \ell)$ for $n - m < \ell$? The fluctuations are determined not by the total number of steps, but by the effective number of independent steps in the random walk.)

The distributions you found in part (b) for the shorter lags should have looked quite close to Gaussian—corresponding nicely to our Green's function analysis of random walks, or more generally to the central limit theorem. Those in mathematical finance, though, are interested in the deviations from the expected behavior. They

---

[41]Data files and hints for the computations can be found at the book website [181].

have noticed that the tails of the distribution deviate from the predicted Gaussian.

(c) *Show that the logarithm of a Gaussian is an inverted parabola. Plot the logarithm of the histogram of the weekly percentage changes from part (b). Are there more large percentage changes than expected from a Gaussian distribution (fat tails) or fewer?* (Hint: Far in the tails the number of measurements starts becoming sparse, fluctuating between zero and one. Focus on the region somewhat closer in to the center, where you have reasonable statistics.)

Some stocks, stock funds, or indices are more risky than others. This is not to say that one on average loses money on risky investments; indeed, they usually on average pay a better return than conservative investments. Risky stocks have a more variable return; they sometimes grow faster than anticipated but sometimes decline steeply. Risky stocks have a high standard deviation in their percentage return. In finance, the standard deviation of the percentage return is called the *volatility*

$$v_\ell = \sqrt{\left\langle \left(P_\ell(t) - \bar{P}_\ell\right)^2 \right\rangle}.$$

(d) *Calculate the daily volatility, the weekly volatility, and the monthly volatility of the inflation-corrected Standard and Poor's 500 data. Plot the volatility as a function of lag, and the volatility squared as a function of lag, for lags from zero to 100 days. Does it behave as a random walk should?*

The volatility of a stock is often calculated from the price fluctuations within a single day, but it is then *annualized* to estimate the fluctuations after a year, by multiplying by the square root of 252.

The individual stocks in the Standard and Poor's 500 index will mostly have significantly higher volatility than the index as a whole.

(e) *Suppose these five hundred stocks had mean annual percentage returns m_i and each had mean volatility σ_i. Suppose they were equally weighted in the index, and their fluctuations were uncorrelated. What would the return and volatility for the index be? Without inside information[42] or insight as to which stocks will have higher mean returns, is the expected return for the index bigger than the average return for buying individual stocks? Which strategy has lower volatility?*

Investment advisers emphasize the importance of *diversification*. The fluctuations of different stocks are not independent, especially if they are in the same industry; one should have investments spread out between different sectors of the economy, and between stocks and bonds and other types of investments, in order to avoid risk and volatility.

(2.12) **Computational finance: pricing derivatives.**[43] (Finance) ②

Suppose you hope to buy a particular house in two years when you get your degree. You are worried about it going way up in price (you have budgeted "only" $100,000), but you do not wish to purchase it now. Furthermore, your plans may change. What you want is a *call option*,[44] where you pay a few thousand dollars to the current owner, who promises (if you choose to exercise your option) to sell the house to you in two years for $100,000. Your mother, who plans to retire in fifteen years, might want a *put* option, which for a fee gives her the option to sell the house at a fixed price in fifteen years. Since these options are not tangible property, but they *derive* their value from something else (here a house), these options are called *derivatives*. Derivatives are not common in housing transactions, but they are big business in stocks and in foreign currencies.[45]

The buyer of the option is shielding themselves from risk; the seller of the option gets cash now in exchange for risking an unusual price rise or drop in the future. What price should the seller of the option charge you for incurring this risk? The rather elegant answer to this question is given by the *Black–Scholes* model [146], and launched a multitrillion dollar industry.

Black and Scholes make several assumptions: no jumps in stock prices, instant trading, etc. These assumed, there is a risk-free strategy and a *fair price* for the derivative, at which no net profit is made. (The 1987 market crash may have been

---

[42] Insider trading is illegal.

[43] This exercise was developed in collaboration with Eric Grannan, based on Hull [91].

[44] Technically, this is a European-style call option; an American-style option would allow you to buy the house at any time in the next two years, not just at the end date.

[45] If you sell widgets for dollars, but pay salaries in pesos, you are likely to want to buy insurance to help out if the dollar falls dramatically with respect to the peso between now and when you are paid for the widgets.

caused by traders using the model, a seeming conspiracy to punish those who think they can eliminate risk.) We treat a special case.

- There are only two investments in the world: a risky asset (which we will call a stock) and cash (a risk-free investment). Initially the stock is worth X_0; cash is worth 1.

- The stock has one of two values at the date of the option (the expiration date), $X_u > X_d$.[46]

- The interest rates are zero, so the cash at the expiration date is still worth 1. (This does not change anything fundamental.)

- We can borrow and lend any amount of cash at the prevailing interest rate (that is, zero) and can buy or sell stock (even if we do not own any; this is called *selling short*). There are no transaction costs.

Let the two possible values of the option at the expiration date be V_u and V_d.[47] Let V_0 be the fair initial price of the derivative that we wish to determine.

Consider a portfolio $\mathcal{P}$ that includes the derivative and a certain amount α of the stock. Initially the value of $\mathcal{P}$ is $P_0 = V_0 + \alpha X_0$. At the expiration date the value will either be $V_u + \alpha X_u$ or $V_d + \alpha X_d$.

(a) *What value of α makes these two final portfolio values equal? What is this common final value P_F?*

(b) *What initial value V_0 of the derivative makes the initial value of the portfolio equal to the final value?* (Express your answer first in terms of P_F, α, and X_0, before substituting in your answers for part (a).) This is the value at which no net profit is made by either buyer or seller of the derivative; on average, the derivative gives the same return as cash.

(c) *Does your answer depend upon the probabilities p_u and p_d of going up or down?*

This portfolio is a weighted average of derivative and stock that makes the owner indifferent as to whether the stock goes up or down. It becomes a risk-free asset, and so its value must increase at the risk-free rate; this is the fundamental insight of *arbitrage* pricing. (An arbitrage is roughly a situation where there is free money to be made; a strategy for updating a portfolio where some final states have positive value and no final states have negative value with respect to risk-free investments. In an *efficient market*, there are no opportunities for arbitrage; large investors have bought and sold until no free money is available.) You can run exactly the same argument for more than one time step, starting at the final state where the values of the derivative are known, and working your way back to the initial state; this is the *binomial tree* method of pricing options. If the market is efficient, the average growth in the value of the stock must also grow at the risk-free rate, so the only unknown is the volatility of the stock (how large the fluctuations are in the stock price, Exercise 2.11). In the continuum limit this tree becomes the famous Black–Scholes partial differential equation.

(2.13) **Building a percolation network.**[48] (Complexity, Computation) ④

Figure 2.12 shows what a large sheet of paper, held at the edges, would look like if small holes were successively punched out at random locations. Here the ensemble averages over the different choices of random locations for the holes; this figure shows the sheet just before it fell apart. Certain choices of hole positions would cut the sheet in two far earlier (a straight line across the center) or somewhat later (checkerboard patterns), but for the vast majority of members of our ensemble the paper will have the same kinds of hole patterns seen here. Again, it is easier to analyze all the possible patterns of punches than to predict a particular pattern.

Percolation theory is the study of the qualitative change in connectivity of a large system as its components are randomly removed. Outside physics, it has become an archetype of criticality at continuous transitions, presumably because the problem is simple to state and the analysis does not demand a background in equilibrium statistical mechanics.[49] In this exercise, we will study bond percolation and site percolation (Figs. 2.12 and 2.13) in two dimensions.

---

[46]Having only two final prices makes the calculation less complicated. The subscripts u and d stand for up and down.

[47]E.g., for a call option allowing the buyer to purchase the stock at X_f with $X_u > X_f > X_d$, the value of the derivative at the expiration date will either be $V_u = X_u - X_f$, or $V_d = 0$ (since the buyer would choose not to exercise the option).

[48]This exercise was developed in collaboration with Christopher Myers. Hints can be found at the book website [181].

[49]Percolation can be mapped onto an equilibrium phase transition—the $q \to 1$ limit of an equilibrium q-state Potts model (where each site has a spin which can take q different states) [39, section 8.4].

Fig. 2.12 Bond percolation network. Each bond on a 10×10 square lattice is present with probability $p = 0.4$. This is below the percolation threshold $p = 0.5$ for the infinite lattice, and indeed the network breaks up into individual clusters (each shaded separately). Note the periodic boundary conditions. Note there are many small clusters, and only a few large ones; here twelve clusters of size $S = 1$, three of size $S = 2$, and one cluster of size $S = 29$ (black). For a large lattice near the percolation threshold the probability distribution of cluster sizes $\rho(S)$ forms a power law (Exercise 12.12).

Bond percolation on a square lattice.

(a) *Define a 2D bond percolation network with periodic boundary conditions on the computer, for size $L \times L$ and bond probability p. For this exercise, the nodes will be represented by pairs of integers (i, j). You will need the method* **GetNeighbors(node)**, *which returns the neighbors of an existing node. Use the bond-drawing software provided to draw your bond percolation network for various p and L, and use it to check that you have implemented the periodic boundary conditions correctly.* (There are two basic approaches. You can start with an empty network and use **AddNode** and **AddEdge** in loops to generate the nodes, vertical bonds, and horizontal bonds (see Exercise 1.7). Alternatively, and more traditionally, you can set up a 2D array of vertical and horizontal bonds, and implement **GetNeighbors(node)** by constructing the list of neighbors from the bond networks when the site is visited.)

The percolation threshold and duality. In most continuous phase transitions, one of the challenges is to find the location of the transition.

We chose bond percolation on the square lattice because one can argue, in the limit of large systems, that the percolation threshold $p_c = 1/2$. The argument makes use of the *dual lattice*.

The nodes of the dual lattice are the centers of the squares between nodes in the original lattice. The edges of the dual lattice are those which do not cross an edge of the original lattice. Since every potential dual edge crosses exactly one edge of the original lattice, the probability p^* of having bonds on the dual lattice is $1 - p$, where p is the probability of bonds for the original lattice. If we can show that the dual lattice percolates if and only if the original lattice does not, then $p_c = 1/2$. This is easiest to see graphically.

(b) *Generate and print a small lattice with $p = 0.4$, picking one where the largest cluster does not span across either the vertical or the horizontal direction (or print Fig. 2.12). Draw a path on the dual lattice spanning the system from top to bottom and from left to right. (You will be emulating a rat running through a maze.) Is it clear for large systems that the dual lattice will percolate if and only if the original lattice does not?*

Finding the clusters.

(c) *Write the following two functions that together find the clusters in the percolation network.*

(1) **FindClusterFromNode(graph, node, visited)**, *which returns the cluster in* **graph** *containing* **node**, *and marks the sites in the cluster as having been* **visited**. *The cluster is the union of* **node**, *the neighbors, the neighbors of the neighbors, etc. The trick is to use the set of* **visited** *sites to avoid going around in circles. The efficient algorithm is a* breadth-first traversal *of the graph, working outward from* **node** *in shells. There will be a* **currentShell** *of nodes whose neighbors have not yet been checked, and a* **nextShell** *which will be considered after the current one is finished (hence breadth first), as follows.*

- *Initialize* **visited[node] = True**, **cluster = [node]**, *and* **currentShell = graph.GetNeighbors(node)**.

- *While there are nodes in the new* **currentShell**:

  * *start a new empty* **nextShell**;

  * *for each node in the current shell, if the node has not been visited,*

· *add the node to the cluster,*

· *mark the node as visited,*

· *and add the neighbors of the node to the* `nextShell`*;*

∗ *set the current shell to* `nextShell`.

— *Return the cluster.*

(2) `FindAllClusters(graph)`, *which sets up the* `visited` *set to be* `False` *for all nodes, and calls* `FindClusterFromNode(graph, node, visited)` *on all nodes that have not been visited, collecting the resulting clusters. Optionally, you may want to order the clusters from largest to smallest, for convenience in the graphics (and in finding the largest cluster).*

Check your code by running it for small L and using the graphics software provided. Are the clusters, drawn in different colors, correct?

Site percolation on a triangular lattice. Universality states that the statistical behavior of the percolation clusters at long length scales should be independent of the microscopic detail. That is, removing bonds from a square lattice should leave the same fractal patterns of holes, near p_c, as punching out circular holes in a sheet just before it falls apart. Nothing about your algorithms from part (c) depended on their being four neighbors of a node, or their even being nodes at all sites. Let us implement site percolation on a triangular lattice (Fig. 2.13); nodes are occupied with probability p, with each node connected to any of its six neighbor sites that are also filled (punching out hexagons from a sheet of paper). The triangular site lattice also has a duality transformation, so again $p_c = 0.5$.

It is computationally convenient to label the site at (x, y) on a triangular lattice by $[i, j]$, where $x = i + j/2$ and $y = (\sqrt{3}/2)j$. If we again use periodic boundary conditions with $0 \leq i < L$ and $0 \leq j < L$, we cover a region in the shape of a $60°$ rhombus.[50] Each site $[i, j]$ has six neighbors, at $[i, j] + e$ with $e = [1, 0], [0, 1], [-1, 1]$ upward and to the right, and minus the same three downward and to the left.

(d) *Generate a site percolation network on a triangular lattice. You can treat the sites one at a time, using* `AddNode` *with probability p, and check*

`HasNode(neighbor)` *to bond to all existing neighbors. Alternatively, you can start by generating a whole matrix of random numbers in one sweep to determine which sites are occupied by nodes, add those nodes, and then fill in the bonds. Check your resulting network by running it for small L and using the graphics software provided. (Notice the shifted periodic boundary conditions at the top and bottom, see Fig. 2.13.) Use your routine from part (c) to generate the clusters, and check these (particularly at the periodic boundaries) using the graphics software.*

Fig. 2.13 Site percolation network. Each site on a 10×10 triangular lattice is present with probability $p = 0.4$, below the percolation threshold for the infinite lattice. Note the periodic boundary conditions at the sides, and the shifted periodic boundaries at the top and bottom.

(e) *Generate a small square-lattice bond percolation cluster, perhaps 30×30, and compare with a small triangular-lattice site percolation cluster. They should look rather different in many ways. Now generate a large[51] cluster of each, perhaps $1,000 \times 1,000$ (or see Fig. 12.7). Stepping back and blurring your eyes, do the two look substantially similar?*

Chapter 12 and Exercise 12.12 will discuss percolation theory in more detail.

(2.14) **Drifting random walk.** @

Random walks with a constant drift term will have a net correlation between steps. This problem can be reduced to the problem without drift by shifting to the moving reference frame. In

---

[50]The graphics software uses the periodic boundary conditions to shift this rhombus back into a rectangle.

[51]Your code, if written properly, should run in a time of order N, the number of nodes. If it seems to slow down more than a factor of 4 when you increase the length of the side by a factor of two, then check for inefficiencies.

particular, suppose we have a random walk with steps independently drawn from a uniform density $\rho(\ell)$ on [0,1), but with a nonzero mean $\langle \ell \rangle = \bar{\ell} \neq \vec{0}$.

Argue that the sums $s'_N = \sum_n^N (\ell_n - \bar{\ell})$ describe random walks in a moving reference frame, with zero mean. Argue that the variance of these random walks (the squared standard deviation) is the same as the variance $\langle (s_N - \bar{s}_N)^2 \rangle$ of the original random walks.

(2.15) **Diffusion of nonconserved particles.** @

Photons diffusing in clouds are occasionally absorbed by the water droplets. Neutrons diffusing in a reactor, or algae diffusing in the sea, may multiply as they move.

How would you modify the derivation of the diffusion equation in eqns 2.9–2.12 to allow for particle non-conservation? Which equation in 2.9 should change? What would the new term in the diffusion equation look like?

(2.16) **Density dependent diffusion.** @

The diffusion constant can be density dependent; for example, proteins diffusing in a cell membrane are so crowded they can get in the way of one another.

What should the diffusion equation be for a conserved particle density ρ diffusing with diffusion constant $D(\rho)$? (Hint: See note 21, p. 29.)

(2.17) **Local conservation.** ⓟ

Tin is deposited on a surface of niobium. At high temperatures, the niobium atoms invade into the tin layer. *Is the number of niobium atoms $\rho_{Nb}(\mathbf{x})$ a locally conserved quantity? (See note 16 on p. 28.)* Republicans perhaps should be concerned about the diffusion of Democrats moving into the Red states of the Southwest. *Is the number of Democrats $\rho_{Dem}(\mathbf{x})$ locally conserved?*

(2.18) **Absorbing boundary conditions.** ⓟ

A particle starting at x' diffuses on the positive x axis for a time t, except that whenever it hits the origin it is absorbed. The resulting probability density gives the Green's function $\rho(x,t) = G(x|x',t) = \int G(x|x',t)\rho(x',0)dx'$.

Solve for G. (Hint: Use the method of images: add a negative δ function at $-x'$.)

(2.19) **Run & tumble.** (Active matter, Biology) ③

Purcell, in an essay *Life at low Reynolds number* [154], describes the strange physical world at the scale of bacteria.[52]

The bacterium *E. coli* swims using roughly five corkscrew-shaped propellers called *flagella*, which spin at 100 revolutions per second. These propellers mesh nicely into a bundle when they rotate counter-clockwise, causing the bacterium to *run* forward. But when they rotate clockwise, the bundle flies apart and the bacterium *tumbles*. Assume that during a tumble the bacterium does not change position, and after a tumble it is pointed in a random direction. Pretend the runs are of fixed duration $T \approx 1\,\mathrm{s}$ and speed $V \approx 20\,\mu\mathrm{m/s}$, and they alternate with tumbles of duration $\tau \approx 0.1\,\mathrm{s}$. (Real cells shift from runs to tumbles with a continuous distribution of run times, and do not completely scramble their orientation after a tumble, Exercise 2.22.)

(a) *What is the mean-square distance $\langle \mathbf{r}^2(t) \rangle$ moved by our bacterium after a time $t = N(T + \tau)$, in terms of V, T, t, and τ? What is the formula for the diffusion constant?* (Hint: Be careful; your formula for the diffusion constant should depend on the fact that the diffusion is in three dimensions.)

Purcell tells us that the cell does not need to swim to get to new food after it has exhausted the local supply. Instead, it can just wait for food molecules to diffuse to it, with a rate he says is $4\pi a N D$ food molecules per second. Here a is the radius of the cell, N is the food concentration at infinity, and $D \approx 10^{-9}\,\mathrm{m^2/s}$ is the food diffusion constant.

(b) *Assume the food is eaten by the bacterium with perfect efficiency at the sphere of radius a. Solve the diffusion equation for the density of food molecules, and confirm Purcell's formula for the rate at which food is eaten.* (Hint: You may want to use the Laplacian in spherical coordinates: $\nabla^2\rho = (1/r^2)\partial(r^2\partial\rho/\partial r)/\partial r$ if the function $\rho(\mathbf{r}) = \rho(r)$ has spherical symmetry.)

The cell lives in an environment which varies in space. It swims to move toward regions with higher concentrations of food, and lower concentrations of poisons (a behavior called *chemotaxis*). Bacteria are too small to sense the concentration gradient from one side of the cell to

---

[52]He also analyzes swimming as *motion generated by a periodic change in shape*. For small, slow things like bacteria where inertia is unimportant, motions like kicking your feet that trace forward and backward along the same path in shape space (*reciprocal motions*) return you back where you started; spinning a corkscrew works. The net motion is an integral of a *curvature* over the loop in shape space [212].

the other. The run-and-tumble strategy is designed to move them far enough to tell if things are getting better. In particular, the cells run for longer times when things are getting better (but not shorter when things are getting worse).

(c) *Model chemotaxis with a one-dimensional run-and-tumble model along a coordinate x. The velocity $\pm V$ is chosen with equal probability at each tumble, with the same velocity and tumble time τ as above. But now the duration T_+ of runs in the positive x direction is larger than the duration T of runs in the negative direction. Compare the run speed V to the average velocity $\langle dx/dt \rangle$ of the bacterium toward a better life.*

Note that this one-dimensional biased random walk is also used to analyze the transcription of DNA to RNA [119, SI.C].

(2.20) **Flocking.**[53] (Active matter) ③

Active matter is a growing field of statistical mechanics—the description of emergent behavior from systems of self-propelled *agents* that maintain themselves out of thermal equilibrium. Applications include flocking birds, schools of fish, collective motion of bacterial colonies, manufactured self-propelled colloidal particles, and bio-polymers like microtubules and actin which actively grow and (together with protein molecular motors) can exert forces and move around. Use the exercise as a motivation to also explore the excellent videos and simulations on the Web in this field.

Boids may have started the field of active matter. A model for the flocking of birds, boids obey slightly complicated rules to avoid collisions, form groups, and aligning their velocities with their neighbors.

(a) *Find a boids simulation on the Web.* (Currently, the most portable ones are written in javascript, as is the mosh pit simulator.) *Compare the behavior to a video of starling murmuration. Describe one feature the boids capture well, and one that the boids fail to mimic properly.*

Later research on active matter focused on velocity alignment. Toner et al. [199] discuss the behavior of wildebeests (also known as gnu), who graze as individuals for months, but will at some point in the season start stirring around, pick a direction, and migrate as a group—ending some

two thousand miles away. How do the wildebeests reach consensus on what direction to go? (Assume a cloudy day on a featureless Serengeti plain. Assume also that they just want to find more water, or better grazing, and do not need to find the Masai Mara where they spend the dry season.)

We shall explore flocking using the mosh pit simulator [32].[54] Inspired by flocking theories, the simulator was developed to model humans at heavy metal concerts, where loud, fast music, flashing lights, and intoxication lead to segregated regions known as *mosh pits* where participants engage in violent collisional behavior [31, 188–190]. The simulation has two types of agents—active (red) and passive (black); both interact via a soft repulsive potential, and have a damping force $-\mu\mathbf{v}$ to absorb kinetic energy. The passive agents prefer to remain stationary, but the active agents are subject to several other types of forces: noise, "flocking", and "speed". In this exercise, we shall explore only the active agents (setting **Fraction Red** to one).

The mosh pit simulator aligns velocities by pulling each agent toward the average heading of its neighbors,

$$F_i^{\text{flock}} = \alpha \sum_{j=1}^{N_i} \mathbf{v}_j \left/ \left| \sum_{j=1}^{N_i} \mathbf{v}_j \right| \right., \qquad (2.36)$$

where the sum runs over the N_i agents within four radii of agent i and the flocking strength α controls the acceleration along the average velocity direction.

Reset the system by reloading the page. Set all agents to active (**Fraction Red** equal to one). Set **Particle count** $N = 40$, **Flocking strength** and **Speed** to zero, **Damping** and **Noise strength** to 0.2, and click **Change**. Let it equilibrate: the agents should mostly jiggle around without collisions, mimicking wildebeests grazing. Change **Flocking strength** to 0.2, and watch the collective behavior of the system. You may increase or decrease the box size and the number of particles (keeping the density fixed), depending on whether your computer is powerful or struggling; if so, report the number and size you used.

(b) *Describe the dynamics. Does the flock end up*

---

[53]This exercise was developed in collaboration with David Hathcock. It makes use of the mosh pit simulator [32] developed by Matt Bierbaum for [190].

[54]The current version of the software sometimes stops responding to the user interface. It seems to be more reliable on computers than on phones and tablets; reducing the number of agents may also help.

going largely in a single direction? If so, is the direction of motion the same for different simulation runs? Identify the slow particles in the animation. How do the slow particles forming this bump differ from the rest? Increase the noise strength. Estimate the noise level at which the particles stop moving collectively. (Note: If you wait long enough, the flocking direction will shift because of finite-size fluctuations.)

The flocking simulations *spontaneously break rotation invariance* (see Chapter 9). This is a surprising result, for technical reasons. Toner [199] illustrates this in a contest. He asks whether physicists standing at random on a featureless, cloudy Serengeti plain could all agree to point in the same direction. In analogy with eqn 2.36, one might imagine each physicist points along the average angle given by its N_i near neighbors plus a small, uncorrelated angular error ξ, with $\langle \xi_j \xi_k \rangle = \epsilon \delta_{ij}$:

$$\theta_i = (1/N_i) \sum_{j=1}^{N_i} \theta_j + \xi_i \qquad (2.37)$$

We shall model the error as a finite-temperature deviation from a minimum energy state with Hamiltonian given by a spring connecting θ_i to each of its neighbors. Let us calculate the behavior of this model for a one-dimensional lattice of physicists:

$$\mathcal{H} = \sum_i \tfrac{1}{2} K (\theta_i - \theta_{i-1})^2. \qquad (2.38)$$

(This is a version of the one-dimensional XY model ignoring phase slip.)

(c) *Calculate the forces on θ_j in eqn 2.38. Show that the energy is a minimum when θ_j points along the average angle of its two near neighbors. Change variables to $\delta_j = \theta_j - \theta_{j-1}$, and use equipartition $\tfrac{1}{2} K \delta^2 = \tfrac{1}{2} k_B T$ (see Sections 3.2.2 and 6.2) to calculate $\langle \delta_j^2 \rangle$ at temperature T.*

(d) *At what temperature will the RMS angle $\sqrt{\langle \delta^2 \rangle}$ between neighboring physicists be one degree? At that temperature, how large must n be before the RMS error $\sqrt{\langle (\theta_n - \theta_0)^2 \rangle}$ is $180°$?* (K has units of energy per radian squared.)

What makes wildebeests surprising is that this result also holds in two dimensions. No matter

how small the noise ξ, physicists standing randomly on a plane cannot break rotational symmetry by cooperating with their neighbors! This was proven by Hohenberg [85] and Mermin and Wagner [131] (see note 12 on p. 260). Toner et al. [198] show that this result does not apply to active matter systems, by developing a systematic continuum hydrodynamic theory of flocking. What gives the wildebeests an advantage? The physicists always observe the same neighbors. The active agents (wildebeests) that are not going with the flow of their neighbors keep bumping into new agents—collecting better information about the error in their way.

(2.21) **Lévy flight.** ③

Lévy flights are random walks which occasionally take unusually large jumps (Fig. 2.14). The probability distributions for their step lengths have *heavy tails*. Just as long ordinary random walks have a Gaussian probability distribution of net lengths, these Lévy flights (in one dimension[55]) have symmetric[56] *stable* probability distributions $S_\alpha(x)$, with the limit $S_2(x) = 1/(2\sqrt{\pi}) \exp(-x^2/4)$ giving the Green's function for ordinary diffusion (eqn 2.32) with $Dt = 1$. The particular case $S_1(x)$ also has an analytic form

$$S_1(x) = 1/(\pi(1 + x^2)) \qquad (2.39)$$

(the *Cauchy* distribution), but the other distributions are written only through their Fourier transforms

$$\widetilde{S}_\alpha(k) = \int_{-\infty}^{\infty} e^{-ikx} S_\alpha(x) dx = \exp(-|k|^\alpha).$$
$$(2.40)$$

where $0 < \alpha \leq 2$. Just as $S_1(x) \to 1/x^2$ for large x, the probability of long steps decays as a power law $S_\alpha(x) \to 1/x^{1+\alpha}$ (except for $\alpha = 2$, where the power-law part vanishes).

(a) *Show by inspection that the symmetric stable distribution $S_\alpha(x)$ is normalized ($\int S_\alpha(x) dx = 1$), using its Fourier transform (eqn 2.40). Why do we call S_α a stable distribution?*

(b) *Given two random steps x and y with distributions $P(x) = S_\alpha(x)$ and $P(y) = S_\alpha(y)$, compute the distribution of distances $P(z)$, where $z = x + y$ is the length of the two-step random walk. Show that the resulting distribution has the*

---

[55]Oddly, the Lévy flight in two dimensions (Fig. 2.14) is usually defined as random angles with one-dimensional stable steps, $P(r, \theta) \propto S_\alpha(r)/(2\pi r)$. This is not a stable multivariate distribution! The appropriate symmetric stable multivariate distribution in 2D would be given by $\widetilde{P}(k_x, k_y) = \exp(-(k_x^2 + k_y^2)^{\alpha/2})$, which probably would give rather similar results.

[56]There is a larger family of stable distributions that are not symmetric. Confusingly, one of the asymmetric ones is called the Lévy distribution.

same shape, just rescaled in width by a factor λ_α. (Hint: Use the fact that convolutions become products in Fourier space, eqn A.23.) For regular random walks, the sum of two steps had standard deviation $\sqrt{2}$ larger than that for a single step. *Check that* $\lambda_2 = \sqrt{2}$.

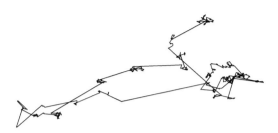

Fig. 2.14 Lévy flight. Trajectory of particle whose steps have random angles and random lengths distributed by a Cauchy distribution $S_1(x)$.

Thus, just as the sum of two Gaussian random steps is a wider Gaussian, the sum of two steps drawn from S_α is a wider S_α distribution. The *shape* of the distribution is stable as we take more steps (see also Exercise 12.11), hence the name *stable distribution*.

For ordinary random walks, we used the fact that the variance of an N-step random walk was N times the variance of one step to argue that the distance scaled as $\sqrt{N}$. This argument fails for our fat-tailed distribution $S_\alpha(x)$ because it has infinite variance for $\alpha < 2$. We can use the stability you found in part (b), though, to determine how the distribution of distances changes with the number of steps.

(c) *Repeat your argument from part (b) m times, to determine the scale factor after $N = 2^m$ steps. (Express your answer in terms of λ_α.) With what power does the distribution of lengths of an N-step walk drawn from S_α vary with N?*

(2.22) **Continuous time walks: Ballistic to diffusive.**[57] ③
Random walks diffuse at long times, with $\langle x^2 \rangle \propto Dt$, with a proportionality constant that depends on dimension. But what happens at short times? In many cases, the motion at short times is *ballistic*—straight line motion with constant speed. Perfume molecules move in straight lines between collisions with molecules in the air.

E. coli and other bacteria travel in fairly straight runs separated by tumbles (Exercise 2.19). Electrons in semiconductors can have long mean-free paths between collisions, and the ballistic to diffusive crossover we study here has practical implications for the behavior of some electronic devices.

Let us consider the trajectory of a single particle moving through vacuum, subject to an external force F and to collisions at random times. The collision at time t_α resets the velocity to a value v_α. Under a constant external force F, the velocity $v(t) = v_\Omega + (F/m)(t - t_\Omega)$, where t_Ω is the time of the collision immediately preceding the current time t. We can write this as an equation:

$$v(t') = \sum_\alpha \left(v_\alpha + (F/m)(t' - t_\alpha) \right) \tag{2.41}$$
$$\Theta(t' - t_\alpha)\Theta(t_{\alpha+1} - t').$$

Here the Heaviside step function Θ (zero for negative argument and one for positive argument), is used to select the time interval between collisions.

We assume the velocities after each collision have mean zero and are uncorrelated with one another, with mean-square velocity σ_v^2, so

$$\langle v_\alpha \rangle = 0, \tag{2.42}$$
$$\langle v_\alpha v_\beta \rangle = \sigma_v^2 \delta_{\alpha\beta}. \tag{2.43}$$

We assume that the scattering events happen at random with a rate $1/\Delta t$, so the probability density that the very next collision happens after a time δ is

$$\rho(\delta) = \exp(-\delta/\Delta t)/\Delta t. \tag{2.44}$$

Note that $\rho(\delta)$ also gives the probability density that the previous collision happened a time δ in the past (see Exercise 1.3).

Let us first calculate the mobility γ. Again, let the collision immediately before the current time t be labeled by Ω.

(a) *Show that the expectation value of the velocity $\langle v(t) \rangle$ is given by F/m times the mean time $\langle t - t_\Omega \rangle$ since the last collision. Your answer should use eqns 2.41 and 2.42, and the probability distribution $\rho(\delta)$. Calculate the mobility in terms of Δt and m. Why is your answer different from the calculation for fixed intervals Δt considered in Section 2.3?* (See Exercise 1.3 and Feynman's discussion in [61, I.43].)

---

[57]Feynman has a nice discussion of continuous-time diffusion [61, I.43].

Now let us explore the crossover from ballistic to diffusive motion, in the absence of an external force. Let $r(t)$ be the distance moved by our random walk since time zero. At times much less than Δt, the particle will not yet have scattered, and the motion is in a straight line. At long times we expect diffusive motion.

How does the crossover happen between these two limits? Let us calculate $\langle dr^2/dt \rangle$ as a function of time t for a continuous-time random walk starting at $t = 0$, in the absence of an external force, giving us the crossover and the diffusion constant D. Assume the random walk has been traveling since $t = -\infty$, but that we measure the distance traveled since $t = 0$.

(b) *Calculate $\langle dr^2/dt \rangle$ as a function of time, where $r(t) = x(t) - x(0)$ is the distance moved since $t = 0$. To avoid getting buried in algebra, we ask you to do this in a particular sequence. (1) Write r as an abstract integral over v. Square it, and differentiate with respect to time. (2) Substitute eqn 2.41 with $F = 0$, and use eqn 2.43 to ensemble average over the collision velocities.* (Hint: Your formula at this point should only involve the last collision time t_Ω, and should be an integral from 0 to t involving a Θ function.) *(3) Then take the ensemble average over the collision times to get $\langle dr^2/dt \rangle$. What is the diffusion constant D? Does it satisfy the relation $D/\gamma = m\bar{v}^2$ of note 21 on p. 29?* (Hint: You may end up with a double integral over t_Ω and t'. Examine the integration region in the (t_Ω, t') plane, and exchange orders of integration.)

The crossover between ballistic motion and diffusion is easily found in the literature, and is a decaying exponential:

$$\langle dr^2/dt \rangle = 2D(1 - \exp(-t/\Delta t)). \quad (2.45)$$

(c) *Use this to check your answer in part (b). Integrate, and provide a log–log plot of $\sqrt{\langle r^2 \rangle/D\Delta t}$ versus $t/\Delta t$, over the range $0.1 < t/\Delta t < 100$. Solve for the asymptotic power-law dependence at short and long times, and plot them as straight lines on your log–log plot. Label your axes.* (Are the power laws tangent to your curves as $t \to 0$ and $t \to \infty$?)

(2.23) **Random walks and generating functions.**[58] (Mathematics) @

Consider a one-dimensional random walk with step-size ± 1 starting at the origin. What is the probability f_t that it *first* returns to the origin at t steps? (See also Exercise 2.18.)

(a) *Argue that the probability is zero unless $t = 2m$ is even. How many total paths are there of length $2m$? Calculate the probability for f_{2m} for up to eight-step hops ($m < 5$) by drawing the different paths that touch the origin only at their endpoints.* (Hint: You can save paper by drawing the paths starting to the right, and multiplying by two. Check your answer by comparing to the results for general m in eqn 2.46.)

This *first return* problem is well studied, and is usually solved using a *generating function*. Generating functions extend a series (here f_{2m}) into a function. The generating function for our one-dimensional random walk first return problem, it turns out, is

$$F(x) = \sum_{m=0}^{\infty} f_{2m} x^m = 1 - \sqrt{1 - x}$$
$$= \sum_m \frac{2^{-2m}}{2m - 1} \binom{2m}{m} x^m. \quad (2.46)$$

(Look up the derivation if you want an example of how it is done.)

(b) *Evaluate the probability of returning to the origin for the first time at $t = 20$ steps ($m = 10$). Use Stirling's formula $n! \sim \sqrt{2\pi n}(n/e)^n$ to give the probability f_{2m} of first returning at time $t = 2m$ for large t.* (Note that the rate per unit time $P_{\text{hop}}(t) = f_{2m}/2$.)

There are lots of other uses for generating functions: partial sums, moments, convolutions...

(c) *Argue that $F(1)$ is the probability that a particle will eventually return to the origin, and that $F'(1)$ is $\langle m \rangle$, half the expected time to return to the origin. What is the probability that our one-dimensional walk will return to the origin? What is the mean time to return to the origin?*

---

[58]This problem was created with the assistance of Archishman Raju.

Temperature and equilibrium

3

We now turn to equilibrium statistical mechanics—a triumph of the nineteenth century. Equilibrium statistical mechanics provided them the fundamental definition for temperature and the laws determining the behavior of all common liquids and gases.[1] We will switch in this chapter between discussing the general theory and applying it to a particular system—the ideal gas. The ideal gas provides a tangible example of the formalism, and its analysis will provide a preview of material coming in the next few chapters.

A system which is not acted upon by the external world[2] is said to approach *equilibrium* if and when it settles down[3] at long times to a state which is independent of the initial conditions (except for conserved quantities like the total energy). Statistical mechanics describes the equilibrium state as an average over all states in *phase space* consistent with the conservation laws; this *microcanonical ensemble* is introduced in Section 3.1. In Section 3.2, we shall calculate the properties of the ideal gas using the microcanonical ensemble. In Section 3.3 we shall define *entropy* and *temperature* for equilibrium systems, and argue from the microcanonical ensemble that heat flows to maximize the entropy and equalize the temperature. In Section 3.4 we will derive the formula for the *pressure* in terms of the entropy, and define the *chemical potential*. Finally, in Section 3.5 we calculate the entropy, temperature, and pressure for the ideal gas, and introduce two refinements to our definitions of phase-space volume.

[1]Quantum statistical mechanics is necessary to understand solids at low temperatures.

[2]If the system is driven (i.e. there are externally imposed forces or currents) we instead call this final condition the *steady state*.

[3]If the system is large, the equilibrium state will also usually be time independent and "calm", hence the name. Small systems will continue to fluctuate substantially even in equilibrium.

3.1 The microcanonical ensemble

Statistical mechanics allows us to solve *en masse* many problems that are impossible to solve individually. In this chapter we address the general equilibrium behavior of N atoms in a box of volume V—any kinds of atoms, in an arbitrary external potential.[4] The walls of our box are rigid, so that energy is conserved when atoms bounce off the walls. This makes our system *isolated*, independent of the world around it.

How can we solve for the behavior of our atoms? We can in principle determine the positions[5] $\mathbb{Q} = (x_1, y_1, z_1, x_2, \ldots, x_N, y_N, z_N) = (q_1, \ldots, q_{3N})$ and momenta $\mathbb{P} = (p_1, \ldots, p_{3N})$ of the particles at any future time given their initial positions and momenta using Newton's

[4]However, we do ignore quantum mechanics until Chapter 7.

[5]The $3N$-dimensional space of positions $\mathbb{Q}$ is called *configuration space*. The $3N$-dimensional space of momenta $\mathbb{P}$ is called *momentum space*. The $6N$-dimensional space $(\mathbb{P}, \mathbb{Q})$ is called *phase space*.

Statistical Mechanics: Entropy, Order Parameters, and Complexity. James P. Sethna, Oxford University Press (2021). © Oxford University Press. DOI:10.1093/oso/9780198865247.003.0003

laws:

$$\dot{\mathbb{Q}} = \mathbf{m}^{-1}\mathbb{P}, \quad \dot{\mathbb{P}} = \mathbb{F}(\mathbb{Q}) \tag{3.1}$$

[6]Here $\mathbf{m}$ is a diagonal matrix if the particles are not all the same mass.

[7]This scrambling is precisely the approach to equilibrium.

[8]In an infinite system, total momentum and angular momentum would also be conserved; the box breaks rotation and translation invariance.

[9]What about quantum mechanics, where the energy levels in a finite system are discrete? In that case (Chapter 7), we will need to keep δE large compared to the spacing between energy eigenstates, but small compared to the total energy.

Fig. 3.1 Energy shell. The shell of energies between E and $E + \delta E$ can have an irregular "thickness". The volume of this shell in $6N$-dimensional phase space, divided by δE, is the definition of $\Omega(E)$. Notice that the microcanonical average weights the thick regions more heavily (see Exercise 3.14). In Section 4.1 we will show that, just as a water drop in a river spends more time in the deep sections where the water flows slowly, so also a trajectory in phase space spends more time in the thick regions.

(where $\mathbb{F}$ is the $3N$-dimensional force due to the other particles and the walls, and $\mathbf{m}$ is the particle mass).[6]

In general, solving these equations is plainly not feasible:

- Many systems of interest involve far too many particles to allow one to solve for their trajectories.

- Most systems of interest exhibit chaotic motion, where the time evolution depends with ever increasing sensitivity on the initial conditions—you cannot know enough about the current state to predict the future.

- Even if it were possible to evolve our trajectory, knowing the solution would for most purposes be useless; we are far more interested in the typical number of atoms striking a wall of the box, say, than the precise time a particular particle hits.

How can we extract the simple, important predictions out of the complex trajectories of these atoms? The chaotic time evolution will rapidly scramble[7] whatever knowledge we may have about the initial conditions of our system, leaving us effectively knowing only the conserved quantities—for our system, just the total energy E.[8] Rather than solving for the behavior of a particular set of initial conditions, let us hypothesize that the energy is all we need to describe the equilibrium state. This leads us to a statistical mechanical description of the equilibrium state of our system as an ensemble of all possible states with energy E—the *microcanonical ensemble*.

We calculate the properties of our ensemble by averaging over states with energies in a shell $(E, E + \delta E)$ taking the limit[9] $\delta E \to 0$ (Fig. 3.1). Let us define the function $\Omega(E)$ to be the phase-space volume of this thin shell, divided by δE:[10]

$$\Omega(E)\,\delta E = \int_{E < \mathcal{H}(\mathbb{P},\mathbb{Q}) < E + \delta E} d\mathbb{P}\,d\mathbb{Q}. \tag{3.5}$$

[10]More formally, one can write the energy shell $E < \mathcal{H}(\mathbb{P},\mathbb{Q}) < E + \delta E$ in terms of the Heaviside step function $\Theta(x)$, where $\Theta(x) = 1$ for $x \geq 0$, and $\Theta(x) = 0$ for $x < 0$. We see that $\Theta(E + \delta E - \mathcal{H}) - \Theta(E - \mathcal{H})$ is one precisely inside the energy shell (Fig. 3.1). In the limit $\delta E \to 0$, we can write $\Omega(E)$ as a derivative:

$$\Omega(E) = \frac{1}{\delta E}\int_{E < \mathcal{H}(\mathbb{P},\mathbb{Q}) < E + \delta E} d\mathbb{P}\,d\mathbb{Q} = \frac{1}{\delta E}\int d\mathbb{P}\,d\mathbb{Q}\,[\Theta(E + \delta E - \mathcal{H}) - \Theta(E - \mathcal{H})] \tag{3.2}$$

$$= \frac{\partial}{\partial E}\int d\mathbb{P}\,d\mathbb{Q}\,\Theta(E - \mathcal{H}) = \int d\mathbb{P}\,d\mathbb{Q}\,\delta(E - \mathcal{H}), \tag{3.3}$$

using the fact that the derivative of the Heaviside function is the Dirac δ-function $\partial\Theta(x)/\partial x = \delta(x)$ (see note 3). Similarly, the expectation of a general operator O is

$$\langle O \rangle = \frac{1}{\Omega(E)\delta E}\int d\mathbb{P}\,d\mathbb{Q}\,[\Theta(E + \delta E - \mathcal{H}) - \Theta(E - \mathcal{H})]\,O(\mathbb{P},\mathbb{Q}) = \frac{1}{\Omega(E)}\frac{\partial}{\partial E}\int d\mathbb{P}\,d\mathbb{Q}\,\Theta(E - \mathcal{H})O(\mathbb{P},\mathbb{Q})$$

$$= \frac{1}{\Omega(E)}\int d\mathbb{P}\,d\mathbb{Q}\,\delta(E - \mathcal{H}(\mathbb{P},\mathbb{Q}))\,O(\mathbb{P},\mathbb{Q}). \tag{3.4}$$

It will be important later to note that the derivatives in eqns 3.2 and 3.4 are at constant N and constant V: $(\partial/\partial E)|_{V,N}$. Thus the microcanonical ensemble can be written as a probability density $\delta(E - \mathcal{H}(\mathbb{P},\mathbb{Q}))/\Omega(E)$ in phase space.

Here $\mathcal{H}(\mathbb{P}, \mathbb{Q})$ is the Hamiltonian for our system.[11] Finding the average $\langle O \rangle$ of a property in the microcanonical ensemble is done by averaging $O(\mathbb{P}, \mathbb{Q})$ over this same energy shell:

$$\langle O \rangle_E \ = \ \frac{1}{\Omega(E)\delta E} \int_{E < \mathcal{H}(\mathbb{P},\mathbb{Q}) < E+\delta E} O(\mathbb{P}, \mathbb{Q}) \, \mathrm{d}\mathbb{P} \, \mathrm{d}\mathbb{Q}. \qquad (3.6)$$

Notice that, by averaging equally over all states in phase space compatible with our knowledge about the system (that is, the conserved energy), we have made a hidden assumption: all points in phase space (with a given energy) are a priori equally likely, so the average should treat them all with equal weight. In Section 3.2, we will see that this assumption leads to sensible behavior, by solving the case of an ideal gas.[12] We will fully justify this equal-weighting assumption in Chapter 4, where we will also discuss the more challenging question of why so many systems actually reach equilibrium.

The fact that the microcanonical distribution describes equilibrium systems should be amazing to you. The long-time equilibrium behavior of a system is precisely the typical behavior of all systems with the same value of the conserved quantities. This fundamental *regression to the mean* is the basis of statistical mechanics.

3.2   The microcanonical ideal gas

We can *talk* about a general collection of atoms, and derive general statistical mechanical truths for them, but to *calculate* specific properties we must choose a particular system. The archetypal statistical mechanical system is the monatomic[13] ideal gas. You can think of helium atoms at high temperatures and low densities as a good approximation to this ideal gas—the atoms have very weak long-range interactions and rarely collide. The ideal gas will be the limit when the interactions between particles vanish.[14]

For the ideal gas, the energy does not depend upon the spatial configuration $\mathbb{Q}$ of the particles. This allows us to study the positions (Section 3.2.1) separately from the momenta (Section 3.2.2).

3.2.1   Configuration space

Since the energy is independent of the position, our microcanonical ensemble must weight all configurations equally. That is to say, it is precisely as likely that all the particles will be within a distance ϵ of the middle of the box as it is that they will be within a distance ϵ of any other particular configuration.

What is the probability density $\rho(\mathbb{Q})$ that the ideal gas particles will be in a particular configuration $\mathbb{Q} \in \mathbb{R}^{3N}$ inside the box of volume V? We know ρ is a constant, independent of the configuration. We know that the gas atoms are in *some* configuration, so $\int \rho \, \mathrm{d}\mathbb{Q} = 1$. The integral

[11] The Hamiltonian $\mathcal{H}$ is the function of $\mathbb{P}$ and $\mathbb{Q}$ that gives the energy. For our purposes, this will always be

$$\frac{\mathbb{P}^2}{2m} + U(\mathbb{Q}) = \sum_{\alpha=1}^{3N} \frac{p_\alpha^2}{2m} + U(q_1, \dots, q_{3N}),$$

where the force in Newton's laws (eqn 3.1) is $F_\alpha = -\partial U/\partial q_\alpha$.

[12] In Section 3.5, we shall add two refinements to our definition of the energy-shell volume $\Omega(E)$.

[13] Air is a mixture of gases, but most of the molecules are diatomic: O_2 and N_2, with a small admixture of triatomic CO_2 and monatomic Ar. The properties of diatomic ideal gases are only slightly more complicated than the monatomic gas; one must keep track of the internal rotational degree of freedom and, at high temperatures, the vibrational degrees of freedom.

[14] With no interactions, how can the ideal gas reach equilibrium? If the particles never collide, they will forever be going with whatever initial velocity we started them. We imagine delicately taking the long-time limit first, before taking the limit of infinitely weak interactions, so we can presume an equilibrium distribution has been established.

over the positions gives a factor of V for each of the N particles, so $\rho(\mathbb{Q}) = 1/V^N$.

It may be counterintuitive that unusual configurations, like all the particles on the right half of the box, have the same probability density as more typical configurations. If there are two noninteracting particles in an $L \times L \times L$ box centered at the origin, what is the probability that both are on the right (have $x > 0$)? The probability that two particles are on the right half is the integral of $\rho = 1/L^6$ over the six-dimensional volume where both particles have $x > 0$. The volume of this space is $(L/2) \times L \times L \times (L/2) \times L \times L = L^6/4$, so the probability is $1/4$, just as one would calculate by flipping a coin for each particle. The probability that N such particles are on the right is 2^{-N}—just as your intuition would suggest. Do not confuse probability density with probability! The unlikely states for molecules are not those with small probability density. Rather, they are states with small net probability, because their allowed configurations and/or momenta occupy insignificant volumes of the total phase space.

Notice that configuration space typically has dimension equal to several times Avogadro's number.[15] Enormous-dimensional vector spaces have weird properties—which directly lead to important principles in statistical mechanics.

As an example of weirdness, most of configuration space has almost exactly half the x-coordinates on the right side of the box. If there are $2N$ noninteracting particles in the box, what is the probability P_m that $N + m$ of them will be in the right half? There are 2^{2N} equally likely ways the distinct particles could sit in the two sides of the box. Of these, $\binom{2N}{N+m} = (2N)!/((N+m)!(N-m)!)$ have m extra particles in the right half.[16] So,

$$P_m = 2^{-2N} \binom{2N}{N+m} = 2^{-2N} \frac{(2N)!}{(N+m)!(N-m)!}. \qquad (3.7)$$

We can calculate the fluctuations in the number on the right using Stirling's formula,[17]

$$n! \sim (n/e)^n \sqrt{2\pi n} \sim (n/e)^n. \qquad (3.8)$$

For now, let us use the second, less accurate form; keeping the factor $\sqrt{2\pi n}$ would fix the prefactor in the final formula (Exercise 3.9), which we will instead derive by normalizing the total probability to one. Using Stirling's formula, eqn 3.7 becomes

$$\begin{aligned}
P_m &\approx 2^{-2N} \left(\frac{2N}{e}\right)^{2N} \Big/ \left(\frac{N+m}{e}\right)^{N+m} \left(\frac{N-m}{e}\right)^{N-m} \\
&= N^{2N}(N+m)^{-(N+m)}(N-m)^{-(N-m)} \\
&= (1+m/N)^{-(N+m)}(1-m/N)^{-(N-m)} \\
&= ((1+m/N)(1-m/N))^{-N}(1+m/N)^{-m}(1-m/N)^m \\
&= (1-m^2/N^2)^{-N}(1+m/N)^{-m}(1-m/N)^m, \qquad (3.9)
\end{aligned}$$

[15]A gram of hydrogen has approximately $N = 6.02 \times 10^{23}$ atoms, known as Avogadro's number. So, a typical $3N$ will be around 10^{24}.

[16]$\binom{p}{q}$ is the number of ways of choosing an unordered subset of size q from a set of size p. There are $p(p-1)\cdots(p-q+1) = p!/(p-q)!$ ways of choosing an ordered subset, since there are p choices for the first member and $p-1$ for the second, ... There are $q!$ different ordered sets for each unordered one, so $\binom{p}{q} = p!/(q!(p-q)!)$.

[17]Stirling's formula tells us that the "average" number in the product $n! = n(n-1)\cdots 1$ is roughly n/e; see Exercise 1.4.

and, since $|m| \ll N$ we may substitute $1 + \epsilon \approx \exp(\epsilon)$, giving us

$$P_m \approx \left(e^{-m^2/N^2}\right)^{-N} \left(e^{m/N}\right)^{-m} \left(e^{-m/N}\right)^m \approx P_0 \exp(-m^2/N), \tag{3.10}$$

where P_0 is the prefactor we missed by not keeping enough terms in Stirling's formula. We know that the probabilities must sum to one, so again for $|m| \ll N$, we have $1 = \sum_m P_m \approx \int_{-\infty}^{\infty} P_0 \exp(-m^2/N)\,\mathrm{d}m = P_0\sqrt{\pi N}$. Hence

$$P_m \approx \sqrt{\frac{1}{\pi N}} \exp\left(-\frac{m^2}{N}\right). \tag{3.11}$$

This is a nice result: it says that the number fluctuations are distributed in a Gaussian or normal distribution[18] $(1/\sqrt{2\pi}\sigma)\exp(-x^2/2\sigma^2)$ with a standard deviation $\sigma = \sqrt{N/2}$. If we have Avogadro's number of particles $N \sim 10^{24}$, then the fractional fluctuations $\sigma/N = 1/\sqrt{2N} \sim 10^{-12} = 0.0000000001\%$. In almost all the volume of a box in $\mathbb{R}^{3N}$, almost exactly half of the coordinates are on the right half of their range. In Section 3.2.2 we will find another weird property of high-dimensional spaces.

In general, the relative fluctuations of most quantities of interest in equilibrium statistical mechanics go as $1/\sqrt{N}$. *For many properties of macroscopic systems, statistical mechanical fluctuations about the average value are very small.*[19]

3.2.2 Momentum space

Working with the microcanonical momentum distribution is more challenging,[20] but more illuminating, than working with the ideal gas configuration space of the last section. Here we must study the geometry of spheres in high dimensions.

The kinetic energy for particles (whether or not they interact) is

$$\sum_{\alpha=1}^{3N} \tfrac{1}{2}m_\alpha v_\alpha^2 = \sum_{\alpha=1}^{3N} \frac{p_\alpha^2}{2m_\alpha} = \frac{\mathbb{P}^2}{2m}, \tag{3.12}$$

where the last form assumes all of our atoms have the same mass m. Hence the condition that a system of equal-mass particles has energy E is that the system lies on a sphere in $3N$-dimensional momentum space of radius $R = \sqrt{2mE}$. Mathematicians[21] call this the $3N-1$ sphere, $\mathbb{S}_R^{3N-1}$. Specifically, if the energy of the system is known to be in a small range between E and $E + \delta E$, what is the corresponding volume of momentum space? The volume $\mu\left(\mathbb{S}_R^{\ell-1}\right)$ of the $\ell - 1$ sphere (in ℓ dimensions) of radius R is[22]

$$\mu\left(\mathbb{S}_R^{\ell-1}\right) = \pi^{\ell/2} R^\ell / (\ell/2)!. \tag{3.13}$$

[18]This is the central limit theorem again. We derived it in Section 2.4.2 using random walks and a continuum approximation, instead of Stirling's formula; the Gaussian was the Green's function for the number of heads in $2N$ coin flips. We will derive it again in Exercise 12.11 using renormalization-group methods.

[19]We will show this in great generality in Section 10.7, see eqn 10.50.

[20]The numerical factors won't be important here: it is just easier to keep them than to explain why we don't need to. Watch the factors of R. In Section 6.2 we will find that these same derivations are far less complicated using the canonical distribution.

[21]Mathematicians like to name surfaces, or manifolds, after the number of dimensions or local coordinates internal to the manifold, rather than the dimension of the space the manifold lives in. After all, one can draw a circle embedded in any number of dimensions (down to two). Thus a basketball is a two sphere $\mathbb{S}^2$, the circle is the one sphere $\mathbb{S}^1$, and the zero sphere $\mathbb{S}^0$ consists of the two points ± 1.

[22]Does this give the area of a circle in $\ell = 2$ dimensions? The factorial function can be defined for non-integers (see Exercise 1.5); $(3/2)! = 3\sqrt{\pi}/4$ and eqn 3.13 imply the correct area of a sphere in three dimensions. The formula in general dimensions is an induction exercise in multiple integration. Hint: It is easiest to do the integrals two dimensions at a time.

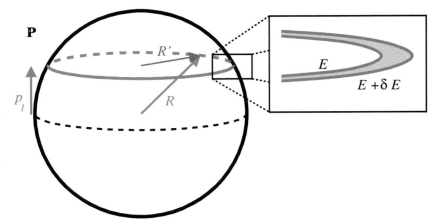

Fig. 3.2 The energy surface in momentum space is the $3N-1$ sphere of radius $R = \sqrt{2mE}$. The conditions that the x-component of the momentum of atom #1 is p_1 restricts us to a circle (or rather a $3N-2$ sphere) of radius $R' = \sqrt{2mE - p_1{}^2}$. The condition that the energy is in the shell $(E, E + \delta E)$ extends the circle into the annular region shown in the inset.

[23]Why is this not the surface area? Because its width is an infinitesimal energy δE, and not an infinitesimal thickness $\delta R \approx \delta E(\partial R/\partial E) = \delta E(R/2E)$. The distinction does not matter for the ideal gas (both would give uniform probability densities over all directions of $\mathbb{P}$) but it is important for interacting systems (where the thickness of the energy shell varies, see Fig. 3.1).

[24]It is a sloppy physics convention to use ρ to denote probability densities of all sorts. Earlier, we used it to denote probability density in $3N$-dimensional configuration space; here we use it to denote probability density in one variable. The argument of the function ρ tells us which function we are considering.

The volume of the thin shell[23] between E and $E + \delta E$ is given by

$$
\begin{aligned}
\frac{\text{shell volume}}{\delta E} &= \frac{\mu\left(\mathbb{S}^{3N-1}_{\sqrt{2m(E+\delta E)}}\right) - \mu\left(\mathbb{S}^{3N-1}_{\sqrt{2mE}}\right)}{\delta E} \\
&= \mathrm{d}\mu\left(\mathbb{S}^{3N-1}_{\sqrt{2mE}}\right)\Big/\mathrm{d}E \\
&= \frac{\mathrm{d}}{\mathrm{d}E}\left(\pi^{3N/2}(2mE)^{3N/2}\big/(3N/2)!\right) \\
&= \pi^{3N/2}(3Nm)(2mE)^{3N/2-1}\big/(3N/2)! \\
&= (3Nm)\pi^{3N/2}R^{3N-2}\big/(3N/2)!.
\end{aligned}
\tag{3.14}
$$

Given our microcanonical ensemble that equally weights all states with energy E, the probability density for having any particular set of particle momenta $\mathbb{P}$ is the inverse of this shell volume.

Let us do a tangible calculation with this microcanonical ensemble. Let us calculate the probability density $\rho(p_1)$ that the x-component of the momentum of the first atom is p_1.[24] The probability density that this momentum is p_1 and the energy is in the range $(E, E + \delta E)$ is proportional to the area of the annular region in Fig. 3.2. The sphere has radius $R = \sqrt{2mE}$, so by the Pythagorean theorem, the circle has radius $R' = \sqrt{2mE - p_1{}^2}$. The volume in momentum space of the annulus is given by the difference in areas inside the two "circles" ($3N-2$ spheres) with momentum p_1 and energies E and $E + \delta E$. We can use eqn 3.13 with $\ell = 3N - 1$:

$$
\begin{aligned}
\frac{\text{annular area}}{\delta E} &= \mathrm{d}\mu\left(\mathbb{S}^{3N-2}_{\sqrt{2mE-p_1{}^2}}\right)\Big/\mathrm{d}E \\
&= \frac{\mathrm{d}}{\mathrm{d}E}\left(\pi^{(3N-1)/2}(2mE - p_1{}^2)^{(3N-1)/2}\big/[(3N-1)/2]!\right) \\
&= \pi^{(3N-1)/2}(3N-1)m(2mE-p_1^2)^{(3N-3)/2}\big/[(3N-1)/2]!. \\
&= (3N-1)m\pi^{(3N-1)/2}R'^{3N-3}\big/[(3N-1)/2]!
\end{aligned}
\tag{3.15}
$$

The probability density $\rho(p_1)$ of being in the annulus at height p_1 is its area divided by the shell volume in eqn 3.14:

$$\rho(p_1) = \text{annular area} \big/ \text{shell volume}$$
$$= \frac{(3N-1)m\pi^{(3N-1)/2}R'^{3N-3}/\left[(3N-1)/2\right]!}{3Nm\pi^{3N/2}R^{3N-2}/(3N/2)!}$$
$$\propto (R^2/R'^3)(R'/R)^{3N}$$
$$= (R^2/R'^3)(1 - p_1{}^2/2mE)^{3N/2}. \tag{3.16}$$

The probability density $\rho(p_1)$ will be essentially zero unless $R'/R = \sqrt{1 - p_1{}^2/2mE}$ is nearly equal to one, since this factor in eqn 3.16 is taken to an enormous power ($3N$, around Avogadro's number). We can thus simplify $R^2/R'^3 \approx 1/R = 1/\sqrt{2mE}$ and $1 - p_1{}^2/2mE = 1 - \epsilon \approx \exp(-\epsilon) = \exp(-p_1{}^2/2mE)$, giving us

$$\rho(p_1) \propto \frac{1}{\sqrt{2mE}} \exp\left(\frac{-p_1{}^2}{2m}\frac{3N}{2E}\right). \tag{3.17}$$

The probability density $\rho(p_1)$ is a Gaussian distribution of standard deviation $\sqrt{2mE/3N}$; we can again set the constant of proportionality to normalize the Gaussian, leading to

$$\rho(p_1) = \frac{1}{\sqrt{2\pi m(2E/3N)}} \exp\left(\frac{-p_1{}^2}{2m}\frac{3N}{2E}\right). \tag{3.18}$$

This is the probability distribution for any momentum component of any of our particles; there was nothing special about particle number one. Our ensemble assumption has allowed us to calculate the momentum distribution explicitly in terms of E, N, and m, without ever considering a particular trajectory; this is what makes statistical mechanics powerful.

Formula 3.18 tells us that most of the surface area of a large-dimensional sphere is very close to the equator! Think of p_1 as the latitude on the sphere. The range of latitudes containing most of the area is $\delta p \approx \pm\sqrt{2mE/3N}$, and the total range of latitudes is $\pm\sqrt{2mE}$; the belt divided by the height is the square root of Avogadro's number. This is true *whatever equator you choose*, even intersections of several equators. Geometry is weird in high dimensions.

In the context of statistical mechanics, this seems much less strange; typical configurations of gases have the kinetic energy divided roughly equally among all the components of momentum; configurations where one atom has most of the kinetic energy (far from its equator) are vanishingly rare.

Formula 3.18 foreshadows four key results that will emerge from our systematic study of equilibrium statistical mechanics in the following few chapters.

(1) **Temperature.** In our calculation, a single momentum component competed for the available energy with the rest of the ideal gas. In

[25]We shall see that temperature is naturally measured in units of energy. Historically we measure temperature in degrees and energy in various other units (Joules, ergs, calories, eV, foot-pounds, ...); Boltzmann's constant $k_B = 1.3807 \times 10^{-23}$ J/K is the conversion factor between units of temperature and units of energy.

[26]This is different from the probability of the subsystem *having* energy E, which is the product of the Boltzmann probability and the number of states with that energy.

[27]Molecular gases will have internal vibration modes that are often not well described by classical mechanics. At low temperatures, these are often frozen out; including rotations and translations but ignoring vibrations is a good approximation for air at room temperature (see note 13 on p. 51).

[28]Relativistic effects, magnetic fields, and quantum mechanics will change the velocity distribution. Equation 3.19 will be reasonably accurate for all gases at reasonable temperatures, all liquids but helium, and many solids that are not too cold. Notice that almost all molecular dynamics simulations are done classically: their momentum distributions are given by eqn 3.19.

Section 3.3 we will study the competition in general between two large subsystems for energy, and will discover that the balance is determined by the *temperature*. The temperature T for our ideal gas will be given (eqn 3.52) by $k_B T = 2E/3N$.[25] Equation 3.18 then gives us the important formula

$$\rho(p_1) = \frac{1}{\sqrt{2\pi m k_B T}} \exp\left(-\frac{p_1{}^2}{2m k_B T}\right). \qquad (3.19)$$

(2) **Boltzmann distribution.** The probability of the x momentum of the first particle having kinetic energy $K = p_1^2/2m$ is proportional to $\exp(-K/k_B T)$ (eqn 3.19). This is our first example of a *Boltzmann distribution*. We shall see in Section 6.1 that the probability of a small subsystem being in a particular state[26] of energy E will in completely general contexts have probability proportional to $\exp(-E/k_B T)$.

(3) **Equipartition theorem.** The average kinetic energy $\langle p_1^2/2m \rangle$ from eqn 3.19 is $k_B T/2$. This is an example of the *equipartition theorem* (Section 6.2): each harmonic degree of freedom in an equilibrium classical system has average energy $k_B T/2$.

(4) **General classical[27] momentum distribution.** Our derivation was in the context of a monatomic ideal gas. But we could have done an analogous calculation for a system with several gases of different masses; our momentum sphere would become an ellipsoid, but the momentum distribution is given by the same formula. More surprising, we shall see (using the canonical ensemble in Section 6.2) that interactions do not matter either, as long as the system is classical:[28] the probability densities for the momenta are still given by the same formula, independent of the potential energies. The momentum distribution of formula 3.19 is correct for nearly all classical equilibrium systems; interactions will affect only the configurations of such particles, not their velocities.

3.3 What is temperature?

Our ordinary experience suggests that heat energy will flow from a hot body into a neighboring cold body until they reach the same temperature. Statistical mechanics insists that the distribution of heat between the two bodies is determined by the microcanonical assumption that all possible states of fixed total energy for the two bodies are equally likely. Can we make these two statements consistent? Can we define the temperature so that two large bodies in equilibrium with one another will have the same temperature?

Consider a general, isolated system of total energy E consisting of two parts, labeled 1 and 2. Each subsystem has a fixed volume and a fixed number of particles, and is energetically weakly connected to the other subsystem. The connection is weak in that we assume we can neglect the dependence of the energy E_1 of the first subsystem on the state s_2 of the second one, and vice versa.[29]

Our microcanonical ensemble then asserts that the equilibrium behavior of the total system is an equal weighting of all possible states of the two subsystems having total energy E. A particular state of the whole system is given by a pair of states (s_1, s_2) with $E = E_1 + E_2$. This immediately implies that a particular configuration or state s_1 of the first subsystem at energy E_1 will occur with probability density[30]

$$\rho(s_1) \propto \Omega_2(E - E_1), \tag{3.20}$$

where $\Omega_1(E_1)\,\delta E_1$ and $\Omega_2(E_2)\,\delta E_2$ are the phase-space volumes of the energy shells for the two subsystems. The volume of the energy surface for the total system at energy E will be given by adding up the product of the volumes of the subsystems for pairs of energies summing to E:

$$\Omega(E) = \int dE_1\, \Omega_1(E_1)\Omega_2(E - E_1), \tag{3.21}$$

as should be intuitively clear.[31] Notice that the integrand in eqn 3.21, normalized by the total integral, is just the probability density[32] for the subsystem to have energy E_1:

$$\rho(E_1) = \Omega_1(E_1)\Omega_2(E - E_1)/\Omega(E). \tag{3.23}$$

If the two subsystems have a large number of particles then it turns out[33] that $\rho(E_1)$ is a very sharply peaked function near its maximum at E_1^*. Hence in equilibrium the energy in subsystem 1 is given (apart from small fluctuations) by the maximum in the integrand $\Omega_1(E_1)\Omega_2(E - E_1)$. The maximum is found when the derivative of the integrand $(d\Omega_1/dE_1)\,\Omega_2 - \Omega_1\,(d\Omega_2/dE_2)$ is zero, which is where

$$\frac{1}{\Omega_1}\frac{d\Omega_1}{dE_1}\bigg|_{E_1^*} = \frac{1}{\Omega_2}\frac{d\Omega_2}{dE_2}\bigg|_{E-E_1^*}. \tag{3.24}$$

It is more convenient not to work with Ω, but rather to work with its logarithm. We define the *equilibrium entropy*

$$S_{\text{equil}}(E) = k_B \log(\Omega(E)) \tag{3.25}$$

for each of our systems.[34] Like the total energy, volume, and number of particles, the entropy in a large system is ordinarily proportional

[29]A macroscopic system attached to the external world at its boundaries is usually weakly connected, since the interaction energy is only important near the surfaces, a negligible fraction of the total volume. More surprising, the momenta and configurations in a nonmagnetic, non quantum system are two uncoupled subsystems: no terms in the Hamiltonian mix them (although the dynamical evolution certainly does).

[30]That is, if we compare the probabilities of two states s_1^a and s_1^b of subsystem 1 with energies E_1^a and E_1^b, and if $\Omega_2(E - E_1^a)$ is 50 times larger than $\Omega_2(E - E_1^b)$, then $\rho(s_1^a) = 50\,\rho(s_1^b)$ because the former has 50 times as many partners that it can pair with to get an allotment of probability.

[32] Warning: Again we are being sloppy; we use $\rho(s_1)$ in eqn 3.20 for the probability density that the subsystem is in a particular state s_1 and we use $\rho(E_1)$ in eqn 3.23 for the probability density that a subsystem is in any of many particular states with energy E_1.

[33]Just as for the configurations of the ideal gas, where the number of particles in half the box fluctuated very little, so also the energy E_1 fluctuates very little from the value E_1^* at which the probability is maximum. We will show this explicitly in Exercise 3.8, and more abstractly in note 37 later.

[34]Again, Boltzmann's constant k_B is a unit conversion factor with units of [energy]/[temperature]; the entropy would be unitless except for the fact that we measure temperature and energy with different scales (note 25 on p. 56).

[31]It is also easy to derive using the Dirac δ-function (Exercise 3.6). We can also derive it, more awkwardly, using energy shells:

$$\Omega(E) = \frac{1}{\delta E}\int_{E<\mathcal{H}_1+\mathcal{H}_2<E+\delta E} d\mathbb{P}_1\,d\mathbb{Q}_1\,d\mathbb{P}_2\,d\mathbb{Q}_2 = \int d\mathbb{P}_1\,d\mathbb{Q}_1\left(\frac{1}{\delta E}\int_{E-\mathcal{H}_1<\mathcal{H}_2<E+\delta E-\mathcal{H}_1} d\mathbb{P}_2\,d\mathbb{Q}_2\right)$$
$$= \int d\mathbb{P}_1\,d\mathbb{Q}_1\,\Omega_2(E - \mathcal{H}_1(\mathbb{P}_1,\mathbb{Q}_1)) = \sum_n \int_{n\delta E<\mathcal{H}_1<(n+1)\delta E} d\mathbb{P}_1\,d\mathbb{Q}_1\,\Omega_2(E - \mathcal{H}_1(\mathbb{P}_1,\mathbb{Q}_1))$$
$$\approx \int dE_1\left(\frac{1}{\delta E}\int_{E_1<\mathcal{H}_1<E_1+\delta E} d\mathbb{P}_1\,d\mathbb{Q}_1\right)\Omega_2(E - E_1) = \int dE_1\,\Omega_1(E_1)\Omega_2(E - E_1), \tag{3.22}$$

where we have converted the sum to an integral $\sum_n f(n\,\delta E) \approx (1/\delta E)\int dE_1 f(E_1)$.

[35]We can see this by using the fact that most large systems can be decomposed into many small, weakly coupled subsystems, for which the entropies add. (For systems with long-range forces like gravitation, breaking the system up into many weakly coupled subsystems may not be possible, and entropy and energy need not be extensive.) This additivity of the entropy for uncoupled systems is exactly true in the canonical ensemble (Section 6.2). It is true for macroscopic systems in the microcanonical ensemble; eqn 3.28 tells us $\Omega(E) \approx \Omega_1(E_1^*)\Omega_2(E_2^*)\int e^{-(E_1-E_1^*)^2/2\sigma_E}\,dE_1 = \Omega_1(E_1^*)\Omega_2(E_2^*)(\sqrt{2\pi}\sigma_E)$, so the entropy of the total system is

$$S_{tot}(E) = k_B \log \Omega(E)$$
$$\approx S_1(E_1^*) + S_2(E - E_1^*)$$
$$+ k_B \log(\sqrt{2\pi}\sigma_E).$$

This is extensive up to the microscopic correction $k_B \log(\sqrt{2\pi}\sigma_E)$ (due to the enhanced energy fluctuations by coupling the two subsystems).

[36]Entropy is a maximum rather than just an extremum because eqn 3.26 is the logarithm of the probability (eqn 3.23) expanded about a maximum.

[38]More correctly, one pays in negative entropy; one must accept entropy $\delta E/T$ when buying energy δE from the heat bath.

to its size.[35] Quantities like these which scale linearly with the system size are called *extensive*. (Quantities like the temperature, pressure, and chemical potential, that stay constant as the system grows are called *intensive*.)

Thus $dS/dE = k_B(1/\Omega)(d\Omega/dE)$, and eqn 3.24 simplifies to the statement

$$\frac{d}{dE_1}\left(S_1(E_1) + S_2(E - E_1)\right) = \left.\frac{dS_1}{dE_1}\right|_{E_1^*} - \left.\frac{dS_2}{dE_2}\right|_{E-E_1^*} = 0 \qquad (3.26)$$

that the total entropy $S_1 + S_2$ is maximized.[36] We want to define the temperature so that it becomes equal when the two subsystems come to equilibrium. We have seen that

$$\frac{dS_1}{dE_1} = \frac{dS_2}{dE_2} \qquad (3.27)$$

in thermal equilibrium. As dS/dE decreases upon increasing energy, we *define the temperature* in statistical mechanics as $1/T = dS/dE$. We have been assuming constant volume and number of particles in our derivation; the formula for a general system is[37]

$$\frac{1}{T} = \left.\frac{\partial S}{\partial E}\right|_{V,N}. \qquad (3.29)$$

The inverse of the temperature is the cost of buying energy from the rest of the world. The lower the temperature, the more strongly the energy is pushed downward. Entropy is the currency being paid. For each unit of energy δE bought, we pay $\delta E/T = \delta E\,(dS/dE) = \delta S$ in reduced entropy of the outside world. Inverse temperature is the cost in entropy to buy a unit of energy.[38]

The "rest of the world" is often called the *heat bath*; it is a source and sink for heat and fixes the temperature. All heat baths are equivalent, depending only on the temperature. More precisely, the equilibrium behavior of a system weakly coupled to the external world is independent of what the external world is made of—it depends only on the world's temperature. This is a deep truth.

[37]Is the probability density $\rho(E_1)$ sharply peaked, as we have assumed? We can Taylor expand the numerator in eqn 3.23 about the maximum $E_1 = E_1^*$, and use the fact that the temperatures balance at E_1^* to remove the terms linear in $E_1 - E_1^*$:

$$\Omega_1(E_1)\Omega_2(E - E_1) = \exp\left(S_1(E_1)/k_B + S_2(E - E_1)/k_B\right)$$
$$\approx \exp\left[\left(S_1(E_1^*) + \frac{1}{2}(E_1 - E_1^*)^2\frac{\partial^2 S_1}{\partial E_1^2} + S_2(E - E_1^*) + \frac{1}{2}(E_1 - E_1^*)^2\frac{\partial^2 S_2}{\partial E_2^2}\right)\Big/k_B\right]$$
$$= \Omega_1(E_1^*)\Omega_2(E_2^*)\exp\left((E_1 - E_1^*)^2\left(\frac{\partial^2 S_1}{\partial E_1^2} + \frac{\partial^2 S_2}{\partial E_2^2}\right)\Big/(2k_B)\right). \qquad (3.28)$$

Thus the energy fluctuations are Gaussian: $\rho(E_1) = (1/\sqrt{2\pi}\sigma_E)\,e^{-(E_1-E_1^*)^2/2\sigma_E^2}$, with standard deviation σ_E given by $1/\sigma_E^2 = -(1/k_B)\left(\partial^2 S_1/\partial E_1^2 + \partial^2 S_2/\partial E_2^2\right)$. (Note the minus sign; $\partial^2 S/\partial E^2 = \partial(1/T)/\partial E$ is typically negative, because $1/T$ decreases as energy increases. This is also the statement that $S(E)$ is convex downward.) Since both S and E are extensive, they are proportional to the number of particles N, and $\sigma_E^2 \propto 1/(\partial^2 S/\partial E^2) \propto N$ (because there is one S in the numerator and two Es in the denominator). Hence, the energy fluctuations per particle σ_E/N are tiny; they scale as $1/\sqrt{N}$. This is typical of fluctuations in statistical mechanics.

3.4 Pressure and chemical potential

The entropy $S(E, V, N)$ is our first example of a *thermodynamic potential*.[39] In thermodynamics, all the macroscopic properties can be calculated by taking derivatives of thermodynamic potentials with respect to their arguments. It is often useful to think of thermodynamic potentials as surfaces; Fig. 3.4 shows the surface in S, E, V space (at constant number of particles N). The energy $E(S, V, N)$ is another thermodynamic potential, completely equivalent to $S(E, V, N)$; it is the same surface with a different direction "up".

In Section 3.3 we defined the temperature using $(\partial S/\partial E)|_{V,N}$. What about the other two first derivatives, $(\partial S/\partial V)|_{E,N}$ and $(\partial S/\partial N)|_{E,V}$? That is, how does the entropy change when volume or particles are exchanged between two subsystems? The change in the entropy for a tiny shift ΔE, ΔV, and ΔN from subsystem 2 to subsystem 1 (Fig. 3.3) is

$$\Delta S = \left(\left.\frac{\partial S_1}{\partial E_1}\right|_{V,N} - \left.\frac{\partial S_2}{\partial E_2}\right|_{V,N}\right)\Delta E + \left(\left.\frac{\partial S_1}{\partial V_1}\right|_{E,N} - \left.\frac{\partial S_2}{\partial V_2}\right|_{E,N}\right)\Delta V$$

$$+ \left(\left.\frac{\partial S_1}{\partial N_1}\right|_{E,V} - \left.\frac{\partial S_2}{\partial N_2}\right|_{E,V}\right)\Delta N. \tag{3.30}$$

The first term is, as before, $(1/T_1 - 1/T_2)\Delta E$; exchanging energy to maximize the entropy sets the temperatures equal. Just as for the energy, if the two subsystems are allowed to exchange volume and number then the entropy will maximize itself with respect to these variables as well, with small fluctuations. Equating the derivatives with respect to volume gives us our statistical mechanics definition of the pressure P:

$$\frac{P}{T} = \left.\frac{\partial S}{\partial V}\right|_{E,N}, \tag{3.31}$$

and equating the derivatives with respect to number gives us the definition of the chemical potential μ:[40]

$$-\frac{\mu}{T} = \left.\frac{\partial S}{\partial N}\right|_{E,V}. \tag{3.32}$$

These definitions are a bit odd; usually we define pressure and chemical potential in terms of the change in energy E, not the change in entropy S. We can relate our definitions to the more usual ones using an important mathematical identity that we derive in Exercise 3.10; if f is a function of x and y, then (see Fig. 3.4)[41]

$$\left.\frac{\partial f}{\partial x}\right|_y \left.\frac{\partial x}{\partial y}\right|_f \left.\frac{\partial y}{\partial f}\right|_x = -1. \tag{3.33}$$

Remember also that if we keep all but one variable fixed, partial derivatives are like regular derivatives, so

$$\left.\frac{\partial f}{\partial x}\right|_y = 1 \bigg/ \left.\frac{\partial x}{\partial f}\right|_y. \tag{3.34}$$

Fig. 3.3 Two subsystems. Two subsystems, isolated from the outside world, may exchange energy (open door through the insulation), volume (piston), or particles (tiny uncorked holes).

[39]Most of the other thermodynamic potentials we will use are more commonly called *free energies*.

[40]These relations are usually summarized in the formula $dS = (1/T)\,dE + (P/T)\,dV - (\mu/T)\,dN$ (see Section 6.4).

[41]Notice that this is exactly minus the result you would have derived by canceling ∂f, ∂x, and ∂y from "numerator" and "denominator"; derivatives are almost like fractions, but not quite.

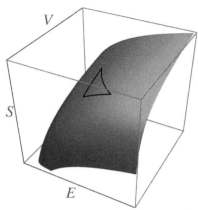

Fig. 3.4 The surface of state. The entropy $S(E, V, N)$ as a function of energy E and volume V (at fixed number N). Viewed sideways, this surface also defines the energy $E(S, V, N)$. The three curves are lines at constant S, E, and V; the fact that they must close yields the relation

$$\left.\frac{\partial S}{\partial E}\right|_{V,N} \left.\frac{\partial E}{\partial V}\right|_{S,N} \left.\frac{\partial V}{\partial S}\right|_{E,N} = -1$$

(see Exercise 3.10).

Using this for $S(E, V)$ and fixing N, we find

$$-1 = \left.\frac{\partial S}{\partial V}\right|_{E,N} \left.\frac{\partial V}{\partial E}\right|_{S,N} \left.\frac{\partial E}{\partial S}\right|_{V,N} = \frac{P}{T}\left(1 \bigg/ \left.\frac{\partial E}{\partial V}\right|_{S,N}\right) T, \qquad (3.35)$$

so

$$\left.\frac{\partial E}{\partial V}\right|_{S,N} = -P. \qquad (3.36)$$

Thus the pressure is minus the energy cost per unit volume at constant entropy. Similarly,

$$-1 = \left.\frac{\partial S}{\partial N}\right|_{E,V} \left.\frac{\partial N}{\partial E}\right|_{S,V} \left.\frac{\partial E}{\partial S}\right|_{N,V} = -\frac{\mu}{T}\left(1 \bigg/ \left.\frac{\partial E}{\partial N}\right|_{S,V}\right) T, \qquad (3.37)$$

so

$$\left.\frac{\partial E}{\partial N}\right|_{S,V} = \mu; \qquad (3.38)$$

the chemical potential is the energy cost of adding a particle at constant entropy.

We can feel pressure and temperature as our bodies exchange volume with balloons and heat with coffee cups. It is not often realized, but we also perceive chemical potentials; our senses of taste and smell are complicated, but they are sensitive to the chemical potential of the molecules comprising the odor and flavor of our foods (Exercise 3.16). You can view chemical potential as a "force" associated with particle number, in the same spirit as pressure is a force associated with volume and temperature is a force associated with energy; differences in μ, P, and T will induce transfers of particles, volume, or energy from one subsystem into another. Chemical potentials are crucial to the study of chemical reactions; whether a reaction will proceed depends in part on the relative cost of the products and the reactants, measured by the differences in their chemical potentials (Section 6.6). They drive the osmotic pressure that holds your cell membranes taut. The chemical potential will also play a central role in calculations involving noninteracting quantum systems, where the number of particles in each quantum state can vary (Chapter 7). Your intuition about chemical potentials will improve as you work with them.

Pressure in mechanics and statistical mechanics (advanced)

Our familiar notion of pressure is from mechanics: the energy of a subsystem increases as the volume decreases, as $\Delta E = -P\Delta V$. Our statistical mechanics definition $(P = -(\partial E/\partial V)|_{S,N}$, eqn 3.36) states that this energy change is measured at fixed entropy—which may not be so familiar.

Not all mechanical volume changes are acceptable for measuring the pressure. A mechanical measurement of the pressure must not exchange heat with the body. (Changing the volume while adding heat to keep the temperature fixed, for example, is a different measurement.) The mechanical measurement must also change the volume slowly. If the volume changes fast enough that the subsystem goes out of equilibrium (typically a piston moving near the speed of sound), then the energy needed to change the volume will include the energy for generating the sound and shock waves—energies that are not appropriate to include in a good measurement of the pressure. We call a process *adiabatic* if it occurs without heat exchange and sufficiently slowly that the system remains in equilibrium.

In this section we will show in a microcanonical ensemble that the mechanical definition of the pressure (the rate of change in average internal energy as one slowly varies the volume, $P_{\mathrm{m}} = -\Delta E/\Delta V$) equals the statistical mechanical definition of the pressure ($P_{\mathrm{stat}} = T(\partial S/\partial V)|_{E,N} = -(\partial E/\partial V)|_{S,N}$, eqns 3.31 and 3.36). Hence, an adiabatic measurement of the pressure is done at constant entropy.

The argument is somewhat technical and abstract, using methods that will not be needed in the remainder of the text. Why is this question important, beyond justifying our definition of pressure? In Chapter 5, the entropy will become our fundamental measure of irreversibility. Since the system remains in equilibrium under adiabatic changes in the volume, its entropy should not change.[42] Our arguments there will work backward from the macroscopic principle that perpetual motion machines should not exist. Our argument here works forward from the microscopic laws, showing that systems that stay in equilibrium (changed adiabatically, thermally isolated and slowly varying) are consistent with a constant entropy.[43]

We must first use statistical mechanics to find a formula for the mechanical force per unit area P_{m}. Consider some general liquid or gas whose volume is changed smoothly from V to $V + \Delta V$, and is otherwise isolated from the rest of the world.

We can find the mechanical pressure if we can find out how much the energy changes as the volume changes. The initial system at $t = 0$ is an microcanonical ensemble at volume V, uniformly filling phase space in an energy range $E < \mathcal{H} < E + \delta E$ with density $1/\Omega(E, V)$. A member of this volume-expanding ensemble is a trajectory $\mathbb{P}(t), \mathbb{Q}(t)$ that evolves in time under the changing Hamiltonian $\mathcal{H}(\mathbb{P}, \mathbb{Q}, V(t))$. The amount this particular trajectory changes in energy under the time-dependent Hamiltonian is

$$\frac{\mathrm{d}\mathcal{H}\big(\mathbb{P}(t), \mathbb{Q}(t), V(t)\big)}{\mathrm{d}t} = \frac{\partial \mathcal{H}}{\partial \mathbb{P}}\dot{\mathbb{P}} + \frac{\partial \mathcal{H}}{\partial \mathbb{Q}}\dot{\mathbb{Q}} + \frac{\partial \mathcal{H}}{\partial V}\frac{\mathrm{d}V}{\mathrm{d}t}. \tag{3.39}$$

A Hamiltonian for particles of kinetic energy $\frac{1}{2}\mathbb{P}^2/m$ and potential energy $U(\mathbb{Q})$ will have $\partial \mathcal{H}/\partial \mathbb{P} = \mathbb{P}/m = \dot{\mathbb{Q}}$ and $\partial \mathcal{H}/\partial \mathbb{Q} = \partial U/\partial \mathbb{Q} = -\dot{\mathbb{P}}$, so the first two terms cancel on the right-hand side of eqn 3.39.[44] Hence the energy change for this particular trajectory is

$$\frac{\mathrm{d}\mathcal{H}\big(\mathbb{P}(t), \mathbb{Q}(t), V(t)\big)}{\mathrm{d}t} = \frac{\partial \mathcal{H}}{\partial V}(\mathbb{P}, \mathbb{Q})\frac{\mathrm{d}V}{\mathrm{d}t}. \tag{3.40}$$

[42]It is the entropy of the entire system, including the mechanical instrument that changes the volume, that cannot decrease. We're using the fact that the instrument can be made with few moving parts that couple to our system; the entropy of a system with only a few degrees of freedom can be neglected (see, however, Exercise 6.23). You will notice that the entropy is always Nk_B times a logarithm. The logarithm is an extremely slowly varying function, so the entropy is always a reasonably small constant times N times k_B. If a system has only a few moving parts N, its entropy is only a few k_B—hence tiny.

[43]Our argument will not use the fact that the parameter V is the volume. Any adiabatic change in the system happens at constant entropy.

[44]Some may recognize these as Hamilton's equations of motion: the cancellation works for general Hamiltonian systems, even those not of the standard Newtonian form.

That is, the energy change of the evolving trajectory is the same as the expectation value of $\partial \mathcal{H} / \partial t$ at the static current point in the trajectory: we need not follow the particles as they zoom around.

We still must average this energy change over the equilibrium ensemble of initial conditions. This is in general not possible, until we make the second assumption involved in the adiabatic measurement of pressure: we assume that the potential energy turns on so slowly that the system remains in equilibrium at the current volume $V(t)$ and energy $E(t)$. This allows us to calculate the ensemble average energy change in terms of an equilibrium average at the fixed, current volume:

$$\frac{\mathrm{d}\langle\mathcal{H}\rangle}{\mathrm{d}t} = \left\langle \frac{\partial\mathcal{H}}{\partial V} \right\rangle_{E(t),V(t)} \frac{\mathrm{d}V}{\mathrm{d}t}. \tag{3.41}$$

Since this energy change must equal $-P_{\mathrm{m}} (\mathrm{d}V/\mathrm{d}t)$, we find (eqn 3.4):

$$- P_{\mathrm{m}} = \left\langle \frac{\partial\mathcal{H}}{\partial V} \right\rangle = \frac{1}{\Omega(E)} \int \mathrm{d}\mathbb{P}\,\mathrm{d}\mathbb{Q}\ \delta(E - \mathcal{H}(\mathbb{P},\mathbb{Q},V)) \frac{\partial\mathcal{H}}{\partial V}. \tag{3.42}$$

We now turn to calculating the derivative of interest for the statistical mechanical definition of pressure:

$$\left.\frac{\partial S}{\partial V}\right|_{E,N} = \frac{\partial}{\partial V} k_B \log(\Omega) = \frac{k_B}{\Omega} \left.\frac{\partial\Omega}{\partial V}\right|_{E,N}. \tag{3.43}$$

Using eqn 3.2 to write Ω in terms of a derivative of the Θ function, we can change orders of differentiation:

$$
\begin{aligned}
\left.\frac{\partial\Omega}{\partial V}\right|_{E,N} &= \left.\frac{\partial}{\partial V}\right|_{E,N} \left.\frac{\partial}{\partial E}\right|_{V,N} \int \mathrm{d}\mathbb{P}\,\mathrm{d}\mathbb{Q}\ \Theta(E - \mathcal{H}(\mathbb{P},\mathbb{Q},V)) \\
&= \left.\frac{\partial}{\partial E}\right|_{V,N} \int \mathrm{d}\mathbb{P}\,\mathrm{d}\mathbb{Q}\ \frac{\partial}{\partial V} \Theta(E - \mathcal{H}(\mathbb{P},\mathbb{Q},V)) \\
&= -\left.\frac{\partial}{\partial E}\right|_{V,N} \int \mathrm{d}\mathbb{P}\,\mathrm{d}\mathbb{Q}\ \delta(E - \mathcal{H}(\mathbb{P},\mathbb{Q},V)) \frac{\partial\mathcal{H}}{\partial V}.
\end{aligned} \tag{3.44}
$$

But the phase-space integral in the last equation is precisely the same integral that appears in our mechanical formula for the pressure, eqn 3.42: it is $\Omega(E)(-P_{\mathrm{m}})$. Thus

$$
\begin{aligned}
\left.\frac{\partial\Omega}{\partial V}\right|_{E,N} &= \left.\frac{\partial}{\partial E}\right|_{V,N} (\Omega(E)P_{\mathrm{m}}) \\
&= \left.\frac{\partial\Omega}{\partial E}\right|_{V,N} P_{\mathrm{m}} + \Omega \left.\frac{\partial P_{\mathrm{m}}}{\partial E}\right|_{V,N},
\end{aligned} \tag{3.45}
$$

so

$$
\begin{aligned}
\left.\frac{\partial S}{\partial V}\right|_{E,N} &= \frac{\partial}{\partial V} k_B \log(\Omega) = \frac{k_B}{\Omega} \left(\left.\frac{\partial\Omega}{\partial E}\right|_{V,N} P_{\mathrm{m}} + \Omega \left.\frac{\partial P_{\mathrm{m}}}{\partial E}\right|_{V,N} \right) \\
&= \left.\frac{\partial k_B \log(\Omega)}{\partial E}\right|_{V,N} P_{\mathrm{m}} + k_B \left.\frac{\partial P_{\mathrm{m}}}{\partial E}\right|_{V,N} \\
&= \left.\frac{\partial S}{\partial E}\right|_{V,N} P_{\mathrm{m}} + k_B \left.\frac{\partial P_{\mathrm{m}}}{\partial E}\right|_{V,N} = \frac{P_{\mathrm{m}}}{T} + k_B \left.\frac{\partial P_{\mathrm{m}}}{\partial E}\right|_{V,N}.
\end{aligned} \tag{3.46}
$$

Now, P and T are both intensive variables, but E is extensive (scales linearly with system size). Hence P/T is of order one for a large system, but $k_B(\partial P/\partial E)$ is of order $1/N$ where N is the number of particles. (For example, we shall see that for the ideal gas, $PV = \frac{2}{3}E = Nk_BT$, so $k_B(\partial P/\partial E) = 2k_B/3V = 2P/3NT = 2P/3NT \ll P/T$ for large N.) Hence the second term, for a large system, may be neglected, giving us the desired relation:

$$\left.\frac{\partial S}{\partial V}\right|_{E,N} = \frac{P_m}{T}. \tag{3.47}$$

The derivative of the entropy $S(E, V, N)$ with respect to V at constant E and N is thus indeed the mechanical pressure divided by the temperature.

3.5 Entropy, the ideal gas, and phase-space refinements

Let us find the temperature and pressure for the ideal gas, using our microcanonical ensemble. We will then introduce two subtle refinements to the phase-space volume (one from quantum mechanics, and one for undistinguished particles) which will not affect the temperature or pressure, but will be important for the entropy and chemical potential.

We derived the volume $\Omega(E)$ of the energy shell in phase space in Section 3.2; it factored[45] into a momentum-shell volume from eqn 3.14 and a configuration-space volume V^N. Before our refinements, we have

[45] It factors only because the potential energy is zero.

$$\Omega_{\mathrm{crude}}(E) = V^N \left(\frac{3N}{2E}\right) \pi^{3N/2}(2mE)^{3N/2} \Big/ (3N/2)!$$

$$\approx V^N \pi^{3N/2}(2mE)^{3N/2} \Big/ (3N/2)!. \tag{3.48}$$

Notice that in the second line of eqn 3.48 we have dropped the term $3N/2E$; it divides the phase-space volume by a negligible factor (two-thirds the energy per particle).[46] The entropy and its derivatives are (before our refinements)

[46] Multiplying $\Omega(E)$ by a factor independent of the number of particles is equivalent to adding a constant to the entropy. The entropy of a typical system is so large (of order Avogadro's number times k_B) that adding a number-independent constant to it is irrelevant. Notice that this implies that $\Omega(E)$ is so large that *multiplying* it by a constant does not significantly change its value (Exercise 3.2).

$$S_{\mathrm{crude}}(E) = k_B \log\left(V^N \pi^{3N/2}(2mE)^{3N/2} \Big/ (3N/2)!\right)$$

$$= Nk_B \log(V) + \frac{3Nk_B}{2} \log(2\pi mE) - k_B \log[(3N/2)!], \tag{3.49}$$

$$\frac{1}{T} = \left.\frac{\partial S}{\partial E}\right|_{V,N} = \frac{3Nk_B}{2E}, \tag{3.50}$$

$$\frac{P}{T} = \left.\frac{\partial S}{\partial V}\right|_{E,N} = \frac{Nk_B}{V}, \tag{3.51}$$

so the temperature and pressure are given by

$$k_BT = \frac{2E}{3N}, \tag{3.52}$$

$$PV = Nk_BT. \tag{3.53}$$

The first line above is the temperature formula we promised in forming eqn 3.19; velocity components of the particles in an ideal gas each have average energy equal to $\frac{1}{2}k_BT$.

The second formula is the *equation of state*[47] for the ideal gas. The equation of state is the relation between the macroscopic variables of an equilibrium system that emerges in the limit of large numbers of particles. The force per unit area on the wall of an ideal gas will fluctuate in time around the pressure $P(T, V, N) = Nk_BT/V$ given by the equation of state, with the magnitude of the fluctuations vanishing as the system size gets large.

In general, our definition for the energy-shell volume in phase space needs two refinements. First, the phase-space volume has dimensions of $([\text{length}][\text{momentum}])^{3N}$; the volume of the energy shell depends multiplicatively upon the units chosen for length, mass, and time. Changing these units will change the corresponding crude form for the entropy by adding a constant times $3N$. Most physical properties, like temperature and pressure, are dependent only on derivatives of the entropy, so the overall constant will not matter; indeed, the zero of the entropy is undefined within classical mechanics. It is suggestive that $[\text{length}][\text{momentum}]$ has units of Planck's constant h, and we shall see in Chapter 7 that quantum mechanics in fact does set the zero of the entropy. We shall see in Exercise 7.3 that dividing[48] $\Omega(E)$ by h^{3N} nicely sets the entropy density to zero in equilibrium quantum systems at absolute zero.

Second, there is an important subtlety in quantum physics regarding *identical* particles. Two electrons, or two helium atoms of the same isotope, are not just hard to tell apart; they really are completely and utterly the same (Fig. 7.3). We shall see in Section 7.3 that the proper quantum treatment of identical particles involves averaging over possible states using Bose and Fermi statistics.

In classical physics, there is an analogous subtlety regarding *undistinguished* particles. Undistinguished classical particles include the case of identical (indistinguishable) particles at high temperatures, where the Bose and Fermi statistics become unimportant (Chapter 7). We use the term *undistinguished* to also describe particles which in principle are not identical, but for which our Hamiltonian and measurement instruments treat identically (pollen grains and colloidal particles, for example). For a system of two undistinguished particles, the phase-space points $(\mathbf{p}_A, \mathbf{p}_B, \mathbf{q}_A, \mathbf{q}_B)$ and $(\mathbf{p}_B, \mathbf{p}_A, \mathbf{q}_B, \mathbf{q}_A)$ should not both be counted; the volume of phase space $\Omega(E)$ should be half that given by a calculation for distinguished particles. For N undistinguished particles, the phase-space volume should be divided by $N!$, the total number of ways the labels for the particles can be permuted.[49]

Unlike the introduction of the factor h^{3N} above, dividing the phase-space volume by $N!$ does change the predictions of classical statistical mechanics in important ways. We will see in Section 5.2.1 that the entropy increase for joining containers of different kinds of particles should

[47]It is rare that the equation of state can be written out as an explicit equation! Only in special cases (e.g., noninteracting systems like the ideal gas) can one solve in closed form for the thermodynamic potentials, equations of state, or other properties.

[48]This is equivalent to using units for which $h = 1$.

[49]This $N!$ is sometimes known as the *Gibbs factor*.

be substantial, while the entropy increase for joining containers filled with undistinguished particles should be near zero. This result is correctly treated by dividing $\Omega(E)$ by $N!$ for each set of N undistinguished particles. We call the resulting ensemble *Maxwell–Boltzmann* statistics, to distinguish it from distinguishable statistics and from the quantum-mechanical Bose and Fermi statistics.

Combining these two refinements gives us the proper energy-shell volume for classical undistinguished particles, replacing eqn 3.5:[50]

$$\Omega(E) = \int_{E < \mathcal{H}(\mathbb{P},\mathbb{Q}) < E + \delta E} \frac{d\mathbb{P}\, d\mathbb{Q}}{N!\, h^{3N}}. \tag{3.54}$$

For the ideal gas, this refines Ω_{crude} of eqn 3.48 to

$$\Omega(E) = (V^N/N!)\, (\pi^{3N/2}(2mE)^{3N/2} / (3N/2)!)\, (1/h)^{3N}, \tag{3.55}$$

$$S(E) = Nk_B \log\left[\frac{V}{h^3}(2\pi mE)^{3/2}\right] - k_B \log\left[N!\,(3N/2)!\right]. \tag{3.56}$$

We can make our equation for the ideal gas entropy more useful by using Stirling's formula $\log(N!) \approx N \log N - N$, valid at large N:

$$S(E, V, N) = \frac{5}{2}Nk_B + Nk_B \log\left[\frac{V}{Nh^3}\left(\frac{4\pi mE}{3N}\right)^{3/2}\right]. \tag{3.57}$$

This is the standard formula for the entropy of an ideal gas. We can put it into a somewhat simpler form by writing it in terms of the particle density $\rho = N/V$:

$$S = Nk_B \left(\frac{5}{2} - \log(\rho\lambda^3)\right), \tag{3.58}$$

where[51]

$$\lambda = h/\sqrt{4\pi mE/3N} \tag{3.59}$$

is called the *thermal de Broglie wavelength*, and will be physically significant for quantum systems at low temperature (Chapter 7).

[50] Note that we also have dropped δE from the left-hand side. Just as in eqn 3.48, where we dropped a factor of $3N/2E$ for the ideal gas to simplify the formulas (e.g., making the entropy extensive), one factor of δE is negligible when the number of particles N is large (see note 46).

[51] De Broglie realized that matter could act as a wave; a particle of momentum p had a wavelength $\lambda_{\text{quantum}} = h/p$. The mean square of one component of momentum in our gas is $p^2 = 2m(E/3N)$, so our particles have a quantum wavelength of $h/\sqrt{2mE/3N} = \sqrt{2\pi}\lambda_{\text{thermal}}$—close enough that we give de Broglie's name to the thermal λ as well.

Exercises

It is fun to notice strange things that arise because things in our world have so many atoms, as in *Temperature and energy, Large and very large numbers*, and *Weirdness in high dimensions*. *Undistinguished particles* explores the Gibbs factor $1/N!$ in our phase-space volume, a preview of the entropy of mixing, and *Ideal gas glass* applies the Gibbs factor to the ideal gas to model the entropy of glass.

Escape velocity explores why the Earth is not a gas giant like Jupiter. *Pressure computation* provides a concrete derivation of the pressure for the ideal gas, directly from a simulation of the molecular impacts on a surface. *Hard sphere gas* calculates a new equation of state, one step more sophisticated than the ideal gas law. *Taste, smell, and chemical potentials* explains how our nose and mouth sense μ just as our fingers sense P and T.

Five exercises explore the statistics and fluctuations of weakly coupled systems. *Entropy maximization and temperature* motivates and *Connecting two macroscopic systems* rederives the product law for the phase-space energy-shell volume. *Gas mixture* shows that the energy fluctuations are tiny in a specific case; *Microcanonical energy fluctuations* shows this is true for any large system. *Gauss and Poisson* explores the dependence of these fluctuations on the outside world.

Pendulum energy shell explores the geometrical richness of the energy surface for systems with potential energy. *Triple product relation*, and *Maxwell relations* introduce some of the tricky partial derivative relations in thermodynamics. *Solving the pendulum* introduces the numerical methods used in molecular dynamics, which can be used to simulate more realistic gases (and liquids and solids), emphasizing the three themes of accuracy, stability, and fidelity. Finally, the *Random energy model* provides a prototype theory for glasses and disordered systems—an advanced application of the microcanonical ensemble.

(3.1) **Temperature and energy.** ⓘ
What units [joules, millijoules, microjoules, nanojoules, ..., zeptojoules (10^{-21} joules), yoctojoules (10^{-24} joules)] would we use to measure temperature if we used energy units instead of introducing Boltzmann's constant $k_B = 1.3807 \times 10^{-23}$ J/K?
Temperature is a typical energy of a single atomic degree of freedom. The 10^{-23} in Boltzmann's constant is small because the 10^{23} in Avogadro's number is large.

(3.2) **Large and very large numbers.** @
The numbers that arise in statistical mechanics can defeat your calculator. A googol is 10^{100} (one with a hundred zeros after it). A googolplex is 10^{googol}.[52]
Consider a monatomic ideal gas with one mole of particles ($N =$ Avogadro's number, 6.02×10^{23}), room temperature $T = 300$ K, and volume $V = 22.4$ liters (at atmospheric pressure).
(a) *Which of the properties (S, T, E, and $\Omega(E)$) of our gas sample are larger than a googol? A googolplex? Does it matter what units you use, within reason?*
If you double the size of a large equilibrium system (say, by taking two copies and weakly coupling them), some properties will be roughly unchanged; these are called *intensive*. Some, like the number N of particles, will roughly double;

they are called *extensive*. Some will grow much faster than the size of the system.
(b) *To which category (intensive, extensive, faster) does each property from part (a) belong?*
For a large system of N particles, one can usually ignore terms which add a constant independent of N to extensive quantities. (Adding 17 to 10^{23} does not change it enough to matter.) For properties which grow even faster, overall *multiplicative* factors often are physically unimportant.

(3.3) **Escape velocity.** ②
The molecules in planetary atmospheres are slowly leaking into interstellar space. The trajectory of each molecule is a random walk, with a step size that grows with height as the air becomes more dilute. Let us consider a crude ideal gas model for this leakage, with no collisions except with the ground.
For this exercise, treat diatomic oxygen as a monatomic ideal gas with twice the mass of an oxygen atom $m_{O_2} = 2m_O$. (This is not an approximation; the rotations and vibrations do not affect the center-of-mass trajectory.) Assume that the probability distribution for the z-component of momentum[53] is that of the ideal gas, given in eqn 3.19: $\rho(p_z) = 1/\sqrt{2\pi m k_B T} \exp(-p_z^2/2m k_B T)$. For your convenience, $k_B = 1.3807 \times 10^{-16}$ erg/K, $m_{O_2} = 5.3 \times 10^{-23}$ g, and $T = 300$ K.
(a) *What is the RMS vertical velocity $\sqrt{\langle v_z^2 \rangle}$ of an O_2 molecule? If a collisionless molecule started at the Earth's surface with this RMS vertical velocity, how high would it reach? How long before it hit the ground?* (Useful constant: $g = 980$ cm/s².)
(b) *Give the probability that an O_2 molecule will have a vertical component of the velocity greater than Earth's escape velocity (about 11 km/s).* (Hint: $\int_x^\infty e^{-t^2} dt \approx e^{-x^2}/2x$ for large x.)
(c) *If we assume that the molecules do not collide with one another, and each thermalizes its velocity each time it collides with the ground (at roughly the time interval from part (a)), about what fraction of the oxygen molecules will we lose per year? Do we need to worry about losing our atmosphere?* (It is useful to know that there happen to be about $\pi \times 10^7$ seconds in a year.)

---

[52]The firm Google is named for the number googol, with a slight spelling change. They named their corporate headquarters the googleplex.
[53]That is, we ignore wind and the rotation of the Earth.

Try this for H_2, using $T = 1,000\,\mathrm{K}$ (the temperature in the upper atmosphere where the last collision occurs). Is this why Jupiter has a hydrogen atmosphere, and Earth does not?

(3.4) **Pressure simulation.**[54] (Computation) ②
Microscopically, pressure comes from atomic collisions onto a surface. Let us calculate this microscopic pressure for an ideal gas, both analytically and using a molecular dynamics simulation. Run a simulation of the ideal gas in a system with reflective walls. Each time an atom collides with a wall, it undergoes *specular reflection*, with the parallel momentum components unchanged and the perpendicular momentum component reversed.

(a) *Remembering that pressure $P = F/A$ is the force per unit area, and that force $F = \mathrm{d}p/\mathrm{d}t = (\sum \Delta P)/\Delta t$ is the net rate of momentum per unit time. Suppose a wall of area A at $x = L$ is holding atoms to values $x < L$ inside a box. Write a formula for the pressure in terms of $\rho_c(p_x)$, the expected number of collisions at that wall per unit time with incoming momentum p_x.* (Hint: Check the factors of two, and limits of your integral. Do negative momenta contribute?)

The simulation provides an *observer*, which records the magnitudes of all impacts on a wall during a given time interval.

(b) *Make a histogram of the number of impacts on the wall during an interval Δt with momentum transfer Δp. By what factor must you multiply $\rho_c(p_x)$ from part (a) to get this histogram?*
Unlike the distribution of momenta in the gas, the probability $\rho_c(p_x)$ of a wall collision with momentum p_x goes to zero as p_x goes to zero; the ideal gas atoms which are not moving do not collide with walls. The density of particles of momentum p_x per unit volume per unit momentum is the total density of particles N/V times the probability that a particle will have momentum p_x (eqn 3.19):

$$\frac{N}{V} \frac{1}{\sqrt{2\pi m k_B T}} \exp\left(-\frac{p_x{}^2}{2m k_B T}\right). \qquad (3.60)$$

(c) *In a time Δt, from how far away will will atoms of incoming momentum p_x collide with*

the wall? What should the resulting formula be for $\rho_c(p_x)$? Does it agree with your histogram of part (b)? What is your resulting equation for the pressure P? Does it agree with the ideal gas law?

(3.5) **Hard sphere gas.** @

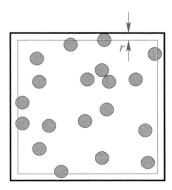

Fig. 3.5 Hard sphere gas.

We can improve on the realism of the ideal gas by giving the atoms a small radius. If we make the potential energy infinite inside this radius (*hard spheres*), the potential energy is simple (zero unless the spheres overlap, which is forbidden). Let us do this in two dimensions; three dimensions is only slightly more complicated, but harder to visualize.

A two-dimensional $L \times L$ box with hard walls contains a gas of N distinguishable[55] hard disks of radius $r \ll L$ (Fig. 3.5). The disks are dilute; the summed area $N\pi r^2 \ll L^2$. Let A be the effective area allowed for the disks in the box (Fig. 3.5): $A = (L - 2r)^2$.

Fig. 3.6 Excluded area around a hard disk.

---

[54]This exercise and the associated software were developed in collaboration with Christopher Myers. Computational hints can be found at the book website [181].
[55]Using indistinguishable particles adds a factor of $1/N!$ to all the estimates for Ω, the volume in configuration space, and $-k_B \log(N!) \approx -k_B(N \log N - N)$ to all the entropy estimates. It does not change the pressure, and does make the calculation slightly less illuminating.

(a) *The area allowed for the second disk is $A - \pi(2r)^2$ (Fig. 3.6), ignoring the small correction when the excluded region around the first disk overlaps the excluded region near the walls of the box. What is the allowed $2N$-dimensional volume in configuration space*[56] *Ω_{HD}^Q of allowed zero-energy configurations of hard disks, in this dilute limit? Leave your answer as a product of N terms.*

Our formula in part (a) expresses Ω_{HD}^Q strangely, with each disk in the product only feeling the excluded area from the former disks. For large numbers of disks and small densities, we can rewrite Ω_{HD}^Q more symmetrically, with each disk feeling the same excluded area $A_{\rm excl}$.

(b) *Use $(1 - \epsilon) \approx \exp(-\epsilon)$ and*[57] *$\sum_{m=0}^{N-1} m = N(N-1)/2 \approx N^2/2$ to approximate $\Omega_{HD}^Q \approx (Ae^{-N\delta})^N$, solving for δ and evaluating any products or sums over disks. Then use $\exp(-\epsilon) \approx (1 - \epsilon)$ again to write $\Omega_{HD}^Q \approx (A - A_{\rm excl})^N$. Interpret your formula for the excluded area $A_{\rm excl}$, in terms of the range of excluded areas you found in part (a) as you added disks.*

(c) *Find the pressure for the hard-disk gas in the large N approximation of part (b), as a function of temperature T, A, r, and N. Does it reduce to the ideal gas law if the disk radius $r = 0$?* (Hint: Constant energy is the same as constant temperature for hard particles, since the potential energy is zero.)

(3.6) **Connecting two macroscopic systems.** ③
An isolated system with energy E is composed of two macroscopic subsystems, each of fixed volume V and number of particles N. The subsystems are weakly coupled, so the sum of their energies is $E_1 + E_2 = E$ (Fig. 3.3 with only the energy door open). We can use the Dirac delta-function $\delta(x)$ (note 3 on p. 7) to define the volume of the energy surface of a system with Hamiltonian $\mathcal{H}$:

$$\Omega(E) = \int \frac{d\mathbb{P}\, d\mathbb{Q}}{h^{3N}} \delta\left(E - \mathcal{H}(\mathbb{P}, \mathbb{Q})\right) \quad (3.61)$$

$$= \int \frac{d\mathbb{P}_1\, d\mathbb{Q}_1}{h^{3N_1}} \frac{d\mathbb{P}_2\, d\mathbb{Q}_2}{h^{3N_2}}$$

$$\times \delta\left(E - (\mathcal{H}_1(\mathbb{P}_1, \mathbb{Q}_1) + \mathcal{H}_2(\mathbb{P}_2, \mathbb{Q}_2))\right).$$

Derive formula 3.21, $\Omega(E) = \int dE_1\, \Omega_1(E_1)\Omega_2(E - E_1)$, for the volume of the energy surface of the whole system using Dirac δ-functions (instead of using energy shells, as in eqn 3.22). (Hint: Insert $\int \delta(E_1 - \mathcal{H}_1(\mathbb{P}_1, \mathbb{Q}_1))\, dE_1 \equiv 1$ into eqn 3.61.)

(3.7) **Gas mixture.** ③
Consider a monatomic gas (He) mixed with a diatomic gas (H_2). We approximate both as ideal gases, so we may treat the helium gas as a separate system, weakly coupled to the hydrogen gas (despite the intermingling of the two gases in space). We showed that a monatomic ideal gas of N atoms has $\Omega_1(E_1) \propto E_1^{3N/2}$. A diatomic molecule has $\Omega_2(E_2) \propto E_2^{5N/2}$.[58]

(a) *Calculate the probability density of system 1 being at energy E_1 (eqn 3.23). For these two gases, which energy $E_1^{\rm max}$ has the maximum probability?*

(b) *Approximate your answer to part (a) as a Gaussian, by expanding the logarithm in a Taylor series $\log(\rho(E_1)) \approx \log(\rho(E_1^{\rm max})) + (E_1 - E_1^{\rm max}) + \dots$ up to second derivatives, and then re-exponentiating. In this approximation, what is the mean energy $\langle E_1 \rangle$? What are the energy fluctuations per particle $\sqrt{\langle (E_1 - E_1^{\rm max})^2 \rangle}/N$? Are they indeed tiny (proportional to $1/\sqrt{N}$)?*

For subsystems with large numbers of particles N, temperature and energy density are well defined because $\Omega(E)$ for each subsystem grows extremely rapidly with increasing energy, in such a way that $\Omega_1(E_1)\Omega_2(E - E_1)$ is sharply peaked near its maximum.

(3.8) **Microcanonical energy fluctuations.** ②
We argued in Section 3.3 that the energy fluctuations between two weakly coupled subsystems are of order $\sqrt{N}$. Let us calculate them explicitly.

Equation 3.28 showed that for two subsystems with energy E_1 and $E_2 = E - E_1$ the probability density of E_1 is a Gaussian with variance (standard deviation squared):

$$\sigma_{E_1}^2 = -k_B / \left(\partial^2 S_1/\partial E_1^2 + \partial^2 S_2/\partial E_2^2\right). \quad (3.62)$$

---

[56] Again, ignore small corrections when the excluded region around one disk overlaps the excluded regions around other disks, or near the walls of the box.

[57] If this formula is not familiar, you can check the exact formula for the first few N, or convert the sum to an integral $\int_0^{N-1} m\, dm \approx \int_0^N m\, dm = N^2/2$.

[58] This is true in the range $\hbar^2/2I \ll k_B T \ll \hbar\omega$, where ω is the vibrational frequency of the stretch mode and I is the moment of inertia. The lower limit makes the rotations classical; the upper limit freezes out the vibrations, leaving us with three classical translation modes and two rotational modes—a total of five harmonic momentum degrees of freedom.

(a) *Show that*

$$\frac{1}{k_B}\frac{\partial^2 S}{\partial E^2} = -\frac{1}{k_B T}\frac{1}{N c_v T}, \qquad (3.63)$$

where c_v is the inverse of the total specific heat at constant volume. (The specific heat c_v is the energy needed per particle to change the temperature by one unit: $N c_v = (\partial E/\partial T)|_{V,N}$.)

The denominator of eqn 3.63 is the product of two energies. The second term $N c_v T$ is a system-scale energy; it is the total energy that would be needed to raise the temperature of the system from absolute zero, if the specific heat per particle c_v were temperature independent. However, the first energy, $k_B T$, is an atomic-scale energy independent of N. The fluctuations in energy, therefore, scale like the geometric mean of the two, summed over the two subsystems in eqn 3.28, and hence scale as $\sqrt{N}$; the total energy fluctuations per particle are thus roughly $1/\sqrt{N}$ times a typical energy per particle.

This formula is quite remarkable; it is a fluctuation-response relation (see Section 10.7). Normally, to measure a specific heat one would add a small energy and watch the temperature change. This formula allows us to measure the specific heat of an object by watching the equilibrium fluctuations in the energy. These fluctuations are tiny for the sample sizes in typical experiments, but can be quite substantial in computer simulations.

(b) *If $c_v^{(1)}$ and $c_v^{(2)}$ are the specific heats per particle for two subsystems of N particles each, show using eqns 3.62 and 3.63 that*

$$\frac{1}{c_v^{(1)}} + \frac{1}{c_v^{(2)}} = \frac{N k_B T^2}{\sigma_{E_1}^2}. \qquad (3.64)$$

We do not even need to couple two systems. The positions and momenta of a molecular dynamics simulation (atoms moving under Newton's laws of motion) can be thought of as two uncoupled subsystems, since the kinetic energy does not depend on the configuration $\mathbb{Q}$, and the potential energy does not depend on the momenta $\mathbb{P}$.

Assume a molecular dynamics simulation of N interacting particles has measured the kinetic energy as a function of time in an equilibrium, constant-energy simulation,[59] and has found a mean kinetic energy $K = \langle E_1 \rangle$ and a standard deviation σ_K.

(c) *Using the equipartition theorem, write the temperature in terms of K. Show that $c_v^{(1)} = 3k_B/2$ for the momentum degrees of freedom. In terms of K and σ_K, solve for the total specific heat of the molecular dynamics simulation (configurational plus kinetic).*

(3.9) **Gauss and Poisson.** ②
The deep truth underlying equilibrium statistical mechanics is that the behavior of large, weakly coupled systems is largely independent of their environment. What kind of heat bath surrounds the system is irrelevant, so long as it has a well-defined temperature, pressure, and chemical potential. This is not true, however, of the *fluctuations* around the average behavior, unless the bath is large compared to the system. In this exercise, we will explore the number fluctuations of a subvolume of a total system K times as large.

Let us calculate the probability of having n particles in a subvolume V, for a box with total volume KV and a total number of particles $T = K N_0$. For $K = 2$ we will derive our previous result, eqn 3.11, including the prefactor. As $K \to \infty$ we will derive the infinite volume result.

(a) *Find the exact formula for this probability; n particles in V, with a total of T particles in KV.* (Hint: What is the probability that the *first* n particles fall in the subvolume V, and the remainder $T - n$ fall outside the subvolume $(K - 1)V$? How many ways are there to pick n particles from T total particles?)

The Poisson probability distribution

$$\rho_n = a^n e^{-a}/n! \qquad (3.65)$$

arises in many applications. It arises whenever there is a large number of possible events T each with a small probability a/T; e.g., the number of cars passing a given point during an hour on a mostly empty street, the number of cosmic rays hitting in a given second, etc.

(b) *Show that the Poisson distribution is normalized: $\sum_n \rho_n = 1$. Calculate the mean of the distribution $\langle n \rangle$ in terms of a. Calculate the variance (standard deviation squared) $\langle (n - \langle n \rangle)^2 \rangle$.*

(c) *As $K \to \infty$, show that the probability that n particles fall in the subvolume V has the Poisson distribution 3.65. What is a?* (Hint: You will need to use the fact that $e^{-a} = (e^{-1/K})^{Ka} \to (1 - 1/K)^{Ka}$ as $K \to \infty$, and the fact that

---

[59] Assume hard walls, so total momentum and angular momentum are not conserved.

$n \ll T$. Here do not assume that n is large; the Poisson distribution is valid even if there are only a few events.)

From parts (b) and (c), you should be able to conclude that the variance in the number of particles found in a volume V inside an infinite system should be equal to N_0, the expected number of particles in the volume:

$$\langle (n - \langle n \rangle)^2 \rangle = N_0. \qquad (3.66)$$

This is twice the squared fluctuations we found for the case where the volume V was half of the total volume, eqn 3.11. That makes sense, since the particles can fluctuate more freely in an infinite volume than in a doubled volume.

If N_0 is large, will the probability P_m that N_0+m particles lie inside our volume still be Gaussian? Let us check this for all K. First, as in Section 3.2.1, let us use the weak form of Stirling's approximation, eqn 3.8 dropping the square root: $n! \sim (n/e)^n$.

(d) *Using your result from part (a), write the exact formula for* $\log(P_m)$*. Apply the weak form of Stirling's formula. Expand your result around* $m = 0$ *to second order in* m*, and show that* $\log(P_m) \approx -m^2/2\sigma_K^2$*, giving a Gaussian form*

$$P_m \sim e^{-m^2/2\sigma_K^2}. \qquad (3.67)$$

What is σ_K*? In particular, what are* σ_2 *and* σ_∞*?* Your result for σ_2 should agree with the calculation in Section 3.2.1, and your result for σ_∞ should agree with eqn 3.66.

Finally, we should address the normalization of the Gaussian. Notice that the ratio of the strong and weak forms of Stirling's formula (eqn 3.8) is $\sqrt{2\pi n}$. We need to use this to produce the normalization $1/\sqrt{2\pi}\sigma_K$ of our Gaussian.

(e) *In terms of* T *and* n*, what factor would the square root term have contributed if you had kept it in Stirling's formula going from part (a) to part (d)? (It should look like a ratio involving three terms like* $\sqrt{2\pi X}$*.) Show from eqn 3.67 that the fluctuations are small,* $m = n - N_0 \ll N_0$ *for large* N_0*. Ignoring these fluctuations, set* $n = N_0$ *in your factor, and give the prefactor multiplying the Gaussian in eqn 3.67.* (Hint: Your answer should be normalized.)

(3.10) **Triple product relation.** (Thermodynamics, Mathematics) @

In traditional thermodynamics, there are many useful formulas like

$$dE = T\,dS - P\,dV \qquad (3.68)$$

(see Section 6.4 and the inside front cover of this text). For example, if V is held constant (and hence $dV = 0$ then $dE = T\,dS$ from eqn 3.68), giving $(\partial S/\partial E)|_V = 1/T$ (the definition of temperature, eqn 3.29).

(a) *Use eqn 3.68 to rederive the traditional formula for the pressure* P.

Let us consider a general formula of this type,

$$A\,dx + B\,dy + C\,df = 0. \qquad (3.69)$$

(b) *What is* $(\partial f/\partial x)|_y$*?* $(\partial f/\partial y)|_x$*?* $(\partial x/\partial y)|_f$*? Use these to derive the triple product relation eqn 3.33,* $(\partial x/\partial y)|_f(\partial y/\partial f)|_x(\partial f/\partial x)|_y = -1$.

The author has always been uncomfortable with manipulating dXs.[60] How can we derive these relations geometrically, with traditional partial derivatives? Our equation of state $S(E, V, N)$ at fixed N is a surface embedded in three dimensions. Figure 3.4 shows a triangle on this surface, which we can use to derive the general triple-product relation between partial derivatives.

(c) *Derive the triple product relation* $(\partial f/\partial x)|_y(\partial x/\partial y)|_f(\partial y/\partial f)|_x = -1$. (Hint: Consider the triangular path in Fig. 3.4, viewing $f = S$, $x = E$, and $y = V$, so the surface is $S(E, V)$. The first side starts at the lower right at (E_0, V_0, S_0) and moves along the hypotenuse at constant S to $V_0 + \Delta V$. The resulting vertex at the upper left will thus be at $(E_0 + (\partial E/\partial V)|_S\Delta V, V_0 + \Delta V, S_0)$. The second side runs at constant E back to $V = V_0$, and the third side runs at constant V back to (E_0, V_0, S_0). The curve must close to make S a single-valued function; the resulting equation should imply the triple-product relation.)

(3.11) **Maxwell relations.** (Thermodynamics) @
Consider the microcanonical formula for the equilibrium energy $E(S, V, N)$ of some general system.[61] One knows that the second derivatives of E are symmetric; at fixed N, we get the same answer whichever order we take partial derivatives with respect to S and V.

---

[60]They are really *differential forms*, which are mathematically subtle (see note 23 on p. 150).
[61]One can derive the formula by solving $S = S(N, V, E)$ for E (see Fig. 3.4).

Use this to show the Maxwell relation

$$\left.\frac{\partial T}{\partial V}\right|_{S,N} = -\left.\frac{\partial P}{\partial S}\right|_{V,N}. \qquad (3.70)$$

(This should take two lines of calculus or less.) Generate two other similar formulæ by taking other second partial derivatives of E. There are many of these relations [76].

(3.12) **Solving the pendulum.**[62] (Computation) ④ Physical systems usually evolve continuously in time; their laws of motion are differential equations. Computer simulations must approximate these differential equations using discrete time steps. In this exercise, we will introduce the most common and important method used for molecular dynamics simulations, together with fancier techniques used for solving more general differential equations.

We will use these methods to solve for the dynamics of a pendulum:

$$\frac{d^2\theta}{dt^2} = \ddot{\theta} = -\frac{g}{L}\sin(\theta). \qquad (3.71)$$

This equation gives the motion of a pendulum with a point mass at the tip of a massless rod[63] of length L. You may wish to rederive it using a free-body diagram.

Download the hints. The animation should show a pendulum oscillating from an initial condition $\theta_0 = 2\pi/3$, $\dot{\theta} = 0$; the equations being solved have $g = 9.8\,\text{m/s}^2$ and $L = 1\,\text{m}$.

There are three independent criteria for picking a good algorithm for solving differential equations: *fidelity*, *accuracy*, and *stability*.

Fidelity. In our time step algorithm, we do not make the straightforward choice—using the current $(\theta(t), \dot{\theta}(t))$ to produce $(\theta(t+\delta), \dot{\theta}(t+\delta))$. Rather, we use a staggered algorithm: $\theta(t)$ determines the acceleration and the update $\dot{\theta}(t) \to \dot{\theta}(t+\delta)$, and then $\dot{\theta}(t+\delta)$ determines the update $\theta(t) \to \theta(t+\delta)$:

$$\dot{\theta}(t+\delta) = \dot{\theta}(t) + \ddot{\theta}(t)\,\delta, \qquad (3.72)$$
$$\theta(t+\delta) = \theta(t) + \dot{\theta}(t+\delta)\,\delta. \qquad (3.73)$$

Would it not be simpler and make more sense to update θ and $\dot{\theta}$ simultaneously from their current values, so that eqn 3.73 would read $\theta(t+\delta) = \theta(t) + \dot{\theta}(t)\,\delta$? This simplest of all time-stepping schemes is called the *Euler method*, and should not be used for ordinary differential equations (although it is sometimes used for solving partial differential equations).

(a) *Try the Euler method. First, see why reversing the order of the updates to θ and $\dot{\theta}$,*

$$\theta(t+\delta) = \theta(t) + \dot{\theta}(t)\,\delta,$$
$$\dot{\theta}(t+\delta) = \dot{\theta}(t) + \ddot{\theta}(t)\,\delta, \qquad (3.74)$$

in the code you have downloaded would produce a simultaneous update. Swap these two lines in the code, and watch the pendulum swing for several turns, until it starts looping the loop. Is the new algorithm as good as the old one? (Make sure you switch the two lines back afterwards.)

The simultaneous update scheme is just as accurate as the one we chose, but it is not as faithful to the physics of the problem; its fidelity is not as good. For subtle reasons that we will not explain here, updating first $\dot{\theta}$ and then θ allows our algorithm to exactly simulate an approximate Hamiltonian system;[64] it is called a *symplectic* algorithm.[65] Improved versions of this algorithm—like the Verlet algorithms below—are often used to simulate systems that conserve energy (like molecular dynamics) because they exactly[66] simulate the dynamics for an approximation to the Hamiltonian—preserving important physical features not kept by just approximately solving the dynamics.

Accuracy. Most computational methods for solving differential equations (and many other continuum problems like integrating functions) involve a step size δ, and become more accurate as δ gets smaller. What is most important is not the error in each time step, but the accuracy of the answer after a fixed time T, which is the accumulated error after T/δ time steps. If this accumulated error varies as δ^n, we say

---

[62]This exercise and the associated software were developed in collaboration with Christopher Myers. See also *Numerical Recipes* [153, chapter 16].

[63]Our current animation emphasizes the rod rather than the mass; the equation for a physical rod without an end mass is similar.

[64]Equation 3.73 is Hamiltonian dynamics for noninteracting particles, and m times eqn 3.72 is the momentum evolution law $\dot{p} = F$ for a system of particles with infinite mass. Alternating these steps gives an effective Hamiltonian.

[65]It conserves a *symplectic form*. In non mathematician's language, this means our time step perfectly simulates a Hamiltonian system satisfying Liouville's theorem and energy conservation, but with an approximation to the true energy.

[66]Up to rounding errors.

that the algorithm has nth order cumulative accuracy. Our algorithm is not very high order!

(b) *Solve eqns 3.72 and 3.73 to give $\theta(t + \delta)$ in terms of $\theta(t)$, $\dot{\theta}(t)$ and $\ddot{\theta}(t)$ for our staggered algorithm. Comparing to the Taylor series $x(t + \tau) = x(t) + v\tau + \frac{1}{2}a\tau^2 + O(\tau^3)$ applied to $\theta(t)$, what order in δ is the error for θ in a single time-step? Looking at eqn 3.73, what is the error in one time step for $\dot{\theta}$? Given that the worst of the two accuracies should determine the overall accuracy, and that the time step error accumulates over more steps as the step size decreases, what order should the cumulative accuracy be for our staggered algorithm?*

(c) *Plot the pendulum trajectory $\theta(t)$ for time steps $\delta = 0.1$, 0.01, and 0.001. Zoom in on the curve at one of the coarse points (say, $t = 1$) and visually compare the values from the three time steps. Does it appear that the trajectory is converging[67] as $\delta \rightarrow 0$? What order cumulative accuracy do you find: is each curve better by a factor of 10, 100, $1{,}000\ldots$?*

A rearrangement of our staggered time-step (eqns 3.72 and 3.73) gives the *velocity Verlet* algorithm:

$$\dot{\theta}(t + \delta/2) = \dot{\theta}(t) + \tfrac{1}{2}\ddot{\theta}(t)\,\delta,$$
$$\theta(t + \delta) = \theta(t) + \dot{\theta}(t + \delta/2)\,\delta, \qquad (3.75)$$
$$\dot{\theta}(t + \delta) = \dot{\theta}(t + \delta/2) + \tfrac{1}{2}\ddot{\theta}(t + \delta)\delta.$$

The trick that makes this algorithm so good is to cleverly split the velocity increment into two pieces, half for the acceleration at the old position and half for the new position. (Initialize $\ddot{\theta}$ once before starting the loop.)

(d) *Show that N steps of our staggered time-step would give the velocity Verlet algorithm, if we shifted the velocities before and afterward by $\mp\frac{1}{2}\delta\ddot{\theta}$.*

(e) *As in part (b), write $\theta(t+\delta)$ for velocity Verlet in terms of quantities at t. What order cumulative accuracy does this suggest?*

(f) *Implement velocity Verlet, and plot the trajectory for time steps $\delta = 0.1$, 0.01, and 0.001. What is the order of cumulative accuracy?*

Stability. In many cases high accuracy is not crucial. What prevents us from taking enormous time steps? In a given problem, there is usually a typical fastest time scale: a vibration or oscillation period (as in our exercise) or a growth or decay rate. When our time step becomes a substantial fraction of this fastest time scale, algorithms like ours usually become *unstable*; the first few time steps may be fairly accurate, but small errors build up exponentially until the errors become unacceptable (indeed, often one's first warning of problems are machine overflows).

(g) *Plot the pendulum trajectory $\theta(t)$ for time steps $\delta = 0.1$, 0.2, $\ldots$, 0.8, using a small-amplitude oscillation $\theta_0 = 0.01$, $\dot{\theta}_0 = 0.0$, up to $t_{\max} = 10$. At about what δ_c does it go unstable? How does δ_c compare with the characteristic time period of the pendulum? At $\delta_c/2$, how accurate is the amplitude of the oscillation? (You will need to observe several periods in order to estimate the maximum amplitude of the solution.)*

In solving the properties of large, nonlinear systems (e.g., partial differential equations (PDEs) and molecular dynamics) stability tends to be the key difficulty. The maximum step-size depends on the local configuration, so e.g. highly energetic collisions can trigger instability before one might expect. The maximum safe stable step-size often has accuracy far higher than needed.

The Verlet algorithms are not hard to code. There are higher-order symplectic algorithms for Hamiltonian systems, but they are mostly used in unusual applications (planetary motion) where high accuracy is demanded, because they are typically significantly less stable. In systems of differential equations where there is no conserved energy or Hamiltonian, or even in Hamiltonian systems (like high-energy collisions) where accuracy at short times is more crucial than fidelity at long times, we use general-purpose methods.

ODE (Ordinary Differential Equation) packages. The general-purpose solvers come in a variety of basic algorithms (Runge–Kutta, predictor–corrector, $\ldots$), and methods for maintaining and enhancing accuracy (variable step size, Richardson extrapolation). There are also *implicit* methods for *stiff* systems. A system is stiff if there is a large separation between the slowest and fastest relevant time scales; implicit methods often allow one to take time steps much larger than the fastest time scale (unlike the explicit Verlet methods you studied in part (d), which go unstable). Large, sophisticated packages have been

---

[67]You may note that the approximate answers seem to extrapolate nicely to the correct answer. One can use this to converge more quickly to the correct answer. This is called Richardson extrapolation and is the basis for the Bulirsch–Stoer methods.

developed over many years for solving differential equations—switching between algorithms and varying the time steps to most efficiently maintain a given level of accuracy. They solve $d\mathbf{y}/dt = \mathbf{dydt}(\mathbf{y}, t)$, where for us $\mathbf{y} = [\theta, \dot{\theta}]$ and $\mathbf{dydt} = [\dot{\theta}, \ddot{\theta}]$. They typically come in the form of subroutines or functions, which need the following as arguments:

- initial conditions $\mathbf{y}_0$,

- the right-hand side $\mathbf{dydt}$, a function of the vector $\mathbf{y}$ and time t, which returns a vector giving the current rate of change of $\mathbf{y}$, and

- the initial and final times, and perhaps intermediate times, at which the trajectory $\mathbf{y}(t)$ is desired.

They often have options that

- ask for desired accuracy goals, typically a relative (fractional) accuracy and an absolute accuracy, sometimes set separately for each component of $\mathbf{y}$,

- ask for and return derivative and time step information from the end of the last step (to allow efficient restarts after intermediate points),

- ask for a routine that computes the derivatives of $\mathbf{dydt}$ with respect to the current components of $\mathbf{y}$ (for use by the stiff integrator), and

- return information about the methods, time steps, and performance of the algorithm.

You will be supplied with one of these general-purpose packages, and instructions on how to use it.

(h) *Write the function* $\mathbf{dydt}$, *and use the general-purpose solver to solve for the motion of the pendulum as in parts (a)–(c), and informally check that the trajectory is accurate.*

(3.13) **Weirdness in high dimensions.** ⓟ
We saw in momentum space that most of the surface area of a high-dimensional sphere is along the equator. Consider the volume enclosed by a high-dimensional sphere.
Is most of the volume near the center, or the surface? How might this relate to statistical mechanics treatments of momentum space, which in some texts approximate the volume of an energy shell with the entire volume enclosed by the sphere?

(3.14) **Pendulum energy shell.** ③
In this exercise, we shall explain why we focus not on the surface of constant energy in phase space, but rather the energy shell. As noted in Fig. 3.1, the energy shell in phase space will typically vary in thickness, causing the microcanonical average to weigh thick regions of the energy surface more heavily than thin regions. (The hyperspheres of the ideal gas are not typical energy shells!) Here we show, for the pendulum, that this weighting by thickness is physically correct. Figure 3.7 shows the phase space of the pendulum with Hamiltonian

$$\begin{aligned}\mathcal{H} &= p^2/2m\ell^2 + mg\ell(1 - \cos(\theta)) \\ &= p^2/2 + (1 - \cos(\theta)),\end{aligned} \tag{3.76}$$

setting $mg\ell = m\ell^2 = 1$. The inner grey ring is an energy shell at a relatively low energy, and the outer ring is at a higher energy where the anharmonic terms in the potential energy are strong. (a) *Why is the inner energy shell roughly circular? For the outer shell, at about what position and momentum is our pendulum at the thinnest point? The thickest?* Note that the outer shell comes close to $E = 2$. *What changes topologically in the energy surface at $E = 2$? Describe physically how our pendulum motion alters when E crosses two.*

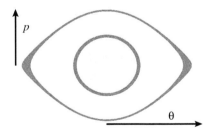

Fig. 3.7 Pendulum energy shells. Two energy shells for the pendulum of eqn 3.76. Both are of thickness $\delta E = 0.1$; the inner one spans from $E_1 = \frac{1}{2}$ to $E_1 + \delta E = 0.6$, the second spans from $E_2 = 1.88$ to 1.98.

In statistical mechanics, we average observables over the energy shell to get typical behaviors. In particular, we show in Chapter 4 that time averages for a typical initial condition are given by averages over the energy surface. Is the varying thickness of the energy shell important for getting the correct energy-shell average?

(b) *Does the pendulum spend extra phase-space time in the regions where the outer energy shell from Fig. 3.7 is thickest? Explain clearly why you say so. Make sure you explain physically how the two energy shells differ in this regard.*
One can use Hamilton's equations 4.1 to show the thickness precisely works to give the correct average in a general Hamiltonian system. Let us check this explicitly for the pendulum.
First calculate the time average. Let us parameterize the phase-space curve of constant energy by its arclength[68] s, where $ds = \sqrt{dp^2 + d\theta^2}$.
(c) *Argue that the time average of an operator over a period T,*

$$\langle O \rangle_t = 1/T \int_0^T dt\, O(p(t), \theta(t)), \qquad (3.77)$$

equals the weighted average

$$1/T \int \frac{ds}{|\mathbf{v}|} O(p(s), \theta(s)), \qquad (3.78)$$

where $\mathbf{v} = (\dot{p}, \dot{\theta})$ is the phase-space velocity around the trajectory. Does $|\mathbf{v}|$ vary significantly around the inner energy shell in Fig. 3.7?
Next calculate the average over the energy shell. To do that, we need to know its local thickness.
(d) *Draw a picture of a piece of the energy shell with varying thickness. Starting from a point (p, θ) at energy E, in what direction must you move to reach the corresponding point at $E + \delta E$? Argue that the thickness of the energy shell at that point (the distance you must move in that direction) is given by $\delta E / |\nabla \mathcal{H}|$, where $\nabla \mathcal{H} = (\partial \mathcal{H}/\partial p, \partial \mathcal{H}/\partial \theta)$. (Do not use Hamilton's equations yet.) Use eqns 3.5 and 3.6 to argue that the microcanonical average of an operator O acting on the pendulum is*

$$\langle O \rangle_{MC} = \frac{\int ds\, O(p(s), \theta(s))/|\nabla \mathcal{H}|}{\int ds/|\nabla \mathcal{H}|}. \qquad (3.79)$$

Now we explicitly relate the phase-space velocity $\mathbf{v}$ to the gradient $\nabla \mathcal{H}$.
(e) *What is the gradient $\nabla \mathcal{H}$ for our pendulum, in terms of p and θ? Does it agree with $\mathbf{v}$? Do the lengths $|\nabla \mathcal{H}|$ and $|\mathbf{v}|$ agree? Use Hamilton's equations of motion (eqn 4.1) to check that this also holds for a general Hamiltonian system.*

(f) *Using your results above, show that $\langle O \rangle_t = \langle O \rangle_{MC}$ for our pendulum.*

(3.15) **Entropy maximum and temperature.** ⓟ
Explain in words why, for two weakly coupled systems, that eqn 3.23

$$\rho(E_1) = \Omega_1(E_1)\Omega_2(E - E_1)/\Omega(E) \qquad (3.80)$$

is intuitive for a system where all states of energy E have equal probability density. Using $S = k_B \log(\Omega)$, show in one step that maximizing the probability of E_1 makes the two temperatures $1/T = \partial S/\partial E$ the same, and hence that maximizing $\rho(E_1)$ maximizes the total entropy.

(3.16) **Taste, smell, and μ.**[69] ⓐ
When we dip our toe into a lake, our skin measures temperature, not heat or energy. What about smell and taste? Do they measure the concentration of flavor and odor molecules, or their chemical potentials? Usually the chemical potential goes up monotonically with concentration—how can we guess what is being measured?
Smell and taste measure the binding of molecules to certain receptors.[70]
(a) *Suppose a flavor molecule binds tightly to alcohol but weakly to water. At a given concentration, would you expect to taste it less in an alcoholic drink? Do your taste buds directly measure concentration?*
Imagine an experiment, which compares a series of liquids at fixed flavor molecule concentration ρ and a series at fixed flavor molecule chemical potential μ. One could then ask the test subjects how strongly they taste the molecule. Can we develop a simplified statistical mechanics model to predict the outcome of such an experiment, within the microcanonical ensemble we use in this chapter?
Let Ω be the energy shell volume for the receptor immersed in the liquid. Let Ω^B be the volume of the portion of the energy shell where a flavor molecule is bound to the receptor, and Ω^U be the portion where no flavor molecules are bound.
(b) *What is the total phase-space volume in terms of Ω^B and Ω^U? What is the ratio of the probabilities P^B/P^U?*

---

[68]One may note that the units of position and momentum are different! Indeed, there is no natural metric in phase space. Our exercise illustrates a particular case using a convenient set of units (eqn 3.76). The general case is the focus of Chapter 4.
[69]Developed in consultation with Robin Dando.
[70]Sweet things are sensed through receptors; sour and salty tastes are measured by *ion channels*. These are indistinguishable for this exercise's modeling purposes.

Model the local region of the tongue as two weakly coupled systems, a fluid with entropy S_f and a receptor with entropy S_r. Let the receptor entropy change by $\Delta S_r = S_r^B - S_r^U$ during binding.

(c) *Write the phase-space volume Ω^B in terms of Ω_r^B and Ω_f^B. Write Ω^B in terms of S_r^B and S_f^B. Doing the same for Ω^U, write the relative probability P^B/P^U in terms of ΔS_r and ΔS_f.*

(d) *Write a formula for the entire energy shell volume Ω in terms of the bound and unbound phase-space volumes for the two subsystems, Ω_r^B, Ω_r^U, Ω_f^B, and Ω_f^U. The energy-shell volume for Ω^U factored into two because the fluid and receptor subsystems were weakly coupled. Can you think of a physical reason why the whole system Ω cannot be factored into a phase-space volume for the fluid and for the receptor, in the microcanonical ensemble?*

We shall find in Chapter 6 that weakly coupled systems that exchange energy and particles can be factored into independent pieces using the *grand canonical ensemble*. Indeed, our final formula, eqn 3.81, follows directly from that ensemble.

Let the receptor energy change by $\Delta E_r = E_r^B - E_r^U < 0$ as it binds the flavor molecule; assume this energy is supplied by the fluid surrounding the receptor. Let the chemical potential of the flavor molecule in the fluid be μ. Our assumption that the two systems are weakly coupled implies that neither ΔE_r nor ΔS_r depend on the alcohol level in the fluid. The fluid can be assumed large and to have many flavor molecules, so its chemical potential and temperature is unchanged during the binding process.

(e) *When one flavor molecule is bound to the receptor, what will the entropy change ΔS_f be in the fluid, in terms of ΔE_r and μ?* (Hint: $\Delta N_f = -1$ and $\Delta E_f = -\Delta E_r$; use eqns 3.29 and 3.32.)

(f) *Show that, in our approximation the receptor binding is unaffected by the alcohol level in the fluid, that[71]*

$$P^B/P^U = e^{-(\Delta E_r - T\Delta S_r - \mu)/k_B T}. \quad (3.81)$$

Does this make physical sense? (Hint: $S =$

$k_B \log \Omega$, so $\Omega = \exp(S/k_B)$. Remember that the change in total entropy is $\Delta S_r + \Delta S_f$.)

Just as your skin measures the temperature and not the heat of the object you touch, your nose and tongue, to this approximation, measure the chemical potential and not the density.

Our sensations of taste and particularly smell are more complicated and nonlinear than our sensation of heat and pressure, sound and light.[72] Partly this is due to the gigantic number of different species of molecules being sensed; partly it might be due to the fact that different species can compete for the same receptor, and several receptors can sense the same species.

(3.17) **Undistinguished particles.** ⓟ
Why should we divide the phase-space volume by $N!$, whenever we do not keep track of the differences between the N particles? Look up "entropy of mixing". In Fig. 5.4, can you see how to extract work from the mixing using methods that cannot tell white from blue particles? If we had a door in the partition wall that let through only orange particles, what would happen? Could we then extract work from the system?

(3.18) **Ideal gas glass.** ⓟ
In this exercise, we explore a physical picture for the position-space entropy of an ideal gas of undistinguished particles, and use this to build an amazingly simple model for the difference in energy between a glass and a crystal made of the same materials (the *residual entropy*, see Exercise 5.11).

If we have N particles in a box of volume V and do not distinguish between the particles, then the effective position-space part of the energy shell volume is $\Omega_\mathbb{Q} = V^N/N!$, where $N!$ is sometimes called the *Gibbs factor*. We saw that this factor keeps the entropy of mixing small for undistinguished particles.

(a) *Use Stirling's formula to show that $\Omega_\mathbb{Q}$ can be rewritten as the phase-space volume of N particles each in a box of effective size roughly V/N, the volume per particle. By what factor is this effective volume different from V/N?*

One can measure the difference in entropy between a glass and a crystal made of the same material. The crystal has only vibrational entropy

---

[71]Equation 3.81 is the change in the *grand canonical free energy*, which we shall study in Section 6.3. Here we ignored the change in volume ΔV occupied by the flavor molecule as it binds to the receptor; adding it in yields the *Gibbs free energy* typically used in chemistry.

[72]Most of our senses are *compressive*; our perceptions depend logarithmically on the sound level / light level / flavor chemical potential. Taste and smell also show fast and strong *adaptation*; prolonged exposure leads to decreased response.

at low temperatures, since all atoms lie close to the sites of a single crystalline lattice. Glasses can freeze into an enormous number of possible configurations, plus similar small vibrations. The *residual* entropy of glasses due to these configurations has been measured using the specific heat (Exercise 5.11) to be around k_B per molecular unit.

Fig. 3.8 Glass vs. Crystal. (a) Ideal gas of N atoms in a big box of size V. (b) N ideal gas atoms each in a small box of size V/N. The ideal gas on the left has an additional entropy due to the $N^N/N!$ configurations of filling the different small boxes with the atoms. This entropy difference is quantitatively similar to the zero-temperature entropy difference between a glass (with many metastable configurations) and a crystal (with a regular lattice of atoms).

Let us model an "ideal gas crystal" as a regular grid of boxes of size V/N, with one ideal gas atom per box (Fig. 3.8). The vibrations of the atoms in the crystal are represented by the motion of the atoms inside their boxes. Let us model the corresponding "ideal gas glass" as the ideal gas in a volume V.

(b) *Using your answer from part (a), calculate the residual entropy of our model glass. Check that you get the same residual entropy using eqn 3.57 for the ideal gas. How well does our model work to predict the experimentally measured residual entropy?*

(3.19) **Random energy model.**[73]   ④

The nightmare of every optimization algorithm is a random landscape; if every new configuration has an energy uncorrelated with the previous ones, no search method is better than systematically examining every configuration. Finding ground states of disordered systems like spin glasses and random-field models, or equilibrating them at nonzero temperatures, is challenging because the energy landscape has many features that are quite random. The random energy model (REM) is a caricature of these disordered systems, where the correlations are completely ignored. While optimization of a single REM becomes hopeless, we shall see that the study of the ensemble of REM problems is quite fruitful and interesting.

The REM has $M = 2^N$ states for a system with N particles (like an Ising spin glass with N spins), each state with a randomly chosen energy. It describes systems in limit when the interactions are so strong and complicated that flipping the state of a single particle completely randomizes the energy. The states of the individual particles then need not be distinguished; we label the states of the entire system by $j \in \{1, \ldots, 2^N\}$. The energies of these states E_j are assumed independent, uncorrelated variables with a Gaussian probability distribution

$$P(E) = \frac{1}{\sqrt{\pi N}} e^{-E^2/N} \tag{3.82}$$

of standard deviation $\sqrt{N/2}$.

Microcanonical ensemble. Consider the states in a small range $E < E_j < E+\delta E$. Let the number of such states in this range be $\Omega(E)\delta E$.

(a) *Calculate the average*

$$\langle \Omega(N\epsilon) \rangle_{\text{REM}} \tag{3.83}$$

over the ensemble of REM systems, in terms of the energy per particle ϵ. For energies per particle near zero, show that this average density of states grows exponentially as the system size N grows. In contrast, show that $\langle \Omega(N\epsilon) \rangle_{\text{REM}}$ decreases exponentially for $\epsilon = E/N < -\epsilon_*$ and for $\epsilon > \epsilon_*$, where the limiting energy per particle

$$\epsilon_* = \sqrt{\log 2}. \tag{3.84}$$

(Hint: As N grows, the probability density $P(N\epsilon)$ decreases exponentially, while the total number of states 2^N grows exponentially. Which one wins?)

What does an exponentially growing number of states mean? Let the entropy per particle be $s(\epsilon) = S(N\epsilon)/N$. Then (setting $k_B = 1$ for notational convenience) $\Omega(E) = \exp(S(E)) = \exp(Ns(\epsilon))$ grows exponentially whenever the entropy per particle is positive.

What does an exponentially decaying number of states for $\epsilon < -\epsilon_*$ mean? It means that, for any particular REM, the likelihood of having *any* states with energy per particle near ϵ vanishes rapidly as the number of particles N grows large.

---

[73]This exercise draws heavily from [133, chapter 5].

How do we calculate the entropy per particle $s(\epsilon)$ of a typical REM? Can we just use the annealed[74] average

$$s_{\text{annealed}}(\epsilon) = \lim_{N \to \infty} (1/N) \log \langle \Omega(N\epsilon) \rangle_{\text{REM}}$$

$$(3.85)$$

computed by averaging over the entire ensemble of REMs?

(b) *Show that* $s_{\text{annealed}}(\epsilon) = \log 2 - \epsilon^2$.

If the energy per particle is above $-\epsilon_*$ (and below ϵ_*), the expected number of states $\Omega(E)\,\delta E$ grows exponentially with system size, so the fractional fluctuations become unimportant as $N \to \infty$. The typical entropy will become the annealed entropy. On the other hand, if the energy per particle is below $-\epsilon_*$, the number of states in the energy range $(E, E + \delta E)$ rapidly goes to zero, so the typical entropy $s(\epsilon)$ goes to minus infinity. (The annealed entropy is not minus infinity because it gets a contribution from exponentially rare REMs that happen to have an energy level far into the tail of the probability distribution.) Hence

$$s(\epsilon) = s_{\text{annealed}}(\epsilon) = \log 2 - \epsilon^2 \quad |\epsilon| < \epsilon_*$$
$$s(\epsilon) = -\infty \qquad\qquad\qquad\quad |\epsilon| > \epsilon_*.$$
$$(3.86)$$

Notice why these arguments are subtle. Each REM model in principle has a different entropy. For large systems as $N \to \infty$, the entropies of different REMs look more and more similar to one another[75] (the entropy is self-averaging) whether $|\epsilon| < \epsilon_*$ or $|\epsilon| > \epsilon_*$. However, $\Omega(N\epsilon)$ is not self-averaging for $|\epsilon| > \epsilon_*$, so the typical entropy is not given by the *annealed* logarithm of $\langle \Omega(E) \rangle_{\text{REM}}$.

The REM has a *glass transition* at the temperature T_c corresponding to ϵ_*. Above T_c the entropy is extensive and the REM acts much like an equilibrium system. Below T_c one can show [133, eqn 5.25] that the REM thermal population condenses onto a finite number of states (i.e., a number that does not grow as the size of the system increases), which goes to zero linearly as $T \to 0$. The mathematical structure of the REM also arises in other, quite different contexts, such as combinatorial optimization and random error correcting codes [133, chapter 6].

---

[74]Annealing a disordered system (like an alloy or a disordered metal with frozen-in defects) is done by heating it to allow the defects and disordered regions to reach equilibrium. By averaging $\Omega(E)$ not only over levels within one REM but also over all REMs, we are computing the result of equilibrating over the disorder—an annealed average.

[75]Mathematically, the entropies per particle of REM models with N particles approach that given by eqn 3.86 with probability one [133, eqn 5.10].

Phase-space dynamics and ergodicity

So far, our justification for using the microcanonical ensemble was simple ignorance; all we know about the complex motion is that energy must be conserved, so we average over all states in phase space of fixed energy.[1] Here we provide a much more convincing argument for the ensemble, and hence for equilibrium statistical mechanics as a whole. In Section 4.1 we will show for classical systems that averaging over the energy surface[2] is consistent with time evolution. Liouville's theorem will tell us that volume in phase space is conserved, so the trajectories only stir the energy surface around—they do not change the relative weights of different parts of the surface. In Section 4.2 we introduce the concept of *ergodicity*; an ergodic system has an energy surface which is well stirred. Using Liouville's theorem and assuming ergodicity will allow us to show[3] that the microcanonical ensemble average gives the long-time average behavior that we call equilibrium.

[1]If you are willing to take this energy-surface average on trust, the rest of this text does not depend on the results in this chapter.

[2]The energy surface should be thought of as the energy shell $E < \mathcal{H}(\mathbb{P}, \mathbb{Q}) < E + \delta E$ in the limit $\delta E \to 0$. We focus here on the energy surface rather than the energy shell because the different energies in the shell do not get intermingled by the time evolution.

[3]We do not aspire to mathematical rigor, but we will provide physical arguments for rigorously known results; see [113].

4.1 Liouville's theorem

In Chapter 3, we saw that treating all states in phase space with a given energy on an equal footing gave sensible predictions for the ideal gas, but we did not show that this democratic treatment was necessarily the correct one. Liouville's theorem, true for all Hamiltonian systems, will tell us that all states are created equal.

Systems of point particles obeying Newton's laws without dissipation are examples of Hamiltonian dynamical systems. These systems conserve the energy, given by the Hamiltonian $\mathcal{H}(\mathbb{P}, \mathbb{Q})$. The laws of motion are given from $\mathcal{H}$ by *Hamilton's equations*:

$$\dot{q}_\alpha = \partial \mathcal{H}/\partial p_\alpha,$$
$$\dot{p}_\alpha = -\partial \mathcal{H}/\partial q_\alpha, \tag{4.1}$$

where as usual $\dot{X}$ is the time derivative of X. The standard example of a Hamiltonian, and the only example we will discuss in this text, is a bunch of particles interacting with a potential energy U:

$$\mathcal{H}(\mathbb{P}, \mathbb{Q}) = \sum_\alpha p_\alpha{}^2/2m_\alpha + U(q_1, \ldots, q_{3N}). \tag{4.2}$$

In this case, one immediately finds the expected Newtonian equations

Statistical Mechanics: Entropy, Order Parameters, and Complexity. James P. Sethna, Oxford University Press (2021). © Oxford University Press. DOI:10.1093/oso/9780198865247.003.0004

of motion:

$$\dot{q}_\alpha = \partial \mathcal{H}/\partial p_\alpha = p_\alpha/m_\alpha,$$
$$\dot{p}_\alpha = -\partial \mathcal{H}/\partial q_\alpha = -\partial U/\partial q_\alpha = f_\alpha(q_1, \ldots, q_{3N}), \tag{4.3}$$

where f_α is the force on coordinate α. More general Hamiltonians[4] arise when studying, for example, the motions of rigid bodies or mechanical objects connected by hinges and joints, where the natural variables are angles or relative positions rather than points in space. Hamiltonians also play a central role in quantum mechanics.[5]

Hamiltonian systems have properties that are quite distinct from general systems of differential equations. They not only conserve energy, but they also have many other unusual properties.[6] Liouville's theorem describes the most important of these properties.

Consider the evolution law for a general probability density in phase space:

$$\rho(\mathbb{P}, \mathbb{Q}) = \rho(q_1, \ldots, q_{3N}, p_1, \ldots, p_{3N}). \tag{4.4}$$

(As a special case, the microcanonical ensemble has ρ equal to a constant in a thin range of energies, and zero outside that range.) This probability density ρ is *locally conserved*: probability cannot be created or destroyed, it can only flow around in phase space. As an analogy, suppose a fluid of mass density $\rho_{3D}(x)$ in three dimensions has a velocity $v(x)$. Because mass is locally conserved, ρ_{3D} must satisfy the continuity equation $\partial \rho_{3D}/\partial t = -\nabla \cdot \mathbf{J}$, where $\mathbf{J} = \rho_{3D}\mathbf{v}$ is the mass current (Fig. 4.1). In the same way, the probability density in $6N$ dimensions has a phase-space probability current $(\rho\dot{\mathbb{P}}, \rho\dot{\mathbb{Q}})$ and hence satisfies a continuity equation

$$\frac{\partial \rho}{\partial t} = -\nabla_{6N}(\rho\mathbf{v}_{6N}) = -\sum_{\alpha=1}^{3N} \left(\frac{\partial(\rho\dot{q}_\alpha)}{\partial q_\alpha} + \frac{\partial(\rho\dot{p}_\alpha)}{\partial p_\alpha} \right)$$
$$= -\sum_{\alpha=1}^{3N} \left(\frac{\partial \rho}{\partial q_\alpha}\dot{q}_\alpha + \rho\frac{\partial \dot{q}_\alpha}{\partial q_\alpha} + \frac{\partial \rho}{\partial p_\alpha}\dot{p}_\alpha + \rho\frac{\partial \dot{p}_\alpha}{\partial p_\alpha} \right). \tag{4.5}$$

Now, it is clear what is meant by $\partial \rho/\partial q_\alpha$, since ρ is a function of the q_αs and p_αs. But what is meant by $\partial \dot{q}_\alpha/\partial q_\alpha$? For our example of point particles, $\dot{q}_\alpha = p_\alpha/m$ has no dependence on q_α; nor does $\dot{p}_\alpha = f_\alpha(q_1, \ldots, q_{3N})$ have any dependence on the momentum p_α. Hence these two mysterious terms in eqn 4.5 both vanish for Newton's laws for point particles. Indeed, using Hamilton's equations 4.1, we find that they cancel one another for a general Hamiltonian system:

$$\partial \dot{q}_\alpha/\partial q_\alpha = \partial(\partial \mathcal{H}/\partial p_\alpha)/\partial q_\alpha = \partial^2 \mathcal{H}/\partial p_\alpha \partial q_\alpha = \partial^2 \mathcal{H}/\partial q_\alpha \partial p_\alpha$$
$$= \partial(\partial \mathcal{H}/\partial q_\alpha)/\partial p_\alpha = \partial(-\dot{p}_\alpha)/\partial p_\alpha = -\partial \dot{p}_\alpha/\partial p_\alpha. \tag{4.6}$$

This leaves us with the equation

$$\frac{\partial \rho}{\partial t} + \sum_{\alpha=1}^{3N} \left(\frac{\partial \rho}{\partial q_\alpha}\dot{q}_\alpha + \frac{\partial \rho}{\partial p_\alpha}\dot{p}_\alpha \right) = \frac{\mathrm{d}\rho}{\mathrm{d}t} = 0. \tag{4.7}$$

[4]You will cover Hamiltonian dynamics in detail in most advanced courses in classical mechanics. For those who do not already know about Hamiltonians, rest assured that we will not use anything other than the special case of Newton's laws for point particles; you can safely ignore the more general case for our purposes.

[5]In Section 7.1 we discuss the quantum version of Liouville's theorem.

[6]For the mathematically sophisticated reader, Hamiltonian dynamics preserves a *symplectic form* $\omega = \mathrm{d}q_1 \wedge \mathrm{d}p_1 + \ldots + \mathrm{d}q_{3N} \wedge \mathrm{d}p_{3N}$; Liouville's theorem follows because the volume in phase space is ω^{3N}.

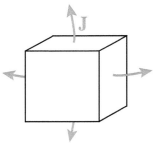

Fig. 4.1 Conserved currents in 3D. Think of the flow in and out of a small volume ΔV in space. The change in the density inside the volume $\partial \rho_{3D}/\partial t \, \Delta V$ must equal minus the flow of material out through the surface $-\int \mathbf{J} \cdot \mathrm{d}S$, which by Gauss' theorem equals $-\int \nabla \cdot \mathbf{J} \, \mathrm{d}V \approx -\nabla \cdot \mathbf{J} \, \Delta V$.

This is Liouville's theorem.

What is $d\rho/dt$, and how is it different from $\partial\rho/\partial t$? The former is called the *total derivative* of ρ with respect to time; it is the evolution of ρ seen by a particle *moving with the flow*. In a three-dimensional flow, $d\rho_{3D}/dt = \partial\rho/\partial t + \mathbf{v}\cdot\nabla\rho = \partial\rho/\partial t + \sum_{i=1}^{3}\dot{x}_i(\partial\rho/\partial x_i)$; the first term is the change in ρ due to the time evolution at fixed position, and the second is the change in ρ that a particle moving with velocity $\mathbf{v}$ would see if the ρ field did not change in time. Equation 4.7 is the same physical situation, but in $6N$-dimensional phase space.

What does Liouville's theorem, $d\rho/dt = 0$, tell us about Hamiltonian dynamics?

Fig. 4.2 Incompressible flow. A small volume in phase space may be stretched and twisted by the flow, but Liouville's theorem shows that the volume stays unchanged.

- **Flows in phase space are incompressible.** In fluid mechanics, if the density $d\rho_{3D}/dt = 0$ it means that the fluid is incompressible. The density of a small element of fluid does not change as it moves around in the fluid; hence the small element is not compressing or expanding. In Liouville's theorem, it means the same thing; a small volume in phase space will evolve into a new shape, perhaps stretched, twisted, or folded, but with exactly the same volume (Fig. 4.2).

- **There are no attractors.** In other dynamical systems, most states of the system are usually transient, and the system settles down onto a small set of states called the *attractor*. A damped pendulum will stop moving; the attractor has zero velocity and vertical angle (Exercise 4.2). A forced, damped pendulum will settle down to oscillate with a particular amplitude; the attractor is a circle in phase space. The decay of these transients in dissipative systems would seem closely related to equilibration in statistical mechanics, where at long times all initial states of a system will settle down into static equilibrium behavior.[7] Perversely, we have just proven that equilibration in statistical mechanics happens by a completely different mechanism! In equilibrium statistical mechanics all states are created equal; transient states are temporary only insofar as they are very unusual, so as time evolves they disappear, to arise again only as rare fluctuations.

- **Microcanonical ensembles are time independent.** An initial uniform density in phase space will stay uniform. More generally, since energy is conserved, a uniform density over a small shell of energies $(E, E+\delta E)$ will stay uniform.

[7]We will return to the question of how irreversibility and damping emerge from statistical mechanics many times in the rest of this book. It will always involve introducing approximations to the microscopic theory.

Liouville's theorem tells us that the energy surface may get stirred around, but the relative weights of parts of the surface are given by their phase-space volumes (Fig. 3.1) and do not change. This property is a necessary condition for our microcanonical ensemble to describe the time-independent equilibrium state.

4.2 Ergodicity

By averaging over the energy surface, statistical mechanics is making a hypothesis, first stated by Boltzmann. Roughly speaking, the hypothesis

Fig. 4.3 KAM tori and nonergodic motion. This is a (Poincaré) cross-section (Fig. 4.9) of Earth's motion in the three-body problem (Exercise 4.4), with Jupiter's mass set at almost 70 times its actual value. The closed loops correspond to trajectories that form tori in phase space, whose cross-sections look like deformed circles in our view. The complex filled region is a single trajectory exhibiting chaotic motion, and represents an ergodic component. The tori, each an ergodic component, can together be shown to occupy nonzero volume in phase space, for small Jovian masses. Note that this system is not ergodic according to either of our definitions. The trajectories on the tori never explore the rest of the energy surface. The region R formed by the chaotic domain is invariant under the time evolution; it has positive volume and the region outside R also has positive volume.

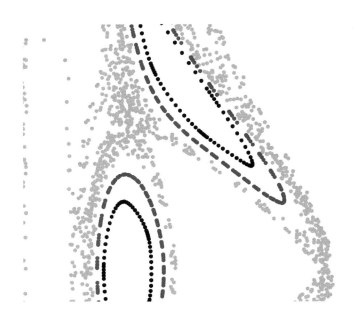

[8]Mathematicians distinguish between ergodic (stirred) and *mixing* (scrambled); we only need to assume ergodicity here. See [113] for more information about ergodicity.

[9]What does *almost every* mean? Technically, it means all but a set of zero volume (measure zero). Basically, the qualification "almost" is there to avoid problems with unusual, specific initial conditions like all the particles moving precisely at the same velocity in neat rows.

[10]Why not just assume that every point on the energy surface gets passed through? Boltzmann originally did assume this. However, it can be shown that a smooth curve (our time-trajectory) cannot fill up a whole volume (the energy surface). In an ergodic system the trajectory covers the energy surface densely, but not completely.

[11]If an ergodic system equilibrates (e.g., does not oscillate forever), the time average behavior will be determined by the equilibrium behavior; ergodicity then implies that the equilibrium properties are equal to the microcanonical averages.

[12]Here S is the energy surface.

is that the energy surface is thoroughly stirred by the time evolution; it is not divided into some kind of components that do not intermingle (see Fig. 4.3). A system which is thoroughly stirred is said to be *ergodic*.[8] The original way of defining ergodicity is due to Boltzmann. Adapting his definition, we have

Definition 1 In an *ergodic* system, the trajectory of almost every[9] point in phase space eventually passes arbitrarily close[10] to every other point (position and momentum) on the surface of constant energy.

The most important consequence of ergodicity is that *time averages are equal to microcanonical averages.*[11] Intuitively, since the trajectory $(\mathbb{P}(t), \mathbb{Q}(t))$ covers the whole energy surface, the average of any property $O(\mathbb{P}(t), \mathbb{Q}(t))$ over time is the same as the average of O over the energy surface.

This turns out to be tricky to prove, though. It is easier mathematically to work with another, equivalent definition of ergodicity. This definition roughly says that the energy surface cannot be divided into components which do not intermingle. Let us define an *ergodic component* R of a set[12] S to be a subset that remains invariant under the flow (so $r(t) \in R$ for all $r(0) \in R$).

Definition 2 A time evolution in a set S is *ergodic* if and only if all the ergodic components R in S either have zero volume or have a volume equal to the volume of S.

We can give an intuitive explanation of why these two definitions are equivalent (but it is hard to prove). A trajectory $r(t)$ must lie within a single ergodic component. If $r(t)$ covers the energy surface densely

(Definition 1), then there is "no more room" for a second ergodic component with nonzero volume (Definition 2).[13] Conversely, if there is only one ergodic component R with volume equal to S (Definition 2), then any trajectory starting in R must get arbitrarily close to all points in R (Definition 1), otherwise the points in R "far" from the trajectory (outside the closure of the trajectory) would be an invariant set of nonzero volume.

Using this second definition of ergodic, we can argue that time averages must equal microcanonical averages. Let us denote the microcanonical average of an observable O as $\langle O \rangle_S$, and let us denote the time average starting at initial condition $(\mathbb{P}, \mathbb{Q})$ as $\overline{O(\mathbb{P}(0), \mathbb{Q}(0))} = \lim_{T \to \infty} (1/T) \int_0^T O(\mathbb{P}(t), \mathbb{Q}(t)) \, dt$.

Showing that the time average $\overline{O}$ equals the ensemble average $\langle O \rangle_S$ for an ergodic system (using this second definition) has three steps.

(1) **Time averages are constant on trajectories.** If O is a nice function (e.g. without any infinities on the energy surface), then

$$\overline{O(\mathbb{P}(0), \mathbb{Q}(0))} = \overline{O(\mathbb{P}(t), \mathbb{Q}(t))}; \tag{4.8}$$

the future time average does not depend on the values of O during the finite time interval $(0, t)$. Thus the time average $\overline{O}$ is constant along the trajectory.[14]

(2) **Time averages are constant on the energy surface.** Now consider the subset R_a of the energy surface where $\overline{O} < a$, for some value a. Since $\overline{O}$ is constant along a trajectory, any point in R_a is sent under the time evolution to another point in R_a, so R_a is an ergodic component. If we have ergodic dynamics on the energy surface, that means the set R_a has either zero volume or the volume of the energy surface. This implies that $\overline{O}$ is a constant on the energy surface (except on a set of zero volume); its value is a^*, the lowest value where R_{a^*} has the whole volume. Thus the equilibrium, time average value of our observable O is independent of initial condition.

(3) **Time averages equal microcanonical averages.** Is this equilibrium value given by the microcanonical ensemble average over S? We need to show that the trajectories do not linger in some regions of the energy surface more than they should (based on the thickness of the energy shell, Fig. 3.1). Liouville's theorem in Section 4.1 told us that the microcanonical ensemble was time independent, so the ensemble average equals its time average, which equals the ensemble average of the time average. But the time average is constant (except on a set of zero volume), so in an ergodic system the ensemble average equals the time average everywhere (except on a set of zero volume).[15]

Can we show that our systems are ergodic? Usually not.[16] Ergodicity has been proven for the collisions of hard spheres, and for geodesic motion on finite surfaces with constant negative curvature,[17] but not for many systems of immediate practical importance. Indeed, several fundamental problems precisely involve systems which are not ergodic.

[13]Mathematicians must be careful in the definitions and proofs to exclude different invariant sets that are infinitely finely intertwined.

[14]If we could show that $\overline{O}$ had to be a continuous function, we would now be able to use the first definition of ergodicity to show that it was constant on the energy surface, since our trajectory comes close to every point on the surface. But it is not obvious that $\overline{O}$ is continuous; for example, is not continuous for Hamiltonian systems that are not ergodic. We can see this from Fig. 4.3; consider two initial conditions at nearby points, one just inside a chaotic region and the other on a KAM torus. The infinite time averages on these two trajectories for most quantities will be quite different; $\overline{O}$ will typically have a jump at the boundary.

[15]In formulæ, $\langle O \rangle_S = \langle O(t) \rangle_S = \langle O(\mathbb{P}(t), \mathbb{Q}(t)) \rangle_S$, where the average $\langle \cdot \rangle_S$ integrates over initial conditions $(\mathbb{P}(0), \mathbb{Q}(0))$ but evaluates O at $(\mathbb{P}(t), \mathbb{Q}(t))$. Averaging over all time, and using the fact that the time average $\overline{O} = a^*$ (almost everywhere), tells us

$$\begin{aligned}
\langle O \rangle_S &= \lim_{T \to \infty} \frac{1}{T} \int_0^T \langle O(\mathbb{P}(t), \mathbb{Q}(t)) \rangle_S \, dt \\
&= \left\langle \lim_{T \to \infty} \frac{1}{T} \int_0^T O(\mathbb{P}(t), \mathbb{Q}(t)) \, dt \right\rangle_S \\
&= \langle \overline{O(\mathbb{P}, \mathbb{Q})} \rangle_S = \langle a^* \rangle_S = a^*.
\end{aligned} \tag{4.9}$$

[16]It is rarely proven for the microscopic dynamics: it is often straightforward to show ergodicity for computer equilibration algorithms (see Section 8.2).

[17]Geodesic motion on a sphere would be motion at a constant speed around great circles. Geodesics are the shortest paths between two points. In general relativity, falling bodies travel on geodesics in space-time.

- **KAM tori and the three-body problem.** Generations of mathematicians and physicists have worked on the gravitational three-body problem.[18] The key challenge was showing that the interactions between the planets do not completely mess up their orbits over long times. One must note that "messing up their orbits" is precisely what an ergodic system must do! (There is just as much phase space at constant energy with Earth and Venus exchanging places, and a whole lot more with Earth flying out into interstellar space.) In the last century the KAM theorem was proven, which showed that (for small interplanetary interactions and a large fraction of initial conditions) the orbits of the planets qualitatively stayed in weakly perturbed ellipses around the Sun (KAM tori, see Fig. 4.3). Other initial conditions, intricately intermingled with the stable ones, lead to chaotic motion. Exercise 4.4 investigates the KAM tori and chaotic motion in a numerical simulation.

  From the KAM theorem and the study of chaos in these systems we learn that Hamiltonian systems with small numbers of particles are often, even usually, not ergodic—tori fill regions of nonzero volume which do not mix with the rest of the energy surface.

- **Fermi, Pasta, Ulam, and KdV.** You might think that this is a peculiarity of having only a few particles. Surely if there are lots of particles, such funny behavior has to go away? On one of the early computers developed for the Manhattan project, Fermi, Pasta, and Ulam tested this [58]. They took a one-dimensional chain of atoms, coupled them with anharmonic potentials, and tried to look for thermalization:

  > Let us say here that the results of our computations were, from the beginning, surprising us. Instead of a continuous flow of energy from the first mode to the higher modes, all of the problems show an entirely different behavior. [...] Instead of a gradual increase of all the higher modes, the energy is exchanged, essentially, among only a certain few. It is, therefore, very hard to observe the rate of "thermalization" or mixing in our problem, and this was the initial purpose of the calculation [58, p. 978].

  It turns out that their system, in the continuum limit, gave a partial differential equation (the Korteweg–de Vries equation) that was even weirder than planetary motion; it had an *infinite family* of conserved quantities, and could be exactly solved using a combination of fronts called *solitons*.

  The kind of nonergodicity found in the Korteweg–de Vries equation was thought to arise in only rather special one-dimensional systems. The discovery of anharmonic localized modes in generic, three-dimensional systems [160, 164, 186] suggests that nonergodicity may arise in rather realistic lattice models.

- **Broken symmetry phases.** Many phases have *broken symmetries*. Magnets, crystals, superfluids, and liquid crystals, for example, violate ergodicity because they only explore one of a variety of equal-energy ground states (see Chapter 9). For example, a liquid may explore all of phase space with a given energy, but an infinite crystal (with a neat grid of atoms aligned in a particular orientation) will never fluctuate to change its orientation, or (in three dimensions) the registry of its grid. That is, a 3D crystal has broken *orientational* and *translational* symmetries. The real system will explore only one ergodic component of the phase space (one crystal position and orientation), and we must do the same when making theories of the system.

- **Glasses.** There are other kinds of breakdowns of the ergodic hypothesis. For example, glasses fall out of equilibrium as they are cooled; they no longer ergodically explore all configurations, but just oscillate about one of many metastable glassy states. Certain models of glasses and disordered systems can be shown to break ergodicity—not just into a small family of macroscopic states as in normal symmetry-breaking phase transitions, but into an infinite number of different, disordered ground states. It is an open question whether real glasses truly break ergodicity when cooled infinitely slowly, or whether they are just sluggish, "frozen liquids".

Should we be concerned that we cannot prove that our systems are ergodic? It is entertaining to point out the gaps in the foundations of statistical mechanics, especially since they tie into so many central problems in mathematics and physics. We emphasize that these gaps are for most purposes purely of academic concern. Statistical mechanics works phenomenally well in most systems with large numbers of interacting degrees of freedom.

Indeed, the level of rigor here is unusual in science. Equilibrium statistical mechanics rests on a remarkably sound foundations, provided by Liouville's theorem and the assumption of ergodicity.

Exercises

No attractors in Hamiltonian systems discusses how stability of fixed points differs between Hamiltonian and dissipative systems. *Perverse initial conditions* explores counterexamples to ergodicity. *Equilibration* checks that realistic molecular dynamics simulations actually do settle down to states predicted by our equilibrium theories.

Liouville vs. the damped pendulum and *Invariant measures* explore analogues of Liouville's theorem in dissipative and chaotic systems. The first investigates how the theorem breaks down when dissipation is added. The second explores the complex, singular ensemble formed by the folding and stretching of a chaotic map. *Crooks and*

Jarzynski discuss recently discovered *exact* results in *non*-equilibrium statistical mechanics—powerful and striking implications of Liouville's theorem for calculating entropy and free energy changes.

2D turbulence and Jupiter's great red spot provides a statistical mechanics theory of hurricanes. *Jupiter! and the KAM theorem* vividly illustrates the breakdown of ergodicity in planetary motion: planetary dynamics is not chaotic enough to equilibrate. An ergodic solar system would be an unpleasant place to live.

(4.1) **Equilibration.**[19] (Computation) ②
Can we verify that realistic systems of atoms equilibrate? As discussed in Section 4.2, we do not know how to prove that systems of realistic atoms are ergodic. Also, phase space is so large we cannot verify ergodicity by checking computationally that a trajectory visits all portions of it.

(a) *For 20 particles in a box of size $L \times L \times L$, could we hope to test if our trajectory came close to all spatial configurations of the atoms? Let us call two spatial configurations "nearby" if the corresponding atoms in the two configurations are in the same $(L/10) \times (L/10) \times (L/10)$ subvolume. How many "distant" spatial configurations are there? On a hypothetical computer that could test 10^{12} such configurations per second, how many years would it take to sample this number of configurations?* (Hint: Conveniently, there are roughly $\pi \times 10^7$ seconds in a year.)
We certainly can solve Newton's laws using molecular dynamics to check the equilibrium *predictions* made possible by assuming ergodicity.
Run a constant-energy (microcanonical) simulation of a fairly dilute gas of Lennard–Jones particles (a pair potential $V(r) = 4\epsilon((r/\sigma)^{-12} - (r/\sigma)^{-6})$ crudely modeling argon or other noble gases). Start the atoms at rest (an atypical, nonequilibrium state), but in a random configuration (except ensure that no two atoms in the initial configuration overlap, less than $|\Delta\mathbf{r}| = \sigma$ apart). The atoms that start close to one another should start moving rapidly, eventually colliding with the more distant atoms until the gas equilibrates into a statistically stable state.
We have derived the distribution of the components of the momenta (p_x, p_y, p_z) for an equilibrium ideal gas (eqn 3.19 of Section 3.2.2),

$$\rho(p_x) = \frac{1}{\sqrt{2\pi m k_B T}} \exp\left(-\frac{p_x{}^2}{2m k_B T}\right) \quad (4.10)$$

This momentum distribution also describes interacting systems such as the one we study here (as we shall show in Chapter 6).
(b) *Plot a histogram of the components of the momentum in your gas for a few time intervals, multiplying the averaging time by four for each new graph, starting with just the first time-step. At short times, this histogram should be peaked around zero, since the atoms start at rest. Do they appear to equilibrate to the Gaussian predic-*
tion of eqn 4.10 at long times? Roughly estimate the equilibration time, measured using the time dependence of the velocity distribution. Estimate the final temperature from your histogram.
These particles, deterministically following Newton's laws, spontaneously evolve to satisfy the predictions of equilibrium statistical mechanics. This equilibration, peculiar and profound from a dynamical systems point of view, seems obvious and ordinary from the perspective of statistical mechanics. But see Fig. 4.3 and Exercise 4.4 for a system of interacting particles (the planets) which does not equilibrate.

(4.2) **Liouville vs. the damped pendulum.** (Mathematics, Dynamical systems) @

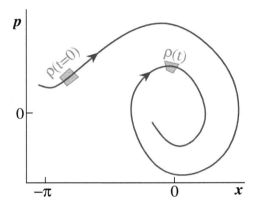

Fig. 4.4 Total derivatives. The total derivative gives the local density as measured by a particle moving with the flow: $d\rho/dt = d/dt\,(\rho(x(t), p(t), t))$. Applying the chain rule gives the definition of the total derivative, $d\rho/dt = \partial\rho/\partial t + \partial\rho/\partial x\,\dot{x} + \partial\rho/\partial p\,\dot{p}$.

The damped pendulum has a force $-\gamma p$ proportional to the momentum slowing down the pendulum (Fig. 4.4). It satisfies the equations

$$\begin{aligned} \dot{x} &= p/M, \\ \dot{p} &= -\gamma p - K\sin(x). \end{aligned} \quad (4.11)$$

At long times, the pendulum will tend to an equilibrium stationary state, zero velocity at $x = 0$ (or more generally at the equivalent positions $x = 2m\pi$, for m an integer); $(p, x) = (0, 0)$ is an attractor for the damped pendulum. An ensemble of damped pendulums is started with initial conditions distributed with probability $\rho(p_0, x_0)$.

---

[19]This exercise and the associated software were developed in collaboration with Christopher Myers. Computational hints can be found at the book website [181].

At late times, these initial conditions are gathered together near the equilibrium stationary state; Liouville's theorem clearly is not satisfied.

(a) *In the steps leading from eqn 4.5 to eqn 4.7, why does Liouville's theorem not apply to the damped pendulum? More specifically, what are $\partial\dot{p}/\partial p$ and $\partial\dot{q}/\partial q$?*

(b) *Find an expression for the total derivative $\mathrm{d}\rho/\mathrm{d}t$ in terms of ρ for the damped pendulum. If we evolve a region of phase space of initial volume $A = \Delta p \, \Delta x$, how will its volume depend upon time?*

(4.3) **Invariant measures.**[20] (Mathematics, Complexity, Computation, Dynamical systems) ④

Liouville's theorem tells us that all available points in phase space are equally weighted when a Hamiltonian system is averaged over all times. What happens for systems that evolve according to laws that are not Hamiltonian? Usually, the system does not continue to explore all points in its state space; at long times it is confined to a subset of the original space known as the *attractor*.

We consider the behavior of the "logistic" mapping from the unit interval $(0, 1)$ into itself:

$$f(x) = 4\mu x(1 - x). \qquad (4.12)$$

We talk of the trajectory of an initial point x_0 as the sequence of points x_0, $f(x_0)$, $f(f(x_0))$, ..., $f^{[n]}(x_0)$, ... Iteration can be thought of as a time step (one iteration of a Poincaré return map of Exercise 4.4 or one step Δt in a time-step algorithm as in Exercise 3.12).

Attracting fixed point. For small μ, our mapping has an attracting fixed point. A fixed point of a mapping is a value $x^* = f(x^*)$; a fixed point is stable if small perturbations shrink after iterating:

$$|f(x^* + \epsilon) - x^*| \approx |f'(x^*)|\epsilon < \epsilon, \qquad (4.13)$$

which happens if the derivative $|f'(x^*)| < 1$.[21]

(a) *Iteration. Set $\mu = 0.2$; iterate f for some initial points $0 < x_0 < 1$ of your choosing, and convince yourself that they are all attracted to zero. Plot f and the diagonal $y = x$ on the same plot.*

Are there any fixed points other than $x = 0$? Repeat for $\mu = 0.4$, and 0.6. What happens?

Analytics. *Find the nonzero fixed point $x^*(\mu)$ of the map 4.12, and show that it exists and is stable for $1/4 < \mu < 3/4$. If you are ambitious or have a computer algebra program, show that there is a stable, period-two cycle for $3/4 < \mu < (1 + \sqrt{6})/4$.*

An attracting fixed point is the antithesis of Liouville's theorem; all initial conditions are transient except one, and all systems lead eventually to the same, time-independent state. (On the other hand, this is precisely the behavior we expect in statistical mechanics *on the macroscopic scale*; the system settles down into a time-independent equilibrium state! All microstates are equivalent, but the vast majority of accessible microstates have the same macroscopic behavior in most large systems.) We could define a rather trivial "equilibrium ensemble" for this system, which consists of the single point x^*; any property $O(x)$ will have the long-time average $\langle O \rangle = O(x^*)$.

For larger values of μ, more complicated things happen. At $\mu = 1$, the dynamics can be shown to fill the entire interval (part b next); the dynamics is ergodic, and the attractor fills the entire set of available states. However, unlike the case of Hamiltonian systems, not all states are weighted equally (i.e., Liouville's theorem does not hold).

We can find time averages for functions of x in two ways: by averaging over time (many iterates of the map) or by weighting an integral over x by the *invariant density* $\rho(x)$, the probability density held fixed under an iteration of the map. The invariant density $\rho(x)\,\mathrm{d}x$ is the probability that a point on a long trajectory will lie between x and $x + \mathrm{d}x$. To find it numerically, we iterate a typical point[22] x_0 a thousand or so times ($N_{\text{transient}}$) to find a point x_a on the attractor, and then collect a long trajectory of perhaps a million points (N_{cycles}). A histogram of this trajectory gives $\rho(x)$. Averaging over this density is manifestly the same as a time average over the trajectory of a million points. We call $\rho(x)$ invariant because it is left the same under the mapping f; iterating our million-point approximation for ρ once under

---

[20]This exercise and the associated software were developed in collaboration with Christopher Myers; see [94]. Computational hints can be found at the book website [181]. There are several other exercises exploring this chaotic logistic map (see Index).

[21]For many-dimensional mappings, a sufficient criterion for stability is that all the eigenvalues of the Jacobian have magnitude smaller than one. A continuous time evolution $\mathrm{d}y/\mathrm{d}t = F(y)$ will be stable if $\mathrm{d}F/\mathrm{d}y$ is smaller than zero, or (for multidimensional systems) if the Jacobian DF has eigenvalues whose real parts are all less than zero.

[22]For example, we must not choose an unstable fixed point or unstable periodic orbit!

f only removes the first point x_a and adds a new final point.

The points in a range dx around a point x map under f to a range $dy = f'(x) dx$ around the image $y = f(x)$. Each iteration maps two points x_a and $x_b = 1 - x_a$ to y, and thus maps all the density $\rho(x_a)|dx_a|$ and $\rho(x_b)|dx_b|$ into dy. Hence the probability $\rho(y) dy$ must equal $\rho(x_a)|dx_a| + \rho(x_b)|dx_b|$, so

$$\rho(f(x_a)) = \rho(x_a)/|f'(x_a)| + \rho(x_b)/|f'(x_b)|. \tag{4.14}$$

(b) Invariant density. *Set $\mu = 1$; iterate f many times, and form a histogram of values giving the density $\rho(x)$ of points along the trajectory. You should find that points x near the boundaries are approached more often than points near the center.*

Analytics. *Using the fact that the long-time average $\rho(x)$ must be independent of time, verify for $\mu = 1$ that the density of points is*[23]

$$\rho(x) = \frac{1}{\pi\sqrt{x(1-x)}}. \tag{4.15}$$

Plot this theoretical curve with your numerical histogram. (Hint: Substitute eqn 4.15 into eqn 4.14. You will need to factor a polynomial.) Mathematicians call this probability density $\rho(x) dx$ the *invariant measure* on the attractor.[24] To get the long-term average of any function $O(x)$, one can use

$$\langle O \rangle = \int O(x)\rho(x) dx. \tag{4.16}$$

To a mathematician, a measure is a way of weighting different regions when calculating integrals— precisely our $\rho(x) dx$. Notice that, for the case of an attracting fixed point, we would have $\rho(x) = \delta(x - x^*)$.

Cusps in the invariant density. At values of μ slightly smaller than one, our mapping has a rather complex invariant density (Fig. 4.5).

(c) *Find the invariant density (as described earlier) for $\mu = 0.9$. Make your trajectory length N_{cycles} big enough and the bin size small enough to see the interesting structures. Notice that the attractor no longer fills the whole range $(0, 1)$; locate roughly where the edges are. Notice also the cusps in $\rho(x)$ at the edges of the attractor, and*

also at places inside the attractor (called bound-*aries, see [94]). Locate some of the more prominent cusps.*

Fig. 4.5 Invariant density in the chaotic region ($\mu = 0.95$).

Analytics of cusps. Notice that $f'(\tfrac{1}{2}) = 0$, so by eqn 4.14 we know that $\rho(f(x)) \geq \rho(x)/|f'(x)|$ must have a singularity near $x = \tfrac{1}{2}$; all the points near $x = \tfrac{1}{2}$ are squeezed together and folded to one side by f. Further iterates of this singularity produce more cusps; the crease after one fold stays a crease after being further stretched and kneaded.

(d) *Set $\mu = 0.9$. Calculate $f(\tfrac{1}{2})$, $f(f(\tfrac{1}{2}))$, $\dots$ and compare these iterates to the locations of the edges and cusps from part (c). (You may wish to include them both on the same plot.)*

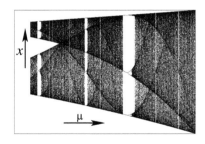

Fig. 4.6 Bifurcation diagram in the chaotic region. Notice the boundary lines threading through the diagram, images of the crease formed by the folding at $x = \tfrac{1}{2}$ in our map (see [94]).

Bifurcation diagram. The evolution of the attractor and its invariant density as μ varies are plotted in the *bifurcation diagram*, which is shown for large μ in Fig. 4.6. One of the striking features in this plot are the sharp boundaries formed by the cusps.

---

[23]You need not derive the factor $1/\pi$, which normalizes the probability density to one.

[24]There are actually many possible invariant measures on some attractors; this one is the SRB measure (John Guckenheimer, private communication).

(e) *Bifurcation diagram. Plot the attractor (duplicating Fig. 4.6) as a function of μ, for $0.9 < \mu < 1$. (Pick regularly spaced $\delta\mu$, run $n_{transient}$ steps, record n_{cycles} steps, and plot. After the routine is working, you should be able to push $n_{transient}$ and n_{cycles} both larger than 100, and $\delta\mu < 0.01$.)*
On the same plot, for the same μs, plot the first eight images of $x = \frac{1}{2}$, that is, $f(\frac{1}{2}), f(f(\frac{1}{2})), \ldots$. Are the boundaries you see just the cusps? What happens in the bifurcation diagram when two boundaries touch? (See [94].)

(4.4) Jupiter! and the KAM theorem.[25] (Astrophysics, Mathematics, Computation, Dynamical systems) ③
The foundation of statistical mechanics is the *ergodic hypothesis*: any large system will explore the entire energy surface. Statistical mechanics focuses on large systems because it is well known that many systems with a few interacting particles are definitely not ergodic.

Fig. 4.7 The Earth's trajectory around the Sun if Jupiter's mass is abruptly increased by about a factor of 100.

The classic example of a nonergodic system is the Solar System. Jupiter has plenty of energy to send the other planets out of the Solar System. Most of the phase-space volume of the energy surface has eight planets evaporated and Jupiter orbiting the Sun alone; the ergodic hypothesis would doom us to one long harsh winter. So, the big question is: why has the Earth not been kicked out into interstellar space?
Mathematical physicists have studied this problem for hundreds of years. For simplicity, they focused on the *three-body problem*: for example, the Sun, Jupiter, and the Earth. The early (failed) attempts tried to do perturbation theory in the strength of the interaction between planets. Jupiter's gravitational force on the Earth is not tiny, though; if it acted as a constant brake or accelerator, our orbit would be seriously perturbed in a few thousand years. Jupiter's effects must cancel out over time rather perfectly...
This exercise is mostly discussion and exploration; only a few questions need to be answered. Open the hints file [181] and run Earth's orbit for 100 years. Check that Jupiter does not seem to send the Earth out of the Solar System. Try increasing Jupiter's mass to 31,000 Earth masses to illustrate our concern.
Reset Jupiter's mass back to 317.83 Earth masses. View Earth's trajectory, run for a while, and zoom in to see the small effects of Jupiter on the Earth. Note that the Earth's position shifts depending on whether Jupiter is on the near or far side of the Sun.

(a) *Estimate the fraction that the Earth's radius from the Sun changes during the first Jovian year (about 11.9 years). How much does this fractional variation increase over the next hundred Jovian years?*
Jupiter thus warps Earth's orbit into a kind of spiral around a tube. This orbit in physical three-dimensional space is a projection of the tube in $6N$-dimensional phase space.
The tube in phase space already exists for massless planets! Let us start in the approximation that Earth and Jupiter have zero mass (the noninteracting planet approximation). Both Earth's orbit and Jupiter's orbit then become circles, or more generally ellipses. The field of topology does not distinguish an ellipse from a circle; any stretched, "wiggled" rubber band is a circle so long as it forms a curve that closes into a loop. Similarly, a torus (the surface of a doughnut) is topologically equivalent to any closed surface with one hole in it (like the surface of a coffee cup, with the handle as the hole). Convince yourself in this noninteracting approximation that Earth's orbit remains topologically a circle in its six-dimensional phase space.[26]
(b) *In the noninteracting planet approximation, what topological surface is it in the eighteen-*

[25]Hints for the computations can be found at the book website [181]. See also the Jupiter web pages [170].
[26]Hint: Plot the orbit in the (x, y), (x, p_x), and other planes. It should look like the projection of a circle along various axes.

dimensional phase space that contains the trajectory of the three bodies? *Choose between*
(i) *sphere*, (ii) *torus*, (iii) *Klein bottle*, (iv) *two-
hole torus*, *and* (v) *complex projective plane.*
(Hint: It is a circle cross a circle, parameterized
by two independent angles—one representing the
time during Earth's year, and one representing
the time during a Jovian year. Feel free to look
at Fig. 4.8 and part (c) before committing yourself, if pure thought is not enough.) *About how
many times does Earth wind around this surface
during each Jovian year?* (This ratio of years is
called the *winding number*.)

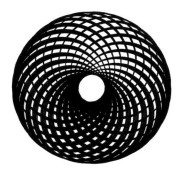

Fig. 4.8 Torus. The Earth's orbit around the Sun after adiabatically increasing the Jovian mass to 50,000
Earth masses.

The mathematical understanding of the three-
body problem was only solved in the past hundred years or so, by Kolmogorov, Arnol'd, and
Moser. Their proof focuses on the topological integrity of this tube in phase space (called now the
KAM torus). They were able to prove stability
if the winding number (Jupiter year over Earth
year) is sufficiently irrational.[27] More specifically,
they could prove in this case that for sufficiently
small planetary masses there is a distorted torus
in phase space, near the unperturbed one, around
which the planets spiral around with the same
winding number (Fig. 4.8).
(c) *About how large can you make Jupiter's mass
before Earth's orbit stops looking like a torus
(Fig. 4.7)?* (Restart each new mass at the same
initial conditions; otherwise, your answer will

depend upon the location of Jupiter in the sky
when you begin.) *Admire the remarkable trajectory when the mass becomes too heavy.*
Thus, for "small" Jovian masses, the trajectory in
phase space is warped and rotated a bit, so that
its toroidal shape is visible looking at Earth's position alone. (The circular orbit for zero Jovian
mass is looking at the torus on edge.)
The fact that the torus is not destroyed immediately is a serious problem for statistical mechanics! The orbit does not ergodically explore the
entire allowed energy surface. This is a counterex-
ample to Boltzmann's ergodic hypothesis. That
means that time averages are not equal to averages over the energy surface; our climate would
be very unpleasant if our orbit were ergodic.

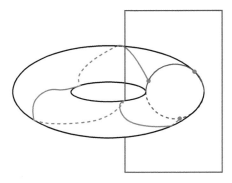

Fig. 4.9 The Poincaré section of a torus is a circle.
The dynamics on the torus becomes a mapping of the
circle onto itself.

Let us use a Poincaré section to explore these
tori, and the chaotic regions between them. If
a dynamical system keeps looping back in phase
space, one can take a cross-section of phase space
and look at the mapping from that cross-section
back into itself (see Fig. 4.9).
The Poincaré section shown in Fig. 4.9 is a planar
cross-section in a three-dimensional phase space.
We can reduce our problem to one with three
phase-space coordinates.[28] This leaves us with
a trajectory in three dimensions (so, for small
Jovian masses, we have a torus embedded in a

---

[27]Almost all numbers are *good irrationals*; they cannot be approximated by rationals p/q to better than $\propto 1/q^2$.

[28] Moving to the center of mass frame reduces eighteen to twelve, restricting to a plane leaves eight. Taking the Earth mass
to zero (the *restricted* planar three-body problem) and changing to frame rotating with Jupiter leaves four: two positions and
two momenta for the Earth. We can remove another variable by confining ourselves to a fixed "energy". The true energy of
the Earth is not conserved (because Earth feels a periodic potential), but there is a conserved quantity which is like an energy
in the rotating frame, see [170] under "Description of the three-body problem".

three-dimensional space). Now we take a Poincaré cross-section; we plot a point of the trajectory every time Earth passes directly between Jupiter and the Sun. We plot the distance to Jupiter along the horizontal axis, and the velocity component towards Jupiter along the vertical axis; the perpendicular component of the velocity (determined by the "energy") is not shown.

Set Jupiter's mass to its true value, and run the Poincaré section for 1,000 years. You should see a nice elliptical cross-section of the torus. As you increase the mass (resetting to the original initial conditions) watch the toroidal cross-sections as they break down. Run for a few thousand years at $M_J = 22,000\, M_e$; notice that the toroidal cross-section has become much more complex.

Fixing the mass at $M_J = 22,000\, M_e$, let us explore the dependence of the planetary orbits on the initial condition. Set up a chaotic trajectory ($M_J = 22,000\, M_e$) and observe the Poincaré section. The software will allow you to launch trajectories at various locations in the Poincaré section. Notice that many initial conditions slowly fill out closed curves. These are KAM tori that have been squashed and twisted like rubber bands. Explore until you find some orbits that seem to fill out whole regions; these represent *chaotic* orbits.[29]

(d) *Print out a Poincaré section with initial conditions both on KAM tori and in chaotic regions; label each.* See Fig. 4.3 for a small segment of the picture you should generate.

It turns out that proving that Jupiter's effects cancel out depends on Earth's smoothly averaging over the surface of the torus. If Jupiter's year is a rational multiple of Earth's year, the orbit closes after a few years and you do not average over the whole torus; only a closed spiral. Rational winding numbers, we now know, lead to chaos when the interactions are turned on; the large chaotic region you found above is associated with an unperturbed orbit with a winding ratio of 3:1 (hence the three circles). Disturbingly, the rational numbers are dense; between any two KAM tori there are chaotic regions, just because between any two irrational numbers there are rational ones.[30] It was amazingly tricky to prove that lots of tori survive nonetheless. You can imagine why this

took hundreds of years to understand (especially without computers to illustrate the tori and chaos graphically).

(4.5) **No Hamiltonian attractors.** (Dynamical systems, Astrophysics) ⓟ
Planetary motion is one of the prime examples of a Hamiltonian dynamical system. In the three-body problem with one body of small mass, Lagrange showed that there were five configurations (L1, L2, L3, L4, and L5) that rotated as a unit—forming fixed points in a rotating frame of reference. Three of these (L1–L3) are unstable, and the other two are stable. In dissipative dynamical systems, a stable fixed point implies that neighboring initial conditions will converge to the fixed point at long times.
Is that true here? What does it mean to be stable, or unstable, in a Hamiltonian system?

(4.6) **Perverse initial conditions.** ⓟ
If we start a gas of classical spherical particles in a square box all in a vertical line, all moving exactly in the vertical direction, they will bounce back and forth vertically forever.
Does that mean a gas of particles in a square box is not ergodic? Why or why not?

(4.7) **Crooks.** ③
Just at the end of the twentieth century, striking new fluctuation theorems were developed for broad classes of systems *driven out of equilibrium*. We can derive a version of Crooks' fluctuation theorem [47] using Liouville's theorem in the microcanonical ensemble. (See also Exercises 4.8 and 5.25.)
Consider a system in microcanonical equilibrium in the energy shell $(E, E + \delta E)$. (Think of a piston of gas, insulated from the outside world). Let the system be subject to an external forcing, giving it some time-dependent potential energy taking the system from an initial state to a final state. In particular, if the system starts at a point in phase-space $(\mathbb{P}, \mathbb{Q})$ when the external force starts, let us denote the final point $U(\mathbb{P}, \mathbb{Q})$ as in Fig. 4.10. (Consider compressing the gas, perhaps very rapidly, sending it out of equilibrium.) The external force will do different amounts of work W on the system depending on

---

[29]Notice that the chaotic orbit does not throw the Earth out of the Solar System. The chaotic regions near infinity and near our initial condition are not connected, because a two-torus separates the three-dimensional energy surface into pieces. In higher dimensional Hamiltonian systems it is conjectured that chaotic regions are typically all joined through *Arnol'd diffusion*.

[30]It is even worse; it turns out that numbers which are extremely close to rational (Liouville numbers like $1 + 1/10 + 1/10^{10} + 1/10^{10^{10}} + \ldots$) may also lead to chaos.

the initial condition. (Consider a system with just a few gas atoms. As the gas is compressed, how strongly the atoms collide with the moving piston will depend on their initial state.)

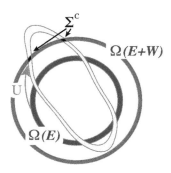

Fig. 4.10 Crooks fluctuation theorem: evolution in time. The inner energy shell $\Omega(E)$ gets mapped by a time-dependent Hamiltonian evolution U into a region spanning a variety of energies (wiggly region); different initial conditions exchange different amounts of work. The region Σ^C denotes the final states that landed in the energy shell $\Omega(E + W)$ (absorbing work W).

(a) *Check that the derivation of Liouville's theorem, eqns 4.5–4.7, also applies for time-dependent potential energies; that they also conserve phase-space volume.*
(b) *The system starts in microcanonical equilibrium in the energy shell $(E, E + \delta E)$, and the probability density evolves under U into the wiggly region in Fig. 4.10. Inside this wiggly region, what is the final phase-space probability density? Conversely, if the system starts in microcanonical equilibrium in the energy shell $(E + W, E + \delta E + W)$, what is the final phase space density under the evolution U^{-1} (Fig. 4.11)?* Express your answer in terms of the function $\Omega(E)$.
The microscopic equations of motion are invariant under time reversal. The volume Σ^C in Fig. 4.10 includes the trajectories that started near energy E and absorbed work W under the time evolution U. Under the reverse evolution U^{-1} these trajectories all map to the volume Σ^D that emitted work W under the reverse evolution U^{-1} to end near E. Liouville's theorem thus tells us that the phase-space volume of Σ^C must therefore be equal to the phase-space volume of Σ^D.
An experiment starts in a microcanonical state in

the energy shell $(E, E + \delta E)$, and measures the probability $p^C(W)$ that the evolution U leaves it in the energy shell $(E + W, E + \delta E + W)$ (roughly doing work W). Another experiment starts in $(E + W, E + \delta E + W)$, and measures the probability $p^D(-W)$ that the evolution U^{-1} leaves it in the energy shell $(E, E + \delta E)$.

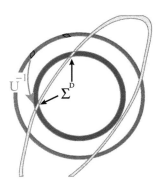

Fig. 4.11 Crooks fluctuation theorem: evolution backward in time. Starting from the outer energy shell $\Omega(E + W)$ transformed under the inverse evolution U^{-1} (wiggly region), the region Σ^C gets mapped back into Σ^D (and vice versa under U).

(c) *Write $p^C(W)$, using only $\Omega(E)$, $\Omega(E + W)$, and Σ^C. Write $p^D(-W)$ in terms of the energy-shell volumes and Σ^D.* (See Figs. 4.10 and 4.11. Hint: the microcanonical ensemble fills its energy shell with uniform density in phase space.)
Now let us use Liouville's theorem to derive a way of using this *nonequilibrium* experiment to measure the *equilibrium* entropy difference $S(E+W) - S(E)$.
(d) *Use your result from part (c) to write $\Omega(E + W)/\Omega(E)$ in terms of the measured $p^C(W)$ and $p^D(-W)$. Use this to derive the entropy difference $S(E+W) - S(E)$.* (Hint: $S(E) = k_B \log(\Omega(E))$.) Who would have thought that anything could be proven about an arbitrary time-dependent system, driven out of equilibrium? Who would have thought that anything physical like $p^C(W)$ would be ever given in terms of the very-large-number $\Omega(E)$ (instead of its log)?
Your result in part (d) is a version of Crooks fluctuation theorem [47], which states that

$$p^D(-W)/p^C(W) = \exp(-\Delta S/k_B). \qquad (4.17)$$

(4.8) Jarzynski. ③

You cannot create a perpetual motion machine—a machine that extracts energy from thermal vibrations to do useful work. Tiny machines, however, fluctuate—sometimes doing work on their environment. This exercise discusses a remarkable result, the Jarzynski equality [93], which quantifies the effects of fluctuations in small nonequilibrium systems. (See also Exercises 4.7 and 4.7.)

Suppose we compress a piston from length L to length L', doing work $W_c \geq 0$ on the gas in the piston, heating the gas. As we allow the gas to expand back to length L, the net work done by the outside world $W_e \leq 0$ is negative, cooling the gas. Perpetual motion is impossible, so no net work can be done extracting kinetic energy from the gas:

$$\overline{W_c + W_e} \geq 0. \qquad (4.18)$$

Note the average in eqn 4.18. Especially for small systems compressed quickly, there will be fluctuations in the work done. Sometimes it will happen that the gas will indeed do net work on the outside world. One version of the Jarzynski equality states that a thermally isolated system, starting in equilibrium at temperature T, evolved under a time-dependent Hamiltonian back to its initial state, will always satisfy

$$\overline{\exp(-(W_c + W_e)/k_B T)} = 1. \qquad (4.19)$$

(a) *Suppose the system is indeed driven out of equilibrium (so $\overline{W_c + W_e} > 0$). Can eqn 4.19 be true if the work $W_c + W_e > 0$ for all initial conditions? Must every system, evolved to a nonequilibrium state under a time-dependent Hamiltonian, sometimes do net work on the outside world?*

Note that the Jarzynski equality eqn 4.19 implies the fact that the average work must be positive, eqn 4.18. Because e^{-x} is convex, $\overline{e^{-x}} \geq e^{-\overline{x}}$. (Roughly, because e^{-x} is curved upward everywhere, its average is bigger than its value at the average x.) Applying this to eqn 4.19 gives us $1 = \overline{e^{-(W_c+W_e)/k_B T}} \geq e^{-\overline{W_c+W_e}/k_B T}$. Taking logs of both sides tells us $0 \geq -\overline{W_c+W_e}/k_B T$ so $0 \leq \overline{W_c + W_e}$; the net work done is not negative. Other forms of the Jarzynski equality give the change in free energy for nonequilibrium systems that start and end in different states.

We shall test a version of this equality in the smallest, simplest system we can imagine—one particle of an ideal gas inside a thermally insulated piston, Fig. 4.12, initially confined to $0 < x < L$, compressed into a region $L - L' < x < L$ of length L' at velocity V, and then immediately allowed to expand back to L at velocity $-V$. For simplicity, we shall assume that the particles which are hit by the piston during compression do not have time to bounce off the far wall and return.[31]

Fig. 4.12 One particle in a piston. The piston is pushed forward at velocity V (not assumed small) a distance ΔL, changing the piston length from L to $L' = L - \Delta L$. The particle is initially at position x and momentum p. If it is hit by the piston, it will recoil to a new velocity, absorbing work $W_c(p)$ from the piston.

First, we carefully work through some elementary mechanics. How much energy is transferred during a collision with the piston? Which initial conditions will collide during contraction? During expansion?

(b) *If a particle with momentum p elastically collides with the compressing piston moving with velocity V, as shown in Fig. 4.12, what is the final kinetic energy after the collision? (Hint: It is easy in the moving reference frame of the piston.) What work $W_c(p)$ is done by the piston on the particle? If a particle with momentum p elastically collides with the expanding piston moving with velocity $-V$, what is the work done $W_e(p)$ by the piston on the particle? Can the piston do negative work during compression? (Hint: What is the direction of the impulsive force during a collision?) Can $W_e(p)$ be positive?*

Many initial conditions (x_0, p_0) for the particle at $t = 0$ will not collide with the piston during compression, and many will not collide with it during the expansion. Fig. 4.13 shows the initial conditions in phase space from which a particle will collide with the piston during the compression and expansion.

(c) *Which boundary describes the initial conditions for a particle that collides with the piston* at the very start of the compression cycle? At the end of the compression cycle? Which boundary describes the initial conditions which collide at the beginning of the expansion cycle? At the end of the expansion cycle? Will any particle collide with the piston more than once, given our assumption that the particle does not have time to bounce off of the wall at $x = L$ and return? Why or why not?

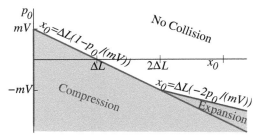

Fig. 4.13 Collision phase diagram. Initial conditions (x_0, p_0) in phase space for the particle, showing regions where it collides with the piston during the compression cycle and during the expansion cycle. The equations determining the boundaries are shown.

Our macroscopic intuition is that the fluctuations that do net negative work should be rare. But these fluctuations, weighted exponentially as in the Jarzynski equality eqn 4.19, must balance those that do positive work.

(d) *Which of the initial conditions of part (c) will do negative net work on the particle? At low temperatures and high velocities V, where $k_B T \ll \frac{1}{2} mV^2$, are these initial conditions rare?* The theoretical argument for the Jarzynski equality is closely related to that for the Crooks thermodynamic relation of Exercise 4.7. Instead of deriving it, we shall test the theorem numerically for our piston. Choose $m = 2$, $k_B T = 8$, $V = 2$, $L = 20$, and $L' = 19$ (so $\Delta L = 1$).

We could calculate the expectation values $\overline{X}$ by doing an integral over phase space (Fig. 4.13) weighted by the Boltzmann distribution. Instead, let us do an ensemble of experiments, starting particles with a Boltzmann distribution of initial conditions in phase space. (This is an example of a Monte-Carlo simulation; see Chapter 8).

(e) *Write a routine that generates N initial conditions (x_0, p_0) with a Boltzmann probability distribution for our ideal gas particle in the piston,*

and calculates $W_c(x_0, p_0) + W_e(x_0, p_0)$ for each. Calculate the average work done by the piston $\overline{W_c + W_e}$. How big must N be before you get a reliable average? Does your average obey eqn 4.18?

(f) *Use your routine with $N = 1,000$ to calculate Jarzynski's exponentially weighted average $\overline{\exp(-W/k_B T)}$ in eqn 4.19, and compare to $\exp(-\overline{W}/k_B T)$. Do this 200 times, and plot a histogram of each. Does the Jarzynski equality appear to hold?*

For big systems, the factor $\exp(-W/k_B T)$ in the Jarzynski equality balances extremely rare fluctuations with low work against the vast majority of instances where the work has small fluctuations about the average.

(4.9) **2D turbulence and Jupiter's great red spot.**[32] (Astrophysics, Computation, Dynamical systems) ③

Fig. 4.14 Jupiter's Red Spot. Jupiter's atmosphere is stormy, with winds that can exceed 400 miles per hour in a complex, turbulent pattern. Jupiter's great red spot is a giant storm comparable to the size of the earth, that has lasted for at least 186 years. (Image from Voyager I, courtesy of NASA/JPL [138]. As of this writing, the Red Spot has been shrinking.)

Fully developed turbulence is one of the outstanding challenges in science. The agitated motion of water spurting through a fire hose has a complex, fluctuating pattern of swirls and eddies spanning many length scales. Turbulence in two-dimensional fluids is much better understood, with the latest advances involving conformal field theories [151]. The inspiration for this exercise

---

[32] This exercise was developed in collaboration with Jaron Kent-Dobias. Hints files can be found at the book website [181].

is an ancient model introduced by Onsager and others (see [135]).

The model describes the motion of point *vortices*, describing local rotational eddies in the fluid. It provides an analogy to Liouville's theorem, an example of a system with an infinite number of conserved quantities, and a system with a negative temperature. And it provides a plausible explanation of Jupiter's Great Red Spot (Fig. 4.14), cyclones, hurricanes, and typhoons on Earth, and experiments on 2D superfluids [66, 95].

The flow of most liquids is almost incompressible; if $\mathbf{u}(\mathbf{x})$ is the velocity field, then to an excellent approximation $\nabla \cdot \mathbf{u} = 0$. Helmholtz's theorem tells us that a smooth field $\mathbf{u}$ is determined by its divergence and curl, if it dies away fast enough at infinity and the region has no holes.[33] So knowing how the *vorticity* $\boldsymbol{\omega} = \nabla \times \mathbf{u}$ evolves is enough to determine the evolution of the fluid. In two dimensions, the vorticity is a scalar $\omega(x) = \partial u_y / \partial x - \partial u_x / \partial y$.

(a) *Show that*

$$\mathbf{u}_0(\mathbf{x}) = \frac{\Gamma}{2\pi} \frac{\hat{\theta}}{r} \qquad (4.20)$$

in two dimensions has both zero curl and zero divergence, except at zero. (This is the velocity field around one of Onsager's vortices.) *Use Stokes' theorem to show that the curl of this velocity field is $\omega(\mathbf{x}) = \Gamma\delta(\mathbf{x}) = \Gamma\delta(x)\delta(y)$. Argue using this solution that the velocity field in an annulus is* not *uniquely defined by its divergence and curl alone (as described by note 33).* (Hint: Consider the annulus bounded by two concentric circles surrounding our vortex.)

We provide a simulation of the dynamics of interacting vortices [181].

(b) *Run the simulation with one vortex with $\Gamma_1 = 1$ and a "tracer vortex" with $\Gamma = 0$. Does the tracer vortex rotate around the test vortex?*

At low velocities, high viscosity, and small distances, fluids behave smoothly as they move; turbulence happens at high Reynolds numbers, where velocities are big, distances small, and viscosities are low. If we reduce the viscosity to zero, the kinetic energy of the fluid is conserved.

We can write $\mathbf{u}$ as a nonlocal function of ω

$$\begin{aligned} u_x(\mathbf{r}) &= -\frac{1}{2\pi} \int d\mathbf{r}' \frac{y - y'}{(\mathbf{r} - \mathbf{r}')^2} \omega(\mathbf{r}') \\ u_y(\mathbf{r}) &= \frac{1}{2\pi} \int d\mathbf{r}' \frac{x - x'}{(\mathbf{r} - \mathbf{r}')^2} \omega(\mathbf{r}') \end{aligned} \qquad (4.21)$$

(the Biot-Savart law).

(c) *Show that the Biot-Savart law agrees with your answer from part (a) for the case of a δ-function vortex.*

We can write the kinetic energy H in terms of the vorticity, which turns out to be a nonlocal convolution

$$\begin{aligned} H &= \int \tfrac{1}{2}\mathbf{u}(\mathbf{r})^2 d\mathbf{r} \qquad\qquad (4.22) \\ &= -\frac{1}{4\pi} \int d\mathbf{r} \int d\mathbf{r}' \omega(\mathbf{r})\omega(\mathbf{r}') \log|\mathbf{r} - \mathbf{r}'| \end{aligned}$$

where we set the density of the fluid equal to one and ignore an overall constant.[34] We can also write the equations of motion for the vorticity ω in terms of itself and the velocity $\mathbf{u}$

$$\frac{\partial \omega}{\partial t} = -\mathbf{u} \cdot \nabla \omega. \qquad (4.23)$$

Note that this continuum vorticity has total derivative zero ($d\omega/dt = \partial\omega/\partial t + \mathbf{u} \cdot \nabla\omega = 0$); the vorticity measured along a packet of fluid remains constant in time. This means that the amount of fluid $\rho(\omega, t)d\omega$ in a small range $(\omega, \omega + d\omega)$ of vorticity stays constant in time, *for each possible* ω—an infinite number of conservation laws.

To set up our simulation, we need to find a numerical representation for the vorticity field. Onsager suggested a discretization into vortices (part (a)), $\omega(\mathbf{r}) = \sum_{i=1}^{N} \Gamma_i \delta(\mathbf{r} - \mathbf{r}_i)$. For point vortices, the energy becomes

$$\mathcal{H} = -\frac{1}{2\pi} \sum_i \sum_{j \ne i} \Gamma_i \Gamma_j \log\left((\mathbf{r}_i - \mathbf{r}_j)^2\right). \quad (4.24)$$

Note that there is no kinetic energy for the vortices. The energy in the velocity field is purely kinetic energy; the "potential" energy of the vortex Hamiltonian is given by the kinetic energy of the fluid.

The vortices move according to the local velocity field,

$$d\mathbf{r}_i/dt = \mathbf{u}(r_i). \qquad (4.25)$$

---

[33] More generally, if the vorticity is defined in a non simply connected region S with n holes, and if $\mathbf{u}$ is parallel to the boundaries, then knowing $\omega(\mathbf{r})$ and the circulations $\int_{\partial S_n} \mathbf{u} \cdot d\boldsymbol{\ell}$ around the edges ∂S_n of the holes specifies the velocity $\mathbf{u}$ inside.

[34] Miller, Weichman, and Cross tell us that $\log(|\mathbf{r} - \mathbf{r}'|/R_0)$ corresponds to free boundary conditions, where R_0 is a constant with dimensions of length.

This just means that the fluid drags the vortices with it, without changing the strength of the vortices. (This agrees with the continuum law that the total derivative of the vorticity is zero, eqn 4.23). Hence

$$\mathrm{d}x_i/\mathrm{d}t = \frac{1}{2\pi}\sum_{j!=i}\Gamma_j(y_i - y_j)/(\mathbf{r}_i - \mathbf{r}_j)^2 \quad (4.26)$$

$$\mathrm{d}y_i/\mathrm{d}t = -\frac{1}{2\pi}\sum_{j!=i}\Gamma_j(x_i - x_j)/(\mathbf{r}_i - \mathbf{r}_j)^2.$$

If there is no kinetic energy for the vortices, what are the "conjugate variables" analogous to x and p for regular particles?

(d) *Check from eqn 4.26 and eqn 4.24 that*

$$\mathrm{d}x_i/\mathrm{d}t = (1/4\Gamma_i)\partial\mathcal{H}/\partial y_i$$
$$\mathrm{d}y_i/\mathrm{d}t = -(1/4\Gamma_i)\partial\mathcal{H}/\partial x_i. \quad (4.27)$$

Thus $2\sqrt{\Gamma_i}x_i$ and $2\sqrt{\Gamma_i}y_i$ are analogous to x and p in a regular phase space.

Launch the simulation.

(e) *Start with $n = 20$ vortices with a random distribution of vortex strengths $\Gamma_i \in [-1, 1]$ and random positions $\mathbf{r}_i$ within the unit circle.[35] Print the original configuration. Run for a time $t = 10$, and print the final configuration. Do you see the spontaneous formation of a giant whirlpool? Are the final positions roughly also randomly arranged? Measure and report the energy H of your vortex configuration.* (Warning: Sometimes the differential equation solver crashes. Just restart again with a new set of random initial conditions.) Note: we are not keeping the vortices inside the unit circle during our dynamical evolution.

Hurricanes and Jupiter's Red Spot can be thought of as large concentrations of vorticity—all the plus or minus vortices concentrated into one area, making a giant whirlpool. What do we need to do to arrange this? Let's consider the effect of the total energy.

(f) *For a given set of vortices of strength Γ_i, would clustering the positive vortices and the negative vortices each into their own clump (counter-rotating hurricanes) be a high-energy or a low energy configuration, as the size of the clumps goes to zero?* (Hint: You will be able to check this with the simulation later.) *Why?*

You chose points at random for part (e), but did

not see the vortices separated into clumps. How did we know this was likely to happen?

(g) *Estimate the entropy difference between a state where twenty vortices are confined within the unit circle, and a state where two circles of radius $R = \frac{1}{2}$ each have ten vortices, one with all negative vorticity and one all positive,[36] as shown in Fig. 4.15. Leave your answer as a multiple of k_B.* (Note: The vortices have different values of Γ, so are distinguishable.) *How many tries would you need to see this clumping happen in an atomic system?*

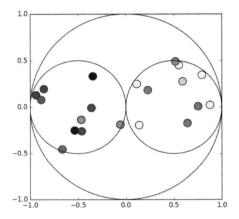

Fig. 4.15 Circles. It is unlikely that the vortices would collect into clumps by accident.

Onsager's problem is one of the best examples of *negative temperature*.[37]

(h) *Using your results from part (f) and part (g), in the energy range where the hurricanes will form, will the change in entropy be positive or negative as the energy increases? Is the temperature negative?*

In most simulations of Onsager's vortices, one selects for states that form a big hurricane by starting with several small ones, and watching them combine. (The small hurricanes must be tightly packed; as they combine they gain entropy because the vortices can spread out more.) Instead, we shall set the energy of our configuration by doing a Metropolis Monte-Carlo simulation at a fixed temperature.

---

[35]You can generate random points in the unit circle by choosing $\theta \in (-\pi, \pi]$ and picking $r = \sqrt{p}$ with p uniformly $\in [0, 1)$.

[36]There are many arrangements of small circles, whose entropy contribution we are not considering here.

[37]Because the phase space is finite, the volume of the energy shell at high energies goes to zero, instead of continuing to increase.

(i) *Thermalize the Monte-Carlo simulation for your $n = 20$ vortices at low temperature $\beta = 1/(k_B T) = 2$, report the final energy, and print out your configuration. Does the thermalized vortex state look similar to the initial conditions you generated in part (e)? Thermalize again at temperature $\beta = -2$, report the energy, and print out your final configuration. Do the vortices separate out into clumps of positive and negative vorticity?*

The time it takes to run a simulation is roughly determined by the minimum distance between vortices. We can use this to keep our simulation from crashing so much.

(j) *Re-run the Monte Carlo simulation with $n = 20$ vortices until you find a configuration with a clear separation of positive and negative vortices, but where the minimum distance between vortices is not too small (say, bigger than 0.01). (This should not take lots of tries, so long as you thermalize with an energy in the right region.) Is the energy you need to thermalize to positive, or negative? Print this initial configuration of vortices. Animate the simulation for $t = 10$ with this state as the initial condition. How many hurricanes do you find? Print out the final configuration of vortices.* (Hint: In Mathematica, one can right-click in the animation window to print the current configuration (Save Graphic As...). In Python, just plot xOft[-1], yOft[-1]; an array evaluated at [-1] gives the last entry. You can copy the axis limits and colors and sizes from the animation, to make a nice plot.)

Entropy

5

Entropy is the most influential concept to arise from statistical mechanics. What does it mean? Can we develop an intuition for it?

We shall see in this chapter that entropy has three related interpretations.[1] *Entropy measures the disorder in a system;* in Section 5.2 we will observe this using the entropy of mixing and the residual entropy of glasses. *Entropy measures our ignorance about a system;* in Section 5.3 we will give examples from nonequilibrium systems and information theory. But we will start in Section 5.1 with the original interpretation, that grew out of the nineteenth-century study of engines, refrigerators, and the end of the Universe. *Entropy measures the irreversible changes in a system.*

[1] Equilibrium is a word with positive connotations, presumably because it allows us to compute properties easily. Entropy and the quantities it measures—disorder, ignorance, uncertainty—are words with negative connotations, presumably because entropy interferes with making efficient heat engines. Notice that these connotations are not always reliable; in information theory, for example, having high Shannon entropy is good, reflecting better compression of data.

5.1 Entropy as irreversibility: engines and the heat death of the Universe

The early 1800s saw great advances in understanding motors and engines. In particular, scientists asked a fundamental question: how efficient can an engine be? The question was made more difficult because there were two relevant principles[2] to be discovered: energy is conserved and entropy always increases.[3]

For some kinds of engines, only energy conservation is important. For example, there are electric motors that convert electricity into mechanical work (running an electric train), and generators that convert mechanical work (from a rotating windmill) into electricity.[4] For these electromechanical engines, the absolute limitation is given by the conservation of energy: the motor cannot generate more energy in mechanical work than is consumed electrically, and the generator cannot generate more electrical energy than is input mechanically. An ideal electromechanical engine can convert all the energy from one form to another.

Steam engines are more complicated. Scientists in the early 1800s were figuring out that heat is a form of energy. A steam engine, running a power plant or an old-style locomotive, transforms a fraction of the heat energy from the hot steam (the "hot bath") into electrical energy or work, but some of the heat energy always ends up wasted—dumped into the air or into the cooling water for the power plant (the "cold bath"). In fact, if the only limitation on heat engines was conservation of energy, one would be able to make a motor using the heat energy from a rock,

[2] These are the first and second laws of thermodynamics, respectively (Section 6.4).

[3] Some would be pedantic and say only that entropy never decreases, since a system in equilibrium has constant entropy. The phrase "entropy always increases" has a ring to it, though.

[4] Electric motors are really the same as generators run in reverse; turning the shaft of an electric motor can generate electricity.

Statistical Mechanics: Entropy, Order Parameters, and Complexity. James P. Sethna, Oxford University Press (2021). © Oxford University Press. DOI:10.1093/oso/9780198865247.003.0005

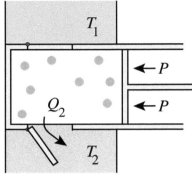

Fig. 5.1 Perpetual motion and Carnot's cycle. How to use an engine which produces Δ more work than the Carnot cycle to build a perpetual motion machine doing work Δ per cycle.

Fig. 5.2 Prototype heat engine. A piston with external exerted pressure P, moving through an insulated cylinder. The cylinder can be put into thermal contact with either of two heat baths: a hot bath at temperature T_1 (say, a coal fire in a power plant) and a cold bath at T_2 (say water from a cold lake). During one cycle of the piston in and out, heat energy Q_1 flows into the piston, mechanical energy W is done on the external world by the piston, and heat energy Q_2 flows out of the piston into the cold bath.

[5]This postulate is one formulation of the second law of thermodynamics. It is equivalent to the more standard version, that entropy always increases.

[6]Carnot did not consider the entropy cost for guiding the piston through the cycle. A calculation of this cost [120] (see Exercise 6.23) argues that it scales as the square root of the number of particles—important for small engines, but unimportant for power plants and refrigerators.

getting both useful work and a very cold rock.

There is something fundamentally less useful about energy once it becomes heat. By spreading out the energy among all the atoms in a macroscopic chunk of material, not all of it can be retrieved again to do useful work. The energy is more useful for generating power when divided between hot steam and a cold lake, than in the form of water at a uniform, intermediate warm temperature. Indeed, most of the time when we use mechanical or electrical energy, the energy ends up as heat, generated from friction or other dissipative processes.

The equilibration of a hot and cold body to two warm bodies in an isolated system is *irreversible*; one cannot return to the original state without inputting some kind of work from outside the system. Carnot, publishing in 1824, realized that the key to producing the most efficient possible engine was to avoid irreversibility. A heat engine run in reverse is a refrigerator; it consumes mechanical work or electricity and uses it to pump heat from a cold bath to a hot one. A reversible heat engine would be able to run forward generating work by transferring heat from the hot to the cold bath, and then run backward using the same work to pump the heat back into the hot bath.

If you had an engine more efficient than a reversible one, you could run it side by side with a reversible engine running as a refrigerator (Fig. 5.1). The pair of engines would together generate work by extracting energy from the hot bath (as from our rock, above) without adding heat to the cold one. After we used this work, we could dump the extra heat from friction back into the hot bath, getting a perpetual motion machine that did useful work without consuming anything. In thermodynamics *we postulate that such perpetual motion machines are impossible.*[5] By calculating the properties of this reversible engine, Carnot placed a fundamental limit on the efficiency of heat engines and discovered what would later be called the entropy.

Carnot considered a prototype heat engine (Fig. 5.2), built from a piston with external pressure P, two heat baths at a hot temperature T_1 and a cold temperature T_2, and some type of gas inside the piston. During one cycle of his engine heat Q_1 flows out of the hot bath, heat Q_2 flows into our cold bath, and net work $W = Q_1 - Q_2$ is done by the piston on the outside world. To make his engine reversible Carnot must avoid (i) friction, (ii) letting hot things touch cold things, (iii) letting systems at high pressure expand into systems at low pressure, and (iv) moving the walls of the container too quickly (emitting sound or shock waves).[6]

Carnot, a theorist, could ignore the practical difficulties. He imagined a frictionless piston which ran through a cycle at arbitrarily low velocities. The piston was used both to extract work from the system and to raise and lower the temperature. Carnot connected the gas thermally to each bath only when its temperature agreed with the bath, so his engine was fully reversible.

The Carnot cycle moves the piston in and out in four steps (Fig. 5.3).

- (a→b) The compressed gas is connected to the hot bath, and the piston moves outward at a varying pressure; heat Q_1 flows in to maintain the gas at temperature T_1.

- (b→c) The piston expands further at varying pressure, cooling the gas to T_2 without heat transfer.

- (c→d) The expanded gas in the piston is connected to the cold bath and compressed; heat Q_2 flows out maintaining the temperature at T_2.

- (d→a) The piston is compressed, warming the gas to T_1 without heat transfer, returning it to the original state.

Energy conservation tells us that the net heat energy flowing into the piston, $Q_1 - Q_2$, must equal the work done on the outside world W:

$$Q_1 = Q_2 + W. \tag{5.1}$$

The work done by the piston is the integral of the force exerted times the distance. The force is the piston surface area times the pressure, and the distance times the piston surface area is the volume change, giving the geometrical result

$$W = \int F \, dx = \int (F/A)(A \, dx) = \int_{cycle} P \, dV = \text{area inside } PV \text{ loop.} \tag{5.2}$$

That is, if we plot P versus V for the four steps of our cycle, the area inside the resulting closed loop is the work done by the piston on the outside world (Fig. 5.3).

Carnot realized that all reversible heat engines working with the same hot and cold bath had to produce exactly the same amount of work for a given heat flow (since they are all perfectly efficient). This allowed him to fill the piston with the simplest possible material (a monatomic ideal gas), for which he knew the relation between pressure, volume, and temperature. We saw in Section 3.5 that the ideal gas equation of state is

$$PV = Nk_BT \tag{5.3}$$

and that its total energy is its kinetic energy, given by the equipartition theorem

$$E = \frac{3}{2}Nk_BT = \frac{3}{2}PV. \tag{5.4}$$

Along a→b where we add heat Q_1 to the system, we have $P(V) = Nk_BT_1/V$. Using energy conservation, we have

$$Q_1 = E_b - E_a + W_{ab} = \frac{3}{2}P_bV_b - \frac{3}{2}P_aV_a + \int_a^b P \, dV. \tag{5.5}$$

But $P_aV_a = Nk_BT_1 = P_bV_b$, so the first two terms cancel; the last term can be evaluated, giving

$$Q_1 = \int_a^b \frac{Nk_BT_1}{V} \, dV = Nk_BT_1 \log\left(\frac{V_b}{V_a}\right). \tag{5.6}$$

Fig. 5.3 Carnot cycle P–V diagram. The four steps in the Carnot cycle: a→b, heat in Q_1 at constant temperature T_1; b→c, expansion without heat flow; c→d, heat out Q_2 at constant temperature T_2; and d→a, compression without heat flow to the original volume and temperature.

Similarly,

$$Q_2 = Nk_BT_2 \log\left(\frac{V_c}{V_d}\right). \tag{5.7}$$

For the other two steps in our cycle we need to know how the ideal gas behaves under expansion without any heat flow in or out. Again, using energy conservation on a small segment of the path, the work done for a small volume change $-P\,dV$ must equal the change in energy dE. Using eqn 5.3, $-P\,dV = -(Nk_BT/V)\,dV$, and using eqn 5.4, $dE = \frac{3}{2}Nk_B\,dT$, so $dV/V = -\frac{3}{2}\,dT/T$. Integrating both sides from b to c, we find

$$\int_b^c \frac{dV}{V} = \log\left(\frac{V_c}{V_b}\right) = \int_b^c -\frac{3}{2}\frac{dT}{T} = -\frac{3}{2}\log\left(\frac{T_2}{T_1}\right), \tag{5.8}$$

so $V_c/V_b = (T_1/T_2)^{3/2}$. Similarly, $V_d/V_a = (T_1/T_2)^{3/2}$. Thus $V_c/V_b = V_d/V_a$, and hence

$$\frac{V_c}{V_d} = \frac{V_c}{V_b}\frac{V_b}{V_d} = \frac{V_d}{V_a}\frac{V_b}{V_d} = \frac{V_b}{V_a}. \tag{5.9}$$

We can use the volume ratios from the insulated expansion and compression (eqn 5.9) to substitute into the heat flow (eqns 5.6 and 5.7) to find

$$\frac{Q_1}{T_1} = Nk_B \log\left(\frac{V_b}{V_a}\right) = Nk_B \log\left(\frac{V_c}{V_d}\right) = \frac{Q_2}{T_2}. \tag{5.10}$$

This was Carnot's fundamental result; his cycle, and hence any reversible engine, satisfies the law

$$\frac{Q_1}{T_1} = \frac{Q_2}{T_2}. \tag{5.11}$$

Later scientists decided to define[7] the *entropy change* to be this ratio of heat flow to temperature:

$$\Delta S_{\text{thermo}} = \frac{Q}{T}. \tag{5.12}$$

For a reversible engine the entropy flow from the hot bath into the piston Q_1/T_1 equals the entropy flow from the piston into the cold bath Q_2/T_2; no entropy is created or destroyed. Any real engine will create[8] net entropy during a cycle; no engine can reduce the net amount of entropy in the Universe.

The irreversible increase of entropy is not a property of the microscopic laws of nature. In particular, the microscopic laws of nature are *time-reversal invariant*: the laws governing the motion of atoms are the same whether time is running backward or forward.[9] The microscopic laws do not tell us the *arrow of time*. The direction of time in which entropy increases is our definition of the *future*.[10]

This confusing point may be illustrated by considering the game of pool or billiards. Neglecting friction, the trajectories of the pool balls are also time-reversal invariant. If the velocities of the balls were reversed halfway through a pool shot, they would retrace their motions, building

---

[7]The thermodynamic entropy is derived with a heat flow $\Delta E = Q$ at a fixed temperature T, so our statistical mechanics definition of temperature $1/T = \partial S/\partial E$ (from eqn 3.29) is equivalent to the thermodynamics definition of entropy $1/T = \Delta S/\Delta E \Rightarrow \Delta S = Q/T$ (eqn 5.12).

[8]For example, a small direct heat leak from the hot bath to the cold bath of δ per cycle would generate

$$\frac{Q_2+\delta}{T_2} - \frac{Q_1+\delta}{T_1} = \delta\left(\frac{1}{T_2} - \frac{1}{T_1}\right) > 0 \tag{5.13}$$

entropy per cycle.

[9]More correctly, the laws of nature are only invariant under CPT: changing the direction of time (T) along with inverting space (P) and changing matter to antimatter (C). Radioactive beta decay and other weak interaction forces are not invariant under time-reversal. The basic conundrum for statistical mechanics is the same, though: we cannot tell from the microscopic laws if we are matter beings living forward in time or antimatter beings living backward in time in a mirror. Time running backward would appear strange macroscopically even if we were made of antimatter.

[10]In electromagnetism, the fact that waves radiate away from sources more often than they converge upon sources is a closely related distinction of past from future.

up all the velocity into one ball that then would stop as it hit the cue stick. In pool, the feature that distinguishes forward from backward time is the greater order at early times; all the momentum starts in one ball, and is later distributed among all the balls involved in collisions. Similarly, the only reason we can resolve the arrow of time—distinguish the future from the past—is that the Universe started in an unusual, low-entropy state, and is irreversibly moving towards equilibrium.[11] (One would think that the Big Bang was high entropy! It was indeed hot (high momentum-space entropy), but it was dense (low position-space entropy), and in total the entropy was lower[12] than it is now.) The temperature and pressure differences we now observe to be moving towards equilibrium as time increases are echoes of this low-entropy state in the distant past.

The cosmic implications of the irreversible increase of entropy were not lost on the intellectuals of the nineteenth century. In 1854, Helmholtz predicted the *heat death of the Universe*: he suggested that as the Universe ages all energy will become heat, all temperatures will become equal, and everything will "be condemned to a state of eternal rest". In 1895, H. G. Wells speculated about the state of the Earth in the distant future:

> ...the sun, red and very large, halted motionless upon the horizon, a vast dome glowing with a dull heat...The earth had come to rest with one face to the sun, even as in our own time the moon faces the earth...There were no breakers and no waves, for not a breath of wind was stirring. Only a slight oily swell rose and fell like a gentle breathing, and showed that the eternal sea was still moving and living. ...the life of the old earth ebb[s] away... *The Time Machine* [210, chapter 11]

This gloomy prognosis has been re-examined recently; it appears that the expansion of the Universe may provide loopholes. While there is little doubt that the Sun and the stars will indeed die, it may be possible—if life can evolve to accommodate the changing environments—that civilization, memory, and thought could continue for an indefinite subjective time (Exercise 5.1).

5.2 Entropy as disorder

A second intuitive interpretation of entropy is as a measure of the disorder in a system. Scientist parents tell their children to lower the entropy by tidying their rooms; liquids have higher entropy than crystals intuitively because their atomic positions are less orderly.[13] We illustrate this interpretation by first calculating the *entropy of mixing*, and then discussing the *zero-temperature entropy of glasses*.

---

[11]Suppose after waiting a cosmologically long time one observed a spontaneous fluctuation of an equilibrium system into a low-entropy, ordered state. Preceding that time, with extremely high probability, all of our laws of macroscopic physics would appear to run backward. The most probable route building up to an ordered state from equilibrium is the time reverse of the most probable decay of that ordered state back to equilibrium.

[12]In the approximation that the Universe remained in equilibrium as it expanded, the entropy would have remained constant; indeed, the photon entropy of the microwave background radiation has remained roughly constant during the expansion of the Universe (Exercise 7.15). The root cause of the low entropy fueling life on Earth is a feature of primordial nucleosynthesis; the formation of heavier nuclei from protons and neutrons stalled after a few light elements [205, 207] (see Exercise 5.24).

[13]There are interesting examples of systems that appear to develop more order as their entropy (and temperature) rises. These are systems where adding order of one, visible type (say, crystalline or orientational order) allows increased disorder of another type (say, vibrational disorder). Entropy is a precise measure of disorder, but is not the only possible or useful measure.

5.2.1 Entropy of mixing: Maxwell's demon and osmotic pressure

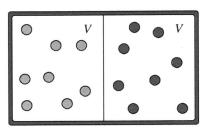

Fig. 5.4 Unmixed atoms. The pre-mixed state: $N/2$ orange atoms on one side, $N/2$ blue atoms on the other.

Fig. 5.5 Mixed atoms. The mixed state: $N/2$ orange atoms and $N/2$ blue atoms scattered through the volume $2V$.

Scrambling an egg is a standard example of irreversibility; you cannot re-separate the yolk from the white. A model for scrambling is given in Figs. 5.4 and 5.5: the mixing of two different types of particles. Here the entropy change upon mixing is a measure of increased disorder.

Consider a volume separated by a partition into two equal volumes of volume V. There are $N/2$ undistinguished ideal gas orange atoms on one side of the partition, and $N/2$ undistinguished ideal gas blue atoms on the other. The position-space entropy of this system (eqn 3.55, ignoring the momentum space parts) is

$$S_{\text{unmixed}} = 2\,k_B \log[V^{N/2}/(N/2)!], \qquad (5.14)$$

just twice the position-space entropy of $N/2$ undistinguished atoms in a volume V. We assume that the blue and orange atoms have the same masses and the same total energy. Now consider the entropy change when the partition is removed, and the two sets of atoms are allowed to mix. Because the temperatures and pressures from both sides are equal, removing the partition does not involve any irreversible sound emission or heat transfer; any entropy change is due to the mixing of the orange and blue atoms. In the mixed state, the entropy has increased to

$$S_{\text{mixed}} = 2k_B \log[(2V)^{N/2}/(N/2)!], \qquad (5.15)$$

twice the entropy of $N/2$ undistinguished atoms in a volume $2V$. Since $\log(2^m x) = m \log 2 + \log x$, the change in entropy due to the mixing is

$$\Delta S_{\text{mixing}} = S_{\text{mixed}} - S_{\text{unmixed}} = k_B \log 2^N = N k_B \log 2. \qquad (5.16)$$

We gain $k_B \log 2$ in entropy every time we place an atom into one of two boxes without looking which box we chose. More generally, we might define a counting entropy:

$$S_{\text{counting}} = k_B \log(\text{number of configurations}) \qquad (5.17)$$

for systems with a discrete number of equally likely configurations.

This kind of discrete choice arises often in statistical mechanics. In equilibrium quantum mechanics (for a finite system) the states are quantized; so adding a new (noninteracting) particle into one of m degenerate states adds $k_B \log m$ to the entropy. In communications theory (Section 5.3.2, Exercises 5.14 and 5.15), each bit transmitted down your channel can be in one of two states, so a random stream of bits of length N has $\Delta S = k_S N \log 2$.[14]

In more general cases, the states available to one particle depend strongly on the configurations of the other particles. Nonetheless, the equilibrium entropy still measures the logarithm of the number of different states that the total system could be in. For example, our equilibrium statistical mechanics entropy $S_{\text{equil}}(E) = k_B \log(\Omega(E))$ (eqn 3.25)

[14]The Shannon constant k_S is defined in Section 5.3.2.

is the logarithm of the number of states of energy E, with phase-space volume h^{3N} allocated to each state.

What would happen if we removed a partition separating $N/2$ blue atoms on one side from $N/2$ blue atoms on the other?[15] The initial entropy is the same as above $S^{\text{BB}}_{\text{unmixed}} = 2\,k_B \log[V^{N/2}/(N/2)!]$, but the final entropy is now $S^{\text{BB}}_{\text{mixed}} = k_B \log((2V)^N/N!)$. Notice we have $N!$ rather than the $((N/2)!)^2$ from eqn 5.15, since all of our particles are now undistinguished. Now $N! = (N)(N-1)(N-2)(N-3)\ldots$ and $((N/2)!)^2 = (N/2)(N/2)[(N-2)/2][(N-2)/2]\ldots$; they roughly differ by 2^N, canceling the entropy change due to the volume doubling. Indeed, expanding the logarithm using Stirling's formula $\log n! \approx n \log n - n$ we find the entropy per atom is unchanged.[16] This is why we introduced the $N!$ term for undistinguished particles in Section 3.2.1; without it the entropy would decrease by $N \log 2$ whenever we split a container into two pieces.[17]

How can we intuitively connect this entropy of mixing with the thermodynamic entropy of pistons and engines in Section 5.1? Can we use our mixing entropy to do work? To do so we must differentiate between the two kinds of atoms. Suppose that the barrier separating the two walls in Fig. 5.4 was a membrane that was impermeable to blue atoms but allowed orange ones to cross. Since both blue and orange atoms are ideal gases, the orange atoms would spread uniformly to fill the entire system, while the blue atoms would remain on one side. This would lead to a pressure imbalance; if the semipermeable wall was used as a piston, work could be extracted as the blue chamber was enlarged to fill the total volume.[18]

Suppose we had a more active differentiation? Maxwell introduced the idea of an intelligent "finite being" (later termed *Maxwell's demon*) that would operate a small door between the two containers. When a blue atom approaches the door from the left or a orange atom approaches from the right the demon would open the door; for the reverse situations the demon would leave the door closed. As time progresses, this active sorting would re-sort the system, lowering the entropy. This is not a concern for thermodynamics, since running a demon is an entropy-generating process! Indeed, one can view this thought experiment as giving a fundamental limit on demon efficiency, putting a lower bound on how much entropy an intelligent being must create in order to engage in this kind of sorting process (Fig. 5.6 and Exercise 5.2).

5.2.2   Residual entropy of glasses: the roads not taken

Unlike a crystal, in which each atom has a set position, a glass will have a completely different configuration of atoms each time it is formed. That is, the glass has a *residual entropy*; as the temperature goes to absolute zero, the glass entropy does not vanish, but rather equals $k_B \log \Omega_{\text{glass}}$, where Ω_{glass} is the number of zero-temperature configurations in which the glass might be trapped.

[15] Alternatively, what if our experiments were color-blind, and the orange and blue atoms were undistinguished.

[16] If you keep Stirling's formula to higher order, you will see that the entropy increases by a microscopic amount when you remove the partition. This is due to the number fluctuations on the two sides that are now allowed. See Exercise 5.2 for a small system where this entropy increase is important.

[17] This is often called the Gibbs paradox.

[18] Such semipermeable membranes are quite common, not for gases, but for dilute solutions of ions in water; some ions can penetrate and others cannot. The resulting force on the membrane is called *osmotic pressure*.

Fig. 5.6 Ion pump as Maxwell's demon. An implementation of Maxwell's demon in biology is Na^+/K^+-ATPase, an enzyme located on the membranes of almost every cell in your body. This enzyme maintains extra potassium (K+) ions inside the cell and extra sodium (Na^+) ions outside the cell. The enzyme exchanges two K^+ ions from outside for three Na^+ ions inside, burning as fuel one ATP (adenosine with three phosphates, the fuel of the cell) into ADP (two phosphates). When you eat too much salt (Na^+Cl^-), the extra sodium ions in the blood increase the osmotic pressure on the cells, draw more water into the blood, and increase your blood pressure. (PDB ID: 4XE5 [75] from rcsb.org [23], figure made using Mol* [165].)

[19]Windows are made from soda-lime glass (silica (SiO_2) mixed with sodium and calcium oxides).

[20]Pyrex[TM] is a borosilicate glass (boron and silicon oxides) with a low thermal expansion, used for making measuring cups that do not shatter when filled with boiling water.

[21]*Hard candy* is an American usage; in British English they are called *boiled sweets*.

[22]One can measure $S_{liquid}(T_\ell)$ by slowly heating a crystal from absolute zero and measuring $\int_0^{T_\ell} dQ/T$ flowing in.

Fig. 5.7 Double-well potential. A model for the potential energy for one coordinate q_i in a glass; two states separated by a barrier V_i and with a small energy difference δ_i. In equilibrium, the atom is $\exp(-\delta_i/k_B T)$ less likely to be in the upper well. For barriers $V_i \gg k_B T$, the molecule will spend most of its time vibrating in one of the wells, and rarely hop across the barrier.

[24]Atomic vibration times are around 10^{-12} seconds, and cooling times are typically between seconds and years, so the cooling rate is indeed slow compared to microscopic times.

What is a glass? Glasses are disordered like liquids, but are rigid like crystals. They are not in equilibrium; they are formed when liquids are cooled too fast to form the crystalline equilibrium state. You are aware of glasses made from silica, like window glass,[19] and Pyrex[TM].[20] You also know of some molecular glasses, like hard candy[21] (a glass made of sugar). Many other materials (even metals) can form glasses when cooled quickly.

How is the residual glass entropy measured? First, one estimates the entropy of the equilibrium liquid;[22] then one measures the entropy flow Q/T out from the glass as it is cooled from the liquid down to absolute zero. The difference

$$S_{residual} = S_{liquid}(T_\ell) - \int \frac{1}{T}\frac{dQ}{dt}dt = S_{liquid}(T_\ell) - \int_0^{T_\ell} \frac{1}{T}\frac{dQ}{dT}dT \quad (5.18)$$

gives the residual entropy.

How big is the residual entropy of a typical glass? The residual entropy is of the order of k_B per molecular unit of the glass (SiO_2 or sugar molecule, for example). This means that the number of glassy configurations e^{S/k_B} is enormous (Exercise 5.11).

How is it possible to measure the number of glass configurations the system did not choose? The glass is, after all, in one particular configuration. How can measuring the heat flow $Q(t)$ out of the liquid as it freezes into one glassy state be used to measure the number Ω_{glass} of possible glassy states? In other words, how exactly is the statistical mechanics definition of entropy $S_{stat} = k_B \log \Omega_{glass}$ related to the thermodynamic definition $\Delta S_{thermo} = Q/T$?

To answer this question, we need a simplified model of how a glass might fall out of equilibrium as it is cooled.[23] We view the glass as a collection of independent molecular units. Each unit has a double-well potential energy: along some internal coordinate q_i there are two minima with an energy difference δ_i and separated by an energy barrier V_i (Fig. 5.7). This internal coordinate might represent a rotation of a sugar molecule in a candy, or a shift in the location of an oxygen atom in a SiO_2 window glass.

Consider the behavior of one of these double-well degrees of freedom. As we cool our system, the molecular unit will be thermally excited over its barrier more and more slowly. So long as the cooling rate Γ_{cool} is small compared to this hopping rate, our unit will remain in equilibrium. However, at the temperature T_i where the two rates cross for our unit the transitions between the wells will not keep up and our molecular unit will freeze into position. If the cooling rate Γ_{cool} is very slow compared to the molecular vibration frequencies (as it almost always is)[24] the

[23]The glass transition is not a sharp phase transition; the liquid grows thicker (more viscous) as it is cooled, with slower and slower dynamics, until the cooling rate becomes too fast for the atomic rearrangements needed to maintain equilibrium to keep up. At that point, there is a smeared-out transition as the viscosity effectively becomes infinite and the glass becomes bonded together. Our model is not a good description of the glass transition, but is a rather accurate model for the continuing thermal rearrangements (β-relaxation) at temperatures below the glass transition, and an excellent model for the quantum dynamics (tunneling centers) which dominate many properties of glasses below a few degrees Kelvin. See Section 12.3.4 for how little we understand glasses.

hopping rate will change rapidly with temperature, and our system will freeze into place in a small range of temperature near T_i.[25]

Our frozen molecular unit has a population in the upper well given by the Boltzmann factor $e^{-\delta_i/k_B T_i}$ times the population in the lower well. Hence, those units with $\delta_i \gg k_B T_i$ will be primarily in the ground state (and hence already roughly in equilibrium). However, consider those N units with barriers high compared to the asymmetry, $\delta_i \ll k_B T_i$; as the glass is cooled, one by one these units randomly freeze into one of two states (Fig. 5.8). For these units, both states will be roughly equally populated when they fall out of equilibrium, so each will contribute about $k_B \log 2$ to the residual entropy. Thus the N units with $T_i > \delta_i/k_B$ will contribute about $k_B \log 2 \sim k_B$ to the statistical residual entropy S_{stat}.[26]

Is the thermodynamic entropy flow $\Delta S = \int dQ/T$ out of the glass also smaller than it would be in equilibrium? Presumably so, since some of the energy remains trapped in the glass within those units left in their upper wells. Is the residue the same as the statistical residual entropy? Those units with $k_B T_i \ll \delta_i$ which equilibrate into the lower well before they freeze will contribute the same amount to the entropy flow into the heat bath as they would in an equilibrium system. On the other hand, those units with $k_B T \gg \delta_i$ will each fail (half the time) to release their energy δ_i to the heat bath, when compared to an infinitely slow (equilibrium) cooling. Since during an equilibrium cooling this heat would be transferred to the bath at a temperature around δ_i/k_B, the missing entropy flow for that unit is $\Delta Q/T \sim \delta_i/(\delta_i/k_B) \sim k_B$. Again, the N units each contribute around k_B to the (experimentally measured) thermodynamic residual entropy S_{thermo}.

Thus the heat flow into a *particular* glass configuration counts the number of roads not taken by the glass on its cooling voyage.

5.3 Entropy as ignorance: information and memory

The most general interpretation of entropy is as a measure of our *ignorance*[27] about a system. The equilibrium state of a system maximizes the entropy because we have lost all information about the initial conditions except for the conserved quantities; maximizing the entropy maximizes our ignorance about the details of the system. The entropy of a glass, or of our mixture of blue and orange atoms ,is a measure of the number of arrangements the atoms could be in, given our ignorance.[28]

This interpretation—that entropy is not a property of the system, but of our knowledge about the system[29] (represented by the ensemble of possibilities)—cleanly resolves many otherwise confusing issues. The atoms in a glass are in a definite configuration, which we could measure using some futuristic X-ray holographic technique. If we did so, our ignorance would disappear, and the residual entropy would become zero for us.[30] We could in principle use our knowledge of the glass atom

Fig. 5.8 Roads not taken by the glass. The branching path of glassy states in our model. The entropy (both statistical and thermodynamic) is proportional to the number of branchings the glass chooses between as it cools. A particular glass will take one trajectory through this tree as it cools; nonetheless the thermodynamic entropy measures the total number of states.

[25]This can be derived straightforwardly using Arrhenius rates (Section 6.6) for the two transitions.

[26]We are losing factors like $\log 2$ because we are ignoring those units with $k_B T_i \sim \delta_i$, which freeze into partly occupied states that are not equally occupied, a case we have not treated yet. The thermodynamic and statistical entropies of course agree quantitatively if calculated more carefully.

[27]In information theory, they use the alternative term *uncertainty*, which has misleading connotations from quantum mechanics; Heisenberg uncertainty has no associated entropy.

[28]Again, entropy is a precise measure of ignorance, but not necessarily a sensible one for all purposes. In particular, entropy does not distinguish the *utility* of the information. Isothermally compressing a mole of gas to half its volume decreases our ignorance by $>10^{23}$ bits—a far larger change in entropy than would be produced by memorizing all the written works of human history.

[29]The entropy of an *equilibrium* system remains purely a property of the composition of the system, because our knowledge is fixed (near zero).

[30]The X-ray holographic process must, naturally, create at least as much entropy during the measurement as the glass loses.

positions to extract extra useful work out of the glass, not available before measuring the positions (Exercises 5.2 and 5.26).

So far, we have confined ourselves to cases where our ignorance is maximal, where all allowed configurations are equally likely. What about systems where we have partial information, where some configurations are more probable than others? There is a powerful generalization of the definition of entropy to general probability distributions, which we will introduce in Section 5.3.1 for traditional statistical mechanical systems. In Section 5.3.2 we will show that this nonequilibrium entropy provides a generally useful measure of our ignorance about a wide variety of systems, with broad applications outside of traditional physics.

5.3.1   Nonequilibrium entropy

The second law of thermodynamics tells us that entropy increases. We want a definition of entropy for systems that are out of equilibrium, so we can watch it grow. In general, we may describe our partial knowledge about such a system as a probability distribution ρ, defining an ensemble of states.

Let us start with a probability distribution among a discrete set of states. We know from Section 5.2.1 that the entropy for M equally likely states (eqn 5.17) is $S(M) = k_B \log M$. In this case, the probability of each state is $p_i = 1/M$. If we write $S(M) = -k_B \log(1/M) = -k_B \langle \log(p_i) \rangle$, we get an appealing generalization for the counting entropy for cases where p_i is not constant:

$$S_{\text{discrete}} = -k_B \langle \log p_i \rangle = -k_B \sum_i p_i \log p_i. \tag{5.19}$$

We shall see in Section 5.3.2 and Exercise 5.17 that this is the correct generalization of entropy to systems out of equilibrium.

What about continuum distributions? Any nonequilibrium state of a classical Hamiltonian system can be described with a probability density $\rho(\mathbb{P}, \mathbb{Q})$ on phase space. The nonequilibrium entropy then becomes

$$\begin{aligned} S_{\text{nonequil}} &= -k_B \langle \log \rho \rangle = -k_B \int \rho \log \rho \\ &= -k_B \int_{E < \mathcal{H}(\mathbb{P}, \mathbb{Q}) < E + \delta E} \frac{d\mathbb{P}\, d\mathbb{Q}}{h^{3N}}\, \rho(\mathbb{P}, \mathbb{Q}) \log \rho(\mathbb{P}, \mathbb{Q}). \end{aligned} \tag{5.20}$$

In the case of the microcanonical ensemble where $\rho_{\text{equil}} = 1/(\Omega(E)\delta E)$, the nonequilibrium definition of the entropy is shifted from our equilibrium definition $S = k_B \log \Omega$ by a negligible amount $k_B \log(\delta E)/N$ per particle:[31]

$$\begin{aligned} S_{\text{micro}} &= -k_B \log \rho_{\text{equil}} = k_B \log(\Omega(E)\delta E) \\ &= k_B \log(\Omega(E)) + k_B \log(\delta E). \end{aligned} \tag{5.21}$$

For quantum systems, the nonequilibrium entropy will be written in terms of the density matrix $\boldsymbol{\rho}$ (Section 7.1):

$$S_{\text{quantum}} = -k_B \text{Tr}(\boldsymbol{\rho} \log \boldsymbol{\rho}). \tag{5.22}$$

[31]The arbitrary choice of the width of the energy shell in the microcanonical ensemble is thus related to the arbitrary choice of the zero for the entropy of a classical system. Unlike the (extensive) shift due to the units of phase space (Section 3.5), this shift is microscopic.

Finally, notice that S_{noneq} and S_{quantum} are defined for the microscopic laws of motion, which (Section 5.1) are time-reversal invariant. We can thus guess that these microscopic entropies will be time independent, since microscopically the system does not know in which direction of time entropy should increase.[32] No information is lost (in principle) by evolving a closed system in time. Entropy (and our ignorance) increases only in theories where we ignore or exclude some degrees of freedom. These degrees of freedom may be external (information flowing into the environment or heat bath) or internal (information flowing into irrelevant microscopic degrees of freedom which are ignored in a coarse-grained theory).

[32]You can show this explicitly in Exercises 5.7 and 7.4.

5.3.2 Information entropy

Understanding ignorance is central to many fields. Entropy as a measure of ignorance has been useful in everything from the shuffling of cards (Exercise 5.13) to reconstructing noisy images. For these other applications, the connection with temperature is unimportant, so we do not need to make use of Boltzmann's constant. Instead, we normalize the entropy with the constant $k_S = 1/\log(2)$:

$$S_S = -k_S \sum_i p_i \log p_i = -\sum_i p_i \log_2 p_i, \qquad (5.23)$$

so that entropy is measured in *bits*.[33]

[33]Each bit doubles the number of possible states Ω, so $\log_2 \Omega$ is the number of bits.

This normalization was introduced by Shannon [183], and the formula 5.23 is referred to as Shannon entropy in the context of *information theory*. Shannon noted that this entropy, applied to the ensemble of possible messages or images, can be used to put a fundamental limit on the amount they can be compressed[34] to efficiently make use of disk space or a communications channel (Exercises 5.14 and 5.15). A low-entropy data set is highly predictable; given the stream of data so far, we can predict the next transmission with some confidence. In language, twins and long-married couples can often complete sentences for one another. In image transmission, if the last six pixels were white the region being depicted is likely to be a white background, and the next pixel is also likely to be white. We need only transmit or store data that violates our prediction. The entropy measures our ignorance, how likely the best predictions about the rest of the message are to be wrong.

[34]Lossless compression schemes (files ending in gif, png, zip, and gz) remove the redundant information in the original files, and their efficiency is limited by the entropy of the ensemble of files being compressed (Exercise 5.15). Lossy compression schemes (files ending in jpg, mpg, and mp3) also remove information that is thought to be unimportant for humans looking at or listening to the files (Exercise 5.14).

Entropy is so useful in these various fields because it is the unique (continuous) function that satisfies three key properties.[35] In this section, we will explain what these three properties are and why they are natural for any function that measures ignorance. We will also show that our nonequilibrium Shannon entropy satisfies these properties. In Exercise 5.17 you can show that this entropy is the only function to do so.

[35]Unique, that is, up to the overall constant k_S or k_B.

To take a tangible example of ignorance, suppose your roommate has lost their keys, and they are asking for your advice. You want to measure the roommate's progress in finding the keys by measuring your ignorance

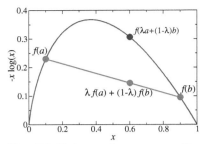

Fig. 5.9 Entropy is concave. For $x \geq 0$, $f(x) = -x \log x$ is strictly convex downward (concave). That is, for $0 < \lambda < 1$, the linear interpolation lies below the curve:

$$f(\lambda a + (1-\lambda)b) \geq \lambda f(a) + (1-\lambda)f(b). \quad (5.24)$$

We know f is concave because its second derivative, $-1/x$, is everywhere negative.

[36]In Exercise 6.6 we will ask you to show that the Shannon entropy S_S is an extremum when all probabilities are equal. Here we provide a stronger proof that it is a global maximum, using the convexity of $x \log x$ (note 37).

with some function S_I. Suppose there are Ω possible sites A_k that they might have left the keys, which you estimate have probabilities $p_k = P(A_k)$, with $\sum_1^\Omega p_i = 1$.

What are the three key properties we want our ignorance function $S_I(p_1, \ldots, p_\Omega)$ to have?

(1) *Entropy is maximum for equal probabilities.* Without further information, surely the best plan is for your roommate to look first at the most likely site, which maximizes p_i. Your ignorance must therefore be maximal if all Ω sites have equal likelihood:

$$S_I\left(\frac{1}{\Omega}, \ldots, \frac{1}{\Omega}\right) > S_I(p_1, \ldots, p_\Omega) \quad \text{unless } p_i = \frac{1}{\Omega} \text{ for all } i. \quad (5.25)$$

Does the Shannon entropy obey eqn 5.25, property (1)?[36] We notice that the function $f(p) = -p \log p$ is concave (convex downward, Fig. 5.9). For a concave function f, the average value of $f(p)$ over a set of points p_k is less than or equal to f evaluated at the average of the p_k:[37]

$$\frac{1}{\Omega} \sum_k f(p_k) \leq f\left(\frac{1}{\Omega} \sum_k p_k\right). \quad (5.27)$$

But this tells us that

$$S_S(p_1, \ldots, p_\Omega) = -k_S \sum p_k \log p_k = k_S \sum f(p_k)$$

$$\leq k_S \Omega f\left(\frac{1}{\Omega} \sum_k p_k\right) = k_S \Omega f\left(\frac{1}{\Omega}\right)$$

$$= -k_S \sum_1^\Omega \frac{1}{\Omega} \log\left(\frac{1}{\Omega}\right) = S_S\left(\frac{1}{\Omega}, \ldots, \frac{1}{\Omega}\right). \quad (5.28)$$

(2) *Entropy is unaffected by extra states of zero probability.* If there is no possibility that the keys are in your shoe (site A_Ω), then your ignorance is no larger than it would have been if you had not included your shoe in the list of possible sites:

$$S_I(p_1, \ldots, p_{\Omega-1}, 0) = S_I(p_1, \ldots, p_{\Omega-1}). \quad (5.29)$$

The Shannon entropy obeys property (2) because $p_\Omega \log p_\Omega \to 0$ as $p_\Omega \to 0$.

[37]Equation 5.27 is Jensen's inequality. It can be proven by induction from the definition of concave (eqn 5.24). For $\Omega = 2$, we use $\lambda = 1/2$, $a = p_1$, and $b = p_2$ to see that $f((p_1 + p_2)/2) \geq (f(p_1) + f(p_2))/2$. For general Ω, we use $\lambda = (\Omega - 1)/\Omega$, $a = (\sum_1^{\Omega-1} p_k)/(\Omega - 1)$, and $b = p_\Omega$ to see that

$$f\left(\frac{\sum_{k=1}^\Omega p_k}{\Omega}\right) = f\left(\frac{\Omega - 1}{\Omega} \frac{\sum_1^{\Omega-1} p_k}{\Omega - 1} + \frac{1}{\Omega} p_\Omega\right) \geq \frac{\Omega - 1}{\Omega} f\left(\frac{\sum_1^{\Omega-1} p_k}{\Omega - 1}\right) + \frac{1}{\Omega} f(p_\Omega)$$

$$\geq \frac{\Omega - 1}{\Omega}\left(\sum_{k=1}^{\Omega-1} \frac{1}{\Omega - 1} f(p_k)\right) + \frac{1}{\Omega} f(p_\Omega) = \frac{1}{\Omega} \sum_{k=1}^\Omega f(p_k), \quad (5.26)$$

where we have used the truth of eqn 5.27 for $\Omega - 1$ to inductively prove it for Ω.

(3) *Entropy change for conditional probabilities.* This last property for our ignorance function demands a new concept, *conditional probability.*[38]

To aid in the search, you are likely to ask the roommate where they were when they last saw the keys. Suppose there are M locations B_ℓ that the roommate may have been (opening the apartment door, driving the car, in the basement laundry room, ...), with probabilities q_ℓ. Surely the likelihood that the keys are currently in a coat pocket is larger if the roommate was outdoors when the keys were last seen. Let $r_{k\ell} = P(A_k \text{ and } B_\ell)$ be the probability the keys are at site k and were last seen at location ℓ. Let[39]

$$c_{k\ell} = P(A_k|B_\ell) = \frac{P(A_k \text{ and } B_\ell)}{P(B_\ell)} = \frac{r_{k\ell}}{q_\ell} \qquad (5.30)$$

be the conditional probability, given that the keys were last seen at B_ℓ, that the keys are now at site A_k. Naturally

$$\sum_k P(A_k|B_\ell) = \sum_k c_{k\ell} = 1; \qquad (5.31)$$

wherever $[\ell]$ they were last seen the keys are now *somewhere* with probability one.

Before you ask your roommate where the keys were last seen, you have ignorance $S_I(A) = S_I(p_1, \ldots, p_\Omega)$ about the site of the keys, and ignorance $S_I(B) = S_I(q_1, \ldots, q_M)$ about the location they were last seen. You have a joint ignorance about the two questions given by the ignorance function applied to all $\Omega \times M$ conditional probabilities:

$$S_I(AB) = S_I(r_{11}, r_{12}, \ldots, r_{1M}, r_{21}, \ldots, r_{\Omega M})$$
$$= S_I(c_{11}q_1, c_{12}q_2, \ldots, c_{1M}q_M, c_{21}q_1, \ldots, c_{\Omega M}q_M). \qquad (5.32)$$

After the roommate answers your question, your ignorance about the location last seen is reduced to zero (decreased by $S_I(B)$). If the location last seen was in the laundry room (site B_ℓ), the probability for the keys being at A_k shifts to $c_{k\ell}$ and your ignorance about the site of the keys is now

$$S_I(A|B_\ell) = S_I(c_{1\ell}, \ldots, c_{\Omega\ell}). \qquad (5.33)$$

So, your combined ignorance has decreased from $S_I(AB)$ to $S_I(A|B_\ell)$.

We can measure the usefulness of your question by the expected amount that it decreases your ignorance about where the keys reside. The expected ignorance after the question is answered is given by weighting the ignorance after each answer B_ℓ by the probability q_ℓ of that answer:

$$\langle S_I(A|B_\ell)\rangle_B = \sum_\ell q_\ell S_I(A|B_\ell). \qquad (5.34)$$

This leads us to the third key property for an ignorance function. If we start with the joint distribution AB, and then measure B, it would be tidy if, on average, your joint ignorance declined by your original ignorance of B:

$$\langle S_I(A|B_\ell)\rangle_B = S_I(AB) - S_I(B). \qquad (5.35)$$

[38] If you find the discussion of conditional probability subtle, and the resulting third property for the ignorance function (eqn 5.35) less self-evident than the first two properties, you are in good company. Generalizations of the entropy that modify this third condition are sometimes useful.

[39] The conditional probability $P(A|B)$ (read as *the probability of A given B*) times the probability of B is the probability of A and B both occurring, so $P(A|B)P(B) = P(A \text{ and } B)$ (or equivalently $c_{k\ell}q_\ell = r_{k\ell}$).

Does the Shannon entropy satisfy eqn 5.35, property (3)? The conditional probability $S_S(A|B_\ell) = -k_S \sum_k c_{k\ell} \log c_{k\ell}$, since $c_{k\ell}$ is the probability distribution for the A_k sites given location ℓ. So,

$$
\begin{aligned}
S_S(AB) &= -k_S \sum_{k\ell} c_{k\ell} q_\ell \log(c_{k\ell} q_\ell) \\
&= -k_S \left(\sum_{k\ell} c_{k\ell} q_\ell \log(c_{k\ell}) + \sum_{k\ell} c_{k\ell} q_\ell \log(q_\ell) \right) \\
&= \sum_\ell q_\ell \left(-k_S \sum_k c_{k\ell} \log(c_{k\ell}) \right) - k_S \sum_\ell q_\ell \log(q_\ell) \left(\sum_k \cancel{c_{k\ell}} \right) \\
&= \sum_\ell q_\ell S_S(A|B_\ell) + S_S(B) \\
&= \langle S_S(A|B_\ell) \rangle_B + S_S(B),
\end{aligned}
\tag{5.36}
$$

and the Shannon entropy does satisfy condition (3).

If A and B are uncorrelated (for example, if they are measurements on uncoupled systems), then the probabilities of A will not change upon measuring B, so $S_I(A|B_\ell) = S_I(A)$. Then our third condition implies $S_I(AB) = S_I(A) + S_I(B)$; our ignorance of uncoupled systems is additive. This is simply the condition that *entropy is extensive*. We argued that the entropy of weakly coupled subsystems in equilibrium must be additive in Section 3.3. Our third condition implies that this remains true for uncorrelated systems in general.

Exercises

The exercises explore four broad aspects of entropy.

Entropy provides fundamental limits. We explore entropic limits to thought in *Life and the heat death of the Universe*, to information in *Burning information and Maxwellian demons*, to aging in *Aging, entropy, and DNA*, to computation in *Reversible computation*, to civilization in *The Dyson sphere*, and to memory in *Black hole thermodynamics*. In the practical world, we explore entropic limits to engine efficiency in *P–V diagram* and *Carnot refrigerator*.

Entropy is an emergent property. We argued from time-reversal invariance that a complete microscopic description of a closed system cannot lose information, and hence the entropy must be a constant; you can show this explicitly in *Does entropy increase?* Entropy increases because the information stored in the initial conditions is rapidly made irrelevant by chaotic motion; this is illustrated pictorially by *The Arnol'd cat map* and *Equili-*

bration in phase space, and numerically (in a dissipative system) in *Chaos, Lyapunov, and entropy increase*. Entropy also increases in coarse-grained theories, which ignore microscopic degrees of freedom; we study our first example of this in *Entropy increases: diffusion*. *Gravity and entropy* shows that our theory of entropy cannot explain the emergent patterns governing the motion of stars in the galaxy. *Phase conjugate mirror* demonstrates that our threshold for declaring ignorance and increasing the entropy in a scrambled system can depend on the information available for future experiments about the scrambling process. Finally, we explore the emergence of the arrow of time by modeling the Universe as a piston in *Nucleosynthesis and the arrow of time*.

Entropy has tangible experimental consequences. In *Entropy of glasses* we explore how an experiment can put upper and lower bounds on the entropy due to our massive ignorance about the zero-temperature atomic configuration in a glass. In *Entropy of socks* we find that the entropy increase due to disorder on human scales is small. In *Rubber band* we find that the entropy in a random walk can exert forces. In *Entropy of the Galaxy* we compare different sinks of entropy in outer space.

Entropy is a general measure of ignorance, with widespread applications to other fields. In *How many shuffles?* we apply it to card shuffling (where ignorance is the goal). In *Information entropy*, *Shannon entropy*, and *Data compression*, we explore the key role entropy plays in communication theory and compression algorithms. In *Fractal dimensions* we find a useful characterization of fractal sets in dissipative systems that is closely related to the entropy. In *Deriving entropy* you can reproduce the famous proof that the Shannon entropy is the unique measure of ignorance with the three key properties explicated in Section 5.3.2.

(5.1) **Life and the heat death of the Universe.** (Astrophysics) @

Freeman Dyson [54] discusses how living things might evolve to cope with the cooling and dimming we expect during the heat death of the Universe.

Normally one speaks of living things as beings that consume energy to survive and proliferate. This is of course not correct; energy is conserved, and cannot be consumed. *Living beings intercept entropy flows*; they use low-entropy sources of energy (e.g., high-temperature solar radiation for plants, candy bars for us) and emit high-entropy forms of the same energy (body heat). Dyson ignores the survival and proliferation issues; he's interested in getting a lot of thinking in before the Universe ends. He presumes that an intelligent being generates a fixed entropy S_{th} per thought.[40]

Energy needed per thought. Assume that the being draws heat Q from a hot reservoir at T_{hot} and radiates it away to a cold reservoir using a heat exchanger at T_{rad}.

(a) *What is the minimum energy Q needed per thought, in terms of S_{th} and T_{rad}?* *You may take T_{hot} very large.* Related formulæ: $\Delta S = Q_2/T_2 - Q_1/T_1$; $Q_1 - Q_2 = W$ (energy is conserved).

Time needed per thought to radiate energy. Dyson shows, using theory not important here, that the power radiated by our intelligent-

being-as-entropy-producer is no larger than CT_{rad}^3, a constant times the cube of the cold temperature.[41]

(b) *Derive the maximum rate of thoughts per unit time dH/dt (the inverse of the time Δt per thought), in terms of S_{th}, C, and T_{rad}.*

Number of thoughts for an ecologically efficient being. Our Universe is expanding; the radius R grows roughly linearly in time t. The microwave background radiation has a characteristic temperature $\Theta(t) \sim R^{-1}$ which is getting lower as the Universe expands; this red-shift is due to the Doppler effect. An ecologically efficient being would naturally try to use as little heat as possible, and so wants to choose T_{rad} as small as possible. It cannot radiate heat at a temperature below $T_{rad} = \Theta(t) = A/t$.

(c) *How many thoughts H can an ecologically efficient being have between now and time infinity, in terms of S_{th}, C, A, and the current time t_0?*

Time without end: greedy beings. Dyson would like his beings to be able to think an infinite number of thoughts before the Universe ends, but consume a finite amount of energy. He proposes that his beings need to be profligate in order to get their thoughts in before the world ends; he proposes that they radiate at a temperature $T_{rad}(t) = Bt^{-3/8}$ which falls with time, but not as fast as $\Theta(t) \sim t^{-1}$.

(d) *Show that with Dyson's cooling schedule, the total number of thoughts H is infinite, but the total energy consumed U is finite.*

We should note that there are many refinements on Dyson's ideas. There are potential difficulties that may arise, like quantum limits to cooling and proton decay. And there are different challenges depending on the expected future of the Universe; a big crunch (where the Universe collapses back on itself) demands that we adapt to heat and pressure, while dark energy may lead to an accelerating expansion and a rather lonely Universe in the end.

(5.2) **Burning information and Maxwellian demons.** (Computer science) ③

If entropy applies both to thermodynamics and to information, can we burn information as fuel (Fig. 5.10)? In this exercise, we shall summarize the work of Szilard [193, 194] as presented

---

[40]Each thought can in principle be an arbitrarily complex computation; the only entropy necessarily generated is believed to be associated with recording the answer (see Exercise 5.3).

[41]C scales with the number of electrons in the being; think of our answer Δt as the time per thought per mole of electrons.

by Bennett [19] and Feynman [60, chapter 5]. (See also [120] and Exercise 6.23 for recent work challenging Szilard's picture.)

Fig. 5.10 Information-burning engine. A memory tape being used to power an engine. If the engine knows or can guess the sequence written on the tape, it can extract useful work in exchange for losing that information.

Consider a really frugal digital memory tape, with one atom used to store each bit (Fig. 5.11). The tape is a series of boxes, with each box containing one ideal gas atom. The box is split into two equal pieces by a removable central partition. If the atom is in the top half of the box, the tape reads one; if it is in the bottom half the tape reads zero. The side walls are frictionless pistons that may be used to push the atom around.

Fig. 5.11 Minimalist digital memory tape. The position of a single ideal gas atom denotes a bit. If it is in the top half of a partitioned box, the bit is one, otherwise it is zero. The side walls of the box are pistons, which can be used to set, reset, or extract energy from the stored bits. The numbers above the boxes are not a part of the tape, they just denote what bit is stored in a given position.

If we know the atom position in the *n*th box, we can move the other side wall in, remove the partition, and gradually retract the piston to its original position (Fig. 5.12)—destroying our information about where the atom is, but extracting useful work.

Fig. 5.12 Expanding piston. Extracting energy from a known bit is a three-step process: compress the empty half of the box, remove the partition, and retract the piston and extract $P\,dV$ work out of the ideal gas atom. (One may then restore the partition to return to an equivalent, but more ignorant, state.) In the process, one loses one bit of information (which side of the the partition is occupied).

(a) Burning the information. *Assuming the gas expands at a constant temperature T, how much work $\int P\,dV$ is done by the atom as the piston retracts?*
This is also the minimum work needed to set a bit whose state is currently unknown. However, a *known* bit can be reset for free (Fig. 5.13).

Fig. 5.13 Pistons changing a zero to a one.

(b) Rewriting a bit. *Give a sequence of partition insertion, partition removal, and adiabatic side-wall motions that will reversibly convert a bit zero (atom on bottom) into a bit one (atom on top), with no net work done on the system.*
Thus the only irreversible act in using this memory tape occurs when one forgets what is written upon it (equivalent to removing and then reinserting the partition).
(c) Forgetting a bit. *Suppose the atom location in the nth box is initially known. What is the change in entropy, if the partition is removed and the available volume doubles? Give both the thermodynamic entropy (involving k_B) and the information entropy (involving $k_S = 1/\log 2$).*
This entropy is the cost of the missed opportunity of extracting work from the bit (as in part (a)).

What prevents a Maxwellian demon from using an atom in an unknown state to extract work? The demon must first measure which side of the box the atom is on. Early workers suggested that there must be a minimum energy cost to take this measurement, equal to the energy gain extractable from the bit (part (a)). Bennett [19] argues that no energy need be expended in the measurement process.[42] Why does this not violate the second law of thermodynamics?

(d) *Demonic states.* *After the bit has been burned, is the demon in a known[43] state? What is its entropy? How much energy would it take to return the demon to its original state, at temperature T? Is the second law violated?*

The demon can extract an unlimited amount of useful work from a tape with an unknown bit sequence if it has enough internal states to store the sequence—basically it can copy the information onto a second, internal tape. But the same work must be expended to re-zero this internal tape, preparing it to be used again.

Szilard's insight might seem self-evident; one can always pay for energy from the bath with an increase in entropy. However, it had deep consequences for the theory of computation (Exercise 5.3).

Recently, Machta [120] has elegantly estimated a minimal entropic cost for controlling machines like pistons, which calls Szilard's basic argument into question (see Exercise 6.23). Time will tell if we can actually exchange thermodynamic and information entropy bit for bit.

(5.3) **Reversible computation.** (Computer science) ③

Is there a minimum energy cost for computation? Again, this exercise is based on the work of Bennett [19] and Feynman [60, chapters 5 and 6].

A general-purpose computer can be built out of certain elementary logic gates. Many of these gates destroy information.

(a) *Irreversible logic gates.* *Consider the XOR gate shown in Fig. 5.14. How many bits of information are lost during the action of this gate? Presuming the four possible inputs are equally probable, use Szilard's argument in Exercise 5.2 to calculate the minimum work needed to perform* this operation, if the gate is run at temperature T.

Early workers in this field thought that these logical operations must have a minimum energy cost. Bennett recognized that they could be done with no energy cost by adding extra outputs to the gates. Naturally, keeping track of these extra outputs involves the use of extra memory storage—he traded an energy cost for a gain in entropy. However, the resulting gates now became *reversible*.

Fig. 5.14 Exclusive-or gate. In logic, an *exclusive or* (XOR, $\oplus$) corresponds to the colloquial English usage of the word: either A or B but not both. An XOR gate outputs a one (true) if the two input bits A and B are different, and outputs a zero (false) if they are the same.

(b) *Reversible gate.* *Add an extra output to the XOR gate, which just copies the state of input A to the first output (C = A, D = A \oplus B). This is the Controlled-Not gate, one of three that Feynman uses to assemble a general-purpose computer, see [60, chapter 6]. Make a table, giving the four possible outputs (C, D) of the resulting gate for the four possible inputs (A, B) = (00, 01, 10, 11). If we run the outputs (C, D) of the gate into the inputs (A', B') of another Controlled-Not gate, what net operation is performed?*

A completely reversible computer can therefore be constructed out of these new gates. The computer performs the calculation, carefully copies the answer into an output buffer, and then performs the reverse computation. In reversing the computation, all the stored information about the extra bits can be reabsorbed, just as in part (b). The only energy or entropy cost for the computation is involved in writing the answer.

---

[42]He modeled the demon as a two-state system (say, another partitioned box). If the demon starts in a known state, one can copy the state of the box into the demon with no cost in energy or entropy, by adiabatically turning on an appropriate coupling. The demon has now measured the atom position, and can extract work from the pistons.

[43]That is, known to the outside world, not just to the demon.

These ideas later led to the subject of *quantum computation*. Quantum computers are naturally reversible, and would also be much more powerful for some kinds of computations than ordinary computers.

(5.4) **Black hole thermodynamics.** (Astrophysics) ③

Astrophysicists have long studied *black holes*: the end state of massive stars which are too heavy to support themselves under gravity (see Exercise 7.16). As the matter continues to fall into the center, eventually the escape velocity reaches the speed of light. After this point, the in-falling matter cannot ever communicate information back to the outside. A black hole of mass M has radius[44]

$$R_s = G \frac{2M}{c^2}, \qquad (5.37)$$

where $G = 6.67 \times 10^{-8} \, \mathrm{cm^3/g \, s^2}$ is the gravitational constant, and $c = 3 \times 10^{10} \, \mathrm{cm/s}$ is the speed of light.

Hawking, by combining methods from quantum mechanics and general relativity, calculated the emission of radiation from a black hole.[45] He found a wonderful result: black holes emit perfect black-body radiation at a temperature

$$T_{\mathrm{bh}} = \frac{\hbar c^3}{8\pi G M k_B}. \qquad (5.38)$$

According to Einstein's theory, the energy of the black hole is $E = Mc^2$.

(a) *Calculate the specific heat of the black hole.* The specific heat of a black hole is *negative*. That is, it gets cooler as you add energy to it. In a bulk material, this would lead to an instability; the cold regions would suck in more heat and get colder. A population of black holes is unstable; the larger ones will eat the smaller ones.[46]

(b) *Calculate the entropy of the black hole, by using the definition of temperature $1/T = \partial S/\partial E$ and assuming the entropy is zero at mass $M = 0$. Express your result in terms of the surface area $A = 4\pi R_s^2$, measured in units of the Planck length $L^* = \sqrt{\hbar G/c^3}$ squared.*

As it happens, Bekenstein had deduced this formula for the entropy somewhat earlier, by thinking about analogies between thermodynamics, information theory, and statistical mechanics. On the one hand, when black holes interact or change charge and angular momentum, one can prove in classical general relativity that the area can only increase. So it made sense to assume that the entropy was somehow proportional to the area. He then recognized that if you had some waste material of high entropy to dispose of, you could ship it into a black hole and never worry about it again. Indeed, given that the entropy represents your lack of knowledge about a system, once matter goes into a black hole one might expect that our knowledge about it completely vanishes.[47] (More specifically, the entropy of a black hole represents the inaccessibility of all information about what it was built out of.) By carefully dropping various physical systems into a black hole (theoretically) and measuring the area increase compared to the entropy increase, he was able to deduce these formulæ purely from statistical mechanics.

We can use these results to provide a fundamental bound on memory storage.

(c) *Calculate the maximum number of bits that can be stored in a sphere of radius one centimeter, in terabytes.* (The Planck length $L^* = \sqrt{\hbar G/c^3} \approx 1.6 \times 10^{-35} \, m.$)

Finally, in perhaps string theory's first physical prediction, your formula for the entropy (part (b)) was derived microscopically for a certain type of black hole.

(5.5) **Pressure–volume diagram.** (Thermodynamics) ②

A monatomic ideal gas in a piston is cycled around the path in the P–V diagram in Fig. 5.15. Leg **a** cools at constant volume by connecting to a heat bath at T_c; leg **b** heats at constant pressure by connecting to a heat bath at T_h; leg

---

[44]This is the Schwarzschild radius of the event horizon for a black hole with no angular momentum or charge.

[45]Nothing can leave a black hole; the radiation comes from vacuum fluctuations just outside the black hole that emit particles.

[46]A thermally insulated glass of ice water also has a negative specific heat. The surface tension at the curved ice surface will decrease the coexistence temperature a slight amount (see Section 11.3); the more heat one adds, the smaller the ice cube, the larger the curvature, and the lower the resulting temperature [145].

[47]Except for the mass, angular momentum, and charge. This suggests that baryon number, for example, is not conserved in quantum gravity. It has been commented that when the baryons all disappear, it will be hard for Dyson to build his progeny out of electrons and neutrinos (Exercise 5.1). Recent work suggests that all information about what fell into the black hole can leak out in later Hawking radiation, albeit in garbled form. This makes black holes quite like refrigerators and steam engines—garbling good information into uselessness, but not destroying it microscopically.

c compresses at constant temperature while remaining connected to the bath at T_h.
Which of the following six statements are true?
(T) (F) *The cycle is reversible; no net entropy is created in the Universe.*
(T) (F) *The cycle acts as a refrigerator, using work from the piston to draw energy from the cold bath into the hot bath, cooling the cold bath.*

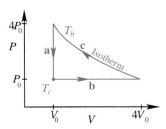

Fig. 5.15 *P–V* **diagram**.

(T) (F) *The cycle acts as an engine, transferring heat from the hot bath to the cold bath and doing positive net work on the outside world.*
(T) (F) *The work done per cycle has magnitude* $|W| = P_0 V_0 |4 \log 4 - 3|$.
(T) (F) *The heat transferred into the cold bath,* Q_c, *has magnitude* $|Q_c| = (9/2) P_0 V_0$.
(T) (F) *The heat transferred from the hot bath,* Q_h, *plus the net work* W *done by the piston onto the gas, equals the heat* Q_c *transferred into the cold bath.*
Related formulæ: $PV = N k_B T$; $U = (3/2) N k_B T$; $\Delta S = Q/T$; $W = -\int P \, dV$; $\Delta U = Q + W$. Notice that the signs of the various terms depend on convention (heat flow out vs. heat flow in); you should work out the signs on physical grounds.

(5.6) **Carnot refrigerator.** (Thermodynamics) ②
Our refrigerator is about $2\,m \times 1\,m \times 1\,m$, and has insulation about $3\,cm$ thick. The insulation is probably polyurethane, which has a thermal conductivity of about 0.02 W/m K. Assume that the refrigerator interior is at $270\,K$, and the room is at $300\,K$.
(a) *How many watts of energy leak from our refrigerator through this insulation?*
Our refrigerator runs at $120\,V$, and draws a maximum of 4.75 amps. The compressor motor turns on every once in a while for a few minutes.
(b) *Suppose* (i) *we do not open the refrigerator door,* (ii) *the thermal losses are dominated by*

the leakage through the foam and not through the seals around the doors, and (iii) *the refrigerator runs as a perfectly efficient Carnot cycle. How much power on average will our refrigerator need to operate? What fraction of the time will the motor run?*

(5.7) **Does entropy increase?** (Mathematics) ③
The second law of thermodynamics says that entropy always increases. Perversely, we can show that in an isolated system, no matter what nonequilibrium condition it starts in, entropy calculated with a complete microscopic description stays constant in time.
Liouville's theorem tells us that the total derivative of the probability density is zero; following the trajectory of a system, the local probability density never changes. The equilibrium states have probability densities that only depend on energy and number. Something is wrong; if the probability density starts nonuniform, how can it become uniform?
Show

$$\frac{\partial f(\rho)}{\partial t} = -\nabla \cdot [f(\rho) \mathbb{V}]$$
$$= -\sum_\alpha \frac{\partial}{\partial p_\alpha} (f(\rho) \dot{p}_\alpha) + \frac{\partial}{\partial q_\alpha} (f(\rho) \dot{q}_\alpha),$$

where f *is any function and* $\mathbb{V} = (\dot{\mathbb{P}}, \dot{\mathbb{Q}})$ *is the* $6N$-*dimensional velocity in phase space. Hence (by Gauss's theorem in* $6N$ *dimensions), show* $\int (\partial f(\rho)/\partial t) \, d\mathbb{P} \, d\mathbb{Q} = 0$, *assuming that the probability density vanishes at large momenta and positions and* $f(0) = 0$. *Show, thus, that the entropy* $S = \int -k_B \rho \log \rho$ *is constant in time.*
We will see that the quantum version of the entropy is also constant for a Hamiltonian system in Exercise 7.4. Some deep truths are not microscopic; the fact that entropy increases is an emergent property.

(5.8) **The Arnol'd cat map.** (Mathematics, Dynamical systems) ③
Why do we suppose equilibrium systems uniformly cover the energy surface? Chaotic motion has *sensitive dependence on initial conditions*; regions on the energy surface get stretched into thin ribbons that twist and fold in complicated patterns. This loses information about the initial conditions and leaves most systems with a uniform distribution of probability over all accessible states. Since energy is the only thing we know is conserved, we average over the energy surface to make predictions.

Arnol'd developed a simple illustration of this stretching and folding using a function taking a two-dimensional square into itself (called the *cat map*, Fig. 5.16). Liouville's theorem tells us that Hamiltonian dynamics preserves the $6N$-dimensional phase-space volume; the cat map preserves area in this 1×1 square, taking $[0, 1) \times [0, 1)$ in the plane onto itself:[48]

$$\Gamma(x, p) = \begin{pmatrix} 2 & 1 \\ 1 & 1 \end{pmatrix} \begin{pmatrix} x \\ p \end{pmatrix} \bmod 1$$
$$= \begin{pmatrix} 2x + p \\ x + p \end{pmatrix} \bmod 1 \qquad (5.39)$$
$$= \text{Mod}_1 \left(M \begin{pmatrix} x \\ p \end{pmatrix} \right)$$

where $M = \begin{pmatrix} 2 & 1 \\ 1 & 1 \end{pmatrix}$ stretches and squeezes our square "energy shell" and Mod_1 cuts and pastes it back back into the square. We imagine a trajectory $(x_{n+1}, p_{n+1}) = \Gamma(x_n, p_n)$ evolving with a discrete time n. This map vividly illustrates how Hamiltonian dynamics leads to the microcanonical ensemble.

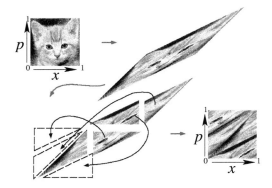

Fig. 5.16 Arnol'd cat map. Arnol'd, in a take-off on Schrödinger's cat, paints a cat on a 2D phase space, which gets warped and twisted as time evolves. By permission from Leon Poon, U Maryland chaos group [152].

The cutting and pasting process confuses our analysis as we repeatedly apply the map $(x_n, p_n) = \Gamma^n(x_0, p_0) = \text{Mod}_1(M(\text{Mod}_1(M(\ldots(x_0, p_0)))))$. Fortunately, we can view the time evolution as many applications of the stretching process, with a final

cut-and-paste:

$$(x_n, p_n) = \text{Mod}_1(M^n(x_0, p_0)). \qquad (5.40)$$

(a) *Show that the last equation 5.40 is true.* (Hint: Work by induction. You may use the fact that $y \bmod 1 = y + n$ for some integer n and that $(z + n) \bmod 1 = z \bmod 1$ for any integer n.) Thus the stretching by M can be done in the entire (x, p) plane, and then the resulting thin strip can be folded back into the unit square. Now we analyze how the map stretches and squeezes the rectangle.

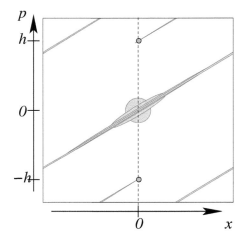

Fig. 5.17 Evolution of an initially concentrated phase-space density. Suppose we know our particle has initial position and momentum confined to a small region $|(x, p)| < r$. This small region is stretched along an irrational angle under the Arnol'd cat map. (This figure shows the origin $x = 0, p = 0$ as the center of the figure; in the map of eqn 5.39 shown in Fig. 5.16 the origin is at the corner.)

(b) *Verify that $(\gamma, 1)$ and $(-1/\gamma, 1)$ are eigenvectors of M, where $\gamma = (\sqrt{5} + 1)/2$ is the Golden Ratio. What are the eigenvalues?* Consider a square box, centered anywhere in the plane, tilted so that its sides are parallel to the two eigenvectors. *How does its height and width change under application of M? Show that its area is conserved.* (Hint: If you are doing this by hand, show that $1/\gamma = (\sqrt{5} - 1)/2$, which can simplify your algebra.)

---

[48]Such maps arise as Poincaré sections of Hamiltonian flows (Fig. 4.9), or as the periodic map given by a Floquet theory. There are many analogies between the behavior of dynamical systems with continuous motion and discrete maps.

The next question explores the map iteration when the initial point is at or near the origin $(x_0, p_0) = (0, 0)$. As illustrated in Fig. 5.17, a small, disk-shaped region of uncertainty, centered on the origin, evolves with time n into a thin strip. When the iteration would make the thin strip hit the square's boundary, it gets cut in two; further iterations stretch and chop our original circle into a bunch of parallel thin lines. (In the version of our map given by eqn 5.40, M^n stretches the initial circle into a growing long thin line, which gets folded in by Mod_1 to a series of parallel line segments.)

(c) *Calculate the momentum h at which the direction of the thin strip (as measured by the expanding eigenvector in part (b)) first crosses the line $x = 0$. Note that h is irrational.* The multiples of an irrational number, taken mod 1, can be shown to be dense in the interval $[0, 1]$. *Using this, argue that the resulting thin strip, in the limit of many iterations, is dense in the unit square (i.e., the maximum gap between line segments goes to zero as time goes to infinity).*

These lines eventually cover the square (our "energy shell") densely and uniformly—intuitively illustrating how the microcanonical ensemble typically describes the long-term behavior. The small circle represents an uncertainty in the initial position. Just as for chaotic motion, Arnol'd's cat map amplifies the uncertainty in the initial position, which leads to the microcanonical ensemble probability distribution spread uniformly over the square. (See also Exercise 5.25.)

Note that there are many initial conditions whose trajectories do not cover the energy surface. The most obvious example is the origin, $x_0 = p_0 = 0$, which maps into itself. In that case, the time average of an operator[49] O, $\lim_{T \to \infty} 1/T \sum_0^T O(x_n, p_n) = O(0, 0)$, clearly will usually be different from its microcanonical average $\int_0^1 dx \int_0^1 dp\, O(x, p)$. The next part of the question asks about a less trivial example of a special initial condition leading to nonergotic behavior.

(d) *If x_0 and p_0 can be written as fractions a/q and b/q for positive integers q, $a \le q$, and $b \le q$, show that x_n and p_n can be written as fractions*

with denominator q as well. Show thus that this trajectory must eventually[50] settle into a periodic orbit, and give an upper bound to the period. Must the time average of an operator O for such a trajectory equal the microcanonical average?

You probably have heard that there are infinitely many more irrational numbers than rational ones. So the probability of landing on one of the periodic orbits is zero (points with rational x and p coordinates are of measure zero). But perhaps there are even more initial conditions which do not equilibrate?

(e) *Find an initial condition with a non-periodic orbit which goes to the origin as $n \to \infty$. What will the time average of O be for this initial condition?*

Arnol'd's argument that the cat map (eqn 5.39) is ergodic shows that, even taking into account points like those you found in parts (d) and (e), the probability of landing on a nonergodic component is zero. For *almost all* initial conditions, or any initial condition with a small uncertainty, the Arnol'd cat map trajectory will be dense and uniformly cover its square phase space at long times.

(5.9) **Chaos, Lyapunov, and entropy increase.**[51] (Mathematics, Complexity, Computation, Dynamical systems) ③
Chaotic dynamical systems have *sensitive dependence on initial conditions*. This is commonly described as the *butterfly effect* (due to Lorenz of the Lorenz attractor): the effects of the flap of a butterfly's wings in Brazil build up with time until months later a tornado in Texas could be launched. In this exercise, we will study this sensitive dependence for a particular system (the logistic map) and measure the sensitivity by defining the *Lyapunov exponents*.
The logistic map takes the interval $(0, 1)$ into itself:

$$f(x) = 4\mu x(1 - x), \qquad (5.41)$$

where the time evolution is given by iterating the map:

$$x_0, x_1, x_2, \ldots = x_0, f(x_0), f(f(x_0)), \ldots . \quad (5.42)$$

In particular, for $\mu = 1$ it precisely folds the unit interval in half, and stretches it (nonuniformly) to cover the original domain.

---

[49]The operator O could be the momentum p^2, the energy, or $\sin^2(2\pi x)(\cos(4\pi p) + 1)$ (with periodic boundary conditions).
[50]Since the cat map of eqn 5.39 is invertible, the orbit will be periodic without an initial transient, but that's harder to argue.
[51]This exercise and the associated software were developed in collaboration with Christopher Myers. Hints for the computations can be found at the book website [181]. There are several other exercises exploring this chaotic logistic map (see Index).

The mathematics community lumps together continuous dynamical evolution laws and discrete mappings as *dynamical systems*.[52] The general stretching and folding exhibited by our map is often seen in driven physical systems without conservation laws.

In this exercise, we will focus on values of μ near one, where the motion is mostly *chaotic*. Chaos is sometimes defined as motion where the final position depends sensitively on the initial conditions. Two trajectories, starting a distance ϵ apart, will typically drift apart in time as $\epsilon e^{\lambda t}$, where λ is the Lyapunov exponent for the chaotic dynamics.

Start with $\mu = 0.9$ and two nearby points x_0 and $y_0 = x_0 + \epsilon$ somewhere between zero and one. Investigate the two trajectories $x_0, f(x_0), f(f(x_0)), \ldots, f^{[n]}(x_0)$ and $y_0, f(y_0), \ldots$. How fast do they separate? Why do they stop separating? Estimate the Lyapunov exponent. (Hint: ϵ can be a few times the precision of the machine, around 10^{-17} for double-precision arithmetic, so long as you are not near the maximum value of f at $x_0 = 0.5$.)

Many Hamiltonian systems are also chaotic. Two configurations of classical atoms or billiard balls, with initial positions and velocities that are almost identical, will rapidly diverge as the collisions magnify small initial deviations in angle and velocity into large ones. It is this chaotic stretching, folding, and kneading of phase space that is at the root of our explanation that entropy increases.

(5.10) **Entropy increases: diffusion.** (Mathematics) @

We can show that entropy microscopically does not increase for a closed system, for any Hamiltonian, either classical (Exercise 5.7) or quantum (Exercise 7.4). However, we can show that entropy increases for most of the coarse-grained effective theories that we use in practice; when we integrate out degrees of freedom, we provide a means for the information about the initial condition to be destroyed. Here you will show that entropy increases for the diffusion equation.

Let $\rho(x, t)$ obey the one-dimensional diffusion equation $\partial\rho/\partial t = D\partial^2\rho/\partial x^2$. Assume that the density ρ and all its gradients die away rapidly at $x = \pm\infty$.[53]

Derive a formula for the time derivative of the entropy $S = -k_B \int \rho(x) \log \rho(x)\, dx$ and show that it strictly increases in time. (Hint: Integrate by parts. You should get an integral of a positive definite quantity.)

(5.11) **Entropy of glasses.**[54] (Condensed matter) ③

Glasses are not really in equilibrium. If you put a glass in an insulated box, it will warm up (very slowly) because of microscopic atomic rearrangements which lower the potential energy. So, glasses do not have a well-defined temperature or specific heat. In particular, the heat flow upon cooling and on heating $(dQ/dT)(T)$ will not precisely match (although their integrals will agree by conservation of energy).

Thomas and Parks in Fig. 5.18 are making the approximation that the specific heat of the glass is dQ/dT, the measured heat flow out of the glass divided by the temperature change of the heat bath. They find that the specific heat defined in this way measured on cooling and heating disagree. Assume that the liquid at $325\,°C$ is in equilibrium both before cooling and after heating (and so has the same liquid entropy S_{liquid}).

(a) *Is the residual entropy, eqn 5.18, experimentally larger on heating or on cooling in Fig. 5.18?* (Hint: Use the fact that the integrals under the curves, $\int_0^{T_\ell}(dQ/dT)\, dT$ give the heat flow, which by conservation of energy must be the same on heating and cooling. The heating curve in Fig. 5.18 shifts weight to higher temperatures; will that increase or decrease the integral $\int(1/T)(dQ/dT)\, dT$?)

(b) *By using the second law (entropy can only increase), show that when cooling and then heating from an equilibrium liquid the residual entropy measured on cooling must always be less than the residual entropy measured on heating.* Your argument should be completely general, applicable to any system out of equilibrium. (Hint: Consider the entropy flow into the outside world upon cooling the liquid into the glass, compared

---

[52]The Poincaré section (Fig. 4.9) takes a continuous, recirculating dynamical system and replaces it with a once-return map, providing the standard motivation for treating maps and continuous evolution laws together. This motivation does not directly apply here, because the logistic map 4.12 is not invertible, so it is not directly given by a Poincaré section of a smooth differential equation. (Remember the existence and uniqueness theorems from math class? The invertibility follows from uniqueness.)

[53]Also, you may assume $\partial^n\rho/\partial x^n \log \rho$ goes to zero at $x = \pm\infty$, even though $\log \rho$ goes to $-\infty$.

[54]This paper is based on work by Stephen Langer [111].

to the entropy flow from the outside world to heat the glass into the liquid again. The initial and final states of the liquid are both in equilibrium.)

Fig. 5.18 Glass transition specific heat. Specific heat of B_2O_3 glass measured while heating and cooling [196]. The glass was cooled from $345\,^\circ C$ to room temperature in 18 hours (diamonds), and then heated from $35\,^\circ C \to 325\,^\circ C$ (crosses). See [109].

The residual entropy of a typical glass is about k_B per molecular unit. It is a measure of how many different glassy configurations of atoms the material can freeze into.

(c) *In a molecular dynamics simulation with one hundred indistinguishable atoms, and assuming that the residual entropy is $k_B \log 2$ per atom, what is the probability that two coolings to zero energy will arrive at equivalent atomic configurations (up to permutations)? In a system with 10^{23} molecular units, with residual entropy $k_B \log 2$ per unit, about how many coolings would be needed for one to duplicate the original configuration again at least once, with probability $1/2$?*

(5.12) **Rubber band.** (Condensed matter) @
Figure 5.19 shows a one-dimensional model for rubber. Rubber is formed from long polymeric molecules, which undergo random walks in the undeformed material. When we stretch the rubber, the molecules respond by rearranging their random walk to elongate in the direction of the external stretch. In our model, the molecule is represented by a set of N links of length d, which with equal energy point either parallel or antiparallel to the previous link. Let the total change in position to the right, from the beginning of the polymer to the end, be L.

As the molecule's extent L increases, the entropy of our rubber molecule decreases.
(a) *Find an exact formula for the entropy of this system in terms of d, N, and L.* (Hint: How many ways can one divide N links into M right-pointing links and $N - M$ left-pointing links, so that the total length is L?)

Fig. 5.19 Rubber band. Simple model of a rubber band with $N = 100$ segments. The beginning of the polymer is at the top; the end is at the bottom; the vertical displacements are added for visualization.

The external world, in equilibrium at temperature T, exerts a force pulling the end of the molecule to the right. The molecule must exert an equal and opposite *entropic* force F.
(b) *Find an expression for the force F exerted by the molecule on the bath in terms of the bath entropy.* (Hint: The bath temperature $1/T = \partial S_{\text{bath}}/\partial E$, and force times distance is energy.) *Using the fact that the length L must maximize the entropy of the Universe, write a general expression for F in terms of the internal entropy S of the molecule.*
(c) *Take our model of the molecule from part (a), the general law of part (b), and Stirling's formula $\log(n!) \approx n \log n - n$, write the force law $F(L)$ for our molecule for large lengths N. What is the spring constant K in Hooke's law $F = -KL$ for our molecule, for small L?*
Our model has no internal energy; this force is entirely entropic. Note how magical this is—we never considered the mechanism of how the segments would generate a force. Statistical mechanics tells us that the force generated by our segmented chain is *independent* of the mechanism. The joint angles in the chain may jiggle from thermal motion, or the constituent polymer monomers may execute thermal motion—so long as the configuration space is segment orientations and the effective potential energy is zero the force will be given by our calculation. For the same reason, the pressure due to compressing an ideal gas is independent of the mechanism.

The kinetic energy of particle collisions for real dilute gases gives the same pressure as the complex water-solvent interactions give for osmotic pressure (Section 5.2.2).

(d) *If we increase the temperature of our rubber band while it is under tension, will it expand or contract? Why?*

In a more realistic model of a rubber band, the entropy consists primarily of our configurational random-walk entropy plus a vibrational entropy of the molecules. If we stretch the rubber band without allowing heat to flow in or out of the rubber, the total entropy should stay approximately constant. (Rubber is designed to bounce well; little irreversible entropy is generated in a cycle of stretching and compression, so long as the deformation is not too abrupt.)

(e) *True or false?*

(T) (F) When we stretch the rubber band, it will cool; the position-space entropy of the random walk will decrease, causing the entropy in the vibrations to decrease, causing the temperature to decrease.

(T) (F) When we stretch the rubber band, it will cool; the position-space entropy of the random walk will decrease, causing the entropy in the vibrations to increase, causing the temperature to decrease.

(T) (F) When we let the rubber band relax after stretching, it will cool; the position-space entropy of the random walk will increase, causing the entropy in the vibrations to decrease, causing the temperature to decrease.

(T) (F) When we let the rubber band relax, there must be no temperature change, since the entropy is constant.

This more realistic model is much like the ideal gas, which also had no position-space energy.

(T) (F) Like the ideal gas, the temperature changes because of the net work done on the system.

(T) (F) Unlike the ideal gas, the work done on the rubber band is positive when the rubber band expands.

You should check your conclusions experimentally; find a rubber band (thick and stretchy is best), touch it to your lips (which are very sensitive to temperature), and stretch and relax it.

(5.13) **How many shuffles?** (Mathematics) @

For this exercise, you will need a deck of cards, either a new box or sorted into a known order (conventionally with an ace at the top and a king at the bottom).[55]

How many shuffles does it take to randomize a deck of 52 cards?

The answer is a bit controversial; it depends on how one measures the information left in the cards. Some suggest that seven shuffles are needed; others say that six are enough.[56] We will follow reference [200], and measure the growing randomness using the information entropy.

We imagine the deck starts out in a known order (say, A♠, 2♠, ..., K♣).

(a) *What is the information entropy of the deck before it is shuffled? After it is completely randomized?*

(b) *Take a sorted deck of cards. Pay attention to the order; (in particular, note the top and bottom cards). Riffle it exactly once, separating it into two roughly equal portions and interleaving the cards in the two portions. Examine the card sequence, paying particular attention to the top few and bottom few cards. Can you tell which cards came from the top portion and which came from the bottom?*

The mathematical definition of a *riffle shuffle* is easiest to express if we look at it backward.[57] Consider the deck after a riffle; each card in the deck either came from the top portion or the bottom portion of the original deck. A riffle shuffle makes each of the 2^{52} patterns $tbbtbttb\ldots$ (denoting which card came from which portion) equally likely.

It is clear that the pattern $tbbtbttb\ldots$ determines the final card order: the number of ts tells us how many cards were in the top portion, and then the cards are deposited into the final pile according to the pattern in order bottom to top. Let us first pretend the reverse is also true: that every pattern corresponds one-to-one with a unique final card ordering.

(c) *Ignoring the possibility that two different riffles could yield the same final sequence of cards, what is the information entropy after one riffle?*

---

[55] Experience shows that those who do not do the experiment in part (b) find it challenging to solve part (c) by pure thought.

[56] More substantively, as the number of cards $N \to \infty$, some measures of information show an abrupt transition near $\frac{3}{2} \log_2 N$, while by other measures the information vanishes smoothly and most of it is gone by $\log_2 N$ shuffles.

[57] The forward definition mimics an expert riffler by splitting the deck into two roughly equal portions by choosing n cards on top with a binomial probability $2^{-52}\binom{52}{n}$. Then the expert drops cards with probability proportional to the number of cards remaining in its portion. This makes each of the 2^{52} choices we found in the backward definition equally likely.

You can convince yourself that the only way two riffles can yield the same sequence is if all the cards in the bottom portion are dropped first, followed by all the cards in the top portion.

(d) *How many riffles drop the entire bottom portion and then the entire top portion, leaving the card ordering unchanged? What fraction of the 2^{52} riffles does this correspond to?* (Hint: $2^{10} = 1,024 \approx 10^3$. Indeed, this approximation underlies measures of computing resources: a gigabyte is not 10^9 bytes, but $(1,024)^3 = 2^{30}$ bytes.) *Hence, what is the actual information entropy after one riffle shuffle?*

We can put a lower bound on the number of riffles needed to destroy all information by assuming the entropy increase stays constant for future shuffles.

(e) *Continuing to ignore the possibility that two different sets of m riffles could yield the same final sequence of cards, how many riffles would it take for our upper bound for the entropy to pass that of a completely randomized deck?*

(5.14) **Information entropy.** (Computer science, Mathematics, Complexity) ⓟ
Entropy is a measure of your ignorance about a system; it is a measure of the lack of information. It has important implications in communication technologies: messages passed across a network communicate information, reducing the information entropy for the receiver.

Your grandparent has sent you an e-mail message. From the header of the message, you know it contains 1,000 characters. You know each character is made of 8 bits, which allows $2^8 = 256$ letters or symbols per character.

(a) *Assuming all possible messages from your grandparent are equally likely (a typical message would then look like G*me'!8V[beep]. . .), how many different messages N could there be? What is the corresponding upper bound $S_{\max}$ for the information entropy $k_S \log N$?*

Your grandparent writes rather dull messages; they all fall into the same pattern. They have a total of 16 equally likely messages.[58] After you read the message, you forget the details of the wording anyhow, and only remember these key points of information.

(b) *What is the actual information entropy change $\Delta S_{\text{Shannon}}$ you undergo when reading the message? If your grandparent writes one message per month, what is the minimum number of 8-bit characters per year that it would take to send your grandparent's messages? (You may split individual characters between messages.)* (Hints: $\Delta S_{\text{Shannon}}$ is your change in entropy from before you read the message to after you read which of 16 messages it was. The length of 1,000 is *not* important for this part.)

This is an extreme form of lossy data compression, like that used in jpeg images, mpeg animations, and mp3 audio files.

(5.15) **Shannon entropy.** (Computer science) @
Natural languages are highly redundant; the number of intelligible fifty-letter English sentences is many fewer than 26^{50}, and the number of distinguishable ten-second phone conversations is far smaller than the number of sound signals that could be generated with frequencies up to $20,000\,\text{Hz}$.[59]
This immediately suggests a theory for signal compression. If you can recode the alphabet so that common letters and common sequences of letters are abbreviated, while infrequent combinations are spelled out in lengthy fashion, you can dramatically reduce the channel capacity needed to send the data. (This is lossless compression, used in zip, gz, and gif file formats.)
An obscure language A'bç! for long-distance communication has only three sounds: a hoot represented by A, a slap represented by B, and a click represented by C. In a typical message, hoots and slaps occur equally often ($p = 1/4$), but clicks are twice as common ($p = 1/2$). Assume the messages are otherwise random.

(a) *What is the Shannon entropy in this language? More specifically, what is the Shannon entropy rate $-k_S \sum p_m \log p_m$, the entropy per sound or letter transmitted?*
(b) *Show that a communication channel transmitting bits (ones and zeros) can transmit no more than one unit of Shannon entropy per bit. (Hint: This should follow by showing that, for*

---

[58]Each message mentions whether they won their bridge hand last week (a fifty-fifty chance), mentions that they wish you would write more often (every time), and speculates who will win the women's college basketball tournament in their region (picking at random one of the eight teams in the league).
[59]Telephones, for example, do not span this whole frequency range: they are limited on the low end at 300–400 Hz, and on the high end at 3,000–3,500 Hz. You can still understand the words, so this crude form of data compression is only losing nonverbal nuances in the communication (Paul Ginsparg, private communication).

$N = 2^n$ messages, the Shannon entropy is maximized by $p_m = 1/N$. We have proved this already in a complicated way in note 37, p. 110; here prove it is a local extremum, either using a Lagrange multiplier or by explicitly setting $p_N = 1 - \sum_{m=1}^{N-1} p_m$.)

(c) *In general, argue that the Shannon entropy gives the minimum number of bits needed to transmit the ensemble of messages.* (Hint: Compare the Shannon entropy of the N original messages with the Shannon entropy of the N (shorter) encoded messages.) *Calculate the minimum number of bits per letter on average needed to transmit messages for the particular case of an A'bç! communication channel.*

(d) *Find a compression scheme (a rule that converts a A'bç! message to zeros and ones, that can be inverted to give back the original message) that is optimal, in the sense that it saturates the bound you derived in part (b).* (Hint: Look for a scheme for encoding the message that compresses one letter at a time. Not all letters need to compress to the same number of bits.)

Shannon also developed a measure of the channel capacity of a noisy wire, discussed error-correction codes, etc.

(5.16) **Fractal dimensions.**[60] (Mathematics, Complexity, Computation, Dynamical systems) ④

There are many strange sets that emerge in science. In statistical mechanics, such sets often arise at continuous phase transitions, where self-similar spatial structures arise (Chapter 12). In chaotic dynamical systems, the attractor (the set of points occupied at long times after the transients have disappeared) is often a fractal (called a *strange attractor*). These sets are often tenuous and jagged, with holes on all length scales; see Figs. 12.2, 12.5, and 12.11.

We often try to characterize these strange sets by a dimension. The dimensions of two extremely different sets can be the same; the path exhibited by a random walk (embedded in three or more dimensions) is arguably a two-dimensional set (note 6 on p. 25), but does not locally look like a surface. However, if two sets have different spatial dimensions (measured in the same way) they are certainly qualitatively different.

There is more than one way to define a dimension. Roughly speaking, strange sets are often spatially inhomogeneous, and what dimension you measure depends upon how you weight different regions of the set. In this exercise, we will calculate the *information dimension* (closely connected to the nonequilibrium entropy), and the *capacity dimension* (originally called the *Hausdorff dimension*, also sometimes called the *fractal dimension*).

To generate our strange set—along with some more ordinary sets—we will use the logistic map

$$f(x) = 4\mu x(1 - x). \tag{5.43}$$

The attractor for the logistic map is a periodic orbit (dimension zero) at $\mu = 0.8$, and a chaotic, cusped density filling two intervals (dimension one)[61] at $\mu = 0.9$. At the onset of chaos at $\mu = \mu_\infty \approx 0.892486418$ (Exercise 12.9) the dimension becomes intermediate between zero and one; this strange, self-similar set is called the *Feigenbaum attractor*.

Both the information dimension and the capacity dimension are defined in terms of the occupation P_n of cells of size ϵ in the limit as $\epsilon \to 0$.

(a) *Write a routine which, given μ and a set of bin sizes ϵ, does the following.*

- *Iterates f hundreds or thousands of times (to get onto the attractor).*

- *Iterates f a large number N_{tot} more times, collecting points on the attractor. (For $\mu \le \mu_\infty$, you could just integrate 2^n times for n fairly large.)*

- *For each ϵ, use a histogram to calculate the probability P_j that the points fall in the jth bin.*

- *Return the set of vectors $P_j[\epsilon]$.*

You may wish to test your routine by using it for $\mu = 1$ (where the distribution should look like $\rho(x) = 1/\pi\sqrt{x(1 - x)}$, Exercise 4.3(b)) and $\mu = 0.8$ (where the distribution should look like two δ-functions, each with half of the points).

The capacity dimension. The definition of the capacity dimension is motivated by the idea that it takes at least

$$N_{\text{cover}} = V/\epsilon^D \tag{5.44}$$

---

[60]This exercise and the associated software were developed in collaboration with Christopher Myers. Hints for the computations can be found at the book website [181]. There are several other exercises exploring this chaotic logistic map (see Index).

[61]See Exercise 4.3. The chaotic region for the logistic map does not have a strange attractor because the map is confined to one dimension; period-doubling cascades for dynamical systems in higher spatial dimensions have fractal, strange attractors in the chaotic region.

bins of size ϵ^D to cover a D-dimensional set of volume V.[62] By taking logs of both sides we find $\log N_{\text{cover}} \approx \log V + D \log \epsilon$. The capacity dimension is defined as the limit

$$D_{\text{capacity}} = \lim_{\epsilon \to 0} \frac{\log N_{\text{cover}}}{\log \epsilon}, \qquad (5.45)$$

but the convergence is slow (the error goes roughly as $\log V / \log \epsilon$). Faster convergence is given by calculating the slope of $\log N$ versus $\log \epsilon$:

$$\begin{aligned} D_{\text{capacity}} &= \lim_{\epsilon \to 0} \frac{\mathrm{d} \log N_{\text{cover}}}{\mathrm{d} \log \epsilon} \\ &= \lim_{\epsilon \to 0} \frac{\log N_{j+1} - \log N_j}{\log \epsilon_{i+1} - \log \epsilon_i}. \end{aligned} \qquad (5.46)$$

(b) *Using your routine from part (a), write a routine to calculate $N[\epsilon]$ by counting nonempty bins. Plot D_{capacity} from the fast convergence eqn 5.46 versus the midpoint $\frac{1}{2}(\log \epsilon_{i+1} + \log \epsilon_i)$. Does it appear to extrapolate to $D = 1$ for $\mu = 0.9$?[63] Does it appear to extrapolate to $D = 0$ for $\mu = 0.8$? Plot these two curves together with the curve for μ_∞. Does the last one appear to converge to $D_1 \approx 0.538$, the capacity dimension for the Feigenbaum attractor gleaned from the literature? How small a deviation from μ_∞ does it take to see the numerical crossover to integer dimensions?*

Entropy and the information dimension. The probability density $\rho(x_j) \approx P_j / \epsilon = (1/\epsilon)(N_j / N_{\text{tot}})$. Converting the entropy formula 5.20 to a sum gives

$$\begin{aligned} S &= -k_B \int \rho(x) \log(\rho(x)) \, \mathrm{d}x \\ &\approx -\sum_j \frac{P_j}{\epsilon} \log\left(\frac{P_j}{\epsilon}\right) \epsilon \\ &= -\sum_j P_j \log P_j + \log \epsilon \end{aligned} \qquad (5.47)$$

(setting the conversion factor $k_B = 1$ for convenience).

You might imagine that the entropy for a fixed point would be zero, and the entropy for a period-m cycle would be $k_B \log m$. But this is incorrect; when there is a fixed point or a periodic limit cycle, the attractor is on a set of dimension zero (a bunch of points) rather than dimension one. The entropy must go to minus infinity—since we have precise information about where the trajectory sits at long times. To estimate the "zero-dimensional" entropy $k_B \log m$ on the computer, we would use the discrete form of the entropy (eqn 5.19), summing over bins P_j instead of integrating over x:

$$S_{d=0} = -\sum_j P_j \log(P_j) = S_{d=1} - \log(\epsilon). \quad (5.48)$$

More generally, the natural measure of the entropy for a set with D dimensions might be defined as

$$S_D = -\sum_j P_j \log(P_j) + D \log(\epsilon). \qquad (5.49)$$

Instead of using this formula to define the entropy, mathematicians solve it for D to define the information dimension:

$$D_{\text{inf}} = \lim_{\epsilon \to 0} \left(\sum P_j \log P_j \right) / \log(\epsilon). \qquad (5.50)$$

The information dimension agrees with the ordinary dimension for sets that locally look like $\mathbb{R}^D$. It is different from the capacity dimension (eqn 5.45), which counts each occupied bin equally; the information dimension counts heavily occupied parts (bins) in the attractor more heavily. Again, we can speed up the convergence by noting that eqn 5.49 says that $\sum_j P_j \log P_j$ is a linear function of $\log \epsilon$ with slope D and intercept S_D. Measuring the slope directly, we find

$$D_{\text{inf}} = \lim_{\epsilon \to 0} \frac{\mathrm{d} \sum_j P_j(\epsilon) \log P_j(\epsilon)}{\mathrm{d} \log \epsilon}. \qquad (5.51)$$

(c) *As in part (b), write a routine that plots D_{inf} from eqn 5.51 as a function of the midpoint $\log \epsilon$, as we increase the number of bins. Plot the curves for $\mu = 0.9$, $\mu = 0.8$, and μ_∞. Does the information dimension agree with the ordinary one for the first two? Does the last one appear to converge to $D_{\text{inf}} \approx 0.517098$, the information dimension for the Feigenbaum attractor from the literature?*

Note that the capacity dimension $D_1 \approx 0.538$ is different from the information dimension $D_{\text{inf}} \approx 0.517098$ in this case. Most "real world" fractals have a whole spectrum of different characteristic spatial dimensions; they are *multifractal*.

---

[62]Imagine covering the surface of a sphere in 3D with tiny cubes; the number of cubes will go as the surface area (2D volume) divided by ϵ^2.

[63]In the chaotic regions, keep the number of bins small compared to the number of iterates in your sample, or you will start finding empty bins between points and eventually get a dimension of zero.

(5.17) **Deriving entropy.** (Mathematics) ③
In this exercise, you will show that there is a unique continuous function S_I (up to the constant k_B) satisfying the three key properties (eqns 5.25, 5.29, and 5.35) for a good measure of ignorance:

$$S_I\left(\frac{1}{\Omega}, \ldots, \frac{1}{\Omega}\right) > S_I(p_1, \ldots, p_\Omega) \quad (5.52)$$

unless $p_i = 1/\Omega$ for all i,

$$S_I(p_1, \ldots, p_{\Omega-1}, 0) = S_I(p_1, \ldots, p_{\Omega-1}), \quad (5.53)$$

and

$$\langle S_I(A|B_\ell)\rangle_B = S_I(AB) - S_I(B). \quad (5.54)$$

Here

$$S_I(A) = S_I(p_1, \ldots, p_\Omega),$$
$$S_I(B) = S_I(q_1, \ldots, q_M),$$
$$\langle S_I(A|B_\ell)\rangle_B = \sum_\ell q_\ell S_I(c_{1\ell}, \ldots, c_{\Omega\ell}),$$
$$S_I(AB) = S_I(c_{11}q_1, \ldots, c_{\Omega M}q_M).$$

You will show, naturally, that this function is our nonequilibrium entropy 5.19. The presentation is based on the proof in the excellent small book by Khinchin [98].
For convenience, define $L(g) = S_I(1/g, \ldots, 1/g)$.
(a) *For any rational probabilities q_ℓ, let g be the least common multiple of their denominators, and let $q_\ell = g_\ell/g$ for integers g_ℓ. Show that*

$$S_I(B) = L(g) - \sum_\ell q_\ell L(g_\ell). \quad (5.55)$$

(Hint: Consider AB to have g possibilities of probability $1/g$, B to measure which group of size g_ℓ, and A to measure which of the g_ℓ members of group ℓ, see Fig. 5.20.)
(b) *If $L(g) = k_S \log g$, show that eqn 5.55 is the Shannon entropy 5.23.*
Knowing that $S_I(A)$ is the Shannon entropy for all rational probabilities, and assuming that $S_I(A)$ is continuous, makes $S_I(A)$ the Shannon entropy. So, we have reduced the problem to showing $L(g)$ is the logarithm up to a constant.
(c) *Show that $L(g)$ is monotone increasing with g.* (Hint: You will need to use both of the first two key properties.)

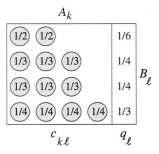

Fig. 5.20 Rational probabilities and conditional entropy. Here the probabilities $q_\ell = (1/6, 1/3, 1/3, 1/2)$ of state B_ℓ are rational. We can split the total probability into $g = 12$ equal pieces (circles, each probability $r_{k\ell} = 1/12$), with $g_k = (2, 3, 3, 4)$ pieces for the corresponding measurement B_ℓ. We can then write our ignorance $S_I(B)$ in terms of the (maximal) equal-likelihood ignorances $L(g) = S_I(1/g, \ldots)$ and $L(g_k)$, and use the entropy change for conditional probabilities property (eqn 5.35) to derive our ignorance $S_I(B)$ (eqn 5.55).

(d) *Show $L(g^n) = nL(g)$.* (Hint: Consider n independent probability distributions each of g equally likely events. Use the third property recursively on n.)
(e) *If $2^m < s^n < 2^{m+1}$, using the results of parts (c) and (d) show*

$$\frac{m}{n} < \frac{L(s)}{L(2)} < \frac{m+1}{n}. \quad (5.56)$$

(Hint: How is $L(2^m)$ related to $L(s^n)$ and $L(2^{m+1})$?) *Show also using the same argument that $m/n < \log(s)/\log(2) < (m+1)/n$. Hence, show that $|L(s)/L(2) - \log(s)/\log(2)| < 1/n$ and thus $L(s) = k \log s$ for some constant k.*
Hence our ignorance function S_I agrees with the formula for the nonequilibrium entropy uniquely (up to an overall constant).

(5.18) **Entropy of socks.** ⓟ
If a billion children neaten their bedrooms, how big is the decrease ΔS_Q of the position-space entropy in their toys and clothing? Assume that they each have one hundred distinguishable items weighing 20 gm, initially randomly placed in a bedroom with $V = 5 m^3$, and that they end up located in their place in a drawer with center-of-mass location confined to a region $1 cm^3$. For this exercise, assume that the internal entropies of the objects are unchanged in the process (atomic vibrations, folding entropy for socks, etc.) Find

the ratio of this entropy change to that of contracting the air in a one liter balloon by 1% in volume at constant temperature. (Hint: An ideal gas at constant temperature has a constant kinetic entropy from the particle momenta, so we compare ΔS_Q for the two.)

(5.19) **Aging, entropy, and DNA.** (Biology) ⓟ
Is human aging inevitable? Does the fact that entropy must increase mean that our cells must run down? In particular, as we get older the DNA in our cells gradually builds up damage—thought to be a significant contribution to the aging process (and a key cause of cancer). Can we use the entropy increase of DNA damage to argue that keeping ourselves young would violate the second law of thermodynamics?

There are roughly thirty trillion (3×10^{13}) cells in the human body, and about three billion (3×10^9) nucleotides in the DNA of each cell. Each nucleotide comes in four types (C, T, A, and G). The damaged DNA of each cell will be different. The repair job for fixing all of our DNA cannot be worse than recreating them all back into exact copies of the correct sequence.

(a) *How many bits of information is it possible to store in the nucleotide sequence of the DNA in an entire human body? How much entropy, in joules/Kelvin, is associated with a completely randomized sequence in every cell?* (Boltzmann's constant is 1.38×10^{-23} J/K.)

Life exists by consuming low-entropy sources and turning them into higher-entropy byproducts. A small cookie has 100 Calories. (Be warned: a Calorie is 1,000 calories. Food calories are measured in kilocalories, and then the kilo is conventionally dropped in favor of a capital C.) Body temperature is about 310 K.

(b) *Calculate the minimum free energy needed to repair a human's DNA if it starts in a completely scrambled state. How many cookies would one need to consume?* (There are 4.184 joules per calorie, and 1,000 calories per Calorie.)

Entropy does not discriminate between important and trivial information. Knowing that air molecules are confined in a balloon is much less useful than knowing that your kid's toys are put away neatly (or knowing the contents of the Library of Congress).

(5.20) **Gravity and entropy.** (Astrophysics) ⓟ
The motion of gravitationally bound collections of particles would seem a natural application of statistical mechanics. The chaotic motion of many stars could maximize a "stellar entropy" of the positions and momenta of the stars treated as particles. But there are challenges.

The gravitational interaction is long range. One of our key assumptions in deriving the entropy was that it should be extensive: the entropy of weakly interacting subsystems adds up to the entropy of the system.

(a) *Can we divide up a galaxy of stars into weakly interacting subsystems? If not, must the entropy of the positions and momenta of the stars in the galaxy be a sum of the stellar entropy of pieces of the galaxy? Do you conclude that Shannon entropy is or is not applicable to stellar dynamics?*

(b) *The Coulomb force is also long range. Why can we describe distributions of atoms using Shannon entropy, when they are composed of electrons and nuclei with long-range forces?*

The $1/r^2$ potential shared by Newton and Coulomb also has a strong singularity at short distances. The Boltzmann distribution for two stars diverges at $r = 0$, so in equilibrium all stars would fall into one another, another problem with applying statistical mechanics to stars. For electrons and nuclei, this divergence is cut off by quantum mechanics.

(5.21) **Data compression.**[64] ③
In this exercise, we compare the performance of different compression algorithms, on a snapshot of the well-studied two-dimensional Ising model. If they worked perfectly, we could measure the entropy of an experimental system or a new simulation by compressing the signal and measuring its length!

The table shows the compressed file sizes (in bits) for 256×256 snapshots of the Ising model at three temperatures (as in Fig. 5.21), using three different storage schemes. It compares them to the predicted Ising model entropy (also in bits)— the entropy $S(T)$ of the probability distribution of Ising snapshots at that temperature. (Note that the Ising entropy for $T = 100$ is omitted; see part (a).) BZ2 and GZIP are general-purpose file compression algorithms, often used for long text files and executables. PNG and GIF are image compression algorithms; the GIF format was the

---

[64]Developed in collaboration with Jaron Kent-Dobias.

lossless compression algorithm most used on the early Web.

Fig. 5.21 Three snapshots of portions of the Ising model. The Ising model is essentially a lattice of ones and zeros, with white signifying zero and black signifying one. (a) A low temperature $T = 1.5$, where all but a few, isolated, random sites are zero. (b) An intermediate temperature, showing some interesting structure. (c) A high temperature, where the pixels are set to zero and one at random. Below: Compressed file sizes, in bits, of the three 256×256 snapshots, using four compression schemes.

	$T = 1.5$	$T \approx 2.27$	$T = 100$
Ising	2,890	28,977	???
BZ2	18,008	63,096	86,240
GZIP	15,992	77,832	125,432
PNG	12,360	85,768	140,648
GIF	7,808	79,376	79,376

Our high-temperature snapshot, Fig. 5.21(c), shows a basically random pattern, with each site set to one or zero with probability $1/2$.

(a) *At absolute minimum, how many bits must it take to compress such an image of size 256×256? That is, what should we insert for the high-temperature Ising entropy in the table?*

Any general purpose algorithm must have some overhead. For example, it must declare somewhere that the image is filled only with ones and zeros, even if the final pattern is then efficiently stored in bits. Let us define the overhead as the difference between the bits used and the Shannon bound on the message size given by the Ising entropy, and define the best algorithm ignoring that overhead.

(b) *Using the table and your calculation from part (a), estimate for our high-temperature images the overhead in bits needed for the best of the four storage formats. Which method is best at low temperatures? Which is best for temperatures near $T_c = 2.27$ where the pictures are most interesting?*

Our low temperature snapshot, Fig. 5.21(a), shows only a few sites set to one. These lone sites are scattered seemingly at random, with only a few clustered regions.

(c) *Presume the sites are independent, and equal to one with small probability p. What would the information entropy of the ensemble of snapshots be, in terms of p?*

The Ising model at low temperatures has probability $p(T) = \exp(-8/T)$ that a site will be flipped with respect to its neighbors. Its free energy at low temperatures is $A(T, V, N) \approx -N\left(2 + T\log(1 + e^{-8/T})\right)$, see Exercise 8.19. (Note that in this formula we set $k = 1$, where in the table we for convenience used $k = k_S$.)

(d) *Compute S from A for the Ising model at low temperatures. Compute S from $p(T)$ and your answer from part (c). Show analytically that they agree at low temperatures and small p (except that the former is measured in "nats" ($k = 1$), rather than bits ($k_S = 1/\log(2)$)).* (Hints: Taylor series will not work directly, since S has a logarithmic singularity at small p—but you can Taylor expand $\log(1+\epsilon)$. You may need approximations like $p \ll 1$, $|\log p| \gg 1$, and corresponding approximations involving $e^{-8/T}$.)

(e) *Verify that your low-temperature formula for the information entropy in part (d) is in rough agreement with the more precise calculation in the first column in the table.*

Consider an alternative compression algorithm, optimized for images with large blocks of a single color. The algorithm will start with a list of colors found in the image, but we shall ignore that overhead. View our image as a one-dimensional line of pixels, and break it up into segments where all the pixels are the same color. Our compressed image will be a list of (color, length) pairs.

(f) *How many bits per segment will our algorithm take, if we know there are only two colors (0 and 1), and each segment will be at most 256^2 long? How many bits on average will it take to compress one of our low temperature images using this algorithm? By what ratio is it worse than the ideal algorithm? How does this algorithm compare to the four compression methods in the table at low temperatures?* (Hint: A number of size at most n can be represented by $\log_2(n)$ bits. Again presume the low temperature images are single sites scattered at random with probability $p = \exp(-8/T)$ with $T = 1.5$. Almost half of our blocks will be of length one, but our algorithm will not take advantage of that.)

(5.22) **The Dyson sphere.** (Astrophysics) @
Life on Earth can be viewed as a heat engine, taking energy from a hot bath (the Sun at temperature $T_S = 6{,}000°$ K) and depositing it into a cold bath (interstellar space, at a microwave background temperature $T_{MB} = 2.725$ K, Exercise 7.15). The outward solar energy flux at the Earth's orbit is $\Phi_S = 1{,}370\,\text{W/m}^2$, and the Earth's radius is approximately $6{,}400\,\text{km}$, $r_E = 6.4 \times 10^6$ m.
(a) *If life on Earth were perfectly efficient (a Carnot cycle with a hot bath at T_S and a cold bath at T_{MB}), how much useful work (in watts) could be extracted from this energy flow? Compare that to the estimated world marketed energy consumption of 4.5×10^{20} J/year. (Useful constant: There are about $\pi \times 10^7$ s in a year.)*
Your answer to part (a) suggests that we have some ways to go before we run out of solar energy. But let's think big.
(b) *If we built a sphere enclosing the Sun at a radius equal to Earth's orbit (about 150 million kilometers, $R_{ES} \approx 1.5 \times 10^{11}$ m), by what factor would the useful work available to our civilization increase?*
This huge construction project is called a *Dyson sphere*, after the physicist who suggested [53] that we look for advanced civilizations by watching for large sources of infrared radiation.
Earth, however, does not radiate at the temperature of interstellar space. It radiates roughly as a black body at near $T_E = 300°$ K $= 23°$ C (see, however, Exercise 7.21).
(c) *How much less effective are we at extracting work from the solar flux, if our heat must be radiated effectively to a $300°$ K cold bath instead of one at T_{MB}, assuming in both cases we run Carnot engines?*
There is an alternative point of view which tracks entropy rather than energy. Living beings maintain and multiply their low-entropy states by dumping the entropy generated into the energy stream leading from the Sun to interstellar space. New memory storage also intrinsically involves entropy generation (Exercise 5.2); as we move into the information age, we may eventually care more about dumping entropy than about generating work. In analogy to the "work effectiveness" of part (c) (ratio of actual work to the Carnot upper bound on the work, given the hot

and cold baths), we can estimate an entropy-dumping effectiveness (the ratio of the actual entropy added to the energy stream, compared to the entropy that could be conceivably added given the same hot and cold baths).
(d) *How much entropy impinges on the Earth from the Sun, per second per square meter cross-sectional area? How much leaves the Earth, per second per cross-sectional square meter, when the solar energy flux is radiated away at temperature $T_E = 300°$ K? By what factor f is the entropy dumped to outer space less than the entropy we could dump into a heat bath at T_{MB}? From an entropy-dumping standpoint, which is more important, the hot-bath temperature T_S or the cold-bath temperature (T_E or T_{MB}, respectively)?*
For generating useful work, the Sun is the key and the night sky is hardly significant. For dumping the entropy generated by civilization, though, the night sky is the giver of life and the realm of opportunity. These two perspectives are not really at odds. For some purposes, a given amount of work energy is much more useful at low temperatures. Dyson later speculated about how life could make efficient use of this by running at much colder temperatures (Exercise 5.1). A hyper-advanced information-based civilization would hence want not to radiate in the infrared, but in the microwave range.
To do this, it needs to increase the area of the Dyson sphere; a bigger sphere can re-radiate the Solar energy flow as black-body radiation at a lower temperature. Interstellar space is a good insulator, and one can only shove so much heat energy through it to get to the Universal cold bath. A body at temperature T radiates the largest possible energy if it is completely black. We will see in Exercise 7.7 that a black body radiates an energy σT^4 per square meter per second, where $\sigma = 5.67 \times 10^{-8}$ J/(s m^2 K^4) is the Stefan–Boltzmann constant.
(e) *How large a radius R_D must the Dyson sphere have to achieve 50% entropy-dumping effectiveness? How does this radius compare to the distance to Pluto ($R_{PS} \approx 6 \times 10^{12}$ m)? If we measure entropy in bits (using $k_S = (1/\log 2)$ instead of $k_B = 1.3807 \times 10^{-23}$ J/K), how many bits per second of entropy can our hyper-advanced civilization dispose of? (You may ignore the relatively small entropy impinging from the Sun onto the Dyson sphere, and ignore both*

the energy and the entropy from outer space.) The Sun would not be bright enough to read by at that distance, but if we had a well-insulated sphere we could keep it warm inside—only the outside need be cold. Alternatively, we could just build the sphere for our computers, and live closer in to the Sun; our re-radiated energy would be almost as useful as the original solar energy.

(5.23) **Entropy of the galaxy.**[65] (Astrophysics) @
What dominates the entropy of our Galaxy? The known sources of entropy include stars, interstellar gas, the microwave background photons inside the Galaxy, and the supermassive black hole at the galactic center.

(a) **Stars.** *Presume there are 10^{11} stars (100 billion) in our galaxy, with mass $m = 2 \times 10^{30}\,kg$, volume $V = 10^{45}\,m^3$, and temperature $T = 10^7\,K$. Assume that they are made of hydrogen, and form an ideal gas (a poor approximation). What is the entropy of one star? Compare to Bekenstein's estimate of $10^{35}\,J/K$ [15]. What is your estimate for the total entropy S_{stars} of the stars in the Galaxy?*

(b) **Gas.** *If the volume of the galaxy is $V = 10^{61}\,m^3$, with interstellar gas of hydrogen atoms of total mass 10% of the mass in the stars, and temperature $10\,K$, what is its entropy S_{gas}?*

(c) **Microwave photons.** *What is the total entropy for the cosmic microwave background radiation at $T_{\text{CMB}} = 2.75\,K$ in the volume V of the galaxy?*

(d) **Galactic black hole.** *What is the entropy for the supermassive black hole at the center of the galaxy, if its mass is $M_{\text{BH}} = 10^{37}\,kg$?*

(e) **Where to live?** *Where would a technological civilization locate, if finding a place to deposit entropy was the bottleneck for their activities?*

(5.24) **Nucleosynthesis and the arrow of time.**[66] (Astrophysics) ③
In this exercise, we shall explain how the first few minutes after the Big Bang set up a low-entropy state that both fuels our life on Earth and provides an arrow of time. To do so, we shall use the statistical mechanics of reaction rates in the

early universe (Exercise 6.26), use the equilibrium number density and entropy of the noninteracting photon gas, and model the Universe as a heat engine (Fig. 5.22).

Fig. 5.22 The Universe as a piston. Imagine the expansion of the Universe as the decompression of a gas of photons and nucleons. The Universe does not exchange heat with the "space-time piston" as it expands, but it can exchange work with the piston. (The image is galaxy cluster MACS 1206 [137]. We shall ignore the formation of galaxies and stars, however, and compare the formation of an H-Universe of hydrogen gas to a Fe-Universe of iron atoms.)

The microscopic laws of physics are invariant under time reversal.[67] It is statistical mechanics that distinguishes future from past—the future is the direction in which entropy increases.

This must arise because the early universe started in a low-entropy state. Experimentally, however, measurements of the cosmic microwave background radiation tell us that the matter in the universe[68] started out as nearly uniform, hot, dense gas (Exercises 7.15 and 10.1)—an *equilibrium state* and hence of *maximum* entropy. How can a state of maximal entropy become our initial low-entropy state?

Some cosmologists suggest that gravitational effects that lead matter to cluster into galaxies, stars, and eventually black holes explain the problem; the initial uniform distribution of matter and energy maximizes the entropy of the particles, but is a low-entropy state for space-time and gravitational degrees of freedom. It is

---

[65]This exercise was developed in collaboration with Katherine Quinn.

[66]This exercise was developed in collaboration with Katherine Quinn.

[67]Technically, the weak interaction is not time-reversal symmetric. But the laws of physics *are* invariant under CPT, where charge-conjugation reverses matter and antimatter and parity reflects spatial coordinates. There are no indications that entropy of antimatter decreases with time—the arrow of time for both appears the same.

[68]We do not have direct evidence for equilibration of neutrinos with the other massive particles, as they must have fallen out of equilibrium before the decoupling of light from matter that formed the microwave background. But estimates suggest that they too started in equilibrium.

certainly true on a cosmological scale that the formation of black holes can dominate the creation of entropy (Exercise 5.23). However, gravity has not been a significant source of free energy[69] (sink for entropy) apart from the vicinity of black holes; a thorough critique is given by Wallace [205].

The dominant source of free energy in the Solar System is not gravitation, but the hydrogen which fuels the Sun. Hydrogen predominates now because the Universe *fell out of equilibrium* as it expanded.

When thermally isolated systems are changed slowly enough to remain in equilibrium, we say the evolution is *adiabatic*.[70] Closed systems evolving adiabatically do not change in entropy. So we might guess that the Universe expanded too fast to stay in equilibrium. Can we produce a low-entropy state in a piston of gas if we expand it quickly?

Consider a piston initially with volume V and filled with N ideal gas atoms of mass m in equilibrium at temperature T. At time $t = 0$, the piston abruptly doubles in volume to $2V$, so quickly that the atoms have unchanged positions or velocities.

(a) *What is the equilibrium entropy at $t = 0^-$, just before the piston moves? What is the (nonequilibrium) entropy at $t = 0^+$? What is the entropy of the final, equilibrium state of the gas as $t \to \infty$?* (Hint: How much work did the gas do on the piston? What is its energy? Its temperature?) *Does the state of the gas at $t = 0^+$ form a low-entropy initial state for our ideal-gas Universe, compared to the final entropy?*

The Universe does not expand into empty space. Rather, space expands between the particles, cooling them in the process (Exercise 7.15). Our Universe fell out of equilibrium because it cooled too fast during the era of nucleosynthesis.

The very early Universe was so hot that any heavier nuclei quickly evaporated into protons and neutrons. Between a few seconds and a couple of minutes after the Big Bang, protons and neutrons began to fuse into light nuclei—mostly helium. The rates for conversion to heavier elements were too slow, however, by the time they

became entropically favorable. Almost all of the heavier elements on Earth were formed later, inside stars (see Exercise 6.26).

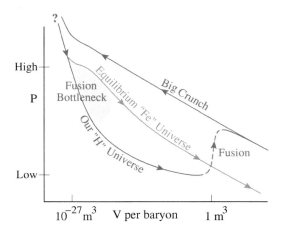

Fig. 5.23 PV diagram for the universe. In our Universe, the baryons largely did not fuse into heavier elements, because the reaction rates were too slow. Only in the eons since have stars been fusing hydrogen into heavier elements, glowing with nonequilibrium flows of low-entropy sunlight. Here we compare a simplified H version of our Universe with an equilibrium Fe-universe where nucleosynthesis finished at early times. This schematic shows the pressure-volume paths as the "piston" of Fig. 5.22 is moved. The H-Universe has nucleosynthesis completely suppressed; the proton gas adiabatically cools into hydrogen atoms, until a time near the present when hydrogen is fused (dashed line). In the Fe-Universe, nucleosynthesis stayed in equilibrium as the universe expanded, leaving only iron atoms. Finally, in the Big Crunch the H-Universe re-compresses, re-forming a hot proton gas.

How would our Universe compare with one with faster nucleosynthesis reactions, which remained in equilibrium? To simplify the problem, we shall ignore gravity (except insofar as it drives the expansion of the Universe), chemistry (no molecule or iron lump formation), and the distinction between protons and neutrons (just keep track of the baryons). We shall compare two universes (see Fig. 5.23). In the Fe-Universe nucleosynthesis is very fast; this universe begins as a

---

[69]Helmholtz and Kelvin in the last half of the 1800's considered whether the Sun could shine from the energy gained by gravitational collapse. They concluded that the Sun could last for a few million years. Even then, geology and biology made it clear that the Earth had been around for billions of years; gravity is not enough.

[70]*Adiabatic* is sometimes used to describe systems that do not exchange heat with their surroundings, and sometimes to describe systems that change slowly enough to stay in equilibrium or in their quantum ground states. Here we mean both.

gas of protons at high temperatures but stays in equilibrium, forming iron atoms early on in the expansion. In the H-Universe we imagine nucleosynthesis is so slow that no fusion takes place; it evolves as an adiabatically[71] expanding gas of hydrogen atoms.[72]

We want to compare the entropy of the H-Universe with that of the Fe-Universe.

(b) *How much does the entropy of the Fe-Universe change as it adiabatically expands from an initial gas of hot protons, through nucleosynthesis, to a gas of iron atoms at the current baryon density? How much does the entropy of the H-Universe change as it expands adiabatically into a gas of hydrogen atoms at the current density?* (No calculations should be needed, but a cogent argument with a precise answer is needed for both entropy changes.)

Some facts and parameters. Wikipedia tells us that nucleosynthesis happened "A few minutes into the expansion, when the temperature was about a billion ... Kelvin and the density was about that of air". The current microwave background radiation temperature is $T_{CMB} = 2.725$ K; let us for convenience set the temperature when nucleosynthesis would happen in equilibrium to a billion times this value, $T_{reaction} = 2.725 \times 10^9$ K. (We estimate the actual temperature in Exercise 6.26.) The thermal photons and the baryons are roughly conserved during the expansion, and the baryon-to-(microwave background photon) ratio is estimated to be $\eta = [Baryons]/[Photons] \sim 5 \times 10^{-10}$ (Exercise 6.26). Hence the temperature, energy density, and pressure are almost completely set by the photon density during this era (see part (c) next). The red shift of the photons increases their wavelength proportional to the expansion, and thus reduces their energy (and hence the temperature) inversely with the expansion rate. Thus the number density of photons and baryons goes as T^3. The current density of baryons is approximately one per cubic meter,[73] and thus we assume that the equilibrium synthesis happens at a density of 10^{27} per cubic meter. Fusing 56 protons into ^{56}Fe releases an energy of

about $\Delta E = 5 \times 10^8$ eV or about $\Delta E/56 = 1.4 \times 10^{-12}$ joules/baryon.

Truths about the photon gas (see Exercise 7.15). The number density for the photon gas is $N/V = (2\zeta(3)/\pi^2)(k_B T)^3/(c^3\hbar^3)$, the internal energy density is[74] $E/V = (1/15)\pi^2(k_B T)^4/(c^3\hbar^3) = 4T^4\sigma/c$, and the pressure is $P = E/3V$. The baryons do not contribute a significant pressure or kinetic energy density.

(c) *Calculate the temperature rise of the Fe-Universe when the hydrogen fuses into iron in equilibrium at high temperature, pressure, and density, releasing thermal photons. Calculate the difference in temperature for the Fe-Universe filled with iron and the H-Universe filled with hydrogen gas, when both reach the current baryon density.* (Hint: How does the temperature change with volume?)

(d) *Calculate the temperature rise in the H-Universe if we fuse all the hydrogen into iron at current densities and pressures (dashed line in Fig. 5.23), using our heat engines to turn the energy into thermal photons. Calculate the pressure rise. Calculate how much entropy would be released per baryon, in units of k_B.*

The key is not that our Universe started in a low-entropy state. It is that it has an untapped source of energy. Why, by postponing our fusion, do we end up with so much more energy? The energy for the matter and photons in an expanding universe is *not conserved*.

(e) *Calculate the difference in work done per baryon by the photon gas during the expansion, between the Fe-Universe and the H-Universe. Compare it to the energy released by nucleosynthesis.* (Hint: Calculate the temperature $T(V)$ and the difference $\Delta T(V)$ as a function of volume. Use that to determine $\Delta P(V)$.)

(f) *Consider a hypothetical Big Crunch, where the universe contracts slowly after expanding (and nothing irreversible happens at low densities). In the post-burn H-Universe, will the pressure and volume return to its original value at volumes per baryon much smaller than 10^{-27}?* (That is, will the P–V loop close in Fig. 5.23?) *If not, why not? What will the Fe-Universe Big*

---

[71]The nuclear degrees of freedom, which do not evolve, can be viewed as decoupled from the photons and the atomic motions.
[72]Without stars, the intelligent beings in the H-Universe will need to scoop up hydrogen with Bussard ramjets, fuse it into heavier elements, and use the resulting heavy elements (and abundant energy) to build their progeny and their technological support system. But they are better off than the inhabitants of the Fe-Universe.
[73]Very roughly.
[74]Here $\sigma = \pi^2 k_B^4/(60\hbar^3 c^2) = 5.67 \times 10^{-8}$ J/m^{2}sK^4 is the Stefan–Boltzmann constant.

Crunch return path be, assuming the return path is also in equilibrium?

(5.25) **Equilibration in phase space.** ③

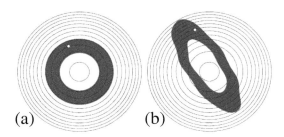

(a)   (b)

Fig. 5.24 Nonequilibrium state in phase space. 2D cartoons illustrating the $6N$-dimensional phase space, with large circles depicting the total energy shells and with the evolution of a random initial configuration depicted by a white dot. (a) Initial microcanonical equilibrium ensemble at energy E_0. (The energy shell is thickened to facilitate later figures.) (b) Nonequilibrium ensemble snapshot immediately following the application of a transient time-dependent Hamiltonian, such as a supersonic compression of the gas in a piston, followed by a rapid re-expansion. The dot represents the evolution of our particular initial state.

Here we discuss the process of equilibration—the smearing of probability until the density is uniform on the energy shell in phase space. Our analysis will be for an abstract system with two coordinates, first forced out of equilibrium by turning on and off a time-dependent Hamiltonian, and then evolved to re-establish an equilibrium. We shall motivate our discussion, however, with a physical example of a piston filled with a gas, subject to a supersonic compression and re-expansion, exhibiting shock waves and ringing as it equilibrates. Our aim is to examine the natural question of exactly when and how entropy increases. This question has been posed sharply by Exercises 5.7 and 7.4 (showing that entropy never increases in the microscopic theory), versus Exercises 5.10 and 8.12 (showing that it does increase in the emergent macroscopic theories). It plagued the early days of statistical mechanics, and inspired the mathematical theory we summarize in Chapter 4.

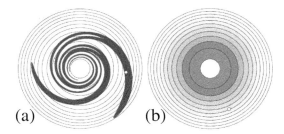

(a)   (b)

Fig. 5.25 Equilibrating the nonequilibrium state. (a) The phase-space volume occupied by the system is stretched and folded by the Hamiltonian time evolution. (b) At long times, the phase-space regions inside and outside the evolved nonequilibrium ensemble become so intertwined that they cannot be distinguished.

Figures 5.24 and 5.25 are two-dimensional cartoons that we shall use to depict the $6N$-dimensional phase space of a nonequilibrium system. The contours shown are total energy contours of constant ΔE, made circular for simplicity. The dark area in Fig. 5.24(a) is the equilibrium microcanonical ensemble energy shell, fattened to aid visualization in the subsequent figures. The white dot represents the initial state of our system (its two coordinates standing in for the positions and momenta of the many particles in the piston just before the supersonic compression). For large N, almost all points in the ensemble depict macroscopic states that are macroscopically indistinguishable.

Figure 5.24(b) denotes the evolution of our ensemble after the application of the time-dependent Hamiltonian.[75] As discussed in Exercise 4.7, Liouville's theorem is applicable also to time-dependent Hamiltonians—phase-space density is only rearranged, so the area and the shade of the dark region remain the same.

In the two-dimensional cartoon of Fig. 5.24(b), the energy of different compressed states varies widely, and the different energy shells (concentric rings) are filled only partially. Also, the energy sometimes decreases (see Exercise 4.8).

(a) *If one supersonically compresses and re-expands a macroscopic piston filled with gas, will the resulting total energy just after the shock fluctuate by a significant percentage? Will the resulting ensemble of states fill the new energy shell?*

---

[75]We apply the map $x' = x + K\sin(\omega y) - \delta,,\ y' = y + x + K\sin(\omega y) + \delta$, with $K = 0.1$, $\omega = 2$, $\delta = -0.3$ to represent the action of our time-dependent Hamiltonian.

Why or why not?

Figure 5.25 depicts the subsequent evolution[76] of a nonequilibrium system as chaotic motion[77] destroys the shock waves and ringing that give evidence of the particular initial conditions of our system. (See Exercise 5.8.)

Presume our 2D phase-space cartoons for the time evolution faithfully reflect the conserved quantities and Hamiltonian dynamics of a real physical system.

(b) *Must the total area of the shaded region in Fig. 5.25(a) (depicting the occupied phase space area after some time elapses) be the same as the area in the shaded region in Fig. 5.24(b) (at the moment the piston stops moving)? Why or why not? Between which phase-space snapshots does the net entropy increase, Figs. 5.24(a)→5.24(b)→5.25(a)→5.25(b)?*

Hence the shock waves in the macroscopic system are represented in our cartoon of Fig. 5.25(a) by the structure in the dark spirals, and equilibration in Fig. 5.25(b) by their blurring into a uniform shade as a function of energy (radius).

At late times, our system equilibrates so that the probability density $\rho(\mathbb{P}, \mathbb{Q})$ in phase space becomes effectively constant in each energy shell. Figure 5.25(b) represents this by shading regions based on the fraction of each energy shell that is occupied by the nonequilibrium ensemble of Fig. 5.24(b); the darker the shade, the larger the probability density ρ in that region of phase space. Our final equilibrated state is not a microcanonical ensemble; the system may be left by the compression in an equilibrated state with a variety of possible energies E (like the canonical ensemble of Chapter 6). Let the probability density for experimentally being left with energy E be $P(E)$.

(c) *Must the area of the dark dark shaded region within each energy shell be preserved? Why or why not? Write $\rho(\mathbb{P}, \mathbb{Q})$, the phase-space density of the final equilibrated state (depicted schematically in Fig. 5.25(b)) as a formula involving $P(E)$, the Hamiltonian $\mathcal{H}(\mathbb{P}, \mathbb{Q})$, and the energy-shell volume $\Omega(E)$ as a function of energy. Explain your reasoning.*

Note that we are using the term equilibrate in seemingly two different ways. We refer to a single system equilibrating, as it evolves in time at fixed energy from a peculiar initial condition to one typical of its energy, number, and volume. We refer to the ensemble equilibrating, as it spreads probability evenly over all states with the same energy. We know both are equivalent, because time averages are the same as ensemble averages

Note the entropy jump in going from Fig. 5.25(a) to Fig. 5.25(b) is an *emergent* law, expressing the fact that the information of which regions are shaded and which are not became useless as the regions become tightly intertwined. Chaos and ergodicity cause us to lose all information about our evolving state except the conserved energy. Furthermore, the theorem that time averages equal the microcanonical average guarantees that the long-time behavior will become typical of states in that energy shell for *any* measured property. Equilibrium statistical mechanics embraces our complete ignorance of the initial state, and uses it to explain our world. Entropy—useless in the underlying microscopic fundamental theory—becomes central to the emergent laws governing our world.

(5.26) **Phase conjugate mirror.** ③

If we start with a gas of particles in a corner of a box and evolve them in time, Liouville's theorem tells us that the phase-space volume is unchanged, but statistical mechanics tells us that entropy increases. We understand this in terms of a loss of useful information—the fact that the particles began in a small volume is encoded in multiparticle correlations between the positions and velocities at late times, but unless one had a magic wand that could reverse all of the velocities, that encoded information is useless, and will not affect future measurements.

Phase conjugate mirrors provide precisely such a magic wand. A phase conjugate mirror performs a *time-reversal* of the incoming wave—reversing the local momentum and phase of whatever wave impinges upon it.

---

[76] We chose a time evolution that rotates by a varying angular velocity $\dot\theta = 1/\sqrt{x^2 + p^2}$ for aesthetic reasons. This corresponds to the Hamiltonian $\mathcal{H} = \sqrt{x^2 + p^2}$. Our depicted energy shells thus ought to have been equally spaced in radius. Ignore these details in working the exercise.

[77] The motion in Fig. 5.25(a) is not chaotic—it stirs and stretches, but does not fold. True chaos, demanding at least three dimensions for continuous time evolution, would equilibrate faster. In four dimensions, chaos can arise even within one energy shell.

Fig. 5.26 Phase conjugate mirror. A phase conjugate mirror time-reverses an incoming beam, allowing the reconstruction of a clouded image by passing the light "backward in time" through the cloud. See [57] for a video.

Light scattered from an image, after passing through a cloud (Fig. 5.26) will be blurred. One normally would describe this blurring as an increase in entropy of the photons due to random scattering in the cloud.

(a) *View the demonstration of a phase conjugate mirror at [57]. Estimate from watching the video how long it takes for the $BaTiO_3$ crystal to "learn" from the blurred beam how to reflect the photons back from whence they came.* The phase conjugate mirror has resurrected the knowledge of how the photons scattered, reversing the presumed increase of entropy!

We can analyze this using a classical analogy to the light beam: a bunch of N noninteracting particles of mass M moving in the (x, z) plane with a nearly uniform, large horizontal velocity $v_z = V$. Our particles enter the cloud each with a phase-space probability distribution

$$\rho_{\text{initial}}(x, p_x) = \frac{1}{2\pi\sigma_x\sigma_p}e^{-\frac{x^2}{2\sigma_x^2} - \frac{p_x^2}{2\sigma_p^2}}. \quad (5.57)$$

We assume the particle velocities remain at small angles θ relative to $\hat{z}$ and that the distributions along the z direction can be ignored, so we may use the small angle approximations that $\theta \approx p_x/p_z$.

(b) *What is the nonequilibrium entropy per particle associated with the initial distribution $\rho_{\text{initial}}(x, p_x)$?* (You may ignore the Gibbs factor $N!$ and the quantum correction h^{2N}).

In beam dynamics, the *emittance* ε of a beam is defined by the volume in phase space occupied for the particles in the bunch. For Gaussian distributions like our initial bunch, the volume is fuzzy; we can use the entropy to generalize the emittance, defining $\varepsilon = \exp(S/k_B)$. Collisions and scattering can increase the emittance, but as long as the beam does not exchange entropy with the outside world the emittance cannot decrease.[78] The emittance is often recast in terms of the spread of angles θ of the beam particles with respect to the direction of motion $\hat{z}$.

(c) *Show that our definition of emittance agrees with the volume in phase space, for the case where $\rho(x, p_x)$ is constant in a volume ε_0 and zero elsewhere. Use the small angle approximation to recast our Gaussian initial distribution of momenta into a distribution $\rho_{\theta 0}(\theta_0)$ of initial angles θ_0 for our beam; show that the latter is also a Gaussian, and calculate its standard deviation $\sigma_{\theta 0}$. Write the entropy per particle S_0 and emittance ε_0 in terms of $\sigma_{\theta 0}$ and σ_x.* (Ignore the constant shift in the entropy[79] involving MV.) Lenses can focus a beam, reducing its σ_x, but this necessarily increases the angular spread σ_θ enough to keep the emittance the same. The emittance (and hence its entropy per particle) is the main measure of the quality of a particle beam.

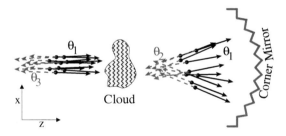

Fig. 5.27 Corner mirror. A bunch of particles (left, solid) passes through a scattering cloud (right, solid), bounces off a mirror made of corner reflectors (right, dashed), and passes back through the cloud (left, dashed). The corner mirror reverses the momentum of each particle as it hits the mirror. Insofar as the small change in position of the particle during the collision may be ignored, each particle trajectory should then retrace its path, reforming the original bunch.

---

[78]Emission of synchrotron radiation can dump entropy into the outside world, as can *stochastic cooling* methods that actively seek to reduce the momentum spread.

[79]The quantum correction h^{2N} that makes the phase-space volume unitless is no longer appropriate if we convert momenta to angles. We have seen in Exercise 7.3 that this correction only shifts the entropy by an overall constant.

We shall examine how $\rho(x,\theta)$, S, and ε evolve as the beam passes through the cloud, reflects off a *corner mirror* (Fig. 5.27) back to the cloud, and returns through the cloud. Our initial ρ_0, S_0, and ε_0 represent the distribution just as it enters the cloud. We shall calculate ρ_1, S_1, and ε_1 just as the particles leave the cloud to the right (before they spread out), ρ_2, S_2, and ε_2 just before they re-enter the cloud moving left, and then ρ_3, S_3, and ε_3 just after they return back through the cloud.

We start by calculating what happens to our beam as it enters a cloud. Let each particle suffer many small-angle collisions with the cloud, so its angle θ undergoes a random walk. For a particle entering the cloud with angle θ_0, the probability distribution for the angle $\theta_1 = \theta_0 + \Delta$ exiting the cloud, where the net scattering angle Δ has a Gaussian distribution

$$\rho_\Delta(\Delta) = \frac{1}{\sqrt{2\pi}\sigma_\Delta}\mathrm{e}^{-\Delta^2/2\sigma_\Delta^2}. \qquad (5.58)$$

Let $\sigma_\Delta = 10\sigma_{\theta 0}$, so the particle velocities leave the cloud much more spread out than they entered. We shall assume for simplicity that the cloud is thin, so the particle displacements x do not change significantly as they pass through. The cloud is also assumed time independent, and time-reversal invariant.[80] The scatterers are stationary on the time scale of the particle motions, and do not recoil when our particles glance off of them.

(d) *What will the probability distribution $\rho_1(x,\theta_1)$ of the angles be as the beam exits the cloud? How much has the entropy S_1 increased, given that $\sigma_\Delta = 10\sigma_{\theta 0}$? By what factor $\varepsilon_1/\varepsilon_0$ has the emittance increased?*

The particles spread out in position as they leave the cloud and approach the mirror. However, the velocities of the particles become closely related to their positions—those with positive angles θ_1 move upward, and with negative angles move downward in x. This spread is thus not fundamentally increasing the entropy: it could be reversed by an ordinary mirror, refocusing the particles back to their exit points.[81]

We use a *corner reflector* as a classical analogy to the phase-conjugate mirror. A particle impinging at any angle to a corner reflector will

be reflected back along its incoming momentum, displaced sideways a distance $\Delta x_{\mathrm{mirror}}$ of order the size L of the corner (Fig. 5.28).

Let us consider a corner mirror made up of tiny corners of size L small enough that we may ignore the errors in retracing the particle trajectories. Thus the particles will re-enter the cloud backward (from the right) at their departing positions x with reversed angle $\theta_1 + \pi$.

Fig. 5.28 Corner reflector. A right-angle corner will reflect incoming particles twice. If the collisions are perfectly elastic, and the corner has sides of length L, it returns particles with $p'_x = -p_x$ and $|x' - x| < \sqrt{2}L$. A 3D corner reflector was put in place on the moon by Apollo 11; a laser beam reflected off of it from Earth has allowed us to measure the distance to the moon to within $3\,\mathrm{cm}$ (one part in 10^8 of the total distance of $384,400\,\mathrm{km}$). Arrays of corner reflectors (as in Fig. 5.27) have been used in road signs, so the light from your headlights reflects backward toward your car.

(e) *Write the distribution $\rho_2(x,\theta_2)$ of the particles in phase space just before the bunch passes back through the cloud, in terms of ρ_0 and/or ρ_1. What is its entropy S_2 and the emittance ε_2? Have we succeeded in reassembling the compact, but velocity-scrambled packet?*

A more challenging task is to restore the velocity distribution of the bunch. Here we must use information stored in the cloud.

If the bunch was focused back through the cloud by a regular mirror, it would get even more clouded by the further random scattering. However, our corner mirror aims particles to re-enter the cloud (in our small L approximation) at the same position and reversed angle as it departed

---

[80]For lasers, frosted glass, etched or sandblasted to pit the surface, would serve as such a thin, time-independent cloud.

[81] In electron beams, Coulomb interactions (*space charge* effects) cause a similar correlation between the spread in positions and the distribution of momenta. Suitable electron lenses can sometimes be used to restore the compact beam—giving an apparent reduction in emittance.

the cloud. Assume the thin cloud scatters a particle coming in from the left at a given incoming angle by $\Delta(x) = \theta_1 - \theta_0$ that depends on the height x at which it hits the cloud. Our cloud is thin, so x does not change as it passes through the cloud. Our cloud is time-reversal invariant, so if the particle enters the cloud backward (from the right) at position x and reversed angle $\theta_1 + \pi$, it must be scattered by an angle $-\Delta(x)$ to $\theta_0 + \pi$—the reverse of the net scattering angle it suffered from the first passage.

(f) *Write the distribution $\rho_3(x, \theta_3)$ for the particles just after they pass back through the cloud, in terms of ρ_0, ρ_1, and/or ρ_2. What is its entropy? Have we violated the second law of thermodynamics, that entropy never decreases?*

The increase of entropy reflects our ignorance about the system. Under ordinary circumstances, scattering off of random disorder is naturally described as an irreversible loss of information. Our calculation of the entropy change in part (d) is perfectly justified, so long as we do not retain detailed information about exactly how the scattering arose. Our corner mirror, and actual phase conjugate mirrors, can usefully make use of the detailed information stored in the scattering potential. In that case, declaring that the entropy of the bunch has increased upon scattering in the cloud is premature.

There is a nice analogy with entanglement entropy, where a quantum system in a pure state can be separated into a subsystem of interest and an environment whose states can no longer be accessed (Think of an experiment on earth emitting photons into the sky, Exercise 7.26.) The density matrix of the subsystem has an entropy which reflects the information lost to the environment. This entropy cannot decrease unless the information in the environment can somehow be accessed (say, using a mirror that reflects the photons back to Earth).

Our calculations often assume that random disorder scrambles information after multiple scatterings have happened. But in some quantum systems, the effects of the time-reversed scattering paths can be of great importance. One example is *localization*: electrons in a disordered environment have quantum states that are not extended, but exponentially decay away from some center. Instead of diffusing outward with time, the localization happens because of *coherent backscattering*, quantum interference with their time-reversed paths. Another example is dirty superconductors, which maintain their performance even with large amounts of nonmagnetic elastic scattering; the Cooper pairs are formed between a state and its time-reversed partner. These important effects would be ignored if we assumed that the quantum particle loses its information as it suffers multiple scattering events in a random environment.

Free energies

6

In this chapter we explain how to study *parts* of statistical mechanical systems. If we ignore most of our system—agreeing not to ask questions about certain degrees of freedom—the statistical mechanical predictions about the remaining parts of our system are embodied in a new statistical ensemble and its associated *free energy*. These free energies usually make calculations easier and the physical behavior more comprehensible. What do we ignore?

We ignore the external world. Most systems are not isolated; they often can exchange energy, volume, or particles with an outside world in equilibrium (often called the *heat bath*). If the coupling to the external world is weak, we can remove it from consideration. The constant-temperature *canonical*[1] ensemble (Section 6.1) and the *Helmholtz* free energy arise from a bath which can exchange energy; the *grand canonical* ensemble (Section 6.3) and the grand free energy arise from baths which also exchange particles at fixed chemical potential. For large systems, these different ensembles predict the same average behavior (apart from tiny fluctuations); we could in principle do most calculations of interest at constant energy and particle number. However, calculations using the appropriate free energy can be much simpler (Section 6.2).

We ignore unimportant internal degrees of freedom. In studying (say) chemical reactions, magnets, or the motion of large mechanical objects, one is normally not interested in the motions of the individual atoms. To ignore them in mechanical systems, one introduces *friction* and *noise* (Section 6.5). By ignoring the atomic motions in chemical reactions, one derives reaction rate theory (Section 6.6).

We coarse-grain. Many systems are not homogeneous, because of initial conditions or boundary conditions; their properties vary in space and/or time. If these systems are locally near equilibrium, we can ignore the internal degrees of freedom in small volumes, *coarse-graining* our description by keeping only the continuum fields which describe the local state.[2] As an example, in Section 6.7 we will calculate the free energy density for the ideal gas, and use it to (again) derive the diffusion equation.

We will calculate free energies explicitly in several important cases in this chapter. Note that free energies are important tools, however, even for systems too complex to solve analytically. We provide these solvable examples in part to motivate later continuum derivations of free energies for systems where microscopic calculations are not feasible.

[1]**Canonical** (*Oxford English dictionary*): ...4. *gen.* Of the nature of a canon or rule: of admitted authority, excellence, or supremacy; authoritative; orthodox; accepted; standard. 5. *Math.* Furnishing, or according to, a general rule or formula.

[2]These are the *order parameter fields* that we will study in Chapter 9.

Statistical Mechanics: Entropy, Order Parameters, and Complexity. James P. Sethna, Oxford University Press (2021). © Oxford University Press. DOI:10.1093/oso/9780198865247.003.0006

6.1 The canonical ensemble

The canonical ensemble governs the equilibrium behavior of a system at fixed temperature. We defined the temperature in Section 3.3 by considering a total system comprised of two weakly coupled parts, with phase-space coordinates $(\mathbb{P}_1, \mathbb{Q}_1)$ and $(\mathbb{P}_2, \mathbb{Q}_2)$ that can exchange energy. We will now focus on the first of these two parts (the *system*); the second part (the *heat bath*) we will assume to be large. We are not interested in measuring any properties that involve the heat bath coordinates; we want a statistical ensemble for the system that averages over the relevant states of the bath (Fig. 6.1).

How does the probability that our system is in a state s depend upon its energy E_s? As we discussed in Section 3.3, the probability density that our system will be in the particular state s is proportional to the volume of the energy shell for our heat bath at bath energy $E - E_s$:

$$\rho(s) \propto \Omega_2(E - E_s) = \exp\left(S_2(E - E_s)/k_B\right) \tag{6.1}$$

since a state s gets a share of the microcanonical probability for each heat-bath partner it can coexist with at the fixed total energy E.

Let us compare the probability of two typical states A and B of our equilibrium system. We know that the energy fluctuations are small, and we assume that the heat bath is large. We can therefore assume that the inverse temperature $1/T_2 = \partial S_2/\partial E_2$ of the heat bath is constant in the range $(E - E_A, E - E_B)$. Hence,

$$\begin{aligned}
\rho(s_B)/\rho(s_A) &= \Omega_2(E - E_B)/\Omega_2(E - E_A) \\
&= e^{(S_2(E-E_B) - S_2(E-E_A))/k_B} = e^{(E_A - E_B)\,(\partial S_2/\partial E)/k_B} \\
&= e^{(E_A - E_B)/k_B T_2}.
\end{aligned} \tag{6.2}$$

This is the general derivation of the Boltzmann distribution; the probability of a particular system state[3] of energy E_s is

$$\rho(s) \propto \exp(-E_s/k_B T). \tag{6.3}$$

We know that the probability is normalized, so

$$\begin{aligned}
\rho(s) &= \exp(-E_s/k_B T) \Big/ \int \frac{d\mathbb{P}_1\, d\mathbb{Q}_1}{h^{3N_1}} \exp(-\mathcal{H}_1(\mathbb{P}_1, \mathbb{Q}_1)/k_B T) \\
&= \exp(-E_s/k_B T) \Big/ \sum_n \exp(-E_n/k_B T) \\
&= \exp(-E_s/k_B T)/Z,
\end{aligned} \tag{6.4}$$

where the normalization factor

$$Z(T, V, N) = \sum_n \exp(-E_n/k_B T) = \int \frac{d\mathbb{P}_1\, d\mathbb{Q}_1}{h^{3N_1}} \exp(-\mathcal{H}_1(\mathbb{P}_1, \mathbb{Q}_1)/k_B T) \tag{6.5}$$

is the *partition function*.[4]

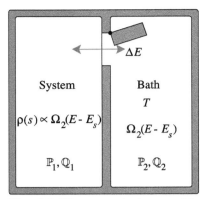

Fig. 6.1 The canonical ensemble describes equilibrium systems which can exchange energy with a heat bath. The bath is at temperature T. The probability of a state s of the system with energy E_s is $\rho(s) = \exp(-E_s/k_B T)/Z$. The thermodynamics of the canonical ensemble is embodied in the Helmholtz free energy $A(T, V, N) = E - TS$.

[3]For a classical system, this is instead the probability density at state s.

[4]To avoid blinding ourselves with integrals, we will write them as a continuous sum; $\int d\mathbb{P}\, d\mathbb{Q}/h^{3N} \to \sum_n$ for the rest of this chapter. This notation foreshadows quantum mechanics (Chapter 7), where for bound systems the energy levels are discrete; it will also be appropriate for lattice systems like the Ising model (Section 8.1), where we have integrated away all the continuous degrees of freedom. No complications arise from translating the sums for the equations in this chapter back into integrals over phase space.

Equation 6.4 is the definition of the canonical ensemble,[5] appropriate for calculating properties of systems which can exchange energy with an external world at temperature T.

The partition function Z is just the normalization factor that keeps the total probability summing to one. It may surprise you to discover that this normalization factor plays a central role in the theory. Indeed, most quantities of interest can be calculated in two different ways: as an explicit sum over states or in terms of derivatives of the partition function. Let us see how this works by using Z to calculate the mean energy, the specific heat, and the entropy of a general system.

Internal energy. To calculate the average internal energy of our system[6] $\langle E \rangle$, we weight each state by its probability. Writing $\beta = 1/(k_B T)$,

$$\langle E \rangle = \sum_n E_n P_n = \frac{\sum_n E_n e^{-\beta E_n}}{Z} = -\frac{\partial Z/\partial \beta}{Z}$$

$$= -\partial \log Z/\partial \beta. \tag{6.11}$$

[6] The angle brackets represent canonical averages.

Specific heat. Let c_v be the specific heat per particle at constant volume. (The specific heat is the energy needed to increase the temperature by one unit, $\partial \langle E \rangle / \partial T$.) Using eqn 6.11, we get

$$N c_v = \frac{\partial \langle E \rangle}{\partial T} = \frac{\partial \langle E \rangle}{\partial \beta} \frac{\mathrm{d}\beta}{\mathrm{d}T} = -\frac{1}{k_B T^2} \frac{\partial \langle E \rangle}{\partial \beta} = \frac{1}{k_B T^2} \frac{\partial^2 \log Z}{\partial \beta^2}. \tag{6.12}$$

---

[5] A formal method of deriving the canonical ensemble is as a *partial trace*, removing the bath degrees of freedom from a microcanonical ensemble. To calculate the expectation of an operator B that depends only on system coordinates $(\mathbb{P}_1, \mathbb{Q}_1)$, we start by averaging over the energy shell in the entire space (eqn 3.5), including both the system coordinates and the bath coordinates $(\mathbb{P}_2, \mathbb{Q}_2)$:

$$\Omega(E) = \frac{1}{\delta E} \int_{E < \mathcal{H}_1 + \mathcal{H}_2 < E + \delta E} \mathrm{d}\mathbb{P}_1 \, \mathrm{d}\mathbb{Q}_1 \, \mathrm{d}\mathbb{P}_2 \, \mathrm{d}\mathbb{Q}_2 = \int \mathrm{d}E_1 \, \Omega_1(E_1) \Omega_2(E - E_1). \tag{6.6}$$

$$\langle B \rangle = \frac{1}{\Omega(E)\delta E} \int_{E < \mathcal{H}_1 + \mathcal{H}_2 < E + \delta E} \mathrm{d}\mathbb{P}_1 \, \mathrm{d}\mathbb{Q}_1 \, B(\mathbb{P}_1, \mathbb{Q}_1) \, \mathrm{d}\mathbb{P}_2 \, \mathrm{d}\mathbb{Q}_2 = \frac{1}{\Omega(E)} \int \mathrm{d}\mathbb{P}_1 \, \mathrm{d}\mathbb{Q}_1 \, B(\mathbb{P}_1, \mathbb{Q}_1) \, \Omega_2(E - \mathcal{H}_1(\mathbb{P}_1, \mathbb{Q}_1)). \tag{6.7}$$

(The indistinguishability factors and Planck's constants in eqn 3.54 complicate the discussion here in inessential ways.) Again, if the heat bath is large the small variations $E_1 - \langle E_1 \rangle$ will not change its temperature. $1/T_2 = \partial S_2 / \partial E_2$ being fixed implies $\partial \Omega_2(E - E_1)/\partial E_1 = -(1/k_B T)\Omega_2$; solving this differential equation gives

$$\Omega_2(E - E_1) = \Omega_2(E - \langle E_1 \rangle) \exp(-(E_1 - \langle E_1 \rangle)/k_B T). \tag{6.8}$$

This gives us

$$\Omega(E) = \int \mathrm{d}E_1 \, \Omega_1(E_1) \Omega_2(E - \langle E_1 \rangle) \, e^{-(E_1 - \langle E_1 \rangle)/k_B T} = \Omega_2(E - \langle E_1 \rangle) \, e^{\langle E_1 \rangle/k_B T} \int \mathrm{d}E_1 \, \Omega_1(E_1) \, e^{-E_1/k_B T}$$

$$= \Omega_2(E - \langle E_1 \rangle) \, e^{\langle E_1 \rangle/k_B T} Z \tag{6.9}$$

and

$$\langle B \rangle = \frac{\int \mathrm{d}\mathbb{P}_1 \, \mathrm{d}\mathbb{Q}_1 \, B(\mathbb{P}_1, \mathbb{Q}_1) \, \Omega_2(E - \langle E_1 \rangle) \, e^{-(\mathcal{H}_1(\mathbb{P}_1, \mathbb{Q}_1) - \langle E_1 \rangle)/k_B T}}{\Omega_2(E - \langle E_1 \rangle) \, e^{\langle E_1 \rangle/k_B T} Z} = \frac{1}{Z} \int \mathrm{d}\mathbb{P}_1 \, \mathrm{d}\mathbb{Q}_1 \, B(\mathbb{P}_1, \mathbb{Q}_1) \exp(-\mathcal{H}_1(\mathbb{P}_1, \mathbb{Q}_1)/k_B T). \tag{6.10}$$

By explicitly doing the integrals over $\mathbb{P}_2$ and $\mathbb{Q}_2$, we have turned a microcanonical calculation into the canonical ensemble (eqn 6.4). Our calculation of the momentum distribution $\rho(p_1)$ in Section 3.2.2 was precisely of this form; we integrated out all the other degrees of freedom, and were left with a Boltzmann distribution for the x-momentum of particle number one. This process is called *integrating out* the degrees of freedom for the heat bath, and is the general way of creating free energies.

We can expand the penultimate form of this formula into a sum, finding the intriguing result

$$
\begin{aligned}
Nc_v &= -\frac{1}{k_B T^2}\frac{\partial \langle E \rangle}{\partial \beta} = -\frac{1}{k_B T^2}\frac{\partial}{\partial \beta}\frac{\sum E_n e^{-\beta E_n}}{\sum e^{-\beta E_n}} \\
&= -\frac{1}{k_B T^2}\left[\frac{\left(\sum E_n e^{-\beta E_n}\right)^2}{Z^2} + \frac{\sum -E_n{}^2 e^{-\beta E_n}}{Z}\right] \\
&= \frac{1}{k_B T^2}\left[\langle E^2 \rangle - \langle E \rangle^2\right] = \frac{\sigma_E{}^2}{k_B T^2},
\end{aligned}
\tag{6.13}
$$

[7]We have used the standard trick $\sigma_E^2 = \langle (E - \langle E \rangle)^2 \rangle = \langle E^2 \rangle - 2\langle E \langle E \rangle\rangle + \langle E \rangle^2 = \langle E^2 \rangle - \langle E \rangle^2$, since $\langle E \rangle$ is just a constant that can be pulled out of the ensemble average.

[8]We will properly introduce susceptibilities (linear responses) and other remarkable relations in Chapter 10.

where σ_E is the RMS fluctuation[7] in the energy of our system at constant temperature. Equation 6.13 is remarkable;[8] it is a relationship between a macroscopic susceptibility (c_v, the energy changes when the temperature is perturbed) and a microscopic fluctuation (σ_E, the energy fluctuation in thermal equilibrium). In general, fluctuations can be related to responses in this fashion. These relations are extremely useful, for example, in extracting susceptibilities from numerical simulations. For example, to measure the specific heat there is no need to make small changes in the temperature and measure the heat flow; just watch the energy fluctuations in equilibrium (Exercises 3.8 and 8.1).

Are results calculated using the canonical ensemble the same as those computed from our original microcanonical ensemble? Equation 6.13 tells us that the energy fluctuations per particle

$$
\sigma_E/N = \sqrt{\langle E^2 \rangle - \langle E \rangle^2}/N = \sqrt{(k_B T)(c_v T)}/\sqrt{N}.
\tag{6.14}
$$

are tiny; they are $1/\sqrt{N}$ times the geometric mean of two microscopic energies: $k_B T$ (two-thirds the kinetic energy per particle) and $c_v T$ (the energy per particle to heat from absolute zero, if the specific heat were temperature independent). These tiny fluctuations will not change the properties of a macroscopic system; the constant energy (microcanonical) and constant temperature (canonical) ensembles predict the same behavior.

Entropy. Using the general statistical mechanical formula[9] for the entropy 5.20, we find

[9]Alternatively, we could use the microcanonical definition of the entropy of the entire system and eqn 6.8 to show

$$
\begin{aligned}
S &= k_B \log \int dE_1\, \Omega_1(E_1)\Omega_2(E - E_1) \\
&= k_B \log \int dE_1\, \Omega_1(E_1)\Omega_2(E - \langle E_1 \rangle) \\
&\qquad \exp(-(E_1 - \langle E_1 \rangle)/k_B T) \\
&= k_B \log \Omega_2(E - \langle E_1 \rangle) \\
&\quad + k_B \log\left(\exp(\langle E_1 \rangle/k_B T)\right) \\
&\quad + k_B \log \int dE_1\, \Omega_1(E_1) \\
&\qquad\qquad \exp(-E_1/k_B T) \\
&= k_B \log \Omega_2(E_2) \\
&\quad + k_B \beta \langle E_1 \rangle + k_B \log Z_1 \\
&= S_2 + \langle E_1 \rangle/T - A_1/T,
\end{aligned}
$$

so

$$
\begin{aligned}
S_1 &= \langle E_1 \rangle/T + k_B \log Z_1 \\
&= \langle E_1 \rangle/T - A_1/T,
\end{aligned}
\tag{6.15}
$$

avoiding the use of the nonequilibrium entropy to derive the same result.

[10]Historically, thermodynamics and the various free energies came before statistical mechanics.

$$
\begin{aligned}
S &= -k_B \sum P_n \log P_n = -k_B \sum \frac{\exp(-\beta E_n)}{Z}\log\left(\frac{\exp(-\beta E_n)}{Z}\right) \\
&= -k_B \frac{\sum \exp(-\beta E_n)(-\beta E_n - \log Z)}{Z} \\
&= k_B \beta \langle E \rangle + k_B \log Z \frac{\sum \exp(-\beta E_n)}{Z} = \frac{\langle E \rangle}{T} + k_B \log Z.
\end{aligned}
\tag{6.16}
$$

Notice that the formulæ for $\langle E \rangle$, c_v, and S all involve $\log Z$ and its derivatives. This motivates us[10] to define a *free energy* for the canonical ensemble, called the *Helmholtz* free energy:

$$
A(T, V, N) = -k_B T \log Z = \langle E \rangle - TS.
\tag{6.17}
$$

The entropy is minus the derivative of A with respect to T. Explicitly,

$$
\left.\frac{\partial A}{\partial T}\right|_{N,V} = -\frac{\partial k_B T \log Z}{\partial T} = -k_B \log Z - k_B T \frac{\partial \log Z}{\partial \beta}\frac{\partial \beta}{\partial T}
$$

$$
= -k_B \log Z - k_B T \langle E \rangle /(k_B T^2) = -k_B \log Z - \langle E \rangle / T
$$

$$
= -S. \tag{6.18}
$$

Why is it called a free energy? First, $k_B T$ gives it the dimensions of an energy. Second, it is the energy available (free) to do work. A heat engine drawing energy $E = Q_1$ from a hot bath that must discharge an entropy $S = Q_2/T_2$ into a cold bath can do work $W = Q_1 - Q_2 = E - T_2 S$; $A = E - TS$ is the energy free to do useful work (Section 5.1).

We see that $\exp(-A(T,V,N)/k_B T) = Z$, quite analogous to the Boltzmann weight $\exp(-E_s/k_B T)$. The former is the phase-space volume or weight contributed by all states of a given N and V; the latter is the weight associated with a particular state s. In general, free energies $F(\mathbb{X})$ will remove all degrees of freedom except for certain constraints $\mathbb{X}$. The phase-space volume consistent with the constraints $\mathbb{X}$ is $\exp(-F(\mathbb{X})/k_B T)$.

6.2 Uncoupled systems and canonical ensembles

The canonical ensemble is typically much more convenient for doing calculations because, for systems in which the Hamiltonian splits into uncoupled components, the partition function factors into pieces that can be computed separately. Let us show this.

Suppose we have a system with two weakly interacting subsystems L and R, both connected to a heat bath at $\beta = 1/k_B T$ (Fig. 6.2). The states for the whole system are pairs of states (s_i^L, s_j^R) from the two subsystems, with energies E_i^L and E_j^R, respectively. The partition function for the whole system is

$$
Z = \sum_{ij} \exp\left(-\beta(E_i^L + E_j^R)\right) = \sum_{ij} e^{-\beta E_i^L} e^{-\beta E_j^R}
$$

$$
= \left(\sum_i e^{-\beta E_i^L}\right)\left(\sum_j e^{-\beta E_j^R}\right)
$$

$$
= Z^L Z^R. \tag{6.19}
$$

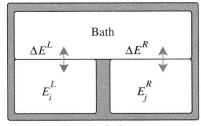

Fig. 6.2 Uncoupled systems attached to a common heat bath. Calculating the properties of two weakly coupled subsystems is easier in the canonical ensemble than in the microcanonical ensemble. This is because energy in one subsystem can be exchanged with the bath, and does not affect the energy of the other subsystem.

Thus partition functions factor for uncoupled subsystems. The Helmholtz free energy therefore adds

$$
A = -k_B T \log Z = -k_B T \log(Z^L \cdot Z^R) = A^L + A^R, \tag{6.20}
$$

as does the entropy, average energy, and other *extensive* properties that one expects to scale with the size of the system.

This is much simpler than the same calculation would be in the microcanonical ensemble! In a microcanonical ensemble, each subsystem would compete with the other for the available total energy. Even though two subsystems are uncoupled (the energy of one is independent of the state of the other) the microcanonical ensemble intermingles them in the calculation. By allowing each to draw energy from a large heat bath, the canonical ensemble allows uncoupled subsystems to become independent calculations.

We can now immediately solve several important examples of uncoupled systems.

Ideal gas. The different atoms in an ideal gas are uncoupled. The partition function for N distinguishable ideal gas atoms of mass m in a cube of volume $V = L^3$ factors into a product over each degree of freedom α:

$$Z_{\text{ideal}}^{\text{dist}} = \prod_{\alpha=1}^{3N} \frac{1}{h} \int_0^L dq_\alpha \int_{-\infty}^\infty dp_\alpha \, e^{-\beta p_\alpha^2 / 2m_\alpha} = \left(\frac{L}{h} \sqrt{\frac{2\pi m}{\beta}} \right)^{3N}$$
$$= (L\sqrt{2\pi m k_B T / h^2})^{3N} = (L/\lambda)^{3N}.$$

Here

$$\lambda = h/\sqrt{2\pi m k_B T} = \sqrt{2\pi \hbar^2 / m k_B T} \tag{6.21}$$

is again the thermal de Broglie wavelength (eqn 3.59).

The mean internal energy in the ideal gas is

$$\langle E \rangle = -\frac{\partial \log Z_{\text{ideal}}}{\partial \beta} = -\frac{\partial}{\partial \beta} \log(\beta^{-3N/2}) = \frac{3N}{2\beta} = \frac{3}{2} N k_B T, \quad (6.22)$$

giving us the equipartition theorem for momentum without our needing to find volumes of spheres in $3N$ dimensions (Section 3.2.2).

For N undistinguished particles, we have counted each real configuration $N!$ times for the different permutations of particles, so we must divide $Z_{\text{ideal}}^{\text{dist}}$ by $N!$ just as we did for the phase-space volume Ω in Section 3.5:

$$Z_{\text{ideal}}^{\text{indist}} = (L/\lambda)^{3N}/N!. \tag{6.23}$$

This does not change the internal energy, but does affect the Helmholtz free energy:

$$\begin{aligned}
A_{\text{ideal}}^{\text{indist}} &= -k_B T \log\left((L/\lambda)^{3N}/N! \right) \\
&= -N k_B T \log(V/\lambda^3) + k_B T \log(N!) \\
&\sim -N k_B T \log(V/\lambda^3) + k_B T (N \log N - N) \\
&= -N k_B T \left(\log(V/N\lambda^3) + 1 \right) \\
&= N k_B T \left(\log(\rho \lambda^3) - 1 \right),
\end{aligned} \tag{6.24}$$

where $\rho = N/V$ is the average density, and we have used Stirling's formula $\log(N!) \approx N \log N - N$.

Finally, the entropy of N undistinguished particles, in the canonical ensemble, is

$$S = -\frac{\partial A}{\partial T} = -Nk_B \left(\log(\rho\lambda^3) - 1\right) - Nk_BT\frac{\partial \log T^{-3/2}}{\partial T}$$
$$= Nk_B \left(5/2 - \log(\rho\lambda^3)\right), \tag{6.25}$$

as we derived (with much more effort) using the microcanonical ensemble (eqn 3.58).

Classical harmonic oscillator. Electromagnetic radiation, the vibrations of atoms in solids, and the excitations of many other systems near their equilibria can be approximately described as a set of uncoupled harmonic oscillators.[11] Using the canonical ensemble, the statistical mechanics of these systems thus decomposes into a calculation for each mode separately.

A harmonic oscillator of mass m and frequency ω has a total energy

$$\mathcal{H}(p, q) = p^2/2m + m\omega^2 q^2/2. \tag{6.29}$$

The partition function for one such oscillator is (using $\hbar = h/2\pi$)

$$Z = \int_{-\infty}^{\infty} dq \int_{-\infty}^{\infty} dp \, \frac{1}{h} \, e^{-\beta(p^2/2m + m\omega^2 q^2/2)} = \frac{1}{h}\sqrt{\frac{2\pi}{\beta m\omega^2}}\sqrt{\frac{2\pi m}{\beta}}$$
$$= \frac{1}{\beta\hbar\omega}. \tag{6.30}$$

Hence the Helmholtz free energy for the classical oscillator is

$$A_\omega(T) = -k_BT \log Z = k_BT \log(\hbar\omega/k_BT), \tag{6.31}$$

the internal energy is

$$\langle E \rangle_\omega(T) = -\frac{\partial \log Z}{\partial \beta} = \frac{\partial}{\partial \beta}(\log \beta + \log \hbar\omega) = 1/\beta = k_BT, \tag{6.32}$$

and hence $c_v = \partial\langle E \rangle/\partial T = k_B$. This is the general statement of the equipartition theorem:[12] each harmonic degree of freedom (p and q count as two) in a classical equilibrium system has mean energy $\frac{1}{2}k_BT$.

Classical velocity distributions. One will notice both for the ideal gas and for the harmonic oscillator that each component of the momentum contributed a factor $\sqrt{2\pi m/\beta}$ to the partition function. As we promised in Section 3.2.2, this will happen in any classical system where the momenta are uncoupled to the positions; that is, where the momentum parts of the energy are of the standard form $\sum_\alpha p_\alpha^2/2m_\alpha$ (non quantum, nonrelativistic, nonmagnetic particles). In these systems the velocity distribution is always Maxwellian (eqn 1.2), independent of what configuration the positions have.

This is a powerful, counterintuitive truth. The equilibrium velocity distribution of atoms crossing barriers in chemical reactions (Section 6.6) or surrounding mountain tops is the same as those in the low-energy valleys.[13] Each atom does slow down as it climbs, but only the formerly energetic ones make it to the top. The population density at the top is thus smaller, but the kinetic energy distribution remains the same (Exercise 6.1).

[11]For example, at temperatures low compared to the melting point a solid or molecule with an arbitrary many-body interaction potential $\mathcal{V}(\mathbb{Q})$ typically only makes small excursions about the minimum $\mathbb{Q}_0$ of the potential. We expand about this minimum, giving us

$$\mathcal{V}(\mathbb{Q}) \approx \mathcal{V}(\mathbb{Q}_0) + \sum_\alpha (\mathbb{Q} - \mathbb{Q}_0)_\alpha \partial_\alpha \mathcal{V}$$
$$+ \sum_{\alpha,\beta} \frac{1}{2} (\mathbb{Q} - \mathbb{Q}_0)_\alpha (\mathbb{Q} - \mathbb{Q}_0)_\beta$$
$$\times \partial_\alpha \partial_\beta \mathcal{V} + \dots . \tag{6.26}$$

Since the potential is a minimum at $\mathbb{Q}_0$, the gradient of the potential must be zero, so second term on the right-hand side must vanish. The third term is a large $3N \times 3N$ quadratic form, which we may diagonalize by converting to *normal modes* q_k. (If the masses of the atoms are not all the same, one must rescale the components of $\mathbb{Q} - \mathbb{Q}_0$ by the square root of the corresponding mass before diagonalizing.) In terms of these normal modes, the Hamiltonian is a set of uncoupled harmonic oscillators:

$$\mathcal{H} = \sum_k p_k^2/2m + m\omega_k^2 q_k^2/2. \tag{6.27}$$

At high enough temperatures that quantum mechanics can be ignored, we can then use eqn 6.30 to find the total partition function for our harmonic system

$$Z = \prod_k Z_k = \prod_k (1/\beta\hbar\omega_k). \tag{6.28}$$

(In Section 7.2 we will consider the quantum harmonic oscillator, which often gives quite an accurate theory for atomic vibrations at all temperatures up to the melting point.)

[12]We saw the theorem for p in Section 3.2.2.

[13]Mountain tops would not be colder if the atmosphere were in equilibrium.

Fig. 6.3 The grand canonical ensemble describes equilibrium systems which can exchange energy and particles with a heat bath. The probability of a state s of the system with N_s particles and energy E_s is $\rho(s) = \exp\left(-(E_s - \mu N_s)/(k_B T)\right)/Z$. The thermodynamics of the grand canonical ensemble is embodied in the grand free energy $\Phi(T, V, \mu)$.

6.3 Grand canonical ensemble

The grand canonical ensemble allows one to decouple the calculations of systems which can exchange both energy and particles with their environment.

Consider our system in a state s with energy E_s and number N_s, together with a bath with energy $E_2 = E - E_s$ and number $N_2 = N - N_s$ (Fig. 6.3). By analogy with eqn 6.3, the probability density that the system will be in state s is proportional to

$$
\begin{aligned}
\rho(s) &\propto \Omega_2(E - E_s, N - N_s) \\
&= \exp\left(\left(S_2(E - E_s, N - N_s)\right)/k_B\right) \\
&\propto \exp\left(\left(-E_s \frac{\partial S_2}{\partial E} - N_s \frac{\partial S_2}{\partial N}\right)/k_B\right) \\
&= \exp\left(-E_s/k_B T + N_s \mu/k_B T\right) \\
&= \exp\left(-(E_s - \mu N_s)/k_B T\right),
\end{aligned}
\tag{6.33}
$$

where

$$
\mu = -T \partial S/\partial N
\tag{6.34}
$$

is the chemical potential. Notice the factor of $-T$; this converts the entropy change into an energy change. Using $\mathrm{d}E = T\,\mathrm{d}S - P\,\mathrm{d}V + \mu\,\mathrm{d}N$, we see that $\mu = (\partial E/\partial N)|_{S,V}$ is precisely the energy change needed to add an additional particle adiabatically and keep the $(N+1)$-particle system in equilibrium. At low temperatures a given system will fill with particles until the energy needed to jam in another particle reaches μ, and then exhibit thermal number fluctuations around that filling.

Again, just as for the canonical ensemble, there is a normalization factor called the grand partition function

$$
\Xi(T, V, \mu) = \sum_n e^{-(E_n - \mu N_n)/k_B T};
\tag{6.35}
$$

the probability density of state s_i is $\rho(s_i) = e^{-(E_i - \mu N_i)/k_B T}/\Xi$. There is a grand free energy

$$
\Phi(T, V, \mu) = -k_B T \log(\Xi) = \langle E \rangle - TS - \mu N
\tag{6.36}
$$

analogous to the Helmholtz free energy $A(T, V, N)$. In Exercise 6.8 you can derive the Euler relation $E = TS - PV + \mu N$, and hence show that $\Phi(T, \mu, V) = -PV$.

Partial traces.[14] Note in passing that we can write the grand canonical partition function as a sum over canonical partition functions. Let us separate the sum over states n of our system into a double sum—an inner restricted sum[15] over states of fixed number of particles M in the

[14]The classical mechanics integrals over phase space become traces over states in Hilbert space in quantum mechanics. Removing some of the degrees of freedom in quantum mechanics is done by a partial trace over the states (Chapter 7 and Exercises 7.26 and 7.27). The name *partial trace* for removing some of the degrees of freedom has also become standard in classical statistical physics (as in note 5 on p. 141).

[15]This restricted sum is said to *integrate over* the internal degrees of freedom ℓ_M.

system and an outer sum over M. Let $s_{\ell_M,M}$ have energy $E_{\ell_M,M}$, so

$$
\begin{aligned}
\Xi(T,V,\mu) &= \sum_M \sum_{\ell_M} e^{-(E_{\ell_M,M}-\mu M)/k_BT} \\
&= \sum_M \left(\sum_{\ell_M} e^{-E_{\ell_M,M}/k_BT} \right) e^{\mu M/k_BT} \\
&= \sum_M Z(T,V,M)\, e^{\mu M/k_BT} \\
&= \sum_M e^{-(A(T,V,M)-\mu M)/k_BT}.
\end{aligned}
\tag{6.37}
$$

Again, notice how the Helmholtz free energy in the last equation plays exactly the same role as the energy plays in eqn 6.35; $\exp(-E_n/k_BT)$ is the probability of the system being in a particular system state n, while $\exp(-A(T,V,M)/k_BT)$ is the probability of the system having any state with M particles.

Using the grand canonical ensemble. The grand canonical ensemble is particularly useful for noninteracting quantum systems (Chapter 7). There each energy eigenstate can be thought of as a separate subsystem, independent of the others except for the competition between eigenstates for energy and particles. A closely related ensemble emerges in chemical reactions (Section 6.6).

For now, to illustrate how to use the grand canonical ensemble, let us compute the number fluctuations. The expected number of particles for a general system is

$$
\langle N \rangle = \frac{\sum_m N_m e^{-(E_m-\mu N_m)/k_BT}}{\sum_m e^{-(E_m-\mu N_m)/k_BT}} = \frac{k_BT}{\Xi}\frac{\partial \Xi}{\partial \mu} = -\frac{\partial \Phi}{\partial \mu}.
\tag{6.38}
$$

Just as the fluctuations in the energy were related to the specific heat (the rate of change of energy with temperature, Section 6.1), the number fluctuations are related to the rate of change of particle number with chemical potential:

$$
\begin{aligned}
\frac{\partial \langle N \rangle}{\partial \mu} &= \frac{\partial}{\partial \mu} \frac{\sum_m N_m e^{-(E_m-\mu N_m)/k_BT}}{\Xi} \\
&= -\frac{1}{\Xi^2} \frac{\left(\sum_m N_m e^{-(E_m-\mu N_m)/k_BT} \right)^2}{k_BT} \\
&\quad + \frac{1}{k_BT} \frac{\sum_m N_m{}^2 e^{-(E_m-\mu N_m)/k_BT}}{\Xi} \\
&= \frac{\langle N^2 \rangle - \langle N \rangle^2}{k_BT} = \frac{\langle (N-\langle N \rangle)^2 \rangle}{k_BT}.
\end{aligned}
\tag{6.39}
$$

6.4   What is thermodynamics?

Thermodynamics and statistical mechanics historically were closely tied, and often they are taught together. What is thermodynamics?

(0) **Thermodynamics** (*Oxford English dictionary*): The theory of the relations between heat and mechanical energy, and of the conversion of either into the other.

(1) **Thermodynamics is the theory that emerges from statistical mechanics in the limit of large systems.** Statistical mechanics originated as a derivation of thermodynamics from an atomistic microscopic theory (somewhat before the existence of atoms was universally accepted). Thermodynamics can be viewed as statistical mechanics in the limit[16] as the number of particles $N \to \infty$. When we calculate the relative fluctuations in properties like the energy or the pressure and show that they vanish like $1/\sqrt{N}$, we are providing a microscopic justification for thermodynamics. Thermodynamics is the statistical mechanics of near-equilibrium systems when one ignores the fluctuations.

[16]The limit $N \to \infty$ is thus usually called the *thermodynamic limit*.

In this text, we will summarize many of the important methods and results of traditional thermodynamics in the exercises (see the index of this book under "Exercises, thermodynamics"). Our discussions of order parameters (Chapter 9) will be providing thermodynamic laws, broadly speaking, for a wide variety of states of matter.

Statistical mechanics has a broader purview than thermodynamics. Particularly in applications to other fields like information theory, dynamical systems, and complexity theory, statistical mechanics describes many systems where the emergent behavior does not have a recognizable relation to thermodynamics.

(2) **Thermodynamics is a self-contained theory.** Thermodynamics can be developed as an axiomatic system. It rests on the so-called three laws of thermodynamics, which for logical completeness must be supplemented by a *zeroth law*. Informally, they are as follows.

(0) *Transitivity of equilibria.* If two systems are in equilibrium with a third, they are in equilibrium with one another.

(1) *Conservation of energy.* The total energy of an isolated system, including the heat energy, is constant.

(2) *Entropy always increases.* An isolated system may undergo irreversible processes, whose effects can be measured by a state function called the entropy.

(3) *Entropy goes to zero at absolute zero.* The entropy per particle of any two large equilibrium systems will approach the same value[17] as the temperature approaches absolute zero.

[17]In classical statistical mechanics, this value is set to zero by measuring phase-space volume in units of h^{3N} (Section 3.5).

The zeroth law (transitivity of equilibria) becomes the basis for defining the temperature. Our statistical mechanics derivation of the temperature in Section 3.3 provides the microscopic justification of the zeroth law: systems that can only exchange heat energy are in equilibrium with one another when they have a common value of $1/T = (\partial S/\partial E)|_{V,N}$.

The first law (conservation of energy) is a fundamental principle of physics. Thermodynamics automatically inherits it from the microscopic theory. Historically, the thermodynamic understanding of how work transforms into heat was important in establishing that energy is conserved. Careful arguments about the energy transfer due to heat flow and mechanical work[18] are central to thermodynamics.

The second law (entropy always increases) is the heart of thermodynamics,[19] and was the theme of Chapter 5.

The third law (entropy goes to zero at $T = 0$, also known as Nernst's theorem) basically reflects the fact that quantum systems at absolute zero are in a ground state. Since the number of ground states of a quantum system typically is small[20] and the number of particles is large, equilibrium systems at absolute zero have zero entropy per particle.

The laws of thermodynamics have been written in many equivalent ways.[21] Carathéodory [37, 38], for example, states the second law as

> There are states of a system, differing infinitesimally from a given state, which are unattainable from that state by any quasi-static adiabatic[22] process.

The axiomatic form of the subject has attracted the attention of mathematicians ever since Carathéodory. In this text, we will not attempt to derive properties axiomatically or otherwise from the laws of thermodynamics; we focus on statistical mechanics.

(3) **Thermodynamics is a zoo of partial derivatives, transformations, and relations.** More than any other field of science, the thermodynamics literature seems filled with partial derivatives and tricky relations between varieties of physical quantities.

This is in part because there are several alternative free energies to choose among. For studying molecular systems one has not only the entropy (or the internal energy), the Helmholtz free energy, and the grand free energy, but also the Gibbs free energy, the enthalpy, and a number of others. There are corresponding free energies for studying magnetic systems, where instead of particles one studies the local magnetization or spin. There appears to be little consensus across textbooks on the symbols or even the names for these various free energies (see the inside front cover of this text).

How do we transform from one free energy to another? Let us write out the Helmholtz free energy in more detail:

$$A(T, V, N) = E - TS(E, V, N). \tag{6.40}$$

The terms on the right-hand side of the equation involve four variables: T, V, N, and E. Why is A not a function also of E? Consider the derivative of $A = E_s - T_b S_s(E_s)$ with respect to the energy E_s of the system, at fixed bath temperature T_b:

$$\partial A/\partial E_s = 1 - T_b \partial S_s/\partial E_s = 1 - T_b/T_s. \tag{6.41}$$

Since A represents the system in equilibrium with the bath, the temperature of the system and the bath must agree, and hence $\partial A/\partial E = 0$; A is independent of E. Physically, energy is transferred until A is a minimum; E is no longer an independent variable. This is an example

[18]As we saw in our analysis of the Carnot cycle in Section 5.1.

[19]In *The Two Cultures*, C. P. Snow suggests being able to describe the second law of thermodynamics is to science as having read a work of Shakespeare is to the arts. (Some in non-English speaking cultures may wish to object.) Remembering which law is which number is not of great import, but the concept of entropy and its inevitable increase is indeed central.

[20]Some systems have multiple degenerate ground states, but the number of such states is typically constant or slowly growing with system size, so the entropy per particle goes to zero. Glasses have large residual entropy, (Section 5.2.2), but are not in equilibrium.

[21]Occasionally you hear the first and second laws stated: (1) you can't win; and (2) you can't break even. Popular versions of the zeroth and third laws are not as compelling.

[22]Carathéodory is using the term *adiabatic* just to exclude heat flow; we use it to also imply infinitely slow (quasi-static) transitions.

of a *Legendre transformation* (Exercise 6.7). Legendre transformations allow one to change from one type of energy or free energy to another, by changing from one set of independent variables (here E, V, and N) to another (T, V, and N).

Thermodynamics seems cluttered in part also because it is so powerful. Almost any macroscopic property of interest can be found by taking derivatives of the free energy. First derivatives of the entropy, energy, or free energy give properties like the temperature and pressure. Thermodynamics introduces a condensed notation to help organize these derivatives. For example,[23]

$$dE = T\,dS - P\,dV + \mu\,dN \tag{6.42}$$

basically asserts that $E(S, V, N)$ satisfies eqns 3.29, 3.36, and 3.38:

$$\left.\frac{\partial E}{\partial S}\right|_{N,V} = T, \quad \left.\frac{\partial E}{\partial V}\right|_{N,S} = -P, \quad \text{and} \quad \left.\frac{\partial E}{\partial N}\right|_{V,S} = \mu. \tag{6.43}$$

The corresponding equation for the Helmholtz free energy $A(T, V, N)$ is

$$\begin{aligned} dA = d(E - TS) &= dE - T\,dS - S\,dT \\ &= -S\,dT - P\,dV + \mu\,dN, \end{aligned} \tag{6.44}$$

which satisfies

$$\left.\frac{\partial A}{\partial T}\right|_{N,V} = -S, \quad \left.\frac{\partial A}{\partial V}\right|_{N,T} = -P, \quad \text{and} \quad \left.\frac{\partial A}{\partial N}\right|_{V,T} = \mu. \tag{6.45}$$

Similarly, systems at constant temperature and pressure (for example, most biological and chemical systems) minimize the *Gibbs free energy* (Fig. 6.4)

$$G(T, P, N) = E - TS + PV, \quad dG = -S\,dT + V\,dP + \mu\,dN. \tag{6.46}$$

Systems at constant entropy and pressure minimize the *enthalpy*

$$H(S, P, N) = E + PV, \quad dH = T\,dS + V\,dP + \mu\,dN, \tag{6.47}$$

and, as noted in Section 6.3, systems at constant temperature, volume, and chemical potential are described by the grand free energy

$$\Phi(T, V, \mu) = E - TS - \mu N, \quad d\Phi = -S\,dT - P\,dV - N\,d\mu. \tag{6.48}$$

There are also many tricky, unintuitive relations in thermodynamics. The first derivatives must agree around a tiny triangle (Fig. 3.4), yielding a relation between their products (eqn 3.33, Exercise 3.10). Second derivatives of the free energy give properties like the specific heat, the bulk modulus, and the magnetic susceptibility. The second derivatives must be symmetric ($\partial^2/\partial x \partial y = \partial^2/\partial y \partial x$), giving Maxwell relations between what naively seem like different susceptibilities (Exercise 3.11). There are further tricks involved with taking derivatives in terms of

[23]These formulæ have precise meanings in differential geometry, where the terms dX are *differential forms*. Thermodynamics distinguishes between *exact* differentials like dS and *inexact* differentials like work and heat which are not derivatives of a state function, but path-dependent quantities. Mathematicians have *closed* and *exact* differential forms, which (in a simply-connected space) both correspond to the exact differentials in thermodynamics. The relation between closed and exact forms gives a type of *cohomology theory*, etc. These elegant topics are not central to statistical mechanics, and we will not pursue them here.

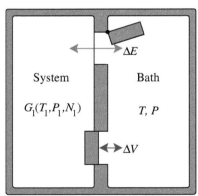

Fig. 6.4 The Gibbs ensemble $G(T, P, N)$ embodies the thermodynamics of systems that can exchange heat and volume with a bath. The **enthalpy** $H(S, P, N)$ is used for systems that only exchange volume.

"unnatural variables",[24] and there are many inequalities that can be derived from stability criteria.

Of course, statistical mechanics is not really different from thermodynamics; it inherits the entire zoo of complex relationships. Indeed, statistical mechanics has its own collection of important relations that connect equilibrium fluctuations to transport and response, like the Einstein relation connecting fluctuations to diffusive transport in Section 2.3 and the fluctuation-dissipation theorem we will derive in Chapter 10. In statistical mechanics, though, the focus of attention is usually not on the general relations between properties, but on calculating the properties of specific systems.

[24]For example, there are useful tricks to take the derivative of $S(E, V, N)$ with respect to P at constant T without re-expressing it in the variables P and T [76].

6.5   Mechanics: friction and fluctuations

A mass m hangs on the end of a spring with spring constant K and unstretched length h_0, subject to a gravitational field of strength g. How far does the spring stretch? We have all solved innumerable statics exercises of this sort in first-year mechanics courses. The spring stretches to a length h^* where $-mg = K(h^* - h_0)$. At h^* the forces balance and the energy is minimized (Fig. 6.5).

What principle of physics is this? In physics, energy is conserved, not minimized! Should we not be concluding that the mass will oscillate with a constant amplitude forever?

We have now come to the point in your physics education where we can finally explain why the mass appears to minimize energy. Here our system (the mass and spring)[25] is coupled to a very large number N of internal atomic or molecular degrees of freedom. The oscillation of the mass is coupled to these other degrees of freedom (friction) and will share its energy with them. The vibrations of the atoms is heat; the energy of the pendulum is dissipated by friction into heat. Indeed, since the spring potential energy is quadratic we can use the equipartition theorem: in equilibrium $\frac{1}{2}K(h - h^*)^2 = \frac{1}{2}k_B T$. For a spring with $K = 10\,\text{N/m}$ at room temperature ($k_B T = 4 \times 10^{-21}$ J), $\sqrt{\langle (h - h^*)^2 \rangle} = \sqrt{k_B T / K} = 2 \times 10^{-11}$ m $= 0.2$ Å. The position indeed minimizes the energy up to thermal fluctuations smaller than an atomic radius.[26]

How do we connect this statistical mechanics picture to the friction coefficient of the damped harmonic oscillator? A careful statistical mechanics treatment (Exercise 10.7) gives a law of motion for the mass of the form

$$\ddot{h} = -\frac{K}{m}(h - h^\star) - \gamma \dot{h} + \xi(t), \qquad (6.49)$$

where γ represents the friction or *dissipation*, and $\xi(t)$ is a random, time-dependent noise force coming from the internal vibrational degrees of freedom of the system. This is an example of a *Langevin equation*. The strength of the noise ξ depends on the dissipation γ and the temperature T so as to guarantee a Boltzmann distribution as the steady state. In general both ξ and γ can be frequency dependent; we will study these issues in detail in Chapter 10.

Fig. 6.5 A mass on a spring in equilibrium sits very close to the minimum of the energy.

[25]We think of the subsystem as being just the macroscopic configuration of mass and spring, and the atoms comprising them as being part of the environment, the rest of the system.

[26]We will return to the fluctuating harmonic oscillator in Exercises 6.18, 10.13, 10.15, 10.16, and 10.18.

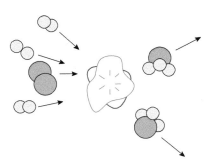

Fig. 6.6 Ammonia collision. The simple motivating argument for the law of mass-action views the reaction as a simultaneous collision of all the reactants.

[27]Experimentally it is more common to work at constant pressure, which makes things more complicated but conceptually no more interesting.

[28]More generally, we can write a reaction as $\sum_i \nu_i A_i = 0$. Here the ν_i are the *stoichiometries*, giving the number of molecules of type A_i changed during the reaction (with $\nu_i < 0$ for reactants and $\nu_i > 0$ for products). The law of mass-action in general states that $\prod_i [A_i]^{\nu_i} = K_{eq}$.

[29]The Haber–Bosch process used industrially for producing ammonia involves several intermediate states. The nitrogen and hydrogen molecules adsorb (stick) onto an iron substrate, and disassociate into atoms. The nitrogen atom picks up one hydrogen atom at a time. Finally, the NH_3 molecule desorbs (leaves the surface) into the vapor. The iron acts as a *catalyst*, lowering the energy barrier and speeding up the reaction without itself being consumed. (Protein catalysts in biology are called *enzymes*.)

6.6 Chemical equilibrium and reaction rates

In studying chemical reactions, one is often interested in the number of molecules of various types as a function of time, and not interested in observing properties depending on the positions or momenta of the molecules. In this section we develop a free energy to derive the laws of chemical equilibrium, and in particular the *law of mass-action*. We will then discuss more carefully the subtle question of exactly when the chemical reaction takes place, and motivate the *Arrhenius law* of thermally activated reaction rates.

Chemical reactions change one type of molecule into another. For example, ammonia (NH_3) can be produced from hydrogen and nitrogen through the reaction

$$3H_2 + N_2 \leftrightharpoons 2NH_3. \tag{6.50}$$

All chemical reactions are in principle reversible, although the backward reaction rate may be very different from the forward rate. In *chemical equilibrium*,[27] the concentrations [X] of the various molecules X (in number per unit volume, say) satisfy the law of mass-action[28]

$$\frac{[NH_3]^2}{[N_2][H_2]^3} = K_{eq}(T). \tag{6.51}$$

The law of mass-action can naively be motivated by imagining a chemical reaction arising from a simultaneous collision of all the reactants (Fig. 6.6). The probability of one nitrogen and three hydrogen molecules colliding in a small reaction region is proportional to the nitrogen concentration and to the cube of the hydrogen concentration, so the forward reaction would occur with some rate per unit volume $K_F[N_2][H_2]^3$; similarly, the backward reaction would occur with a rate per unit volume $K_B[NH_3]^2$ proportional to the probability that two NH_3 molecules will collide. Balancing these two rates to get a steady state gives us eqn 6.51 with $K_{eq} = K_F/K_B$.

This naive motivating argument becomes unconvincing when one realizes that the actual reaction may proceed through several short-lived intermediate states—at no point is a multiple collision required.[29] How can we derive the law of mass-action soundly from statistical mechanics?

Since we are uninterested in the positions and momenta, at fixed volume and temperature our system is described by a Helmholtz free energy $A(T, V, N_{H_2}, N_{N_2}, N_{NH_3})$. When the chemical reaction takes place, it changes the number of the three molecules, and changes the free energy of the system:

$$\Delta A = \frac{\partial A}{\partial N_{H_2}} \Delta N_{H_2} + \frac{\partial A}{\partial N_{N_2}} \Delta N_{N_2} + \frac{\partial A}{\partial N_{NH_3}} \Delta N_{NH_3}$$
$$= -3\mu_{H_2} - \mu_{N_2} + 2\mu_{NH_3}, \tag{6.52}$$

where $\mu_X = \partial A/\partial X$ is the chemical potential of molecule X. The reaction will proceed until the free energy is at a minimum, so

$$-3\mu_{H_2} - \mu_{N_2} + 2\mu_{NH_3} = 0 \tag{6.53}$$

in equilibrium.

To derive the law of mass-action, we must now make an assumption: that the molecules are uncorrelated in space.[30] This makes each molecular species into a separate ideal gas. The Helmholtz free energies of the three gases are of the form

$$A(N, V, T) = Nk_BT \left[\log((N/V)\lambda^3) - 1\right] + NF_0, \tag{6.54}$$

where $\lambda = h/\sqrt{2\pi m k_B T}$ is the thermal de Broglie wavelength. The first two terms give the contribution to the partition function from the positions and momenta of the molecules (eqn 6.24); the last term NF_0 comes from the internal free energy of the molecules.[31] So, the chemical potential is

$$\begin{aligned}
\mu(N, V, T) = \frac{\partial A}{\partial N} &= k_BT \left[\log((N/V)\lambda^3) - 1\right] + Nk_BT(1/N) + F_0 \\
&= k_BT \log((N/V)\lambda^3) + F_0 \\
&= k_BT \log(N/V) + c + F_0, \tag{6.55}
\end{aligned}$$

where the constant $c = k_BT \log(\lambda^3)$ is independent of density. Using eqn 6.55 in eqn 6.53, dividing by k_BT, writing concentrations $[X] = N_X/V$, and pulling terms independent of concentrations to the right, we find the law of mass-action:

$$-3\log[H_2] - \log[N_2] + 2\log[NH_3] = \log(K_{eq}),$$

$$\implies \quad \frac{[NH_3]^2}{[H_2]^3[N_2]} = K_{eq}. \tag{6.56}$$

We also find that the equilibrium constant depends exponentially on the net internal free energy difference $\Delta F_{net} = -3F_0^{H_2} - F_0^{N_2} + 2F_0^{NH_3}$ between reactants and products:

$$K_{eq} = K_0 \exp(-\Delta F_{net}/k_BT) \tag{6.57}$$

with a prefactor

$$K_0 = \frac{\lambda_{H_2}^9 \lambda_{N_2}^3}{\lambda_{NH_3}^6} = \frac{h^6 m_{NH_3}^3}{8\pi^3 k_B^3 T^3 m_{H_2}^{9/2} m_{N_2}^{3/2}} \propto T^{-3} \tag{6.58}$$

that depends weakly on the temperature. The factor $e^{-\Delta F_{net}/k_BT}$ represents the Boltzmann factor favoring a final state with molecular free energy ΔF_{net} lower than the initial state. The temperature dependence of the prefactor (four molecules have more momentum-space entropy than two), and the temperature dependence of the molecular free energies F_0 (see note 31), are usually weak compared to the exponential dependence on the difference in molecular ground-state energies ΔE_{net}:

$$K_{eq} \propto \exp(-\Delta E_{net}/k_BT), \tag{6.59}$$

[30] This assumption is also often valid for ions and atoms in solution; if the ion–ion interactions can be neglected and the solute (water) is not important, the ions are well described as ideal gases, with corrections due to integrating out the solvent degrees of freedom.

[31] For atoms in their ground state, F_0 is the ground-state energy E_0. For small spinless molecules near room temperature, the internal free energy F_0 is usually dominated by the molecular ground-state energy E_0, with a classical contribution from rotational motion; vibrations and internal electronic excitations are frozen out. In this regime, the homonuclear diatomic molecules H_2 and N_2 have $F_0 = E_0 - k_BT \log(IT/\hbar^2)$, where I is the moment of inertia; ammonia with three moments of inertia has $F_0 = E_0 - k_BT \log(\sqrt{8\pi T^3 I_1 I_2 I_3}/3\hbar^3)$ (see [105, sections 47–51]).

[32]Ammonia synthesis is *exothermic*, releasing energy. The reverse reaction, consuming ammonia to make nitrogen and hydrogen is *endothermic*.

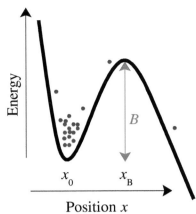

Position x

Fig. 6.7 Barrier-crossing potential. Energy E as a function of some reaction coordinate X for a chemical reaction. The dots schematically represent how many atoms are at each position. The reactants (left) are separated from the products (right) by an energy barrier of height B. One can estimate the rate of the reaction by calculating the number of reactants crossing the top of the barrier per unit time.

[33]The picture 6.7 applies to barrier-crossing problems in extended systems, like diffusion in crystals and atomic transitions between metastable states (glasses, Section 5.2.2; biomolecules, Exercise 6.4; and nanojunctions, Exercise 10.5). In each of these systems, this partial trace leaves us with a discrete set of states.

[34]Crowding could change this. For example, a surface catalyst where the product does not leave the surface could stop reacting as the product covers the active sites.

[35]Re-crossing is a *dynamical correction* to transition-state theory; see [79].

[36]At low temperatures, it is mainly important that this surface be perpendicular to the unstable "downward" eigendirection of the Hessian (second derivative matrix) for the potential energy at the transition state.

with $\Delta E_{net} = 92.4\,\text{kJ/mole}$ for the ammonia synthesis reaction.[32] Equilibrium constants usually grow in the thermally activated form suggested by eqn 6.57.

We now go beyond chemical equilibrium, to consider the rates of the forward and backward reactions. To do so, we need a more precise definition of which atoms belong to which molecules. Exactly when during the trajectory do we say that the reaction has occurred?

An M-atom chemical reaction (classically) is a trajectory in a $3M$-dimensional configuration space. It is traditional in chemistry to pick out one *reaction coordinate* X, and plot the energy (minimized with respect to the other $3M-1$ coordinates) versus X. Figure 6.7 shows this energy plot. Notice the energy barrier B separating the reactants from the products; the atomic configuration at the top of the barrier is called the *transition state*. (This barrier, in $3M$-dimensional configuration space, is actually a saddlepoint; dividing the reactants from the products demands the identification of a $(3M-1)$-dimensional transition-state *dividing surface*.) Our free energy $A(T, V, N_{H_2}, N_{N_2}, N_{NH_3})$ is properly a partial trace, with all configurations to the left of the transition state B contributing to the free energy of the reactants and all configurations to the right of B contributing as products.[33]

How fast does our chemical reaction proceed, if we start out of equilibrium with extra reactants? In dilute systems where the mass-action law holds, the forward reaction rate is to a good approximation independent of the concentration of the product.[34] If our reactions occur slowly enough so that the molecules remain in equilibrium at the current concentrations, we can estimate the nonequilibrium reaction rate by studying the equilibrium transitions from reactant to product at the same reactant concentration.

The reaction rate cannot be larger than the total number of atoms in equilibrium crossing past the energy barrier from reactant to product. (It can be smaller if trajectories which do cross the barrier often immediately re-cross backward before equilibrating on the other side.[35] Such re-crossings are minimized by choosing the transition-state dividing surface properly.[36]) The density of particles at the top of the barrier is smaller than the density at the bottom of the well by a Boltzmann factor of $\exp(-B/k_B T)$. The rate of the reaction will therefore be of the thermally activated, or *Arrhenius*, form

$$\Gamma = \Gamma_0 \exp(-B/k_B T), \tag{6.60}$$

with some prefactor Γ_0 which will be proportional to the mass-action formula (e.g., $\Gamma_0 \propto [H_2]^2[N_2]$ for our ammonia formation reaction). By carefully calculating the population near the bottom of the well and the population and velocities near the top of the barrier, one can derive a formula for the constant of proportionality (see Exercise 6.11).

This Arrhenius law for thermally activated motion governs not only chemical reaction rates, but also diffusion constants and more macroscopic phenomena like nucleation rates (Section 11.3).[37]

6.7   Free energy density for the ideal gas

We began our text (Section 2.2) studying the diffusion equation. How is it connected with free energies and ensembles? Broadly speaking, inhomogeneous systems out of equilibrium can also be described by statistical mechanics, if the gradients in space and time are small enough that the system is close to a *local equilibrium*. We can then represent the local state of the system by *order parameter fields*, one field for each property (density, temperature, magnetization) needed to characterize the state of a uniform, macroscopic body. We can describe a spatially varying, inhomogeneous system that is nearly in equilibrium using a *free energy density*, typically depending on the order parameter fields and their gradients. The free energy of the inhomogeneous system will be given by integrating the free energy density.[38]

We will be discussing order parameter fields and free energy densities for a wide variety of complex systems in Chapter 9. There we will use symmetries and gradient expansions to derive the form of the free energy density, because it will often be too complex to compute directly. In this section, we will directly derive the free energy density for an inhomogeneous ideal gas, to give a tangible example of the general case.[39]

Remember that the Helmholtz free energy of an ideal gas is nicely written (eqn 6.24) in terms of the density $\rho = N/V$ and the thermal de Broglie wavelength λ:

$$A(N, V, T) = Nk_BT \left[\log(\rho\lambda^3) - 1\right]. \qquad (6.61)$$

Hence the free energy density for $n_j = \rho(\mathbf{x}_j)\Delta V$ atoms in a small volume ΔV is

$$\mathcal{F}^{\text{ideal}}(\rho(\mathbf{x}_j), T) = \frac{A(n_j, \Delta V, T)}{\Delta V} = \rho(\mathbf{x}_j)k_BT \left[\log(\rho(\mathbf{x}_j)\lambda^3) - 1\right]. \qquad (6.62)$$

The probability of observing particle density $\rho(x)$ is

$$P\{\rho\} = e^{-\beta \int dV \, \mathcal{F}^{\text{ideal}}(\rho(\mathbf{x}))}/Z. \qquad (6.63)$$

The free energy $F\{\rho\} = \int \mathcal{F}(\rho(\mathbf{x}))\,d\mathbf{x}$ acts just like the energy in the Boltzmann distribution. We have *integrated out* the microscopic degrees of freedom (positions and velocities of the individual particles, Fig. 6.8) and replaced them with a coarse-grained field $\rho(x)$ (Fig. 6.9). The free energy density of eqn 6.62 can be used to determine any equilibrium property of the system that can be written in terms of the density $\rho(x)$.[40]

The free energy density also provides a framework for discussing the evolution laws for nonuniform densities. A system prepared with some nonuniform density will evolve in time $\rho(x, t)$. If in each small volume ΔV the system is close to equilibrium, then one may expect that its time evolution can be described by equilibrium statistical mechanics

[37]There are basically three ways in which slow processes arise in physics. (1) Large systems can respond slowly to external changes because communication from one end of the system to the other is sluggish; examples are the slow decay at long wavelengths in the diffusion equation (Section 2.2) and Goldstone modes (Section 9.3). (2) Systems like radioactive nuclei can respond slowly—decaying with lifetimes of billions of years—because of the slow rate of quantum tunneling through barriers. (3) Systems can be slow because they must thermally activate over barriers (with the Arrhenius rate of eqn 6.60).

[38]Properly, given an order parameter field $s(\mathbf{x})$ there is a functional $F\{s\}$ which gives the system free energy. (A *functional* is a mapping from a space of functions into the real numbers.) Writing this functional as an integral over a free energy density (as we do) can be subtle, not only due to long-range fields, but also due to total divergence terms (Exercise 9.3).

[39]We will also use the free energy density of the ideal gas when we study correlation functions in Section 10.3.

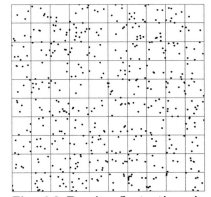

Fig. 6.8 Density fluctuations in space. If n_j is the number of points in a box of size ΔV at position $\mathbf{x}_j$, then $\rho(\mathbf{x}_j) = n_j/\Delta V$.

[40]In Chapter 10, for example, we will use it to calculate correlation functions $\langle\rho(x)\rho(x')\rangle$, and will discuss their relationship with susceptibilities and dissipation.

even though it is not globally in equilibrium. A nonuniform density will have a force which pushes it towards uniformity; the total free energy will decrease when particles flow from regions of high particle density to low density. We can use our free energy density to calculate this force, and then use the force to derive laws (depending on the system) for the time evolution.

The chemical potential for a uniform system is

$$\mu = \frac{\partial A}{\partial N} = \frac{\partial A/V}{\partial N/V} = \frac{\partial \mathcal{F}}{\partial \rho},$$

i.e., the change in free energy for a change in the average density ρ. For a nonuniform system, the local chemical potential at a point x is

$$\mu(x) = \frac{\delta}{\delta \rho(x)} \int \mathcal{F}(y) \mathrm{d}y \tag{6.64}$$

the variational derivative[41] of the free energy density with respect to $\rho(x)$. Because our ideal gas free energy has no terms involving gradients of ρ, the variational derivative $\delta \int \mathcal{F}/\delta\rho$ equals the partial derivative $\partial \mathcal{F}/\partial \rho$:

$$\begin{aligned}
\mu(x) &= \frac{\delta \int \mathcal{F}^{\mathrm{ideal}}}{\delta \rho} = \frac{\partial}{\partial \rho} \left(\rho k_B T \left[\log(\rho \lambda^3) - 1 \right] \right) \\
&= k_B T \left[\log(\rho \lambda^3) - 1 \right] + \rho k_B T / \rho \\
&= k_B T \log(\rho \lambda^3).
\end{aligned} \tag{6.65}$$

The chemical potential is like a number pressure for particles: a particle can lower the free energy by moving from regions of high chemical potential to low chemical potential. The gradient of the chemical potential $-\partial \mu / \partial x$ is thus a pressure gradient, effectively the statistical mechanical force on a particle.

How will the particle density ρ evolve in response to this force $\mu(x)$? This depends upon the problem. If our density were the density of the entire gas, the atoms would accelerate under the force—leading to sound waves.[42] There momentum is conserved as well as particle density. If our particles could be created and destroyed, the density evolution would include a term $\partial \rho / \partial t = -\eta \mu$ not involving a current. In systems that conserve (or nearly conserve) energy, the evolution will depend on Hamilton's equations of motion for the free energy density; in magnets, the magnetization responds to an external force by precessing; in superfluids, gradients in the chemical potential are associated with winding and unwinding the phase of the order parameter field (vortex motion)...

Let us focus on the case of a small amount of perfume in a large body of still air. Here particle density is locally conserved, but momentum is strongly damped (since the perfume particles can scatter off of the air molecules). The velocity of our particles will be proportional to the effective force on them $\mathbf{v} = -\gamma(\partial \mu / \partial x)$, with γ the mobility.[43] Hence

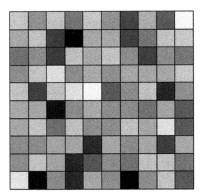

Fig. 6.9 Coarse-grained density in space. The density field $\rho(\mathbf{x})$ represents all configurations of points $\mathbb{Q}$ consistent with the average. Its free energy density $\mathcal{F}^{\mathrm{ideal}}(\rho(\mathbf{x}))$ contains contributions from all microstates $(\mathbb{P}, \mathbb{Q})$ with the correct number of particles per box. The probability of finding this particular density is proportional to the integral of all ways $\mathbb{Q}$ that the particles in Fig. 6.8 can be arranged to give this density.

[41]The variational derivative is the derivative in function space, giving the linear approximation to $\int (\mathcal{F}\{\rho + \delta\rho\} - \mathcal{F}\{\rho\})$ (see note 7 on p. 290). We will seldom touch upon variational derivatives in this text.

[42]In that case, we would need to add the local velocity field into our description of the local environment.

[43]This is *linear response*. Systems that are nearly in equilibrium typically have currents proportional to gradients in their properties. Examples include Ohm's law where the electrical current is proportional to the gradient of the electromagnetic potential $I = V/R = (1/R)(\mathrm{d}\phi/\mathrm{d}x)$, thermal conductivity where the heat flow is proportional to the gradient in temperature $\mathbf{J} = \kappa \nabla T$, and viscous fluids, where the shear rate is proportional to the shear stress. We will study linear response with more rigor in Chapter 10.

the current $\mathbf{J} = \rho v$ of particles will be

$$\mathbf{J} = \gamma \rho(x) \left(-\frac{\partial \mu}{\partial x} \right) = -\gamma \rho(x) \frac{\partial k_B T \log(\rho \lambda^3)}{\partial x}$$
$$= -\gamma \rho(x) \frac{k_B T}{\rho} \frac{\partial \rho}{\partial x} = -\gamma k_B T \frac{\partial \rho}{\partial x}, \tag{6.66}$$

and thus the rate of change of ρ is given by the diffusion equation

$$\frac{\partial \rho}{\partial t} = -\nabla \cdot J = \gamma k_B T \frac{\partial^2 \rho}{\partial x^2}. \tag{6.67}$$

Notice,

- we have again derived the diffusion equation (eqn 2.7) $\partial \rho / \partial t = D \, \partial^2 \rho / \partial x^2$, this time by starting with a free energy density from equilibrium statistical mechanics, and assuming a linear law relating velocity to force;
- we have rediscovered the Einstein relation (eqn 2.22) $D = \gamma k_B T$;
- we have asserted that $-\partial \mu / \partial x$ acts just like an external force, even though μ comes from the ideal gas, which itself has no potential energy (see Exercises 5.12 and 6.13).

Our free energy density for the ideal gas is simpler than the free energy density of a general system because the ideal gas has no stiffness (no free energy cost for gradients in the density). In general, the boxed regions in Fig. 6.8 will not be independent systems, and there will be a free energy difference that depends on the change in the coarse-grained fields between boxes (e.g., the (free) energy cost to bend an elastic rod) leading to terms involving gradients of the field. Free energy densities will play a large role in Chapters 9–12.

Exercises

Free energies are the work-horses of statistical mechanics and thermodynamics. There are an immense number of practical applications of the various free energies in chemistry, condensed matter physics, and engineering. These exercises do not explore these applications in depth; rather, they emphasize the central themes and methods that span across fields.

Exponential atmosphere, *Two-state system*, *Negative temperature*, and *Gas vs. rubber band* explore the prototypical examples of canonical ensembles. *Molecular motors* provides a biophysics example of how to construct the appropriate free energies in systems exchanging things other than energy, volume, and number, and introduces

us to *telegraph noise* in bistable systems.

Laplace, *Lagrange*, and *Legendre* explore mathematical tools important in thermodynamics, and their links to statistical mechanics. *Lagrange* can be viewed as an alternative mathematical derivation of much of statistical mechanics. *Euler*, *Gibbs–Duhem*, and *Clausius–Clapeyron* introduce the corresponding fundamental thermodynamic relations.

Barrier crossing introduces the quantitative methods used to calculate the rates of chemical reactions. *Michaelis–Menten and Hill* provides a microscopic example of how one integrates out degrees of freedom leading to effective theories; it derives two forms for chemical reaction rates commonly seen in molecular biology. *Pollen and hard squares* derives depletion forces in colloidal systems as a purely entropic free energy. *Rubber band free energy* and *Rubber band formalism* solve the entropic model of rubber introduced in Exercise 5.12 to illustrate the nature and advantages of working in the canonical ensemble. And *Langevin dynamics* and *Langevin simulation* introduce one method used for simulating systems in the canonical ensemble.

Bayesian statistics and statistical mechanics share a mathematical framework of information geometry applied to probability distributions (see exercises in Chapter 1). *Statistical mechanics and statistics* introduces Bayesian analysis. *Gibbs for pistons* introduces the ensemble as a probability distribution. *Pistons in probability space* analyzes the Gibbs piston ensemble as a surface (model manifold) in prediction space; *FIM for Gibbs* analyzes the metric for this manifold in a general Gibbs ensemble. And *Can we burn information?* is a recent advance which estimates the entropy cost of controlling a system— challenging Szilard's argument that information entropy and thermodynamic entropy can be converted to one another (as in Exercise 5.2).

We conclude with three cool applications of free energies. *Word frequency: Zipf's law* argues that our language optimizes ease of communication. *Epidemics and zombies* uses popular culture to illustrate both basic models of epidemiology, and the stochastic underpinnings of chemical reaction rate equations. And *Nucleosynthesis as a chemical reaction* sets up the fusion reaction of the early Universe that led to the arrow of time (Exercise 5.24).

(6.1) **Exponential atmosphere.**[44] (Computation) ②

As you climb a mountain, the air becomes thin and cold, and typically rather windy. Are any of these effects due to equilibrium statistical mechanics? The wind is not; it is due to nonuniform heating and evaporation in far distant regions. We have determined that equilibrium statistical mechanics demands that two equilibrium bodies in contact must share the same temperature, even when one of them is above the other. But gas molecules fall down under gravity, ...

This example is studied in [61, I.40], where Feynman uses it to deduce much of classical equilibrium statistical mechanics. Let us reproduce his argument. Simulate an ideal gas in a box with reflecting walls, under the influence of gravity. Since the ideal gas has no internal equilibration, the simulation will start in an equilibrium ensemble at temperature T.

(a) *Does the distribution visually appear statistically stationary? How is it possible to maintain a static distribution of heights, even though all the atoms are continuously accelerating downward? After running for a while, plot a histogram of the height distribution and velocity distribution. Do these distributions remain time independent, apart from statistical fluctuations? Do their forms agree with the predicted equilibrium Boltzmann distributions?*

The equilibrium thermal distribution is time independent even if there are no collisions to keep things in equilibrium. The number of atoms passing a plane at constant z from top to bottom must match the number of atoms passing from bottom to top. There are more atoms at the bottom, but many of them do not have the vertical kinetic energy to make it high enough. Macroscopically, we can use the ideal gas law ($PV = Nk_BT$, so $P(z) = n(z)k_BT$, where $n = N/V$ is the number density of atoms) to deduce the Boltzmann distribution giving the density dependence on height.[45]

(b) *The pressure increases with depth due to the increasing weight of the air above. What is the force due to gravity on a slab of thickness Δz and area A? What is the change in pressure from z to $z - \Delta z$? Use this, and the ideal gas law, to find the density dependence on height. Does it agree with the Boltzmann distribution?*

Feynman then deduces the momentum distribution of the particles from the balancing of upward and downward particle fluxes we saw in part (a). He starts by arguing that the equilibrium probability ρ_v that a given atom has a particular vertical velocity v_z is independent of height. (The atoms at different heights are all at the same temperature, and only differ in their overall density; since they do not interact, they do not know the density, hence the atom's velocity distribution cannot depend on z.)

(c) *If the unknown velocity distribution is $\rho_v(v_z)$, use it and the Boltzmann height distribution deduced in part (b) to write the joint equilibrium*

---

[44]This exercise and the associated software were developed in collaboration with Christopher Myers. Computational hints can be found at the book website [181].

[45]Feynman [61, I.40] uses this to derive the Boltzmann distribution for an ideal gas in any potential $U(\mathbf{x})$. The change in pressure $dP = dnk_BT = k_BT\,dn/dx$ must balance the force on the particles $Fn = -n\,dU/dx$, so $(1/n)dn/dx = d\log n/dx = -dU/dx/k_BT$, $\log n = C - U(x)/k_BT$ for some constant C, and hence $n = \exp(-U(x)/k_BT)/Z$ with $Z = \exp(C)$.

probability distribution $\rho(v_z, z, t)$.

Now consider[46] the atoms with vertical velocity v_z in a slab of gas of area A between z and $z + \Delta z$ at time t. Their probability density (per unit vertical velocity) is $\rho(v_z, z, t) A \Delta z$. After a time Δt, this slab will have accelerated to $v_z - g\Delta t$, and risen a distance $h + v_z \Delta t$, so

$$\rho(v_z, z, t) = \rho(v_z - g\Delta t, z + v_z\Delta t, t + \Delta t). \quad (6.68)$$

(d) *Using the fact that $\rho(v_z, z, t)$ is time independent in equilibrium, write a relation between $\partial\rho/\partial v_z$ and $\partial\rho/\partial z$. Using your result from part (c), derive the equilibrium velocity distribution for the ideal gas.*

Feynman then argues that interactions and collisions will not change the velocity distribution.

(e) *Simulate an interacting gas in a box with reflecting walls, under the influence of gravity. Use a temperature and a density for which there is a layer of liquid at the bottom (just like water in a glass). Plot the height distribution (which should show clear interaction effects) and the momentum distribution. Use the latter to determine the temperature; do the interactions indeed not distort the momentum distribution?*

What about the atoms which evaporate from the fluid? Only the very most energetic atoms can leave the liquid to become gas molecules. They must, however, use up every bit of their extra energy (on average) to depart; their kinetic energy distribution is precisely the same as that of the liquid.[47]

Feynman concludes his chapter by pointing out that the predictions resulting from the classical Boltzmann distribution, although they describe many properties well, do not match experiments on the specific heats of gases, foreshadowing the need for quantum mechanics.[48]

(6.2) **Two-state system.** @
Consider the statistical mechanics of a tiny object with only two discrete states:[49] one of energy E_1 and the other of higher energy $E_2 > E_1$.

(a) Boltzmann probability ratio. *Find the ratio of the equilibrium probabilities ρ_2/ρ_1 to find our system in the two states, when weakly coupled to a heat bath of temperature T. What is the limiting probability as $T \to \infty$? As $T \to 0$?* Related formula: Boltzmann probability $= Z(T)\exp(-E/kT) \propto \exp(-E/kT)$.

(b) Probabilities and averages. *Use the normalization of the probability distribution (the system must be in one or the other state) to find ρ_1 and ρ_2 separately. (Hint: Solve for $Z(T)$ in the "related formula" for part (a).) What is the average value of the energy E?*

(6.3) **Negative temperature.** ③
Consider a system of N atoms, each with two electronic states at energies $\pm\varepsilon/2$. The atoms are isolated from the outside world. There are only weak couplings between the atoms, sufficient to bring them into internal equilibrium but without significantly affecting the energy of the system.

(a) Microcanonical entropy. *If the net energy is E (corresponding to a number of excited atoms $m = E/\varepsilon + N/2$), what is the microcanonical entropy $S_{\text{micro}}(E)$ of our system? Simplify your expression using Stirling's formula, $\log n! \sim n\log n - n$.*

(b) Negative temperature. *Find the temperature, using your simplified expression from part (a). What happens to the temperature when $E > 0$?* Having the energy $E > 0$ is a kind of population inversion. Population inversion is the driving mechanism for lasers.

For many quantities, the thermodynamic derivatives have natural interpretations when viewed as sums over states. It is easiest to see this in small systems.

(c) Canonical ensemble. (i) *Take one of our atoms and couple it to a heat bath of temperature $k_B T = 1/\beta$. Write explicit formulæ for Z_{canon}, E_{canon}, and S_{canon} in the canonical ensemble, as a trace (or sum) over the two states of the atom.* (E should be the energy of each state multiplied by the probability ρ_n of that state, and S should

---

[46]Feynman gives a complicated argument avoiding partial derivatives and gently introducing probability distributions, which becomes cleaner if we just embrace the math.

[47]Ignoring quantum mechanics.

[48]Quantum mechanics is important for the internal vibrations within molecules, which absorb energy as the gas is heated. Quantum effects are not so important for the pressure and other properties of gases, which are dominated by the molecular center-of-mass motions.

[49]Visualize this as a tiny biased coin, which can be in the "heads" or "tails" state but has no other internal vibrations or center of mass degrees of freedom. Many systems are well described by large numbers of these two-state systems: some paramagnets, carbon monoxide on surfaces, glasses at low temperatures, ...

be the trace of $-k_B \rho_n \log \rho_n$.) (ii) *Compare the results with what you get by using the thermodynamic relations. Using Z from the trace over states, calculate the Helmholtz free energy A, S as a derivative of A, and E from $A = E - TS$. Do the thermodynamically derived formulæ you get agree with the statistical traces?*

(d) *What happens to E in the canonical ensemble as $T \to \infty$? Can you get into the negative-temperature regime discussed in part (b)?*

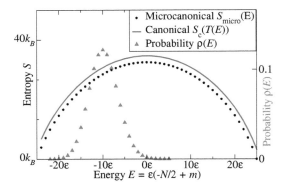

Fig. 6.10 Negative temperature. Entropies and energy fluctuations for this problem with $N = 50$. The canonical probability distribution for the energy is for $\langle E \rangle = -10\varepsilon$, and $k_B T = 1.207\varepsilon$. For $E > 0$, the entropy goes down as E goes up, so $T(E) < 0$.

(e) *Canonical–microcanonical correspondence. Find the entropy in the canonical distribution for N of our atoms coupled to the outside world, from your answer to part (c). Explain the value of $S(T = \infty) - S(T = 0)$ by counting states. Using the approximate form of the entropy from part (a) and the temperature from part (b), show that the canonical and microcanonical entropies agree, $S_{micro}(E) = S_{canon}(T(E))$ for large N.* (Perhaps useful: $\text{arctanh}(x) = \frac{1}{2} \log((1 + x)/(1 - x))$.) Notice that the two are not equal in Fig. 6.10; the form of Stirling's formula we used in part (a) is not very accurate for $N = 50$. *Explain in words why the microcanonical entropy is smaller than the canonical entropy.*

(f) *Fluctuations. Calculate the RMS energy fluctuations in our system in the canonical ensemble.*

Evaluate it at $T(E)$ from part (b). For large N, are the fluctuations in E small compared to E?

(6.4) **Molecular motors and free energies.**[50] (Biology) @
Figure 6.11 shows the set-up of an experiment on the molecular motor RNA polymerase that transcribes DNA into RNA.[51] Choosing a good ensemble for this system is a bit involved. It is under two constant forces (F and pressure), and involves complicated chemistry and biology. Nonetheless, you know some things based on fundamental principles. Let us consider the optical trap and the distant fluid as being part of the external environment, and define the system as the local region of DNA, the RNA, motor, and the fluid and local molecules in a region immediately enclosing the region, as shown in Fig. 6.11.

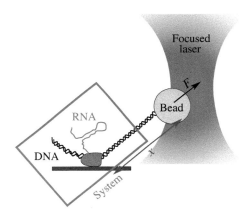

Fig. 6.11 RNA polymerase molecular motor attached to a glass slide is pulling along a DNA molecule (transcribing it into RNA). The opposite end of the DNA molecule is attached to a bead which is being pulled by an optical trap with a constant external force F. Let the distance from the motor to the bead be x; thus the motor is trying to move to decrease x and the force is trying to increase x.

(a) *Without knowing anything further about the chemistry or biology in the system, which of the following must be true on average, in all cases? (T) (F) The total entropy of the Universe (the system, bead, trap, laser beam, ...) must increase or stay unchanged with time.*

[50]This exercise was developed with the help of Michelle Wang.
[51]RNA, ribonucleic acid, is a long polymer like DNA, with many functions in living cells. It has four monomer units (A, U, C, and G: Adenine, Uracil, Cytosine, and Guanine); DNA has T (Thymine) instead of Uracil. Transcription just copies the DNA sequence letter for letter (except for this substitution) into one strand of RNA.

(T) (F) The entropy S_s of the system cannot decrease with time.

(T) (F) The total energy E_T of the Universe must decrease with time.

(T) (F) The energy E_s of the system cannot increase with time.

(T) (F) $G_s - Fx = E_s - TS_s + PV_s - Fx$ cannot increase with time, where G_s is the Gibbs free energy of the system.

Related formula: $G = E - TS + PV$.

(Hint: Precisely two of the answers are correct.) The sequence of monomers on the RNA can encode information for building proteins, and can also cause the RNA to fold into shapes that are important to its function. One of the most important such structures is the hairpin (Fig. 6.12). Experimentalists study the strength of these hairpins by pulling on them (also with laser tweezers). Under a sufficiently large force, the hairpin will unzip. Near the threshold for unzipping, the RNA is found to jump between the zipped and unzipped states, giving *telegraph noise*[52] (Fig. 6.13). Just as the current in a telegraph signal is either on or off, these systems are bistable and make transitions from one state to the other; they are a two-state system.

Fig. 6.12 Hairpins in RNA. (Reprinted with permission from Liphardt et al. [117], ©2001 AAAS.) A length of RNA attaches to an inverted, complementary strand immediately following (just as two DNA strands attach to form their double helix). This forms a hairpin fold.

The two RNA configurations presumably have different energies (E_z, E_u), entropies (S_z, S_u), and volumes (V_z, V_u) for the local region around the zipped and unzipped states, respectively. The environment is at temperature T and pressure P. Let $L = L_u - L_z$ be the extra length of RNA in the unzipped state. Let ρ_z be the fraction of the time our molecule is zipped at a given external force F, and $\rho_u = 1 - \rho_z$ be the unzipped fraction of time.

(b) *Of the following statements, which are true, assuming that the pulled RNA is in equilibrium?*

(T) (F) $\rho_z/\rho_u = \exp((S_z^{tot} - S_u^{tot})/k_B)$, where S_z^{tot} and S_u^{tot} are the total entropy of the Universe when the RNA is in the zipped and unzipped states, respectively.

(T) (F) $\rho_z/\rho_u = \exp(-(E_z - E_u)/k_BT)$.

(T) (F) $\rho_z/\rho_u = \exp(-(G_z - G_u)/k_BT)$, where $G_z = E_z - TS_z + PV_z$ and $G_u = E_u - TS_u + PV_u$ are the Gibbs energies in the two states.

(T) (F) $\rho_z/\rho_u = \exp(-(G_z - G_u + FL)/k_BT)$, where L is the extra length of the unzipped RNA and F is the applied force.

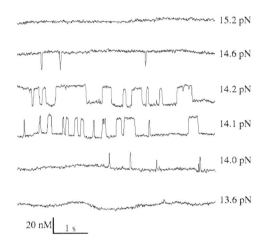

Fig. 6.13 Telegraph noise in RNA unzipping. (Reprinted with permission from Liphardt et al. [117], ©2001 AAAS.) As the force increases, the fraction of time spent in the zipped state decreases.

(6.5) **Laplace.**[53] (Thermodynamics) @
The Laplace transform of a function $f(t)$ is a function of x:

$$\mathcal{L}\{f\}(x) = \int_0^\infty f(t)e^{-xt}\,dt. \qquad (6.69)$$

---

[52]The dots and dashes of telegraph signals are thought analogous to the two states of the molecule.
[53]Pierre-Simon Laplace (1749–1827). See [128, section 4.3].

Show that the canonical partition function $Z(\beta)$ can be written as the Laplace transform of the microcanonical volume of the energy shell $\Omega(E)$.

(6.6) **Lagrange.**[54] (Thermodynamics) @
In this exercise, we explore a statistical mechanics from an information-theory perspective. We shall derive the canonical and grand canonical perspective by maximizing the entropy subject to constraints on the energy and the number of particles. In the process, we introduce the mathematics of *Lagrange multipliers*.
Lagrange multipliers allow one to find the extremum of a function $f(\mathbf{x})$ given a constraint $g(\mathbf{x}) = g_0$. One sets the derivative of

$$f(\mathbf{x}) + \lambda(g(\mathbf{x}) - g_0) \qquad (6.70)$$

with respect to λ and $\mathbf{x}$ to zero. The derivatives with respect to components x_j of $\mathbf{x}$ then include terms involving λ which push x_j. Setting the derivative with respect to λ to zero determines the value λ needed to enforce the constraint.
We will use Lagrange multipliers to find the maximum of the nonequilibrium entropy

$$S = -k_B \sum p_i \log p_i$$

constraining the normalization, energy, and number.
(a) Microcanonical. *Using a Lagrange multiplier to enforce the normalization $\sum_i p_i = 1$, show that the probability distribution that extremizes the entropy is a constant (the microcanonical distribution).*
(b) Canonical. *Add another Lagrange multiplier to fix the mean energy $\langle E \rangle = \sum_i E_i p_i$. Show that the canonical distribution maximizes the entropy given the constraints of normalization and fixed energy.*
(c) Grand canonical. *Summing over different numbers of particles N and adding the constraint that the average number is $\langle N \rangle = \sum_i N_i p_i$, show that you get the grand canonical distribution by maximizing the entropy.*
(d) Links to ensembles. *Write the grand partition function Ξ, the temperature T, and the chemical potential μ in terms of your three Lagrange multipliers in part (c).*

(6.7) **Legendre.**[55] (Thermodynamics) @
The Legendre transform of a function $f(t)$ is given by minimizing $f(x) - xp$ with respect to x, so that p is the slope ($p = \partial f/\partial x$):

$$g(p) = \min_x \{f(x) - xp\}. \qquad (6.71)$$

Equation 6.17, shows that the Helmholtz free energy $A(T, V, N) = \langle E(S, V, N) \rangle - TS$ is the Legendre transform of the energy $E(S, V, N)$, eliminating the entropy as a variable. Equation 6.36 uses Legendre to remove N in exchange for μ to give the grand free energy; eqns 6.46 and 6.47 do this for the Gibbs free energy and the enthalpy. Eqns 6.40 and 6.41 show that the Helmholtz free energy $A(T, V, N)$ is given by a similar transformation of the entropy $S(E, V, N)$, removing energy in favor of temperature:

$$A(T, V, N) = \min_E \{E - TS(E, V, N)\}. \qquad (6.72)$$

But close examination shows a subtle difference. The entropy from which one is eliminating a variable is multiplied by temperature, with a minus sign.
This is a historical artifact. Entropy is an oddball, both in that it does not have units of energy, and that it is maximized in equilibrium while free energies are minimized.[56]
Factoring out T from eqn 6.72, show that the Helmholtz free energy is given by a Legendre transform of the negative of the entropy, followed by a change of variables from p to a function of T. What is $p(T)$?

(6.8) **Euler.** (Thermodynamics, Chemistry) @
(a) *Using the fact that the entropy $S(N, V, E)$ is extensive for large systems, show that*

$$N \left.\frac{\partial S}{\partial N}\right|_{V,E} + V \left.\frac{\partial S}{\partial V}\right|_{N,E} + E \left.\frac{\partial S}{\partial E}\right|_{N,V} = S. \qquad (6.73)$$

Show from this that in general

$$S = (E + PV - \mu N)/T \qquad (6.74)$$

and hence

$$E = TS - PV + \mu N. \qquad (6.75)$$

---

[54] Joseph-Louis Lagrange (1736–1813). See [128, section 12, p. 331].
[55] Adrien-Marie Legendre (1752–1833).
[56] Some texts introduce *negentropy* to make things more consistent. Note that the units of entropy (and negentropy) are also different from those of the free energies.

This is Euler's relation.[57]

(b) *Test this explicitly for the ideal gas. Use the ideal gas entropy (eqn 3.57)*

$$S(N, V, E) = \frac{5}{2} N k_B \tag{6.76}$$
$$+ N k_B \log \left[\frac{V}{N h^3} \left(\frac{4 \pi m E}{3N} \right)^{3/2} \right],$$

to derive formulæ for T, P, and μ in terms of E, N, and V, and verify eqn 6.74.

(6.9) **Gibbs–Duhem.** (Thermodynamics, Chemistry) @

As a state function, E is supposed to depend only on S, V, and N. But eqn 6.75 seems to show explicit dependence on T, P, and μ as well; how can this be?

Another answer is to consider a small shift of all six variables. We know that $dE = T\, dS - P\, dV + \mu\, dN$, but if we shift all six variables in Euler's equation we get $dE = T\, dS - P\, dV + \mu\, dN + S\, dT - V\, dP + N\, d\mu$. This implies the *Gibbs–Duhem relation*

$$0 = S\, dT - V\, dP + N\, d\mu. \tag{6.77}$$

This relation implies that the intensive variables T, P, and μ are not all independent; the change in μ is determined given a small change in T and P.

(a) *Write μ as a suitable derivative of the Gibbs free energy $G(T, P, N)$.*

This makes μ a function of the three variables T, P, and N. The Gibbs–Duhem relation says it must be independent of N.

(b) *Argue that changing the number of particles in a large system at fixed temperature and pressure should not change the chemical potential.* (Hint: Doubling the number of particles at fixed T and P doubles the size and energy of the system as well.)

The fact that both $G(T, P, N)$ and N are extensive means that G must be proportional to N. We used this extensivity to prove the Euler relation in Exercise 6.8; we can thus use the Euler relation to write the formula for G directly.

(c) *Use the Euler relation (eqn 6.75) to write a formula for $G = E - TS + PV$. Is it indeed proportional to N? What about your formula for μ from part (a); will it be dependent on N?*

(6.10) **Clausius–Clapeyron.** (Thermodynamics, Chemistry) @

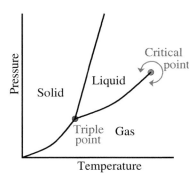

Fig. 6.14 Generic phase diagram, showing the coexistence curves for solids, liquids, and gases.

Consider the phase diagram in Fig. 6.14. Along an equilibrium phase boundary, the temperatures, pressures, and chemical potentials of the two phases must agree; otherwise a flat interface between the two phases would transmit heat, shift sideways, or leak particles, respectively (violating the assumption of equilibrium).

(a) *Apply the Gibbs–Duhem relation 6.77 to both phases, for a small shift by ΔT along the phase boundary. Let s_1, v_1, s_2, and v_2 be the molecular entropies and volumes ($s = S/N$, $v = V/N$ for each phase); derive the* Clausius–Clapeyron *equation for the slope of the coexistence line on the phase diagram*

$$dP/dT = (s_1 - s_2)/(v_1 - v_2). \tag{6.78}$$

It is hard to experimentally measure the entropies per particle; we do not have an entropy thermometer. But, as you will remember, the entropy difference upon a phase transformation $\Delta S = Q/T$ is related to the heat flow Q needed to induce the phase change. Let the latent heat L be the heat flow per molecule.

(b) *Write a formula for dP/dT that does not involve the entropy.*

---

[57]Leonhard Euler (1707–1783). More specifically, it is one of many fundamental relations named after Euler; other Euler relations involve the number of faces, edges, and vertices for a polygonalization of a surface and the polar representation of complex numbers.

(6.11) **Barrier crossing.** (Chemistry) ②
In this exercise, we will derive the Arrhenius law (eqn 6.60)

$$\Gamma = \Gamma_0 \exp(-E/k_B T), \qquad (6.79)$$

giving the rate at which chemical reactions cross energy barriers. The important exponential dependence on the barrier height E is the relative Boltzmann probability that a particle is near the top of the barrier (and hence able to escape). Here we will do a relatively careful job of calculating the prefactor Γ_0.

Fig. 6.15 Well probability distribution. The approximate probability distribution for the atoms still trapped inside the well.

Consider a system having an energy $U(X)$, with an energy well with a local minimum at $X = X_0$ having energy $U(X_0) = 0$. Assume there is an energy barrier of height $U(X_B) = B$ across which particles can escape.[58] Let the temperature of the system be much smaller than B/k_B. To do our calculation, we will make some approximations. (1) We assume that the atoms escaping across the barrier to the right do not scatter back into the well. (2) We assume that the atoms deep inside the well are in local equilibrium. (3) We assume that the particles crossing to the right across the barrier are given by the equilibrium distribution inside the well.
(a) *Let the probability density that a particle has position X be $\rho(X)$. What is the ratio of probability densities $\rho(X_B)/\rho(X_0)$ if the particles near the top of the barrier are assumed to be in equilibrium with those deep inside the*

well? Related formula: Boltzmann distribution $\rho \propto \exp(-E/k_B T)$.

Fig. 6.16 Crossing the barrier. The range of positions for which atoms moving to the right with velocity v will cross the barrier top in time Δt.

If the barrier height $B \gg k_B T$, then most of the particles in the well stay near the bottom of the well. Often, the potential near the bottom is accurately described by a quadratic approximation $U(X) \approx \frac{1}{2} M \omega^2 (X - X_0)^2$, where M is the mass of our system and ω is the frequency of small oscillations in the well.
(b) *In this approximation, what is the probability density $\rho(X)$ near the bottom of the well? (See Fig. 6.15.) What is $\rho(X_0)$, the probability density of being precisely at the bottom of the well?* Related formula: Gaussian probability distribution $(1/\sqrt{2\pi\sigma^2}) \exp(-x^2/2\sigma^2)$.
Knowing the answers from (a) and (b), we know the probability density $\rho(X_B)$ at the top of the barrier.[59] We also need to know the probability that particles near the top of the barrier have velocity V, because the faster-moving parts of the distribution of velocities contribute more to the flux of probability over the barrier (see Fig. 6.16). As usual, because the total energy is the sum of the kinetic energy and potential energy, the total Boltzmann probability factors; in equilibrium the particles will always have a velocity probability distribution $\rho(V) = 1/\sqrt{2\pi k_B T/M} \exp(-\frac{1}{2} M V^2/k_B T)$.
(c) *First give a formula for the decay rate Γ (the probability per unit time that a given particle crosses the barrier towards the right), for an unknown probability density $\rho(X_B)\rho(V)$ as an integral over the velocity V. Then, using your*

[58]This potential could describe a chemical reaction, with X being a reaction coordinate. At the other extreme, it could describe the escape of gas from a moon of Jupiter, with X being the distance from the moon in Jupiter's direction.
[59]Or rather, we have calculated $\rho(X_B)$ in equilibrium, half of which (the right movers) we assume will also be crossing the barrier in the nonequilibrium reaction.

*formulæ from parts (a) and (b), give your esti-
mate of the decay rate for our system.* Related
formula: $\int_0^\infty x \exp(-x^2/2\sigma^2)\,\mathrm{d}x = \sigma^2$.

How could we go beyond this one-dimensional
calculation? In the olden days, Kramers studied other one-dimensional models, changing the
ways in which the system was coupled to the
external bath (see Exercise 12.22). On the computer, one can avoid a separate heat bath and
directly work with the full multidimensional configuration space, leading to *transition-state theory*. The transition-state theory formula is very
similar to the one you derived in part (c), except
that the prefactor involves the product of all the
frequencies at the bottom of the well and all the
positive frequencies at the saddlepoint at the top
of the barrier (see [79]). Other generalizations
arise when crossing multiple barriers [92] or in
nonequilibrium systems [122].

(6.12) **Michaelis–Menten and Hill.** (Biology, Computation) ③
Biological reaction rates are often saturable; the
cell needs to respond sensitively to the introduction of a new chemical S, but the response should
not keep growing indefinitely as the new chemical concentration $[S]$ grows.[60] Other biological reactions act as switches; a switch not only
saturates, but its rate or state changes sharply
from one value to another as the concentration
of a chemical S is varied. These reactions give
tangible examples of how one develops effective
dynamical theories by removing degrees of freedom; here, instead of coarse-graining some large
statistical mechanics system, we remove a single
enzyme E from the equations to get an effective
reaction rate.

The rate of a chemical reaction,

$$NS + B \rightarrow C, \tag{6.80}$$

where N substrate molecules S combine with a B
molecule to make a C molecule, will occur with
a reaction rate given by the law of mass-action:

$$\frac{\mathrm{d}[C]}{\mathrm{d}t} = k[S]^N[B]. \tag{6.81}$$

Saturation and the Michaelis–Menten equation.
Saturation is not seen in ordinary chemical reaction kinetics. Notice that the reaction rate goes
as the Nth power of the concentration $[S]$; far

from saturating, the reaction rate grows linearly
or faster with concentration.

The archetypal example of saturation in biological systems is the *Michaelis–Menten* reaction
form. A reaction of this form converting a chemical S (the substrate) into P (the product) has a
rate given by the formula

$$\frac{\mathrm{d}[P]}{\mathrm{d}t} = \frac{V_{\max}[S]}{K_M + [S]}, \tag{6.82}$$

where K_M is called the Michaelis constant
(Fig. 6.17). This reaction at small concentrations acts like an ordinary chemical reaction with
$N = 1$ and $k = V_{\max}/K_M$, but the rate saturates
at $V_{\max}$ as $[S] \rightarrow \infty$. The Michaelis constant K_M
is the concentration $[S]$ at which the rate is equal
to half of its saturation rate (Fig. 6.17).

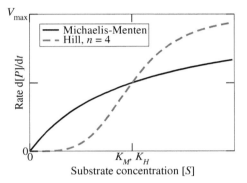

Fig. 6.17 Michaelis–Menten and Hill equation
forms.

We can derive the Michaelis–Menten form by hy-
pothesizing the existence of a catalyst or enzyme
E, which is in short supply. The enzyme is presumed to be partly free and available for binding (concentration $[E]$) and partly bound to the
substrate (concentration[61] $[E:S]$), helping it to
turn into the product. The total concentration
$[E] + [E:S] = E_{\text{tot}}$ is fixed. The reactions are
as follows:

$$E + S \; \underset{k_1}{\overset{k_{-1}}{\leftrightarrows}} \; E:S \; \overset{k_{\text{cat}}}{\rightarrow} \; E + P. \tag{6.83}$$

We must then assume that the supply of substrate is large, so its concentration changes
slowly with time. We can then assume that the
concentration $[E:S]$ is in steady state, and remove it as a degree of freedom.

---

[60]$[S]$ is the concentration of S (number per unit volume). S stands for *substrate*.
[61]The colon denotes the bound state of the two, called a *dimer*.

(a) *Assume the binding reaction rates in eqn 6.83 are of traditional chemical kinetics form (eqn 6.81), with constants k_1, k_{-1}, and k_{cat}, and with $N = 1$ or $N = 0$ as appropriate. Write the equation for $\mathrm{d}[E:S]/\mathrm{d}t$, set it to zero, and use it to eliminate $[E]$ in the equation for $\mathrm{d}P/\mathrm{d}t$. What are V_{max} and K_M in the Michaelis–Menten form (eqn 6.82) in terms of the ks and E_{tot}?*

We can understand this saturation intuitively: when all the enzyme is busy and bound to the substrate, adding more substrate cannot speed up the reaction.

Cooperativity and sharp switching: the Hill equation. Hemoglobin (Hb) is what makes blood red; this iron-containing protein can bind up to four molecules of oxygen in the lungs, and carries them to the tissues of the body where it releases them. If the binding of all four oxygens were independent, the $[O_2]$ concentration dependence of the bound oxygen concentration would have the Michaelis–Menten form; to completely de-oxygenate the Hb would demand a very low oxygen concentration in the tissue.

What happens instead is that the Hb binding of oxygen looks much more sigmoidal—a fairly sharp transition between nearly four oxygens bound at high $[O_2]$ (lungs) to nearly none bound at low oxygen concentrations. This arises because the binding of the oxygens is enhanced by having other oxygens bound. This is not because the oxygens somehow stick to one another; instead, each oxygen deforms the Hb in a nonlocal *allosteric*[62] fashion, changing the configurations and affinity of the other binding sites. The Hill equation was introduced for hemoglobin to describe this kind of cooperative binding. Like the Michaelis–Menten form, it is also used to describe reaction rates, where instead of the carrier Hb we have an enzyme, or perhaps a series of transcription binding sites (see Exercise 8.11). We will derive the Hill equation not in terms of allosteric binding, but in the context of a saturable reaction involving n molecules binding simultaneously. As a reaction rate, the Hill equation is

$$\frac{\mathrm{d}[P]}{\mathrm{d}t} = \frac{V_{max}[S]^n}{K_H^n + [S]^n} \qquad (6.84)$$

(see Fig. 6.17). For Hb, the concentration of the n-fold oxygenated form is given by the right-

hand side of eqn 6.84. In both cases, the transition becomes much more of a switch, with the reaction turning on (or the Hb accepting or releasing its oxygen) sharply at a particular concentration (Fig. 6.17). The transition can be made more or less sharp by increasing or decreasing n. The Hill equation can be derived using a simplifying assumption that n molecules bind in a single reaction:

$$E + nS \quad \underset{k_b}{\overset{k_u}{\leftrightharpoons}} \ E:(nS), \qquad (6.85)$$

where E might stand for hemoglobin and S for the O_2 oxygen molecules. Again, there is a fixed total amount $E_{tot} = [E] + [E:nS]$.

(b) *Assume that the two reactions in eqn 6.85 have the chemical kinetics form (eqn 6.81) with $N = 0$ or $N = n$ as appropriate. Write the equilibrium equation for $E:(nS)$, and eliminate $[E]$ using the fixed total E_{tot}. What are V_{max} and K_H in terms of k_b, k_u, and E_{tot}?*

Usually, and in particular for hemoglobin, this cooperativity is not so rigid; the states with one, two, and three O_2 molecules bound also compete with the unbound and fully bound states. This is treated in an approximate way by using the Hill equation, but allowing n to vary as a fitting parameter; for Hb, $n \approx 2.8$.

Both Hill and Michaelis–Menten equations are often used in biological reaction models even when there are no explicit mechanisms (enzymes, cooperative binding) known to generate them.

(6.13) **Pollen and hard squares.**[63] @

Objects embedded in a gas will have an effective attractive force at short distances, when the gas molecules can no longer fit between the objects. This is called the *depletion force*, and is a common tool in physics to get micron-scale particles to clump together. One can view this force as a pressure imbalance (no collisions from one side) or as an entropic attraction.

Let us model the entropic attraction between a pollen grain and a wall using a two-dimensional ideal gas of classical indistinguishable particles as the fluid. For convenience, we imagine that the pollen grain and the fluid are formed from square particles lined up with the axes of the box, of lengths B and b, respectively (Fig. 6.18). We assume *no interaction* between the ideal gas

---

[62]Allosteric comes from *Allo* (other) and *steric* (structure or space). Allosteric interactions can be cooperative, as in hemoglobin, or inhibitory.

[63]This problem was inspired by conversations with Ard Louis.

molecules (unlike in Exercise 3.5), but the potential energy is infinite if the gas molecules overlap with the pollen grain or with the wall. The container as a whole has one pollen grain, N gas molecules, and total area $L \times L$.

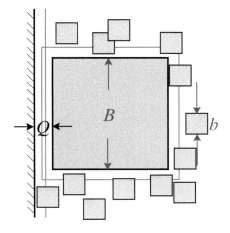

Fig. 6.18 Square pollen grain in fluid of oriented square molecules, next to a wall. The thin lines represents the exclusion region around the pollen grain and away from the wall.

Assume the pollen grain is close to only one wall. Let the distance from the surface of the wall to the closest face of the pollen grain be Q. (A similar square-particle problem with interacting small molecules is studied in [65].) Because the molecules are noninteracting, it is useful to start by considering the configuration space area $A(Q)$ available for one molecule.

(a) *Plot the area $A(Q)$ available for each gas molecule, in units of $(length)^2$. One can proceed in three steps: (i) Plot $A(Q \gg 0)$ when the pollen grain is far from the wall. At what distance Q_0 does this change? (ii) Plot $A(0)$, the area available when the pollen grain is touching the wall. (iii) Finally, how does the area vary with Q (linearly, quadratically, step function ...) between zero and Q_0? (See Fig. 6.18.) Plot this, and give formulæ for $A(Q)$ as a function of Q for the two relevant regions, $Q < Q_0$ and $Q > Q_0$.*

(b) *What is the configuration-space volume $\Omega(Q)$ for the gas, in units of $(length)^{2N}$? What is the position-space entropy of the ideal gas, $S(Q)$? (Write your answers here in terms of $A(Q)$.)*

Your answers to part (b) can be viewed as giving a free energy for the pollen grain after integrating over the gas degrees of freedom (also known as a partial trace, or coarse-grained free energy).

(c) *What is the resulting coarse-grained free energy of the pollen grain, $\mathcal{F}(Q) = E - T\,S(Q)$, in the two regions $Q > b$ and $Q < b$? Use $\mathcal{F}(Q)$ to calculate the force on the pollen grain for $Q < b$. Is the force positive (away from the wall) or negative? Why?*

(d) *Directly calculate the force due to the ideal gas pressure on the far side of the pollen grain, in terms of $A(Q)$. Compare it to the force from the partial trace in part (c). Why is there no balancing force from the other side? Effectively how "long" is the far side of the pollen grain?*

(6.14) **Statistical mechanics and statistics.**[64] (Statistics) ③

Consider the problem of fitting a theoretical model to experimentally determined data. Let our model predict a time-dependent function $y^\theta(t)$, where $\boldsymbol{\theta}$ are the model parameters. Let there be N experimentally determined data points d_i at times t_i with errors of standard deviation σ. We assume that the experimental errors for the data points are independent and Gaussian distributed, so that the probability that a given model produced the observed data points (the probability $P(D|\boldsymbol{\theta})$ of the data given the model) is

$$P(D|\boldsymbol{\theta}) = \prod_{i=1}^{N} \frac{1}{\sqrt{2\pi}\sigma} e^{-\left(y^\theta(t_i)-d_i\right)^2/2\sigma^2}. \quad (6.86)$$

(a) *True or false: This probability density corresponds to a Boltzmann distribution with energy H and temperature T, with $H = \sum_{i=1}^{N}(y^\theta(t_i) - d_i)^2/2$ and $k_B T = \sigma^2$.*

There are two approaches to statistics. Among a family of models, the *frequentists* will pick the parameters $\boldsymbol{\theta}$ with the largest value of $P(D|\boldsymbol{\theta})$ (the *maximum likelihood estimate*); the ensemble of best-fit models is then deduced from the range of likely input data (deduced from the error bars σ). The *Bayesians* take a different point of view. They argue that there is no reason to believe a priori that all models have the same probability. (In model parameter space, there is no analogue of Liouville's theorem, Section 4.1.) Suppose the probability of the model (the *prior*)

[64]This exercise was developed with the help of Robert Weiss.

is $P(\boldsymbol{\theta})$. They use the theorem

$$P(\boldsymbol{\theta}|D) = P(D|\boldsymbol{\theta})P(\boldsymbol{\theta})/P(D). \qquad (6.87)$$

(b) *Prove Bayes' theorem (eqn 6.87) using the fact that $P(A$ and $B) = P(A|B)P(B)$ (see note 39 on p. 111).*

The Bayesians will often pick the maximum of $P(\boldsymbol{\theta}|D)$ as their model for the experimental data. But, given their perspective, it is even more natural to consider the entire *ensemble* of models, weighted by $P(\boldsymbol{\theta}|D)$, as the best description of the data. This ensemble average then naturally provides error bars for the parameters as well as for the predictions of various quantities.

Consider the problem of fitting a line to two data points. Suppose the experimental data points are at $t_1 = 0$, $d_1 = 1$ and $t_2 = 1$, $d_2 = 2$, where both y-values have uncorrelated Gaussian errors with standard deviation $\sigma = 1/2$, as assumed in eqn 6.86 above. Our model $M(m,b)$, with parameters $\boldsymbol{\theta} = (m,b)$, is $y(t) = mt + b$. Our Bayesian statistician has prior knowledge that m and b both lie between zero and two, and assumes that the probability density is otherwise uniform; $P(m,b) = 1/4$ for $0 < m < 2$ and $0 < b < 2$.

(c) *Which of the contour plots shown accurately represent the probability distribution $P(\boldsymbol{\theta}|D)$ for the model, given the observed data? (The spacing between the contour lines is arbitrary.)*

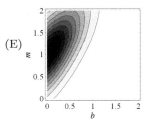

(6.15) **Gas vs. rubber band.** ⓟ

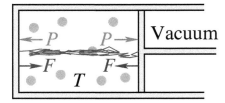

Fig. 6.19 Piston with rubber band. A piston in vacuum is held closed by a rubber band. The gas in the piston exerts an outward force. The rubber band stretches across the piston, exerting an inward tensile force. The piston adjusts to its thermal equilibrium distance, with no external force applied.

In Fig. 6.19, we see a piston filled with a gas which exerts a pressure P. The volume outside the piston is empty (vacuum, $P_{\text{ext}} = 0$), and no external forces are applied. A rubber band attached to the piston and the far wall resists the outward motion with an inward force F. The piston moves to its equilibrium position.

Assume the gas and the rubber band are modeled as both having zero potential energy $E_{\text{int}} = 0$ for all accessible states. (So, for example, perhaps the gas was modeled as hard spheres, similar to Exercise 3.5, and the rubber band as zero-energy random walk chain as in Exercise 5.12). *What will happen to the position of the piston if the temperature is increased by a factor of two? Fundamentally, why does this occur? Give an elegant and concise reason for your answer.*

(6.16) **Rubber band free energy.** (Condensed matter) @

This exercise illustrates the convenience of choosing the right ensemble to decouple systems into independent subunits. Consider again the random-walk model of a rubber band in Exercise 5.12—N segments of length d, connected by hinges that had zero energy both when straight and when bent by 180°. We there calculated its entropy $S(L)$ as a function of the folded length L, used Stirling's formula to simplify the combinatorics, and found its spring constant K at $L = 0$.

Here we shall do the same calculation, without the combinatorics. Instead of calculating the entropy $S(L)$ at fixed L,[65] we shall work at fixed temperature and fixed force, calculating an appropriate free energy $\chi(T, F)$. View our model rubber band as a collection of segments $s_n = \pm 1$ of length d, so $L = d\sum_{n=1}^{N} s_n$. Let F be the force exerted by the band on the world (negative for positive L).

(a) *Write a Hamiltonian for our rubber band under an external force.[66] Show that it can be written as a sum of uncoupled Hamiltonians $\mathcal{H}_n$, one for each link of the rubber band model. Solve for the partition function X_n of one link, and then use this to solve for the partition function $X(T, F)$.* (Hint: Just as for the canonical ensemble, the partition function for a sum of uncoupled Hamiltonians is the product of the partition functions of the individual Hamiltonians.)

(b) *Calculate the associated thermodynamic potential $\chi(T, F)$. Derive the abstract formula for* $\langle L \rangle$ *as a derivative of χ, in analogy with the calculation in eqn 6.11. Find the spring constant $K = \partial F/\partial L$ in terms of χ. Evaluate it for our rubber band free energy at $F = 0$.*

(6.17) **Rubber band formalism.** (Condensed matter) ⓟ

Consider the rubber band of Exercise 5.12. View it as a system that can exchange length $L_M = d(2M - N)$ with the external world, with the force F being the restoring force of the rubber band (negative for positive L). Let $X(T, F)$ be the analog of the partition function for the ensemble which keeps both temperature and force constant.

(a) *Use the number of configurations $\Omega(L_M)$ at fixed length to write a formal expression for $X(T, F)$ as a sum over M. (See the analogous third line in eqn 6.37.)*

(b) *Write the corresponding thermodynamic potential $\chi = -k_B T \log(X(T, F))$ in terms of E, T, S, F, and L. (For example, eqn 6.17 says that a system connected to an external bath is described by $\langle E \rangle - TS$; what analogous expression applies here?)*

(6.18) **Langevin dynamics.** (Computation, Dynamical systems) ⓟ

Even though energy is conserved, macroscopic objects like pendulums and rubber balls tend to minimize their potential and kinetic energies unless they are externally forced. Section 6.5 explains that these energies get transferred into heat—they get lost into the $6N$ internal degrees of freedom. (See also Exercise 10.7.) Equation 6.49 suggests that these microscopic degrees of freedom can produce both friction and noise. First consider the equation $m\ddot{h} = \xi(t)$ for the position $h(t)$ of a particle of mass m. Assume the noise $\xi(t)$ gives a kick to the particle at regular intervals separated by Δt: $\xi(t) = \sum_{j=-\infty}^{\infty} \xi_j \delta(t - j\Delta t)$, with $\langle \xi_i^2 \rangle = \sigma^2$ and $\langle \xi_i \xi_j \rangle = 0$ for $i \neq j$.

(a) *Assume the particle starts at rest at time $t = +\epsilon$ (just after the kick at $t = 0$). Integrate $m\ddot{h}$ to a time $t_n = n\Delta t + \epsilon$ to argue that this leads to a random walk in momentum space. Calculate the expectation of the energy $\langle p_n^2/2m \rangle$ at this time. Will noise alone lead this system to equilibrate at late times?*

---

[65] Usually we would write $S(E, L)$, but our model rubber band is purely entropic—$E = 0$ for all states.

[66] Our model rubber band has no internal energy. This Hamiltonian includes the energy exchanged with the outside world when the length of the rubber band changes, just as the Gibbs free energy $G(T, P) = E - TS + PV$ includes the energy needed to borrow volume from the external world.

At late times, can we generate a thermal velocity distribution for this system? Let us introduce a damping force $F = -\dot{h}/\gamma$, to remove some of the energy that we insert by kicking the system. (Here γ is the mobility.) Using $\dot{p} = F = -p/\gamma$ over the interval Δt, we find we multiply the momentum by $\exp(-\Delta t/m\gamma)$ during each time step, so $p_j = \exp(-\Delta t/m\gamma)p_{j-1} + \xi_j$.

(b) *For this part, assume the momentum has been evolving since $t = -\infty$. By summing the geometrical series, find $\langle p_j^2 \rangle$ in terms of the mass m, the noise σ, the damping γ, and the time step Δt.*

(c) *What is the relation between the noise and the damping needed to make $\langle p^2/2m \rangle = \frac{1}{2}k_B T$, as needed for thermal equilibrium? How does the relation simplify in the small time-step limit $\Delta t \to 0$?*

This is *Langevin dynamics*, which both is a way of modeling the effects of the environment on macroscopic systems and a numerical method for simulating systems at fixed temperature (i.e., in the canonical ensemble).

(6.19) **Langevin simulation.**[67] ③

This exercises leverages the mosh pit simulator [32], designed to model humans at heavy metal concerts [31, 188–190]. Here we shall use the simulation to explore the equilibrium behavior of interacting particles. View the simulator as a kind of experimental system; explore how the system responds to changing the control parameters.

Launch the mosh pit simulator [32]. In this exercise, we shall use only the active (red) agents subject to noise, and will not use the active matter forces "flocking", and "speed" that lead to nonequilibrium behavior. Instead, we shall explore whether and how noise and damping induce equilibration.

Return the simulation to its defaults by reloading the page. Turn all the agents active (**Fraction Red** to 1), and reduce their number N to 100 (set **Particle count** and click **Change**). Set **Flock strength** and **Speed** to zero, **Damping** to 0.05 and **Noise strength** to 0.2, and click **Change**. Verify that the the particles are moving in noisy paths between collisions. Adjust the number of frames skipped to speed up the visualization.

The noise and damping implement Langevin dynamics, used for finite-temperature simulations (Exercises 6.18 and 10.7).

The particles interact via a soft repulsive potential, and have a damping force $-\mu\mathbf{v}$ to absorb kinetic energy. Noise adds an uncorrelated stochastic force $\boldsymbol{\eta}(t)$ in the interval (t_n, t_{n+1}) with strength $t_n = n\Delta t$ with $\langle \eta_\alpha(t_i)\eta_\beta(t_j) \rangle = \sigma^2 \delta_{ij}\delta_{\alpha\beta}$. This noise might represent the agitation of our agents, who randomly thrash around—or it could represent equilibrium buffeting of red pollen grains by surrounding water molecules. In Exercise 10.7, we argued that this dynamics could lead to a thermal distribution for the momentum of one particle.

In the first part, we shall check whether our system exhibits a speed distribution compatible with equilibrium statistical mechanics. Return to the settings above (reload, all active, $N = 100$, flocking = speed = 0, damping = 0.05, noise = 0.2, and **Change**). Show the graph of the speed histogram for the particles.[68]

(a) *Examine the shape of the distribution of speeds $\rho(s)$ for the particles. Derive what the probability distributions of speeds should take for an equilibrium thermal ensemble, and show your work. Does your prediction roughly agree with the speed histogram shown? What qualitative features of the observed distribution does your prediction explain that a Gaussian distribution or the traditional Maxwellian distribution (eqn 1.2) cannot?*

Langevin dynamics uses the balance between damping and noise to maintain a constant temperature. Exercise 6.18 derives the well known relation $k_B T = \sigma^2/(2\mu\Delta t)$ (which in turn is related to the fluctuation-dissipation theorem, see Exercise 10.7). But perhaps our definitions of σ and μ differ from the adjustable parameters **Noise strength** and **Damping** used in the simulation (which may differ from the σ and μ in the paper).

In the guise of an experimentalist probing the effects of different controls, we shall in the next part check our formula by observing how the speed distribution changes as we change param-

---

[67] This exercise was developed in collaboration with David Hathcock. It makes use of the mosh pit simulator [32] developed by Matt Bierbaum for [190].

[68] In the current implementation, the speeds are shown on the lower right; the horizontal axis is not labeled, but stays fixed as parameters change. The vertical axis is set by the peak of the distribution.

eters. Return to the settings above.

(b) *How much should the speeds increase if we quadruple σ? How much should they change if we quadruple μ? Double Δt? Make a table of the measurements you take to check whether these three variables correspond directly to the three experimental controls.* (Measuring the peak of a distribution is noisy; try time-averaging the point where the distribution crosses half of the peak value. I used a ruler placed against the screen. Pick values of the parameters allowing for good measurements. If you are really ambitious, you can examine the source Javascript code on github.)

In the third part, we shall explore what happens to the speed distribution when the interactions between particles becomes strong.

Return to the settings above.

(c) *Alter the number of particles to 350, where the density is typical of a liquid. Does the final distribution of speeds change from that at lower density? Alter it to 500. Does the distribution change in the crystal formed at this density? Is this a surprise? Explain.*

(6.20) **Gibbs for pistons.** (Thermodynamics) ④
The degrees of freedom in a piston are $\mathbf{X} = \{\mathbb{P}, \mathbb{Q}, V\}$, where $\mathbb{P}$ and $\mathbb{Q}$ are the $3N$ positions and momenta of the particles, and V is the current volume of the piston. The Gibbs ensemble for a piston is the probability density

$$\rho = (1/\Gamma) \exp(-\beta \mathcal{H}(\mathbb{P}, \mathbb{Q}) - \beta PV). \quad (6.88)$$

Here Γ is the partition function for the Gibbs ensemble, normalizing the distribution to one.

Let our piston be filled with an ideal gas of particles of mass m. What is the partition function $Z(V, \beta)$ for the canonical ensemble? (Be sure to include the Gibbs factor N!; the quantum phase-space refinements are optional.) Show that the partition function for the Gibbs ensemble is

$$\Gamma(P, \beta) = (2\pi m/\beta)^{3N/2} (\beta P)^{-(N+1)}, \quad (6.89)$$

Show that the joint probability density for finding the N particles with $3N$ dimensional momenta $\mathbb{P}$, the piston with volume V, and the $3N$ dimensional positions $\mathbb{Q}$ inside V (eqn 6.88), is

$$\rho_{\text{Gibbs}}(\mathbb{P}, \mathbb{Q}, V | P, \beta) = (1/\Gamma(P, \beta)) e^{-\beta \mathbb{P}^2/2m - \beta PV}. \quad (6.90)$$

(6.21) **Pistons in probability space.**[69] (Mathematics, Information geometry) ④
Fig. 5.3 shows the Carnot cycle as a path in the P–V space of pressure and volume—parameters varied from the outside. One could draw a similar diagram in the space of pressure and temperature, or volume and temperature. Here we shall explore how to describe the path in the space of *probability distributions*. In the process, we shall compute the *model manifold* of the ideal gas, and show that it is a two-dimensional plane.

As discussed in Exercise 1.15, there is a natural distance, or metric, in the space of probability distributions:

$$g_{\mu\nu} = -\left\langle \frac{\partial^2 \log(\rho)}{\partial \theta_\mu \partial \theta_\nu} \right\rangle, \quad (6.91)$$

the *Fisher information metric*. So, a system in the Gibbs ensemble is described in terms of two parameters, usually P and T. We shall instead use the "natural" parameters $\theta_1 = p = \beta P$ and $\theta_2 = \beta$, where $\beta = 1/k_B T$ (see Exercise 6.22). The squared distance in probability space between two systems with tiny changes in pressure and temperature is then

$$d^2(\rho(\mathbf{X}|\boldsymbol{\theta}), \rho(\mathbf{X}|\boldsymbol{\theta} + d\boldsymbol{\theta})) = g_{\mu\nu} d\theta_\mu d\theta_\nu. \quad (6.92)$$

(a) *Compute $g_{\mu\nu}^{(p,\beta)} = -\langle \partial^2 \log(\rho)/\partial \theta_\mu \partial \theta_\nu \rangle$ using eqn 6.90 from Exercise 6.20.*
The metric tensor $g^{(p,\beta)}$ for the Gibbs ensemble of the piston tells us the distance in probability space between neighboring pressures and temperatures. What kind of surface (the *model manifold*) is formed by this two-parameter family of probability distributions? Does it have an intrinsic curvature?

(b) *Show that one can turn the metric tensor into the identity $g_{\mu\nu}^{(x,y)} = \delta_{\mu\nu}$ by a coordinate transformation $(p, \beta) \to (x = A \log(p), y = B \log(\beta))$. What are the necessary scale factors A and B?*
Hence the model manifold of the piston in the Gibbs ensemble is a plane! We can draw our control paths in the (x, y) plane. We label the four steps of the Carnot cycle as in Fig. 5.3.

(c) *Draw the Carnot cycle path in as a parameterized curve in (x, y), with $P_a = 1$, $P_b = 0.5$, $T_1 = 1$ and $T_2 = 0.8$, for $N = 1$. (Hint: eqn 5.8 will be helpful in finding the adiabatic parts of the path $p(\beta)$.) Is the length of the expansion at fixed pressure the same as you calculated in Exercise 6.23?*

---

[69]This exercise was developed in collaboration with Ben Machta, Archishman Raju, Colin Clement, and Katherine Quinn

(6.22) **FIM for Gibbs.**[70] (Mathematics, Thermodynamics, Information geometry) ④
In this exercise, we study the geometry in the space of probability distributions defined by the Gibbs ensemble[71] of a general equilibrium system. We compute the Fisher Information Metric (FIM, Exercises 1.15 and 6.21)

$$g_{\mu\nu} = -\left\langle \frac{\partial^2 \log(\rho)}{\partial\theta_\mu \partial\theta_\nu} \right\rangle, \qquad (6.93)$$

of the Gibbs phase space ensemble $\rho(\mathbb{P}, \mathbb{Q})$ in terms of thermodynamic properties of the system.
In Exercise 6.21 we calculated $g_{\mu\nu}$ for the ideal gas, using the "natural" variables $\theta_1 = p = \beta P$ and $\theta_2 = \beta$, rather than P and T. Why are these coordinates special? The log of the Gibbs probability distribution for an arbitrary interacting collection of particles with Hamiltonian $\mathcal{H}$ (eqn 6.88) is

$$\begin{aligned} \log(\rho) &= -\beta\mathcal{H}(\mathbb{P}, \mathbb{Q}) - \beta PV - \log\Gamma \\ &= -\beta\mathcal{H}(\mathbb{P}, \mathbb{Q}) - pV - \log\Gamma. \end{aligned} \qquad (6.94)$$

This is the logarithm of the partition function Γ plus terms *linear* in $p = \beta P$ and β.[72] So the second derivatives with respect to p and β only involve $\log(\Gamma)$. We know that the Gibbs free energy $G(p, \beta) = -k_B T \log(\Gamma) = -(1/\beta)\log(1/\Gamma(p,\beta))$, so $\log(\Gamma) = -\beta G(p, \beta)$. The first derivatives of the Gibbs free energy $dG = -SdT + VdP + \mu dN$ are related to things like volume and entropy and chemical potential; our metric is given by the second derivatives (compressibility, specific heat, ...)

(a) *For a collection of particles interacting with Hamiltonian $\mathcal{H}$, relate the four terms $g_{\mu\nu}^{(\mathrm{p},\beta)}$ in terms of physical quantities given by the second derivatives of G. Write your answer in terms of N, p, β, the particle density $\rho = N/\langle V \rangle$, the isothermal compressibility $\kappa = -(1/\langle V \rangle)(\partial\langle V \rangle/\partial P)|_T$, the thermal expansion coefficient $\alpha = (1/\langle V \rangle)(\partial\langle V \rangle/\partial T)|_P$, and the specific heat per particle at constant pressure, $c_\mathrm{P} = (T/N)(\partial S/\partial T)|_P$.* (Hint: $G(P,T) = G(p/\beta, 1/\beta)$. Your answer will be a bit less

complicated if you pull out an overall factor of $N/(\rho\beta^2)$.)
The metric tensor for a general Hamiltonian is a bit simpler in the more usual coordinates (P, β) or (P, T).
(b) *Show that*

$$g^{(\mathrm{P},\beta)} = N\begin{pmatrix} \beta\kappa/\rho & \alpha/\beta\rho \\ \alpha/\beta\rho & c_\mathrm{P}/\beta^2 \end{pmatrix}$$

and

$$g^{(\mathrm{P},\mathrm{T})} = N\begin{pmatrix} \kappa/\rho T & -\alpha/\rho T \\ -\alpha/\rho T & c_\mathrm{P}/T^2 \end{pmatrix}.$$

(c) *Calculate $g^{(\mathrm{p},\beta)}$ for the ideal gas using your answer from part (a). Compare with your results calculating $g^{(\mathrm{p},\beta)}$ directly from the probability distribution in Exercise 6.21. Is the difference significant for macroscopic systems?* (Hint: If you use $G = A + PV$ directly from eqn 6.24, remember that the thermal de Broglie wavelength λ depends on temperature.)
The standard formulas for an ideal gas do not include the piston wall as a degree of freedom, so part (c) has one fewer positional degree of freedom than in Exercise 6.21. That is, the macroscopic calculation neglects the entropic contribution of the fluctuations in volume (the position of the piston inside the cylinder).

(6.23) **Can we burn information?**[73] (Mathematics, Thermodynamics, Information geometry) ④
The use of entropy to measure information content has been remarkably fruitful in computer science, communications, and even in studying the efficiency of signaling and sensing in biological systems. The Szilard engine (Exercise 5.2) was a key argument that thermodynamic entropy and information entropy could be exchanged for one another—that one could burn information. We ask here—can they be exchanged? Are information and entropy fungible?

---

[70]This exercise was developed in collaboration with Ben Machta, Archishman Raju, Colin Clement, and Katherine Quinn
[71]The Fisher information distance is badly defined except for changes in *intensive* quantities. In a microcanonical ensemble, for example, the energy E is constant and so the derivative $\partial\rho/\partial E$ would be the derivative of a δ function. So we study pistons varying P and $\beta = 1/k_B T$, rather than at fixed volume or energy.
[72]In statistics, log probability distributions which depend on parameters in this linear fashion are called *exponential families*. Many common distributions, including lots of statistical mechanical models like ours, are exponential families.
[73]This exercise was developed in collaboration with Ben Machta, Archishman Raju, Colin Clement, and Katherine Quinn

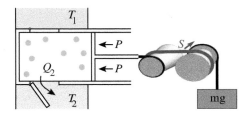

Fig. 6.20 Piston control. Machta [120] studies a piston plus a control system to extract work during expansion. To change the pressure, a continuously variable transmission, controlled by a gradient of the entropy S, connects the piston to a mass under a gravitational force. Minimizing the control cost plus the entropy cost due to fluctuations in the gear ratio lead to a minimum entropy cost for control.

Szilard stores a bit of information as an atom on one side of a piston, and extracts PdV work $k_B T \log 2$ as the piston expands—the same work needed to store a bit. Machta [120] argues that there is a fundamental bound on the entropy cost for extracting this work. He considers a system consisting of the piston plus a control mechanism to slowly decrease the pressure and extract the work, Fig. 6.20. (See Feynman's *Ratchet and pawl* discussion [61, I.46], discussing fluctuations in a similar system.)

Machta argues that this cost is given by a *path length in parameter space*. To be specific, Machta argues that to guide a system through a change in pressure from P_i to P_f should cost an entropy[74]

$$\langle \Delta S_{\text{control}} \rangle = 2 \int_{P_i}^{P_f} \sqrt{g_{PP}} |dP|. \qquad (6.95)$$

The metric in this space, as discussed in Exercise 1.15, is

$$g_{PP} = -\langle \partial^2 \log(\rho)/\partial P^2 \rangle, \qquad (6.96)$$

the *Fisher information metric*, giving the natural distance between two nearby probability distributions.

For example, in a Gibbs ensemble at constant pressure $P = \theta_1$, the squared distance between two nearby pressures at the same temperature is

$$d^2(\rho(\mathbf{X}|P), \rho(\mathbf{X}|P + dP)) = g_{PP}(dP)^2, \qquad (6.97)$$

leading directly to eqn 6.95.

Let us compute the metric g_{PP} in the coordinates for the ideal gas in a piston (Exercise 6.20), and then analyze the cost for thermodynamic control for Szilard's burning information engine in Exercise 5.2.

(a) *Using eqns 6.96 and 6.90, show that $g_{PP} = (1 + N)/P^2$.*

(b) *What is the entropy cost to expand a piston containing a single atom at constant temperature by a factor of two? What is the work done by the piston? How does this affect Szilard's argument about burning information in Exercise 5.2?*

Machta's result thus challenges Szilard's argument that information entropy and thermodynamic entropy can be exchanged. It also gives a (subextensive) cost for the Carnot cycle (see Exercise 6.21).

(6.24) **Word frequencies: Zipf's law.**[75] (Linguistics) ③

The words we use to communicate with one another show a fascinating pattern. Naturally, we use some words more frequently than others. In English[76] **the** is the most frequently occurring word (about 7% of all words), followed by **of** (3.5%), followed by **and** (about 3%). Following down the list, the word ranked j in usage has frequency roughly proportional to $1/j$—Zipf's law (Fig. 6.21). This law holds approximately for all human languages. There are many explanations offered for this, but no clear consensus.

Here we explore a version of one of many explanations [142, 150] for Zipf's law—that it maximizes the communication needed for a given effort. This will build on our understanding of Shannon entropy and communications, and will provide an interesting analogue to our derivation in Exercise 6.6 of different statistical ensembles as maximizing entropy subject to constraints on probability, energy, and number.

First, how much information is communicated by a common word? Surely the word **primeval** tells us more than the word **something**—the former immediately tells us our message is about primitive times or traits, even though the latter is longer. How can we measure the information in a word? Shannon tells us that a communication channel used to transmit message #i

---

[74]We shall follow Machta and set $k_B = 1$ in this exercise, writing it explicitly only when convenient.
[75]This exercise was developed with the help of Colin Clement.
[76]American English, using the Brown corpus. Surprisingly, in the Oxford English corpus sampling international usage, the highest ranked words are **the**, **be**, **to**, ...

with probability p_i has an information capacity of $-k_s \sum p_i \log p_i$ bits/message.

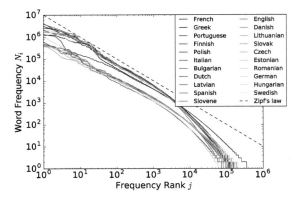

Fig. 6.21 Zipf's law for word frequencies. Frequency of the jth most used word, versus rank j (courtesy Colin Clement). Data is from the Proceedings of the European Parliament, 1996–2011 [100].

(a) *Argue that the information from message #i is given by $k_S \log 1/p_i$. This quantity is called the surprise: the number of bits of information learned from the message. Using each word of our language as a message, argue that*

$$I = k_S \sum_j N_j \log(1/p_j) \qquad (6.98)$$

is the total information transmitted in a series of N words with word #j being transmitted with probability p_j.

Second, how much effort is needed to communicate a given word? Here the presumption is that an unusual word (like **primeval**, $j = 17{,}627$) will take more effort to transmit and receive than a simple word (like **something**, $j = 287$). Shannon would say that a good encoding scheme will use a short string to compress the common word **something** and a longer string to compress **primeval**. Even though the two words are of similar length in letters, we imagine that our brains engage in a kind of compression algorithm too. If we restrict our compression to whole words, and ignore correlations between words, then Shannon would say that the optimal encoding would take the smallest words into the shortest strings, so **the** would be stored as

the string "0", **something** would be stored as "100011111" (287 base 2, 9 bits), and **primeval** would be "100010011011011" in our brain (15 bits).

We are not interested in maximizing the information per word, but maximizing the information per unit of effort.

Let C_j be the effort needed to speak and hear the word of rank j (the number of bits of effort needed to retrieve and identify the word by the brain). Thus the effort for a series of N words with frequency N_j of word j is

$$E = \sum_i N_i C_i. \qquad (6.99)$$

We may assume the brain stores words also in a tree-like fashion, with a few words at the bottom of the tree and more and more words as one goes up the tree. Suppose the branches divides into two at each level, with one word stored at each branching point. Suppose there is a fixed cost c_0 to recognize that a new word has started, and that it costs one additional bit of effort for the brain to climb each level of the tree. There are two words at height 1, four words at height 2, eight words at height 3,... so the cost of the jth most costly word is $C_j^{tree} = \lfloor \log_2(j+2) \rfloor + c_0$ where $\lfloor x \rfloor = \text{floor}(x)$ is the largest integer less than x. We shall approximate this as

$$C_j = k_s \log j + c_0, \qquad (6.100)$$

so in our examples, for **something** $C_{287} = 9 + c_0$. (Remember $k_S = 1/\log(2)$.)

We shall study the optimal probability distribution for the words in our language by maximizing the information I transmitted with respect to the N_j, given a total effort E. We assume our message is long enough that N_j is well approximated by Np_j.[77] So, we shall maximize I using two constraints: eqn 6.99 to fix the total effort and $N = \sum_j N_j$ to fix the normalization.

Remember that we can derive the Boltzmann distribution (Exercise 6.6) by optimizing the entropy $-k_B \sum p_i \log p_i$ subject to two constraints: $E = \sum E_i p_i$ to fix the total energy and $\sum p_i = 1$ to fix the normalization. We used Lagrange multipliers to enforce these constraints. We shall use the same strategy here to examine the optimal word frequency distribution. We thus want to

[77]This is analogous to, say, pressure P in thermodynamics, which is well approximated in a given large system by $\langle P \rangle$.

find the extremum of

$$\mathcal{L} = k_S \sum_i N_i \log(N/N_i)$$

$$- \beta(\sum_i N_i C_i - E) \qquad (6.101)$$

$$- \lambda(\sum_i N_i - N)$$

with respect to N_j, β, and λ.

(b) *Give the equations found by varying $\mathcal{L}$ with respect to N_j. Solve for N_j in terms of λ, β, c_0 and N.*

You should get a power law in j. Our job now is to estimate what power law maximizes the information per unit effort.

We know that λ and β must take values that satisfy the two constraints. First, let us use the normalization condition $N = \sum_j N_j$ to solve for λ.

(c) *Calculate $\sum_j N_j$, and solve for λ.* (Hint: You will want to use the Riemann ζ function, $\zeta(\beta) = \sum_{j=1}^{\infty} j^{-\beta}$.) *Plugging in your formula for λ, find a simpler formula for N_j.*

Now let us calculate the effort in terms of β and c_0.

(d) *Calculate the effort E from eqn 6.99, using β, our assumption 6.100 for the cost C_j of figuring out word j, and your new form for N_j.* (Hint: You will need the derivative $\zeta'(\beta)$ of the the Riemann ζ function. Note that $\mathrm{d}j^{-\beta}/d\beta = -\log(j)j^{-\beta}$.)

We can use this same trick with ζ functions to get a formula for the information $I = -k_S \sum_j N_j \log(N_j/N)$ of eqn 6.98.

(e) *Calculate the information I.*

We are interested in finding the value of β that optimizes the information transfer per unit effort. Let us start with the case where there is no extra effort to recognize (or form) a new word; $c_0 = 0$.

(f) *Notice that $N_j \propto j^{-\beta}$ cannot be normalized for $\beta \leq 1$. Plot $I(\beta)/E(\beta)$ for $c_0 = 0$ in the range β from one to four. Where does it reach a maximum?* In transmitting computer files, we use use only two "words", zero and one, and we do not bother to add spaces between them. *Does this choice jive with your results for $c_0 = 0$?*

(g) *Plot $I(\beta)/E(\beta)$ for $c_0 = 1$ and $c_0 = 2$. Do you find values of β that agree better with the* value assumed in Zipf's law (Fig. 6.21).

This is one of many possible motivations for Zipf's law...[78]

(6.25) **Epidemics and zombies.**[79] (Biology, Epidemiology, Computation) ③

This exercise is based on Alemi and Bierbaum's class project, published in *You can run, you can hide: The epidemiology and statistical mechanics of zombies* [2]. See also the Zombietown site [29] and simulator [30].

Epidemics are studied by disease control specialists using statistical methods, modeling the propagation of the disease as susceptible people are infected, infect others, and recover (e.g., to study measles outbreaks as anti-vaccine sentiments change with time [148]). The SIR model is the simplest commonly studied model (see Exercise 12.33), with three coupled differential equations: $\dot{S} = -\beta SI$ reflects the rate β at which each infected person I infects each susceptible person S, and $\dot{R} = \kappa I$ reflects the rate κ that each infected person joins the recovered population R.

(a) *What is the equation for $\dot{I}$ implied by these first two equations, assuming no infected people die or shift groups other than by new infections or recoveries?*

We shall use a less-common, but even simpler SZR model [2], designed to predict the evolution of a zombie outbreak.

$$\dot{S} = -\beta SZ$$
$$\dot{Z} = (\beta - \kappa)SZ \qquad (6.102)$$
$$\dot{R} = \kappa SZ.$$

Here the zombie population Z never recovers, but if it is destroyed by a member of the surviving population S, it joins the removed population R. The bite parameter β describes the rate at which a zombie bites a human it encounters, and the kill parameter κ gives the rate that a human may destroy a zombie it finds.

The SZR model is even simpler than the SIR model, in that one can write an explicit solution for the evolution as a function of time. We do so in two steps.

(b) *Argue that the only stationary states have all zombies or all humans. Both $\dot{S}$ and $\dot{Z}$ are linear*

---

[78]For example, $\beta \to 1$ generates the largest vocabulary. Perhaps this enhances reproductive success? Communicating with patterns of two types of grunts might not compete in the social arena, no matter how efficient.

[79]Hints for the computations can be found at the book website [181].

in SZ, so there must be a linear combination of the two that has no time dependence. *Show that $P = Z + (1 - \kappa/\beta)S$ satisfies $\dot{P} = 0$. Argue from these two facts that for $P < 0$ the zombies lose.*
(c) *Show that $\chi = S/Z$ satisfies $\dot{\chi} = \gamma\chi$, and so $\chi(t) = \chi_0 \exp(\gamma t)$. Show that $\gamma = -\beta P$.[80] Check that this answer concurs with your criterion for human survival in part (b).*

So the fraction of the doomed species exponentially decays, and the population of the surviving species is determined by P. If desired, one could use the added equation $S(t) + Z(t) + R(t) = N$ with your answers to parts (b) and (c) to solve analytically for the explicit time evolution of S, Z, and R. We shall solve these equations numerically instead (parts e-g).

Suppose now that we start with a single zombie $Z_0 = 1$, and the number of humans S_0 is large. It would seem from our equation for our invariant P from part (b) that if the bite parameter β is greater than the kill parameter κ that the humans are doomed. But surely there is some chance that we will be lucky, and kill the zombie before it bites any of us? This will happen with probability $\kappa/(\beta + \kappa)$. If we fail the first time, we can hope to destroy two zombies before either bites again...

Here is where the statistical fluctuations become important. The differential eqns 6.102 are a continuum approximation to the discrete transitions between integer numbers of the three species. These three equations are similar to reaction rate equations in chemistry (as in eqn 6.50) with molecules replaced by people:

$$\begin{aligned} S + Z &\xrightarrow{\beta SZ} 2Z \\ S + Z &\xrightarrow{\kappa SZ} S + R. \end{aligned} \quad (6.103)$$

Just as for disease outbreaks, if the number of molecules is small then chemical reactions exhibit important statistical fluctuations. These fluctuations are important, for example, for the biology inside cells, where the numbers of a given species of RNA or protein can be small, and the number of DNA sites engaging in creating RNA is usually either zero or one (see Exercises 8.10 and 8.11).
We can simulate each individual bite and kill event for a population of S humans and Z zombies. (Indeed, this can be done rather efficiently for the entire population of the USA;

see [2, 29, 30].) The time to the next event is an exponential random variable given by the total event rate. Which event happens next is then weighted by the individual rates for the different events.

It is well known that decay rates add. Let us nonetheless derive this. Let the probability density of event type #n be $\rho_n(t)$, with survival probability $S_n(t) = 1 - \int_0^t \rho_n(\tau)d\tau$. Let there be two types of events.
(d) *Write the probability density of the first event, $\rho_{\text{tot}}(t)$, in terms of ρ_1, ρ_2, S_1, and S_2. (The final answer should involve no integrals or derivatives.) Specialize to systems with constant rates, which have an exponentially decaying survival time, $S_n(t) = \exp(-\gamma_n t)$. Show that the total event rate γ_{tot} is the sum $\gamma_1 + \gamma_2$. Show that the probability of the next event being of type #1 is $\gamma_1/\gamma_{\text{tot}}$.*

To simulate one step of our discrete SZR model, we

- find the total rate of events γ_{tot},
- increment t by Δt, a random number pulled from an exponential distribution with decay rate γ_{tot},
- choose to bite or to kill by choosing a random number uniform in $(0, \gamma_{\text{tot}})$, and checking if it is less than γ_{bite}, the total rate of bites.
- change S, Z, and R appropriately for the event, and
- perform any observations needed.

This is a simple example of the *Gillespie* algorithm, discussed in more detail in Exercises 8.10 and 8.11.
(e) *Write a routine to use the Gillespie algorithm to solve the discrete SZR model for eqns 6.103, keeping track of t, S, Z, and R for each event, from $0 < t < t_{\max}$ or until $S = 0$ or $Z = 0$; add an extra point at $t_{\max}$ if the system terminates early. Write a routine to numerically solve the continuum equations 6.102. Use $\beta = 0.001$, $\kappa = 0.0008$, and $t_{\max} = 5$. (Should the zombies win?) Plot the two for initial conditions $Z_0 = 100$, $S_0 = 9,900$, and $R_0 = 0$ (a simultaneous outbreak of a hundred zombies). Is the continuum limit faithfully describing the behavior for large numbers of zombies and humans?*
Now let us examine the likelihood of zombies being stamped out despite their advantage in bit-

---

[80]Note that some of the solutions in reference [2] have typographical errors.

ing, if we start with only one zombie.

(f) *Plot the zombie simulation for twenty initial conditions using $Z_0 = 1$, $S_0 = 9,999$, and $R_0 = 0$, together with the continuum solution for the same initial conditions. Are the simulations suffering from more fluctuations than they did for larger number of zombies? Do you see evidence for zombie extinction early in the outbreak? What fraction of the initial outbreaks appear to have killed off all the humans? Zoom in and plot at early times ($Z \leq 5$, $t < 0.5$) and note a few trajectories where the first zombie is killed before biting, and trajectories where the zombie population goes extinct after reaching a peak population of two or three.*

We can write a formula for the probability P_{ext}^{∞} that a single initial zombie will be defeated. As described in reference [2], the probability P_{ext} that the zombies go extinct, in the limit of many humans, is equal to the probability that the first one is destroyed, plus the probability that it bites first times the probability that both zombie lines go extinct:

$$P_{\text{ext}}^{\infty} = \frac{\kappa}{\beta + \kappa} + \frac{\beta}{\beta + \kappa}(P_{\text{ext}}^{\infty})^2 \qquad (6.104)$$

where we assume a large initial human population $S_0 \to \infty$. This can be solved to show $P_{\text{ext}}^{\infty} = \kappa/\beta$.

Exactly this same argument holds for regular disease outbreaks. Similar arguments can be used to determine the likelihood that an advantageous gene mutation will take over a large population.

(g) *Was the fraction of extinctions you observed in part (f) roughly given by the calculation above? Write a simpler routine that simulates the Gillespie algorithm over n epidemics, reporting the fraction in which zombies go extinct (observing only whether $Z = 0$ happens before $S = 0$, ignoring the time and the trajectory). For $S_0 = 9,999$ and 1,000 epidemics, how good is the prediction?*

(6.26) **Nucleosynthesis as a chemical reaction.**[81]
(Astrophysics) ③
As discussed in Exercise 5.24, the very early Universe was so hot that any heavier nuclei quickly

evaporated into protons and neutrons. Between a few seconds and a couple of minutes after the Big Bang, protons and neutrons began to fuse into light nuclei—mostly helium. The Universe contracted too fast, however, to allow the heavier elements to form. Almost all of the heavier elements on Earth were formed later, inside stars. In this exercise, we calculate how and when the nucleons would have fused into heavy elements if the energy barriers to the nuclear reactions were low enough, or the expansion rate were slow enough, for the Universe to have stayed in equilibrium.

The nucleons for the equilibrated Universe will all fuse into their most stable form[82] long before it cools to the present temperature. The binding energy ΔE released by fusing nucleons into ^{56}Fe is about 5×10^8 eV (about half a proton mass). In this exercise, we shall ignore the difference between protons and neutrons,[83] ignore nuclear excitations (assuming no internal entropy for the nuclei, so ΔE is the free energy difference), and ignore the electrons (so protons are the same as hydrogen atoms, etc.)

The creation of ^{56}Fe from nucleons involves a complex cascade of reactions. We argued in Section 6.6 that however complicated the reactions, they must in net be described by the overall reaction, here

$$56\text{p} \to {}^{56}\text{Fe}, \qquad (6.105)$$

releasing an energy of about $\Delta E = 5 \times 10^8$ eV or about $\Delta E/56 = 1.4 \times 10^{-12}$ joules/baryon. We used the fact that most chemical species are dilute (and hence can be described by an ideal gas) to derive the corresponding law of mass-action, here

$$[\text{Fe}]/[\text{p}]^{56} = K_{\text{eq}}(T). \qquad (6.106)$$

(Note that this equality, true in equilibrium no matter how messy the necessary reactions pathways, can be rationalized using the oversimplified picture that 56 protons must simultaneously collect in a small region to form an iron nucleus.) We then used the Helmholtz free energy for an ideal gas to calculate the reaction constant $K_{\text{eq}}(T)$ for the reaction, in the ideal gas limit.

---

[81]This exercise was developed in collaboration with Katherine Quinn.

[82]It is usually said that ^{56}Fe is the most stable nucleus, but actually ^{62}Ni has a higher nuclear binding energy per nucleon. Iron-56 is favored by nucleosynthesis pathways and conditions inside stars, and we shall go with tradition and use it for our calculation.

[83]The entropy cost for our reaction in reality depends upon the ratio of neutrons to protons. Calculations imply that this ratio falls out of equilibrium a bit earlier, and is about 1/7 during this period. By ignoring the difference between neutrons and protons, we avoid considering reactions of the form $M\text{p} + (56 - M)\text{n} + (M - 26)\text{e} \to {}^{56}\text{Fe}$.

(a) *Give symbolic formulas for $K_{eq}(T)$ in eqn 6.106, assuming the nucleons form an approximate ideal gas. Reduce it to expressions in terms of ΔE, $k_B T$, h, m_p, and the atomic number $A = 56$. (For convenience, assume $m_{Fe} = A m_p$, ignoring the half-proton-mass energy release. Check the sign of ΔE: should it be positive or negative for this reaction?) Evaluate $K_{eq}(T)$ in MKS units, as a function of T.* (Note: Do not be alarmed by the large numbers.) Wikipedia tells us that nucleosynthesis happened "A few minutes into the expansion, when the temperature was about a billion . . . Kelvin and the density was about that of air". To figure out when half of the protons would have fused ([Fe] = [p]/A), we need to know the temperature and the net baryon density [p] + A[Fe]. We can use eqn 6.106 to give one equation relating the temperature to the baryon density.

(b) *Give the symbolic formula for the net baryon density[84] at which half of the protons would have fused, in terms of K_{eq} and A.*

What can we use for the other equation relating temperature to baryon density? One of the fundamental constants in the Universe as it evolves is the baryon to photon number ratio. The total number of baryons has been conserved in accelerator experiments of much higher energy than those during nucleosynthesis. The cosmic microwave background (CMB) photons have mostly been traveling in straight lines[85] since the decoupling time, a few hundred thousand years after the Big Bang when the electrons and nucleons combined into atoms and became transparent (decoupled from photons). The ratio of baryons to CMB photons $\eta \sim 5 \times 10^{-10}$ has been constant since that point: two billion photons per nucleon.[86] Between the nucleosynthesis and decoupling eras, photons were created and scattered by matter, but there were so many more photons than baryons that the number of photons (and hence the baryon to photon ratio η) was still approximately constant.

We know the density of black-body photons at a temperature T; we integrate eqn 7.66 for the number of photons per unit frequency to get

$$\rho_{photons}(T) = (2\zeta(3)/\pi^2)(k_B T/\hbar c)^3; \quad (6.107)$$

the current density $\rho_{photons}(T_{CMB})$ of microwave background photons at temperature $T_{CMB} = 2.725\,\text{K}$ is thus a bit over 400 per cubic centimeter, and hence the current density of a fraction of a baryon per cubic meter.

(c) *Numerically matching your formula for the net baryon density at the halfway point in part (b) to $\eta\rho_{photons}(T_{reaction})$, derive the temperature $T_{reaction}$ at which our reaction would have occurred. (You can do this graphically, if needed.) Check Wikipedia's assertions. Is $T_{reaction}$ roughly a billion degrees Kelvin? Is the nucleon density roughly equal to that of air?* (Hint: Air has a density of about 1 kg/m^3.)

---

[84]Here we ask you to assume, unphysically, that all baryons are either free protons or in iron nuclei—ignoring the baryons forming helium and other elements. These other elements will be rare at early (hot) times, and rare again at late (cold) times, but will spread out the energy release from nucleosynthesis over a larger range than our one-reaction estimate would suggest.

[85]More precisely, they have traveled on geodesics in space-time.

[86]It is amusing to note that this ratio is estimated using the models of nucleosynthesis that we are mimicking in this exercise.

Quantum statistical mechanics

7

Quantum statistical mechanics governs most of solid-state physics (metals, semiconductors, and glasses) and parts of molecular physics and astrophysics (white dwarfs, neutron stars). Statistical mechanics spawned the origin of quantum mechanics (Planck's theory of the black-body spectrum). Quantum statistical mechanics forms the framework for our understanding of other exotic quantum phenomena (Bose condensation, superfluids, and superconductors). Applications of quantum statistical mechanics are significant components of courses in these various subjects. We condense our treatment of this important subject into this one chapter in order to avoid overlap with other physics and chemistry courses, and also in order to keep our treatment otherwise accessible to those uninitiated into the quantum mysteries.

In this chapter we assume the reader has some background in quantum mechanics. We will proceed from the abstract to the concrete, through a series of simplifications. We begin (Section 7.1) by introducing *mixed states* for quantum ensembles, and the advanced topic of *density matrices* (for non stationary quantum systems which are necessarily not mixtures of energy eigenstates). We illustrate mixed states in Section 7.2 by solving the finite-temperature quantum harmonic oscillator. We discuss the statistical mechanics of identical particles (Section 7.3). We then make the vast simplification of presuming that the particles are noninteracting (Section 7.4), which leads us to the Bose–Einstein and Fermi distributions for the filling of single-particle eigenstates. We contrast Bose, Fermi, and Maxwell–Boltzmann statistics in Section 7.5. We illustrate how amazingly useful the noninteracting particle picture is for quantum systems by solving the classic problems of black-body radiation and Bose condensation (Section 7.6), and for the behavior of metals (Section 7.7).

7.1 Mixed states and density matrices

Classical statistical ensembles are probability distributions $\rho(\mathbb{P}, \mathbb{Q})$ in phase space. How do we generalize them to quantum mechanics? Two problems immediately arise. First, the Heisenberg uncertainty principle tells us that one cannot specify both position and momentum for a quantum system at the same time. The states of our quantum sys-

[1]Quantum systems with many particles
have wavefunctions that are functions
of all the positions of all the particles
(or, in momentum space, all the momenta of all the particles).

tem will not be points in phase space. Second, quantum mechanics
already has probability densities; even for systems in a definite state[1]
$\Psi(\mathbb{Q})$ the probability is spread among different configurations $|\Psi(\mathbb{Q})|^2$
(or momenta $|\widetilde{\Psi}(\mathbb{P})|^2$). In statistical mechanics, we need to introduce a
second level of probability, to discuss an ensemble that has probabilities
p_n of being in a variety of quantum states $\Psi_n(\mathbb{Q})$. Ensembles in quantum mechanics are called *mixed states*; they are not superpositions of
different wavefunctions, but incoherent mixtures.[2]

Suppose we want to compute the ensemble expectation of an operator
$\mathbf{A}$. In a particular state Ψ_n, the quantum expectation is

$$\langle \mathbf{A} \rangle_{\text{pure}} = \int \Psi_n^*(\mathbb{Q}) \mathbf{A} \Psi_n(\mathbb{Q}) \, \mathrm{d}^{3N}\mathbb{Q}. \tag{7.1}$$

So, in the ensemble the expectation is

$$\langle \mathbf{A} \rangle = \sum_n p_n \int \Psi_n^*(\mathbb{Q}) \mathbf{A} \Psi_n(\mathbb{Q}) \, \mathrm{d}^{3N}\mathbb{Q}. \tag{7.2}$$

[2]So, for example, if $|V\rangle$ is a vertically
polarized photon, and $|H\rangle$ is a horizon-
tally polarized photon, then the super-
position $(1/\sqrt{2})\,(|V\rangle + |H\rangle)$ is a diag-
onally polarized photon, while the un-
polarized photon is a mixture of half
$|V\rangle$ and half $|H\rangle$, described by the den-
sity matrix $\frac{1}{2}(|V\rangle\langle V| + |H\rangle\langle H|)$. The
superposition is in *both* states, the mix-
ture is in perhaps one or perhaps the
other (see Exercise 7.5).

Except for selected exercises, for the rest of the book we will use mixtures
of states (eqn 7.2). Indeed, for all of the equilibrium ensembles, the
Ψ_n may be taken to be the energy eigenstates, and the p_n either a
constant in a small energy range (for the microcanonical ensemble), or
$\exp(-\beta E_n)/Z$ (for the canonical ensemble), or $\exp\left(-\beta(E_n - N_n\mu)\right)/\Xi$
(for the grand canonical ensemble). For most practical purposes you may
stop reading this section here, and proceed to the quantum harmonic
oscillator.

Density matrices (advanced)

What do we gain from going beyond mixed states? First, there are
lots of time-dependent systems that cannot be described as mixtures
of energy eigenstates. (Any such mixture will be time independent.)
Second, although one can define a general, time-dependent ensemble in
terms of more general bases Ψ_n, it is useful to be able to transform between a variety of bases. For example, superfluids and superconductors
show an exotic *off-diagonal long-range order* when looked at in position
space (Exercise 9.8). Third, we will see that the proper generalization of
Liouville's theorem demands the more elegant, operator-based approach.

Our goal is to avoid carrying around the particular states Ψ_n. Instead,
we will write the ensemble average (eqn 7.2) in terms of $\mathbf{A}$ and an
operator $\boldsymbol{\rho}$, the *density matrix*. For this section, it is convenient to
use Dirac's bra–ket notation, in which the mixed-state ensemble average
can be written[3]

[3]In Dirac's notation, $\langle \Psi | \mathbf{M} | \Phi \rangle = \int \Psi^* \mathbf{M} \Phi$.

$$\langle \mathbf{A} \rangle = \sum_n p_n \langle \Psi_n | \mathbf{A} | \Psi_n \rangle. \tag{7.3}$$

Pick any complete orthonormal basis Φ_α. Then the identity operator is

$$\mathbf{1} = \sum_\alpha |\Phi_\alpha\rangle\langle\Phi_\alpha| \tag{7.4}$$

and, substituting the identity (eqn 7.4) into eqn 7.3 we find

$$
\begin{aligned}
\langle \mathbf{A} \rangle &= \sum_n p_n \langle \Psi_n | \left(\sum_\alpha |\Phi_\alpha\rangle\langle\Phi_\alpha| \right) \mathbf{A} |\Psi_n\rangle \\
&= \sum_n p_n \sum_\alpha \langle \Phi_\alpha | \mathbf{A} \Psi_n\rangle\langle\Psi_n|\Phi_\alpha\rangle \\
&= \sum_\alpha \langle \Phi_\alpha \mathbf{A} | \left(\sum_n p_n |\Psi_n\rangle\langle\Psi_n| \right) |\Phi_\alpha\rangle \\
&= \mathrm{Tr}(\mathbf{A}\boldsymbol{\rho}),
\end{aligned}
\tag{7.5}
$$

where[4]

$$
\boldsymbol{\rho} = \left(\sum_n p_n |\Psi_n\rangle\langle\Psi_n| \right)
\tag{7.6}
$$

is the *density matrix*.

There are several properties we can now deduce about the density matrix.

Sufficiency. In quantum mechanics, all measurement processes involve expectation values of operators. Our density matrix therefore suffices to embody everything we need to know about our quantum system.

Pure states. A pure state, with a definite wavefunction Φ, has $\boldsymbol{\rho}_{\mathrm{pure}} = |\Phi\rangle\langle\Phi|$. In the position basis $|\mathbb{Q}\rangle$, this pure-state density matrix has matrix elements $\boldsymbol{\rho}_{\mathrm{pure}}(\mathbb{Q}, \mathbb{Q}') = \langle\mathbb{Q}|\boldsymbol{\rho}_{\mathrm{pure}}|\mathbb{Q}'\rangle = \Phi^*(\mathbb{Q}')\Phi(\mathbb{Q})$. Thus in particular we can reconstruct[5] the wavefunction from a pure-state density matrix, up to a physically unmeasurable overall phase. Since our wavefunction is normalized $\langle\Phi|\Phi\rangle = 1$, we note also that the square of the density matrix for a pure state equals itself: $\boldsymbol{\rho}_{\mathrm{pure}}^2 = |\Phi\rangle\langle\Phi||\Phi\rangle\langle\Phi| = |\Phi\rangle\langle\Phi| = \boldsymbol{\rho}_{\mathrm{pure}}$.

Normalization. The trace of a pure-state density matrix $\mathrm{Tr}\boldsymbol{\rho}_{\mathrm{pure}} = 1$, since we can pick an orthonormal basis with our wavefunction Φ as the first basis element, making the first term in the trace sum one and the others zero. The trace of a general density matrix is hence also one, since it is a probability distribution of pure-state density matrices:

$$
\mathrm{Tr}\boldsymbol{\rho} = \mathrm{Tr}\left(\sum_n p_n |\Psi_n\rangle\langle\Psi_n| \right) = \sum_n p_n \mathrm{Tr}\left(|\Psi_n\rangle\langle\Psi_n| \right) = \sum_n p_n = 1.
\tag{7.7}
$$

Canonical distribution. The canonical distribution is a mixture of the energy eigenstates $|E_n\rangle$ with Boltzmann weights $\exp(-\beta E_n)$. Hence the density matrix $\boldsymbol{\rho}_{\mathrm{canon}}$ is diagonal in the energy basis:[6]

$$
\boldsymbol{\rho}_{\mathrm{canon}} = \sum_n \frac{\exp(-\beta E_n)}{Z} |E_n\rangle\langle E_n|.
\tag{7.8}
$$

We can write the canonical density matrix in a basis-independent form using the Hamiltonian operator $\mathcal{H}$. First, the partition function is given[7]

---

[4]The trace of a matrix is the sum of its diagonal elements, and is independent of the basis in which you write it. The same is true of operators; we are summing the diagonal elements $\mathrm{Tr}(\mathbf{M}) = \sum_\alpha \langle\Phi_\alpha|\mathbf{M}|\Phi_\alpha\rangle$.

[5]Since Φ is normalized, $|\Phi(\mathbb{Q}')|^2 = \int d\widetilde{\mathbb{Q}}\,|\rho(\widetilde{\mathbb{Q}}, \mathbb{Q}')|^2$. Pick any point $\mathbb{Q}'$ where $\Phi(\mathbb{Q}') = |\Phi(\mathbb{Q}')|\exp(i\theta(\mathbb{Q}'))$ is not zero. Then

$$
\Phi^*(\mathbb{Q}') = \sqrt{\int d\widetilde{\mathbb{Q}}\,|\rho(\widetilde{\mathbb{Q}}, \mathbb{Q}')|^2}\,e^{-i\theta(\mathbb{Q}')}
$$

and thus

$$
\Phi(\mathbb{Q}) = \frac{\rho(\mathbb{Q}, \mathbb{Q}')}{\sqrt{\int d\widetilde{\mathbb{Q}}\,|\rho(\widetilde{\mathbb{Q}}, \mathbb{Q}')|^2}}\,e^{i\theta(\mathbb{Q}')},
$$

a function of ρ times the single phase $\exp(i\theta(\mathbb{Q}'))$.

[6]Notice that the states Ψ_n in a general mixture need not be eigenstates or even orthogonal.

[7]What is the exponential of a matrix M? We can define it in terms of a power series, $\exp(M) = \mathbf{1} + M + M^2/2! + M^3/3! + \ldots$, but it is usually easier to change basis to diagonalize M. In that basis, any function $f(M)$ is given by

$$
f(\boldsymbol{\rho}) = \begin{pmatrix} f(\boldsymbol{\rho}_{11}) & 0 & 0 \ldots \\ 0 & f(\boldsymbol{\rho}_{22}) & 0 \ldots \\ \ldots & & \end{pmatrix}.
\tag{7.9}
$$

At the end, change back to the original basis. This procedure also defines $\log M$ (eqn 7.13).

by the trace

$$Z = \sum_n \exp(-\beta E_n) = \sum_n \langle E_n | \exp(-\beta \mathcal{H}) | E_n \rangle = \text{Tr}\left(\exp(-\beta \mathcal{H})\right). \tag{7.10}$$

Second, the numerator

$$\sum_n |E_n\rangle \exp(-\beta E_n)\langle E_n| = \exp(-\beta \mathcal{H}), \tag{7.11}$$

since $\mathcal{H}$ (and thus $\exp(-\beta \mathcal{H})$) is diagonal in the energy basis. Hence

$$\boldsymbol{\rho}_{\text{canon}} = \frac{\exp(-\beta \mathcal{H})}{\text{Tr}\left(\exp(-\beta \mathcal{H})\right)}. \tag{7.12}$$

Entropy. The entropy for a general density matrix will be

$$S = -k_B \text{Tr}(\boldsymbol{\rho} \log \boldsymbol{\rho}). \tag{7.13}$$

Time evolution for the density matrix. The time evolution for the density matrix is determined by the time evolution of the pure states composing it:[8]

$$\frac{\partial \boldsymbol{\rho}}{\partial t} = \sum_n p_n \left(\frac{\partial |\Psi_n\rangle}{\partial t} \langle \Psi_n| + |\Psi_n\rangle \frac{\partial \langle \Psi_n|}{\partial t} \right). \tag{7.14}$$

Now, the time evolution of the ket wavefunction $|\Psi_n\rangle$ is given by operating on it with the Hamiltonian:

$$\frac{\partial |\Psi_n\rangle}{\partial t} = \frac{1}{i\hbar} \mathcal{H} |\Psi_n\rangle, \tag{7.15}$$

and the time evolution of the bra wavefunction $\langle \Psi_n|$ is given by the time evolution of $\Psi_n^*(\mathbb{Q})$:

$$\frac{\partial \Psi_n^*}{\partial t} = \left(\frac{\partial \Psi_n}{\partial t} \right)^* = \left(\frac{1}{i\hbar} \mathcal{H} \Psi_n \right)^* = -\frac{1}{i\hbar} \mathcal{H} \Psi_n^*, \tag{7.16}$$

so since $\mathcal{H}$ is Hermitian, we have

$$\frac{\partial \langle \Psi_n|}{\partial t} = -\frac{1}{i\hbar} \langle \Psi_n| \mathcal{H}. \tag{7.17}$$

Hence[9]

$$\frac{\partial \boldsymbol{\rho}}{\partial t} = \sum_n p_n \frac{1}{i\hbar} \left(\mathcal{H} |\Psi_n\rangle \langle \Psi_n| - |\Psi_n\rangle \langle \Psi_n| \mathcal{H} \right) = \frac{1}{i\hbar} \left(\mathcal{H} \boldsymbol{\rho} - \boldsymbol{\rho} \mathcal{H} \right)$$

$$= \frac{1}{i\hbar} [\mathcal{H}, \boldsymbol{\rho}]. \tag{7.18}$$

Quantum Liouville theorem. This time evolution law 7.18 is the quantum version of Liouville's theorem. We can see this by using the equations of motion 4.1, $\dot{q}_\alpha = \partial \mathcal{H}/\partial p_\alpha$, and $\dot{p}_\alpha = -\partial \mathcal{H}/\partial q_\alpha$ and the definition of Poisson brackets

$$\{A, B\}_P = \sum_\alpha \frac{\partial A}{\partial q_\alpha} \frac{\partial B}{\partial p_\alpha} - \frac{\partial A}{\partial p_\alpha} \frac{\partial B}{\partial q_\alpha}. \tag{7.19}$$

[8]The p_n are the probability that one *started* in the state Ψ_n, and thus manifestly do not change with time.

[9]The *commutator* of two matrices is $[A, B] = AB - BA$. Notice that eqn 7.18 is *minus* the formula one uses for the time evolution of operators in the Heisenberg representation.

Liouville's theorem tells us that the total time derivative of the phase-space probability density is zero (eqn 4.7). We can turn this into a statement about the partial time derivative:

$$0 = \frac{d\rho}{dt} = \frac{\partial \rho}{\partial t} + \sum_\alpha \frac{\partial \rho}{\partial q_\alpha} \dot{q}_\alpha + \frac{\partial \rho}{\partial p_\alpha} \dot{p}_\alpha$$
$$= \frac{\partial \rho}{\partial t} + \sum_\alpha \left(\frac{\partial \rho}{\partial q_\alpha} \frac{\partial \mathcal{H}}{\partial p_\alpha} - \frac{\partial \rho}{\partial p_\alpha} \frac{\partial \mathcal{H}}{\partial q_\alpha} \right). \qquad (7.20)$$

Hence

$$\partial \rho / \partial t = \{\mathcal{H}, \boldsymbol{\rho}\}_P. \qquad (7.21)$$

Using the classical↔quantum correspondence between the Poisson brackets and the commutator $\{\ \}_P \leftrightarrow (1/i\hbar)[\ ]$ the time evolution law 7.18 is precisely the analogue of Liouville's theorem 7.21.

Quantum Liouville and statistical mechanics. The classical version of Liouville's equation is far more compelling an argument for statistical mechanics than is the quantum version. The classical theorem, you remember, states that $d\rho/dt = 0$; the density following a point on the trajectory is constant, hence any time-independent density must have ρ constant along the trajectories. If the trajectory covers the energy surface (ergodicity), then the probability density has to be constant on the energy surface, justifying the microcanonical ensemble.

For an isolated quantum system, this argument breaks down. The condition that an equilibrium state must be time independent is not very stringent. Indeed, $\partial \boldsymbol{\rho}/\partial t = [\mathcal{H}, \boldsymbol{\rho}] = 0$ for any mixture of many-body energy eigenstates. In principle, isolated quantum systems are very nonergodic, and one must couple them to the outside world to induce transitions between the many-body eigenstates needed for equilibration.[10]

7.2 Quantum harmonic oscillator

The harmonic oscillator is a great example of how statistical mechanics works in quantum systems. Consider an oscillator of frequency ω. The energy eigenvalues are $E_n = (n + \frac{1}{2})\hbar\omega$ (Fig. 7.1). Hence its partition function is a geometric series $\sum x^n$, which we can sum to $1/(1-x)$:

$$Z_{\text{qho}} = \sum_{n=0}^{\infty} e^{-\beta E_n} = \sum_{n=0}^{\infty} e^{-\beta\hbar\omega(n+1/2)}$$
$$= e^{-\beta\hbar\omega/2} \sum_{n=0}^{\infty} \left(e^{-\beta\hbar\omega} \right)^n = e^{-\beta\hbar\omega/2} \frac{1}{1 - e^{-\beta\hbar\omega}}$$
$$= \frac{1}{e^{\beta\hbar\omega/2} - e^{-\beta\hbar\omega/2}} = \frac{1}{2\sinh(\beta\hbar\omega/2)}. \qquad (7.22)$$

[10]This may seem less of a concern when one realizes just how peculiar many-body eigenstates are. What would a many-body eigenstate look like, corresponding to an atom in an excited state, inside a large box at zero temperature? The normal analysis considers the atom to be in an energy eigenstate which decays after some time into a ground-state atom plus some photons. Clearly the atom was only in an approximate eigenstate (or it would not decay); it is in a *resonance*, which has an imaginary part to its energy. The joint photon-atom many-body quantum state is a weird superposition of states with photons being absorbed by the atom and the atom emitting photons, carefully crafted to produce a stationary state.

Many-body eigenstates of systems with a nonzero energy per unit volume often are quantitatively described by equilibrium statistical mechanics. If the quantum dynamics is ergodic enough, any local region has a density matrix corresponding to a thermal ensemble (see Exercises 7.17, 7.26, and 7.27). Thus these energy eigenstates are time independent *because the quantum system is already in equilibrium.*

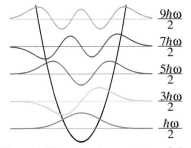

Fig. 7.1 The quantum states of the harmonic oscillator are at equally spaced energies.

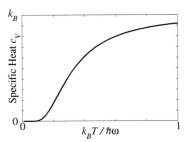

Fig. 7.2 The specific heat for the quantum harmonic oscillator.

[11]We call it the energy gap in solid-state physics; it is the minimum energy needed to add an excitation to the system. In quantum field theory, where the excitations are particles, we call it the *mass* of the particle mc^2.

[12]In three dimensions, this phase change must be ±1. In two dimensions one can have any phase change, so one can have not only fermions and bosons but *anyons*. Anyons, with fractional statistics, arise as excitations in the fractional quantized Hall effect.

[13]Examples of bosons include mesons, He^4, phonons, photons, gluons, W$^\pm$ and Z bosons, and (presumably) gravitons. The last four mediate the fundamental forces—the electromagnetic, strong, weak, and gravitational interactions. The spin-statistics theorem (not discussed here) states that bosons have integer spins.

[14]Most of the common elementary particles are fermions: electrons, protons, neutrons, neutrinos, quarks, etc. Fermions have half-integer spins. Particles made up of even numbers of fermions are bosons.

[15]A permutation $\{P_1, P_2, \ldots, P_N\}$ is just a reordering of the integers $\{1, 2, \ldots, N\}$. The sign $\sigma(P)$ of a permutation is $+1$ if P is an even permutation, and -1 if P is an odd permutation. Swapping two labels, keeping all the rest unchanged, is an odd permutation. One can show that composing two permutations multiplies their signs, so odd permutations can be made by odd numbers of pair swaps, and even permutations are composed of even numbers of pair swaps.

The average energy is

$$\langle E \rangle_{\text{qho}} = -\frac{\partial \log Z_{\text{qho}}}{\partial \beta} = \frac{\partial}{\partial \beta}\left[\frac{1}{2}\beta\hbar\omega + \log\left(1 - e^{-\beta\hbar\omega}\right)\right]$$
$$= \hbar\omega\left(\frac{1}{2} + \frac{e^{-\beta\hbar\omega}}{1 - e^{-\beta\hbar\omega}}\right) = \hbar\omega\left(\frac{1}{2} + \frac{1}{e^{\beta\hbar\omega} - 1}\right), \qquad (7.23)$$

which corresponds to an average excitation level

$$\langle n \rangle_{\text{qho}} = \frac{1}{e^{\beta\hbar\omega} - 1}. \qquad (7.24)$$

The specific heat is thus

$$c_V = \frac{\partial E}{\partial T} = k_B \left(\frac{\hbar\omega}{k_B T}\right)^2 \frac{e^{-\hbar\omega/k_B T}}{\left(1 - e^{-\hbar\omega/k_B T}\right)^2} \qquad (7.25)$$

(Fig. 7.2). At high temperatures, $e^{-\hbar\omega/k_B T} \approx 1 - \hbar\omega/k_B T$, so $c_V \to k_B$ as we found for the classical harmonic oscillator (and as given by the equipartition theorem). At low temperatures, $e^{-\hbar\omega/k_B T}$ becomes exponentially small, so the specific heat goes rapidly to zero as the energy asymptotes to the zero-point energy $\frac{1}{2}\hbar\omega$. More specifically, there is an energy gap[11] $\hbar\omega$ to the first excitation, so the probability of having any excitation of the system is suppressed by a factor of $e^{-\hbar\omega/k_B T}$.

7.3 Bose and Fermi statistics

In quantum mechanics, identical particles are not just hard to tell apart—their quantum wavefunctions must be the same, up to an overall phase change,[12] when the coordinates are swapped (see Fig. 7.3). In particular, for bosons[13] the wavefunction is unchanged under a swap, so

$$\Psi(\mathbf{r}_1, \mathbf{r}_2, \ldots, \mathbf{r}_N) = \Psi(\mathbf{r}_2, \mathbf{r}_1, \ldots, \mathbf{r}_N) = \Psi(\mathbf{r}_{P_1}, \mathbf{r}_{P_2}, \ldots, \mathbf{r}_{P_N}) \qquad (7.26)$$

for any permutation P of the integers $1, \ldots, N$. For fermions[14]

$$\Psi(\mathbf{r}_1, \mathbf{r}_2, \ldots, \mathbf{r}_N) = -\Psi(\mathbf{r}_2, \mathbf{r}_1, \ldots, \mathbf{r}_N) = \sigma(P)\,\Psi(\mathbf{r}_{P_1}, \mathbf{r}_{P_2}, \ldots, \mathbf{r}_{P_N}), \qquad (7.27)$$

where $\sigma(P)$ is the sign of the permutation P.[15]

The eigenstates for systems of identical fermions and bosons are a subset of the eigenstates of distinguishable particles with the same Hamiltonian:

$$\mathcal{H}\Psi_n = E_n\Psi_n; \qquad (7.28)$$

in particular, they are given by the distinguishable eigenstates which obey the proper symmetry properties under permutations. A non-symmetric eigenstate Φ with energy E may be symmetrized to form a Bose eigenstate by summing over all possible permutations P:

$$\Psi_{\text{sym}}(\mathbf{r}_1, \mathbf{r}_2, \ldots, \mathbf{r}_N) = (\text{normalization})\sum_P \Phi(\mathbf{r}_{P_1}, \mathbf{r}_{P_2}, \ldots, \mathbf{r}_{P_N}) \qquad (7.29)$$

or antisymmetrized to form a fermion eigenstate

$$\Psi_{\mathrm{asym}}(\mathbf{r}_1, \mathbf{r}_2, \ldots, \mathbf{r}_N) = (\text{normalization}) \sum_P \sigma(P)\Phi(\mathbf{r}_{P_1}, \mathbf{r}_{P_2}, \ldots, \mathbf{r}_{P_N}) \tag{7.30}$$

if the symmetrization or antisymmetrization does not make the sum zero. These remain eigenstates of energy E, because they are combinations of eigenstates of energy E.

Quantum statistical mechanics for identical particles is given by restricting the ensembles to sum over symmetric wavefunctions for bosons or antisymmetric wavefunctions for fermions. So, for example, the partition function for the canonical ensemble is still

$$Z = \mathrm{Tr}\left(\mathrm{e}^{-\beta H}\right) = \sum_n \mathrm{e}^{-\beta E_n}, \tag{7.31}$$

but now the trace is over a complete set of many-body symmetric (or antisymmetric) states, and the sum is over the symmetric (or antisymmetric) many-body energy eigenstates.

7.4 Noninteracting bosons and fermions

Many-body quantum statistical mechanics is hard. We now make a huge approximation: we will assume our quantum particles do not interact with one another. Just as for the classical ideal gas, this will make our calculations straightforward.

The noninteracting Hamiltonian is a sum of single-particle quantum Hamiltonians H:

$$\mathcal{H}^{\mathrm{NI}} = \sum_{j=1}^N H(\mathbf{p}_j, \mathbf{r}_j) = \sum_{j=1}^N \frac{\hbar^2}{2m}\nabla_j^2 + V(\mathbf{r}_j). \tag{7.32}$$

Let ψ_k be the single-particle eigenstates of H, then

$$H\psi_k(\mathbf{r}) = \varepsilon_k \psi_k(\mathbf{r}). \tag{7.33}$$

For distinguishable particles, the many-body eigenstates can be written as a product of orthonormal single-particle eigenstates:

$$\Psi_{\mathrm{dist}}^{\mathrm{NI}}(\mathbf{r}_1, \mathbf{r}_2, \ldots, \mathbf{r}_N) = \prod_{j=1}^N \psi_{k_j}(\mathbf{r}_j), \tag{7.34}$$

where particle j is in the single-particle eigenstate k_j. The eigenstates for noninteracting bosons are given by symmetrizing over the coordinates r_j:

$$\Psi_{\mathrm{boson}}^{\mathrm{NI}}(\mathbf{r}_1, \mathbf{r}_2, \ldots, \mathbf{r}_N) = (\text{normalization}) \sum_P \prod_{j=1}^N \psi_{k_j}(\mathbf{r}_{P_j}), \tag{7.35}$$

Fig. 7.3 Feynman diagram: identical particles. In quantum mechanics, two electrons (or two atoms of the same isotope) are fundamentally identical. We can illustrate this with a peek at an advanced topic mixing quantum field theory and relativity. Here is a scattering event of a photon off an electron, viewed in two reference frames; time is vertical, a spatial coordinate is horizontal. On the left we see two "different" electrons, one which is created along with an anti-electron or positron e^+, and the other which later annihilates the positron. On the right we see the same event viewed in a different reference frame; here there is only one electron, which scatters two photons. (The electron is *virtual*, moving faster than light, between the collisions; this is allowed in intermediate states for quantum transitions.) The two electrons on the left are not only indistinguishable, they are the *same particle*! The antiparticle is also the electron, traveling backward in time.

This antisymmetrization can be written as

$$\frac{1}{\sqrt{N!}} \begin{vmatrix} \psi_{k_1}(\mathbf{r_1}) & \dots & \psi_{k_1}(\mathbf{r_N}) \\ \psi_{k_2}(\mathbf{r_1}) & \dots & \psi_{k_2}(\mathbf{r_N}) \\ \dots & & \dots \\ \psi_{k_N}(\mathbf{r_1}) & \dots & \psi_{k_N}(\mathbf{r_N}) \end{vmatrix}$$ (7.36)

called the *Slater determinant*.

[17]Notice that the normalization of the boson wavefunction depends on how many single-particle states are multiply occupied.

[18]Because the spin of the electron can be in two directions $\pm 1/2$, this means that two electrons can be placed into each single-particle spatial eigenstate.

and naturally the fermion eigenstates are given by antisymmetrizing over all $N!$ possible permutations, and renormalizing to one,[16]

$$\Psi_{\text{fermion}}^{\text{NI}}(\mathbf{r}_1, \mathbf{r}_2, \dots, \mathbf{r}_N) = \frac{1}{\sqrt{N!}} \sum_P \sigma(P) \prod_{j=1}^{N} \psi_{k_j}(\mathbf{r}_{P_j}).$$ (7.37)

Let us consider two particles in orthonormal single-particle energy eigenstates ψ_k and ψ_ℓ. If the particles are distinguishable, there are two eigenstates $\psi_k(\mathbf{r_1})\psi_\ell(\mathbf{r_2})$ and $\psi_k(\mathbf{r_2})\psi_\ell(\mathbf{r_1})$. If the particles are bosons, the eigenstate is $(1/\sqrt{2})(\psi_k(\mathbf{r_1})\psi_\ell(\mathbf{r_2}) + \psi_k(\mathbf{r_2})\psi_\ell(\mathbf{r_1}))$. If the particles are fermions, the eigenstate is $(1/\sqrt{2})(\psi_k(\mathbf{r_1})\psi_\ell(\mathbf{r_2}) - \psi_k(\mathbf{r_2})\psi_\ell(\mathbf{r_1}))$.

What if the particles are in the same single-particle eigenstate ψ_ℓ? For bosons, the eigenstate $\psi_\ell(\mathbf{r_1})\psi_\ell(\mathbf{r_2})$ is already symmetric and normalized.[17] For fermions, antisymmetrizing a state where both particles are in the same state gives zero: $\psi_\ell(\mathbf{r_1})\psi_\ell(\mathbf{r_2}) - \psi_\ell(\mathbf{r_2})\psi_\ell(\mathbf{r_1}) = 0$. This is the Pauli exclusion principle: you cannot have two fermions in the same quantum state.[18]

How do we do statistical mechanics for noninteracting fermions and bosons? Here it is most convenient to use the grand canonical ensemble (Section 6.3); in this ensemble we can treat each eigenstate as being populated independently from the other eigenstates, exchanging particles directly with the external bath (analogous to Fig. 6.2). The grand partition function hence factors:

$$\Xi^{\text{NI}} = \prod_k \Xi_k.$$ (7.38)

The grand canonical ensemble thus allows us to separately solve the case of noninteracting particles one eigenstate at a time.

Bosons. For bosons, all fillings n_k are allowed. Each particle in eigenstate ψ_k contributes energy ε_k and chemical potential $-\mu$, so

$$\Xi_k^{\text{boson}} = \sum_{n_k=0}^{\infty} e^{-\beta(\varepsilon_k-\mu)n_k} = \sum_{n_k=0}^{\infty} \left(e^{-\beta(\varepsilon_k-\mu)} \right)^{n_k} = \frac{1}{1-e^{-\beta(\varepsilon_k-\mu)}}$$ (7.39)

and the boson grand partition function is

$$\Xi_{\text{boson}}^{\text{NI}} = \prod_k \frac{1}{1-e^{-\beta(\varepsilon_k-\mu)}}.$$ (7.40)

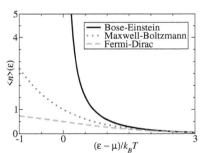

Fig. 7.4 **Bose–Einstein, Maxwell–Boltzmann, and Fermi–Dirac** distributions, $\langle n \rangle(\varepsilon)$. Occupation number for single-particle eigenstates as a function of energy ε away from the chemical potential μ. The Bose–Einstein distribution diverges as μ approaches ε; the Fermi–Dirac distribution saturates at one as μ gets small.

The grand free energy ($\Phi = -k_B T \log \Xi$, eqn 6.36) is a sum of single-state grand free energies:

$$\Phi_{\text{boson}}^{\text{NI}} = \sum_k \Phi_k^{\text{boson}} = \sum_k k_B T \log \left(1 - e^{-\beta(\varepsilon_k-\mu)} \right).$$ (7.41)

Because the filling of different states is independent, we can find out the expected number of particles in state ψ_k. From eqn 6.38,

$$\langle n_k \rangle = -\frac{\partial \Phi_k^{\text{boson}}}{\partial \mu} = -k_B T \frac{-\beta e^{-\beta(\varepsilon_k-\mu)}}{1-e^{-\beta(\varepsilon_k-\mu)}} = \frac{1}{e^{\beta(\varepsilon_k-\mu)}-1}.$$ (7.42)

This is called the *Bose–Einstein* distribution (Fig. 7.4)

$$\langle n \rangle_{\text{BE}} = \frac{1}{e^{\beta(\varepsilon - \mu)} - 1}. \tag{7.43}$$

The Bose–Einstein distribution describes the filling of single-particle eigenstates by noninteracting bosons. For states with low occupancies, where $\langle n \rangle \ll 1$, $\langle n \rangle_{\text{BE}} \approx e^{-\beta(\varepsilon - \mu)}$, and the boson populations correspond to what we would guess naively from the Boltzmann distribution.[19] The condition for low occupancies is $\varepsilon_k - \mu \gg k_B T$, which usually arises at high temperatures[20] (where the particles are distributed among a larger number of states). Notice also that $\langle n \rangle_{\text{BE}} \to \infty$ as $\mu \to \varepsilon_k$ since the denominator vanishes (and becomes negative for $\mu > \varepsilon_k$); systems of noninteracting bosons always have μ less than or equal to the lowest of the single-particle energy eigenvalues.[21]

Notice that the average excitation $\langle n \rangle_{\text{qho}}$ of the quantum harmonic oscillator (eqn 7.24) is given by the Bose–Einstein distribution (eqn 7.43) with $\mu = 0$. We will use this in Exercise 7.2 to argue that one can treat excitations inside harmonic oscillators (vibrations) as particles obeying Bose statistics (phonons).

Fermions. For fermions, only $n_k = 0$ and $n_k = 1$ are allowed. The single-state fermion grand partition function is

$$\Xi_k^{\text{fermion}} = \sum_{n_k=0}^{1} e^{-\beta(\varepsilon_k - \mu)n_k} = 1 + e^{-\beta(\varepsilon_k - \mu)}, \tag{7.44}$$

so the total fermion grand partition function is

$$\Xi_{\text{fermion}}^{\text{NI}} = \prod_k \left(1 + e^{-\beta(\varepsilon_k - \mu)} \right). \tag{7.45}$$

For summing over only two states, it is hardly worthwhile to work through the grand free energy to calculate the expected number of particles in a state:

$$\langle n_k \rangle = \frac{\sum_{n_k=0}^{1} n_k \exp(-\beta(\varepsilon_k - \mu)n_k)}{\sum_{n_k=0}^{1} \exp(-\beta(\varepsilon_k - \mu)n_k)} = \frac{e^{-\beta(\varepsilon_k - \mu)}}{1 + e^{-\beta(\varepsilon_k - \mu)}} = \frac{1}{e^{\beta(\varepsilon_k - \mu)} + 1}, \tag{7.46}$$

leading us to the *Fermi–Dirac* distribution

$$f(\varepsilon) = \langle n \rangle_{\text{FD}} = \frac{1}{e^{\beta(\varepsilon - \mu)} + 1}, \tag{7.47}$$

where $f(\varepsilon)$ is also known as the Fermi function (Fig. 7.5). Again, when the mean occupancy of state ψ_k is low, it is approximately given by the Boltzmann probability distribution, $e^{-\beta(\varepsilon - \mu)}$. Here the chemical potential can be either greater than or less than any given eigenenergy ε_k. Indeed, at low temperatures the chemical potential μ separates filled states $\varepsilon_k < \mu$ from empty states $\varepsilon_k > \mu$; only states within roughly $k_B T$ of μ are partially filled.

[19]We will derive this from Maxwell–Boltzmann statistics in Section 7.5.

[20]This may seem at odds with the formula, but as T gets large μ gets large and negative even faster. This happens (at fixed total number of particles) because more states at high temperatures are available for occupation, so the pressure μ needed to keep them filled decreases.

[21]Chemical potential is like a pressure pushing atoms into the system. When the river level gets up to the height of the fields, your farm gets flooded.

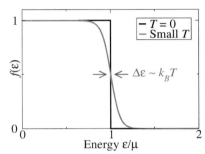

Fig. 7.5 The Fermi distribution $f(\varepsilon)$ of eqn 7.47. At low temperatures, states below μ are occupied, states above μ are unoccupied, and states within around $k_B T$ of μ are partially occupied.

[22] Just in case you have not heard, neutrinos are quite elusive. A lead wall that can stop half of the neutrinos would be light-years thick.

[23] Landau's insight was to describe interacting systems of fermions (e.g. electrons) at temperatures low compared to the Fermi energy by starting from the noninteracting Fermi gas and slowly "turning on" the interaction. (The Fermi energy $\varepsilon_F = \mu(T = 0)$, see Section 7.7.) Excited states of the noninteracting gas are electrons excited into states above the Fermi energy, leaving holes behind. They evolve in two ways when the interactions are turned on. First, the excited electrons and holes push and pull on the surrounding electron gas, creating a *screening cloud* that *dresses* the bare excitations into *quasiparticles*. Second, these quasiparticles develop lifetimes; they are no longer eigenstates, but *resonances*. Quasiparticles are useful descriptions so long as the interactions can be turned on slowly enough for the screening cloud to form but fast enough so that the quasiparticles have not yet decayed; this occurs for electrons and holes near the Fermi energy, which have long lifetimes because they can only decay into energy states even closer to the Fermi energy [13, p. 345]. Later workers fleshed out Landau's ideas into a systematic perturbative calculation, where the quasiparticles are poles in a quantum Green's function (see Exercises 7.24 and 10.9 for classical examples of how this works). More recently, researchers have found a renormalization-group interpretation of Landau's argument, whose coarse-graining operation removes states far from the Fermi energy, and which flows to an effective noninteracting Fermi gas (see Chapter 12 and Exercise 12.8).

The chemical potential μ is playing a large role in these calculations. How do you determine it? You normally know the expected number of particles N, and must vary μ until you reach that value. Hence μ directly plays the role of a particle pressure from the outside world, which is varied until the system is correctly filled.

The amazing utility of noninteracting bosons and fermions. The classical ideal gas is a great illustration of statistical mechanics, and does a good job of describing many gases, but nobody would suggest that it captures the main features of solids and liquids. The noninteracting approximation in quantum mechanics turns out to be far more powerful, for quite subtle reasons.

For bosons, the noninteracting approximation is quite accurate in three important cases: photons, phonons, and the dilute Bose gas. In Section 7.6 we will study two fundamental problems involving noninteracting bosons: black-body radiation and Bose condensation. The behavior of superconductors and superfluids shares some common features with that of the Bose gas.

For fermions, the noninteracting approximation would rarely seem to be useful. Electrons are charged, and the electromagnetic repulsion between the electrons in an atom, molecule, or material is always a major contribution to the energy. Neutrons interact via the strong interaction, so nuclei and neutron stars are also poor candidates for a noninteracting theory. Neutrinos are hard to pack into a box.[22] There are experiments on cold, dilute gases of fermion atoms, but noninteracting fermions would seem a model with few applications.

The truth is that the noninteracting Fermi gas describes all of these systems (atoms, metals, insulators, nuclei, and neutron stars) remarkably well. The emergent low-energy theory for interacting Fermi systems is often noninteracting screened *quasielectrons* and *quasiholes* in a modified potential (Exercise 7.24). The approximation is so powerful that in most circumstances we ignore the interactions; whenever we talk about exciting a "1S electron" in an oxygen atom, or an "electron–hole" pair in a semiconductor, we are using this effective noninteracting electron approximation. The explanation for this amazing fact is called Landau Fermi-liquid theory.[23]

7.5 Maxwell–Boltzmann "quantum" statistics

In classical statistical mechanics, we treated indistinguishable particles as distinguishable ones, except that we divided the phase-space volume, (or the partition function, in the canonical ensemble) by a factor of $N!$:

$$\Omega_N^{\text{MB}} = \frac{1}{N!} \Omega_N^{\text{dist}},$$
$$Z_N^{\text{MB}} = \frac{1}{N!} Z_N^{\text{dist}}. \tag{7.48}$$

This was important to get the entropy to be extensive (Section 5.2.1). This approximation is also sometimes used in quantum statistical mechanics, although we should emphasize that it does not describe either bosons, fermions, or any physical system. These bogus particles are said to obey *Maxwell–Boltzmann* statistics.[24]

What is the canonical partition function for the case of N noninteracting distinguishable quantum particles? If the partition function for one particle is

$$Z_1 = \sum_k e^{-\beta \varepsilon_k} \tag{7.49}$$

then the partition function for N noninteracting, distinguishable (but otherwise similar) particles is

$$Z_N^{\mathrm{NI,dist}} = \sum_{k_1, k_2, \ldots, k_n} e^{-\beta(\varepsilon_{k_1} + \varepsilon_{k_2} + \cdots + \varepsilon_{k_N})} = \prod_{j=1}^{N} \left(\sum_{k_j} e^{-\beta \varepsilon_{k_j}} \right) = Z_1^N. \tag{7.50}$$

So, the Maxwell–Boltzmann partition function for noninteracting particles is

$$Z_N^{\mathrm{NI,MB}} = Z_1^N / N!. \tag{7.51}$$

Let us illustrate the relation between these three distributions by considering the canonical ensemble of two noninteracting particles in three possible states of energies ε_1, ε_2, and ε_3. The Maxwell–Boltzmann partition function for such a system would be

$$\begin{aligned}
Z_2^{\mathrm{NI,MB}} &= \frac{1}{2!} \left(e^{-\beta \varepsilon_1} + e^{-\beta \varepsilon_2} + e^{-\beta \varepsilon_3} \right)^2 \\
&= \frac{1}{2} e^{-2\beta \varepsilon_1} + \frac{1}{2} e^{-2\beta \varepsilon_2} + \frac{1}{2} e^{-2\beta \varepsilon_3} \\
&\quad + e^{-\beta(\varepsilon_1 + \varepsilon_2)} + e^{-\beta(\varepsilon_1 + \varepsilon_3)} + e^{-\beta(\varepsilon_2 + \varepsilon_3)}.
\end{aligned} \tag{7.52}$$

The $1/N!$ fixes the weights of the singly occupied states[25] nicely; each has weight one in the Maxwell–Boltzmann partition function. But the doubly occupied states, where both particles have the same wavefunction, have an unintuitive suppression by $\frac{1}{2}$ in the sum.

There are basically two ways to fix this. One is to stop discriminating against multiply occupied states, and to treat them all democratically. This gives us noninteracting bosons:

$$Z_2^{\mathrm{NI,boson}} = e^{-2\beta \varepsilon_1} + e^{-2\beta \varepsilon_2} + e^{-2\beta \varepsilon_3} + e^{-\beta(\varepsilon_1 + \varepsilon_2)} + e^{-\beta(\varepsilon_1 + \varepsilon_3)} + e^{-\beta(\varepsilon_2 + \varepsilon_3)}. \tag{7.53}$$

The other way is to "squelch" multiple occupancy altogether. This leads to fermions:

$$Z_2^{\mathrm{NI,fermion}} = e^{-\beta(\varepsilon_1 + \varepsilon_2)} + e^{-\beta(\varepsilon_1 + \varepsilon_3)} + e^{-\beta(\varepsilon_2 + \varepsilon_3)}. \tag{7.54}$$

Thus the Maxwell–Boltzmann distribution treats multiple occupancy of states in an unphysical compromise between democratic bosons and exclusive fermions.

[24] Sometimes it is said that distinguishable particles obey Maxwell–Boltzmann statistics. Many properties are independent of the $N!$ in the denominator of eqn 7.48, such as the occupancy $\langle n \rangle$ of noninteracting single-particle eigenstates (eqn 7.59). But this factor does matter for other properties, like the entropy of mixing and the Helmholtz free energy, so we reserve the term Maxwell–Boltzmann for undistinguished particles (Section 3.5).

[25] More precisely, we mean those many-body states where the single-particle states are all singly occupied or vacant.

Here we have been comparing the different distributions within the canonical ensemble. What about the grand canonical ensemble, which we actually use for calculations?[26] The grand partition function for Maxwell–Boltzmann statistics is

$$\Xi^{\mathrm{NI,MB}} = \sum_M Z_M^{\mathrm{NI,MB}} \, \mathrm{e}^{M\beta\mu} = \sum_M \frac{1}{M!} \left(\sum_k \mathrm{e}^{-\beta\varepsilon_k} \right)^M \mathrm{e}^{M\beta\mu}$$

$$= \sum_M \frac{1}{M!} \left(\sum_k \mathrm{e}^{-\beta(\varepsilon_k - \mu)} \right)^M = \exp\left(\sum_k \mathrm{e}^{-\beta(\varepsilon_k - \mu)} \right)$$

$$= \prod_k \exp\left(\mathrm{e}^{-\beta(\varepsilon_k - \mu)} \right). \tag{7.55}$$

The grand free energy is

$$\Phi^{\mathrm{NI,MB}} = -k_B T \log \Xi^{\mathrm{NI,MB}} = \sum_k \Phi_k, \tag{7.56}$$

with the single-particle grand free energy

$$\Phi_k = -k_B T \mathrm{e}^{-\beta(\varepsilon_k - \mu)}. \tag{7.57}$$

[27]It is amusing to note that noninteracting particles fill single-particle energy states according to the same law

$$\langle n \rangle = \frac{1}{\mathrm{e}^{\beta(\varepsilon - \mu)} + c}, \tag{7.58}$$

with $c = -1$ for bosons, $c = 1$ for fermions, and $c = 0$ for Maxwell–Boltzmann statistics.

Finally, the expected[27] number of particles in a single-particle eigenstate with energy ε is

$$\langle n \rangle_{\mathrm{MB}} = -\frac{\partial\Phi}{\partial\mu} = \mathrm{e}^{-\beta(\varepsilon - \mu)}. \tag{7.59}$$

This is precisely the Boltzmann factor for filling the state that we expect for noninteracting distinguishable particles; the indistinguishability factor $N!$ does not alter the filling of the noninteracting single-particle states.

7.6 Black-body radiation and Bose condensation

7.6.1 Free particles in a box

Fig. 7.6 Particle in a box. The quantum states of a particle in a one-dimensional box with periodic boundary conditions are sine and cosine waves ψ_n with n wavelengths in the box, $k_n = 2\pi n/L$. With a real box (zero boundary conditions at the walls) one would have only sine waves, but at half the spacing between wavevectors $k_n = \pi n/L$, giving the same net density of states.

[28]That is, the value of ψ at the walls need not be zero (as for an infinite square well), but rather must agree on opposite sides, so $\psi(0, y, z) \equiv \psi(L, y, z)$, $\psi(x, 0, z) \equiv \psi(x, L, z)$, and $\psi(x, y, 0) \equiv \psi(x, y, L)$. Periodic boundary conditions are not usually seen in experiments, but are much more natural to compute with, and the results are unchanged for large systems.

For this section and the next section on fermions, we shall simplify even further. We consider particles which are not only noninteracting and identical, but are also free. That is, they are subject to no external potential, apart from being confined in a box of volume $L^3 = V$ with *periodic boundary conditions* (Fig. 7.6).[28] The single-particle quantum eigenstates of such a system are products of sine and cosine waves along the three directions—for example, for any three nonnegative integers n_i,

$$\psi = \left(\frac{2}{L} \right)^{3/2} \cos\left(\frac{2\pi n_1}{L} x \right) \cos\left(\frac{2\pi n_2}{L} y \right) \cos\left(\frac{2\pi n_3}{L} z \right). \tag{7.60}$$

There are eight such states with the same energy, substituting sine for cosine in all possible combinations along the three directions. These are more conveniently organized if we use the complex exponential instead of sine and cosine:

$$\psi_{\mathbf{k}} = (1/L)^{3/2} \exp(i\mathbf{k} \cdot \mathbf{r}), \qquad (7.61)$$

with $\mathbf{k} = (2\pi/L)(n_1, n_2, n_3)$ and the n_i can now be any integer.[29] The allowed single-particle eigenstates form a regular square grid in the space of wavevectors k, with an average density $(L/2\pi)^3$ per unit volume of k-space:

$$\text{density of plane waves in } \mathbf{k}\text{-space} = V/8\pi^3. \qquad (7.62)$$

For a large box volume V, the grid is extremely fine, and one can use a continuum approximation that the number of states falling into a $\mathbf{k}$-space region is given by its volume times the density (eqn 7.62).[30]

7.6.2 Black-body radiation

Our first application is to electromagnetic radiation. Electromagnetic radiation has plane-wave modes similar to eqn 7.61. Each plane wave travels at the speed of light c, so its frequency is $\omega_k = c|\mathbf{k}|$. There are two modes per wavevector $\mathbf{k}$, one for each polarization. When one quantizes the electromagnetic field, each mode becomes a quantum harmonic oscillator.

Before quantum mechanics, people could not understand the equilibration of electromagnetic radiation. The equipartition theorem predicted that if you could come to equilibrium, each mode would have $k_B T$ of energy. Since there are immensely more wavevectors in the ultraviolet and X-ray ranges than in the infrared and visible,[31] opening your oven door would theoretically give you a suntan or worse (the so-called *ultraviolet catastrophe*). Experiments saw a spectrum which looked compatible with this prediction for small frequencies, but was (fortunately) cut off at high frequencies.

Let us calculate the equilibrium energy distribution inside our box at temperature T. The number of single-particle plane-wave eigenstates $g(\omega)\,\mathrm{d}\omega$ in a small range $\mathrm{d}\omega$ is[32]

$$g(\omega)\,\mathrm{d}\omega = (4\pi k^2)\left(\frac{\mathrm{d}|\mathbf{k}|}{\mathrm{d}\omega}\mathrm{d}\omega\right)\left(\frac{2V}{(2\pi)^3}\right), \qquad (7.63)$$

where the first term is the surface area of the sphere of radius k, the second term is the thickness of the spherical shell for a small $\mathrm{d}\omega$, and the last is the density of single-particle plane-wave eigenstate wavevectors times two (because there are two photon polarizations per wavevector). Knowing $\mathbf{k}^2 = \omega^2/c^2$ and $\mathrm{d}|\mathbf{k}|/\mathrm{d}\omega = 1/c$, we find the density of plane-wave eigenstates per unit frequency:

$$g(\omega) = \frac{V\omega^2}{\pi^2 c^3}. \qquad (7.64)$$

Now, the number of photons is not fixed; they can be created or destroyed, so their chemical potential μ is zero.[33] Their energy $\varepsilon_k = \hbar\omega_k$. Finally, they are to an excellent approximation identical, noninteracting

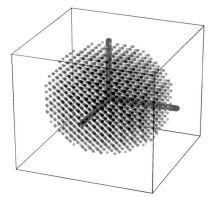

Fig. 7.7 k-sphere. The allowed **k**-space points for periodic boundary conditions form a regular grid. The points of equal energy lie on a sphere.

[29]The eight degenerate states are now given by the choices of sign for the three integers.

[30]Basically, the continuum limit works because the shape of the box (which affects the arrangements of the allowed **k** vectors) is irrelevant to the physics so long as the box is large. For the same reason, the energy of the single-particle eigenstates is independent of direction; it will be proportional to $|\mathbf{k}|$ for massless photons, and proportional to $\mathbf{k}^2$ for massive bosons and electrons (Fig. 7.7). This makes the calculations in the following sections tractable.

[31]There are a thousand times more wavevectors with $|k| < 10k_0$ than for $|k| < k_0$. The optical frequencies and wavevectors span roughly a factor of two, so there are eight times as many optical modes as there are radio and infrared modes.

[32]We are going to be sloppy and use $g(\omega)$ as eigenstates per unit frequency for photons, and later we will use $g(\varepsilon)$ as single-particle eigenstates per unit energy. Be warned: $g_\omega(\omega)\,\mathrm{d}\omega = g_\varepsilon(\hbar\omega)\,\mathrm{d}\,\hbar\omega$, so $g_\omega = \hbar g_\varepsilon$.

[33]See Exercise 7.2 to derive this from the quantum harmonic oscillator.

Fig. 7.8 The Planck black-body radiation power spectrum, with the Rayleigh–Jeans approximation, valid for low frequency ω.

[34]Why is this called *black-body* radiation? A black surface absorbs all radiation at all frequencies. In equilibrium, the energy it absorbs at a given frequency must equal the energy it emits, otherwise it would push the system out of equilibrium. (This is called *detailed balance*, Section 8.2.) Hence, ignoring the small surface cooling due to radiation, a black body emits a thermal distribution of photons (see Exercise 7.7).

bosons, so the number of photons per eigenstate with frequency ω is $\langle n \rangle = 1/(e^{\hbar\omega/k_B T} - 1)$. This gives us a number of photons:

$$(\text{\# of photons}) \; \mathrm{d}\omega = \frac{g(\omega)}{e^{\hbar\omega/k_B T} - 1} \, \mathrm{d}\omega \tag{7.65}$$

and an electromagnetic (photon) energy per unit volume $u(\omega)$ given by

$$
\begin{aligned}
V u(\omega) \, \mathrm{d}\omega &= \frac{\hbar\omega g(\omega)}{e^{\hbar\omega/k_B T} - 1} \, \mathrm{d}\omega \\
&= \frac{V\hbar}{\pi^2 c^3} \frac{\omega^3 \, \mathrm{d}\omega}{e^{\hbar\omega/k_B T} - 1}
\end{aligned} \tag{7.66}
$$

(Fig. 7.8). This is Planck's famous formula for black-body radiation.[34] At low frequencies, we can approximate $e^{\hbar\omega/k_B T} - 1 \approx \hbar\omega/k_B T$, yielding the Rayleigh–Jeans formula

$$
\begin{aligned}
V u_{\mathrm{RJ}}(\omega) \, \mathrm{d}\omega &= V \left(\frac{k_B T}{\pi^2 c^3} \right) \omega^2 \, \mathrm{d}\omega \\
&= k_B T g(\omega),
\end{aligned} \tag{7.67}
$$

just as one would expect from equipartition: $k_B T$ per classical harmonic oscillator.

For modes with frequencies high compared to $k_B T/\hbar$, equipartition no longer holds. The energy gap $\hbar\omega$, just as for the low-temperature specific heat from Section 7.2, leads to an excitation probability that is suppressed by the exponential Boltzmann factor $e^{-\hbar\omega/k_B T}$ (eqn 7.66, approximating $1/(e^{\hbar\omega/k_B T} - 1) \approx e^{-\hbar\omega/k_B T}$). Planck's discovery that quantizing the energy averted the ultraviolet catastrophe was the origin of quantum mechanics, and led to his name being given to $\hbar$.

7.6.3 Bose condensation

How does our calculation change when the noninteracting free bosons cannot be created and destroyed? Let us assume that our bosons are spinless, have mass m, and are nonrelativistic, so their energy is $\varepsilon = p^2/2m = -\hbar^2\nabla^2/2m$. If we put them in our box with periodic boundary conditions, we can make the same continuum approximation to the density of states as we did in the case of black-body radiation. In eqn 7.62, the number of plane-wave eigenstates per unit volume in **k**-space is $V/8\pi^3$, so the density in momentum space $\mathbf{p} = \hbar\mathbf{k}$ is $V/(2\pi\hbar)^3$. For our massive particles $\mathrm{d}\varepsilon/\mathrm{d}|\mathbf{p}| = |\mathbf{p}|/m = \sqrt{2\varepsilon/m}$, so the number of plane-wave eigenstates in a small range of energy $\mathrm{d}\varepsilon$ is

$$
\begin{aligned}
g(\varepsilon) \, \mathrm{d}\varepsilon &= (4\pi p^2) \left(\frac{\mathrm{d}|\mathbf{p}|}{\mathrm{d}\varepsilon} \, \mathrm{d}\varepsilon \right) \left(\frac{V}{(2\pi\hbar)^3} \right) \\
&= (4\pi(2m\varepsilon)) \left(\sqrt{\frac{m}{2\varepsilon}} \, \mathrm{d}\varepsilon \right) \left(\frac{V}{(2\pi\hbar)^3} \right) \\
&= \frac{V m^{3/2}}{\sqrt{2}\pi^2\hbar^3} \sqrt{\varepsilon} \, \mathrm{d}\varepsilon,
\end{aligned} \tag{7.68}
$$

where the first term is the surface area of the sphere in **p**-space, the second is the thickness of the spherical shell, and the third is the density of plane-wave eigenstates per unit volume in **p**-space.

Now we fill each of these single-particle plane-wave eigenstates with an expected number given by the Bose–Einstein distribution at chemical potential μ, $1/(e^{(\varepsilon-\mu)/k_BT}-1)$. The total number of particles N is then given by

$$N(\mu) = \int_0^\infty \frac{g(\varepsilon)}{e^{(\varepsilon-\mu)/k_BT}-1} \, d\varepsilon. \tag{7.69}$$

We must vary μ in this equation to give us the correct number of particles N. For bosons, as noted in Section 7.4, μ cannot be larger than the lowest single-particle eigenenergy (here $\varepsilon_0 = 0$), so μ will always be negative. For larger numbers of particles we raise μ up from below, forcing more particles into each of the single-particle states. There is a limit, however, to how hard we can push; when $\mu = 0$ the ground state gets a diverging number of particles.

For free bosons in three dimensions, the integral for $N(\mu = 0)$ converges to a finite value.[35] Thus the largest number of particles N_{max}^{cont} we can fit into our box within our continuum approximation for the density of states is the value of eqn 7.69 at $\mu = 0$:

$$\begin{aligned}
N_{max}^{cont} &= \int \frac{g(\varepsilon)}{e^{\varepsilon/k_BT}-1} \, d\varepsilon \\
&= \frac{Vm^{3/2}}{\sqrt{2}\pi^2\hbar^3} \int_0^\infty d\varepsilon \frac{\sqrt{\varepsilon}}{e^{\varepsilon/k_BT}-1} \\
&= V\left(\frac{\sqrt{2\pi mk_BT}}{h}\right)^3 \frac{2}{\sqrt{\pi}} \int_0^\infty \frac{\sqrt{z}}{e^z-1} \, dz \\
&= \left(\frac{V}{\lambda^3}\right) \zeta(3/2). \tag{7.70}
\end{aligned}$$

Here ζ is the Riemann zeta function,[36] with $\zeta(\tfrac{3}{2}) \approx 2.612$ and $\lambda = h/\sqrt{2\pi mk_BT}$ is again the thermal de Broglie wavelength (eqn 3.59). Something new has to happen at a critical density:

$$\frac{N_{max}^{cont}}{V} = \frac{\zeta(3/2)}{\lambda^3} = \frac{2.612 \text{ particles}}{\text{de Broglie volume}}. \tag{7.71}$$

This has an elegant interpretation: the quantum statistics of the particles begin to dominate the behavior when they are within around a thermal de Broglie wavelength of one another.

What happens when we try to cram more particles in? Our approximation of the distribution of eigenstates as a continuum breaks down. Figure 7.9 shows a schematic illustration of the first few single-particle eigenvalues. When the distance between μ and the bottom level ε_0 becomes significantly smaller than the distance between the bottom and the next level ε_1, the continuum approximation (which approximates the filling of ε_0 using an integral halfway to ε_1) becomes qualitatively wrong. The low-energy states, viewed as a continuum, cannot accommodate the extra bosons. Instead, the lowest state absorbs all the extra particles

[35]At $\mu = 0$, the denominator of the integrand in eqn 7.69 is approximately ε/k_BT for small ε, but the numerator goes as $\sqrt{\varepsilon}$, so the integral converges at the lower end: $\int_0^X \varepsilon^{-1/2} \sim (\tfrac{1}{2}\varepsilon^{1/2})|_0^X = \sqrt{X}/2$.

[36]The Riemann ζ function $\zeta(s) = [1/(s-1)!] \int_0^\infty z^{s-1}/(e^z-1) \, dz$ is famous for many reasons. It is related to the distribution of prime numbers. It is the subject of the famous unproven *Riemann hypothesis*, that its zeros in the complex plane, apart from those at the negative even integers, all have real part equal to $\tfrac{1}{2}$.

Fig. 7.9 Bose condensation. The chemical potential μ is here so close to the ground state energy ε_0 that the continuum approximation to the density of states breaks down. The ground state is macroscopically occupied (that is, filled by a nonzero fraction of the total number of particles N).

[37]The next few states have quantitative corrections, but the continuum approximation is only off by small factors.

added to the system beyond $N_{\max}^{\text{cont}}$.[37] This is called *Bose–Einstein condensation*.

Usually we do not add particles at fixed temperature; instead we lower the temperature at fixed density N/V. Bose condensation then occurs at temperature

$$k_B T_c^{\text{BEC}} = \frac{h^2}{2\pi m}\left(\frac{N}{V\zeta(3/2)}\right)^{2/3}. \tag{7.72}$$

Bose condensation was first accomplished experimentally in 1995 (see Exercise 7.14).

Bose condensation has also long been considered the underlying principle behind superfluidity. Liquid He^4 undergoes an unusual transition at about 2.176 K to a state without viscosity; it will swirl round a circular tube for as long as your refrigeration lasts. The quantitative study of the superfluid transition involves the interactions between the helium atoms, and uses the scaling methods that we will introduce in Chapter 12. But it is interesting to note that the Bose condensation temperature for liquid He^4 (with $m = 6.65 \times 10^{-24}$ g and volume per particle $V/N = 27.6\,\text{cm}^3/\text{mole}$) is 3.13 K—quite close to the superfluid transition temperature.

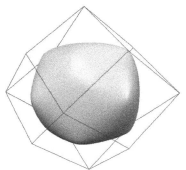

Fig. 7.10 The Fermi surface for lithium, from [43]. The Fermi energy for lithium is 4.74 eV, with one conduction electron outside a helium closed shell. As for most metals, the Fermi energy in lithium is much larger than k_B times its melting point (4.74 eV = 55,000 K, melting point 453 K). Hence it is well described by this $T = 0$ Fermi surface, slightly smeared by the Fermi function (Fig. 7.5).

7.7   Metals and the Fermi gas

We claimed in Section 7.4 that many systems of strongly interacting fermions (metals, neutron stars, nuclei) are surprisingly well described by a model of noninteracting fermions. Let us solve for the properties of N *free* noninteracting fermions in a box.

Let our particles be nonrelativistic and have spin 1/2. The single-particle eigenstates are the same as those for bosons except that there are two states (spin up, spin down) per plane wave. Hence the density of states is given by twice that of eqn 7.68:

$$g(\varepsilon) = \frac{\sqrt{2}Vm^{3/2}}{\pi^2\hbar^3}\sqrt{\varepsilon}. \tag{7.73}$$

[38]Equation 7.76 has an illuminating derivation in **k**-space, where we fill all states with $|\mathbf{k}| < k_F$. Here the Fermi wavevector k_F has energy equal to the Fermi energy, $\hbar k_F^2/2m = p_F^2/2m = \varepsilon_F$, and hence $k_F = \sqrt{2\varepsilon_F m}/\hbar$. The resulting sphere of occupied states at $T = 0$ is called the *Fermi sphere*. The number of fermions inside the Fermi sphere is thus the **k**-space volume of the Fermi sphere times the **k**-space density of states,

$$N = ((4/3)\pi k_F^3)\left(\frac{2V}{(2\pi)^3}\right) = \frac{k_F^3}{3\pi^2}V, \tag{7.75}$$

equivalent to eqn 7.76.

The number of fermions at chemical potential μ is given by integrating $g(\varepsilon)$ times the expected number of fermions in a state of energy ε, given by the Fermi function $f(\varepsilon)$ of eqn 7.47:

$$N(\mu) = \int_0^\infty g(\varepsilon)f(\varepsilon)\,\mathrm{d}\varepsilon = \int_0^\infty \frac{g(\varepsilon)}{\mathrm{e}^{(\varepsilon-\mu)/k_BT}+1}\,\mathrm{d}\varepsilon. \tag{7.74}$$

What chemical potential will give us N fermions? At nonzero temperature, one must do a self-consistent calculation, but at $T = 0$ one can find N by counting the number of states below μ. In the zero-temperature limit (Fig. 7.5) the Fermi function is a step function $f(\varepsilon) = \Theta(\mu - \varepsilon)$; all states below μ are filled, and all states above μ are empty. The zero-temperature value of the chemical potential is called the *Fermi energy* ε_F. We can find the number of fermions by integrating up to $\mu = \varepsilon_F$:[38]

$$N = \int_0^{\varepsilon_F} g(\varepsilon)\,\mathrm{d}\varepsilon = \frac{\sqrt{2}m^{3/2}}{\pi^2\hbar^3}V\int_0^{\varepsilon_F}\sqrt{\varepsilon}\,\mathrm{d}\varepsilon = \frac{(2\varepsilon_F m)^{3/2}}{3\pi^2\hbar^3}V. \qquad (7.76)$$

We mentioned earlier that the independent fermion approximation was startlingly useful even though the interactions are not small. Ignoring the Coulomb repulsion between electrons in a metal, or the strong interaction between neutrons in a neutron star, gives an excellent description of their actual behavior. However, our calculation above also assumed that the electrons are free particles, experiencing no external potential. This approximation is not particularly accurate in general; the interactions with the atomic nuclei are important, and is primarily what makes one material different from another. In particular, the atoms in a crystal will form a periodic potential for the electrons.[39] One can show that the single-particle eigenstates in a periodic potential are periodic functions times $\exp(i\mathbf{k}\cdot\mathbf{r})$, with exactly the same wavevectors $\mathbf{k}$ as in the free fermion case. The filling of the Fermi surface in $\mathbf{k}$-space is changed only insofar as the energies of these single-particle states are no longer isotropic. Some metals (particularly the alkali metals, like lithium in Fig. 7.10) have roughly spherical Fermi surfaces; many (see Fig. 7.11 for aluminum) are quite intricate, with several pieces to them [13, chapters 9–11].

Fig. 7.11 The Fermi surface for aluminum, also from [43]. Aluminum has a Fermi energy of 11.7 eV, with three conduction electrons outside a Ne closed shell.

[39]Rather than using the Coulomb potential for the nucleus, a better approximation is given by incorporating the effects of the *inner shell* electrons into the periodic potential, and filling the Fermi sea with the remaining *conduction* electrons.

Exercises

We use *Phase-space units and the zero of entropy* to explain the factors of Planck's constant in the microcanonical ensemble volume. *Crystal defects*, *Photon density matrices*, and *Spin density matrix* use mixed states (equilibrium and nonequilibrium) to calculate practical properties of quantum systems. *Eigenstate thermalization*, *Does entropy increase in quantum systems?*, *Entanglement of two spins*, *Heisenberg entanglement*, and *Quantum measurement and entropy* address the profound connections between entropy growth, the "wavefunction collapse" associated with macroscopic objects, thermalization, and the use of entropy to characterize the information lost to an environment.

Phonons and photons are bosons and *Phonons on a string* draw connections between the simple harmonic oscillator to quantum fields of particles. *Drawing wavefunctions* and *Many-fermion wavefunction nodes* prompt one to address wavefunctions of interacting particles. *Ensembles and quantum statistics* and *Bosons are gregarious: superfluids and lasers* use the noninteracting particle approximation to illuminate the effects of quantum statistics.

The photon-dominated Universe, *Light emission and absorption*, *Cooling coffee*, and *The greenhouse effect* explore ramifications of the Planck law for black-body radiation, and *Einstein's A and B* discusses the interaction of matter with electromagnetism. *Light baryon superfluids*, *Bose condensation in a band*, and *Bose condensation: the experiment* explore the condensation of noninteracting bosons at low temperature; *Fermions in semiconductors* and *White dwarfs, neutron stars, and black holes* use noninteracting fermions to describe modern electronics and the stability of dying stars. *Is sound a quasiparticle* provides insight into why noninteracting approximations work. Finally, *Why are atoms classical?* addresses why we can ignore quantum mechanics and still describe much of our world.

(7.1) **Ensembles and quantum statistics.** (Quantum) ③

A system has two single-particle eigenfunctions, with energies (measured in degrees Kelvin) $E_0/k_B = -10$ and $E_2/k_B = 10$. Experiments are performed by adding three noninteracting particles to these two states, either identical spin-1/2 fermions, identical spinless bosons, distinguishable particles, or spinless identical particles obeying Maxwell–Boltzmann statistics. Please make a table for this exercise, giving your answers for the four cases (Fermi, Bose, Distinguishable, and Maxwell–Boltzmann) for each of the three parts. Substantive calculations may be needed.

(a) *The system is first held at constant energy. In Fig. 7.12 which curve represents the entropy of the fermions as a function of the energy? Bosons? Distinguishable particles? Maxwell–Boltzmann particles?*

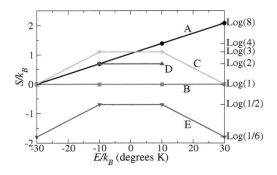

Fig. 7.12 Microcanonical three particles.

(b) *The system is now held at constant temperature. In Fig. 7.13 which curve represents the mean energy of the fermions as a function of temperature? Bosons? Distinguishable particles? Maxwell–Boltzmann particles?*

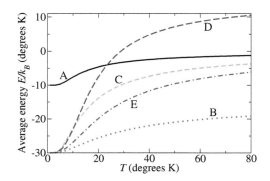

Fig. 7.13 Canonical three particles.

(c) *The system is now held at constant temperature, with chemical potential set to hold the average number of particles equal to three. In Fig. 7.14, which curve represents the chemical potential of the fermions as a function of temperature? Bosons? Distinguishable? Maxwell–Boltzmann?*

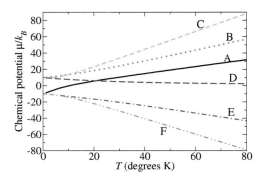

Fig. 7.14 Grand canonical three particles.

(7.2) **Phonons and photons are bosons.** (Quantum) @

Photons are elementary particles. How do they add together to make electromagnetic waves? Sound waves in a crystal are normal modes of vibration. Why do we talk about quantized sound in terms of phonons—as if sound were made up of particles?

Phonons and photons are the elementary, harmonic excitations of the elastic and electromagnetic fields. We have seen in Exercise 7.11 that phonons are decoupled harmonic oscillators, with a distribution of frequencies ω. A similar analysis shows that the Hamiltonian of the electromagnetic field can be decomposed into harmonic normal modes.

This exercise will explain why we think of phonons and photons as particles, instead of excitations of harmonic modes. In the process, we discover that both obey noninteracting Bose statistics, in the harmonic limit.

(a) *Show that the canonical partition function for a quantum harmonic oscillator of frequency ω is the same as the grand canonical partition function for bosons multiply filling a single state with energy $\hbar\omega$, with $\mu = 0$ (apart from a shift in the arbitrary zero of the total energy of the system).*

The Boltzmann filling of a harmonic oscillator is therefore the same as the Bose–Einstein filling of bosons into a single quantum state, except for an extra shift in the energy of $\hbar\omega/2$. This extra shift is called the *zero-point energy*. The excitations within the harmonic oscillator are thus often considered as particles with Bose statistics: the nth excitation is n bosons occupying the oscillator's quantum state.

This particle analogy becomes even more compelling for systems like phonons and photons where there are many harmonic oscillator states, each labeled by a wavevector k (see Exercise 7.11). Real, massive Bose particles like He^4 in free space have single-particle quantum eigenstates with a dispersion relation[40] $\varepsilon_k = \hbar^2 k^2/2m$. Phonons and photons have one harmonic oscillator for every k, with an excitation energy $\varepsilon_k = \hbar\omega_k$. If we treat them, as in part (a), as bosons filling these as single-particle states, we find that they are completely analogous to ordinary massive particles. (Photons even have the dispersion relation of a massless boson. If we take the mass to zero of a relativistic particle, the invariant mass relation $m_0^2 c^2 = (\varepsilon/c)^2 - \mathbf{p}^2$ implies $\varepsilon = \sqrt{m_0^2 c^4 + \mathbf{p}^2 c^2}$, which goes to $|\mathbf{p}|c = \hbar c |\mathbf{k}|$ as $m_0 \to 0$.)

(b) *Do phonons or photons Bose condense at low temperatures? Why or why not? Can you think of a nonequilibrium Bose condensation of photons, where a macroscopic occupation of a single frequency and momentum state occurs?*

You might think that photons are real bosons, while phonons are somehow a collective excitation of atomic vibrations that happen to obey Bose statistics. Instead, one should ask what kind of statistics the elementary excitations of an interacting system obeys! For quasielectrons and quasiholes in metals and insulators, the excitations are fermions (even including the screening cloud). But for the fractional quantum Hall state, the elementary excitations have *fractional* statistics—they are neither fermions nor bosons, but *anyons*.

(7.3) **Phase-space units and the zero of entropy.** (Quantum) ③
In classical mechanics, the entropy $S = k_B \log \Omega$ goes to minus infinity as the temperature is lowered to zero. In quantum mechanics the entropy

per particle goes to zero,[41] because states are quantized and the ground state is the only one populated. This is Nernst's theorem, the third law of thermodynamics.

The classical phase-space shell volume $\Omega(E)\,\delta E$ (eqn 3.5) has units of $((\text{momentum}) \times (\text{distance}))^{3N}$. It is a little perverse to take the logarithm of a quantity with units. The natural candidate with these dimensions is Planck's constant h^{3N}; if we measure phase-space volume in units of h per dimension, $\Omega(E)\,\delta E$ will be dimensionless. Of course, the correct dimension could be a constant times h, like $\hbar$.

(a) *Arbitrary zero of the classical entropy. Show that the width of the energy shell δE in the definition of $\Omega(E)$ does not change the microcanonical entropy per particle $S/N = k_B \log(\Omega(E))/N$ in a large system. Show that the choice of units in phase space does change the classical entropy per particle.*

We want to choose the units of classical phase-space volume so that the entropy agrees with the quantum entropy at high temperatures. How many quantum eigenstates per unit volume of classical phase space should we expect at high energies? We will fix these units by matching the quantum result to the classical one for a particular system, and then check it using a second system. Let us start with a free particle.

(b) *Phase-space density of states for a particle in a one-dimensional box. Show, or note, that the quantum momentum-space density of states for a free quantum particle in a one-dimensional box of length L with periodic boundary conditions is L/h. (An infinite square well gives the same answer, but is slightly messier.) Draw a picture of the classical phase space of this box (p, x), and draw a rectangle of length L for each quantum eigenstate. Is the phase-space area per eigenstate equal to h, as we assumed in Section 3.5?*

This works also for N particles in a three-dimensional box.

(c) *Phase-space density of states for N particles in a box. Show that the density of states for N free particles in a cubical box of volume V with periodic boundary conditions is V^N/h^{3N}, and hence that the phase-space volume per state is h^{3N}.*

---

[40]The *dispersion relation* is the relationship between energy and wavevector, here ε_k.

[41]If the ground state is degenerate, the entropy does not go to zero, but it typically stays finite as the number of particles N gets big, so for large N the entropy per particle goes to zero.

Let us see if this choice of units[42] also works for the harmonic oscillator.

(d) *Phase-space density of states for a harmonic oscillator. Consider a harmonic oscillator with Hamiltonian $\mathcal{H} = p^2/2m + \frac{1}{2}m\omega^2 q^2$. Draw a picture of the energy surface with energy E, and find the volume (area) of phase space enclosed.* (Hint: The area of an ellipse is $\pi r_1 r_2$ where r_1 and r_2 are the largest and smallest radii, corresponding to the major and minor axes.) *What is the volume per energy state, the volume between E_n and E_{n+1}, for the eigenenergy $E_n = (n + \frac{1}{2})\hbar\omega$?*

(7.4) **Does entropy increase in quantum systems?** (Mathematics, Quantum) @

We saw in Exercise 5.7 that in classical Hamiltonian systems the nonequilibrium entropy $S_{\text{nonequil}} = -k_B \int \rho \log \rho$ is constant in a classical mechanical Hamiltonian system. Here you will show that in the microscopic evolution of an isolated quantum system, the entropy is also time independent, even for general, time-dependent density matrices $\rho(t)$.

Using the evolution law (eqn 7.18) $\partial\rho/\partial t = [\mathcal{H}, \rho]/(i\hbar)$, prove that $S = \text{Tr}(\rho \log \rho)$ is time independent, where ρ is any density matrix. (Two approaches: (1) Go to an orthonormal basis ψ_i that diagonalizes ρ. Show that $\psi_i(t)$ is also orthonormal, and take the trace in that basis. (2) Let $U(t) = \exp(-i\mathcal{H}t/\hbar)$ be the unitary operator that time evolves the wave function $\psi(t)$. Show that $\rho(t) = U(t)\rho(0)U^\dagger(t)$. Write a general function $F(\rho)$ as a formal power series[43] in $\rho(t)$. Show, term-by-term in the series, that $F(t) = U(t)F(0)U^\dagger(t)$. Then use the cyclic invariance of the trace.)

Again, entropy must be constant in time for the underlying microscopic theory, because the microscopic theory is invariant under time-reversal symmetry. The increase of entropy in quantum systems is closely related to the problem of quantum measurement.

In classical theories, entropy measures the loss of information about the microstate of a macroscopic object into uninterpretable multiparticle correlations (Chapter 4, Exercises 5.8 and 5.25).

Quantum mechanics suffers a closely related *collapse of the wavefunction* when a quantum system interacts with a macroscopic object. That collapse entangles the quantum state with a multiparticle quantum wavefunction, (Exercises 7.26 and 7.27), destroying coherence and leading to an emergent entropy increase (Exercise 7.25).

(7.5) **Photon density matrices.** (Quantum, Optics) ⓟ

Write the density matrix for a vertically polarized photon $|V\rangle$ in the basis where $|V\rangle = \binom{1}{0}$ and a horizontal photon $|H\rangle = \binom{0}{1}$. Write the density matrix for a diagonally polarized photon, $\binom{1/\sqrt{2}}{1/\sqrt{2}}$, and the density matrix for unpolarized light (note 2 on p. 180). Calculate $\text{Tr}(\rho)$, $\text{Tr}(\rho^2)$, and $S = -k_B \text{Tr}(\rho \log \rho)$. Interpret the values of the three traces physically. (One is a check for pure states, one is a measure of information, and one is a normalization. Hint: A bra $\langle V|$ can be written $(\,1\;\;0\,)$.)

(7.6) **Spin density matrix.**[44] (Quantum) ③

Let the Hamiltonian for a spin be

$$\mathcal{H} = -\frac{\hbar}{2}\mathbf{B} \cdot \vec{\sigma}, \qquad (7.77)$$

where $\vec{\sigma} = (\sigma_x, \sigma_y, \sigma_z)$ are the three Pauli spin matrices, and $\mathbf{B}$ may be interpreted as a magnetic field, in units where the gyromagnetic ratio is unity. Remember that $\sigma_i\sigma_j - \sigma_j\sigma_i = 2i\epsilon_{ijk}\sigma_k$. Show that any 2×2 density matrix may be written in the form

$$\rho = \frac{1}{2}(\mathbf{1} + \mathbf{p} \cdot \vec{\sigma}). \qquad (7.78)$$

Show that the equations of motion for the density matrix $i\hbar\partial\rho/\partial t = [\mathcal{H}, \rho]$ can be written as $d\mathbf{p}/dt = -\mathbf{B} \times \mathbf{p}$.

---

[42]You show here that ideal gases should calculate entropy using phase-space units with $h = 1$. To argue this directly for interacting systems usually involves semiclassical quantization [104, chapter 48, p. 170] or path integrals [59]. But it must be true. We could imagine measuring the entropy difference between the interacting system and an ideal gas, by slowly and reversibly turning off the interactions between the particles, measuring the entropy flow into or out of the system. Thus, setting the zero of entropy for the ideal gas sets it for all systems.

[43]One must note that $S(\rho) = -k_B \rho \log \rho$ is singular as $\rho \to 0$ and does not have a series expansion.

[44]Adapted from exam question by Bert Halperin, Harvard University, 1976.

(7.7) **Light emission and absorption.** (Quantum, Optics) ②

The experiment that Planck was studying did not directly measure the energy density per unit frequency (eqn 7.66) inside a box. It measured the energy radiating out of a small hole, of area A. Let us assume the hole is on the upper face of the cavity, perpendicular to the z axis.

What is the photon distribution just inside the boundary of the hole? Few photons come into the hole from the outside, so the distribution is depleted for those photons with $v_z < 0$. However, the photons with $v_z > 0$ to an excellent approximation should be unaffected by the hole—since they were emitted from far distant walls of the cavity, where the existence of the hole is a negligible perturbation. So, presuming the relevant photons just inside the hole are distributed in the same way as in the box as a whole (eqn 7.66), how many leave in a time dt?

Fig. 7.15 Photon emission from a hole. The photons leaving a cavity in a time dt are those within $v_z\,dt$ of the hole.

As one can see geometrically (Fig. 7.15), those photons within $v_z\,dt$ of the boundary will escape in time dt. The vertical velocity $v_z = c\cos(\theta)$, where θ is the photon velocity angle with respect to the vertical. The Planck distribution is isotropic, so the probability that a photon will be moving at an angle θ is the perimeter of the θ circle on the sphere divided by the area of the sphere, $2\pi\sin(\theta)\,d\theta/(4\pi) = \frac12\sin(\theta)\,d\theta$.

(a) *Show that the probability density*[45] *$\rho(v_z)$ for a particular photon to have velocity v_z is independent of v_z in the range $(-c,c)$, and thus is $1/2c$.* (Hint: $\rho(v_z)\,\Delta v_z = \rho(\theta)\,\Delta\theta$.)

An upper bound on the energy emitted from a hole of area A is given by the energy in the box as a whole (eqn 7.66) times the fraction $Ac\,dt/V$ of the volume within $c\,dt$ of the hole.

(b) *Show that the actual energy emitted is $1/4$ of this upper bound.* (Hint: You will need to

integrate $\int_0^c \rho(v_z)v_z\,dv_z$.)

Hence the power per unit area emitted from the small hole in equilibrium is

$$P_{\text{black}}(\omega, T) = \left(\frac{c}{4}\right)\frac{\hbar}{\pi^2 c^3}\frac{\omega^3\,d\omega}{e^{\hbar\omega/k_B T} - 1}. \quad (7.79)$$

Why is this called black-body radiation? Certainly a small hole in a large (cold) cavity looks black—any light entering the hole bounces around inside until it is absorbed by the walls. Suppose we placed a black object—a material that absorbed radiation at all frequencies and angles—capping the hole. This object would absorb radiation from the cavity, rising in temperature until it came to equilibrium with the cavity—emitting just as much radiation as it absorbs. Thus the overall power per unit area emitted by our black object in equilibrium at a given temperature must equal that of the hole. This must also be true if we place a selective filter between the hole and our black body, passing through only particular types of photons. Thus the emission and absorption of our black body must agree with the hole for every photon mode individually, an example of the *principle of detailed balance* we will discuss in more detail in Section 8.2.

How much power per unit area $P_{\text{colored}}(\omega, T)$ is emitted in equilibrium at temperature T by a red or maroon body? A white body? A mirror? These objects are different in the fraction of incident light they absorb at different frequencies and angles $a(\omega, \theta)$. We can again use the principle of detailed balance; we place our colored object next to a black body and match the power emitted and absorbed for each angle and frequency:

$$P_{\text{colored}}(\omega, T, \theta) = P_{\text{black}}(\omega, T)a(\omega, \theta). \quad (7.80)$$

Returning to black bodies, we should calculate $Q_{\text{tot}}(T)$, the total power per unit area emitted from a black body at temperature T, by integrating eqn 7.79 over frequency.

(c) *Using the fact that $\int_0^\infty x^3/(e^x - 1)\,dx = \pi^4/15$, show that*

$$Q_{\text{tot}}(T) = \int_0^\infty P_{\text{black}}(\omega, T)\,d\omega = \sigma T^4 \quad (7.81)$$

and give a formula for the Stefan–Boltzmann constant σ. ($\sigma = 5.67 \times 10^{-5}$ erg cm^{-2} K^{-4} s^{-1}; use this to check your answer.)

---

[45]We are being sloppy again, using the same name ρ for the probability densities per unit velocity and per unit angle.

(7.8) **Einstein's A and B.** (Quantum, Optics, Mathematics) ③

Einstein used statistical mechanics to deduce basic truths about the interaction of light with matter very early in the development of quantum mechanics. In particular, he established that *stimulated emission* was demanded for statistical mechanical consistency, and found formulæ determining the relative rates of absorption, spontaneous emission, and stimulated emission. (See Feynman [61, I.42–5].)

Consider a system consisting of noninteracting atoms weakly coupled to photons (electromagnetic radiation), in equilibrium at temperature $k_B T = 1/\beta$. The atoms have two energy eigenstates E_1 and E_2 with average populations N_1 and N_2; the relative population is given as usual by the Boltzmann distribution

$$\left\langle \frac{N_2}{N_1} \right\rangle = e^{-\beta(E_2 - E_1)}. \qquad (7.82)$$

The energy density in the electromagnetic field is given by the Planck distribution (eqn 7.66):

$$u(\omega) = \frac{\hbar}{\pi^2 c^3} \frac{\omega^3}{e^{\beta \hbar \omega} - 1}. \qquad (7.83)$$

An atom in the ground state will absorb electromagnetic energy from the photons at a rate that is proportional to the energy density $u(\omega)$ at the excitation energy $\hbar\omega = E_2 - E_1$. Let us define this absorption rate per atom to be $2\pi B u(\omega)$.[46] An atom in the excited state E_2, with no electromagnetic stimulation, will decay into the ground state with a rate A, emitting a photon. Einstein noted that neither A nor B should depend upon temperature.

Einstein argued that using just these two rates would lead to an inconsistency.

(a) *Compute the long-time average ratio N_2/N_1 assuming only absorption and spontaneous emission. Even in the limit of weak coupling (small A and B), show that this equation is incompatible with the statistical distributions 7.82 and 7.83.* (Hint: Write a formula for dN_1/dt, and set it equal to zero. Is the resulting B/A temperature independent?)

Einstein fixed this by introducing stimulated emission. Roughly speaking, an atom experiencing an oscillating electromagnetic field is more likely to emit photons into that mode. Einstein found that the stimulated emission rate had to be a constant $2\pi B'$ times the energy density $u(\omega)$.

(b) *Write the equation for dN_1/dt, including absorption (a negative term) and spontaneous and stimulated emission from the population N_2. Assuming equilibrium, use this equation and eqns 7.82 and 7.83 to solve for B, and B' in terms of A.* These are generally termed the Einstein A and B coefficients.

Let us express the stimulated emission rate in terms of the number of excited photons per mode (see Exercise 7.9(a) for an alternative derivation).

(c) *Show that the rate of decay of excited atoms $A + 2\pi B' u(\omega)$ is enhanced by a factor of $\langle n \rangle + 1$ over the zero temperature rate, where $\langle n \rangle$ is the expected number of photons in a mode at frequency $\hbar\omega = E_2 - E_1$.*

(7.9) **Bosons are gregarious: superfluids and lasers.** (Quantum, Optics, Atomic physics) ③

Adding a particle to a Bose condensate. Suppose we have a noninteracting system of bosonic atoms in a box with single-particle eigenstates ψ_n. Suppose the system begins in a Bose-condensed state with all N bosons in a state ψ_0, so

$$\Psi_N^{[0]}(\mathbf{r}_1, \ldots, \mathbf{r}_N) = \psi_0(\mathbf{r}_1) \cdots \psi_0(\mathbf{r}_N). \qquad (7.84)$$

Suppose a new particle is gently injected into the system, into an equal superposition of the M lowest single-particle states.[47] That is, if it were injected into an empty box, it would start in state

$$\phi(\mathbf{r}_{N+1}) = \frac{1}{\sqrt{M}} \big(\psi_0(\mathbf{r}_{N+1}) + \psi_1(\mathbf{r}_{N+1}) + \ldots + \psi_{M-1}(\mathbf{r}_{N+1}) \big). \qquad (7.85)$$

The state $\Phi(\mathbf{r}_1, \ldots \mathbf{r}_{N+1})$ after the particle is inserted into the noninteracting Bose condensate

---

[46]The literature uses $u_{\text{cycles}}(f)$ where $f = \omega/2\pi$ is in cycles per second, and has no factor of 2π. Since $u_{\text{cycles}}(f) \, df = u(\omega) \, d\omega$, the absorption rate $B u_{\text{cycles}}(f) = B u(\omega) \, d\omega/df = 2\pi B u(\omega)$.

[47]For free particles in a cubical box of volume V, injecting a particle at the origin $\phi(\mathbf{r}) = \delta(\mathbf{r})$ would be a superposition of *all* plane-wave states of equal weight, $\delta(\mathbf{r}) = (1/V) \sum_{\mathbf{k}} e^{i\mathbf{k} \cdot \mathbf{x}}$. (In second-quantized notation, $a^\dagger(\mathbf{x} = 0) = (1/V) \sum_{\mathbf{k}} a_{\mathbf{k}}^\dagger$.) We "gently" add a particle at the origin by restricting this sum to low-energy states. This is how quantum tunneling into condensed states (say, in Josephson junctions or scanning tunneling microscopes) is usually modeled.

is given by symmetrizing the product function $\Psi_N^{[0]}(\mathbf{r}_1,\ldots,\mathbf{r}_N)\phi(\mathbf{r}_{N+1})$ (eqn 7.29).

(a) *Symmetrize the initial state of the system after the particle is injected. (For simplicity, you need not calculate the overall normalization constant C.) Calculate the ratio of the probability that all particles in the symmetrized state are in the state ψ_0 to the probability that one of the particles is in another particular state ψ_m for $0 < m < M$. (Hint: First do it for $N = 1$. The enhancement should be by a factor $N + 1$.)*

So, if a macroscopic number of bosons are in one single-particle eigenstate, a new particle will be much more likely to add itself to this state than to any of the microscopically populated states.

Notice that nothing in your analysis depended on ψ_0 being the lowest energy state. If we started with a macroscopic number of particles in a single-particle state with wavevector $\mathbf{k}$ (that is, a superfluid with a supercurrent in direction $\mathbf{k}$), new added particles, or particles scattered by inhomogeneities, will preferentially enter into that state. This is an alternative approach to understanding the persistence of supercurrents, complementary to the topological approach (Exercise 9.7).

Adding a photon to a laser beam. This chummy behavior between bosons is also the principle behind lasers.[48] A laser has N photons in a particular mode. An atom in an excited state emits a photon. The photon it emits will prefer to join the laser beam than to go off into one of its other available modes by a factor $N+1$. Here the N represents *stimulated emission*, where the existing electromagnetic field pulls out the energy from the excited atom, and the $+1$ represents *spontaneous emission* which occurs even in the absence of existing photons.

Imagine a single atom in a state with excitation energy energy E and decay rate Γ, in a cubical box of volume V with periodic boundary conditions for the photons. By the energy-time uncertainty principle, $\langle \Delta E\, \Delta t \rangle \geq \hbar/2$, the energy of the atom will be uncertain by an amount $\Delta E \propto \hbar\Gamma$. Assume for simplicity that, in a cubical box without pre-existing photons, the atom would decay at an equal rate into any mode in the range $E - \hbar\Gamma/2 < \hbar\omega < E + \hbar\Gamma/2$.

(b) *Assuming a large box and a small decay rate Γ, find a formula for the number of modes M per*
unit volume V competing for the photon emitted from our atom. Evaluate your formula for a laser with wavelength $\lambda = 619$ nm and the linewidth $\Gamma = 10^4$ rad/s. (Hint: Use the density of states, eqn 7.64.)

Assume the laser is already in operation, so there are N photons in the volume V of the lasing material, all in one plane-wave state (a *single-mode* laser).

(c) *Using your result from part (a), give a formula for the number of photons per unit volume N/V there must be in the lasing mode for the atom to have 50% likelihood of emitting into that mode.*

The main task in setting up a laser is providing a population of excited atoms. Amplification can occur if there is a *population inversion*, where the number of excited atoms is larger than the number of atoms in the lower-energy state (definitely a nonequilibrium condition). This is made possible by *pumping* atoms into the excited state by using one or two other single-particle eigenstates.

(7.10) **Crystal defects.** (Quantum, Condensed matter) @

A defect in a crystal has one on-center configuration with energy zero, and M off-center configurations with energy ϵ, with no significant quantum tunneling between the states. The Hamiltonian can be approximated by the $(M + 1) \times (M + 1)$ matrix

$$\mathcal{H} = \begin{pmatrix} 0 & 0 & 0 & \cdots \\ 0 & \epsilon & 0 & \cdots \\ 0 & 0 & \epsilon & \cdots \\ & \cdots & & \end{pmatrix}. \tag{7.86}$$

There are N defects in the crystal, which can be assumed stuck in position (and hence distinguishable) and assumed not to interact with one another.

Write the canonical partition function $Z(T)$, the mean energy $E(T)$, the fluctuations in the energy, the entropy $S(T)$, and the specific heat $C(T)$ as a function of temperature. Plot the specific heat per defect $C(T)/N$ for $M = 6$; set the unit of energy equal to ϵ and $k_B = 1$ for your plot. Derive a simple relation between M and the change in entropy between zero and infinite temperature. Check this relation using your formula for $S(T)$.

[48]Laser is an acronym for "light amplification by the stimulated emission of radiation".

The bump in the specific heat for a two-state system is called a *Schottky anomaly*.

(7.11) **Phonons on a string.** (Quantum, Condensed matter) @

A continuum string of length L with mass per unit length μ under tension τ has a vertical, transverse displacement $u(x,t)$. The kinetic energy is $\int (\mu/2)(\partial u/\partial t)^2 dx$ and the potential energy is $\int (\tau/2)(\partial u/\partial x)^2 dx$. The string has fixed boundary conditions at $x = 0$ and $x = L$. *Write the kinetic energy and the potential energy in new variables, changing from $u(x,t)$ to normal modes $q_k(t)$ with $u(x,t) = \sum_n q k_n(t) sin(k_n x)$, $k_n = n\pi/L$. Show in these variables that the system is a sum of decoupled harmonic oscillators with mass $m = \mu L/2$. Calculate the density of normal modes per unit frequency $g(\omega)$ for a long string L. Calculate the specific heat of the string $c(T)$ per unit length in the limit $L \to \infty$, treating the oscillators quantum mechanically. What is the specific heat of the classical string?* (Hint: $\int_0^\infty x/(e^x - 1)dx = \pi^2/6$.)

Almost the same calculation, in three dimensions, gives the low-temperature specific heat of crystals. This is also an example of a *quantum field theory*, in this case for a massless phonon mode (zero frequency as $k \to \infty$).

(7.12) **Semiconductors.** (Quantum, Condensed matter) ③

Let us consider a caricature model of a doped semiconductor [13, chapter 28]. Consider a crystal of phosphorus-doped silicon, with $N - M$ atoms of silicon and M atoms of phosphorus. Each silicon atom contributes one electron to the system, and has two states at energies $\pm\Delta/2$, where $\Delta = 1.16$ eV is the energy gap. Each phosphorus atom contributes *two* electrons and two states, one at $-\Delta/2$ and the other at $\Delta/2 - \epsilon$, where $\epsilon = 0.044$ eV is much smaller than the gap.[49] (Our model ignores the quantum mechanical hopping between atoms that broadens the levels at $\pm\Delta/2$ into the conduction band and the valence band.[50] It also ignores spin and chemistry; each silicon really contributes four electrons and four levels, and each phosphorus five

electrons and four levels.) To summarize, our system has $N + M$ spinless electrons (maximum of one electron per state), N valence band states at energy $-\Delta/2$, M impurity band states at energy $\Delta/2 - \epsilon$, and $N - M$ conduction band states at energy $\Delta/2$.

(a) *Derive a formula for the number of electrons as a function of temperature T and chemical potential μ for the energy levels of our system.*

(b) *In a one centimeter cubed sample, there are $M = 10^{16}$ phosphorus atoms; silicon has about $N = 5 \times 10^{22}$ atoms per cubic centimeter. Find μ at room temperature $(1/40 \, \mathrm{eV})$ from the formula you derived in part (a).* (Probably trying various μ is easiest; set up a program on your calculator or computer.) *At this temperature, what fraction of the phosphorus atoms are ionized (have their upper energy state empty)? Why are so few left unionized?* (Hint: Entropy versus energy.) *What is the density of holes (empty states at energy $-\Delta/2$)?*

Phosphorus is an *electron donor*, and our sample is doped *n*-type, since the dominant carriers are electrons; *p*-type semiconductors are doped with holes.

The behavior of a semiconductor is extremely sensitive to small concentrations of impurities. There are few electrons and holes in the pure crystal, because the gap is large compared to $k_B T$. (We live at temperatures much lower than electronic excitations.) Impurities, however, are easily ionized—their energy levels cluster near to one edge of the energy band, at energies comparable to $k_B T$.[51] So the impurities can add their electrons to the conduction band (or holes to the valence band), dramatically changing the conductivity for small doping.

(c) *What is the limiting occupation probability for the states as $T \to \infty$, where entropy is maximized and all states are equally likely? Using this, find a formula for $\mu(T)$ valid at large T, not involving Δ or ϵ.*

(d) *Draw an energy level diagram showing the filled and empty states at $T = 0$. Find a formula for $\mu(T)$ in the low-temperature limit $T \to 0$,*

[49]The phosphorus atom is neutral when both of its states are filled; the upper state can be thought of as an electron bound to a phosphorus positive ion. The energy shift ϵ represents the Coulomb attraction of the electron to the phosphorus ion; it is small because the dielectric constant is large [13, chapter 28].

[50]In solid state physics, you use Bloch's theorem to start from free electrons and produce bands. Here we start from atoms and could add hopping to reach the same end.

[51]This happens because the dielectric constant is large; look up *hydrogenic impurities*. The dielectric constant is large because the gap of a semiconductor is small (small compared to the electronic energy scales, but still large compared to $k_B T$). If the gap were zero, the material would be a metal, with infinite polarizability and hence infinite dielectric constant.

not involving the variable T. (Hint: Balance the number of holes in the impurity band with the number of electrons in the conduction band. Why can you ignore the valence band?)

(7.13) Bose condensation in a band. (Atomic physics, Quantum) ②
The density of single-particle eigenstates $g(E)$ of a system of noninteracting bosons forms a band; the eigenenergies are confined to a range $E_{min} < E < E_{max}$, so $g(E)$ is nonzero in this range and zero otherwise. The system is filled with a finite density of bosons.
Which of the following is necessary for the system to undergo Bose condensation at low temperatures?

(i) $g(E)/(e^{\beta(E-E_{min})} + 1)$ is finite as $E \to E_{min}^-$.

(ii) $g(E)/(e^{\beta(E-E_{min})} - 1)$ is finite as $E \to E_{min}^-$.

(iii) $E_{min} \geq 0$.

(iv) $\int_{E_{min}}^{E} g(E')/(E'-E_{min}) \, dE'$ is a convergent integral at the lower limit E_{min}.

(v) Bose condensation cannot occur in a system whose states are confined to an energy band.

(7.14) Bose condensation: the experiment. (Quantum, Atomic physics) ④
Anderson, Ensher, Matthews, Wieman, and Cornell in 1995 were able to get a dilute gas of rubidium-87 atoms to Bose condense [7].

(a) *Is rubidium-87 (37 protons and electrons, 50 neutrons) a boson or a fermion?*

(b) *At their quoted maximum number density of 2.5×10^{12}/cm³, at what temperature $T_c^{predict}$ do you expect the onset of Bose condensation in free space? They claim that they found Bose condensation starting at a temperature of $T_c^{measured} = 170$ nK. Is that above or below your estimate?* (Useful constants: $h = 6.6262 \times 10^{-27}$ erg s, $m_n \sim m_p = 1.6726 \times 10^{-24}$ g, $k_B = 1.3807 \times 10^{-16}$ erg/K.)
The trap had an effective potential energy that was harmonic in the three directions, but anisotropic with cylindrical symmetry. The frequency along the cylindrical axis was $f_0 = 120$ Hz so $\omega_0 \sim 750$ Hz, and the two other frequencies were smaller by a factor of $\sqrt{8}$: $\omega_1 \sim 265$ Hz. The Bose condensation was observed

by abruptly removing the trap potential,[52] and letting the gas atoms spread out; the spreading cloud was imaged 60 ms later by shining a laser on them and using a CCD to image the shadow.

Fig. 7.16 Bose–Einstein condensation at 400, 200, and 50 nano-Kelvin. The pictures are spatial distributions 60 ms after the potential is removed; the field of view of each image is $200 \,\mu\text{m} \times 270 \,\mu\text{m}$. The left picture is roughly spherically symmetric, and is taken before Bose condensation; the middle has an elliptical Bose condensate superimposed on the spherical thermal background; the right picture is nearly pure condensate. From [7]. Thanks to the Physics 2000 team for permission to reprint this figure.

For your convenience, the ground state of a particle of mass m in a one-dimensional harmonic oscillator with frequency ω is $\psi_0(x) = (m\omega/\pi\hbar)^{1/4} e^{-m\omega x^2/2\hbar}$, and the momentum-space wavefunction is $\tilde{\psi}_0(p) = (1/(\pi m\hbar\omega))^{1/4} e^{-p^2/2m\hbar\omega}$. In this 3D problem the solution is a product of the corresponding Gaussians along the three axes.

(c) *Will the momentum distribution be broader along the high-frequency axis (ω_0) or one of the low-frequency axes (ω_1)? Assume that you may ignore the small width in the initial position distribution, and that the positions in Fig. 7.16 thus reflect the velocity distribution times the time elapsed. Which axis, x or y in Fig. 7.16, corresponds to the high-frequency cylinder axis? What anisotropy would one expect in the momentum distribution at high temperatures (classical statistical mechanics)? Does this high-temperature anisotropy agree with the experimental measurement in Fig. 7.16?*
Their Bose condensation is not in free space; the

---

[52] Actually, they first slowly reduced it by a factor of 75 and then abruptly reduced it from there; let us ignore that complication.

atoms are in a harmonic oscillator potential. In the calculation in free space, we approximated the quantum states as a continuum density of states $g(E)$. That is only sensible if $k_B T$ is large compared to the level spacing near the ground state.

(d) *Compare $\hbar\omega$ to $k_B T$ at the Bose condensation point T_c^{measured} in their experiment.* ($\hbar = 1.05459 \times 10^{-27}\,\text{erg}\,\text{s}$; $k_B = 1.3807 \times 10^{-16}\,\text{erg/K}$.)

For bosons in a one-dimensional harmonic oscillator of frequency ω_0, it is clear that $g(E) = 1/(\hbar\omega_0)$; the number of states in a small range ΔE is the number of $\hbar\omega_0$s it contains.

(e) *Compute the density of single-particle eigenstates*

$$g(E) = \int_0^\infty d\varepsilon_1\, d\varepsilon_2\, d\varepsilon_3\, g_1(\varepsilon_1) g_2(\varepsilon_2) g_3(\varepsilon_3)$$
$$\times\, \delta\left(E - (\varepsilon_1 + \varepsilon_2 + \varepsilon_3)\right) \quad (7.87)$$

for a three-dimensional harmonic oscillator, with one frequency ω_0 and two of frequency ω_1. (Hint: What is the shape of the energy shell in ε space?)

Their experiment has $N = 2 \times 10^4$ atoms in the trap as it condenses.

(f) *By working in analogy with the calculation in free space and your density of states from part (e), find the maximum number of atoms that can occupy the three-dimensional harmonic oscillator potential (without Bose condensation) at temperature T. (Hint: $\int_0^\infty z^2/(e^z - 1)\,dz = 2\zeta(3) = 2.40411$.) According to your calculation, at what temperature T_c^{HO} should the real experimental trap have Bose condensed?*

(7.15) **The photon-dominated Universe.**[53] (Astrophysics) ③

The Universe is currently not in equilibrium. However, in the microwave frequency range it is filled with radiation that is precisely described by a Planck distribution at $2.725 \pm 0.001\,\text{K}$ (Fig. 7.17).

The microwave background radiation is a window back to the *decoupling time*, about 380,000 years after the Big Bang,[54] when the temperature dropped low enough for the protons and electrons to combine into hydrogen atoms. Light does not travel far in ionized gases; it accelerates

the charges and scatters from them. Hence, before this time, our Universe was very close to an equilibrium soup of electrons, nuclei, and photons.[55] The neutral atoms after this time were transparent enough that almost all of the photons traveled for the next 13 billion years directly into our detectors.

These photons in the meantime have greatly increased in wavelength. This is due to the subsequent expansion of the Universe. The initial Planck distribution of photons changed both because of the Doppler effect (a red-shift because the distant gas that emitted the photon appears to be moving away from us) and because the photons are diluted into a larger volume. The Doppler shift both reduces the photon energy and squeezes the overall frequency range of the photons (increasing the number of photons per unit frequency).

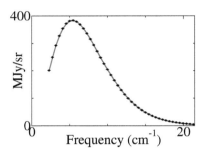

Fig. 7.17 Planck microwave background radiation, as measured by the COBE satellite [63]. The units on the axes are those used in the original paper: inverse centimeters instead of frequency (related by the speed of light) on the horizontal axis and Mega-Janskys/steridian for the vertical axis (1 MegaJansky $= 10^{-20}\,\text{W}\,\text{m}^{-2}\,\text{Hz}^{-1}$). The curve is the Planck distribution at $2.725\,\text{K}$.

One might ask why the current microwave background radiation is thermal (Fig. 7.17), and why it is at such a low temperature . . .

(a) *If the side of the box L and the wavelengths of the photons in the box are all increased by a factor f, what frequency ω' will result from a photon with initial frequency ω? If the original density of photons is $n(\omega)$, what is the density of photons $n'(\omega')$ after the expansion? Show that*

---

[53]This exercise was developed with the help of Dale Fixsen and Eanna Flanagan.
[54]See [207] for a history of the early Universe.
[55]The neutrinos fell out of equilibrium somewhat earlier.

Planck's form for the number of photons per unit frequency per unit volume

$$\frac{\omega^2}{\pi^2 c^3 (e^{\hbar\omega/k_B T} - 1)} \qquad (7.88)$$

(from eqn 7.65) is preserved, except for a shift in temperature. What is the new temperature T', in terms of the original temperature T and the expansion factor f?

This is as expected; an adiabatic expansion leaves the system in equilibrium, but at a different temperature.

(b) *How many microwave background photons are there per cubic centimeter? How does this compare to the average atomic density in the Universe ($n_{\text{matter}} \sim 2.5 \times 10^{-7}$ atoms/cm^3)? (Note $\int_0^\infty x^2/(e^x - 1)\, dx = 2\zeta(3) \approx 2.404$. Useful constants: $\hbar = 1.05 \times 10^{-27}$ erg s, $c = 3 \times 10^{10}$ cm/s, and $k_B = 1.38 \times 10^{-16}$ erg/K.)*

Cosmologists refer to the current Universe as *photon dominated*, because there are currently many more photons than atoms.

We can also roughly estimate the relative contributions of photons and atoms to other properties of the Universe.

(c) *Calculate formulæ for the entropy S, the internal energy E, and the pressure P for the photon gas in a volume V and temperature T. For simplicity, write them in terms of the Stefan–Boltzmann constant*[56] $\sigma = \pi^2 k_B^4/60\hbar^3 c^2$. *Ignore the zero-point energy in the photon modes*[57] *(which otherwise would make the energy and pressure infinite, even at zero temperature).*

(Hint: You will want to use the grand free energy Φ for the photons. For your information, $\int_0^\infty x^3/(e^x - 1)\, dx = \pi^4/15 = -3\int_0^\infty x^2 \log(1 - e^{-x})\, dx$, where the last integral can be integrated by parts to get the first integral.)

(d) *Calculate formulæ for the entropy, mass-energy*[58] *density, and pressure for an ideal gas of hydrogen atoms at density n_{matter} and the same volume and temperature. Can we ignore quantum mechanics for the atomic gas? Assemble your results from parts (c) and (d) into a table comparing photons to atoms, with four columns giving the two analytical formulæ and*

then numerical values for $V = 1\,\text{cm}^3$, the current microwave background temperature, and the current atom density. Which are dominated by photons? By atoms? (Hint: You will want to use the Helmholtz free energy A for the atoms. More useful constants: $\sigma = 5.67 \times 10^{-5}\,\text{erg cm}^{-2}\,\text{K}^{-4}\,\text{s}^{-1}$, and $m_H \approx m_p = 1.673 \times 10^{-24}$ g.)

Before the decoupling time, the coupled light-and-matter soup satisfied a wave eqn [89]:

$$\rho \frac{\partial^2 \Theta}{\partial t^2} = B \nabla^2 \theta. \qquad (7.89)$$

Here Θ represents the local temperature fluctuation $\Delta T/T$. The constant ρ is the sum of three contributions: the matter density, the photon energy density E/V divided by c^2, and a contribution P/c^2 due to the photon pressure P (this comes in as a component in the stress-energy tensor in general relativity).

(e) *Show that the sum of the two photon contributions to the mass density is proportional to $E/(c^2 V)$. What is the constant of proportionality?*

The constant B in our wave eqn 7.89 is the bulk modulus: $B = -V(\partial P/\partial V)|_S$.[59] At decoupling, the dominant contribution to the pressure (and to B) comes from the photon gas.

(f) *Write P as a function of S and V (eliminating T and E), and calculate B for the photon gas. Show that it is proportional to the photon energy density E/V. What is the constant of proportionality?*

Let R be the ratio of ρ_{matter} to the sum of the photon contributions to ρ from part (e).

(g) *What is the speed of sound in the Universe before decoupling, as a function of R and c?* (Hint: Compare with eqn 10.78 in Exercise 10.1 as a check for your answer to parts (e)–(g).)

Exercise 10.1 and the ripples-in-fluids animation at [139] show how this wave equation explains much of the observed fluctuations in the microwave background radiation.

---

[56]The Stefan–Boltzmann law says that a black body radiates power σT^4 per unit area, where σ is the Stefan–Boltzmann constant; see Exercise 7.7.

[57]Treat them as bosons (eqn 7.41) with $\mu = 0$ rather than as harmonic oscillators (eqn 7.22).

[58]That is, be sure to include the mc^2 for the hydrogen atoms into their contribution to the energy density.

[59]The fact that one must compress adiabatically (constant S) and not isothermally (constant T) is subtle but important (Isaac Newton got it wrong). Sound waves happen too fast for the temperature to equilibrate. Indeed, we can assume at reasonably long wavelengths that there is no heat transport (hence we may use the adiabatic modulus). All this is true both for air and for early-Universe photon gases.

(7.16) **White dwarfs, neutron stars, and black holes.** (Astrophysics, Quantum) ③

As the energy sources in large stars are consumed, and the temperature approaches zero, the final state is determined by the competition between gravity and the chemical or nuclear energy needed to compress the material.

A simplified model of ordinary stellar matter is a Fermi sea of noninteracting electrons, with enough nuclei to balance the charge. Let us model a white dwarf (or black dwarf, since we assume zero temperature) as a uniform density of He^4 nuclei and a compensating uniform density of electrons. Assume Newtonian gravity. Assume the chemical energy is given solely by the energy of a gas of noninteracting electrons (filling the levels to the Fermi energy).

(a) *Assuming nonrelativistic electrons, calculate the energy of a sphere with N zero-temperature noninteracting electrons and radius R.* You may assume that the single-particle eigenstates have the same energies and **k**-space density in a sphere of volume V as they do for a cube of volume V; just like fixed versus periodic boundary conditions, the boundary does not matter to bulk properties. *Calculate the Newtonian gravitational energy of a sphere of He^4 nuclei of equal and opposite charge density. At what radius is the total energy minimized?*

A more detailed version of this model was studied by Chandrasekhar and others as a model for white dwarf stars. Useful numbers: $m_p = 1.6726 \times 10^{-24}$ g, $m_n = 1.6749 \times 10^{-24}$ g, $m_e = 9.1095 \times 10^{-28}$ g, $\hbar = 1.05459 \times 10^{-27}$ erg s, $G = 6.672 \times 10^{-8}$ cm^3/(g s^2), 1 eV $= 1.60219 \times 10^{-12}$ erg, $k_B = 1.3807 \times 10^{-16}$ erg/K, and $c = 3 \times 10^{10}$ cm/s.

(b) *Using the nonrelativistic model in part (a), calculate the Fermi energy of the electrons in a white dwarf star of the mass of the Sun, 2×10^{33} g, assuming that it is composed of helium.* (i) *Compare it to a typical chemical binding energy of an atom. Are we justified in ignoring the electron–electron and electron–nuclear interactions (i.e. chemistry)?* (ii) *Compare it to the temperature inside the star, say 10^7 K. Are we justified in assuming that the electron gas is degenerate (roughly zero temperature)?* (iii) *Compare it to the mass of the electron. Are we roughly justified in using a nonrelativistic theory?* (iv) *Compare it to the mass difference between a proton and a neutron.*

The electrons in large white dwarf stars are relativistic. This leads to an energy which grows more slowly with radius, and eventually to an upper bound on the stellar mass.

(c) *Assuming extremely relativistic electrons with $\varepsilon = pc$, calculate the energy of a sphere of noninteracting electrons. Notice that this energy cannot balance against the gravitational energy of the nuclei above a special value of the mass, M_0. Calculate M_0. How does your M_0 compare with the mass of the Sun, above?*

A star with mass larger than M_0 continues to shrink as it cools. The electrons (see (iv) in part (b) above) combine with the protons, staying at a constant density as the star shrinks into a ball of almost pure neutrons (a *neutron star*, often forming a *pulsar* because of trapped magnetic flux). Recent speculations [156] suggest that the neutronium will further transform into a kind of quark soup with many strange quarks, forming a transparent insulating material.

For an even higher mass, the Fermi repulsion between quarks cannot survive the gravitational pressure (the quarks become relativistic), and the star collapses into a black hole. At these masses, general relativity is important, going beyond the purview of this text. But the basic competition, between degeneracy pressure and gravity, is the same.

(7.17) **Eigenstate thermalization.** (Quantum) ⓟ

Footnote 10 on p. 183 discusses many-body eigenstates as "weird superpositions of states with photons being absorbed by the atom and the atom emitting photons". (See also Exercise 7.27.) Many-body eigenstates with a *finite energy density* can be far more comprehensible—they often correspond to *equilibrium* quantum systems.

Look up "eigenstate thermalization". Find a discussion that discusses a single pure state of a large system $A + B$, tracing out the bath B and leaving a density matrix for a small subsystem A.

(7.18) **Drawing wavefunctions.** (Quantum) @

Here we explore the wavefunctions of Bose, Fermi, and distinguishable particles in position space.

(a) *Draw a typical first excited state $\Psi(x, y)$ for two particles at positions x and y in a one-dimensional potential well, given that they are (i) distinguishable, (ii) bosons, and (iii) fermions. For the fermion wavefunction, assume* the two spins are aligned. Do the plots in the 2D region $-L/2 < x, y < L/2$, emphasizing the two regions where the wavefunction is positive and negative, and the nodal curve where $\Psi(x, y) = 0$.

Real particles interact with one another and their environment. We explore what wavefunction features are common to all first excited states.

(b) *Which nodal lines in your plots for part (a) are fixed? Which can vary depending on the many-body wavefunction? (In particular, for the Bose excited state, must the nodal curve be straight?) Which nodal lines in the fermion excited state must remain in position independent of the Hamiltonian? Would you expect that the number of nodal lines be independent of external and interparticle potentials, for the first excited state?*

(7.19) **Many-fermion wavefunction nodes.** (Quantum) @

Consider an N-electron atom or molecule in three dimensions, treating the nuclei as fixed (and hence the Coulomb attraction to the electrons as a fixed external potential), with electron ground-state wavefunction $\Psi(\mathbf{R})$, $\mathbf{R} = \{\mathbf{r}_1, \ldots, \mathbf{r}_N\} \in \mathbb{R}^{3N}$. We shall explore the structure of the zeros of this $3N$-dimensional wave function.

In high-dimensional spaces, one often measures the *co-dimension* of hypersurfaces by counting down from the dimension of the embedding space. Thus the unit hypersphere is co-dimension one, whatever the dimension of space is.

Assume that there are no magnetic fields, and that all spins are aligned, so that $\Psi(\mathbf{r})$ can be assumed to be both real and totally antisymmetric under interchange of particles.

(a) *What is the co-dimension of the sets where $\Psi(\mathbf{R}) = 0$?*

(b) *What zeros are pinned in place by the Fermi statistics of the electrons? What is the co-dimension of this lattice of pinned nodes?*

I view the nodal surfaces as being like a soap bubble, pinned on the fixed nodes and then smoothly interpolating to minimize the energy of the wavefunction. Diffusion quantum Monte Carlo, in fact, makes use of this. One starts with a carefully constructed variational wavefunction

$\Psi_0(\mathbf{r})$, optimizes the wavefunction in a *fixed-node* approximation (finding the energy in a path integral restricted to one local connected region where $\Psi_0(\mathbf{r}) > 0$), and then at the end relaxes the fixed-node approximation.

(c) *Consider a connected region in configuration space surrounded by nodal surfaces where $\Psi(\mathbf{r}) > 0$. Do you expect that the ground state Ψ in this region will define, through antisymmetry, Ψ everywhere in space?* (Hint: Why does the ground state wavefunction for one particle have no nodes?)

(7.20) **Cooling coffee.** ②

Vacuum is an excellent insulator. This is why the surface of the Sun can remain hot ($T_S = 6{,}000°$ K) even though it faces directly onto outer space at the microwave background radiation temperature $T_{MB} = 2.725$ K (Exercise 7.15). The main way[60] in which heat energy can pass through vacuum is by thermal electromagnetic radiation (photons). A black body radiates an energy σT^4 per square meter per second, where $\sigma = 5.67 \times 10^{-8}$ J/(s m^2 K^4) (Exercise 7.7).

A vacuum flask or Thermos^{TM} bottle keeps coffee warm by containing the coffee in a *Dewar*—a double-walled glass bottle with vacuum between the two walls.

Coffee at an initial temperature $T_H(0) = 100°$ C of volume $V = 150$ mL is stored in a vacuum flask with surface area $A = 0.1$ m^2 in a room of temperature $T_C = 20°$ C. Write down symbolically the differential equation determining how the difference between the coffee temperature and the room temperature $\Delta(t) = T_H(t) - T_C$ decreases with time, assuming the vacuum surfaces of the dewar are black and remain at the current temperatures of the coffee and room. Solve this equation symbolically in the approximation that Δ is small compared to T_c (by approximating $T_H^4 = (T_C + \Delta)^4 \approx T_C^4 + 4\Delta T_C^3$). What is the exponential decay time (the time it take for the coffee to cool by a factor of e), both symbolically and numerically in seconds? (Useful conversion: $0°$ C = 273.15° K.)

Real Dewars are not painted black! They are coated with shiny metals in order to minimize this radiative heat loss. (White or shiny materials not only absorb less radiation, but they also emit less radiation, see Exercise 7.7.)

---

[60]The sun and stars can also radiate energy by emitting neutrinos. This is particularly important during a supernova.

(7.21) **The greenhouse effect.** (Astrophysics, Ecology) ②

The outward solar energy flux at the Earth's orbit is $\Phi_S = 1{,}370 \, \mathrm{W/m^2}$, and the Earth's radius is approximately $6{,}400 \, \mathrm{km}$, $r_E = 6.4 \times 10^6 \, \mathrm{m}$. The Earth reflects about 30% of the radiation from the Sun directly back into space (its *albedo* $\alpha \approx 0.3$). The remainder of the energy is eventually turned into heat, and radiated into space again. Like the Sun and the Universe, the Earth is fairly well described as a black-body radiation source in the infrared. A black body radiates an energy σT^4 per square meter per second, where $\sigma = 5.67 \times 10^{-8} \, \mathrm{J/(s\,m^2\,K^4)}$ (Exercise 7.7).

What temperature T_A does the Earth radiate at, in order to balance the energy flow from the Sun after direct reflection is accounted for? Is that hotter or colder than you would estimate from the temperatures you've experienced on the Earth's surface? (Warning: The energy flow in is proportional to the Earth's cross-sectional area, while the energy flow out is proportional to its surface area.)

The reason the Earth is warmer than would be expected from a simple radiative energy balance is the *greenhouse effect*.[61] The Earth's atmosphere is opaque in most of the infrared region in which the Earth's surface radiates heat. (This frequency range coincides with the vibration frequencies of molecules in the Earth's upper atmosphere. Light is absorbed to create vibrations, collisions can exchange vibrational and translational (heat) energy, and the vibrations can later again emit light.) Thus it is the Earth's atmosphere which radiates at the temperature T_A you calculated; the upper atmosphere has a temperature intermediate between that of the Earth's surface and interstellar space.

The vibrations of oxygen and nitrogen, the main components of the atmosphere, are too symmetric to absorb energy (the transitions have no dipole moment), so the main greenhouse gases are water, carbon dioxide, methane, nitrous oxide, and chlorofluorocarbons (CFCs). The last four have significantly increased due to human activities; CO_2 by ~30% (due to burning of fossil fuels and clearing of vegetation), CH_4 by ~150% (due to cattle, sheep, rice farming, escape of natural gas, and decomposing garbage), N_2O by

~15% (from burning vegetation, industrial emission, and nitrogen fertilizers), and CFCs from an initial value near zero (from former aerosol sprays, now banned to spare the ozone layer). Were it not for the Greenhouse effect, we'd all freeze (like Mars)—but we could overdo it, and become like Venus (whose deep and CO_2-rich atmosphere leads to a surface temperature hot enough to melt lead).

(7.22) **Light baryon superfluids.** (Quantum) ⓟ

Presume a parallel universe where neutrons and protons were as light as electrons ($\sim 1{,}840$ times lighter than they are now).

Estimate the superfluid transition temperature T_c for water in the parallel universe. Will it be superfluid at room temperature? (Presume the number of molecules of water per unit volume stays constant, and assume the superfluid transition is roughly at the Bose condensation temperature.)

(7.23) **Why are atoms classical?** (Quantum) ②

What is the point of doing classical statistical mechanics, if everything is really quantum? Most things, from atoms on up, have very small quantum fluctuations. Why are atoms so well described by classical mechanics? (Electrons, in contrast, are thoroughly quantum mechanical.)

We saw in Section 7.4 that bosons and fermions behave as classical, Maxwell–Boltzmann particles at high temperatures. The temperature at which quantum effects can become important is called the *degeneracy temperature* T_{dg}. This happens when the cube of the thermal de Broglie wavelength $\lambda = h/\sqrt{2\pi M k_B T}$ (eqn 3.59) equals the volume per particle V/N:

$$k_B T_{\mathrm{dg}} = (2\pi\hbar^2/M)(N/V)^{2/3}. \tag{7.90}$$

We saw in eqn 7.71 that Bose condensation happens at $T_{\mathrm{dg}}/\zeta(3/2)^{2/3}$, certainly an important effect of quantum mechanics. What about fermions? The main quantum effect for fermions is the Pauli exclusion principle, which causes the formation of a Fermi sea of filled states at high densities and low temperatures.

(a) *Show that the degeneracy temperature is equal to the Fermi energy ε_F (eqn 7.76) up to constants of order one. What would the degeneracy temperature be for the fermion ^3He (two protons, one neutron, two electrons)?* (^3He has

---

[61]The glass in greenhouses also is transparent in the visible and opaque in the infrared. This, it turns out, is not why it gets warm inside; the main insulating effect of the glass is to forbid the warm air from escaping. The greenhouse effect is in that sense poorly named.

a mass $M \sim 3M_p = 5{,}520 m_e$, and a density of 0.081 gm/cm^3, making the number density $\sim 1.6 \, 10^{28}$m^{-3}.)

Indeed, ^3He is degenerate (has a well-formed Fermi surface) as soon as it forms a liquid (at 3.19 K). Cooled to a millikelvin, the atoms form Cooper pairs and condense into an exotic superfluid.

But why is the degeneracy temperature so low for atoms? The masses of atoms are determined by the protons and neutrons, but the sizes and binding energies are determined by the electrons. For example, the hydrogen atom has a diameter of about two Bohr radii $2a_0 = 2\hbar^2/(m_e e^2) \sim 1.06$ Å, depending on the electron mass m_e (with a tiny *reduced mass* correction due to the heavy proton nucleus). Similarly, the excited states of the hydrogen atom are simple fractions of the ionization energy, one Rydberg $R_H = e^2/2a_0 \approx 13.6$ eV $\approx 1.6 \, 10^5$ K. Atoms are all a few Bohr radii (from helium at ~ 0.6 Å to monster cesium at ~ 6 Å), and their covalent binding energies are a few eV.

(b) *Calculate the degeneracy temperature for atoms of mass M and a typical atomic density $N/V = 1/(2a_0)^3$. What function of m_e/M makes the degeneracy temperature $k_B T$ so low compared to the characteristic Rydberg energy R_H?*

So it is the heaviness of the proton compared to the electron which makes atoms classical. (See also Exercise 7.22).

The quantum fluctuations of atoms are suppressed when they bind together to form crystals. Covalent crystals will bind together at a reasonable fraction of the atomic energy scale R_H. Molecular crystals like water freeze at lower temperatures, because the electronic interactions are well optimized in the molecules, leaving less incentive to bind into crystals. Only helium remains liquid to low enough temperatures to form a quantum Fermi or Bose fluid.

What about other quantum effects, like quantum tunneling? Quantum tunneling of electrons is pretty common, but atoms hardly ever tunnel. We can explain this with the same kind of argument [203]. Tunneling rates are suppressed by a WKB factor $\exp(-A\sqrt{2MV}Q/\hbar)$, with V the barrier height, Q the distance, and A a constant of order one depending on the shape of the barrier. If we set the tunneling distance and barrier to electronic scales $Q = a_0$, $V = R_H$, and

the mass to a baryon mass $M = M_p$, we find $\sqrt{2M_p R_H} a_0/\hbar = \sqrt{M_p/m_e} \sim 43$, suppressing tunneling by a factor of $\approx e^{-43} \approx 10^{-19}$.

(7.24) **Is sound a quasiparticle?** (Condensed matter) ⓟ

Sound waves in the harmonic approximation are noninteracting—a general solution is given by a linear combination of the individual frequency modes. Landau's Fermi liquid theory (footnote 23, p. 188) describes how the noninteracting electron approximation can be effective even though electrons are strongly coupled to one another. The quasiparticles are electrons with a screening cloud; they develop long lifetimes near the Fermi energy; they are described as poles of Green's functions.

(a) *Do phonons have lifetimes? Do their lifetimes get long as the frequency goes to zero?* (Look up *ultrasonic attenuation* and Goldstone's theorem.)

(b) *Are they described as poles of a Green's function? (See Section 9.3 and Exercise 10.9.)*

Are there analogies for phonons to the screening cloud around a quasiparticle? A phonon screening cloud would be some kind of collective, nonlinear movement of atoms that behaved at long distances and low frequencies like an effective, harmonic interaction. In particular, the effective scattering between these quasiphonons should be dramatically reduced from that one would calculate assuming that the phonons are harmonic vibrations.

At low temperatures in perfect crystals (other than solid helium), anharmonic interactions are small except at high sound amplitudes. But as we raise the temperature, anharmonic effects lead to thermal expansion and changes in elastic constants. We routinely model sound in crystals in terms of the density and elastic constants at the current temperature (phonons "screened" by thermal vibrations), not in terms of the bare phonon normal modes of the zero-temperature crystal.

Thermal expansion is due to anharmonic terms in lattice vibrations. Roughly speaking, atoms bump into one another more at high temperatures—they have an anharmonic wall in the energy when they get too close. This expansion leads to a lower elastic constant in most cases. At larger average spacing between atoms, the bond strengths are weaker and the elastic moduli drop, so the phonon frequencies at a given wavevector also decrease.

(c) *Describe qualitatively the challenges one would have in describing lattice vibrations at high temperatures using "bare" phonons. Briefly describe the way in which one might absorb these anharmonic terms into an emergent theory for finite-temperature vibrations.*

Particle-like excitations occur in almost all emergent theories in physics. *Quasielectrons* and *quasiholes* emerge in the Fermi liquid theory of metals, as described in Section 7.7. *Polarons* are electrons decorated with nearby lattice deformations in insulators. *Excitons* are bound electron-hole pairs in semiconductors. *Polaritons* are photons bound to optical phonons or excitons. Polyacetylene is a precursor to topological insulators; its *solitons* have reversed spin-charge statistics (neutral spins and spinless charged excitations). *Anyons*, with fractional statistics, arise in the fractional quantum Hall effect (and may enable topologically protected quantum computation). Dislocations in crystals are now being rebranded as *fractons*—topological defects with mobility constraints. . .

(7.25) **Quantum measurement and entropy.** (Quantum) ③

Here we explore deep analogies between the collapse of the wavefunction during a quantum measurement and the increase of entropy in statistical mechanics.

Entropy is an emergent property in statistical mechanics. There is no place in the microscopic laws where entropy increases (Exercises 5.7 and 7.4). For large numbers of particles, chaos and ergodicity leads to a loss of information that emerges as equilibration and entropy growth (Exercises 5.8 and 5.25).

In quantum mechanics, the process of measurement is also an emergent property. Schrödinger's time evolution has no place where the wavefunction collapses. When a quantum subsystem interacts with a macroscopic object[62] enough to change its state, the quantum system will subsequently behave according to the Born rule. The Born rule describes the interaction as the application of an operator $\mathbf{O}$; if the initial quantum state ϕ is decomposed into eigenstates $\phi = \sum c_o \xi_o$ of $\mathbf{O}$ then with probability $|c_o|^2$ the quantum subsystem will behave as if it were in state ξ_o all along (*collapsing* into that state).

It is natural to describe this measurement without picking one of the observations, treating the quantum subsystem after the interaction as a mixed state. Suppose we begin with a photon in a diagonally polarized state $\phi_i = \begin{pmatrix} 1/\sqrt{2} \\ 1/\sqrt{2} \end{pmatrix}$, and apply an operator $\mathbf{O} = \begin{pmatrix} 1 & 0 \\ 0 & -1 \end{pmatrix}$. that measures whether the polarization is vertical $|\phi\rangle = \begin{pmatrix} 1 \\ 0 \end{pmatrix}$ or horizontal $\begin{pmatrix} 0 \\ 1 \end{pmatrix}$.

(a) *What is the density matrix $\boldsymbol{\rho}_i$ for the photon before the measurement? What is the density matrix $\boldsymbol{\rho}_f$ after the measurement, according to the Born rule? How much has the entropy changed?* (If you write the bra $\langle \phi_i |$ as $\begin{pmatrix} 1/\sqrt{2} & 1/\sqrt{2} \end{pmatrix}$ then ρ naturally forms a 2×2 matrix.)

The Born rule asserts that the measurement process alters the density matrix, changing it from a pure state to a mixed, unpolarized state. It is unsatisfying in two ways. First, it does not keep track of the effect of the quantum state on the measuring device: there is no record of the measurement. Second, it provides no intuition as to why quantum evolution should produce a collapse of this form.

Let us consider including the object's many-body wavefunction $\Psi(\mathbf{x})$ into the analysis.[63] Before the measurement, the photon is not interacting with the object, so the system (photon plus object) is in a product state

$$\Phi_i = \phi_i \Psi_i(\mathbf{x}) = \begin{pmatrix} 1/\sqrt{2} \\ 1/\sqrt{2} \end{pmatrix} \Psi_i(\mathbf{x}). \qquad (7.91)$$

We then perform a measurement—turning on a time-dependent Hamiltonian coupling the photon to the object, turning it off again, and waiting until the object as a whole received the information. After the measurement, the object is left in one of two distinct many-body wavefunctions, $\Psi_v(\mathbf{x})$ or $\Psi_h(\mathbf{x})$, depending on whether the photon was measured to be vertically or horizontally polarized. The photon's state is unchanged (we assume the object performed a *quantum nondemolition* measurement, [208, section 3.7]).

---

[62]This object could be a material which exchanges energy with the quantum subsystem, as for a solid whose phonons cause a qubit to suffer decoherence. It could be a measuring instrument, recording a value stored in the qubit. Or it could be Geiger counter, rigged to a jar of poison gas in a box with a cat.

[63]We make the unrealistic assumption that the object starts in a pure state for simplicity: the same argument works for a thermal state, or indeed any macroscopic density matrix for the object.

Thus the system after the measurement is in the state

$$\Phi_f = \begin{pmatrix} 1/\sqrt{2} \\ 0 \end{pmatrix} \Psi_v(\mathbf{x}) + \begin{pmatrix} 0 \\ 1/\sqrt{2} \end{pmatrix} \Psi_h(\mathbf{x}). \quad (7.92)$$

(b) *Write the density matrix for the system, both before and after the measurement, as a 2×2 matrix of functions.* (That is, we use a position-space basis for the object. For example, the density matrix for the *object* before the measurement is $\rho = |\Psi_i\rangle\langle\Psi_i|$, which in position space is $\rho(\mathbf{x}', \mathbf{x}) = \langle\mathbf{x}'|\Psi_i\rangle\langle\Psi_i|\mathbf{x}\rangle = \Psi_i(\mathbf{x}')\Psi_i^*(\mathbf{x})$. Hint: Both matrices will have off-diagonal terms.)
Since our system obeys Schrödinger's equation, it ought to be still in a pure state. Remember that, in position space, the trace of an operator $M(x', x)$ is given by integrating over the diagonal $\text{Tr}(M) = \int M(x, x)\mathrm{d}x$, and the square of the density matrix is $\rho^2(x', x) = \int \rho(x', x'')\rho(x'', x)\mathrm{d}x''$.
(c) *Show that your initial density matrix is in a pure state by computing $\text{Tr}(\boldsymbol{\rho}_i^2)$. Show that your final density matrix $\boldsymbol{\rho}_f$ is also in a pure state by computing the trace of its square.*
(d) *What is the entropy change after the measurement, including both the photon and the object?* (Hint: You need not calculate anything, given your results of part (c).)
Our calculation so far has followed the microscopic rules—evolving the wavefunctions of the photon and the object via Schrödinger's equation. We now must make the same kind of macroscopic approximations we use in explaining the increase of entropy. The information about the *coherence* between the two polarizations stored in the object becomes unusable if the object is large and its response to the interaction is complex.
Specifically, the time-dependent Hamiltonian, in making the measurement, has left an indelible imprint on the object. The vector $\mathbf{x}$ represents the configuration of gigantic numbers of atoms, each of which has shifted in a way that depends upon whether the photon was horizontally or vertically polarized. By the definition of a good measuring apparatus, if the final positions of the atoms $\mathbf{x}$ has nonzero probability density of arising for a vertical polarization (i.e., $|\Psi_v(\mathbf{x})|^2 > 0$), then it must have no probability of arising for

a horizontal polarization (so $|\Psi_h(\mathbf{x})|^2$ must be zero). Otherwise, those shared configurations represent the likelihood us that the object has forgotten which state it measured—that every trace of memory is removed on the atomic level. It is more drastic even than this. One cannot act on Ψ_v to make it overlap with Ψ_h with any sensible, local operator. (Think of the object as including an observer writing down the measurement. What quantum operator[64] could erase that information?) Indeed, for any operator $\mathbf{B}$ acting on the object,

$$\langle\Psi_v|\mathbf{B}|\Psi_h\rangle = \int \mathrm{d}\mathbf{x}'\mathrm{d}\mathbf{x}\langle\Psi_v|\mathbf{x}'\rangle\langle\mathbf{x}'|B|\mathbf{x}\rangle\langle\mathbf{x}|\Psi_h\rangle$$
$$= \int \Psi_v^*(\mathbf{x}')B(\mathbf{x}', \mathbf{x})\Psi_h(\mathbf{x})\mathrm{d}\mathbf{x}'\mathrm{d}\mathbf{x} \equiv 0. \quad (7.93)$$

The two wavefunctions are not just orthogonal. They are not just with zero overlap. It is sometimes said that the two wavefunctions *live in different Hilbert spaces*.[65]
How does this allow us to simplify the final density matrix you derived in part (c)? Suppose we subject our photon and our object to a second observation operator $\mathbf{Q}$, which we represent in complete generality as a 2×2 matrix of operators in the polarization space

$$\mathbf{Q} = \begin{pmatrix} A(\mathbf{x}', \mathbf{x}) & B(\mathbf{x}', \mathbf{x}) \\ B^*(\mathbf{x}', \mathbf{x}) & C(\mathbf{x}', \mathbf{x}) \end{pmatrix}. \quad (7.94)$$

We know from eqn 7.5 that $\langle\mathbf{Q}\rangle = \text{Tr}(\mathbf{Q}\rho)$. We now demonstrate that the pure-state density matrix $\boldsymbol{\rho}_f$, if the object is a good measuring instrument, is equivalent to a mixed state $\widehat{\boldsymbol{\rho}}_f$.
(e) *Using eqn 7.93 and your final density matrix from part (b), show that $\langle\mathbf{Q}\rangle$ is equal to $\text{Tr}(\mathbf{Q}\widehat{\boldsymbol{\rho}}_f)$, where*

$$\widehat{\boldsymbol{\rho}}_f = \begin{pmatrix} \frac{1}{2}\Psi_v(\mathbf{x}')\Psi_v^*(\mathbf{x}) & 0 \\ 0 & \frac{1}{2}\Psi_h(\mathbf{x}')\Psi_h^*(\mathbf{x}) \end{pmatrix}. \quad (7.95)$$

What terms changed between $\boldsymbol{\rho}_f$ from part (c) and $\widehat{\boldsymbol{\rho}}_f$? How do these changes represent the loss of coherence between the two polarizations, stored in the object? Explain in words how $\widehat{\boldsymbol{\rho}}_f$ represents a photon and an object which has recorded the polarization.
(f) *How much has the entropy changed after the measurement, using the emergent density*

---

[64]The operator $\mathbf{U}(-t)$ that evolves the photon and the object backward in time would get them to overlap perfectly, erasing all memory. Of course, reversing time would also allow entropy to decrease. Such operators are declared not sensible.
[65]For the sophisticated, one could build an entire Fock space by applying (sensible, local) creation and annihilation operators to the two states; the resulting Hilbert spaces would never overlap, in the limit of an infinite-sized object.

matrix $\hat{\rho}_f$ (eqn 7.95) that reflects the loss of coherence? (Warning: The entropy of a state described by a wavefunction $\psi(x)$ is zero (since it is a pure state). The entropy is *not* $-k_B \int |\psi(x)|^2 \log |\psi(x)|^2$. That would be the entropy of the mixture generated the observation of the position x. Hint: Ψ_v and Ψ_h are pure states describing the object.)

We can now connect our discussion to the Born rule, by considering an effective theory for the photon valid for any observable not also involving the object. This allows us to "integrate out" the object's degrees of freedom to get an effective *reduced density matrix* just for the photon, as we do in classical statistical mechanics to derive free energies (see note 5 on p. 141, and Section 6.7). After the measurement, the object no longer interacts with the photon. Equation 7.94 still describes a general operator for our system. Operators which do not involve the object will be independent of $\mathbf{x}$ and $\mathbf{x}'$, the coordinates describing the object's degrees of freedom.

(g) *Show that our density matrix $\hat{\rho}_f$ reflecting decoherence is equivalent to the unpolarized density matrix ρ_f given by the Born rule, for any operator that does not involve the object.* (Hint: Integrate out $\mathbf{x}$ and $\mathbf{x}'$ in the trace; see Exercise 7.26.)

Thus the collapse of the wavefunction emerges naturally from the complexity and size of macroscopic objects interacting with microscopic quantum subsystems. There remain deep questions about quantum measurement (see Weinberg [208, section 3.7]), but the wavefunction collapse is an emergent property of macroscopic observing systems, much like the entropy.

(7.26) **Entanglement of two spins.**[66] (Quantum, Computation) ③

How does entropy increase in a quantum system? The typical way is through the loss of coherence—information lost to the environment (Exercise 7.25). A measurement through an operator $\mathbf{O}$ can cause a pure state wavefunction (entropy zero) to split into an ensemble[67] of eigenstates of $\mathbf{O}$ (entropy $-k_B \sum_o |c_o|^2 \log |c_o|^2$).

Here we focus on a different mechanism; *Entropy can increase when we ignore or throw away information.* In this exercise, we will concentrate on a quantum system with two spins, an example of *entanglement* and entropy. In Exercise 7.27 we shall discuss the entanglement of larger systems, and explore both entanglement entropy and *eigenstate thermalization*.

Consider first the spin singlet[68] state of positronium: an electron–positron pair with opposite spins in an antisymmetric spin wave function

$$(1/\sqrt{2})\,(|\uparrow_e\rangle|\downarrow_p\rangle - |\downarrow_e\rangle|\uparrow_p\rangle). \tag{7.96}$$

(a) *What is the entropy of this spin singlet wavefunction?*

What happens if we separate the two particles with an electric field (without disturbing their spins), and throw away[69] the electron?

To study this, we introduce the *reduced density matrix*. Suppose ρ is the density matrix of a system composed of two subsystems A and B with bases $\{|\psi_a\rangle\}$ and $\{|\phi_b\rangle\}$. Suppose we consider experiments solely involving A (since we are ignoring B or have thrown it away). Then these experiments will involve observables $\mathbf{O}$ that do not act on the variables in B, so $\mathbf{O}|\phi_b\rangle = |\phi_b\rangle\mathbf{O}$ and $\langle\phi_b|\mathbf{O} = \mathbf{O}\langle\phi_b|$. We can write the expectation value of $\mathbf{O}$ in terms of the density matrix for the entire system with the usual trace

$$\begin{aligned}
\mathrm{Tr}_{a,b}(\mathbf{O}\rho) &= \sum_{a,b}\langle\psi_a|\langle\phi_b|\mathbf{O}\rho|\phi_b\rangle|\psi_a\rangle \\
&= \sum_a \langle\psi_a|\mathbf{O}\left(\sum_b\langle\phi_b|\rho|\phi_b\rangle\right)|\psi_a\rangle \\
&= \mathrm{Tr}_a\mathbf{O}\rho^{[A]},
\end{aligned} \tag{7.97}$$

where the reduced density matrix

$$\rho^{[A]} = \sum_b \langle\phi_b|\rho|\phi_b\rangle \tag{7.98}$$

is the trace over the basis states in B. The reduced density matrix is thus a partial trace. Indeed, one often talks of partial traces in classical systems where one has integrated out over some degrees of freedom to get an effective free energy (see note 5 on p. 141).

---

[66] This exercise was developed in collaboration with Jaron Kent-Dobias. Computer hints can be found at the book website [181].
[67] Exercise 7.25 describes microscopically how this arises; the wavefunctions of the environment split into pieces for each possible observation that differ so much that no operator can connect them.
[68] This is a classic system for exhibiting "spooky action at a distance" and disproving hidden variables theories [61, III.18-3].
[69] Or, more realistically, what happens if we let the positronium decay into two photons of correlated polarization, and let one escape into outer space [61, III.18-3].

(b) *Calculate the reduced density matrix ρ_P for the positron portion of the spin-singlet wavefunction of eqn 7.96, which can be used to describe subsequent experiments on the positron not involving the electron. Show your steps. What is the entropy if we ignore or discard the electron?*

(c) *Now consider the positronium in a triplet state $|\uparrow_p\rangle|\uparrow_e\rangle$. After discarding the electron, what is its entropy?*

We say that the singlet state is *entangled*, while the triplet state is not. Some speculate the growth of quantum entanglement with time is equivalent to the growth of entropy with time; that all loss of information rests upon quantum entanglement with unobservable degrees of freedom (either information escaping to far places as in part (b), or information escaping into many-body correlations in quantum wavefunctions of macroscopic objects, as in Exercise 7.25). This speculation would seem at odds, however, with the observation that chaotic classical systems also lose information and have entropies that grow with time.

(7.27) **Heisenberg entanglement.**[70] (Quantum, Computation) ④

Here we introduce the quantum Heisenberg antiferromagnet, and use it to explore how entropy, temperature, and equilibration can emerge through the entanglement of two portions of a large system—closely related to the *eigenstate thermalization hypothesis* (see Exercise 7.17). We saw in Exercise 7.26 that ignoring part of a system can take a quantum pure state into a mixture of states on the remaining subsystem. This should remind you of the way we derived the canonical ensemble by splitting a system in the microcanonical ensemble into a subsystem and a bath, and ignoring the bath (Section 6.1, Fig. 6.1). We can make this analogy much more powerful by using a larger quantum system, here a one-dimensional chain of spin $\frac{1}{2}$ particles.

The one-dimensional Heisenberg antiferromagnet has Hamiltonian

$$\mathcal{H}_{N_{\text{spins}}} = \sum_{m=1}^{N_{\text{spins}}-1} \mathbf{S}_m \cdot \mathbf{S}_{m+1}, \quad (7.99)$$

where we have set the strength of the coupling J to 1—positive, and hence favoring antiparallel spins. Here the quantum spins $\mathbf{S} = (\sigma_X, \sigma_Y, \sigma_Z)$ have spin $\frac{1}{2}$, and are written in terms of the Pauli matrices

$$\sigma_x = \begin{pmatrix} 0 & 1 \\ 1 & 0 \end{pmatrix} \quad \sigma_y = \begin{pmatrix} 0 & -\mathrm{i} \\ \mathrm{i} & 0 \end{pmatrix} \quad \sigma_z = \begin{pmatrix} 1 & 0 \\ 0 & -1 \end{pmatrix}. \quad (7.100)$$

Let us begin with an analytical calculation of the Hamiltonian and the eigenstates for $N_{\text{spins}} = 2$, considered already in Exercise 7.26. We work in the four-dimensional σ_z basis

$$\begin{pmatrix} |\uparrow_1\rangle|\uparrow_2\rangle \\ |\uparrow_1\rangle|\downarrow_2\rangle \\ |\downarrow_1\rangle|\uparrow_2\rangle \\ |\downarrow_1\rangle|\downarrow_2\rangle \end{pmatrix}. \quad (7.101)$$

(a) *Show analytically that*

$$\mathcal{H}_2 = \begin{pmatrix} 1 & 0 & 0 & 0 \\ 0 & -1 & 2 & 0 \\ 0 & 2 & -1 & 0 \\ 0 & 0 & 0 & 1 \end{pmatrix}. \quad (7.102)$$

Find the eigenvalues and eigenstates for this Hamiltonian. Is the ground state the triplet or the singlet? Does this make sense for an antiferromagnet? (Hint: The spin $\mathbf{S}_1$ commutes with the kets $|\uparrow_2\rangle$ and $|\downarrow_2\rangle$ and vice versa. The tensor discussion below may also be helpful.)

Implementing this calculation elegantly on the computer demands that we understand how the single-spin σ operators and the dot product $\mathbf{S}_m \cdot \mathbf{S}_{m+1}$ act on the entire $2^{N_{\text{spins}}}$-dimensional Hilbert space. The fact that they commute with the parts of the wavefunction that only involves other spins says that they act as identity matrices on those parts of the Hilbert space. That is, $\sigma_x[1]$ for the first spin needs to be promoted to $\sigma_x[1] \otimes \mathbf{1}_{2^{N_{\text{spins}}-1}}$, and $\sigma_x[2]$ for the second needs to be turned into $\mathbf{1}_2 \otimes \sigma_x[1] \otimes \mathbf{1}_{2^{N_{\text{spins}}-2}}, \ldots$

(b) *Implement this numerically for the two-spin system. Calculate the Heisenberg Hamiltonian, and verify the answer of part (a).* (Hint: Many modern programming languages have support for tensor data structures. These efficient routines will be important in later steps, so use them here.[71] See the hints files at [181].)

---

[70]Suggested by Chao-Ming Jian, and developed with Jaron Kent-Dobias. Computer hints can be found at [181].

[71]One subtle point. In combining an operator L_{ij} acting on subsystem A with $M_{\alpha\beta}$ acting on subsystem B, we want an operator O which labels rows using $i\alpha$ and columns with $j\beta$ $O_{i\alpha\,j\beta}$. We can then use O as a two-index matrix to compute eigenvectors and eigenvalues. In some implementations, this demands that we swap the two inner axes of the naive product $L_{ij}M_{\alpha\beta}$.

In this exercise, we shall discuss how pure energy eigenstates states in a system AB become mixed states when we split the system into a subsystem A and a bath B, and study the properties of these mixed states. We shall index operators acting on the subsystem A with Latin letters i, j, operators on the bath B with Greek letters α, β, and operators on the total system AB with capital letters I, J, or sometimes with pairs of indices $i\alpha$, $j\beta$.

(c) *If $\rho_{i\alpha,j\beta}$ is the density matrix for the whole system AB, show analytically that the sum $\sum_\alpha \rho_{i\alpha j\alpha}$ gives the reduced density matrix for the subsystem (e.g., as defined in Exercise 7.26).*

We can use the two-spin problem of part (a) to preview the rest of the exercise, in a context where you know the answer from Exercise 7.26. Here we view the first spin as the the "subsystem" A, and the second spin as the "bath" B.

(d) *Select the singlet eigenstate, and normalize it if necessary. Generate the pure-state density matrix, and reshape it into the four index tensor $\rho_{i\alpha,j\beta}$. Trace over the bath as in part (c), and verify that the reduced density matrix ρ^A_{ij} describes an unpolarized spin. Calculate the entropy by taking the suitable matrix trace.*

To generate the Heisenberg Hamiltonian for multiple spins, we can save steps by noting that we already know the Hamiltonian for two spins from eqn 7.102. So the term $\mathbf{S}_m \cdot \mathbf{S}_{m+1}$ in eqn 7.99 becomes[72]

$$\mathbf{1}_{2^{m-1}} \otimes \mathcal{H}_2 \otimes \mathbf{1}_{2^{N_{\text{spins}}-(m+1)}}. \qquad (7.103)$$

(e) *Use this to write a function that returns the Heisenberg Hamiltonian $\mathcal{H}_{N_{\text{spins}}}$ (eqn 7.99) as a $2^{N_{\text{spins}}} \times 2^{N_{\text{spins}}}$ matrix. Check, for $N_{\text{spins}} = 2$ it returns $\mathcal{H}_2$ (eqn 7.102). Check also for $N_{\text{spins}} = 3$ its eigenvalues are $(-4, -4, 2, 2, 2, 2, 0, 0)$, and for $N_{\text{spins}} = 4$ that its distinct eigenvalues are $\{-3 - 2\sqrt{3}, -1 - 2\sqrt{2}, 3, -1 + 2\sqrt{2}, -1, -3 + 2\sqrt{3}\} \approx \{-6.46, -3.8, 3, 1.8, -1, 0.46\}$.*

We shall work with a system of $N_{\text{spins}} = N_{AB} = 10$ spins in the chain; we shall primarily study a subsystem with $N_A = 4$ spins, so the bath has $N_B = N_{AB} - N_A = 6$ spins. We shall use a particular eigenstate ψ of $\mathcal{H}_{N_{AB}}$ in a sequence of four steps: *(i)* calculate the reduced density matrix $\boldsymbol{\rho}_A$ for N_A, *(ii)* investigate the entanglement between A and the bath B, *(iii)* calculate the entanglement entropy, and *(iv)* illustrate *eigenstate thermalization*. For the last, we want to choose ψ with an energy that is lower than average, but not near zero.

(f) *Create $\mathcal{H}_{AB} = \mathcal{H}_{10}$. Find its energy eigenvalues and eigenstates, and (if necessary) sort them in increasing order of their energy. Pick the energy eigenstate ψ of the full system that is $1/4$ the way from the bottom (eigenvector number $K = 2^{N_{AB}-3}$). Calculate the pure density matrix $\boldsymbol{\rho}^{\text{pure}}$, reshape it into the four index tensor $\rho^{AB}_{i\alpha,j\beta}$, and trace over the bath to give the reduced density matrix ρ^A_{ij}. Check that $\boldsymbol{\rho}^A$ has trace one (as it must), and calculate $\text{Tr}[(\boldsymbol{\rho}^A)^2]$. Is it is a mixed state?*

The entanglement entropy between A and B for a pure state ψ of AB is the entropy of the reduced density matrix of A.

(g) *Calculate the entanglement entropy $S = -\text{Tr}\boldsymbol{\rho}_A \log \boldsymbol{\rho}_A$.[73] Check that it has the same entropy as subsystem B. See how the entanglement entropy changes with the size of the subsystem, by looping over N_A from zero to N_{AB}. Plot S as a function of N_A for our particular eigenstate ψ. Where is the entanglement entropy largest? Explain why it goes to zero for the two endpoints.*

The term entanglement sounds mutual; A and B are entangled together, rather than the bath B has perturbed A. This nomenclature is not an accident. As you checked numerically, the entanglement entropies of the two subsystems is the same, as can be shown using the *Schmidt decomposition* (not described here).

In statistical mechanics, a large ergodic system AB in the microcanonical ensemble at energy E, when restricted to a relatively small subsystem A, will generate an equilibrium thermal ensemble at the corresponding temperature. The *eigenstate thermalization hypothesis* says that many systems[74] take this to an extreme: for each eigenstate ψ, the reduced density matrix $\boldsymbol{\rho}_A$ of the subsystem will converge to a Boltz-

---

[72]In the C and Python convention where indices start with zero, this would be $\mathbf{1}_{2^m} \otimes \mathcal{H}_2 \otimes \mathbf{1}_{2^{N_{\text{spins}}-(m+2)}}$.

[73]Note: $\boldsymbol{\rho}_A$ will have some zero eigenvalues, which may confuse a matrix logarithm. Remember that diagonalizing ρ also diagonalizes $\log \rho$, and the trace is invariant under a change of basis. Hence, you can define your own function pLogp which returns zero if $p \leq 0$ and $p \log p$ otherwise, and sum it over the eigenvalues of $\boldsymbol{\rho}_A$.

[74]Systems exhibiting, for example, many-body localization (of considerable current interest) do not exhibit eigenstate thermalization.

mann equilibrium thermal ensemble

$$\rho^{\beta}_{jk} = \delta_{jk} e^{-\beta E^A_k} \bigg/ \sum_{\ell} e^{-\beta E^A_{\ell}} \qquad (7.104)$$

as the system size goes to infinity.

Let us calculate the probability p_k that our subsystem is in eigenstate ψ^A_k, $p_k = \mathrm{Tr}(|\psi^A_k\rangle\langle\psi^A_k|\rho_A)$. We are simulating a rather small system, so fluctuations will be large.

(h) *Make a log plot of p_k vs. E^A_k. Do a fit to the predicted form eqn 7.104 to find β, and plot the result with your data.*

In particular, the reduced density matrix is predicted to be at the temperature of the microcanonical ensemble at the energy E of the original pure state ψ.

(i) *Write a function* EbarAB(β) *returning the average energy of the entire system as a function of β. Take a sampling of eigenstates ψ_K of the total system, fit p_k vs. E^A_k as in part (h), and plot β vs. E along with your prediction β(EbarAB). Do you achieve a rough agreement?*

Calculation and computation

Most statistical mechanical systems cannot be solved explicitly.[1] Statistical mechanics does provide general relationships and organizing principles (temperature, entropy, free energy, thermodynamic relations) even when a solution is not available. But there are times when specific answers about specific models or experiments are needed.

There are two basic tools for extracting answers out of statistical mechanics for realistic systems. The first is *simulation*. Sometimes one simply mimics the microscopic theory. For example, a molecular dynamics simulation will move the atoms according to Newton's laws. We will not discuss such methods in this chapter.[2] If one is not interested in the detailed dynamical trajectories of the system, one can use *Monte Carlo* simulation methods to extract the equilibrium properties from a model. We introduce these methods in Section 8.1 in the context of the *Ising model*, the most well studied of the lattice models in statistical mechanics. The theory underlying the Monte Carlo method is the mathematics of *Markov chains*, which arise in many other applications; we discuss them in Section 8.2.

The second tool is to use *perturbation theory*. Starting from a solvable model, one can calculate the effects of small extra terms; for a complex system one can extrapolate from a limit (like zero or infinite temperature) where its properties are known. Section 8.3 briefly discusses perturbation theory, and the deep connection between its convergence and the existence of *phases*.

[1] No tidy formula for the equation of state, entropy, or partition function can typically be found.

[2] Often direct simulation methods also involve sophisticated ideas from statistical mechanics. For example, to emulate a microscopic system connected to a heat bath, one adds friction and noise to the microscopic theory in the correct proportions so as to lead to proper thermal equilibration (the Einstein relation, eqn 2.22; see also Section 10.8, Exercises 6.18, 6.19, 10.7, and [4, section 3.8]).

8.1 The Ising model

Lattice models are a big industry within statistical mechanics. These models have a variable at each site of a regular grid, and a Hamiltonian or evolution law for these variables. Critical phenomena and phase transitions (Chapter 12), lattice QCD[3] and quantum field theories, quantum magnetism and models for high-temperature superconductors, phase diagrams for alloys (Section 8.1.2), the behavior of systems with dirt or disorder, and nonequilibrium systems exhibiting avalanches and crackling noise (Chapter 12) all make important use of lattice models.

In this section, we will introduce the Ising model[4] and three physical systems (among many) to which it has been applied: magnetism, bi-

[3] QCD, quantum chromodynamics, is the theory of the strong interaction that binds the nucleus together.

[4] Ising's name is pronounced *Eesing*, but the model is usually pronounced *Eyesing* with a long I sound.

Statistical Mechanics: Entropy, Order Parameters, and Complexity. James P. Sethna, Oxford University Press (2021). © Oxford University Press. DOI:10.1093/oso/9780198865247.003.0008

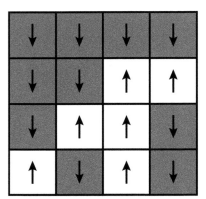

Fig. 8.1 The 2D square-lattice Ising model. It is traditional to denote the values $s_i = \pm 1$ as up and down, or as two different shades.

[5]In simulations of finite systems, we will avoid special cases at the edges of the system by implementing *periodic boundary conditions*, where corresponding sites on opposite edges are also neighbors.

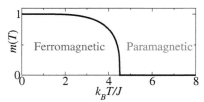

Fig. 8.2 Ising magnetization. The magnetization $m(T)$ per spin for the 3D cubic lattice Ising model. At low temperatures there is a net magnetization, which vanishes at temperatures $T > T_c \approx 4.5$.

[6]*Ferromagnetic* is named after iron (Fe), the most common material that has a spontaneous magnetization.

nary alloys, and the liquid–gas transition. The Ising model is the most extensively studied lattice model in physics. Like the ideal gas in the previous chapters, the Ising model will provide a tangible application for many topics to come: Monte Carlo (this section), low- and high-temperature expansions (Section 8.3), relations between fluctuations, susceptibility, and dissipation (Chapter 10), nucleation of abrupt transitions (Section 11.3), coarsening and phase separation (Section 11.4.1), and self-similarity at continuous phase transitions (Chapter 12).

The Ising model has a lattice of N sites i with a single, two-state degree of freedom s_i on each site that may take values ± 1. We will be primarily interested in the Ising model on square and cubic lattices (in 2D and 3D, Fig. 8.1). The Hamiltonian for the Ising model is

$$\mathcal{H} = -\sum_{\langle ij \rangle} J s_i s_j - H \sum_i s_i. \tag{8.1}$$

Here the sum $\langle ij \rangle$ is over all pairs of nearest-neighbor sites,[5] and J is the *coupling* between these neighboring sites. (For example, there are four neighbors per site on the square lattice.)

8.1.1 Magnetism

The Ising model was originally used to describe magnets. Hence the degree of freedom s_i on each site is normally called a *spin*, H is called the *external field*, and the sum $M = \sum_i s_i$ is termed the *magnetization*.

The energy of two neighboring spins $-J s_i s_j$ is $-J$ if the spins are parallel, and $+J$ if they are antiparallel. Thus if $J > 0$ (the usual case) the model favors parallel spins; we say that the interaction is *ferromagnetic*.[6] At low temperatures, the spins will organize themselves to either mostly point up or mostly point down, forming a *ferromagnetic phase*. If $J < 0$ we call the interaction *antiferromagnetic*; the spins will tend to align (for our square lattice) in a checkerboard *antiferromagnetic phase* at low temperatures. At high temperatures, independent of the sign of J, we expect entropy to dominate; the spins will fluctuate wildly in a *paramagnetic phase* and the magnetization per spin $m(T) = M(T)/N$ is zero (see Fig. 8.2).[7]

---

[7]The Ising model parameters are rescaled from the microscopic ones. The Ising spin $s_i = \pm 1$ represents twice the z-component of a spin-1/2 atom in a crystal, $\sigma_i^z = s_i/2$. The Ising interactions between spins, $J s_i s_j = 4 J \sigma_i^z \sigma_j^z$, is thus shifted by a factor of four from the z–z coupling between spins. The coupling of the spin to the external magnetic field is microscopically $g \mu_B H \cdot \sigma_i^z$, where g is the gyromagnetic ratio for the spin (close to two for the electron) and $\mu_B = e\hbar/2m_e$ is the Bohr magneton. Hence the Ising external field is rescaled from the physical one by $g \mu_B/2$. Finally, the interaction between spins in most materials is not so anisotropic as to only involve the z-component of the spin; it is usually better approximated by the dot product $\sigma_i \cdot \sigma_j = \sigma_i^x \sigma_j^x + \sigma_i^y \sigma_j^y + \sigma_i^z \sigma_j^z$, used in the more realistic *Heisenberg model*. (See Exercise 7.27. Unlike the Ising model, where σ_i^z commutes with $\mathcal{H}$ and the spin configurations are the energy eigenstates, the quantum and classical Heisenberg models differ.) Some materials have anisotropic crystal structures, which make the Ising model at least approximately valid.

8.1.2   Binary alloys

The Ising model is quite a convincing model for binary alloys. A realistic alloy might mix roughly half copper and half zinc to make β-brass. At low temperatures, the copper and zinc atoms each sit on a cubic lattice, with the zinc sites in the middle of the copper cubes, together forming an *antiferromagnetic* phase on the body-centered cubic (bcc) lattice. At high temperatures, the zincs and coppers freely interchange, analogous to the Ising paramagnetic phase. The transition temperature is about 733 °C [214, section 3.11].

Let us consider as an example a square lattice of atoms, which can be either of type A or B (Fig. 8.3). We set the spin values $A = +1$ and $B = -1$. Let the number of the two kinds of atoms be N_A and N_B, with $N_A + N_B = N$, let the interaction energies (bond strengths) between two neighboring atoms be E_{AA}, E_{BB}, and E_{AB}, and let the total number of nearest-neighbor bonds of the three possible types be N_{AA}, N_{BB} and N_{AB}. Then the Hamiltonian for our binary alloy is

$$\mathcal{H}_{\text{binary}} = -E_{AA}N_{AA} - E_{BB}N_{BB} - E_{AB}N_{AB}. \qquad (8.2)$$

Since each site interacts only with its nearest neighbors, this must be the Ising model in disguise. Indeed, one finds[8] $J = \frac{1}{4}(E_{AA} + E_{BB} - 2E_{AB})$ and $H = E_{AA} - E_{BB}$.

To make this a quantitative model, one must include atomic relaxation effects. (Surely if one kind of atom is larger than the other, it will push neighboring atoms off their sites. We simply include this relaxation into the energies in our Hamiltonian 8.2.) We must also incorporate thermal position fluctuations into the Hamiltonian, making it a free energy.[9] More elaborate Ising models (with three-site and longer-range interactions, for example) are commonly used to compute realistic phase diagrams for alloys [216]. Sometimes, though, the interactions introduced by relaxations and thermal fluctuations have important long-range pieces, which can lead to qualitative changes in the behavior—for example, they can change the ordering transition from continuous to abrupt.

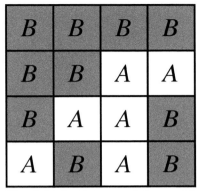

Fig. 8.3 The Ising model as a binary alloy. Atoms in crystals naturally sit on a lattice. The atoms in alloys are made up of different elements (here, types A and B), which can arrange in many configurations on the lattice.

---

[8]Adding an overall shift $-CN$ to the Ising Hamiltonian, one can see that

$$\mathcal{H}_{\text{Ising}} = -J\sum_{\langle ij \rangle} s_i s_j - H\sum_i s_i - CN = -J\left(N_{AA} + N_{BB} - N_{AB}\right) - H\left(N_A - N_B\right) - CN. \qquad (8.3)$$

We solve for N, N_A, and N_B using the facts that on a square lattice there are twice as many bonds as spins ($N_{AA} + N_{BB} + N_{AB} = 2N$), and that for every A atom there must be four bonds ending at each site (so $4N_A = 2N_{AA} + N_{AB}$ and $4N_B = 2N_{BB} + N_{AB}$). Rearranging into the binary alloy form (eqn 8.2), we find the values for J and H above and $C = \frac{1}{2}(E_{AA} + E_{BB} + 2E_{AB})$.

[9]To incorporate thermal fluctuations, we must do a partial trace, integrating out the vibrations of the atoms around their equilibrium positions (as in Section 6.6). This leads to an effective free energy for each pattern of lattice occupancy $\{s_i\}$:

$$\mathcal{F}\{s_i\} = -k_B T \log\left(\int d\mathbb{P}\int_{\text{atom } r_i \text{ of type } s_i \text{ near site } i} d\mathbb{Q}\, \frac{e^{-\mathcal{H}(\mathbb{P}, \mathbb{Q})/k_B T}}{h^{3N}}\right) = \mathcal{H}\{s_i\} - TS\{s_i\}. \qquad (8.4)$$

The entropy $S\{s_i\}$ due to these vibrations will depend upon the particular atomic configuration s_i, and can often be calculated explicitly (Exercise 6.11(b)). $\mathcal{F}\{s_i\}$ can now be used as a lattice Hamiltonian, except with temperature-dependent coefficients; those atomic configurations with more freedom to vibrate will have larger entropy and will be increasingly favored at higher temperature.

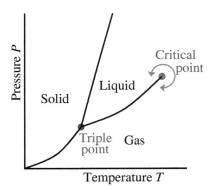

Fig. 8.4 *P–T* **phase diagram** for a typical material. The solid–liquid phase boundary corresponds to a change in symmetry, and cannot end. The liquid–gas phase boundary typically does end; one can go continuously from the liquid phase to the gas phase by increasing the pressure above P_c, increasing the temperature above T_c, and then lowering the pressure again.

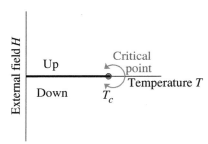

Fig. 8.5 *H–T* **phase diagram for the Ising model**. Below the critical temperature T_c, there is an an up-spin and a down-spin "phase" separated by a jump in magnetization at $H = 0$. Above T_c the behavior is smooth as a function of H.

[10]This is a typical homework exercise in a textbook like ours; with a few hints, you too can do it.

[11]Or do high-temperature expansions, low-temperature expansions, transfer-matrix methods, exact diagonalization of small systems, $1/N$ expansions in the number of states per site, $4 - \epsilon$ expansions in the dimension of space, ...

[12]Monte Carlo is a gambling center in Monaco. Lots of random numbers are generated there.

8.1.3 Liquids, gases, and the critical point

The Ising model is also used as a model for the liquid–gas transition. In this *lattice gas* interpretation, up-spins ($s_i = +1$) count as atoms and down-spins count as a site without an atom. The gas is the phase with mostly down-spins (negative "magnetization"), with only a few up-spin atoms in the vapor. The liquid phase is mostly atoms (up-spins), with a few vacancies.

The Ising model description of the gas phase seems fairly realistic. The liquid, however, seems much more like a crystal, with atoms sitting on a regular lattice. Why do we suggest that this model is a good way of studying transitions between the liquid and gas phase?

Unlike the binary alloy problem, the Ising model is not a good way to get quantitative phase diagrams for fluids. What it is good for is to understand the properties near the *critical point*. As shown in Fig. 8.4, one can go continuously between the liquid and gas phases; the phase boundary separating them ends at a critical point T_c, P_c, above which the two phases blur together seamlessly, with no jump in the density separating them.

The Ising model, interpreted as a lattice gas, also has a line $H = 0$ along which the density (magnetization) jumps, and a temperature T_c above which the properties are smooth as a function of H (the paramagnetic phase). The phase diagram in Fig. 8.5 looks only topologically like the real liquid–gas coexistence line in Fig. 8.4, but the behavior near the critical point in the two systems is remarkably similar. Indeed, we will find in Chapter 12 that in many ways the behavior at the liquid–gas critical point is described *exactly* by the three-dimensional Ising model.

8.1.4 How to solve the Ising model

How do we solve for the properties of the Ising model?

(1) Solve the one-dimensional Ising model, as Ising did.[10]

(2) Have an enormous brain. Onsager solved the two-dimensional Ising model in a bewilderingly complicated way. Since Onsager, many great minds have found simpler, elegant solutions, but all would take at least a chapter of rather technical and unilluminating manipulations to duplicate. Nobody has solved the three-dimensional Ising model.

(3) Perform the Monte Carlo method on a computer.[11]

The Monte Carlo[12] method involves doing a kind of random walk through the space of lattice configurations. We will study these methods in great generality in Section 8.2. For now, let us just outline the heat-bath Monte Carlo method.

Heat-bath Monte Carlo for the Ising model:

- Pick a site $i = (x, y)$ at random.

- Check how many neighbor spins are pointing up:

$$m_i = \sum_{j:\langle ij \rangle} s_j = \begin{cases} 4 & (\text{4 neighbors up}), \\ 2 & (\text{3 neighbors up}), \\ 0 & (\text{2 neighbors up}), \\ -2 & (\text{1 neighbor up}), \\ -4 & (\text{0 neighbors up}). \end{cases} \qquad (8.5)$$

- Calculate $E_+ = -Jm_i - H$ and $E_- = +Jm_i + H$, the energy for spin i to be $+1$ or -1 given its current environment.
- Set spin i up with probability $e^{-\beta E_+}/(e^{-\beta E_+} + e^{-\beta E_-})$ and down with probability $e^{-\beta E_-}/(e^{-\beta E_+} + e^{-\beta E_-})$.
- Repeat.

The heat-bath algorithm just thermalizes one spin at a time; it sets the spin up or down with probability given by the thermal distribution given that its neighbors are fixed. Using it, we can explore statistical mechanics with the Ising model on the computer, just as we have used pencil and paper to explore statistical mechanics with the ideal gas.

8.2 Markov chains

The heat-bath Monte Carlo algorithm is not the most efficient (or even the most common) algorithm for equilibrating the Ising model. Monte Carlo methods in general are examples of *Markov chains*. In this section we develop the mathematics of Markov chains and provide the criteria needed to guarantee that a given algorithm converges to the equilibrium state.

Markov chains are an advanced topic which is not necessary for the rest of this text. Our discussion does introduce the idea of *detailed balance* and further illustrates the important concept of ergodicity. Markov methods play important roles in other topics (in ways we will not pursue here). They provide the mathematical language for studying random walks and other random evolution laws in discrete and continuum systems. Also, they have become important in bioinformatics and speech recognition, where one attempts to deduce the *hidden Markov model* which describes the patterns and relations in speech or the genome.

In this chapter we will consider Markov chains with a finite set of states $\{\alpha\}$, through which the system evolves in a discrete series of steps n.[13] The probabilities of moving to different new states in a Markov chain depend *only* on the current state. In general, systems which lack memory of their history are called *Markovian*.

For example, an N-site Ising model has 2^N states $\mathbb{S} = \{s_i\}$. A Markov chain for the Ising model has a transition rule, which at each step shifts the current state $\mathbb{S}$ to a state $\mathbb{S}'$ with probability $P_{\mathbb{S}' \Leftarrow \mathbb{S}}$.[14] For the heat-bath algorithm, $P_{\mathbb{S}' \Leftarrow \mathbb{S}}$ is equal to zero unless $\mathbb{S}'$ and $\mathbb{S}$ are the same except for at most one spin flip. There are many problems outside of

---

[13]There are analogues of Markov chains which have an infinite number of states, and/or are continuous in time and/or space (see Exercises 8.22 and 12.22).

[14]Some texts will order the subscripts in the opposite direction $P_{\mathbb{S} \Rightarrow \mathbb{S}'}$. We use this convention to make our time evolution correspond to multiplication on the left by $P_{\alpha\beta}$ (eqn 8.6).

mainstream statistical mechanics that can be formulated in this general way. For example, Exercise 8.4 discusses a model with 1,001 states (different numbers α of red bacteria), and transition rates $P_{\alpha+1\Leftarrow\alpha}$, $P_{\alpha-1\Leftarrow\alpha}$, and $P_{\alpha\Leftarrow\alpha}$.

Let the probabilities of being in various states α at step n be arranged in a vector $\boldsymbol{\rho}_\alpha(n)$. Then the rates $P_{\beta\alpha}$ for moving from α to β (dropping the arrow) form a matrix, which when applied to the probability vector $\boldsymbol{\rho}$ takes it from one time to the next (eqn 8.6).

In general, we want to understand the probability of finding different states after long times. Under what circumstances will an algorithm, defined by our matrix P, take our system into thermal equilibrium? To study this, we need to understand some properties of the transition matrix P, its eigenvalues, and its eigenvectors. $P_{\beta\alpha}$ in general must have the following properties:

- **Time evolution.** The probability vector at step $n+1$ is

$$\boldsymbol{\rho}_\beta(n+1) = \sum_\alpha P_{\beta\alpha}\boldsymbol{\rho}_\alpha(n), \qquad \boldsymbol{\rho}(n+1) = P \cdot \boldsymbol{\rho}(n). \quad (8.6)$$

- **Positivity.** The matrix elements are probabilities, so

$$0 \le P_{\beta\alpha} \le 1. \quad (8.7)$$

- **Conservation of probability.** The state α must go somewhere, so

$$\sum_\beta P_{\beta\alpha} = 1. \quad (8.8)$$

- **Not symmetric!** Typically $P_{\beta\alpha} \ne P_{\alpha\beta}$.

This last point is not a big surprise; it should be much more likely to go from a high-energy state to a low one than from low to high. However, this asymmetry means that much of our mathematical intuition and many of our tools, carefully developed for symmetric and Hermitian matrices, will not apply to our transition matrix $P_{\alpha\beta}$. In particular, we cannot assume in general that we can diagonalize our matrix.

What do we know about the Markov chain and its asymmetric matrix P? We will outline the relevant mathematics, proving what is convenient and illuminating and simply asserting other truths.

It is true that our matrix P will have eigenvalues. Also, it is true that for each distinct eigenvalue there will be at least one right eigenvector:[15]

$$P \cdot \boldsymbol{\rho}^\lambda = \lambda\boldsymbol{\rho}^\lambda \quad (8.9)$$

and one left eigenvector:

$$\boldsymbol{\sigma}^{\lambda\top} \cdot P = \lambda\boldsymbol{\sigma}^{\lambda\top}. \quad (8.10)$$

However, for degenerate eigenvalues there may not be multiple eigenvectors, and the left and right eigenvectors usually will not be equal to one another.[16]

[15]For example, the matrix $\begin{pmatrix} 0 & 1 \\ 0 & 0 \end{pmatrix}$ has a double eigenvalue of zero, but only one left eigenvector $(0\ 1)$ and one right eigenvector $\begin{pmatrix} 1 \\ 0 \end{pmatrix}$ with eigenvalue zero.

[16]This follows from a more specific theorem that we will not prove. A general matrix M can be put into *Jordan canonical form* by a suitable change of basis S: $M = SJS^{-1}$. (There is no connection with the canonical ensemble.) The matrix J is block diagonal, with one eigenvalue λ associated with each block (but perhaps multiple blocks per λ). A single block for an eigenvalue λ with multiplicity three would be

$$\begin{pmatrix} \lambda & 1 & 0 \\ 0 & \lambda & 1 \\ 0 & 0 & \lambda \end{pmatrix}. \quad (8.11)$$

The block has only one left and right eigenvector (proportional to the first column and last row).

For the particular case of our transition matrix P, we can go further. If our Markov chain reaches an equilibrium state $\boldsymbol{\rho}^*$ at long times, that state must be unchanged under the time evolution P. That is, $P \cdot \boldsymbol{\rho}^* = \boldsymbol{\rho}^*$, and thus the equilibrium probability density is a right eigenvector with eigenvalue one. We can show that our Markov chain transition matrix P has such a right eigenvector.

Theorem 8.1 *The matrix P has at least one right eigenvector $\boldsymbol{\rho}^*$ with eigenvalue one.*

Proof (sneaky) P has a left eigenvector $\boldsymbol{\sigma}^{*\top} = (1, 1, 1, \ldots, 1)$, which has eigenvalue one because of conservation of probability:

$$(\boldsymbol{\sigma}^{*\top} \cdot P)_\alpha = \sum_\beta \sigma_\beta^* P_{\beta\alpha} = \sum_\beta P_{\beta\alpha} = 1 = \sigma_\alpha^*. \qquad (8.12)$$

Hence P must have an eigenvalue equal to one, and hence it must also have a right eigenvector with eigenvalue one. $\qquad \square$

We can also show that all the other eigenvalues have right eigenvectors that sum to zero, since P conserves probability.[17]

Theorem 8.2 *Any right eigenvector $\boldsymbol{\rho}^\lambda$ with eigenvalue λ different from one must have components that sum to zero.*

Proof $\boldsymbol{\rho}^\lambda$ is a right eigenvector, $P \cdot \boldsymbol{\rho}^\lambda = \lambda \boldsymbol{\rho}^\lambda$. Hence

$$\lambda \sum_\beta \rho_\beta^\lambda = \sum_\beta (\lambda \rho_\beta^\lambda) = \sum_\beta \left(\sum_\alpha P_{\beta\alpha} \rho_\alpha^\lambda \right) = \sum_\alpha \left(\sum_\beta P_{\beta\alpha} \right) \rho_\alpha^\lambda$$
$$= \sum_\alpha \rho_\alpha^\lambda. \qquad (8.13)$$

This implies that either $\lambda = 1$ or $\sum_\alpha \rho_\alpha^\lambda = 0$. $\qquad \square$

Markov chains can have more than one stationary probability distribution.[18] They can have *transient* states, which the system eventually leaves, never to return.[19] They can also have *cycles*, which are probability distributions which, like a clock $1 \to 2 \to 3 \to \cdots \to 12 \to 1$, shift through a finite number of distinct classes of states before returning to the original one. All of these are obstacles in our quest for finding the equilibrium states in statistical mechanics. We can bypass all of them by studying *ergodic* Markov chains.[20] A finite-state Markov chain is ergodic if it does not have cycles and it is *irreducible*: that is, one can get from every state α to every other state β in a finite sequence of moves. We use the following famous theorem, without proving it here.

Theorem 8.3. (Perron–Frobenius theorem) *Let A be a matrix with all nonnegative matrix elements such that A^n has all positive elements. Then A has a positive eigenvalue λ_0, of multiplicity one, whose corresponding right and left eigenvectors have all positive components. Furthermore any other eigenvalue λ of A must be smaller, $|\lambda| < \lambda_0$.*

[17]One can also view Theorem 8.2 as saying that all the right eigenvectors except $\boldsymbol{\rho}^*$ are orthogonal to the left eigenvector $\boldsymbol{\sigma}^*$.

[18]A continuum example of this is given by the KAM theorem of Exercise 4.4. There is a probability density confined to each KAM torus which is time independent.

[19]Transient states are important in dissipative dynamical systems, where they consist of all states not on the attractors.

[20] We are compromising here between the standard Markov chain usage in physics and in mathematics. Physicists usually ignore cycles, and call algorithms which can reach every state ergodic. Mathematicians use the term ergodic to exclude cycles and exclude probability running to infinity (not important here, where we have a finite number of states). However, they allow ergodic chains to have transient states; only the "attractor" need be connected. Our definition of ergodic for finite Markov chains corresponds to a transition matrix P for which some power P^n has all positive (nonzero) matrix elements; mathematicians call such matrices *regular*.

For an ergodic Markov chain, we can use Theorem 8.2 to see that the Perron–Frobenius eigenvector with all positive components must have eigenvalue $\lambda_0 = 1$. We can rescale this eigenvector to sum to one, proving that an ergodic Markov chain has a unique time-independent probability distribution ρ^*.

What is the connection between our definition of ergodic Markov chains and our earlier definition of ergodic (Section 4.2) involving trajectories in phase space? Ergodic in phase space meant that we eventually come close to all states on the energy surface. For finite Markov chains, ergodic is the stronger condition that we have nonzero probability of getting between any two states in the chain after some finite time.

It is possible to show that an ergodic Markov chain will take any initial probability distribution $\rho(0)$ and converge to equilibrium, but the proof in general is rather involved. We can simplify it by specializing one more time, to Markov chains that satisfy *detailed balance*.

A Markov chain satisfies *detailed balance* if there is some probability distribution ρ^* such that[21]

$$P_{\alpha\beta}\rho_\beta^* = P_{\beta\alpha}\rho_\alpha^* \tag{8.14}$$

[21] Note that we do *not* sum over the repeated indices here: this equation holds for each pair of states α and β. Also, there is an elegant equivalent definition of detailed balance directly in terms of P and not involving the equilibrium probability distribution ρ^*; see Exercise 8.5.

for each state α and β. In words, the probability flux from state α to β (the rate times the probability of being in α) balances the probability flux back, in detail (i.e. for every pair of states).

Detailed balance is a sufficient, but not necessary condition for a sensible statistical mechanical system. A system whose states and equilibrium distribution are both time-reversal invariant will automatically satisfy detailed balance, since the difference $P_{\alpha\beta}\rho_\beta^* - P_{\beta\alpha}\rho_\alpha^*$ is thus unchanged under time-reversal while the net flux it represents between β and α changes sign. Magnetic fields break time-reversal invariance (Exercise 9.13), and flows in phase space have states $(\mathbb{P}, \mathbb{Q}) \to (-\mathbb{P}, \mathbb{Q})$ that are not invariant under time-reversal, and yet both describe systems which obey statistical mechanics with sensible equilibrium states.

Detailed balance allows us to find a complete set of right eigenvectors for our transition matrix P. One can see this with a simple transformation. If we divide both sides of eqn 8.14 by $\sqrt{\rho_\beta^* \rho_\alpha^*}$, we create a symmetric matrix $Q_{\alpha\beta}$:

$$Q_{\alpha\beta} = P_{\alpha\beta}\sqrt{\frac{\rho_\beta^*}{\rho_\alpha^*}} = P_{\alpha\beta}\,\rho_\beta^*\Big/\sqrt{\rho_\alpha^* \rho_\beta^*}$$

$$= P_{\beta\alpha}\,\rho_\alpha^*\Big/\sqrt{\rho_\alpha^* \rho_\beta^*} = P_{\beta\alpha}\sqrt{\frac{\rho_\alpha^*}{\rho_\beta^*}} = Q_{\beta\alpha}. \tag{8.15}$$

[22] This works in reverse to get the right eigenvectors of P from Q. One multiplies τ_α^λ by $\sqrt{\rho_\alpha^*}$ to get ρ_α^λ, and divides to get σ_α^λ, so if detailed balance holds, $\sigma_\alpha^\lambda = \rho_\alpha^\lambda\big/\rho_\alpha^*$. In particular, $\sigma^1 = \sigma^* = (1, 1, 1, \dots)^\top$, as we saw in Theorem 8.1.

This particular symmetric matrix has eigenvectors $Q \cdot \tau^\lambda = \lambda\tau^\lambda$ which can be turned into right eigenvectors of P when rescaled[22] by $\sqrt{\rho^*}$:

$$\rho_\alpha^\lambda = \tau_\alpha^\lambda\sqrt{\rho_\alpha^*}; \tag{8.16}$$

$$\sum_\alpha P_{\beta\alpha}\rho_\alpha^\lambda = \sum_\alpha P_{\beta\alpha}(\tau_\alpha^\lambda \sqrt{\rho_\alpha^*}) = \sum_\alpha \left(Q_{\beta\alpha} \sqrt{\frac{\rho_\beta^*}{\rho_\alpha^*}} \right) (\tau_\alpha^\lambda \sqrt{\rho_\alpha^*})$$
$$= \sum_\alpha \left(Q_{\beta\alpha} \tau_\alpha^\lambda \right) \sqrt{\rho_\beta^*} = \lambda \left(\tau_\beta^\lambda \sqrt{\rho_\beta^*} \right) = \lambda \rho_\beta^\lambda. \qquad (8.17)$$

Now we turn to the main theorem underlying the algorithms for equilibrating lattice models in statistical mechanics.

Theorem 8.4. (main theorem) *A discrete dynamical system with a finite number of states can be guaranteed to converge to an equilibrium distribution ρ^* if the computer algorithm*

- *is Markovian (has no memory),*
- *is ergodic (can reach everywhere and is acyclic), and*
- *satisfies detailed balance.*

Proof Let P be the transition matrix for our algorithm. Since the algorithm satisfies detailed balance, P has a complete set of eigenvectors ρ^λ. Since our algorithm is ergodic there is only one right eigenvector ρ^1 with eigenvalue one, which we can choose to be the stationary distribution ρ^*; all the other eigenvalues λ have $|\lambda| < 1$. Decompose the initial condition $\rho(0) = a_1\rho^* + \sum_{|\lambda|<1} a_\lambda \rho^\lambda$. Then[23]

$$\rho(n) = P \cdot \rho(n-1) = P^n \cdot \rho(0) = a_1\rho^* + \sum_{|\lambda|<1} a_\lambda \lambda^n \rho^\lambda. \qquad (8.18)$$

Since the (finite) sum in this equation decays to zero, the density converges to $a_1\rho^*$. This implies both that $a_1 = 1$ and that our system converges to ρ^* as $n \to \infty$. □

Thus, to develop a new equilibration algorithm (Exercises 8.6, 8.8), one need only ensure that it is Markov, ergodic, and satisfies detailed balance.

8.3   What is a phase? Perturbation theory

What is a phase? We know some examples. Water is a liquid phase, which at atmospheric pressure exists between $0\,°C$ and $100\,°C$; the equilibrium density of H_2O jumps abruptly downward when the water freezes or vaporizes. The Ising model is ferromagnetic below T_c and paramagnetic above T_c. Figure 8.6 plots the specific heat of a noninteracting gas of fermions and of bosons. There are many differences between fermions and bosons illustrated in this figure,[24] but the fundamental difference is that *the Bose gas has two different phases*. The specific heat has a cusp at the Bose condensation temperature, which separates the normal phase and the condensed phase.

How do we determine in general how far a phase extends? Inside phases the properties do not shift in a singular way; one can smoothly

[23]The eigenvalue closest to one will correspond to the eigenstate that is the slowest to decay. This decay rate need not, however, be the slowest time scale in the Markov chain [80].

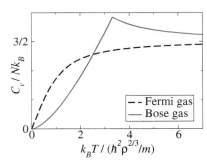

Fig. 8.6 Bose and Fermi specific heats. The specific heats for the ideal Bose and Fermi gases. Notice the cusp at the Bose condensation temperature T_c. Notice that the specific heat of the Fermi gas shows no such transition.

[24]The specific heat of the Fermi gas falls as the temperature decreases; at low temperatures, only those single-particle eigenstates within a few $k_B T$ of the Fermi energy can be excited. The specific heat of the Bose gas initially grows as the temperature decreases from infinity. Both the Fermi and Bose gases have $C_v/N \to 0$ as $T \to 0$, as is always true (otherwise the entropy $\int_0^T C_v/T \, dT$ would diverge).

Fig. 8.7 Perturbation theory. (a) Low-temperature expansions for the cubic Ising model magnetization (Fig. 8.2) with successively larger numbers of terms. (b) The high- and low-temperature expansions for the Ising and other lattice models are sums over (Feynman diagram) clusters. At low T, Ising configurations are small clusters of up-spins in a background of down-spins (or vice versa). This cluster of four sites on the cubic lattice contributes to the term of order x^{20} in eqn 8.19, because flipping the cluster breaks 20 bonds.

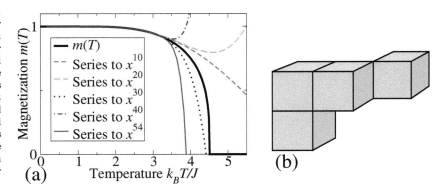

[25]This heroic calculation (27 terms) was not done to get really accurate low-temperature magnetizations. Various clever methods can use these expansions to extrapolate to understand the subtle phase transition at T_c (Chapter 12). Indeed, the $m(T)$ curve shown in both Figs. 8.2 and 8.7(a) was not measured directly, but was generated using eqn 8.19 in a 9, 10 Padé approximant [51].

extrapolate the behavior inside a liquid or magnetic phase under small changes in external conditions. *Perturbation theory works inside phases.* More precisely, inside a phase the properties are analytic (have convergent Taylor expansions) as functions of the external conditions.

Much of statistical mechanics (and indeed of theoretical physics) is devoted to calculating high-order perturbation theories around special solvable limits. (We will discuss *linear* perturbation theory in space and time in Chapter 10.) Lattice theories at high and low temperatures T have perturbative expansions in powers of $1/T$ and T, with Feynman diagrams involving all ways of drawing clusters of lattice points (Fig. 8.7(b)). Gases at high temperatures and low densities have *virial expansions*. Metals at low temperatures have *Fermi liquid theory*, where the electron–electron interactions are perturbatively incorporated by dressing the electrons into quasiparticles. Properties of systems near continuous phase transitions can be explored by perturbing in the dimension of space, giving the ϵ-*expansion*. Some of these perturbation series have zero radius of convergence; they are *asymptotic series* (see Exercise 1.5).

For example the low-temperature expansion [52, 149] (Exercises 8.19 and 8.18) of the magnetization[25] per spin of the cubic-lattice three-dimensional Ising model (Section 8.1) starts out [24]

$$
\begin{aligned}
m = {} & 1 - 2x^6 - 12x^{10} + 14x^{12} - 90x^{14} + 192x^{16} - 792x^{18} + 2148x^{20} \\
& - 7716x^{22} + 23262x^{24} - 79512x^{26} + 252054x^{28} \\
& - 846628x^{30} + 2753520x^{32} - 9205800x^{34} \\
& + 30371124x^{36} - 101585544x^{38} + 338095596x^{40} \\
& - 1133491188x^{42} + 3794908752x^{44} - 12758932158x^{46} \\
& + 42903505303x^{48} - 144655483440x^{50} \\
& + 488092130664x^{52} - 1650000819068x^{54} + \ldots,
\end{aligned} \tag{8.19}
$$

where $x = \mathrm{e}^{-2J/k_B T}$ is the probability to break a bond (parallel energy $-J$ to antiparallel energy $+J$). This series was generated by carefully considering the probabilities of low-energy spin configurations, formed by flipping combinations of clusters of spins (Fig. 8.7(b)). This expansion smoothly predicts the magnetization at low temperatures using the properties at zero temperature; in Fig. 8.7(a) we see that the magnetization is well described by our series for $k_B T \lesssim 3J$. Another power series about $T \sim 3J/k_B$ would converge up to a higher temperature.[26] No finite power series, however, can extrapolate past the temperature T_c at which the magnetization goes to zero. This is easiest to see in the opposite direction; $m(T) \equiv 0$ above T_c, so any extrapolation below T_c must continue to have zero magnetization. Much of the glory in perturbation theory involves summing infinite families of terms to extrapolate through critical points.

Phase boundaries occur at parameter values where the properties are not smooth—where the continuation of the properties on one side does not predict the behavior on the other. We could almost define phases as regions where perturbation theory works—except for the awkward problem that we do not want liquids and gases to be called part of the same fluid "phase", even though they are connected by paths going around the critical point (Fig. 8.4).

This leads to an important experimental method. Suppose you have invented a new exotic liquid crystal. How can you tell if it is in an already known phase? You look for an experimental path, mixing materials and changing external conditions, for smoothly changing your phase to the known one. For example, are oil and water both in the same (liquid) phase? Can we go from one to the other smoothly, without passing through a phase transition?[27] You cannot mix oil and water, but you can mix oil and alcohol, and certainly can mix alcohol and water. Changing the concentrations smoothly starting from oil, going through pure alcohol, and ending at water demonstrates that these two fluids are part of the same phase (see Fig. 8.8). This is often used, for example, to determine to which exotic phase a new liquid crystal should be assigned. This argument is also the basis for much of theoretical physics. If you can go smoothly from A (your theory) to B (the experiment) by adding corrections, then A and B are in the same phase; publish![28]

[26] The radius of convergence of the series is less than T_c because there is another closer singularity in the *complex temperature plane*. This is analogous to the function $1/(1+x^2)$, which is smooth on the real axis but whose Taylor series $1 - x^2 + x^4 - x^6 + \dots$ converges only for $-1 < x < 1$; the poles at $\pm i$ set the radius of convergence even though the function is analytic for all real x.

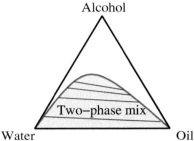

Fig. 8.8 Oil, water, and alcohol. A schematic ternary phase diagram for mixtures of oil, water, and alcohol. Each point in the triangle represents a mixture of percentages of the three, with the corners being pure water, oil, and alcohol. The shaded region shows where phase separation occurs; relative concentrations in the shaded region will separate into a two-phase mixture given by the endpoints of the tie-line passing through that point. Oil and water basically do not dissolve in one another; a mixture of the two will separate into the two separate fluids. You can go smoothly from one to the other, though, by first adding alcohol.

Exercises

The Ising model explores its continuous and abrupt phase transitions as we vary temperature and field. *Ising fluctuations and susceptibilities* studies its linear response to external fields, the connection between its fluctuations and response, and its energy gap and Curie law at low and high temperatures. *Ising hard discs* illustrates how to map a general two-state lattice model onto an Ising model, as we do for binary alloys.

[27] This process is sometimes called *adiabatic continuity* [11]. Phases can also be thought of as universality classes for attracting renormalization-group fixed points; see Chapter 12.

[28] Some unperturbed theories are better than others, even if they are in the same phase. The correct theory of superconductors is due to Bardeen, Cooper, and Schrieffer (BCS), despite the fact that earlier theories involving Bose condensation of electron pairs are not separated from BCS theory by a phase transition. The Cooper pairs in most superconductors are large compared to their separation, so they overlap many other pairs; this makes BCS theory almost exact.

Coin flips and Markov chains, Unicycle, and *Red and green bacteria* discuss the behavior and peculiarities of nonequilibrium Markov chains. *Detailed balance* reformulates this condition in an elegant fashion involving only the transition rates. *Metropolis* explores the most commonly applied Monte Carlo method, and in *Implementing Ising* you write your own heat-bath and Metropolis algorithms. *Wolff* and *Implementing Wolff* analyze a powerful and subtle cluster-flip algorithm.

In small systems like biological cells, the numbers of reacting molecules can be so small that number fluctuations can be important; *Stochastic cells* and *Repressilator* develop Monte Carlo methods (the Gillespie algorithm) for stochastic simulations of chemical reactions in these systems. *Entropy increases! Markov chains* shows that the free energy of a Markov chain decreases with time. *Metastability and Markov* introduces a Markov process (continuous time and space) to describe escape over a barrier.

Ising low temperature expansion and *2D Ising cluster expansion* introduce the technology of calculating free energies and magnetizations via sums of cluster diagrams.

Hysteresis and avalanches introduces the nonequilibrium noise of magnets exposed to external fields, and *Hysteresis algorithms* introduces an efficient $O(N \log N)$ algorithm for evolving the model. *Fruit flies and Markov* explores equilibrium and nonequilibrium models of insect behavior, and *Kinetic proofreading* implements an explanation for the high-fidelity transcription of DNA in our cells. Finally, *NP-completeness and satisfiability* explores the most challenging class of problems in computer science, and find a phase transition at which the truly difficult cases congregate.

(8.1) **The Ising model.**[29] (Computation) @

You will need a two-dimensional square-lattice Ising model simulation.

The Ising Hamiltonian is (eqn 8.1):

$$\mathcal{H} = -J \sum_{\langle ij \rangle} S_i S_j - H \sum_i S_i, \qquad (8.20)$$

where $S_i = \pm 1$ are spins on a square lattice, and the sum $\sum_{\langle ij \rangle}$ is over the four nearest-neighbor bonds (each pair summed once). It is conventional to set the coupling strength $J = 1$ and Boltzmann's constant $k_B = 1$, which amounts to measuring energies and temperatures in units of J. The constant H is called the external field, and $\mathbf{M} = \sum_i S_i$ is called the magnetization. Our simulation does not conserve the number of spins

up, so it is not a natural simulation for a binary alloy. You can think of it as a grand canonical ensemble, or as a model for extra atoms on a surface exchanging with the vapor above.

Play with the simulation. At high temperatures, the spins should not be strongly correlated. At low temperatures the spins should align all parallel, giving a large magnetization.

Roughly locate T_c, the largest temperature where there is a net magnetization (distant spins remain parallel on average) at $H = 0$. Explore the behavior by gradually lowering the temperature from just above T_c to just below T_c; does the behavior gradually change, or jump abruptly (like water freezing to ice)? Explore the behavior at $T = 2$ (below T_c) as you vary the external field $H = \pm 0.1$ up and down through the phase boundary at $H = 0$ (Fig. 8.5). Does the behavior vary smoothly in that case?

(8.2) **Ising fluctuations and susceptibilities.**[30] (Computation) ③

The partition function for the Ising model is $Z = \sum_n \exp(-\beta E_n)$, where the states n run over all 2^N possible configurations of the Ising spins (eqn 8.1), and the free energy $F = -kT \log Z$.

(a) *Show that the average of the magnetization M equals $-(\partial F/\partial H)|_T$. (Hint: Write out the sum for the partition function and take the derivative.) Derive the formula for the susceptibility $\chi_0 = (\partial M/\partial H)|_T$ in terms of $\langle (M - \langle M \rangle)^2 \rangle = \langle M^2 \rangle - \langle M \rangle^2$. (Hint: Remember our derivation of formula 6.13 $\langle (E - \langle E \rangle)^2 \rangle = k_B T^2 C$.)*

Now test this using the Ising model simulation. Notice that the program outputs averages of several quantities: $\langle |m| \rangle$, $\langle (m - \langle m \rangle)^2 \rangle$, $\langle e \rangle$, $\langle (e - \langle e \rangle)^2 \rangle$. In simulations, it is standard to measure $e = E/N$ and $m = M/N$ *per spin* (so that the plots do not depend upon system size; you will need to rescale properties appropriately to make comparisons with formulæ written for the energy and magnetization of the *system as a whole*. You can change the system size and decrease the graphics refresh rate (number of sweeps per draw) to speed your averaging. Make sure to equilibrate before starting to average!

(b) *Correlations and susceptibilities: numerical. Check the formulæ for C and χ from part (a) at $H = 0$ and $T = 3$, by measuring the fluctu-*

---

[29] Ising simulation software can be found at [28].

[30] Ising simulation software can be found at [28].

ations and the averages, and then changing by $\Delta H = 0.02$ or $\Delta T = 0.1$ and measuring the averages again. Check them also for $T = 2$, where $\langle M \rangle \neq 0$.[31]

There are systematic series expansions for the Ising model at high and low temperatures, using Feynman diagrams (see Section 8.3). The first terms of these expansions are both famous and illuminating.

Low-temperature expansion for the magnetization. At low temperatures we can assume all spins flip alone, ignoring clusters.

(c) *What is the energy for flipping a spin antiparallel to its neighbors? Equilibrate at a relatively low temperature $T = 1.0$, and measure the magnetization. Notice that the primary excitations are single spin flips. In the low-temperature approximation that the flipped spins are dilute (so we may ignore the possibility that two flipped spins touch or overlap), write a formula for the magnetization. (Remember, each flipped spin changes the magnetization by 2.) Check your prediction against the simulation.* (Hint: See eqn 8.19.)

The magnetization (and the specific heat) are exponentially small at low temperatures because there is an *energy gap* to spin excitations in the Ising model,[32] just as there is a gap to charge excitations in a semiconductor or an insulator.

High-temperature expansion for the susceptibility. At high temperatures, we can ignore the coupling to the neighboring spins.

(d) *Calculate a formula for the susceptibility of a free spin coupled to an external field. Compare it to the susceptibility you measure at high temperature $T = 100$ for the Ising model, say, $\Delta M / \Delta H$ with $\Delta H = 1$. (Why is $H = 1$ a small field in this case?)*

Your formula for the high-temperature susceptibility is known more generally as Curie's law.

(8.3) **Coin flips and Markov.** (Mathematics) ⓟ

A physicist, testing the laws of chance, flips a coin repeatedly until it lands tails.

(a) *Treat the two states of the physicist ("still flipping" and "done") as states in a Markov chain. The current probability vector then is $\vec{\rho} = \begin{pmatrix} \rho_{\text{flipping}} \\ \rho_{\text{done}} \end{pmatrix}$. Write the transition matrix $\mathcal{P}$, giving the time evolution $\mathcal{P} \cdot \vec{\rho}_n = \vec{\rho}_{n+1}$, assuming that the coin is fair.*

(b) *Find the eigenvalues and right eigenvectors of $\mathcal{P}$. Which eigenvector is the steady state ρ^*? Call the other eigenvector $\widetilde{\rho}$. For convenience, normalize $\widetilde{\rho}$ so that its first component equals one.*

(c) *Assume an arbitrary initial state is written $\rho_0 = A\rho^* + B\widetilde{\rho}$. What are the conditions on A and B needed to make ρ_0 a valid probability distribution? Write ρ_n as a function of A, B, ρ^*, and $\widetilde{\rho}$.*

(8.4) **Red and green bacteria.**[33] (Mathematics, Biology) ②

A growth medium at time $t = 0$ has 500 red bacteria and 500 green bacteria. Each hour, each bacterium divides in two. A color-blind predator eats exactly 1,000 bacteria per hour.[34]

(a) *After a very long time, what is the probability distribution for the number α of red bacteria in the growth medium?*

(b) *Roughly how long will it take to reach this final state? (Assume either that one bacterium reproduces and then one is eaten every $1/1,000$ of an hour, or that at the end of each hour all the bacteria reproduce and then 1,000 are consumed. My approach was to treat it as a random walk with an α-dependent step size.)*

(c) *Assume that the predator has a 1% preference for green bacteria (implemented as you choose). Roughly how much will this change the final distribution?*

(8.5) **Detailed balance.** @

In an equilibrium system, for any two states α and β with equilibrium probabilities ρ_α^* and ρ_β^*, detailed balance states (eqn 8.14) that

$$P_{\beta \Leftarrow \alpha} \rho_\alpha^* = P_{\alpha \Leftarrow \beta} \rho_\beta^*, \qquad (8.21)$$

---

[31] Be sure to wait until the state is equilibrated before you start! Below T_c this means the state should not have red and black domains, but be all in one ground state. You may need to apply a weak external field for a while to remove stripes at low temperatures.

[32] Not all real magnets have a gap; if there is a spin rotation symmetry, one can have gapless *spin waves*, which are like sound waves except twisting the magnetization rather than wiggling the atoms.

[33] Adapted from author's graduate admission to candidacy exam, Princeton University, Fall 1977.

[34] This question is purposely open-ended, and rough answers to parts (b) and (c) within a factor of two are perfectly acceptable. Numerical and analytical methods are both feasible.

that is, the equilibrium flux of probability from α to β is the same as the flux backward from β to α. (Again, here and in eqn 8.22, we do not sum over repeated indices.) It is both possible and elegant to reformulate the condition for detailed balance so that it does not involve the equilibrium probabilities. Consider three states of the system, α, β, and γ.

(a) *Assume that each of the three types of transitions among the three states satisfies detailed balance. Eliminate the equilibrium probability densities to derive*[35]

$$P_{\alpha \Leftarrow \beta} P_{\beta \Leftarrow \gamma} P_{\gamma \Leftarrow \alpha} = P_{\alpha \Leftarrow \gamma} P_{\gamma \Leftarrow \beta} P_{\beta \Leftarrow \alpha}. \quad (8.22)$$

Viewing the three states α, β, and γ as forming a circle, you have derived a relationship between the rates going clockwise and the rates going counter-clockwise around the circle. Can we show the converse to part (a), that if every triple of states in a Markov chain satisfies the condition 8.22 then it satisfies detailed balance? Given the transition matrix P, can we construct a probability density $\boldsymbol{\rho}^*$ which makes the probability fluxes between all pairs of states equal?

In most problems, most of the rates are zero: the matrix P is sparse, connecting only nearby states. This makes the cyclic condition for detailed balance stricter than Equation 8.22. (Detailed balance demands that the forward and reversed products for cycles of *all* lengths must be equal, the *Kolmogorov criterion*.) There are N^3 equations in Equation 8.22 (α, β, and γ each running over all N states), and we need to solve for N unknowns $\boldsymbol{\rho}^*$. But in a sparse matrix most of these "triangle" conditions tell us only that $0 = 0$. However, if we assume that the transition matrix has enough positive entries, the three-state cyclic eqn 8.22 is enough to construct the stationary state $\boldsymbol{\rho}^*$ from P, satisfying detailed balance (eqn 8.21).

(b) *Suppose P is the transition matrix for some Markov chain satisfying the condition 8.22 for every triple of states α, β, and γ. Assume for simplicity that there is a state α_0 with nonzero transition rates from all other states δ.*[36] *Construct a probability density $\boldsymbol{\rho}^*$ that demonstrates that P satisfies detailed balance (eqn 8.21).*

(Hint: Assume you know $\rho^*_{\alpha_0}$; use some of the eqns 8.21 to write a formula for each of the other $N - 1$ elements ρ^*_δ that ensures detailed balance for the pair. Then solve for $\rho^*_{\alpha_0}$ to make the probability distribution normalized. Use the cyclic condition eqn 8.22 to show that this candidate stationary state $\boldsymbol{\rho}^*$ satisfies detailed balance for any two states β and δ.)

(8.6) **Metropolis.** (Mathematics, Computation) @
The heat-bath algorithm described in the text thermalizes one spin at a time. Another popular choice is the Metropolis algorithm, which also flips a single spin at a time:

(1) Pick a spin at random;
(2) Calculate the energy $\Delta\mathbf{E}$ for flipping the spin;
(3) If $\Delta\mathbf{E} < 0$ flip it; if $\Delta\mathbf{E} > 0$, flip it with probability $e^{-\beta\Delta\mathbf{E}}$.

Show that Metropolis satisfies detailed balance. Note that it is ergodic and Markovian (no memory), and hence that it will lead to thermal equilibrium. Is Metropolis more efficient than the heat-bath algorithm (fewer random numbers needed to get to equilibrium)?

(8.7) **Implementing Ising.**[37] (Computation) ④
In this exercise, we will implement a simulation of the two-dimensional Ising model on a square lattice using the heat-bath and Metropolis algorithms.
The heat-bath algorithm flips spins one at a time, putting them into equilibrium with their neighbors: it is described in detail in Section 8.1.

(a) *Implement the heat-bath algorithm for the Ising model. Do not recalculate the exponentials for the transition probabilities for each spin flip! Instead, create an array* `heatBathProbUp[nUp]` *to store the probability that a spin will be set to $+1$ given that* `nUp` *of its neighbors are currently pointing up (equal to $+1$); recalculate it whenever the external field or temperature is changed. Explore the resulting behavior (say, as in Exercise 8.1).*
The Metropolis algorithm also flips one spin at a time, but it always flips spins if the net energy decreases: it is described in detail in Exercise 8.6.

---

[35]Note that, as in eqn 8.21, we do *not* sum over repeated indices here; this equation must hold for all triples of states α, β, and γ. Hence these form N^3 equations for the N^2 variables $P_{\alpha \Leftarrow \beta}$.

[36]See *regular matrices*, note 20 on p. 223.

[37]Computational hints can be found at the book website [181].

(b) *Implement the Metropolis algorithm for the Ising model. Here you will want to set up an array* MetropolisProbUp[s,nUp] *storing the probability that a spin which currently has value* s *will be set to* +1 *if* nUp *of its neighbors are currently up. Is Metropolis noticeably faster than the heat-bath algorithm?*

The Metropolis algorithm is always faster to equilibrate than the heat-bath algorithm, but is never a big improvement. Other algorithms can be qualitatively faster in certain circumstances (see Exercises 8.8 and 8.9).

(8.8) **Wolff.**[38] (Mathematics, Computation) @

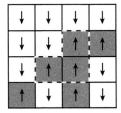

Fig. 8.9 Cluster flip: before. The region inside the dotted line is flipped in one Wolff move. Let this configuration be A. Let the cluster being flipped be C (bounded by the dotted line). Notice that the boundary of C has $n\uparrow= 2$, $n\downarrow= 6$.

Near the critical point T_c where the system develops a magnetization, any single-spin-flip dynamics becomes very slow (the *correlation time* diverges). Wolff [213], building on ideas of Swendsen and Wang [192], came up with a clever method to flip whole clusters of spins.

Find a Wolff simulation (e.g., [28]). Run at $T_c = 2/\log(1 + \sqrt{2})$ using the Metropolis algorithm for a 512×512 system or larger; watch the slow growth and long persistence times of the larger clusters. This is *critical slowing down*. Now change to the Wolff algorithm, and see how much faster the large clusters rearrange. Take single Wolff steps: many will almost completely rearrange the pattern.

The *correlation time* is roughly the time it takes a system, begun at one particular configuration in the equilibrium ensemble, to move to an uncorrelated configuration. We measure the correlation time in *sweeps*. A sweep in a $N \times N$ system will attempt to flip N^2 spins.

Fig. 8.10 Cluster flip: after. Let this configuration be B. Notice that the cluster has been flipped.

(a) *Change to a smaller system size (perhaps 128×128), and visually estimate the correlation time for the Metropolis algorithm. (For small sizes, periodic boundary conditions will make the Ising model appear magnetized. Estimate the number of sweeps you need to scramble the largest islands, not to flip the magnetization. Doing many sweeps between redraws will speed things up, and the pauses will help you to count.) Estimate the correlation time for Wolff (taking single steps). Is the speedup significant?*

How does the Wolff cluster flip method work?

(1) Pick a spin at random, remember its direction $D = \pm 1$, and flip it.

(2) For each of the four neighboring spins, *if* it is in the direction D, flip it with probability p.

(3) For each of the new flipped spins, recursively flip their neighbors as in (2).

Because with finite probability you can flip any spin, the Wolff algorithm is ergodic. As a cluster flip it is Markovian. Let us see that it satisfies detailed balance, when we pick the right value of p for the given temperature. Define $n_\uparrow$ and $n_\downarrow$ to count the number of cluster edges (dotted lines in Figs. 8.9 and 8.10) that are adjacent to external spins pointing up and down, respectively.

(b) *Show for the two configurations in Figs. 8.9 and 8.10 that $E_B - E_A = 2(n_\uparrow - n_\downarrow)J$. Argue that this will be true for flipping any cluster of up-spins to down-spins.*

The cluster flip can start at any site α in the cluster C. The ratio of rates $\Gamma_{A\to B}/\Gamma_{B\to A}$ depends upon the number of times the cluster chose *not* to grow on the boundary. Let P_α^C be the probability that the cluster grows internally from site

α to the cluster C (ignoring the moves which try to grow outside the boundary). Then

$$\Gamma_{A \to B} = \sum_{\alpha} P_{\alpha}^{C} (1-p)^{n_{\uparrow}}, \qquad (8.23)$$

$$\Gamma_{B \to A} = \sum_{\alpha} P_{\alpha}^{C} (1-p)^{n_{\downarrow}}, \qquad (8.24)$$

since the cluster must refuse to grow $n_{\uparrow}$ times when starting from the up-state A, and $n_{\downarrow}$ times when starting from B.

(c) *What value of p lets the Wolff algorithm satisfy detailed balance at temperature T?*

See [143, sections 4.2–3] for more details on the Wolff algorithm, and [97] for the Wolff algorithm for other models including external fields.

(8.9) **Implementing Wolff.**[39] (Computation) ④

In this exercise, we will implement the Wolff algorithm of Exercise 8.8.

Near the critical temperature T_c for a magnet, the equilibration becomes very sluggish: this is called *critical slowing-down*. This sluggish behavior is faithfully reproduced by the single-spin-flip heat-bath and Metropolis algorithms. If one is interested in equilibrium behavior, and not in dynamics, one can hope to use fancier algorithms that bypass this sluggishness, saving computer time.

(a) *Run the two-dimensional Ising model (either from the book website or from your solution to Exercise 8.7) near $T_c = 2/\log(1 + \sqrt{2})$ using a single-spin-flip algorithm. Start in a magnetized state, and watch the spins rearrange until roughly half are pointing up. Start at high temperatures, and watch the up- and down-spin regions grow slowly. Run a large enough system that you get tired of waiting for equilibration.*

The Wolff algorithm flips large clusters of spins at one time, largely bypassing the sluggishness near T_c. It is described in detail in Exercise 8.8. The Wolff algorithm can be generalized to systems with external fields [97].

(b) *Implement the Wolff algorithm. A recursive implementation works only for small system sizes on most computers. Instead, put the spins that are destined to flip on a list* toFlip. *You will also need to keep track of the sign of the original triggering spin.*

While there are spins toFlip,
   if the first spin remains parallel to the original,

   flip it, and
   for each neighbor of the flipped spin,
      if it is parallel to the original spin,
         add it to toFlip with probability p.

(c) *Estimate visually how many Wolff cluster flips it takes to reach the equilibrium state at T_c. Is Wolff faster than the single-spin-flip algorithms? How does it compare at high temperatures?*

(d) *Starting from a random configuration, change to a low temperature $T = 1$ and observe the equilibration using a single-spin flip algorithm. Compare with your Wolff algorithm. (See also Exercise 12.3.) Which reaches equilibrium faster? Is the dynamics changed qualitatively?*

(8.10) **Stochastic cells.**[40] (Biology, Computation) ④

Living cells are amazingly complex mixtures of a variety of complex molecules (RNA, DNA, proteins, lipids, ...) that are constantly undergoing reactions with one another. This complex of reactions has been compared to computation; the cell gets input from external and internal sensors, and through an intricate series of reactions produces an appropriate response. Thus, for example, receptor cells in the retina "listen" for light and respond by triggering a nerve impulse. The kinetics of chemical reactions are usually described using differential equations for the concentrations of the various chemicals, and rarely are statistical fluctuations considered important. In a cell, the numbers of molecules of a given type can be rather small; indeed, there is (often) only one copy of the relevant part of DNA for a given reaction. It is an important question whether and when we may describe the dynamics inside the cell using continuous concentration variables, even though the actual numbers of molecules are always integers.

Consider a dimerization reaction; a molecule M (called the *monomer*) joins up with another monomer and becomes a dimer D: $2M \longleftrightarrow D$. Proteins in cells often form dimers; sometimes (as here) both proteins are the same (homodimers) and sometimes they are different proteins (heterodimers). Suppose the forward reaction rate is k_d and the backward reaction rate is k_u. Figure 8.11 shows this as a Petri net [73]

---

[39] Hints for the computations can be found at the book website [181].

[40] This exercise and the associated software were developed in collaboration with Christopher Myers. Computational hints can be found at the book website [181].

with each reaction shown as a box, with incoming arrows showing species that are consumed by the reaction, and outgoing arrows showing species that are produced by the reaction; the number consumed or produced (the *stoichiometry*) is given by a label on each arrow. There are thus two reactions: the backward unbinding reaction rate per unit volume is $k_u[D]$ (each dimer disassociates with rate k_u), and the forward binding reaction rate per unit volume is $k_b[M]^2$ (since each monomer must wait for a collision with another monomer before binding, the rate is proportional to the monomer concentration squared).

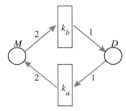

Fig. 8.11 Dimerization reaction. A Petri net diagram for a dimerization reaction, with dimerization rate k_b and dimer dissociation rate k_u.

The brackets [.] denote concentrations. We assume that the volume per cell is such that one molecule per cell is $1\,\mathrm{nM}$ (10^{-9} moles per liter). For convenience, we shall pick nanomoles as our unit of concentration, so $[M]$ is also the number of monomers in the cell. Assume $k_b = 1\,\mathrm{nM^{-1}s^{-1}}$ and $k_u = 2\,\mathrm{s^{-1}}$, and that at $t = 0$ all N monomers are unbound.

(a) Continuum dimerization. *Write the differential equation for* $\mathrm{d}M/\mathrm{d}t$ *treating M and D as continuous variables.* (Hint: Remember that two M molecules are consumed in each reaction.) *What are the equilibrium concentrations for* $[M]$ *and* $[D]$ *for* $N = 2$ *molecules in the cell, assuming these continuous equations and the values above for* k_b *and* k_u? *For* $N = 90$ *and* $N = 10{,}100$ *molecules? Numerically solve your differential equation for* $M(t)$ *for* $N = 2$ *and* $N = 90$, *and verify that your solution settles down to the equilibrium values you found.*

For large numbers of molecules in the cell, we expect that the continuum equations may work well, but for just a few molecules there surely will be relatively large fluctuations. These fluctuations are called *shot noise*, named in early studies of electrical noise at low currents due to individual electrons in a resistor. We can implement a Monte Carlo algorithm to simulate this shot noise.[41] Suppose the reactions have rates Γ_i, with total rate $\Gamma_{\mathrm{tot}} = \sum_i \Gamma_i$. The expected time to the next reaction is $1/\Gamma_{\mathrm{tot}}$, and the probability that the next reaction will be j is $\Gamma_j/\Gamma_{\mathrm{tot}}$. To simulate until a final time t_f:

(1) Calculate a list of the rates of all reactions in the system.

(2) Find the total rate Γ_{tot}.

(3) Pick a random time t_{wait} with probability distribution $\rho(t) = \Gamma_{\mathrm{tot}} \exp(-\Gamma_{\mathrm{tot}} t)$.

(4) If the current time t plus t_{wait} is bigger than t_f, we are finished; return.

(5) Otherwise:

  − increment t by t_{wait};

  − pick a random number r uniformly distributed in the range $[0, \Gamma_{\mathrm{tot}})$;

  − pick the reaction j for which $\sum_{i<j} \Gamma_i \leq r < \sum_{i<j+1} \Gamma_i$ (that is, r lands in the jth interval of the sum forming Γ_{tot});

  − execute that reaction, by incrementing each chemical by its stoichiometry.

(6) Repeat.

There is one important additional change:[42] the binding reaction rate for M total monomers binding is no longer $k_b M^2$ for discrete molecules; it is $k_b M(M-1)$.[43]

(b) Stochastic dimerization. *Implement this algorithm for the dimerization reaction of part (a). Simulate for* $N = 2$, $N = 90$, *and* $N = 10{,}100$ *and compare a few stochastic realizations with the continuum solution. How large a value of* N *do you need for the individual reactions to be well described by the continuum equations (say, fluctuations less than* $\pm 20\%$ *at late times)?*

Measuring the concentrations in a single cell is often a challenge. Experiments often average over many cells. Such experiments will measure

[41] In the context of chemical simulations, this algorithm is named after Gillespie [67]; the same basic approach was used just a bit earlier for the Ising model by Bortz, Kalos, and Lebowitz [33], and is called *continuous-time Monte Carlo* in that context.

[42] Without this change, if you start with an odd number of cells your concentrations can go negative!

[43] Again $[M] = M$, because we assume one molecule per cell gives a concentration of $1\,\mathrm{nM}$.

a smooth time evolution even though the individual cells are noisy. Let us investigate whether this ensemble average is well described by the continuum equations.

(c) *Average stochastic dimerization. Find the average of many realizations of your stochastic dimerization in part (b), for $N = 2$ and $N = 90$, and compare with your deterministic solution. How much is the long-term average shifted by the stochastic noise? How large a value of N do you need for the ensemble average of $M(t)$ to be well described by the continuum equations (say, shifted by less than 5% at late times)?*

(8.11) **Repressilator.**[44] (Biology, Computation) ④

TetR

λ CI —| LacI

Fig. 8.12 Biology repressilator. The biologist's view of the repressilator network. The T-shapes are blunt arrows, signifying that the protein at the tail (bottom of the T) suppresses the production of the protein at the head. Thus LacI (pronounced lack-eye) suppresses TetR (tet-are), which suppresses λ CI (lambda-see-one). This condensed description summarizes a complex series of interactions (see Fig. 8.13).

The "central dogma" of molecular biology is that the flow of information is from DNA to RNA to proteins; DNA is *transcribed* into RNA, which then is *translated* into protein.

Now that the genome is sequenced, it is thought that we have the parts list for the cell. All that remains is to figure out how they work together! The proteins, RNA, and DNA form a complex network of interacting chemical reactions, which governs metabolism, responses to external stimuli, reproduction (*proliferation*), *differentiation* into different cell types, and (when the cell perceives itself to be breaking down in dangerous ways) programmed cell death, or *apoptosis*.

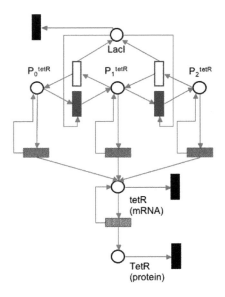

Fig. 8.13 Computational repressilator. The Petri net version [73] of *one-third* of the repressilator network (the LacI repression of TetR). The biologist's shorthand (Fig. 8.12) hides a lot of complexity! We have implemented these equations for you, so studying this figure is optional. The solid lighter vertical rectangles represent binding reactions $A + B \rightarrow C$, with rate $k_b[A][B]$; the open vertical rectangles represent unbinding $C \rightarrow A + B$, with rate $k_u[C]$. The horizontal rectangles represent catalyzed synthesis reactions $C \rightarrow C + P$, with rate $\gamma[C]$; the darker ones represent transcription (formation of mRNA), and the lighter one represent translation (formation of protein). The black vertical rectangles represent degradation reactions, $A \rightarrow$ nothing with rate $k_d[A]$. The LacI protein (top) can bind to the DNA in two *promoter sites* ahead of the gene coding for tetR; when bound, it largely blocks the transcription (formation) of tetR mRNA. P_0 represents the promoter without any LacI bound; P_1 represents the promoter with one site blocked, and P_2 represents the doubly bound promoter. LacI can bind to one or both of the promoter sites, changing P_i to P_{i+1}, or correspondingly unbind. The unbound P_0 state transcribes tetR mRNA quickly, and the bound states transcribe it slowly (leaky repression). The tetR mRNA then catalyzes the formation of the TetR protein.

Our understanding of the structure of these interacting networks is growing rapidly, but our understanding of the dynamics is still rather

---

[44]This exercise draws heavily on Elowitz and Leibler [55]; it and the associated software were developed in collaboration with Christopher Myers. Computational hints can be found at the book website [181].

primitive. Part of the difficulty is that the cellular networks are not neatly separated into different modules; a given protein may participate in what would seem to be several separate regulatory pathways. In this exercise, we will study a model gene regulatory network, the *repressilator*. This experimental system involves three proteins, each of which inhibits the formation of the next. They were added to the bacterium *E. coli*, with hopefully minimal interactions with the rest of the biological machinery of the cell. We will implement the stochastic model that the authors used to describe their experimental system [55]. In doing so, we will:

- Implement in a tangible system an example both of the central dogma and of *transcriptional regulation*: the control by proteins of DNA expression into RNA;

- Introduce sophisticated Monte Carlo techniques for simulations of stochastic reactions;

- Introduce methods for automatically generating continuum descriptions from reaction rates; and

- Illustrate the *shot noise* fluctuations due to small numbers of molecules and the *telegraph noise* fluctuations due to finite rates of binding and unbinding of the regulating proteins.

Figure 8.12 shows the biologist's view of the repressilator network. Three proteins (TetR, λCI, and LacI) each repress the formation of the next. We shall see that, under appropriate circumstances, this can lead to spontaneous oscillations; each protein peaks in turn, suppressing the suppressor of its suppressor, leading to its own later decrease.

The biologist's notation summarizes a much more complex picture. The LacI protein, for example, can bind to one or both of the *transcriptional regulation* or *operator* sites ahead of the gene that codes for the tetR mRNA.[45] When bound, it largely blocks the translation of DNA into tetR. [46] The level of tetR will gradually decrease as it degrades; hence less TetR protein will be translated from the tetR mRNA. The resulting network of ten reactions is depicted

in Fig. 8.13, showing one-third of the total repressilator network. The biologist's shorthand (Fig. 8.12) does not specify the details of how one protein represses the production of the next. The larger diagram, for example, includes two operator sites for the repressor molecule to bind to, leading to three states (P_0, P_1, and P_2) of the promoter region depending upon how many LacI proteins are bound.

(a) *Run the simulation for at least 6,000 seconds and plot the protein, RNA, and promoter states as a function of time. Notice that:*

- *the protein levels do oscillate, as in [55, Fig. 1(c)];*

- *there are significant noisy-looking fluctuations;*

- *there are many more proteins than RNA.*

To see how important the fluctuations are, we should compare the stochastic simulation to the solution of the continuum reaction rate equations (as we did in Exercise 8.10). In [55], the authors write a set of six differential equations giving a continuum version of the stochastic simulation. These equations are simplified; they both "integrate out" or coarse-grain away the promoter states from the system, deriving a Hill equation (Exercise 6.12) for the mRNA production, and they also rescale their variables in various ways. Rather than typing in their equations and sorting out these rescalings, it is convenient and illuminating to write a routine to generate the continuum differential equations directly from our reaction rates.

(b) *Write a* `DeterministicRepressilator`, *derived from* `Repressilator` *just as* `StochasticRepressilator` *was. Write a routine* `dcdt(c,t)` *that does the following:*

- *Sets the chemical amounts in the reaction network to the values in the array* `c`.

- *Sets a vector* `dcdt` *(of length the number of chemicals) to zero.*

- *For each reaction:*
  - *compute its rate;*
  - *for each chemical whose stoichiometry is changed by the reaction, add the stoi-*

---

[45]Messenger RNA (mRNA) codes for proteins. Other forms of RNA can serve as enzymes or parts of the machinery of the cell. Proteins in *E. coli* by convention have the same names as their mRNA, but start with capitals where the mRNA start with small letters.

[46]*RNA polymerase*, the molecular motor responsible for transcribing DNA into RNA, needs to attach to the DNA at a *promoter site*. By binding to the adjacent operator sites, our repressor protein inhibits this attachment and hence partly blocks transcription.

chiometry change times the rate to the corresponding entry of `dcdt`.

Call a routine to integrate the resulting differential equation (as described in the last part of Exercise 3.12, for example), and compare your results to those of the stochastic simulation.
The stochastic simulation has significant fluctuations away from the continuum equation. Part of these fluctuations are due to the fact that the numbers of proteins and mRNAs are small; in particular, the mRNA numbers are significantly smaller than the protein numbers.

(c) *Write a routine that creates a stochastic repressilator network that multiplies the mRNA concentrations by* `RNAFactor` *without otherwise affecting the continuum equations. (That is, multiply the initial concentrations and the transcription rates by* `RNAFactor`*, and divide the translation rate by* `RNAFactor`*.) Try boosting the* `RNAFactor` *by ten and one hundred. Do the RNA and protein fluctuations become significantly smaller?* This noise, due to the discrete, integer values of chemicals in the cell, is analogous to the *shot noise* seen in electrical circuits due to the discrete quantum of electric charge. It scales, as do most fluctuations, as the square root of the number of molecules.

A continuum description of the binding of the proteins to the operator sites on the DNA seems particularly dubious; a variable that must be zero or one is replaced by a continuous evolution between these extremes. (Such noise in other contexts is called *telegraph noise*—in analogy to the telegraph, which is either silent or sending as the operator taps the key.) The continuum description is accurate in the limit where the binding and unbinding rates are fast compared to all of the other changes in the system; the protein and mRNA variations then see the average, local equilibrium concentration. On the other hand, if the rates are slow compared to the response of the mRNA and protein, the latter can have a switching appearance.

(d) *Incorporate a* `telegraphFactor` *into your stochastic repressilator routine, that multiplies the binding and unbinding rates. Run for 1,000 seconds with* `RNAFactor` = 10 *(to suppress the shot noise) and* `telegraphFactor` = 0.001. *Do you observe features in the mRNA curves that appear to switch as the relevant proteins unbind*

and bind?

(8.12) **Entropy increases! Markov chains.** (Mathematics) ③
In Exercise 5.7 you noticed that, formally speaking, entropy does not increase in Hamiltonian systems. Let us show that it does increase for Markov chains.[47]
Convexity arguments are a basic tool in formal statistical mechanics. The function $f(x) = -x \log x$ is strictly concave (convex downward) for $x \geq 0$ (Fig. 5.9); this is easily shown by noting that its second derivative is negative in this region.

(a) Convexity for sums of many terms. *If* $\sum_\alpha \mu_\alpha = 1$, *and if for all* α *both* $\mu_\alpha \geq 0$ *and* $x_\alpha \geq 0$, *show by induction on the number of states* M *that if* $g(x)$ *is concave for* $x \geq 0$, *then*

$$g\left(\sum_{\alpha=1}^{M} \mu_\alpha x_\alpha\right) \geq \sum_{\alpha=1}^{M} \mu_\alpha g(x_\alpha). \qquad (8.25)$$

This is a generalization of Jensen's inequality (eqn 5.27), which was the special case of equal μ_α. *(Hint: In the definition of concave,* $f(\lambda a + (1 - \lambda)b) \geq \lambda f(a) + (1 - \lambda)f(b)$, *take* $(1 - \lambda) = \mu_{M+1}$ *and* $b = x_{M+1}$. *Then* a *is a sum of* M *terms, rescaled from their original values. Do the coefficients of* x_α *in the variable* a *sum to one? Can we apply induction?)*
The Markov chain is implicitly exchanging energy with a heat bath at the temperature T. Thus to show that the entropy for the world as a whole increases, we must show that $\Delta S - \Delta E/T$ increases, where ΔS is the entropy of our system and $\Delta E/T$ is the entropy flow from the heat bath. Hence, showing that entropy increases for our Markov chain is equivalent to showing that the free energy $E - TS$ decreases.
Let $P_{\alpha\beta}$ be the transition matrix for a Markov chain, satisfying detailed balance with energy E_α at temperature T. The current probability of being in state α is ρ_α. The free energy

$$F = E - TS = \sum_\alpha \rho_\alpha E_\alpha + k_B T \sum_\alpha \rho_\alpha \log \rho_\alpha. \qquad (8.26)$$

(b) *Show that the free energy decreases for a Markov chain. In particular, using eqn 8.25, show that the free energy for* $\rho_\beta^{(n+1)} = \sum_\alpha P_{\beta\alpha}\rho_\alpha^{(n)}$ *is less than or equal to the free energy for* $\rho^{(n)}$. *You may use the properties of the*

---

[47]We know that the Markov chain eventually evolves to the equilibrium state, and we argued that the latter minimizes the free energy. What we are showing here is that the free energy goes continuously downhill for a Markov chain.

Markov transition matrix P, $(0 \leq P_{\alpha\beta} \leq 1$ *and* $\sum_\alpha P_{\alpha\beta} = 1)$, *and detailed balance* $(P_{\alpha\beta}\rho_\beta^* = P_{\beta\alpha}\rho_\alpha^*,$ *where* $\rho_\alpha^* = \exp(-E_\alpha/k_B T)/Z)$.

(Hint: You will want to use $\mu_\alpha = P_{\alpha\beta}$ in eqn 8.25, but the entropy will involve $P_{\beta\alpha}$, which is *not* the same. Use detailed balance to convert from one to the other.)

(8.13) **Hysteresis and avalanches.**[48] (Complexity, Computation) ④

Fig. 8.14 Barkhausen noise experiment. By increasing an external magnetic field $H(t)$ (bar magnet approaching), the magnetic domains in a slab of iron flip over to align with the external field. The resulting magnetic field jumps can be turned into an electrical signal with an inductive coil, and then listened to with an ordinary loudspeaker. Barkhausen noise from our computer experiments can be heard on the Internet [102].

A piece of magnetic material exposed to an increasing external field $H(t)$ (Fig. 8.14) will magnetize (Fig. 8.15) in a series of sharp jumps, or *avalanches* (Fig. 8.16). These avalanches arise as magnetic domain walls in the material are pushed by the external field through a rugged potential energy landscape due to irregularities and impurities in the magnet. The magnetic signal resulting from these random avalanches is called *Barkhausen noise*.

We model this system with a nonequilibrium lattice model, the *random field Ising model*. The Hamiltonian or energy function for our system is

$$\mathcal{H} = -\sum_{\langle i,j \rangle} J s_i s_j - \sum_i \left(H(t) + h_i\right) s_i, \quad (8.27)$$

where the spins $s_i = \pm 1$ lie on a square or cubic lattice with periodic boundary conditions. The coupling and the external field H are as in the traditional Ising model (Section 8.1). The disorder in the magnet is incorporated using the *random field* h_i, which is independently chosen at each lattice site from a Gaussian probability distribution of standard deviation R:

$$P(h) = \frac{1}{\sqrt{2\pi}R} e^{-h^2/2R^2}. \quad (8.28)$$

We are not interested in thermal equilibrium; there would be no hysteresis! We take the opposite extreme; we set the temperature to zero. We start with all spins pointing down, and adiabatically (infinitely slowly) increase $H(t)$ from $-\infty$ to ∞.

Fig. 8.15 Hysteresis loop with subloops for our model. As the external field is raised and lowered (vertical), the magnetization lags behind—this is called *hysteresis*. The magnetization curves here look macroscopically smooth.

Fig. 8.16 Tiny jumps: Barkhausen noise. Blowing up a small portion of Fig. 8.15, we see that the magnetization is growing in a series of sharp jumps, or avalanches.

---

[48]This exercise is largely drawn from [103]. It and the associated software were developed in collaboration with Christopher Myers. Computational hints can be found at the book website [181].

Our rules for evolving the spin configuration are simple: each spin flips over when doing so would decrease the energy. This occurs at site i when the local field at that site

$$J \sum_{j \text{ nbr to } i} s_j + h_i + H(t) \qquad (8.29)$$

changes from negative to positive. A spin can be pushed over in two ways. It can be triggered when one of its neighbors flips (by participating in a propagating avalanche) or it can be triggered by the slow increase of the external field (starting a new avalanche).

(a) *Set up lattices* s[m][n] *and* h[m][n] *on the computer. (If you do three dimensions, add an extra index to the arrays.) Fill the former with down-spins (-1) and the latter with random fields (real numbers chosen from the distribution 8.28). Write a routine* FlipSpin *for the lattice, which given i and j flips the spin from $s = -1$ to $s = +1$ (complaining if it is already flipped). Write a routine* NeighborsUp *which calculates the number of up-neighbors for the spin (implementing the periodic boundary conditions).*

On the computer, changing the external field infinitely slowly is easy. To start a new avalanche (or the first avalanche), one searches for the unflipped spin that is next to flip, jumps the field H to just enough to flip it, and propagates the avalanche, as follows:

(1) Find the triggering spin i for the next avalanche, which is the unflipped site with the largest internal field $J \sum_{j \text{ nbr to } i} s_j + h_i$ from its random field and neighbors.

(2) Increment the external field H to minus this internal field, and push the spin onto a first-in–first-out queue (Fig. 8.17, right).

(3) Pop the top spin off the queue.

(4) If the spin has not been flipped,[49] flip it and push all unflipped neighbors with positive local fields onto the queue.

(5) While there are spins on the queue, repeat from step (3).

(6) Repeat from step (1) until all the spins are flipped.

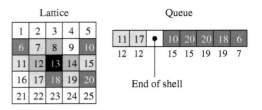

Fig. 8.17 Avalanche propagation in the hysteresis model. Left: a propagating avalanche. Spin 13 triggered the avalanche. It triggered the first shell of spins 14, 8, and 12, which then triggered the second shell 15, 19, 7, 11, and 17, and finally the third shell 10, 20, 18, and 6. Right: the first-in–first-out queue, part way through flipping the second shell. (The numbers underneath are the triggering spins for the spins on the queue, for your convenience.) The spin at the left of this queue is next to flip. Notice that spin 20 has been placed on the queue twice (two neighbors in the previous shell). By placing a marker at the end of each shell in the queue, we can measure the number of spins flipping per shell (unit "time") during an avalanche (Fig. 8.18).

(b) *Write a routine* BruteForceNextAvalanche *for step (1), which checks the local fields of all of the unflipped spins, and returns the location of the next to flip.*

(c) *Write a routine* PropagateAvalanche *that propagates an avalanche given the triggering spin, steps (3)–(5), coloring the spins in the display that are flipped. Run a 300×300 system at $R = 1.4$, 0.9, and 0.7 (or a 50^3 system at $R = 4$, $R = 2.16$, and $R = 2$) and display the avalanches.* If you have a fast machine, you can run a larger size system, but do not overdo it; the sorted list algorithm of Exercise 8.14 can dramatically speed up your simulation.

There are lots of properties that one might wish to measure about this system: avalanche sizes, avalanche correlation functions, hysteresis loop shapes, average pulse shapes (Fig. 8.18) during avalanches,... It can get ugly if you put all of these measurements inside the inner loop of your code. Instead, we suggest that you try the *subject–observer* design pattern: each time a spin is flipped, and each time an avalanche is finished, the subject (our simulation) notifies the list of observers.

---

[49]You need to check if the spin is flipped again after popping it off the queue; spins can be put onto the queue more than once during an avalanche (Fig. 8.17).

$dM/dt \sim$ voltage $V(t)$

Time t (in avalanche shells)

Fig. 8.18 Avalanche time series. Number of domains flipped per time step for the avalanche shown in Fig. 12.5. Notice how the avalanche almost stops several times; if the forcing were slightly smaller compared to the disorder, the avalanche would have separated into smaller ones. The fact that the disorder is just small enough to keep the avalanche growing is the criterion for the phase transition, and the cause of the self-similarity. At the critical point, a partial avalanche of size S will on average trigger another one of size S.

(d) *Build a* `MagnetizationObserver`, *which stores an internal magnetization starting at* $-N$, *adding two to it whenever it is notified. Build an* `AvalancheSizeObserver`, *which keeps track of the growing size of the current avalanche after each spin flip, and adds the final size to a histogram of all previous avalanche sizes when the avalanche ends. Set up* `NotifySpinFlip` *and* `NotifyAvalancheEnd` *routines for your simulation, and add the two observers appropriately. Plot the magnetization curve* $M(H)$ *and the avalanche size distribution histogram* $D(S)$ *for the three systems you ran for part (c).*

(8.14) **Hysteresis algorithms.**[50] (Complexity, Computation) ④

As computers increase in speed and memory, the benefits of writing efficient code become greater and greater. Consider a problem on a system of size N; a complex algorithm will typically run more slowly than a simple one for small N, but if its time used scales proportional to N and the simple algorithm scales as N^2, the added complexity wins as we can tackle larger, more ambitious questions.

In the hysteresis model (Exercise 8.13), the brute-force algorithm for finding the next avalanche for a system with N spins takes a time of order N per avalanche. Since there are roughly N avalanches (a large fraction of all avalanches are of size one, especially in three dimensions) the time for the brute-force algorithm scales as N^2. Can we find a method which does not look through the whole lattice every time an avalanche needs to start?

Lattice   Sorted list

$H = 1.1$

Fig. 8.19 Using a sorted list to find the next spin in an avalanche. The shaded cells have already flipped. In the sorted list, the arrows on the right indicate the `nextPossible[nUp]` pointers—the first spin that would not flip with nUp neighbors at the current external field. Some pointers point to spins that have already flipped, meaning that these spins already have more neighbors up than the corresponding nUp. (In a larger system the unflipped spins will not all be contiguous in the list.)

We can do so using the *sorted list* algorithm: we make[51] a list of the spins in order of their random fields (Fig. 8.19). Given a field range $(H, H + \Delta)$ in a lattice with z neighbors per site, only those spins with random fields in the range $JS + H < -h_i < JS + (H + \Delta)$ need to be checked, for the $z + 1$ possible fields $JS = (-Jz, -J(z-2), \dots, Jz)$ from the neighbors. We can keep track of the locations in the sorted list of the $z + 1$ possible next spins to flip. The spins can be sorted in time $N \log N$, which is practically indistinguishable from linear in N, and a big improvement over the brute-force algorithm.

---

[50]This exercise is also largely drawn from [103], and was developed with the associated software in collaboration with Christopher Myers.

[51]Make sure you use a packaged routine to sort the list; it is the slowest part of the code. It is straightforward to write your own routine to sort lists of numbers, but not to do it efficiently for large lists.

To create a sorted list algorithm:

(1) Define an array `nextPossible[nUp]`, which points to the location in the sorted list of the next spin that would flip if it had `nUp` neighbors. Initially, all the elements of `nextPossible[nUp]` point to the spin with the largest random field h_i.

(2) From the $z + 1$ spins pointed to by `nextPossible`, choose the one `nUpNext` with the largest internal field in `nUp - nDown` + $h_i = 2$ `nUp` $- z + h_i$. Do not check values of `nUp` for which the pointer has fallen off the end of the list; use a variable `stopNUP`.

(3) Move the pointer `nextPossible[nUpNext]` to the next spin on the sorted list. If you have fallen off the end of the list, decrement `stopNUP`.[52]

(4) If the spin `nUpNext` has exactly the right number of up-neighbors, flip it, increment the external field $H(t)$, and start the next avalanche. Otherwise go back to step (2).

Implement the sorted list algorithm for finding the next avalanche. Notice the pause at the beginning of the simulation; most of the computer time ought to be spent sorting the list. Compare the timing with your brute-force algorithm for a moderate system size, where the brute-force algorithm is slightly painful to run. Run some fairly large systems—$2,000^2$ at $R = (0.7, 0.8, 0.9)$ or 200^3 at $R = (2.0, 2.16, 3.0)$ should run quickly—and explore the avalanche shapes and size distribution.

To do really large simulations of billions of spins without needing gigabytes of memory, there is yet another algorithm we call *bits*, which stores the spins as bits and never generates or stores the random fields (see [103] for implementation details).

(8.15) **NP-completeness and kSAT.**[53] (Computer science, Computation, Mathematics) ④
In this exercise you will numerically investigate a phase transition in an ensemble of problems in mathematical logic, called **kSAT** [12, 136]. In particular, you will examine how the computational difficulty of the problems grows near the critical point. This exercise ties together a number of fundamental issues in critical phenomena, computer science, and mathematical logic.

The **kSAT** problem we study is one in a class of problems called **NP**–complete. In other exercises, we have explored how the speed of algorithms for solving computational problems depends on the size N of the system. (Sorting a list of N elements, for example, can be done using of order $N \log N$ size comparisons between elements.) Computer scientists categorize problems into *complexity classes*. For example, a problem is in **P** if it can guarantee a solution[54] in a time that grows no faster than a polynomial in the size N. Sorting lists is in **P** (the time grows more slowly than N^2, for example, since $N \log N < N^2$ for large N); telling whether an N digit number is prime has recently been shown also to be in **P**. A problem is in **NP**[55] if a proposed solution can be *verified* in polynomial time. For example, factoring an integer with N digits is not known to be in **P** (since there is no known algorithm for finding the factors[56] of an N-digit integer that runs in a time polynomial in N), but it is in **NP**.

(a) *Given two proposed factors of an N digit integer, argue that the number of computer operations needed to verify whether their product is correct is less than a constant times N^2.*

There are many problems in **NP** that have no known polynomial-time solution algorithm. A large family of them, the **NP**–complete problems, have been shown to be maximally difficult, in the sense that they can be used to efficiently solve any other problem in **NP**. Specifically, any

---

[52]Either this spin is flipped (move to the next), or it will start the next avalanche (flip and move to the next), or it has too few spins to flip (move to the next, flip it when it has more neighbors up).

[53]This exercise and the associated software were developed in collaboration with Christopher Myers, with help from Bart Selman and Carla Gomes. Computational hints can be found at the book website [181].

[54]**P** and **NP**–complete are defined for deterministic, single-processor computers. On a quantum computer, for example, there are faster algorithms for solving certain problems like factorization into primes.

[55]**NP** does not stand for "not polynomial", but rather for *nondeterministic polynomial* time. **NP** problems can be solved in polynomial time on a hypothetical *nondeterministic* parallel computer—a machine with an indefinite number of CPUs that can be each run on a separate sub-case.

[56]The difficulty of factoring large numbers is the foundation of some of our *public-key cryptography* methods, used for ensuring that your credit card number on the web is available to the merchant without being available to anyone else listening to the traffic. Factoring large numbers is not known to be **NP**–complete.

problem in **NP** can be translated (using an algorithm that runs in time polynomial in the size of the problem) into any one of the **NP**–complete problems, with only a polynomial expansion in the size N. A polynomial-time algorithm for any one of the **NP**–complete problems would allow one to solve all **NP** problems in polynomial time.

- The *traveling salesman problem* is a classic example. Given N cities and a cost for traveling between each pair and a budget K, find a round-trip path (if it exists) that visits each city with cost $< K$. The best known algorithm for the traveling salesman problem tests a number of paths that grows exponentially with N—faster than any polynomial.

- In statistical mechanics, the problem of finding the ground state of a spin glass[57] in 3D is also **NP**–complete (Section 12.3.4).

- Another **NP**–complete problem is 3-colorability (Exercise 1.8). Can the N nodes of a graph be colored red, green, and blue so that no two nodes joined by an edge have the same color?

One of the key challenges in computer science is determining whether **P** is equal to **NP**—that is, whether all of these problems can be solved in polynomial time. It is generally believed that the answer is negative, that in the worst cases **NP**–complete problems require exponential time to solve.

Proving a new type of problem to be **NP**–complete usually involves translating an existing **NP**–complete problem into the new type (expanding N at most by a polynomial). In Exercise 1.8, we introduced the problem of determining *satisfiability* (**SAT**) of Boolean logical expressions. Briefly, the **SAT** problem is to find an assignment of N logical variables (true or false) that makes a given logical expression true, or to determine that no such assignment is possible. A logical expression is made from the variables using the operations OR ($\vee$), AND ($\wedge$), and NOT ($\neg$). We introduced in Exercise 1.8 a particular subclass of logical expressions called **3SAT**, which demand simultaneous satisfaction of M clauses in N variables each an OR of three literals (where a literal is a variable or its negation). For example, a **3SAT** expression might start out

$$[(\neg X_{27}) \vee X_{13} \vee X_3] \wedge [(\neg X_2) \vee X_{43} \vee (\neg X_2 1)] \dots . \tag{8.30}$$

We showed in that exercise that **3SAT** is **NP**–complete by translating a general 3-colorability problem with N nodes into a **3SAT** problem with $3N$ variables. As it happens, **SAT** was the first problem to be proven to be **NP**–complete; *any* **NP** problem can be mapped onto **SAT** in roughly this way. **3SAT** is also known to be **NP**–complete, but **2SAT** (with clauses of only two literals) is known to be **P**, solvable in polynomial time.

Numerics. Just because a problem is **NP**–complete does not make a typical instance of the problem numerically challenging. The classification is determined by worst-case scenarios, not by the ensemble of typical problems. If the difficult problems are rare, the average time for solution might be acceptable even though some problems in the ensemble will take exponentially long times to run. (Most map coloring problems with a few hundred nodes can be either quickly 3-colored or quickly shown to need four; there exist particular maps, though, which are fiendishly complicated.) Statistical mechanics methods are used to study the average time and distribution of times for solving these hard problems.

In the remainder of this exercise we will implement algorithms to solve examples of **kSAT** problems, and apply them to the ensemble of random **2SAT** and **3SAT** problems with M clauses. We will see that, in the limit of large numbers of variables N, the fraction of satisfiable **kSAT** problems undergoes a *phase transition* as the number M/N of clauses per variable grows. Each new clause reduces the scope for possible solutions. The random **kSAT** problems with few clauses per variable are almost always satisfiable, and it is easy to find a solution; the random **kSAT** problems with many clauses per variable are almost always not satisfiable, and it is easy to find a contradiction. Only near the critical point where the mean number of solutions vanishes as $N \to \infty$ is determining satisfiability typically a challenge.

A logical expression in conjunctive normal form with N variables X_m can conveniently be represented on the computer as a list of sublists of

---

[57]Technically, as in the traveling salesman problem, we should phrase this as a decision problem. Find a state (if it exists) with energy less than E.

nonzero integers in the range $[-N, N]$, with each integer representing a literal ($-m$ representing $\neg X_m$) each sublist representing a disjunction (OR) of its literals, and the list as a whole representing the conjunction (AND) of its sublists. Thus $[[-3, 1, 2], [-2, 3, -1]]$ would be the expression $((\neg X_3) \vee X_1 \vee X_2) \wedge ((\neg X_2) \vee X_3 \vee (\neg X_1))$.
(b) *Do Exercise 1.8, part (b). Generate on the computer the conjunctive normal form for the 3-colorability of the two graphs in Fig. 1.8.* (Hint: There should be $N = 12$ variables, three for each node.)

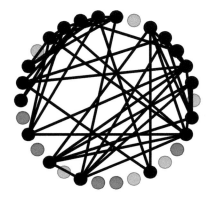

Fig. 8.20 D–P algorithm. A visualization of the Davis–Putnam algorithm during execution. Black circles are unset variables, the other shades are true and false, and bonds denote clauses whose truth is not established.

The DP (Davis–Putnam) algorithm (Fig. 8.20) for determining satisfiability is recursive. Tentatively set a variable to true, reduce the clauses involving the variable, and apply DP to the remainder. If the remainder is satisfiable, return satisfiable. Otherwise set the variable to false, again reduce the clauses involving the variable, and return DP applied to the remainder.
To implement DP, you will want to introduce (i) a data structure that connects a variable to the clauses that contain it, and to the clauses that contain its negation, and (ii) a record of which clauses are already known to be true (because one of its literals has been tentatively set true). You will want a *reduction routine* which tentatively sets one variable, and returns the variables and clauses changed. (If we reach a dead end—a contradiction forcing us to unset the variable—we'll need these changes in order

to back up.) The *recursive solver* which calls the reduction routine should return not only whether the network is satisfiable, and the solution if it exists, but also the number of dead ends that it reached.
(c) *Implement the DP algorithm. Apply it to your 3-colorability expressions from part (b).*
Let us now explore how computationally challenging a typical, random **3SAT** problem is, as the number M/N of clauses per variable grows.
(d) *Write a routine, given k, N, and M, that generates M random* **kSAT** *clauses using N variables. Make sure that no variable shows up twice in the same clause (positive or negative). For $N = 5$, 10, and 20 measure the fraction of* **2SAT** *and* **3SAT** *problems that are satisfiable, as a function of M/N. Does the fraction of unsatisfiable clusters change with M/N? Around where is the transition from mostly satisfiable to mostly unsatisfiable? Make plots of the time (measured as number of dead ends) you found for each run, versus M/N, plotting both mean and standard deviation, and a scatter plot of the individual times. Is the algorithm slowest near the transition?*
The DP algorithm can be sped up significantly with a few refinements. The most important is to remove singletons ("length one" clauses with all but one variable set to values presumed false, hence determining the value of the remaining variable).
(e) *When reducing the clauses involving a tentatively set variable, notice at each stage whether any singletons remain; if so, set them and reduce again. Try your improved algorithm on larger problems. Is it faster?*

Heavy tails and random restarts. The DP algorithm will eventually return either a solution or a judgment of unsatisfiability, but the time it takes to return an answer fluctuates wildly from one run to another. You probably noticed this in your scatter plots of the times—a few were huge, and the others small. You might think that this is mainly because of the rare, difficult cases. Not so. The time fluctuates wildly even with repeated DP runs on the same satisfiability problem [72].
(f) *Run the DP algorithm on a* **2SAT** *problem many times on a single network with $N = 40$ variables and $M = 40$ clauses, randomly shuffling the order in which you select variables to flip. Estimate the power law $\rho(t) \sim t^x$ giv-*

ing the probability of the algorithm finishing after time t. Sort your variables so that the next one chosen (to be tentatively set) is the one most commonly arising (positive or negative) in the clauses. Does that speed up the algorithm? Try also reversing the order, choosing always the least used variable. Does that dramatically slow down your algorithm?

Given that shuffling the order of which spins you start with can make such a dramatic difference in the run time, why persist if you are having trouble? The discovery of the heavy tails motivates adding appropriate *random restarts* to the algorithm [72]; by throwing away the effort spent exploring the neighborhood of one spin choice, one can both improve the average behavior and avoid the heavy tails.

It is known that **2SAT** has a continuous phase transition at $M/N = 1$, and that **3SAT** has an abrupt phase transition (albeit with critical fluctuations) near $M/N = 4.25$. **3SAT** is thought to have severe critical slowing-down near the phase transition, whatever algorithm used to solve it. Away from the phase transition, however, the fiendishly difficult cases that take exponentially long for DP to solve are exponentially rare; DP typically will converge quickly.

(g) *Using your best algorithm, plot the fraction of **2SAT** problems that are **SAT** for values of $N = 25$, 50, and 100. Does the phase transition appear to extrapolate to $M/N = 1$, as the literature suggests? For **3SAT**, try $N = 10$, 20, and 30, and larger systems if your computer is fast. Is your phase transition near $M/N \approx 4.25$? Sitting at the phase transition, plot the mean time (dead ends) versus N in this range. Does it appear that **2SAT** is in **P**? Does **3SAT** seem to take a time which grows exponentially?*

Other algorithms. In the past decade, the methods for finding satisfaction have improved dramatically. **WalkSAT** [166] starts not by trying to set one variable at a time, but starts with a random initial state, and does a zero-temperature Monte Carlo, flipping only those variables which are in unsatisfied clauses. For years, the best known algorithm was **SP**, developed by physicists [71, 134] using techniques developed to study the statistical mechanics of spin-glasses.

(8.16) **Ising hard disks.** ⓟ

In Section 8.1, we noted that the Ising model can be used as a model for alloys, liquids, and gases in addition to magnets. Here we shall see that we can create a one-dimensional lattice model for a hard-disk gas using the Ising model. (See the hard disk gases in Exercises 3.5, 6.13, and 10.11.)

Fig. 8.21 Ising model for 1D hard disks.

Consider the Hamiltonian

$$\mathcal{H} = B \sum_j (s_j + 1)(s_{j+1} + 1) \qquad (8.31)$$

where as usual $s_i = \pm 1$ are Ising spins.

(a) *If we view the up-spins as gas particles, centered at the spin and extending a distance far enough to overlap a neighbor (Fig. 8.21), what is the relation between the energy in eqn 8.31 and the number of disk overlaps?*

(b) *Rewrite $\mathcal{H}$ in eqn 8.31 in the traditional form $\mathcal{H} = -\sum_j J s_j s_{j+1} - H \sum_j s_j + CN$, where C is an overall constant term in the energy, proportional to the number of spins N. What are J and H in terms of B?* (See note 8 on p. 219.)

(8.17) **Ising parallel updates.** (Computation) ⓟ

Our description of the heat-bath and Metropolis algorithm equilibrates one spin at a time. This is inefficient even in Fortran and C++ (memory nonlocality); in interpreted languages like Python or Mathematica conditional loops like this are really slow.

Could we update all the spins at once? If not, can you figure out a way of bringing a large fraction of the spins to local equilibrium with their neighbors in one vector operation (without looping over spins)? No implementation is needed—just a description of the method.

(8.18) **Ising low temperature expansion.** ⓟ

Consider the low-temperature cluster expansion of eqn 8.19. Which zero-temperature ground state (spin-up or spin-down) is it perturbing about? What cluster gives the first term? Explain the power x^6 and the coefficient -2. What cluster(s) contribute to the second term, proportional to x^{10}?

(8.19) 2D Ising cluster expansions. @

In this exercise, we shall derive the first two terms of the low-temperature expansion for 2D square-lattice Ising free energy, corresponding to the 3D magnetization expansion (eqn 8.19). The lattice is finite, having periodic boundary conditions and dimensions L columns by L rows, so that there are $N = L \times L$ sites.

(a) *Let E_g be the energy per spin of the ground state at zero temperature, and δ be the extra energy needed to flip one spin away from the ground state. What is the partition function Z at temperatures low enough that only these two types of states need to be considered?* (Hint: How many states share the energy $NE_g + \delta$?) *For later use, what is the ground state energy E_g per spin and the spin-flip energy δ, as a function of the nearest-neighbor bond strength J and the external field H?*

At low temperatures, it will be a rare event that a particular spin will flip. But for $N \gg \exp(-\delta/k_B T)$, a typical state will have many spins flipped—a *dilute gas* of flipped spins.[58] Can we find the partition function in that limit? Flipping m spins costs m times the energy, but they have many more configurations.

(b) *Argue that the number of configurations with two isolated spins flipped is $N(N - 5)/2 = N^2/2 + O(N)$.* (The two spins flipped cannot be the same spin, and if two neighboring spins flip the energy is not 2δ. Both corrections can be incorporated into higher-order terms in the expansion, analogous to the term $14x^{12}$ in eqn 8.19.) *Argue that the number of configurations with m isolated spins is $N^m/m! + O(N^{m-1})$. Knowing the energy to flip m isolated spins is $m\delta$, write the low-temperature partition function as a sum over m, in the dilute gas approximation where all spins flipping are isolated. Show that $Z = \exp(-NE_g/k_B T + N \exp(-\delta/k_B T))$. What is the free energy f per spin?*

This fact, that summing over an infinite number of clusters for Z is equal to a single cluster calculation for $\log Z$, is an example of the *linked cluster theorem*, used also to simplify Feynman diagram calculations. According to the linked-cluster theorem, Fig. 8.7(b) shows what should likely be considered two clusters in the expansion for Z that happen to be adjacent.

(c) *Using your answer for E_g and δ, find the magnetization per spin $m(T)$ at zero field by evaluating the appropriate derivative of the free energy per spin $f(T, H)$. Compare to the first two terms of eqn 8.19.*

(8.20) Unicycle. ③

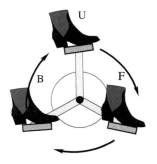

Fig. 8.22 Three–state unicycle.

We model a person pedaling a unicycle as a three-state Markov chain (Fig. 8.22). Let the three states of the Markov chain depict the pedal when the right foot is **U**p, **F**ront, or **B**ack in the cycle, 120° apart. Please use the convention that the vector of probabilities of her being in the three states is $\boldsymbol{\rho} = \begin{pmatrix} \rho_U \\ \rho_F \\ \rho_B \end{pmatrix}$. Assume she pedals steadily, at 120° per time step, as shown by the arrows in Fig. 8.22.

(a) *Write the Markov matrix P for a single time step. Find its eigenvalues and eigenvectors. If she starts cycling in state U, how long will it take to reach a time-independent steady state? Explain how this relates to the eigenvalues of P.*

The Markov matrix for our unicycle in part (a) thus has a *cycle* (Section 8.2).

The Markov matrix is a linear evolution law for the probability distribution. Schrödinger's equation, the diffusion equation, and the wave equation are other examples of linear systems. If a linear evolution law has a symmetry, then its solutions can be chosen from the eigenstates of that symmetry operator (see Section A.4). All three of our other examples all have a translation invariance $x \to x + \Delta$, and thus have solutions of the form $f_k(t) \exp(ikx)$, where $\exp(ikx)$ for various k are the eigenstates of the translation operator.

---

[58]This dilute gas approximation is quite common: for example, it is used in nucleation theory (Section 11.3) to turn the free energy cost of forming a raindrop into a rate per unit volume of raindrop formation, and in quantum mechanics it turns the instanton action into a tunneling rate per unit time.

(b) *What symmetry does our unicycle Markov chain have? Show that your Markov eigenstates are also eigenstates of the unicycle's symmetry.* Our three examples with translation symmetry also have real (non complex) solutions of the form $f(t)\sin(kx)$. *Can your eigenstates be combined into real solutions of this form? What extra symmetry possessed by the three other linear systems guarantees that they have sine and cosine solutions?*

Now our cyclist starts up a steep hill. With probability p she succeeds in pushing uphill at each time step; with probability $1-p$ she stays in the same state.

(c) *Write the transition matrix P^{Hill}. Does it have the same symmetry as P? What are the new eigenvectors and eigenvalues? Will it approach a stationary state $\boldsymbol{\rho}^*$ as time goes to infinity?* (Solving for the motion is not an efficient method for deriving the answer.) *Does P^{Hill} satisfy detailed balance? Why or why not?*

Our probability distribution may be in a stationary state, but our cyclist is not—she is moving uphill, doing work. A Markov chain satisfying detailed balance evolves into a statistical mechanical equilibrium state (Exercise 8.12) that can do no work. One might think that the converse is true: if the steady state has a net flow of probability between two states, one could extract work from the imbalance in the transitions as one might use a paddle wheel in flowing water to make electricity or mill grain. But magnetic fields and such can break detailed balance (Exercise 9.13) without allowing work to be done in equilibrium.

(8.21) **Fruit flies and Markov.**[59] (Biology) ③
Fruit flies exhibit several stereotyped behaviors. A Markov model for the transitions between these states has been used to describe the transitions between behaviors [21, 22] (Fig. 8.23). Let us simplify the fruit fly behavior into three states: idle ($\alpha = 1$), grooming ($\alpha = 2$), and locomotion ($\alpha = 3$). Assume that measurements are done at regular time intervals (say one minute), and let $P_{\beta\alpha}$ be the probability that a fly in state α in time interval n goes into state β in the next interval $n + 1$, so the probability distribution $\rho_\beta(n+1) = \sum_\alpha P_{\beta\alpha}\rho_\alpha(n)$.

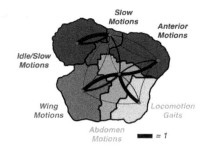

Fig. 8.23 Fruit fly behavior map, from "Hierarchy and predictability in *Drosophilia* behavior" [20]. Fruit fly stereotyped behaviors were analyzed using an unsupervised machine-learning platform and projected into a two-dimensional space to preserve local similarity. The resulting behavior clusters were labeled by hand. Black lines represent transitions between states, with right-handed curvatures indicating direction of transmission, and line thickness proportional to the transition rates. Anterior, abdomen, and wing motions are basically all grooming. We lump together slow motions (consisting of small leg twitches and proboscis extension) with the idle state. The transition matrix here does not faithfully represent the actual behavior. Courtesy of G. J. Berman.

Assume that the transition matrix is

$$P = \begin{pmatrix} 29/32 & 1/8 & 1/32 \\ 1/12 & 55/64 & 1/32 \\ 1/96 & 1/64 & 15/16 \end{pmatrix}. \qquad (8.32)$$

(a) *Show that probability is conserved under the transition matrix P, both directly and by checking the appropriate left eigenvector and eigenvalue.*
(b) *Is the dynamics given by the transition matrix P Markov? Ergodic? Must the dynamics of a fruit fly be Markov? Why or why not? Is real fruit fly dynamics ergodic? If so, why? If not, give a state of the fly (not included in our model) that can never return to the locomotion state.* (Hint: Flies are not immortal.)
(c) *Find the stationary state giving the probability distribution $\rho_\alpha^* = \rho_\alpha(\infty)$ of fruit fly behavior at long times. Is it a left or a right eigenvector? What is its eigenvalue?*
(d) *Does the transition matrix P satisfy detailed balance? First check this directly by demanding that no net probability flows around cycles*

$$P_{\alpha\Leftarrow\beta}P_{\beta\Leftarrow\gamma}P_{\gamma\Leftarrow\alpha} = P_{\alpha\Leftarrow\gamma}P_{\gamma\Leftarrow\beta}P_{\beta\Leftarrow\alpha}. \quad (8.33)$$

---

[59]This problem was created with the assistance of Gordon Berman.

(eqn 8.22 in Exercise 8.5). Then check it using your stationary solution in part (c) by demanding that the stationary state has equal probability flows forward and backward

$$P_{\beta \Leftarrow \alpha} \rho^*_\alpha = P_{\alpha \Leftarrow \beta} \rho^*_\beta, \qquad (8.34)$$

(eqn 8.14).

(e) *Must fruit flies obey detailed balance in their transition rates? If so, why? If not, discuss how the fruit fly might plausibly have a net probability flow around a cycle from idle to grooming to locomotion.*

In the morning, when the lights are turned on, 60% of the fruit flies are idle ($\alpha = 1$), 10% are grooming ($\alpha = 2$), and 30% are undergoing locomotion ($\alpha = 3$).

(f) *What is the net entropy per fly of this initial probability distribution? What is the entropy per fly of the stationary state ρ^*? Does fly entropy increase or decrease from morning to its stationary state? Plot the entropy per fly for the first hour, by iterating our Markov transition matrix directly on the computer. Is the entropy change monotonic?*

Fruit flies are not isolated systems: they can exchange energy with their environment. We would like to check if the fruit fly evolves to minimize its free energy $F = E - TS$.

(g) *What quantities are exchanged between the fruit fly and its environment? Is it feasible to describe the real fruit fly with a free energy, integrating out the rest of the world as a heat bath? Must fruit flies minimize their free energy as time progresses? Why or why not?*

How can we estimate the energy for each fly state?

(h) *Assume that the stationary state of the flies is a Boltzmann distribution $\rho^*_\alpha \propto \exp(-E_\alpha/k_B T)$. Calculate $E_\alpha/k_B T$ for the three fly states.* (Note: Only the differences are determined; choose the zero of energy to make the partition function equal to one.) *What is the average fly energy first thing in the morning? In the stationary state? Does the fly energy increase or decrease from morning to its stationary state? Plot the energy for the first hour; is it monotonic?*

(i) *Let the free energy be $F = E - TS$ for the flies in our simple model. Plot $F/k_B T$ for the first hour. Does it monotonically decrease?*

(8.22) **Metastability and Markov.**[60] ③

This exercise will introduce you to many technologies. You will study thermally activated escape from a barrier (Exercise 6.11) as a Markov process—a generalization of a Markov chain to continuous time and space. You will describe the motion using Langevin dynamics (Exercises 6.18 and 10.7) and the corresponding Fokker–Planck equation. You will provide an alternative estimate of the Arrhenius rate for crossing the barrier—central to chemical reaction rate theory. To do so, you will convert the Fokker–Planck equation into a self-adjoint analogue to a Hamiltonian, and calculate the decay rate and the shape of the slowest-decaying mode—the "probability density in the well" at late times. (See also Exercise 12.22, where we apply renormalization-group methods to calculate the escape rate.)

Fig. 8.24 shows the metastable probability distribution for the position of a particle trapped in a well $V(x)$. If the temperature is low compared to the barrier needed to escape the well, we expect that the probability distribution inside the well will be well equilibrated and approximately given by a Boltzmann factor times an exponential decay

$$\rho^*(x, t) \propto \exp(-\lambda_0 t) \exp(-V(x)/k_B T). \quad (8.35)$$

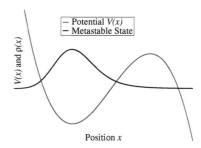

Fig. 8.24 Metastable state in a barrier. A potential energy $V(x)$ with a metastable state with $x < 0$ and an unbound state with $V(x) \to -\infty$ as $x \to \infty$. We use a Markov process to describe thermal motion over the barrier, and derive the metastable probability distribution $\rho^*(x)$ inside the well and the decay rate λ_0.

---

[60]This exercise was developed in collaboration with David Hathcock, and is closely related to his manuscript [80]. Hints for the computational portions are available at the book website [181].

For a noisy overdamped system, the motion of a particle in potential $V(x)$ satisfies the Langevin equation (Exercise 12.22)

$$\dot{x} = -\eta^{-1}\mathrm{d}V/\mathrm{d}x + \sqrt{2k_BT/\eta}\,\xi(t), \quad (8.36)$$

where η is the damping constant, $\sqrt{2k_BT/\eta}$ is the noise amplitude, and $\xi(t)$ is white noise with zero mean $\langle\xi(t)\rangle = 0$ and no time correlations $\langle\xi(t)\xi(t')\rangle = \delta(t - t')$. The probability density satisfied the diffusion equation

$$\frac{\partial\rho}{\partial t} = \frac{\partial}{\partial x}\left(\frac{1}{\eta}\frac{\partial V}{\partial x}\rho\right) + (k_BT/\eta)\frac{\partial^2\rho}{\partial^2 x} = \mathcal{L}[\rho],$$
$$(8.37)$$

an example of a Fokker–Planck equation. This equation should look familiar. It can be written in terms of the current $\partial\rho/\partial t = -\partial J/\partial x$, where the current $J(x) = -\rho/\eta\,\partial V/\partial x - k_BT/\eta\,\partial\rho/\partial x$, includes a contribution from the deterministic force $-\partial V/\partial x$ in addition to the diffusive current we found in Section 2.2. Here $D = k_BT/\eta$ is the Einstein relation again, with η equal to the inverse mobility γ^{-1}.

If we took a finite time step, the noise ξ would become a random step in a random walk, and we could derive the Fokker–Planck equation with the same methods we used to derive the diffusion equation in Section 2.2. Let us instead derive it properly, in continuous time. Given a particular time-dependent noise $\xi(t)$, by definition of $\dot{x}$ in eqn 8.36,

$$x' = x(t + \Delta t) = x + \int_t^{t+\Delta t}\dot{x}\,\mathrm{d}t = x + \Delta x.$$
$$(8.38)$$

Let $\langle X\rangle_\xi$ be the average of a quantity X over noise functions $\xi(t)$ with mean zero and the δ correlations described above.

(a) *Show, to first order in Δt, that $\langle\Delta x\rangle_\xi = -(\Delta t/\eta)\partial V/\partial x$ and that $\langle(\Delta x)^2\rangle_\xi = 2k_BT\Delta t/\eta$.*
(b) *Argue that*

$$\rho(x', t + \Delta t) \quad\quad\quad\quad (8.39)$$
$$= \left\langle\int\rho(x,t)\delta\left(x' - (x + \Delta x)\right)\mathrm{d}x\right\rangle_\xi.$$

Taylor expand the δ function to second order in Δx. Evaluate the expectation values from part (a), integrate by parts to remove the derivatives in the δ function, and evaluate the integral over x.

We will be interested in the metastable state in the well—the near-equilibrium probability distribution formed in the time before the rare escape over the barrier. Suppose $V(x)$ had a true minimum. Would the Fokker–Planck evolution (eqn 8.37) converge to the Boltzmann distribution $\rho^*(x) \propto \exp(-V(x)/k_BT)$?
(c) *Show that the Boltzmann distribution $\rho^*(x)$ is a stationary state under eqn 8.37. Is this distribution normalizable for our potential (Fig. 8.24, with $V(x) \to -\infty$ as $x \to \infty$)?*
In eqn 8.37, we denote $\mathcal{L}$ as the operator acting on $\rho(x)$ that gives $\partial\rho/\partial t$. For a discrete-time Markov chain,[61] all eigenvalues must have $|\Lambda| \leq 1$; similarly, for our continuous time Markov process all eigenvalues must be negative. Thus let us denote the eigenvalues of $\mathcal{L}$ by $-\lambda_n$.

If our potential had a true minimum, then the equilibrium state $\rho^*(x)$ would be an eigenstate of eigenvalue $-\lambda = 0$ of $\mathcal{L}$ (because it does not decay). But ρ^* violates the boundary condition $\rho(x) \to 0$ as $x \to \infty$, so is not a legal eigenfunction. Hence $\mathcal{L}$'s eigenvalues are all negative.

We are particularly interested in the smallest λ_0, corresponding to the slowest decaying mode. If $\rho(x, t = 0)$ is expanded in eigenvectors of $\mathcal{L}$, then the eigenfunction $\rho_0(x)$ will survive longest—the probability density will converge to a multiple of ρ_0 that decays exponentially, and λ_0 is the rate at which particles escape after the initial transient dies away.

Can we find ρ_0 and λ_0? We need to solve the eigenvalue problem $\partial\rho_0/\partial t = \mathcal{L}[\rho_0] = -\lambda_0\rho_0$. Just as most Markov matrices P are not symmetric (with left and right eigenvectors and other unfamiliar behavior), so $\mathcal{L}$ is *non–self-adjoint*.

We shall use the same trick used in eqn 8.15, to "symmetrize" our Markov process into a familiar problem. There we found the eigenvalues of a system satisfying detailed balance by solving for the eigenvalues of a modified linear operator (eqn 8.15) describing the evolution of a probability density $\sigma = \rho/\sqrt{\rho^*}$. (The big advantage is that the new operator is symmetric, making its right and left eigenvectors equal.) Applied to our Fokker–Planck equation 8.37, this yields

$$\frac{\partial\sigma}{\partial t} = \left(\frac{k_BT}{\eta}\frac{\partial^2}{\partial x^2} + \frac{1}{2\eta}V''(x)\right.$$
$$\left. - \frac{1}{4\eta k_BT}\left(V'(x)\right)^2\right)\sigma(x)$$
$$(8.40)$$

---

[61]If a system modeled with a discrete-time Markov chain with step Δt has an eigenvalue Λ, then the same state modeled with a continuous time Markov process will have eigenvalue $-\lambda$, with $\Lambda = \exp(-\lambda\Delta t)$.

(d) *Use eqn 8.37 to derive the evolution equation for $\sigma(x) = \rho/\sqrt{\rho^*}$, eqn 8.40. Up to a factor of* i*, what famous equation (from quantum mechanics) is satisfied by σ? What is the Hamiltonian and effective quantum potential in terms of $V(x)$, k_BT, and η? (Hint: careful with the sign for the last part. Remember that the kinetic energy is proportional to $-\partial^2/\partial x^2$.) Does the effective quantum potential grow as $x \to \infty$ for the cubic potential $V(x)$?*

If we solve for the eigenstates σ_n and eigenvalues λ_n of the Hamiltonian, then $\rho_n = \sqrt{\rho^*}\sigma_n$ are eigenstates of the diffusion equation, which decay at rate λ_n. The barrier crossing time is approximately the inverse of the decay rate of the slowest decaying mode, $\tau \approx \lambda_0^{-1}$.

Now let us specialize to the cubic potential,

$$V(x) = -x^3/3 + x \qquad (8.41)$$

with $k_BT = \frac{1}{2}$ and $\eta = 1$. Notice there is a local minimum "well" at $x_{\min} = -1$ and an energy barrier that peaks at $x_{\max} = 1$. The barrier crossing time can be approximated using the quadratic expansions around these points: $V(x) \approx V_0 + \frac{1}{2}K(x - x_{\min})^2$ and $V(x) \approx V_0 + E - \frac{1}{2}\widetilde{K}(x - x_{\max})^2$.

(e) *For the cubic potential (eqn 8.41), numerically compute the eigenstate of the transformed diffusion equation with smallest eigenvalue. What does the eigenvalue predict for the lifetime? How nearly does it agree with the classic calculation of Kramers, $\lambda_0 \approx \sqrt{K\widetilde{K}}/(2\pi\eta)\exp(-E/k_BT)$ (eqn 12.65)?[62]*

(f) *Using the corresponding eigenstate ρ_0, plot the slowest decaying mode $\rho_0(x) = \sqrt{\rho^*}\sigma_0$, normalized to one, along with the Boltzmann distribution $\rho^*(x)/Z$ and the Boltzmann probability distribution in the approximation that the well is quadratic. Explain how ρ_0 differs from the quadratic approximation, and why this is physically sensible. Explain how ρ_0 differs from ρ^*, and how it resolves an important question about how to determine the metastable probability "in the well".*

(8.23) **Kinetic proofreading in cells.**[63] (Biology) ③
Many biological processes in our body require that we distinguish between two different pathways with very high accuracy. Dur-

ing DNA replication, for example, DNA polymerase (Fig. 8.25), an enzyme that synthesizes the copy of the existing DNA sequence, can sometimes make mistakes. Our body needs an error-correcting mechanism to make sure that incorrect base pairs are not added to the copied DNA sequence and replication happens with high fidelity.

Fig. 8.25 DNA polymerase, guiding replication of DNA. First, helicases break the hydrogen bonds between base pairs and unwind the DNA strand. Then DNA polymerase catalyzes the reaction which adds complementary base pairs to the nucleotide sequence. The process of matching these base pairs is done with extremely high accuracy. The mechanism with which this is achieved is called proofreading. Kinetic proofreading [87] is one explanation for the high accuracy of replication. (Image from https://en.wikipedia.org/wiki/DNA_polymerase reproduced here under the CC BY-SA 3.0 license).

Typically, there is an energy difference between a correct base pair and an incorrect base pair being added to the copy of the DNA strand. However, these energy differences are usually not enough to explain the extremely high accuracy with which DNA replication takes place. Thus, the copying is followed by some activity which adds a further correction. Here, we will explore one particular mechanism for this proposed by Hopfield called *kinetic proofreading* [25,87].
DNA replication is shown in Fig. 8.26. You

---

[62]This has a different prefactor from the rate calculated in Exercise 6.11; here we consider overdamped systems, where there we were considering underdamped barrier escape.

[63]This exercise was developed by Archishman Raju in collaboration with Colin Clement and the author.

imagine the lower end of the DNA (the parent strand) extends up and to the right to a Y-junction, at which another protein (a helicase) holds it apart from its original spousal complementary strand. The upper strand is being assembled one nucleotide at a time, with A matched to T and G matched to C for the four possible DNA nucleotides Adenine, Thymine, Guanine, and Cytosine. This matching is done with high fidelity, with error ratios (incorrect/correct) in yeast and humans quoted between $f = 10^{-7}$ to less than 10^{-9}. The challenge for the enzyme is to attach A to a parent T with that accuracy, and then also to attach G to C with a similar accuracy (Fig. 8.26).

Suppose the enzyme-DNA complex E is bound to a parent strand with the next parent nucleotide T desiring a partner A, and wanting to avoid pairing with G. Clearly the complex must associate the potential partner $X \in \{A, G\}$ with T to find out which matches better. Let this associated bound state be denoted $E'X$, and let the free energies of the three be $\mathcal{E}_E$, $\mathcal{E}_{E'A}$, and $\mathcal{E}_{E'G}$. We may assume that the rate of attachment $E \to E'A$ and $E \to E'G$ are both the same: in a time-step[64] Δt, an unbound complex will bind either A or G with probability α. Because $E'A$ is more tightly bound than $E'G$, the fraction of unbinding μ_A is smaller than μ_G (Fig. 8.27).

Throughout this exercise, we shall take limits to simplify our analysis. In Fig. 8.27 as drawn, we shall assume the final reaction of rate ω completing the nucleotide insertion is small, so we can assume the populations of E, $E'A$, and $E'G$ come to steady state before the complex moves on to other tasks. This makes the error ratio equal to the steady state $f = \rho^*_{E'G}/\rho^*_{E'A}$, where ρ^* is the steady state concentrations of the various components. Similarly, A and G compete in Fig. 8.27 for the enzyme complex E; a complex bound to G is not available to bind to A, and will reduce the overall concentration of $E'A$. If we assume $\alpha \ll \mu_X$ for both reactions, we may assume E is rarely bound to either, and assume ρ^*_E stays close to one. This allows us to analyze the upper and lower branches of Fig. 8.27 separately.[65]

Fig. 8.27 Naive DNA replication reaction diagram. Dark arrows are reactions whose rates are assumed to depend on the nucleotide. Notation from [118].

(a) *Calculate $\rho^*_{E'A}$ and $\rho^*_{E'G}$ in terms of $\mathcal{E}_E$, $\mathcal{E}_{E'A}$, and $\mathcal{E}_{E'G}$, assuming the system comes to equilibrium and given the assumptions discussed. What must be true for most of E to remain unbound? Write a formula f_{ρ^*} for the failure rate in terms of the free energies. Calculate separate formulae for the steady state $\rho^*_{E'A}$ and $\rho^*_{E'G}$ in terms of α, μ_A, and μ_G, again using the approximations discussed above. Write the failure rate f_{rate} also in terms of the reaction rates. Use the two-state detailed balance condition, eqn 8.14, to show that your reaction-rate failure ratio is correct even if ρ_E is not approximately one. (The answers should not be complicated.)*

As a rough estimate, the difference in the binding energy between a correct and incorrect nucleotide in DNA is of the order of 5 kcal/mol. At body temperature, $k_B T \approx 0.616$ kcal/mol.

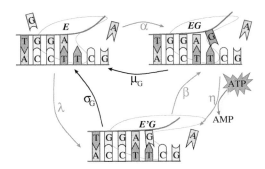

Fig. 8.26 DNA proofreading during replication, here showing a potential copying error, a G rather than A attached to a T. The proofreading involves two states of the enzyme/DNA/nucleotide complex, E'G and EG, each of which can abandon the error and return to the unbound enzyme/DNA complex E.

---

[64]We shall use a discrete time step Δt to model these reaction networks as Markov chains. Taking the time step to zero leads to a continuous time Markov process (see Exercises 8.22 and 12.22).

[65]The failure ratio, eqn 8.44, is unchanged by this decoupling approximation, since the competition reduces all steady-state probabilities by the same factor.

(b) *What is f_{naive} for our naive model at body temperature, assuming $\mathcal{E}_{E'G} - \mathcal{E}_{E'A} = 5$ kcal/mol? How much would the enzyme need to magnify this free energy difference to make the naive error ratio equal to the experimental value 10^{-9}?*

Now let us consider the more elaborate reaction diagram, shown in Fig. 8.28. The basic idea is to add an intermediate state EA between the free enzyme E and the final bound state $E'A$ before nucleotide incorporation. Can this give us a lower error rate, for a given free energy difference $\mathcal{E}_{E'G} - \mathcal{E}_{E'A}$?

(c) *Consider the reaction network shown in Fig. 8.28, including an intermediate state EA. (The reaction network for incorporation of the wrong nucleotide G is similar, with only the labeled rates μ and σ changed.)* Argue, if the system is in equilibrium, that the failure ratio remains f_{ρ^*} in terms of the free energies, as in part (a). Argue, if the reaction rates all satisfy the two-state detailed balance condition (eqn 8.14) with $\rho_X^* = \exp(-\beta\mathcal{E}_X)$, then the error ratio is still f_{rate} from part (a). Use the two-state detailed balance condition between the (slow) reaction rate λ and the two forward reactions σ_A and σ_G to write the error ratio in terms of σ_A and σ_G. Finally, use the cyclic detailed balance condition (eqn 8.22 in Exercise 8.5)*

$$P_{a \Leftarrow c}P_{c \Leftarrow b}P_{b \Leftarrow a} = P_{a \Leftarrow b}P_{b \Leftarrow c}P_{c \Leftarrow a} \quad (8.42)$$

to show that $\sigma_A/\sigma_G = \mu_A/\mu_G$. (Hint: In equilibrium, the probability of being in a state is independent of the dynamics—given only by the energy of the state. This is a profound truth in statistical mechanics.)

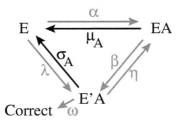

Fig. 8.28 Equilibrium DNA replication reaction diagram. Dark arrows are reactions whose rates are assumed to depend on the binding nucleotide—those that change the association of A or G with the parent nucleotide T. Adding the reaction rates for the alternative pathway (incorrect addition of G) would double the diagram, as in Fig. 8.27.

The key is to make use of a fuel source that drives the reaction out of equilibrium (Fig. 8.29). For example, the reaction $EA \to E'A$ now becomes $EA + \text{ATP} \to E'A + \text{AMP} + PP_i$. Converting ATP to AMP releases a large amount of free energy, both because ATP as a single molecule has a larger free energy, and because the concentration of ATP in a cell is kept much higher than AMP and ADP by the action of mitochondria (which use glucose to convert AMP and ADP to ATP.) In particular, the free energy released in the reaction $EA + \text{ATP} \to E'A + \text{AMP}$ is

$$\delta\Xi = \delta G_0 + kT \log \frac{[\text{ATP}][EA]}{[\text{AMP}][\text{PP}_i][E'A]}. \quad (8.43)$$

Detailed balance between the new reactions with rates β and η now tell us that $\beta/\eta = \exp(-(\mathcal{E}_{E'A} - \mathcal{E}_{EA} - \delta\Xi)/k_BT)$.

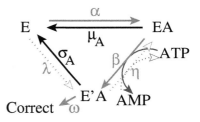

Fig. 8.29 Kinetic proofreading reaction diagram. Here the reactions connecting EA to $E'A$ are driven by the fuel source of the cell, converting ATP (adenosine triphosphate) to AMP (adenosine monophosphate) by removing (hydrolysing) two phosphate groups in the form of PP_i. Dark arrows are reactions whose rates are assumed to depend on the nucleotide.

The typical energy shifts $\delta\Xi$ associated with the removal of phosphate groups is large (20–30 k_BT even for a single phosphate). This inspires another approximation: we may assume $(\mathcal{E}_{E'A} - \mathcal{E}_E) \gg k_BT$ (so the back reaction rate λ is small), and yet we may assume $(\mathcal{E}_{EA} - \mathcal{E}_{E'A} - \delta\Xi) \gg k_BT$ (so the back reaction rate η is small).

(d) *Does the cyclic condition for detailed balance, eqn 8.42, relating the product of clockwise and counter-clockwise reaction rates, hold for our kinetic proofreading reaction diagram? Why not? (See Fig. 8.30.) What changes in the environment when one goes clockwise around the reaction, back to the "same" initial state?*

We shall make another approximation—that the conversion rate β from EA to $E'A$ is slow enough that it does not significantly change the concentration of EA. Just as making α slow compared to μ allowed the correct and incorrect nucleotide reactions to decouple, this approximation allows the reactions $E \Leftrightarrow EA$ to decouple from the reactions $EA \rightarrow E'A \rightarrow E$.

Fig. 8.30 Impossible stair. The addition of the free energy of ATP hydrolysis to one of the reactions is like providing an elevator connecting the ground floor to one two floors up. The effective staircase clockwise goes downhill forever; the system no longer satisfies detailed balance. (Thanks to Escher and the Penrose stair.)

(e) *Use your answers from part (a) to solve for the steady-state behavior of the kinetic proofreading reactions $E \Leftrightarrow EA$ in Fig. 8.29, in these approximations. Solve for the steady state behavior of $\rho^*_{E'G}/\rho^*_{E'A}$ and use your answers from part (c)*

to simplify it. What is the error ratio of the kinetic proofreading reaction diagram?

Kinetic proofreading can *square* the error fraction, at the expense of using many ATPs (one for each error-correcting cycle around the triangle before the nucleotide is accepted at rate ω). Finally, let us use the machinery of Markov chains to solve the problem in greater generality. We shall still ignore the competition between the two nucleotides for E and assume that the fixation rate ω is small.

(f) *Explicitly write the 3×3 Markov transition matrix for Figs. 8.28 and 8.29, evolving the systems by a small time step Δt. (See note 64 on p. 249.)*

By explicitly solving for the steady-state probabilities for this matrix, the error ratio can be calculated and is given by Hopfield's formula [87, Equation 5]

$$f = \frac{(\alpha\beta + \beta\lambda + \lambda\mu_G)(\eta\mu_A + \beta\sigma_A + \mu_A\sigma_A)}{(\alpha\beta + \beta\lambda + \lambda\mu_A)(\eta\mu_G + \beta\sigma_G + \mu_G\sigma_G)}. \tag{8.44}$$

(g) *Look back to part (c) and use the cyclic condition of detailed balance (eqn 8.42) to show that the error rate f is equal to the reaction-rate form of f_{naive} if detailed balance is satisfied.* (Hint: Write an expression for σ_G and σ_A in terms of the other coefficients and substitute them in f. The calculation is a bit complicated.)

(h) *What is the error ratio when we set $\lambda = \eta = 0$? Does it approach your answer from part (e) when β is small?*

Order parameters, broken symmetry, and topology

9

In elementary school, we were taught that there were three states of matter: solid, liquid, and gas. The ancients thought that there were four: earth, water, air, and fire, which was considered sheer superstition. In junior high, the author remembers reading a book as a kid called *The Seven States of Matter* [74]. At least one was "plasma", which made up stars and thus most of the Universe,[1] and which sounded rather like fire.

The original three, by now, have become multitudes. In important and precise ways, magnets are a distinct form of matter. Metals are different from insulators. Superconductors and superfluids are striking new states of matter. The liquid crystal in your wristwatch is one of a huge family of different liquid crystalline states of matter [50] (nematic, cholesteric, blue phase I, II, and blue fog, smectic A, B, C, C*, D, I, ...). There are over 200 qualitatively different types of crystals, not to mention the quasicrystals (Fig. 9.1). There are disordered states of matter like spin glasses, and states like the fractional quantum Hall effect with excitations of charge $e/3$ like quarks. Particle physicists tell us that the vacuum we live within has in the past been in quite different states; in the last vacuum before this one, there were four different kinds of light [46] (mediated by what is now the photon, the W^+, the W^-, and the Z particle).

When there were only three states of matter, we could learn about each one and then turn back to learning long division. Now that there are multitudes, though, we have had to develop a system. Our system is constantly being extended and modified, because we keep finding new phases which do not fit into the old frameworks. It is amazing how the 500th new state of matter somehow screws up a system which worked fine for the first 499. Quasicrystals [116], the fractional quantum hall effect, and spin glasses all really stretched our minds until (1) we understood why they behaved the way they did, and (2) we understood how they fit into the general framework.

In this chapter we are going to tell you the system. It consists of four basic steps [130]. First, you must identify the broken symmetry (Section 9.1). Second, you must define an order parameter (Section 9.2). Third, you are told to examine the elementary excitations (Section 9.3). Fourth, you classify the topological defects (Section 9.4). Most of what we say in this chapter is taken from Mermin [130], Coleman [46], and deGennes and Prost [50], which are heartily recommended.

This chapter is slightly modified from a lecture given at the Santa Fe Institute [168].

[1]They had not heard of dark matter.

Fig. 9.1 Quasicrystals. Crystals are surely the oldest known of the broken-symmetry phases of matter. In the past few decades, we have uncovered an entirely new class of *quasicrystals*, here [62] with icosahedral symmetry. Note the five-fold facets; this rotational symmetry was forbidden in our old categories. (Courtesy I. R. Fisher.)

Statistical Mechanics: Entropy, Order Parameters, and Complexity. James P. Sethna, Oxford University Press (2021). © Oxford University Press. DOI:10.1093/oso/9780198865247.003.0009

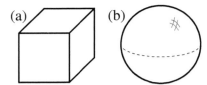

Fig. 9.2 Which is more symmetric? Cube and sphere. (a) The cube has many symmetries. It can be rotated by 90°, 180°, or 270° about any of the three axes passing through the faces. It can be rotated by 120° or 240° about the corners and by 180° about an axis passing from the center through any of the 12 edges. (b) The sphere, though, can be rotated by *any* angle. The sphere respects rotational invariance: all directions are equal. The cube is an object which breaks rotational symmetry: once the cube is there, some directions are more equal than others.

Fig. 9.3 Which is more symmetric? Ice and water. At first glance, water seems to have much less symmetry than ice. (a) The picture of "two-dimensional" ice clearly breaks the rotational invariance; it can be rotated only by 120° or 240°. It also breaks the translational invariance; the crystal can only be shifted by certain special distances (whole number of lattice units). (b) The picture of water has no symmetry at all; the atoms are jumbled together with no long-range pattern at all. However, water as a phase has a complete rotational and translational symmetry; the pictures will look the same if the container is tipped or shoved.

[2]This is not to say that different phases always differ by symmetries! Liquids and gases have the same symmetry, and some fluctuating phases in low dimensions do not break a symmetry. It is safe to say, though, that if the two materials have different symmetries, they are different phases.

9.1 Identify the broken symmetry

What is it that distinguishes the hundreds of different states of matter? Why do we say that water and olive oil are in the same state (the liquid phase), while we say aluminum and (magnetized) iron are in different states? Through long experience, we have discovered that most phases differ in their symmetry.[2]

Consider Figs. 9.2, showing a cube and a sphere. Which is more symmetric? Clearly, the sphere has many more symmetries than the cube. One can rotate the cube by 90° in various directions and not change its appearance, but one can rotate the sphere by any angle and keep it unchanged.

In Fig. 9.3, we see a two-dimensional schematic representation of ice and water. Which state is more symmetric here? Naively, the ice looks much more symmetric; regular arrangements of atoms forming a lattice structure. Ice has a discrete rotational symmetry: one can rotate Fig. 9.3(a) by multiples of 60°. It also has a discrete translational symmetry: it is easy to tell if the picture is shifted sideways, unless one shifts by a whole number of lattice units. The water looks irregular and disorganized. On the other hand, if one rotated Fig. 9.3(b) by an arbitrary angle, it would still look like water! Water is not a snapshot; it is better to think of it as a combination (or ensemble) of all possible snapshots. While the snapshot of the water shown in the figure has no symmetries, water as a phase has complete rotational and translational symmetry.

9.2 Define the order parameter

Particle physics and condensed matter physics have quite different philosophies. Particle physicists are constantly looking for the building blocks. Once pions and protons were discovered to be made of quarks, the focus was on quarks. Now quarks and electrons and photons seem to be made of strings, and strings are hard to study experimentally (so far). Condensed matter physicists, on the other hand, try to understand why messy combinations of zillions of electrons and nuclei do such interesting simple things. To them, the fundamental question is not discovering the underlying quantum mechanical laws, but in understanding and explaining the new laws that emerge when many particles interact.[3]

As one might guess, we do not always keep track of all the electrons and protons. We are always looking for the important variables, the important degrees of freedom. In a crystal, the important variables are the motions of the atoms away from their lattice positions. In a magnet, the important variable is the local direction of the magnetization (an arrow pointing to the "north" end of the local magnet). The local magnetization comes from complicated interactions between the electrons,

[3]The particle physicists also use order parameter fields; their quantum fields also hide lots of details about what their quarks and gluons are composed of. The main difference is that they do not know what their fields are composed of. It ought to be reassuring to them that we do not always find our greater knowledge very helpful.

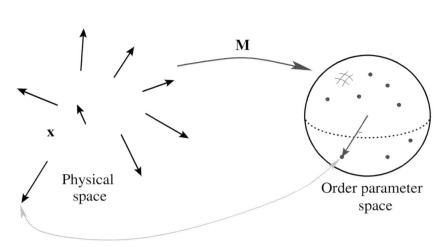

Fig. 9.4 Magnetic order parameter. For a magnetic material at a given temperature, the local magnetization $|\mathbf{M}| = M_0$ will be pretty well fixed, but the energy is often nearly independent of the direction $\widehat{M} = \mathbf{M}/M_0$ of the magnetization. Often, the magnetization changes directions in different parts of the material. (That is why not all pieces of iron are magnetic!) We take the magnetization as the order parameter for a magnet; you can think of it as an arrow pointing to the north end of each atomic magnet. The current state of the material is described by an order parameter field $\mathbf{M}(\mathbf{x})$. It can be viewed either as an arrow at each point in space. or as a function taking points in space $\mathbf{x}$ into points on the sphere. This sphere $\mathbb{S}^2$ is the order parameter space for the magnet.

and is partly due to the little magnets attached to each electron and partly due to the way the electrons dance around in the material; these details are for many purposes unimportant.

The important variables are combined into an *order parameter field*. In Fig. 9.4, we see the order parameter field for a magnet.[4] At each position $\mathbf{x} = (x, y, z)$ we have a direction for the local magnetization $\mathbf{M}(\mathbf{x})$. The length of $\mathbf{M}$ is pretty much fixed by the material, but the direction of the magnetization is undetermined. By becoming a magnet, this material has broken the rotational symmetry. The order parameter $\mathbf{M}$ labels which of the various broken symmetry directions the material has chosen.

The order parameter is a field; at each point in our magnet, $\mathbf{M}(\mathbf{x})$ tells the local direction of the field near $\mathbf{x}$. Why would the magnetization point in different directions in different parts of the magnet? Usually, the material has lowest energy when the order parameter field is uniform, when the symmetry is broken in the same way throughout space. In practice, though, the material often does not break symmetry uniformly. Most pieces of iron do not appear magnetic, simply because the local magnetization points in different directions at different places. The magnetization is already there at the atomic level; to make a magnet, you pound the different domains until they line up. We will see in this chapter that much of the interesting behavior we can study involves the way the order parameter varies in space.

The order parameter field $\mathbf{M}(\mathbf{x})$ can be usefully visualized in two different ways. On the one hand, one can think of a little vector attached to each point in space. On the other hand, we can think of it as a mapping from real space into order parameter space. That is, $\mathbf{M}$ is a function which takes different points in the magnet onto the surface of a sphere (Fig. 9.4). As we mentioned earlier, mathematicians call the

[4]Most magnets are crystals, which already have broken the rotational symmetry. For some *Heisenberg* magnets, the effects of the crystal on the magnetism is small. Magnets are really distinguished by the fact that they break time-reversal symmetry: if you reverse the arrow of time, the magnetization changes sign.

Fig. 9.5 Nematic order parameter space. (a) Nematics are made up of long, thin molecules that prefer to align with one another. (Liquid crystal watches are made of nematics.) Since they do not care much which end is up, their order parameter is not a vector $\hat{n}$ along the axis of the molecules, but is instead a unit vector up to the equivalence $\hat{n} \equiv -\hat{n}$. (b) The nematic order parameter space is a half-sphere, with antipodal points on the equator identified. Thus, for example, the route shown over the top of the hemisphere is a closed path; the two intersections with the equator correspond to the same orientations of the nematic molecules in space.

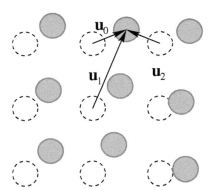

Fig. 9.6 Two-dimensional crystal. A crystal consists of atoms arranged in regular, repeating rows and columns. At high temperatures, or when the crystal is deformed or defective, the atoms will be displaced from their lattice positions. The displacement $\mathbf{u}$ is shown for one of the atoms. Even better, one can think of $\mathbf{u}(\mathbf{x})$ as the local translation needed to bring the ideal lattice into registry with atoms in the local neighborhood of $\mathbf{x}$.

Also shown is the ambiguity in the definition of u. Which ideal atom should we identify with a given real one? This ambiguity makes the order parameter u equivalent to $u + m a \hat{x} + n a \hat{y}$. Instead of a vector in two dimensions, the order parameter space is a square with periodic boundary conditions.

sphere $\mathbb{S}^2$, because locally it has two dimensions. (They do not care what dimension the sphere is embedded in.)

Choosing an order parameter is an art. Usually we are studying a new system which we do not understand yet, and guessing the order parameter is a piece of figuring out what is going on. Also, there is often more than one sensible choice. In magnets, for example, one can treat $\mathbf{M}$ as a fixed-length vector in $\mathbb{S}^2$, labeling the different broken symmetry states. This *topological order parameter* is the best choice at low temperatures, where we study the elementary excitations and topological defects. For studying the transition from low to high temperatures, when the magnetization goes to zero, it is better to consider $\mathbf{M}$ as a "soft-spin" vector of varying length (a vector in $\mathbb{R}^3$, Exercise 9.5). Finding the simplest description for your needs is often the key to the problem.

Before varying our order parameter in space, let us develop a few more examples. The liquid crystals in LCD displays (like those in old digital watches) are nematics. Nematics are made of long, thin molecules which tend to line up so that their long axes are parallel. Nematic liquid crystals, like magnets, break the rotational symmetry. Unlike magnets, though, the main interaction is not to line up the north poles, but to line up the axes. (Think of the molecules as American footballs, the same up and down.) Thus the order parameter is not a vector $\mathbf{M}$ but a headless vector $\vec{n} \equiv -\vec{n}$. The order parameter space is a hemisphere, with opposing points along the equator identified (Fig. 9.5(b), Exercise 9.10). This space is called $\mathbb{RP}^2$ by the mathematicians (the projective plane), for obscure reasons.

For a crystal, the important degrees of freedom are associated with the broken translational order. Consider a two-dimensional crystal which has lowest energy when in a square lattice, but which is deformed away from that configuration (Fig. 9.6). This deformation is described by an arrow connecting the undeformed ideal lattice points with the actual positions of the atoms. If we are a bit more careful, we say that $\mathbf{u}(\mathbf{x})$

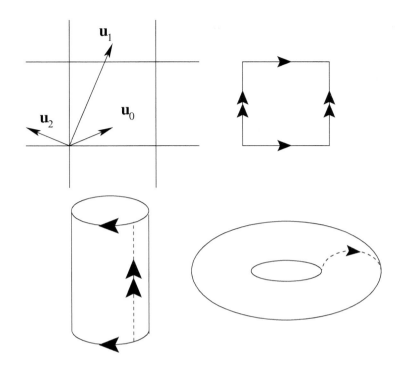

Fig. 9.7 Crystal order parameter space. Here we see that gluing the opposite edges of a square together (giving it periodic boundary conditions) yields a torus. (A torus is a surface of a doughnut, inner tube, or bagel, depending on your background.)

is that displacement needed to align the ideal lattice in the local region onto the real one. By saying it this way, $\mathbf{u}$ is also defined between the lattice positions; there still is a best displacement which locally lines up the two lattices.

The order parameter $\mathbf{u}$ is not really a vector; there is a subtlety. In general, which ideal atom you associate with a given real one is ambiguous. As shown in Fig. 9.6, the displacement vector $\mathbf{u}$ changes by a multiple of the lattice constant a when we choose a different reference atom:

$$\mathbf{u} \equiv \mathbf{u} + a\widehat{x} = \mathbf{u} + ma\widehat{x} + na\widehat{y}. \tag{9.1}$$

The set of distinct order parameters forms a square with periodic boundary conditions. As Fig. 9.7 shows, a square with periodic boundary conditions has the same topology as a torus, $\mathbb{T}^2$.

Finally, let us mention that guessing the order parameter (or the broken symmetry) is not always so straightforward. For example, it took many years before anyone figured out that the order parameter for superconductors and superfluid helium 4 is a complex number ψ.[5] The magnitude of the complex number ψ is a fixed function of temperature, so the topological order parameter space is the set of complex numbers of magnitude $|\psi|$. Thus the order parameter space for superconductors and superfluids is a circle $\mathbb{S}^1$.

Now we examine small deformations away from a uniform order parameter field.

[5]The order parameter field $\psi(\mathbf{x})$ represents the *condensate wavefunction*, which (extremely loosely) is a single quantum state occupied by a large fraction of the Cooper pairs or helium atoms in the material. The corresponding broken symmetry is closely related to the number of particles. In "symmetric", normal liquid helium, the local number of atoms is conserved; in superfluid helium, the local number of atoms becomes indeterminate! (This is because many of the atoms are condensed into that delocalized wavefunction; see Exercise 9.8.)

9.3 Examine the elementary excitations

$u(x)$

Fig. 9.8 One-dimensional sound wave. The order parameter field for a one-dimensional crystal is the local displacement $u(x)$. Long-wavelength waves in $u(x)$ have low frequencies, and cause sound.

Crystals are rigid because of the broken translational symmetry. (Glasses are quite rigid, but fundamentally we do not understand why [182].) Because they are rigid, they fight displacements. Because there is an underlying continuous translational symmetry, a uniform displacement costs no energy. A nearly uniform displacement, thus, will cost little energy, and therefore will have a low frequency. These low-frequency elementary excitations are the sound waves in crystals.

[6]We argue here that low-frequency excitations come from spontaneously broken symmetries. They can also come from conserved quantities; since air cannot be created or destroyed, a long-wavelength density wave cannot relax quickly.

[7]At finite temperatures, we mean a free energy cost.

[8]See Exercises 9.5 and 9.6.

[9]Terms with high derivatives become small when you look on long length and time scales; the nth derivative $\partial^n u/\partial x^n \sim 1/D^n$ for a function with variations on a length D. (Test this; take the 400th derivative of $u(x) = \cos(2\pi x/D)$.) Powers of gradients $(\partial u/\partial x)^n \sim 1/D^n$ are also small.

It is amazing how slow human beings are. The atoms inside your eyelash collide with one another a million million times during each time you blink your eye. It is not surprising, then, that we spend most of our time in condensed matter physics studying those things in materials that happen slowly. Typically only vast conspiracies of immense numbers of atoms can produce the slow behavior that humans can perceive.

A good example is given by sound waves. We will not talk about sound waves in air; air does not have any broken symmetries, so it does not belong in this chapter.[6] Consider instead sound in the one-dimensional crystal shown in Fig. 9.8. We describe the material with an order parameter field $u(x)$, where here x is the position of the reference atom in the ideal crystal and $x + u(x)$ is the actual position of the atom in the material.

Now, there must be an energy cost[7] for deforming the ideal crystal. There will not be any cost, though, for a uniform translation; $u(x) \equiv u_0$ has the same energy as the ideal crystal. (Shoving all the atoms to the right does not cost any energy.) So, the energy will depend only on derivatives of the function $u(x)$. The simplest energy that one can write looks like[8]

$$\mathcal{E} = \int \mathrm{d}x \, \frac{1}{2}\kappa \left(\frac{\partial u}{\partial x}\right)^2. \qquad (9.2)$$

Higher derivatives will not be important for the low frequencies that humans can hear.[9] Now, you may remember Newton's law $F = ma$. The force here is given by the derivative of the energy $F = -\delta\mathcal{E}/\delta u$. The mass is represented by the density of the material ρ. Working out the math (a variational derivative and an integration by parts, for those who are interested) gives us the *wave equation*

$$\rho \, \partial^2 u/\partial t^2 = \kappa \, \partial^2 u/\partial x^2. \qquad (9.3)$$

The solutions to this equation

$$u(x,t) = u_0 \cos(kx - \omega_k t) \qquad (9.4)$$

represent phonons or sound waves. The wavelength of the sound waves is $\lambda = 2\pi/k$, and the frequency is ω in radians per second. Substituting eqn 9.4 into eqn 9.3 gives us the relation

$$\omega = \sqrt{\kappa/\rho}\, k. \qquad (9.5)$$

The frequency gets small only when the wavelength gets large. This is the vast conspiracy: only huge sloshings of many atoms can happen slowly. *Why does the frequency get small?* Well, there is no cost to a uniform translation, which is what eqn 9.4 looks like for infinite wavelength. *Why is there no energy cost for a uniform displacement?* Well, there is a translational symmetry: moving all the atoms the same amount does not change their interactions. *But have we not broken that symmetry?* That is precisely the point.

Long after phonons were understood, Jeffrey Goldstone started to think about broken symmetries and order parameters in the abstract. He found a rather general argument that, whenever a continuous symmetry (rotations, translations, $SU(3)$, ...) is broken, long-wavelength modulations in the symmetry direction should have low frequencies (see Exercise 9.14). The fact that the lowest energy state has a broken symmetry means that the system is stiff; modulating the order parameter will cost an energy rather like that in eqn 9.2. In crystals, the broken translational order introduces a rigidity to shear deformations, and low-frequency phonons (Fig. 9.8). In magnets, the broken rotational symmetry leads to a magnetic stiffness and spin waves (Fig. 9.9). In nematic liquid crystals, the broken rotational symmetry introduces an orientational elastic stiffness (they pour, but resist bending!) and rotational waves (Fig. 9.10).

In superfluids, an exotic broken *gauge symmetry*[10] leads to a stiffness which results in the superfluidity. Superfluidity and superconductivity really are not any more amazing than the rigidity of solids. Is it not amazing that chairs are rigid? Push on a few atoms on one side and, 10^9 atoms away, atoms will move in lock-step. In the same way, decreasing the flow in a superfluid must involve a cooperative change in a macroscopic number of atoms, and thus never happens spontaneously any more than two parts of the chair ever drift apart.

The low-frequency Goldstone modes in superfluids are heat waves! (Do not be jealous; liquid helium has rather cold heat waves.) This is often called second sound, but is really a periodic temperature modulation, passing through the material like sound does through a crystal.

Just to round things out, what about superconductors? They have also got a broken gauge symmetry, and have a stiffness that leads to superconducting currents. What is the low-energy excitation? It does not have one. But what about Goldstone's theorem?

Goldstone of course had conditions on his theorem which excluded superconductors. (Actually, Goldstone was studying superconductors when he came up with his theorem.) It is just that everybody forgot the extra conditions, and just remembered that you always got a low-frequency mode when you broke a continuous symmetry. We condensed matter physicists already knew why there is no Goldstone mode for superconductors; P. W. Anderson had shown that it was related to the long-range Coulomb interaction, and its absence is related to the Meissner effect. We call the loophole in Goldstone's theorem the Anderson–Higgs mechanism.[11]

We end this section by bringing up another exception to Goldstone's theorem; one we have known about even longer, but which we do not have a nice explanation for. What about the orientational order in crystals? Crystals break both the continuous translational order and the continuous orientational order. The phonons are the Goldstone modes

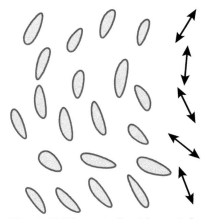

Fig. 9.9 Magnets: spin waves. Magnets break the rotational invariance of space. Because they resist twisting the magnetization locally, but do not resist a uniform twist, they have low-energy spin wave excitations.

[10]See Exercises 9.8 and 9.20.

Fig. 9.10 Nematic liquid crystals: rotational waves. Nematic liquid crystals also have low-frequency rotational waves.

[11]In condensed matter language, the Goldstone mode produces a density wave in the electron gas whose electric fields are independent of wavelength. This gives it a finite frequency (the plasma frequency) even at long wavelength. In high-energy language, the photon eats the Goldstone boson, and gains a mass.

for the translations, but *there are no orientational Goldstone modes in crystals.*[12] Rotational waves analogous to those in liquid crystals (Fig. 9.10) are basically not allowed in crystals; at long distances they tear up the lattice. We understand this microscopically in a clunky way, but do not have an elegant, macroscopic explanation for this basic fact about solids.

9.4 Classify the topological defects

When the author was in graduate school, the big fashion was topological defects. Everybody was studying homotopy groups, and finding exotic systems to write papers about. It was, in the end, a reasonable thing to do.[13] It is true that in a typical application you will be able to figure out what the defects are without homotopy theory. You will spend forever drawing pictures to convince anyone else, though. Most importantly, homotopy theory helps you to think about defects.

A defect is a tear in the order parameter field. A topological defect is a tear that cannot be patched. Consider the piece of two-dimensional crystal shown in Fig. 9.11. Starting in the middle of the region shown, there is an extra row of atoms. (This is called a dislocation.) Away from the middle, the crystal locally looks fine; it is a little distorted, but there is no problem seeing the square grid and defining an order parameter. Can we rearrange the atoms in a small region around the start of the extra row, and patch the defect?

No. The problem is that we can tell there is an extra row without ever coming near to the center. The traditional way of doing this is to traverse a large loop surrounding the defect, and count the net number of rows crossed by the loop. For the loop shown in Fig. 9.12, there are two rows going up and three going down; no matter how far we stay from the center, there will always be an extra row on the right.

How can we generalize this basic idea to other systems with broken symmetries? Remember that the order parameter space for the two-dimensional square crystal is a torus (see Fig. 9.7), and that the order parameter at a point is that translation which aligns a perfect square grid to the deformed grid at that point. Now, what is the order parameter far to the left of the defect (a), compared to the value far to the right (d)? The lattice to the right is shifted vertically by half a lattice constant; the order parameter has been shifted halfway around the torus. As shown in Fig. 9.12, as you progress along the top half of a clockwise loop the order

[13]The next fashion, catastrophe theory, never became particularly important.

Fig. 9.11 Dislocation in a crystal. Here is a topological defect in a crystal. We can see that one of the rows of atoms on the right disappears halfway through our sample. The place where it disappears is a defect, because it does not locally look like a piece of the perfect crystal. It is a topological defect because it cannot be fixed by any local rearrangement. No reshuffling of atoms in the middle of the sample can change the fact that five rows enter from the right, and only four leave from the left! The Burger's vector of a dislocation is the net number of extra rows and columns, combined into a vector (columns, rows).

[12]In two dimensions, crystals provide another loophole in a well-known result, known as the Mermin–Wagner theorem. Hohenberg, Mermin, and Wagner [85,131] proved in the 1960s that two-dimensional systems with a continuous symmetry cannot have a broken symmetry at finite temperature. At least, that is the English phrase everyone quotes when they discuss the theorem; they actually prove it for several particular systems, including superfluids, superconductors, magnets, and translational order in crystals. Indeed, crystals in two dimensions do not break the translational symmetry; at finite temperatures, the atoms wiggle enough so that the atoms do not sit in lock-step over infinite distances (their translational correlations decay slowly with distance). But the crystals do have a broken orientational symmetry: the crystal axes point in the same directions throughout space (as discussed by Mermin [130]). The residual translational correlations (the local alignment into rows and columns of atoms) introduce long-range forces which force the crystalline axes to align, breaking the continuous rotational symmetry.

Fig. 9.12 Loop around the dislocation mapped onto order parameter space. Consider a closed loop around the defect. The order parameter field u changes as we move around the loop. The positions of the atoms around the loop with respect to their local "ideal" lattice drift upward continuously as we traverse the loop. This precisely corresponds to a path around the order parameter space; the path passes once around the hole of the torus. A path *through* the hole corresponds to an extra column of atoms.

Moving the atoms slightly will deform the path, but will not change the number of times the path winds through or around the hole. Two paths which traverse the torus the same number of times through and around are equivalent.

parameter (position of the atom within the unit cell) moves upward, and along the bottom half again moves upward. All in all, the order parameter circles once around the torus. The winding number around the torus is the net number of times the torus is circumnavigated when the defect is orbited once.

Why do we call dislocations topological defects? *Topology* is the study of curves and surfaces where bending and twisting is ignored. An order parameter field, no matter how contorted, which does not wind around the torus can always be smoothly bent and twisted back into a uniform state. If along any loop, though, the order parameter winds either around the hole or through it a net number of times, then enclosed in that loop is a defect which cannot be bent or twisted flat; the winding number (an integer) cannot change in a smooth and continuous fashion.

How do we categorize the defects for two-dimensional square crystals? Well, there are two integers: the number of times we go around the central hole, and the number of times we pass through it. In the traditional description, this corresponds precisely to the number of extra rows and columns of atoms we pass by. This was named the Burger's vector in the old days, and nobody needed to learn about tori to understand it. We now call it the first homotopy group of the torus:

$$\Pi_1(\mathbb{T}^2) = \mathbb{Z} \times \mathbb{Z}, \tag{9.6}$$

where $\mathbb{Z}$ represents the integers. That is, a defect is labeled by two integers (m, n), where m represents the number of extra rows of atoms

Fig. 9.13 Hedgehog defect. Magnets have no line defects (you cannot lasso a basketball), but do have point defects. Here is shown the hedgehog defect, $\mathbf{M}(\mathbf{x}) = M_0 \hat{x}$. You cannot surround a point defect in three dimensions with a loop, but you can enclose it in a sphere. The order parameter space, remember, is also a sphere. The order parameter field takes the enclosing sphere and maps it onto the order parameter space, wrapping it exactly once. The point defects in magnets are categorized by this *wrapping number*; the second homotopy group of the sphere is $\mathbb{Z}$, the integers.

Some paper clips also work harden, but less dramatically (partly because they have already been bent). After several bendings, they will stop hardening and start to weaken dramatically; that is because they are beginning to break in two.

[15]This again is the mysterious lack of rotational Goldstone modes in crystals, (note 12 on p. 260), which would otherwise mediate the bend. See also Exercise 11.5.

Fig. 9.14 Defect line in a nematic liquid crystal. You cannot lasso the sphere, but you can lasso a hemisphere! Here is the defect corresponding to the path shown in Fig. 9.5(b). As you pass clockwise around the defect line, the order parameter rotates counter-clockwise by 180°. This path on Fig. 9.5(b) would actually have wrapped around the right-hand side of the hemisphere. Wrapping around the left-hand side would have produced a defect which rotated clockwise by 180°. The path in Fig. 9.5(b) is halfway in between, and illustrates that these two defects really are not different topologically (Exercise 9.1).

[16]The zeroth homotopy group classifies domain walls. The third homotopy group, applied to defects in three-dimensional materials, classifies what the condensed matter people call *textures* and the particle people sometimes call *skyrmions*. The fourth homotopy group, applied to defects in space-time path integrals, classifies types of *instantons*.

on the right-hand part of the loop, and n represents the number of extra columns of atoms on the bottom.

This is where we show the practical importance of topological defects. Unfortunately for you, we cannot enclose a soft copper tube for you to play with, the way the author does in lectures.[14] They are a few cents each, and machinists on two continents have been quite happy to cut them up for demonstrations, but they do not pack well into books. Anyhow, copper and most metals exhibit what is called *work hardening*. It is easy to bend the tube, but it is amazingly tough to bend it back. The soft original copper is relatively defect free. To bend, the crystal has to create lots of line dislocations, which move around to produce the bending[15] The line defects get tangled up, and get in the way of any new defects. So, when you try to bend the tube back, the metal becomes much stiffer. Work hardening has had a noticeable impact on the popular culture. The magician effortlessly bends the metal bar, and the strongman cannot straighten it ... Superman bends the rod into a pair of handcuffs for the criminals ...

Before we explain why these paths form a group, let us give some more examples of topological defects and how they can be classified. Figure 9.13 shows a "hedgehog" defect for a magnet. The magnetization simply points straight out from the center in all directions. How can we tell that there is a defect, always staying far away? Since this is a point defect in three dimensions, we have to surround it with a sphere. As we move around on this sphere in ordinary space, the order parameter moves around the order parameter space (which also happens to be a sphere, of radius $|\mathbf{M}|$). In fact, the order parameter space is covered exactly once as we surround the defect. This is called the *wrapping number*, and does not change as we wiggle the magnetization in smooth ways. The point defects of magnets are classified by the wrapping number:

$$\Pi_2(\mathbb{S}^2) = \mathbb{Z}. \tag{9.7}$$

Here, the subscript 2 says that we are studying the second homotopy group. It represents the fact that we are surrounding the defect with a two-dimensional spherical surface, rather than the one-dimensional curve we used in the crystal.[16]

You might get the impression that a defect with topological strength seven is really just seven strength 1 defects, stuffed together. You would be quite right; occasionally defects do bunch up, but usually big ones decompose into small ones. This does not mean, though, that adding two defects always gives a bigger one. In nematic liquid crystals, two line defects are as good as none! Magnets do not have any line defects; a loop in real space never surrounds something it cannot smooth out. Formally, we show this by noting that the first homotopy group of the sphere is zero; any closed path on a basketball can be contracted to a point (Exercise 9.16). For a nematic liquid crystal, though, the order parameter space was a hemisphere. There is a path on the hemisphere in Fig. 9.5(b) that you cannot get rid of by twisting and stretching. It does not look like a closed path, but you have to remember that the

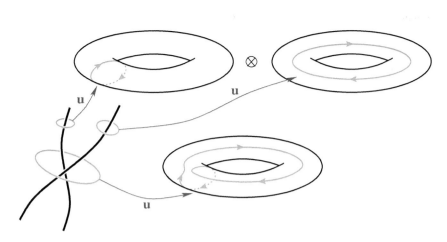

two opposing points on the equator really represent the same nematic orientation. The corresponding defect has a director field n that rotates $180°$ as the defect is orbited; Fig. 9.14 shows one typical configuration (called an $s = -1/2$ defect). Now, if you put two of these defects together, they cancel (Exercise 9.1). Nematic line defects add modulo 2, like "clock arithmetic" with two hours in a day:[17]

$$\Pi_1(\mathbb{RP}^2) = \mathbb{Z}_2. \qquad (9.8)$$

Two parallel defects can coalesce and heal, even though each one individually is stable; each goes halfway around the sphere, and the whole path can be shrunk to zero. Figure 9.16 shows both $s = \pm\frac{1}{2}$ defects and the thick "healed" $s = \pm1$ defects formed when they merge (Exercise 9.16).

Finally, why are these defect categories a group? A group is a set with a multiplication law, not necessarily commutative, and an inverse for each element. For the first homotopy group, the elements of the group are equivalence classes of paths; two paths are equivalent if one can be stretched and twisted onto the other, staying in the order parameter space at all times.[18] For example, any path going through the hole from the top (as in the top right-hand torus in Fig. 9.15) is equivalent to any other one. To multiply a path u and a path v, one must first make sure that they meet at some point (by dragging them together, probably). Then one defines a new path $u \otimes v$ by traversing first the path u and then v.[19]

The inverse of a path u is just the path which runs along the same path in the reverse direction. The identity element consists of the equivalence class of paths which do not enclose a hole; they can all be contracted smoothly to a point (and thus to one another). Finally, the multiplication law has a direct physical implication: encircling two defect lines of strength u and v is completely equivalent to encircling one defect of strength $u \otimes v$ (Fig. 9.15).

This all seems pretty abstract; maybe thinking about order parameter

Fig. 9.15 Multiplying two paths. The product of two paths is given by starting from their intersection, traversing the first path, and then traversing the second. The inverse of a path is the same path traveled backward; compose the two and one can shrink them continuously back to nothing. This definition makes the homotopy classes into a group.

This multiplication law has a physical interpretation. If two defect lines coalesce, their homotopy class must be given by the loop enclosing both. This large loop can be deformed into two little loops, so the homotopy class of the coalesced line defect is the product of the homotopy classes of the individual defects.

Fig. 9.16 Escape into 3D: nematics. Nematic line defects, courtesy Liquid Crystal Institute, Kent State University. The thin lines have winding numbers $s = \pm\frac{1}{2}$, the thick lines wind by $s = \pm1$, and are thick because they *escape into the third dimension* (Exercise 9.16).

[17]In this analogy, we ignore the re-use of hour names in the afternoon.

[18]A path is a continuous mapping from the circle into the order parameter space: $\theta \rightarrow u(\theta)$, $0 \leq \theta < 2\pi$. When we encircle the defect with a loop, we get a path in order parameter space as shown in Fig. 9.4; $\theta \rightarrow \mathbf{x}(\theta)$ is the loop in real space, and $\theta \rightarrow \mathbf{u}(\mathbf{x}(\theta))$ is the path in order parameter space. Two paths are equivalent if there is a continuous one-parameter family of paths connecting one to the other: $u \equiv v$ if there exists $u_t(\theta)$ continuous both in θ and in $0 \leq t \leq 1$, with $u_0 \equiv u$ and $u_1 \equiv v$.

[19]That is, if u and v meet at $\theta = 0 \equiv 2\pi$, we define $u \otimes v(\theta) \equiv u(2\theta)$ for $0 \leq \theta \leq \pi$, and $u \otimes v(\theta) \equiv v(2\theta)$ for $\pi \leq \theta \leq 2\pi$.

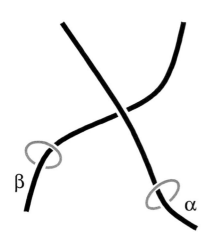

Fig. 9.17 Two defects: can they cross? Can a defect line of class α pass by a line of class β, without getting topologically entangled?

Fig. 9.18 Pulling off a loop around two defects. We see that we can pass by if we leave a trail; is the connecting double line topologically equal to the identity (no defect)? Encircle the double line by a loop. The loop can be wiggled and twisted off the double line, but it still circles around the two legs of the defects α and β.

spaces and paths helps one think more clearly, but are there any real uses for talking about the group structure? Let us conclude this chapter with an amazing, physically interesting consequence of the multiplication laws we described.

Can two defect lines cross one another? Figure 9.17 shows two defect lines, of strength (homotopy type) α and β, which are not parallel. Suppose there is an external force pulling the α defect past the β one. Figure 9.18 shows the two line defects as we bend and stretch one to pass by the other. There is a trail left behind of two parallel defect lines. α can really leave β behind only if it is topologically possible to erase the trail. Can the two lines annihilate one another? Only if their net strength is zero, as measured by the loop in 9.18.

Now, get two wires and some string (Exercise 9.19). Bend the wires into the shape found in Fig. 9.18. Tie the string into a fairly large loop, surrounding the doubled portion. Wiggle the string around, and try to get the string out from around the doubled section. You will find that you cannot completely remove the string (It is against the rules to pull the string past the cut ends of the defect lines!), but that you can slide it downward into the configuration shown in Fig. 9.19.

Now, in this figure we see that each wire is encircled once clockwise and once counter-clockwise. Do they cancel? Not necessarily! If you look carefully, the order of traversal is such that the net homotopy class is $\beta\alpha\beta^{-1}\alpha^{-1}$, which is only the identity if β and α *commute*. Thus the physical entanglement problem for defects is directly connected to the group structure of the paths; commutative defects can pass through one another, noncommutative defects entangle.

It would be tidy if we could tell you that the work hardening in copper is due to topological entanglements of defects. It would not be true. The homotopy group of dislocation lines in fcc copper is commutative. (It is rather like the two-dimensional square lattice; if $\alpha = (m, n)$ and $\beta = (o, p)$ with m, n, o, p the number of extra horizontal and vertical lines of atoms, then $\alpha\beta = (m + o, n + p) = \beta\alpha$.) The reason dislocation lines in copper do not pass through one another is energetic, not topological. The two dislocation lines interact strongly with one another, and energetically get stuck when they try to cross. Remember, at the beginning of the chapter we said that there were gaps in the system; the topological theory can only say when things are impossible to do, not when they are difficult to do.

It would be nice to tell you that this beautiful connection between the commutativity of the group and the entanglement of defect lines is nonetheless important in lots of other contexts. That too would not be true. There are two types of materials known which are supposed to suffer from defect lines which topologically entangle. The first are biaxial nematics, which were thoroughly analyzed theoretically before anyone found one. The other are the metallic glasses, where David Nelson has a theory of defect lines needed to relieve the frustration. These defects do not commute, and the entanglement of defect lines was conjectured to be important for glass formation; this has not been fleshed out into

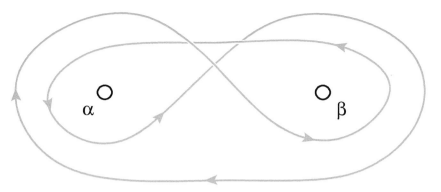

Fig. 9.19 Noncommuting defects.
The homotopy class for the loop is precisely $\beta\alpha\beta^{-1}\alpha^{-1}$, which is the identity (no defect) precisely when $\beta\alpha = \alpha\beta$. Thus two defect lines can pass by one another if their homotopy classes commute!

a practical calculation.

This basic construction, however, is used to study the statistics of *anyons*. Figure 9.18 can represent the 2D space-time paths of two quantum defects (time being vertical). Two fermions will gain a minus sign as one orbits the other; two bosons will give a plus sign; anyons can have other changes in phase, sometimes mixing multiple components. If these changes are noncommutative in a sufficiently complicated way as the paths *braid* around one another, the system can be used as a quantum computer.

Exercises

Ising order parameter checks your understanding of topological order parameters. *Nematic order parameter* explores the weird space $\mathbb{RP}^2$. *Pentagonal order parameter* challenges you to discover the order parameter for a model glass.

Rigidity of crystals emphasizes how rigidity is tied to the broken symmetry. The use of symmetry to develop new laws of matter without microscopic input is explored in *Symmetries and wave equations*, *Chiral wave equation*, and *Landau theory for the Ising model*. *Sound and Goldstone's theorem* and *Superfluid second sound* analyze the slow, undamped elementary excitations at long wavelengths.

Topological defects are entertaining and challenging to visualize and understand. *Can't lasso a basketball* explores the "escape into the third dimension". *Nematic defects* provides a physical visualization of homotopy theory and an integral form for the topological charge. *XY defects* explores the defect composition law. *Fingerprints* explores broken rotation and translational symmetry in a familiar context, *Defects in crystals* simulates both topological and other important defects in crystals, and *Defect entanglement* fleshes out the discussion of how noncommutative defects cannot cross one another.

The energetics and internal structure of a point defect is studied in *Defects and total divergences*, and studied for a surface defect in *Domain walls in magnets*.

Where do the order parameters come from? In simple liquid crystals and magnets, one has an intuitive feel for the broken symmetry state. In *Superfluid order and vortices*, *Number and phase in superfluids*, and the challenging exercise *Superfluids and ODLRO* we derive in increasing levels of rigor the complex order parameter, dynamics, and broken gauge symmetry characteristic of superfluidity.

(9.1) **Nematic defects.** (Mathematics, Condensed matter) @
The winding number S of a defect is $\theta_{\mathrm{net}}/2\pi$, where θ_{net} is the net angle of rotation that the order parameter makes as you circle the defect.

The winding number is positive if the order parameter rotates in the same direction as the traversal (Fig. 9.20(a)), and negative if it rotates in the opposite directions (Fig. 9.20(b)).

Fig. 9.20 Defects in nematic liquid crystals. (a) $S = \frac{1}{2}$ disclination line. (b) $S = -\frac{1}{2}$ disclination. The dots are not physical, but are a guide to help you trace the orientations (starting on the left and moving clockwise); nematic liquid molecules often have a head and tail, but there is no long-range correlation in which direction the heads lie.

As you can deduce topologically (Fig. 9.5(b) on p. 256), the winding number is *not* a topological invariant in general. It is for superfluids $\mathbb{S}^1$ and crystals $\mathbb{T}^D$, but not for Heisenberg magnets (Exercise 9.16) or nematic liquid crystals (Fig. 9.16). If we treat the plane of the figure as the equator of the hemisphere, you can see that the $S = 1/2$ defect rotates around the sphere around the left half of the equator, and the $S = -1/2$ defect rotates around the right half of the equator. These two paths can be smoothly deformed into one another; the path shown on the order parameter space figure (Fig. 9.5(b)) is about halfway between the two.

(a) *Which figure (A–E) represents the defect configuration in real space halfway between $S = 1/2$ and $S = -1/2$, corresponding to the intermediate path shown in Fig. 9.5(b) on p. 256?* (The changing shapes denote rotations into the third dimension.)

(b) *Obtain full-page versions of Fig. 9.20 and Fig. 9.5(b) from the book website [181]. Place the $S = +\frac{1}{2}$ path onto a flat surface, and hold the order parameter hemisphere beside it. Draw the oriented path (with arrows) traversed on the order parameter space as one follows counterclockwise around the $S = +\frac{1}{2}$ defect.* (Your answer will depend on how you orient the hemisphere.) *Draw the oriented path for the $S = -\frac{1}{2}$ defect. Finally, draw the intermediate path from your chosen answer above. Describe how one would smoothly and continuously change from the $+\frac{1}{2}$ to the $-\frac{1}{2}$ path through your intermediate path.*

(9.2) **XY defects.** (Mathematics, Condensed matter) @

Let the order parameter field $\mathbf{m}(x, y)$ of a two-dimensional XY model be a unit vector field in the plane (order parameter space $\mathbb{S}^1$). The topological defects are characterized by their *winding number s*, since $\Pi_1(\mathbb{S}^1) = \mathbb{Z}$. The winding number is the number of counter-clockwise orbits around the order parameter space for a single counter-clockwise path around the defect.

(a) *What are the winding numbers of the two defects surrounded by paths A and B in Fig. 9.21? What should the winding number be around path C, according to the group multiplication law for $\Pi_1(\mathbb{S}^1)$?* (Hint: "+" is the multiplication operation for the group of integers. Warning: Many will intuitively guess the wrong answer.)

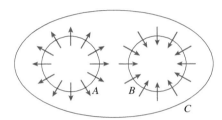

Fig. 9.21 XY defect pair. Two topological defects, circumscribed by loops A and B running counter-clockwise; the pair of defects is circumscribed by a path C also going counter-clockwise. The path in order parameter space mapped to from C, as a homotopy group element, is the group product of the paths from A and B.

(b) *Copy the figure onto a separate sheet of paper, and fill in the region around A and B past C with a smooth, nonsingular, nonvanishing order parameter field of unit vectors.* (Hint: You can use your answer for (b) to check your answer for (a).)

We can find a formula for the winding number as an integral around the defect in the two-dimensional XY model. Let D encircle a defect counter-clockwise once (Fig. 9.22).

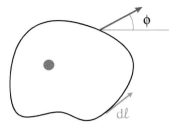

Fig. 9.22 Looping the defect. The curve D encircles the defect once; $d\ell$ is a unit vector tangent to D running counter-clockwise. Define ϕ to be the angle between the unit vector $\mathbf{m}$ and the x-axis.

(c) *Show that the winding number is given by the line integral around the curve D:*

$$s = \frac{1}{2\pi} \oint \sum_{j=1}^{2} (m_1 \partial_j m_2 - m_2 \partial_j m_1) \, d\ell_j, \quad (9.9)$$

where the two coordinates are x_1 and x_2, $\partial_j = \partial / \partial x_j$, and ℓ_j is the tangent vector to the contour being integrated around (so the integral is of the form $\oint \mathbf{v} \cdot d\ell$). (Hints: Write $\mathbf{m} = (\cos(\phi), \sin(\phi))$; the integral of a directional derivative $\nabla f \cdot d\ell$ is the difference in f between the two endpoints.)

There are other useful formulæ of this kind. For example, the wrapping number of a vector order parameter around the sphere $\Pi_2(\mathbb{S}^2)$ is given by an integral of the determinant of the Jacobian of the order parameter field.[20]

(9.3) **Defects and total divergences.** ③

A hypothetical liquid crystal is described by a unit-vector order parameter $\hat{\mathbf{n}}$, representing the orientation of the long axis of the molecules. (Think of it as a nematic liquid crystal where the heads of the molecules all line up as well.[21]) The free energy density is normally written

$$\mathcal{F}_{\text{bulk}}[\hat{\mathbf{n}}] = \frac{K_{11}}{2} (\text{div}\,\hat{\mathbf{n}})^2 + \frac{K_{22}}{2} (\hat{\mathbf{n}} \cdot \text{curl}\,\hat{\mathbf{n}})^2$$
$$+ \frac{K_{33}}{2} (\hat{\mathbf{n}} \times \text{curl}\,\hat{\mathbf{n}})^2. \quad (9.10)$$

Assume a spherical droplet of radius R_0 contains a hedgehog defect (Fig. 9.13, p. 261) in its center, with order parameter field $\hat{\mathbf{n}}(\mathbf{r}) = \hat{\mathbf{r}} = \mathbf{r}/|r| = (x, y, z)/\sqrt{x^2 + y^2 + z^2}$. The hedgehog

[20]Such formulæ are used, for example, in path integrals to change the weights of different topological sectors. Examples include Chern–Simons terms, the θ term in quantum gauge theories, the Wess–Zumino term in the nonlinear sigma model, . . .

[21]The order parameter is the same as the Heisenberg antiferromagnet, but the latter has a symmetry where the order parameter can rotate independently from the spatial rotations, which is not true of liquid crystals.

is a topological defect, which wraps around the sphere once.

(a) *Show that* curl $\hat{\mathbf{n}} = 0$ *for the hedgehog. Calculate the free energy of the hedgehog, by integrating* $\mathcal{F}[\hat{\mathbf{n}}]$ *over the sphere. Compare the free energy to the energy in the same volume with* $\hat{\mathbf{n}}$ *constant (say, in the* $\hat{x}$ *direction).*

There are other terms allowed by symmetry that are usually not considered as part of the free energy density, because they are total divergence terms. Any term in the free energy which is a divergence of a vector field, by Gauss's theorem, is equivalent to the flux of the vector field out of the boundary of the material. For periodic boundary conditions such terms vanish, but our system has a boundary. For large systems these terms scale like the surface area, where the other terms in the free energy can grow proportional to the volume—but they do not always do so.

(b) *Consider the effects of an additional term* $\mathcal{F}_{\mathrm{div}}[\hat{n}] = K_0(\mathrm{div}\,\hat{\mathbf{n}})$, *allowed by symmetry, in the free energy* $\mathcal{F}[\hat{\mathbf{n}}]$. *Calculate its contribution to the energy of the hedgehog, both by integrating it over the volume of the sphere and by using Gauss's theorem to calculate it as a surface integral. Compare the total energy* $\int \mathcal{F}_{\mathrm{bulk}} + \mathcal{F}_{\mathrm{div}}\,\mathrm{d}^3r$ *with that of the uniform state with* $\hat{\mathbf{n}} = \hat{x}$, *and with the anti-hedgehog,* $\hat{\mathbf{n}}(\mathbf{r}) = -\hat{\mathbf{r}}$. *Which is lowest, for large* R_0? *How does the ground state for large* R_0 *depend on the sign of* K_0?

The term K_0 from part (b) is definitely not negligible! Liquid crystals in many cases appear to have *strong pinning* boundary conditions, where the relative angle of the order parameter and the surface is fixed by the chemical treatment of the surface. Some terms like K_0 are not included in the bulk energy because they become too *large*; they rigidly constrain the boundary conditions and become otherwise irrelevant.

(9.4) **Domain walls in magnets.** (Condensed matter) ③

The free energy density of an Ising magnet below T_c can be roughly approximated as a double-well potential (eqn 9.18), with two minima at $\pm m_0$:

$$\mathcal{F} = \tfrac{1}{2}K(\nabla m)^2 + (g/4!)(m^2 - m_0^2)^2. \quad (9.11)$$

This exercise studies the structure of the domain wall separating a region of positive magnetization from one of negative magnetization.

Suppose this soft-spin order parameter $m(x)$ varies only along the x direction, with $m(-\infty) = -m_0$ and $m(\infty) = m_0$. In between, it must pass through a barrier region with $m \approx 0$. The stiffness K penalizes sharp gradients in the magnetization; g penalizes regions with magnetization away from the equilibria at $\pm m_0$.

Let us start by doing a dimensional analysis calculation of the width of the domain wall and its energy. The first term in $\mathcal{F}$ is a stiffness of the order parameter against rapid changes in m. If the wall between $m = -m_0$ and $m = +m_0$ is of thickness Δ, it has roughly energy per unit area $\sim K\Delta(m_0/\Delta)^2$, favoring wide walls. The second term in $\mathcal{F}$ is a double-well potential, with a barrier B separating two wells, with units energy per unit volume. Our interface will have an energy cost $\sim B\Delta$ per unit area due to the barrier, which wants our wall to be as narrow as possible.

(a) *Find* B. *Vary the width of the wall to minimize the sum. Roughly estimate the energy per unit area of the domain wall in terms of* K, m_0, *and* g.

The rest of this exercise will lead you through a variational calculation of the shape of the domain wall (see [128, chapter 12] for information about the calculus of variations).

(b) *Find the equation satisfied by that* $m(x)$ *which minimizes* $F = \int \mathcal{F}\,\mathrm{d}x$, *given the boundary conditions.* (This is the Euler–Lagrange equation from the calculus of variations.)

(c) *Show that the solution* $m(x)$ *has the property that the combination*

$$E = (K/2)(\partial m/\partial x)^2 - (g/4!)(m^2 - m_0^2)^2 \quad (9.12)$$

is independent of x. (Hint: What is $\partial E/\partial x$?)

E is analogous to the energy of a particle in an inverted potential well, with x playing the role of time. The double well becomes a potential with two hills at $\pm m_0$. ("Energy" conservation comes from the symmetry of the system under sideways translations of the wall.) Solving for the minimum $m(x)$ is finding the classical trajectory in the inverted potential; it rolls down one hill and rolls back up the second one.

(d) *Argue from the boundary conditions that* $E = 0$. *Using that, find the minimum free energy path* $m(x)$ *satisfying the boundary conditions* $m(\pm\infty) = \pm m_0$. *Was your wall thickness estimate of part (a) roughly correct?* (Hint: If you know $\mathrm{d}y/\mathrm{d}x = f(y)$, you know $\int \mathrm{d}y/f(y) = \int \mathrm{d}x$.)

(9.5) **Landau theory for the Ising model.** (Condensed matter) ③

This chapter has focused on the topological order parameter, which labels the different ground states of the system when there is a spontaneously broken symmetry. To study the defect cores, interfaces, and high temperatures near phase transitions, one would like an order parameter which can vary in magnitude as well as direction.

In Section 6.7, we explicitly computed a free energy for the ideal gas as a function of the density. Can we use symmetry and gradient expansions to derive free energy densities for more realistic systems—even systems that we do not understand microscopically? Lev Landau used the approach we discuss here to develop theories of magnets, superconductors, and superfluids—before the latter two were understood in microscopic terms.[22] In this exercise, you will develop a Landau theory for the Ising model.[23]

Here we outline the general recipe, and ask you to implement the details for the Ising model. Along the way, we will point out places where the assumptions made in Landau theory can break down—often precisely in the cases where the theory is most useful. There are five steps.

(1) **Pick an order parameter field.**

Remember that the Ising model had a high-temperature paramagnetic phase with zero magnetization per spin m, and a low-temperature ferromagnetic phase with a net magnetization per spin $\pm m(T)$ that went to one at $T = 0$. The Ising model picks one of the two equilibrium states (up or down); we say it spontaneously breaks the up–down symmetry of the Hamiltonian

Hence the natural[24] order parameter is the scalar $m(\mathbf{x})$, the local magnetization averaged over some volume ΔV. This can be done by averaging the magnetization per spin in small boxes, as in Section 6.7.

(a) *What value will $m(\mathbf{x})$ take at temperatures*

high compared to the interaction J in the Ising model? What values will it take at temperatures very low compared to J?

(2) **Write a general local[25] free energy density, for long wavelengths and translational symmetry.**

A local free energy is one which depends on the order parameter field and its gradients:

$$\mathcal{F}^{\text{Ising}}\{m, T\} = \mathcal{F}(\mathbf{x}, m, \partial_j m, \partial_j \partial_k m, \dots). \tag{9.13}$$

As in Section 9.3, we Taylor expand in gradients.[26] Keeping terms with up to two gradients of m (and, for simplicity, no gradients of temperature), we find

$$\begin{aligned}
\mathcal{F}^{\text{Ising}}\{m, T\} = {} & A(m, T) + V_i(m, T)\partial_i m \\
& + B_{ij}(m, T)\partial_i \partial_j m \\
& + C_{ij}(m, T)(\partial_i m)(\partial_j m).
\end{aligned} \tag{9.14}$$

(b) *What symmetry tells us that the unknown functions A, B, C, and V do not explicitly depend on position $\mathbf{x}$?* If a is the natural microscopic length of the problem (the size of the molecules, or lattice, or perhaps the correlation length), all the terms in the expansion should be of the same order in units where $a = 1$. *If the magnetization varies on a large length scale $D \gg a$, how much smaller would a term involving three derivatives be than the terms B and C that we have kept?*

(3) **Impose the other symmetries of the problem.**

The Ising model has an up–down symmetry[27] so the free energy density $\mathcal{F}^{\text{Ising}}\{m\} = \mathcal{F}\{-m\}$. Hence the coefficients A and C are functions of m^2, and the functions $V_i(m, T) = m v_i(m^2, T)$ and $B_{ij}(m) = m\, b_{ij}(m^2, T)$.

The two-dimensional Ising model on a square lattice is symmetric under 90° rotations. This tells us that $v_i = 0$ because no vector is invariant under 90° rotations. Similarly, b and C must commute with these rotations, and so must be mul-

---

[22]Physicists call this Landau theory (or, more properly, Ginzburg–Landau theory); similar methods with different names are used in engineering, liquid crystal physics, and other fields.

[23]See also Exercise 12.5, which derives this form using mean-field theory.

[24]Landau has a more systematic approach for defining the order parameter, based on group representation theory, which can be quite useful in more complex systems.

[25]You can add fields to make long-range interactions local (like electric fields for Coulomb interactions). A complete order parameter should include broken symmetries, long-range fields, and conserved quantities.

[26]A gradient expansion will not be valid at sharp interfaces and in defect cores where the order parameter varies on microscopic length scales. Landau theory is often used anyhow, as a solvable if uncontrolled approximation to the real behavior.

[27]The equilibrium state may not have up–down symmetry, but the model—and hence the free energy density—certainly does.

tiples of the identity matrix.[28] Hence we have

$$\mathcal{F}^{\mathrm{Ising}}\{m, T\} = A(m^2, T) + m\,b(m^2, T)\nabla^2 m \\ + C(m^2, T)(\nabla m)^2. \quad (9.15)$$

Many systems are isotropic: the free energy density is invariant under all rotations. For isotropic systems, the material properties (like the functions A, B_{ij}, and C_{ij} in eqn 9.14) must be invariant under rotations. All terms in a local free energy for an isotropic system must be writable in terms of dot and cross products of the gradients of the order parameter field.

(c) *Would the free energy density of eqn 9.15 change for a magnet that had a continuous rotational symmetry?*

(4) **Simplify using total divergence terms.**
Free energy densities are intrinsically somewhat arbitrary. If one adds to $\mathcal{F}$ a gradient of any smooth vector function $\nabla \cdot \boldsymbol{\xi}(m)$, the integral will differ only by a surface term $\int \nabla \cdot \boldsymbol{\xi}(m)\,\mathrm{d}V = \int \boldsymbol{\xi}(m) \cdot \mathrm{d}S$.
In many circumstances, surface terms may be ignored. (i) If the system has periodic boundary conditions, then the integral $\int \boldsymbol{\xi}(m) \cdot \mathrm{d}S = 0$ because the opposite sides of the box will cancel. (ii) Large systems will have surface areas which are small compared to their volumes, so the surface terms can often be ignored, $\int \nabla \cdot \boldsymbol{\xi}(m)\,\mathrm{d}V = \int \boldsymbol{\xi}(m) \cdot \mathrm{d}S \sim L^2 \ll \int \mathcal{F}\,\mathrm{d}V \sim L^3$. (iii) Total divergence terms can be interchanged for changes in the surface free energy, which depends upon the orientation of the order parameter with respect to the boundary of the sample.[29]
This allows us to integrate terms in the free energy by parts; by subtracting a total divergence $\nabla(uv)$ from the free energy we can exchange a term $u\,\nabla v$ for a term $-v\,\nabla u$. For example, we can subtract a term $-\nabla \cdot \left(m\,b(m^2, T)\nabla m\right)$ from the free energy 9.15, replacing $\left(m\,b(m^2, T)\right)(\nabla^2 m)$ with the equivalent

term $-\nabla\left(m\,b(m^2, T)\right) \cdot \nabla m$:

$$\mathcal{F}^{\mathrm{Ising}}\{m, T\} \\ = A(m^2, T) + m\,b(m^2, T)\nabla^2 m \\ \quad + C(m^2, T)(\nabla m)^2 \\ \quad - \nabla\left(m\,b(m^2, T) \cdot \nabla m\right) \\ = A(m^2, T) + C(m^2, T)(\nabla m)^2 \\ \quad - \nabla\left(m\,b(m^2, T)\right) \cdot \nabla m \\ = A(m^2, T) + C(m^2, T)(\nabla m)^2 \\ \quad - \left(b(m^2, T) + 2m^2 b'(m^2, T)\right)(\nabla m)^2 \\ = A(m^2, T) \\ \quad + \left(C(m^2, T) - b(m^2, T)\right. \\ \quad \left. - 2m^2 b'(m^2, T)\right)(\nabla m)^2. \quad (9.16)$$

Thus we may absorb the b term proportional to $\nabla^2 m$ into an altered $c = C(m^2, T) - b(m^2, T) - 2m^2 b'(m^2, T)$ term times $(\nabla m)^2$:

$$\mathcal{F}^{\mathrm{Ising}}\{m, T\} = A(m^2, T) + c(m^2, T)(\nabla m)^2. \quad (9.17)$$

(5) **(Perhaps) assume the order parameter is small.**[30]
If we assume m is small, we may Taylor expand A and c in powers of m^2, yielding $A(m^2, T) = f_0 + (\mu(T)/2)m^2 + (g/4!)m^4$ and $c(m^2, T) = \frac{1}{2}K$, leading to the traditional Landau free energy for the Ising model:

$$\mathcal{F}^{\mathrm{Ising}} = \tfrac{1}{2}K(\nabla m)^2 + f_0 + (\mu(T)/2)m^2 + (g/4!)m^4, \quad (9.18)$$

where f_0, g, and K can also depend upon T. (The factors of $1/2$ and $1/4!$ are traditional.)
The free energy density of eqn 9.18 is one of the most extensively studied models in physics. The field theorists use ϕ instead of m for the order parameter, and call it the ϕ^4 model. Ken Wilson added fluctuations to this model in developing the renormalization group (Chapter 12).
Notice that the Landau free energy density has a qualitative change at $\mu = 0$ (Fig. 9.23). For positive μ it has a single minimum at $m = 0$; for negative μ it has two minima at $m = \pm\sqrt{-6\mu/g}$. Is this related to the transition in the Ising model from the paramagnetic phase ($m = 0$) to the ferromagnetic phase at T_c?

---

[28] Under a 90° rotation $R = \left(\begin{smallmatrix} 0 & 1 \\ -1 & 0 \end{smallmatrix}\right)$, a vector $\left(\begin{smallmatrix} v_1 \\ v_2 \end{smallmatrix}\right)$ goes to $R \cdot \mathbf{v} = \left(\begin{smallmatrix} v_2 \\ -v_1 \end{smallmatrix}\right)$, so only $\mathbf{v} = \mathbf{0}$ is invariant. Similarly $RCR^{-1} = \left(\begin{smallmatrix} C_{22} & -C_{12} \\ -C_{12} & C_{11} \end{smallmatrix}\right)$ is invariant only if $C_{22} = C_{11}$ and $C_{12} = 0$, so C is a multiple of the identity.
[29] See Exercise 9.3 and [110]. Total divergence terms also change the energy of topological defects (internal surfaces) [167].
[30] The order parameter is small near a continuous phase transition, but Landau theory fails there. In Chapter 12 we show that large, scale-invariant fluctuations lead to an average order parameter that scales as $(T_c - T)^\beta$. Landau theory, ignoring these fluctuations, predicts $\beta = \frac{1}{2}$, where the 3D Ising model has $\beta = 0.3264\ldots$.

Fig. 9.23 Landau free energy density for the Ising model 9.18, at positive, zero, and negative values of the quadratic term μ.

The free energy density already incorporates (by our assumptions) fluctuations in m on length scales small compared to the coarse-graining length W. *If we ignored fluctuations on scales larger than W* then the free energy of the whole system[31] would be given by the volume times the free energy density, and the magnetization at a temperature T would be given by minimizing the free energy density. The quadratic term $\mu(T)$ would vanish at T_c, and if we expand $\mu(T) \sim a(T - T_c) + \ldots$ we find $m = \pm\sqrt{6a/g}\sqrt{T_c - T}$ near the critical temperature.

This is qualitatively correct, but quantitatively wrong. The magnetization does vanish at T_c with a power law $m \sim (T_c - T)^\beta$, but the exponent β is not generally $1/2$; in two dimensions it is $\beta_{2D} = 1/8$ and in three dimensions it is $\beta_{3D} \approx 0.32641\ldots$. These exponents (particularly the presumably irrational one in 3D) cannot be fixed by keeping more or different terms in the analytic Landau expansion.

(d) *Show that the power law $\beta_{Landau} = 1/2$ is not changed in the limit $T \to T_c$ even when one adds another term $(h/6!)\, m^6$ into eqn 9.18. (That is, show that $m(T)/(T_c - T)^\beta$ goes to a constant as $T \to T_c$.)* (Hint: You should get a quadratic equation for m^2. Keep the root that vanishes at $T = T_c$, and expand in powers of h.) *Explore also the alternative phase transition where $g \equiv 0$ but $h > 0$; what is β for that transition?*

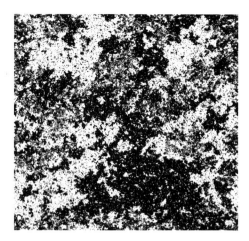

Fig. 9.24 Fluctuations on all scales. A snapshot of the Ising model at T_c. Notice that there are fluctuations on all length scales.

As we see in Fig. 9.24 there is no length W above which the Ising model near T_c looks smooth and uniform. The Landau free energy density gets corrections on all length scales; for the infinite system the free energy has a singularity at T_c (making our power-series expansion for $\mathcal{F}^{Ising}$ inadequate). The Landau free energy density is only a starting-point for studying continuous phase transitions;[32] we must use the renormalization-group methods of Chapter 12 to explain and predict these singularities.

(9.6) **Symmetries and wave equations.** @
We can use symmetries and gradient expansions not only for deriving new free energies (Exercise 9.5), but also for directly deriving equations of motion. This approach (sometimes including fluctuations) has been successful in a number of systems that are strongly out of equilibrium [84, 96, 198]. (See Exercise 9.20 for a more microscopic approach.) In this exercise, you will derive the equation of motion for a scalar order parameter $y(x, t)$ in a one-dimensional system. Our order parameter might represent the height of a string vibrating perpendicular to the x axis,

[31]The total free energy is convex (Fig. 11.2(a)). The free energy density $\mathcal{F}$ in Fig. 9.23 can have a barrier if a boundary between the phases is thicker than the coarse-graining length. The total free energy also has singularities at phase transitions. $\mathcal{F}$ can be analytic because it is the free energy of a finite region; thermal phase transitions do not occur in finite systems.
[32]An important exception to this is (low T_c) superconductivity, where the Cooper pairs are large compared to their separation. Because they overlap so many neighbors, the fluctuations in the order parameter field are suppressed, and Landau theory is valid even very close to the phase transition.

or the displacement of a one-dimensional crystal along x, or the density of particles in a one-dimensional gas.

Write the most general possible law. We start by writing the most general possible evolution law. Such a law might give the time derivative $\partial y / \partial t = \ldots$ like the diffusion equation, or the acceleration $\partial^2 y / \partial t^2 = \ldots$ like the wave equation, or something more general. If we take the left-hand side minus the right-hand side, we can write any equation of motion in terms of some (perhaps nonlinear) function $\mathcal{G}$ involving various partial derivatives of the function $y(x, t)$:

$$\mathcal{G}\left(y, \frac{\partial y}{\partial x}, \frac{\partial y}{\partial t}, \frac{\partial^2 y}{\partial x^2}, \frac{\partial^2 y}{\partial t^2}, \frac{\partial^2 y}{\partial x \partial t}, \ldots, \frac{\partial^7 y}{\partial x^3 \partial t^4}, \ldots\right) = 0. \tag{9.19}$$

Notice that we have already assumed that our system is homogeneous and time independent; otherwise $\mathcal{G}$ would explicitly depend on x and t as well.

First, let us get a tangible idea of how a function $\mathcal{G}$ can represent an equation of motion, say the diffusion equation.

(a) *What common equation of motion results from the choice $\mathcal{G}(a_1, a_2, \ldots) = a_3 - D a_4$ in eqn 9.19?*

Restrict attention to long distances and times: gradient expansion. We are large and slow creatures. We will perceive only those motions of the system that have long wavelengths and low frequencies. Every derivative with respect to time (space) divides our function by a characteristic time scale (length scale). By specializing our equations to long length and time scales, let us drop all terms with more than two derivatives (everything after the dots in eqn 9.19). We will also assume that $\mathcal{G}$ can be written as a sum of products of its arguments—that it is an *analytic* function of y and its gradients. This implies that

$$f + g \frac{\partial y}{\partial t} + h \frac{\partial y}{\partial x} + i \left(\frac{\partial y}{\partial t}\right)^2 + \cdots + n \frac{\partial^2 y}{\partial t \partial x} = 0, \tag{9.20}$$

where $f, g, \ldots, n$ are general analytic functions of y.

(b) *Give the missing terms, multiplying functions $j(y), k(y), \ldots, m(y)$.*

Apply the symmetries of the system.

We will assume that our system is like waves on a string, or one-dimensional phonons, where an overall shift of the order parameter $y \to y + \Delta$ is a symmetry of the system (Fig. 9.25). This implies that $\mathcal{G}$, and hence $f, g, \ldots, n$, are independent of y.

$y(x,t) + \Delta$

$y(x,t)$

Fig. 9.25 Shift symmetry. We assume our system is invariant under overall shifts in the order parameter field. Hence, if $y(x, t)$ is a solution, so is $y(x, t) + \Delta$.

Let us also assume that our system is invariant under flipping the sign of the order parameter $y \to -y$, and to spatial inversion, taking $x \to -x$ (Fig. 9.26). More specifically, we will keep all terms in eqn 9.20 which are *odd* under flipping the sign of the order parameter and *even* under inversion.[33]

$y(x,t)$ $-y(x,t)$ $y(x,t)$ $y(-x,t)$

Fig. 9.26 Flips and inversion. We assume our system is invariant under flipping ($y \to -y$) and inversion ($x \to -x$). Hence, if $y(x, t)$ is a solution, so are $y(-x, t)$ and $-y(x, t)$.

(c) *Which three terms in eqn 9.20 are left after imposing these symmetries? Which one is not part of the wave equation $\partial^2 y / \partial t^2 = c^2 \partial^2 y / \partial x^2$?*

This third term would come from a source of friction. For example, if the vibrating string was embedded in a fluid (like still air), then slow vibrations (low Reynolds numbers) would be damped by a term like the one allowed by symmetry in part (c). Systems with *time inversion* symmetry cannot have dissipation, and you can check that your term changes sign as $t \to -t$, where the other terms in the wave equation do not.

---

[33] All terms in the equation of motion must have the same dependence on a symmetry of the system. One could concoct inversion-symmetric physical systems whose equation of motion involved terms odd under inversion.

This third term would not arise if the vibrating string is in a vacuum. In particular, it is not *Galilean invariant*. A system has Galilean invariance if it is unchanged under *boosts*: for any solution $y(x,t)$, $y(x,t) + vt$ is also a solution.[34] The surrounding fluid stays at rest when our vibrating string gets boosted, so the resulting friction is not Galilean invariant. On the other hand, internal friction due to bending and flexing the string is invariant under boosts. This kind of friction is described by *Kelvin damping* (which you can think of as a dashpot in parallel with the springs holding the material together, Exercises 9.14 and 10.9).

(d) *Show that your third term is not invariant under boosts. Show that the* Kelvin damping *term $\partial^3 y/\partial t \partial x^2$ is invariant under boosts and transforms like the terms in the wave equation under shifts, flips, and inversion.*

(9.7) **Superfluid order and vortices.** (Quantum, Condensed matter) ③

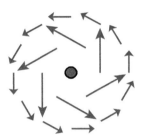

Fig. 9.27 Superfluid vortex line. Velocity flow $\mathbf{v}(\mathbf{x})$ around a superfluid vortex line.

Superfluidity in helium is closely related to Bose condensation of an ideal gas; the strong interactions between the helium atoms quantitatively change things, but many properties are shared. In particular, we describe the superfluid in terms of a complex number $\psi(\mathbf{r})$, which we think of as a wavefunction which is occupied by a large fraction of all the atoms in the fluid.

(a) *If N noninteracting bosons reside in a state $\chi(\mathbf{r})$, write an expression for the net current*

density $J(\mathbf{r})$.[35] *Write the complex field $\chi(\mathbf{r})$ in terms of an amplitude and a phase, $\chi(\mathbf{r}) = |\chi(\mathbf{r})| \exp(i\phi(\mathbf{r}))$. We write the superfluid density as $n_s = N|\chi|^2$. Give the current J in terms of ϕ and n_s. What is the resulting superfluid velocity, $v = J/n_s$, Fig. 9.27? (It should be independent of n_s.)*

The Landau order parameter in superfluids $\psi(\mathbf{r})$ is traditionally normalized so that the amplitude is the square root of the superfluid density; in part (a), $\psi(\mathbf{r}) = \sqrt{N}\chi(\mathbf{r})$. The supercurrent in a strongly interacting quantum system like Helium can be shown to be given by this same formula, except with the mass replaced by an effective mass m^* (Exercise 9.20).

In equilibrium statistical mechanics, the macroscopically occupied state is always the ground state, which is real and hence has no current. We can form nonequilibrium states, however, which macroscopically occupy other quantum states. For example, an experimentalist might cool a container filled with helium while it is moving; the ground state in the moving reference frame has a current in the unmoving laboratory frame. More commonly, the helium is prepared in a rotating state.

(b) *Consider a torus filled with an ideal Bose gas at $T = 0$ with the hole along the vertical axis; the superfluid is condensed into a state which is rotating around the hole. Using your formula from part (a) and the fact that $\phi + 2n\pi$ is indistinguishable from ϕ for any integer n, show that the circulation $\oint v \cdot dr$ around the hole is quantized. What is the quantum of circulation?*

Superfluid helium cannot swirl except in quantized units! Notice that you have now explained why superfluids have no viscosity. The velocity around the torus is quantized, and hence it cannot decay continuously to zero; if it starts swirling with nonzero n around the torus, it must swirl forever.[36] This is why we call them superfluids.

In bulk helium this winding number labels line defects called *vortex lines*.

(c) *Treat $\phi(\mathbf{r})$, the phase of the superconducting wavefunction, as the topological order parameter*

---

[34]This is a nonrelativistic version of *Lorentz invariance*.

[35]You can use the standard quantum mechanics single-particle expression $J = (i\hbar/2m)(\psi\nabla\psi^* - \psi^*\nabla\psi)$ and multiply by the number of particles, or you can use the many-particle formula $J(\mathbf{r}) = (i\hbar/2m) \int d^3\mathbf{r}_1 \cdots d^3\mathbf{r}_N \sum_\ell \delta(\mathbf{r}_\ell - \mathbf{r})(\Psi\nabla_\ell\Psi^* - \Psi^*\nabla_\ell\Psi)$ and substitute in the condensate wavefunction $\Psi(\mathbf{r}_1, \ldots, \mathbf{r}_N) = \prod_n \chi(\mathbf{r}_n)$.

[36]Or at least until a dramatic event occurs which changes n, like a vortex line passing across the torus, demanding an activation energy proportional to the width of the torus. See also Exercise 7.9.

of the superfluid. Is the order parameter space a closed loop, $\mathbb{S}^1$? Classify the types of vortex lines in a superfluid. (That is, either give the first homotopy group of the order parameter space, or give the allowed values of the quantized circulation around a vortex.)

(9.8) **Superfluids and ODLRO.** (Condensed matter, Quantum) ⑤

This exercise develops the quantum theory of the order parameters for superfluids and superconductors, following classic presentations by Anderson [8, 9]. We introduce the reduced density matrix, off-diagonal long-range order, broken gauge symmetry, and deduce that a subvolume of a superfluid is best described as a superposition of states with different numbers of particles. The exercise is challenging; it assumes more quantum mechanics than the rest of the text, it involves technically challenging calculations, and the concepts it introduces are deep and subtle...

Density matrices. We saw in Exercise 9.7 that a Bose-condensed ideal gas can be described in terms of a complex number $\psi(\mathbf{r})$ representing the eigenstate which is macroscopically occupied. For superfluid helium, the atoms are in a strongly interacting liquid state when it goes superfluid. We can define the order parameter $\psi(\mathbf{r})$ even for an interacting system using the *reduced density matrix*.

Suppose our system is in a mixture of many-body states Ψ_α with probabilities P_α. The full density matrix in the position representation, you will remember, is

$$\widehat{\rho}(\mathbf{r}'_1, \ldots, \mathbf{r}'_N, \mathbf{r}_1, \ldots, \mathbf{r}_N) \qquad (9.21)$$
$$= \sum_\alpha P_\alpha \Psi^*(\mathbf{r}'_1, \ldots, \mathbf{r}'_N) \Psi(\mathbf{r}_1, \ldots, \mathbf{r}_N).$$

(Properly speaking, these are the matrix elements of the density matrix in the position representation; rows are labeled by $\{\mathbf{r}'_i\}$, columns are labeled by $\{\mathbf{r}_j\}$.) The reduced density matrix $\widehat{\rho}(\mathbf{r}', \mathbf{r})$ (which I will call the density matrix hereafter) is given by setting $\mathbf{r}'_j = \mathbf{r}_j$ for all but one of the particles and integrating over all possible positions, multiplying by N:

$$\widehat{\rho}_2(\mathbf{r}', \mathbf{r}) =$$
$$N \int \mathrm{d}^3 r_2 \cdots \mathrm{d}^3 r_N \qquad (9.22)$$
$$\times \widehat{\rho}(\mathbf{r}', \mathbf{r}_2 \ldots, \mathbf{r}_N, \mathbf{r}, \mathbf{r}_2, \ldots, \mathbf{r}_N).$$

(For our purposes, the fact that it is called a matrix is not important; think of $\widehat{\rho}_2$ as a function of two variables.)

(a) *What does the reduced density matrix $\rho_2(\mathbf{r}', \mathbf{r})$ look like for a zero-temperature Bose condensate of noninteracting particles, condensed into a normalized single-particle state $\chi(\mathbf{r})$?*

An alternative, elegant formulation for this density matrix is to use second-quantized creation and annihilation operators instead of the many-body wavefunctions. These operators $a^\dagger(\mathbf{r})$ and $a(\mathbf{r})$ add and remove a boson at a specific place in space. They obey the commutation relations

$$[a(\mathbf{r}), a^\dagger(\mathbf{r}')] = \delta(\mathbf{r} - \mathbf{r}'),$$
$$[a(\mathbf{r}), a(\mathbf{r}')] = [a^\dagger(\mathbf{r}), a^\dagger(\mathbf{r}')] = 0; \qquad (9.23)$$

since the vacuum has no particles, we also know

$$a(\mathbf{r})|0\rangle = 0,$$
$$\langle 0|a^\dagger(\mathbf{r}) = 0. \qquad (9.24)$$

We define the ket wavefunction as

$$|\Psi\rangle = (1/\sqrt{N!}) \int \mathrm{d}^3 r_1 \cdots \mathrm{d}^3 r_N \qquad (9.25)$$
$$\times \Psi(\mathbf{r}_1, \ldots, \mathbf{r}_N) a^\dagger(\mathbf{r}_1) \ldots a^\dagger(\mathbf{r}_N)|0\rangle.$$

(b) *Show that the ket is normalized if the symmetric Bose wavefunction Ψ is normalized.* (Hint: Use eqn 9.23 to pull the as to the right through the $a^\dagger$s in eqn 9.25; you should get a sum of $N!$ terms, each a product of N δ-functions, setting different permutations of $\mathbf{r}_1 \cdots \mathbf{r}_N$ equal to $\mathbf{r}'_1 \cdots \mathbf{r}'_N$.) *Show that $\langle \Psi|a^\dagger(\mathbf{r}')a(\mathbf{r})|\Psi\rangle$, the overlap of $a(\mathbf{r})|\Psi\rangle$ with $a(\mathbf{r}')|\Psi\rangle$ for the pure state $|\Psi\rangle$ gives the the reduced density matrix 9.22.* Since this is true of all pure states, it is true of mixtures of pure states as well; hence the reduced density matrix is the same as the expectation value $\langle a^\dagger(\mathbf{r}')a(\mathbf{r})\rangle$.

In a nondegenerate Bose gas, in a system with Maxwell–Boltzmann statistics, or in a Fermi system, one can calculate $\widehat{\rho}_2(\mathbf{r}', \mathbf{r})$ and show that it rapidly goes to zero as $|\mathbf{r}' - \mathbf{r}| \to \infty$. This makes sense; in a big system, $a(\mathbf{r})|\Psi(\mathbf{r})\rangle$ leaves a state with a missing particle localized around $\mathbf{r}$, which will have no overlap with $a(\mathbf{r}')|\Psi\rangle$ which has a missing particle at the distant place $\mathbf{r}'$.

ODLRO and the superfluid order parameter. This is no longer true in superfluids; just as in the condensed Bose gas of part (a), interacting, finite-temperature superfluids have a reduced density matrix with off-diagonal long-range order (ODLRO);

$$\widehat{\rho}_2(\mathbf{r}', \mathbf{r}) \to \psi^*(\mathbf{r}')\psi(\mathbf{r}) \quad \text{as } |\mathbf{r}' - \mathbf{r}| \to \infty. \quad (9.26)$$

It is called long-range order because there are correlations between distant points; it is called off-diagonal because the diagonal of this density matrix in position space is $\mathbf{r} = \mathbf{r}'$. The order parameter for the superfluid is $\psi(\mathbf{r})$, describing the long-range piece of this correlation.

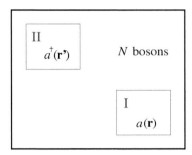

Fig. 9.28 Delocalization and ODLRO. Particles in superfluids are delocalized: the number of particles in a subvolume is not well defined. Annihilating a boson at $\mathbf{r}$ in region I, insofar as the boson comes out of the condensate, is equivalent to annihilating it at $\mathbf{r}'$. The probability overlap between these two states is precisely $\widehat{\rho}_2(\mathbf{r}', r) = \psi^*(\mathbf{r}')\psi(\mathbf{r})$.

(c) *What is $\psi(\mathbf{r})$ for the noninteracting Bose condensate of part (a), in terms of the condensate wavefunction $\chi(\mathbf{r})$?*

This reduced density matrix is analogous in many ways to the density–density correlation function for gases $C(\mathbf{r}', \mathbf{r}) = \langle \rho(\mathbf{r}')\rho(\mathbf{r}) \rangle$ and the correlation function for magnetization $\langle M(\mathbf{r}')M(\mathbf{r}) \rangle$ (Chapter 10). The fact that $\widehat{\boldsymbol{\rho}}_2$ is long range is analogous to the fact that $\langle M(\mathbf{r}')M(\mathbf{r}) \rangle \sim \langle M \rangle^2$ as $\mathbf{r}' - \mathbf{r} \to \infty$; the long-range order in the direction of magnetization is the analog of the long-range phase relationship in superfluids.

Number conservation and ψ. Figure 9.28 illustrates the fact that the local number of particles in a subvolume of a superfluid is indeterminate. Our ground state locally violates conservation of particle number.[37] If the number of particles in a local region is not well defined, perhaps we can

think of the local state as some kind of superposition of states with different particle number? Then we could imagine factoring the off-diagonal long-range order $\langle a^\dagger(\mathbf{r}')a(\mathbf{r}) \rangle \sim \psi^*(\mathbf{r}')\psi(\mathbf{r})$ into $\langle a^\dagger(\mathbf{r}') \rangle \langle a(\mathbf{r}) \rangle$, with $\psi(\mathbf{r}) = \langle a \rangle$. (This is zero in a closed system, since $a(\mathbf{r})$ changes the total number of particles.) The immediate question is how to set the relative phases of the parts of the wavefunction with differing numbers of particles. Let us consider a region small enough that we can ignore the spatial variations.

(d) *Consider a zero-temperature Bose condensate of N noninteracting particles in a local region. Let the state into which the bosons condense, $\chi(\mathbf{r}) = \chi = |\chi| \exp(\mathrm{i}\phi)$, be spatially uniform. What is the phase of the N-particle Bose-condensed state?*

The phase $\exp(\mathrm{i}\phi(\mathbf{r}))$ is the relative phase between the components of the local Bose condensates with N and $N - 1$ particles. The superfluid state is a *coherent superposition of states with different numbers of particles* in local regions. How odd!

Momentum conservation comes from translational symmetry; energy conservation comes from time translational symmetry; angular momentum conservation comes from rotational symmetry. What symmetry leads to number conservation?

(e) *Consider the Hamiltonian $\mathcal{H}$ for a system that conserves the total number of particles, written in second quantized form (in terms of creation and annihilation operators). Argue that the Hamiltonian is invariant under a global symmetry which multiplies all of the creation operators by $\exp(\mathrm{i}\zeta)$ and the annihilation operators by $\exp(-\mathrm{i}\zeta)$. (This amounts to changing the phases of the N-particle parts of the wavefunction by $\exp(\mathrm{i}N\zeta)$. Hint: Note that all terms in $\mathcal{H}$ have an equal number of creation and annihilation operators.)*

The magnitude $|\psi(\mathbf{r})|^2$ describes the superfluid density n_s. As we saw earlier, n_s is the whole density for a zero-temperature noninteracting Bose gas; it is about 1% of the density for superfluid helium, and about 10^{-8} for superconductors. If we write $\psi(\mathbf{r}) = \sqrt{n_s(\mathbf{r})} \exp(\mathrm{i}\phi(\mathbf{r}))$, then the phase $\phi(\mathbf{r})$ (the topological order parameter) labels which of the broken-symmetry

---

[37]This is not just the difference between canonical and grand canonical ensembles. Grand canonical ensembles are probability mixtures between states of different numbers of particles; superfluids have a coherent superposition of wavefunctions with different numbers of particles.

ground states we reside in.

Broken gauge invariance. We can draw a deep connection with quantum electromagnetism by promoting this global symmetry into a local symmetry. Consider the effects of shifting ψ by a spatially dependent phase $\zeta(x)$. It will not change the potential energy terms, but will change the kinetic energy terms because they involve gradients. Consider the case of a single-particle pure state. Our wavefunction $\chi(x)$ changes into $\widetilde{\chi} = \exp(i\zeta(x))\chi(x)$, and $[p^2/2m]\widetilde{\chi} = [((\hbar/i)\nabla)^2/2m]\widetilde{\chi}$ now includes terms involving $\nabla\zeta$.

(f) *Show that this single-particle Hamiltonian is invariant under a transformation which changes the phase of the wavefunction by* $\exp(i\zeta(x))$ *and simultaneously replaces p with* $p - \hbar\nabla\zeta$.

This invariance under multiplication by a phase is closely related to gauge invariance in electromagnetism. Remember in classical electromagnetism the vector potential $\mathbf{A}$ is arbitrary up to adding a gradient of an arbitrary function Λ: changing $\mathbf{A} \to \mathbf{A} + \nabla\Lambda$ leaves the magnetic field unchanged, and hence does not change anything physical. There choosing a particular Λ is called choosing a *gauge*, and this arbitrariness is called *gauge invariance*. Also remember how we incorporate electromagnetism into the Hamiltonian for charged particles: we change the kinetic energy for each particle of charge q to $(p - (q/c)A)^2/2m$, using the *covariant derivative* $(\hbar/i)\nabla - (q/c)A$.

In quantum electrodynamics, particle number is not conserved, but charge is conserved. Our local symmetry, stemming from number conservation, is analogous to the symmetry of electrodynamics when we multiply the wavefunction by $\exp(i(q/c)\zeta(x))$, where $q = -e$ is the charge on an electron.

(g) *Consider the Hamiltonian for a charged particle in a vector potential* $H = ((\hbar/i)\nabla - (q/c)A)^2/2m + V(x)$. *Show that this Hamiltonian is preserved under a transformation which multiplies the wavefunction by* $\exp(i(q/e)\zeta(x))$ *and performs a suitable gauge transformation on* A. *What is the required gauge transformation?*

To summarize, we found that superconductivity leads to a state with a local indeterminacy in the number of particles. We saw that it is natural to describe local regions of superfluids

as coherent superpositions of states with different numbers of particles. The order parameter $\psi(\mathbf{r}) = \langle a(\mathbf{r}) \rangle$ has amplitude given by the square root of the superfluid density, and a phase $\exp(i\phi(\mathbf{r}))$ giving the relative quantum phase between states with different numbers of particles. We saw that the Hamiltonian is symmetric under uniform changes of ϕ; the superfluid ground state breaks this symmetry just as a magnet might break rotational symmetry. Finally, we saw that promoting this global symmetry to a local one demanded changes in the Hamiltonian completely analogous to gauge transformations in electromagnetism; number conservation comes from a gauge symmetry. Superfluids spontaneously break gauge symmetry!

See Exercise 9.20 for Anderson's derivation [8,9] of the emergent superconducting equation of motion from the observation that particle number N and phase ϕ are conjugate variables, like momentum and position in quantum mechanics.

(9.9) **Ising order parameter.** ⓟ
What would the topological order parameter space be for the Ising model? (Hint: what is the space of possible ground states?)

(9.10) **Nematic order parameter.**[38] ⓐ
Let us develop some intuition for the projective plane, the order parameter space of a nematic.
(a) *If one cuts the order parameter space for a two-dimensional crystal (Fig. 9.7) around the edge, it becomes a cylinder. What topological object is formed by cutting the order parameter space for a nematic along the curved path shown in Fig. 9.5(b)?*

This provides us an excuse to exercise our skills with scissors and tape. Unfortunately, the projective plane cannot be pasted together in ordinary three dimensions: we will need to improvise.

(b) *Get a printed, elongated version of Fig. 9.29, suitable for cutting (download from [181]). Tape the short ends together, aligning the arrows. What object do you form? Cut along the dotted line. Can you now tape the long ends together?* (Remember, in higher dimensions one can pass one paper strip through another (moving it into the fourth dimension). You can mimic this by cutting a strip and passing it (without twisting) past another before taping it back together.) *What object do you form in the end?*

---

[38] A printable version of Fig. 9.29 can be found at the book website [181].

Fig. 9.29 Projective Plane. Just like the torus, the order parameter space $\mathbb{RP}^2$ for a nematic can be described as a square with opposite sides identified, but now with a twist. (Think of the square as a top-view of the hemisphere, when the latter is pinched outward along the four diagonals.) The dashed line, for example, connects two points on the edge of the square that correspond to the same nematic orientation. What geometrical figure (disk, cylinder, Möbius strip, sphere with a hole, two disks) is formed when you cut along the dashed line?

Fig. 9.30 Frustrated soccer ball. A material that prefers to form alternating pentagons and hexagons cannot fill two-dimensional flat space, but does live happily on the soccer ball sphere $\mathbb{S}^2$ in three dimensions. Note the distortion of the pentagons and hexagons becomes more and more severe as one works outward from the origin.

(9.11) **Pentagonal order parameter.**[39] (Mathematics) @

Glasses are rigid, like crystals, but have disordered atomic positions, like liquids. We do not have a definitive theory of glasses, but one feature appears to be frustration—good glass-forming systems can form local low energy structures that are difficult to arrange into crystalline patterns. The classic example is the packing of spheres. Four spheres can neatly pack into a low-energy regular tetrahedron, but tetrahedra cannot fill space (Fig. 12.19).

The regular tetrahedron has six edges, each with an dihedral angle of $\arccos(1/3)=70.5°$.[40] Undistorted tetrahedra cannot fit together around an edge without leaving a gap.

(a) *How many tetrahedra are trying to meet along an edge (maybe with a gap) in the frustrated icosahedron of Fig. 12.19(a)? What is the angle of the gap formed by this many tetrahedra meeting at an edge?*

We can create a frustrated order parameter for this kind of glassy system, by considering an *ideal glass* formed by curving space to close the gap. Consider the analogous problem—a two-dimensional material whose lowest-energy local configurations form the faces of a soccer ball, Fig. 9.30.

Figure 9.30 shows a low energy region. To describe the order at a point **x** on the plane, we must both identify the orientations of the polygons and the translational positions within the polygons. How can we combine both into a single order parameter field?

(b) *Get a soccer ball. Generate a copy of Fig. 9.30 blown up so that the polygons roughly match the size of the faces of the soccer ball. Pick a point **x** on the sheet, and align the ball atop the sheet so as to match the local material structure to the ideal soccer ball template. What is the order parameter space you need to describe the local order?* (Hint: This is a challenging question. You need a three-dimensional rotation matrix to define the orientation of the soccer ball tangent to **x**. Is there more than one rotation matrix that can give an equivalent local fit to the structure?)

[39] A printable version of Figs. 9.30 and 9.31 can be found at the book website [181].
[40] A dihedral angle is the angle between two intersecting planes—the opening angle of the edge.

Fig. 9.31 Cutting and sewing: disclinations. The natural defect needed to flatten the soccer ball is a disclination, introducing a wedge of material at the center of a pentagon. This produces a net change in angle as one follows the order parameter around a loop containing the pentagon's center.

Constructing this order parameter, and exploring the consequences, consumed a large portion of the author's postdoc. In particular, the natural low energy structure in Fig. 9.30 is given by rolling the sphere without slipping (a kind of parallel transport).

One cannot fill 2D space by flattening the sphere without stretching (a longstanding complaint of map-makers). One must introduce a disclinations, by adding a wedge of material chosen to sew naturally into the gaps created by flattening the ideal template (Fig. 12.19(a) and Fig. 9.31). For our soccer ball material, one disclination is needed at the center of each pentagon.

(c) *Get a copy of Fig. 9.31 blown up so that the polygons match the faces on the soccer ball (download from [181]). Place a pentagon of the soccer ball atop the central black polygon, tilted so that one edge on the ball lies atop the corresponding edge in the material. Roll the soccer ball around the central polygon so as to trace out its perimeter. As you traverse a path* $\mathbf{x}(\theta)$ *an angle* $\theta = 2\pi$ *around the center of a pentagon in the plane, what is the net angle* ϕ *that the soccer ball rotates? If we define the strength of the defect line as* $s = \phi/2\pi$ *in analogy with the nematic defect line of Fig. 9.20, what is* s?

(9.12) **Rigidity of crystals.** (Order parameters) Ⓟ
Rocks, bridges, and our bones are rigid; unlike fluids, they resist shear. This is because bridges

and most rocks are polycrystalline. The free energy of a crystal rises if the atoms near one surface are displaced relative to those on a distant surface (Fig. 9.32(a)), because the atoms in a crystal have a broken translational symmetry.[41] In an unstressed, equilibrium crystal the atoms on average lie on a regular grid of crystalline lattice positions. Shifting the top layer with respect to the bottom layer forces the grid to distort; this elastic deformation increases the energy. Rigidity of the order parameter is a hallmark of broken symmetry states of matter.

Mountains are built of rocks, and they hold up under gravity. Is this fundamentally because they are rigid? Or is a rock placed on your desk flowing into a puddle like a liquid? It turns out that gravity can transport mass *through* a crystalline lattice without shearing planes of atoms past one another. In most crystals, the dominant mechanism is through *vacancy diffusion* (Fig. 9.32(b)).

Fig. 9.32 Crystal rigidity [176]. (a) Crystals are rigid to shears σ that couple to the lattice (i.e., the broken translational symmetry). (b) Under a force like gravity that couples to the mass density, an equilibrium crystal *can* flow like a liquid (at a rate linear in the gravitational force), via vacancy diffusion. (This is true at temperatures above the *roughening transition* [45], where the equilibrium surfaces all have steps that can absorb and emit vacancies.)

Because gravity couples to the atoms and *not* to the lattice formed by the broken symmetry, we shall calculate below that the crystal will flow at a rate like a very viscous fluid. Apart from flowing slowly, are there important differences in the behavior between the flow of our silicon crystal and than of an extremely viscous liquid?

(a) *Will the crystalline axes be preserved under the flow? Would an single-crystalline ice cube, placed with its hexagonal axis vertical on*

[41]The breakdown of rigidity in crystals is also discussed in Exercise 11.5 (dislocation-mediated plasticity) and Exercise 11.15 (fracture of crystals).

a surface kept below freezing, flow into a circular, squashed droplet (like a liquid) after reaching its equilibrium shape? Into a snowflake, with complex dendritic branches? Roughly what shape would you expect? (Hint: The shape of a water drop on a surface is set by the balance of gravity and surface tension. Is the interfacial tension between a crystal and gas isotropic?)

Let us estimate how slowly the crystal will ooze into this odd puddle. As vacancies move upward, there is a net current of atoms downward. The rate will be some prefactor times the probability $\exp(-\beta f_{trans})$ per site that the atoms in the crystal have a vacancy poised at the transition-state saddle-point[42] between two lattice sites (see Exercise 6.11). It is traditional to think of this probability as the product of the density of vacancies times the probability that a given vacancy will move to its saddle-point. Let the free energy cost of a vacancy be $F_V = E_V - TS_V$. Let the free energy barrier for the motion of a vacancy be $F_B = E_B - TS_B$. Then we see that $f_{trans} = F_V + F_B$.

The experimentalists [191] estimate that $S_{trans} = S_V + S_B = 10.9k_B$. A commonly considered contribution to the entropy of a defect, or the entropy of a transition state, is the change in the vibrational frequencies of nearby atoms. For simplicity, let us model the entropy change for each of these two as due to a change of frequency of one local normal mode. So, e.g. for S_V, the atoms might vibrate radially outward from a vacancy more slowly (ω_V) than they did when an atom occupied the site (ω_{noV}), with all the other modes unchanged.

(b) *Assume the temperature is high compared to $\hbar\omega$ for all the oscillators, so that the entropies are all equal to the classical entropies of the harmonic oscillators. Calculate S_V in terms of ω_V and ω_{noV}. How much would the frequency need to change to account for half of the experimentally measured entropy? Does this frequency change seem plausible? Discuss.*

The free energy of a vacancy, or the barrier to its motion, can depend on temperature for reasons other than the entropy of local phonons. For example, the thermal expansion of our silicon crystal will lower the energy needed for the companion atom to squeeze between neighbors across the barrier. It is likely that silicon vacancy diffusion has a large effective prefactor partly due to this mechanism.

How long will it take for, say, an equilibrium silicon crystal to flow into a puddle? The self-diffusion constant for silicon is experimentally measured to be $D = D_0 \exp(-E_{trans}/k_BT)$, with $D_0 \approx 0.04 \mathrm{m}^2/\mathrm{s}$ and the activation barrier at the transition state $E_{trans} \approx 4.7\mathrm{eV}$ [191].

(c) *Using the Einstein relation relating the mobility to the diffusion constant, give a formula for the upward current J of vacancies due to gravity, as a function of the diffusion constant D, the temperature k_BT, the mass m_{Si} of a silicon atom, the number density ρ_{Si} of silicon atoms per unit volume, and the force g due to gravity. Give the formula involving J giving the velocity v at which the top surface of silicon will shrink downward due to gravity. Evaluate v for silicon just below its melting point, $1,414°C = 1,687°K$. How big is it in Ångstroms/year?* (Useful: $1\mathrm{eV} = 1.6\,10^{-19}\mathrm{J}$; $k_B = 1.38\,10^{-23}\mathrm{J/K}$; $m_{Si} = 28.085\mathrm{au} = 4.66\,10^{-26}\mathrm{kg}$; $g = 9.8\mathrm{m}^2/\mathrm{s}$, and the density of silicon atoms is $\rho_{Si} = 5\,10^{28}\mathrm{atoms/m}^3$.)

The current you derived in (c) should be linear in the applied force—the same linear response one would find in a liquid (as we promised).

Vacancy diffusion is usually the dominant mechanism for linear response bulk flow in perfect crystals, but there are usually much faster mechanisms for moving atoms in solids. Ice in your freezer sublimates (evaporates directly from the solid to gas), explaining why your ice cubes become rounded with time, especially with self-defrosting freezers that occasionally raise their temperature above freezing. Diffusion on surfaces is much faster than bulk diffusion, but scales differently with system size L (since the current downward is proportional to the perimeter L and not the area L^2). Nonequilibrium defects in the crystal (grain boundaries and dislocations) also often have much higher diffusion rates, and also can move to change the crystal shape. (Note that these defects do not exist in an unstrained equilibrium crystal, and their nucleation rate is not linear in the applied force, Exercise 11.5.)

(9.13) **Chiral wave equation.** ②
The evolution of a physical system is described by a field Ξ, obeying a partial differential equa-

---

[42] As an atom falls one lattice site into a vacancy below it, the saddle-point is crossed at the midpoint of the transition.

tion

$$\partial \Xi / \partial t = A \, \partial \Xi / \partial x. \qquad (9.27)$$

(a) Symmetries. *Give the numbers corresponding to ALL the symmetries that this physical system appears to have:*

(1) *Spatial inversion ($x \rightarrow -x$).*

(2) *Time-reversal symmetry ($t \rightarrow -t$).*

(3) *Order parameter inversion ($\Xi \rightarrow -\Xi$).*

(4) *Homogeneity in space ($x \rightarrow x + \Delta$).*

(5) *Time translational invariance ($t \rightarrow t + \Delta$).*

(6) *Order parameter shift invariance ($\Xi \rightarrow \Xi + \Delta$).*

(b) Traveling waves. *Show that our equation $\partial \Xi / \partial t = A \, \partial \Xi / \partial x$ has a traveling wave solution. If $A > 0$, which directions can the waves move?* This chiral wave equation describes the only gapless excitations for the quantum Hall effect—unidirectional waves that move around the edge of a cold, two-dimensional electron gas in a large magnetic field. (What symmetries are broken by an external magnetic field?) There are similar gapless excitations along the surfaces of topological insulators and other recently discovered systems of great current research interest. (See also the conclusion of Exercise 7.24).

In the litany of states of matter in the introduction, the quantum Hall effect is *not* fully described by the system of order parameters, broken symmetries, and topology. One needs to supplement them with Chern numbers, edge states, and topological protection. Some day texts like this one will need another chapter, describing the system my colleagues are developing to to understand these new topological states of matter.

(9.14) **Sound and Goldstone's theorem.** (Condensed matter) ③

Sound waves are the fundamental excitations (or Goldstone modes) associated with *translational symmetry* (see Exercise 7.24). If a system is invariant under shifts sideways by a constant displacement u, they must have low-frequency excitations associated with long-wavelength displacement fields $u(x, t)$ (Section 9.3).

Sound waves have a slight damping or dissipation, called *ultrasonic attenuation*. If our system

is also Galilean invariant,[43] this dissipation becomes small as the wavelength goes to infinity. We illustrate this fact by calculating this dissipation for a particular model of sound propagation. Let our material have speed of sound c and density ρ. Suppose we subject it to an external force $f(x, t)$. We model the dissipation of energy into heat using *Kelvin damping*, with damping constant d^2:

$$\frac{\partial^2 u}{\partial t^2} = c^2 \frac{\partial^2 u}{\partial x^2} + d^2 \frac{\partial}{\partial t} \frac{\partial^2 u}{\partial x^2}. \qquad (9.28)$$

We noted in Exercise 9.6 that Kelvin damping is the dissipative term with fewest derivatives allowed by Galilean invariance.

In the absence of dissipation and forcing $d = 0 = f(x, t)$, the wave equation has plane-wave solutions $u(x, t) = \exp(\mathrm{i}(kx - \omega_k t))$ with the *dispersion relation* $\omega_k = \pm ck$. The damping causes these plane-wave solutions to decay with time, giving the frequency an imaginary part $\omega_k \rightarrow \omega_k - \mathrm{i}\Gamma$. Define the *quality factor* Q_k to be 2π times the number of periods of oscillation needed for the energy in the wave (proportional to u^2) to decay by a factor of $1/e$.

(a) *Show that eqn 9.28 has solutions in the form of damped plane waves,*

$$u(x, t) = \exp(\mathrm{i}(kx - \omega_k t)) \exp(-\Gamma_k t)$$
$$= \exp(\mathrm{i}(kx - \Omega_k t)), \qquad (9.29)$$

with complex frequency $\Omega_k = \omega_k - \mathrm{i}\Gamma_k$. Solve for Ω_k (quadratic formula). What is the new dispersion relation ω_k? What is the damping Γ_k? Show at long wavelengths that the frequency $\omega_k \approx \pm ck$. With what power of k does the quality factor diverge as $k \rightarrow 0$?

Thus the lower the frequency, the smaller the damping per oscillation. (This is probably why it is called *ultrasonic* attenuation—if the sound is not at ultra-high frequencies the attenuation is small.) The Goldstone modes become dissipationless as $\omega \rightarrow 0$. We can use the (undamped) wave equation to describe sound waves in air only because we are large and slow compared to atoms, and so hear sound waves that are long wavelength and low frequency compared to the scales at which dissipation is dominant.

(b) *At what wavelength does the real part of the frequency ω_k vanish?* (This is where the sound

---

[43]A Galilean transformation is a boost by a velocity v: $u(x, t) \rightarrow u(x, t) + vt$ (a nonrelativistic Lorentz boost). A Galilean-invariant system does not notice whether it is moving with a steady velocity.

wave begins to be overdamped; at shorter wavelengths displacements relax diffusively, without oscillations.)

The wave equation for light in outer space has no damping—energy is conserved. There is no fundamental reason why energy should be conserved for sound waves. What feature of everyday life suggests that something like Kelvin damping is occurring for sound in air?

The thunder associated with a nearby lightning strike is of course louder than that for the same lightning strike farther away.

(c) *Is that in itself good evidence for attenuation of sound? How would the power in undamped sound fall off with distance L?* Thunder from a nearby lightning bolt sounds different (a thunderclap) from thunder striking from a far distance L. To a good approximation, all sound frequencies take the same time $\Delta t = L/c$ to arrive. *If frequency ω is damped by a factor of three in power at a distance L, how much will the sound at 2ω (one octave higher) be damped?*

This is why the crack! of a nearby lightning bolt becomes a low frequency rumble at long distances.

(9.15) **Superfluid second sound.** ⓟ

There are two different order parameters for superfluids. The "soft-spin" Landau order parameter is a complex number as a function of position. (Remember the Bose condensate $\Psi(\mathbf{r}) = \psi(x_1)\psi(x_2)\ldots$ of noninteracting bosons; the quantum single-particle state $\psi(x)$ gives the Landau order parameter.) The "topological" order parameter is the phase $\phi(x)$ of this complex function $\psi(x) = \psi_0 \exp(i\phi(x))$. Superfluids break gauge symmetry (making a particular choice for ϕ).

Under most circumstances, for every continuous broken symmetry there is an elementary excitation consisting of a long wavelength, sinusoidal variation of the broken symmetry direction of the corresponding order parameter (Goldstone's theorem).

Argue that the elementary excitation for the superfluid order is a slowly varying small amplitude oscillation in the phase, as in $\phi(x) = \phi_0 + a\cos(kx)$. Using the single-particle formula $\mathbf{J} = (\hbar/2mi)(\psi^*\nabla\psi - \psi\nabla\psi^*)$ (treating the superfluid as a Bose condensate as in Exercise 9.7(a)), argue that the oscillation $\phi(x)$ will change the density of superfluid bosons periodically in space. The supercurrent in a strongly interacting quantum system like helium can be shown to be given by this same formula, except with $\hbar/m$ replaced by the superfluid stiffness per particle (Exercise 9.20).

There is already another density wave in fluids—the longitudinal phonon that gives ordinary sound in liquids and gases. One often uses a *two-fluid model* with a superfluid density and a density of *normal fluid* (quasiparticles); sound waves are an in-phase oscillation of the two fluids (with net density changes), and *second sound* has the normal and superfluid components oscillating out of phase (with little net density change). Since the superfluid carries no heat, the normal fluid component of second sound forms an oscillation of heat entropy...

Second sound has been recently observed in graphite at temperatures above $100\,\mathrm{K}$ [90].

(9.16) **Can't lasso a basketball.**[44] (Mathematics) @ Here we explores why the $s = 1$ lines in Fig. 9.16 are "thick". They are not true singularities—they *escape into the third dimension*.

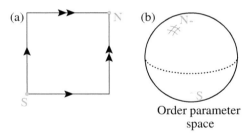

Fig. 9.33 Sphere. The sphere (b) can be topologically constructed by gluing together the edges of a square (a) as shown.

The magnetization $\mathbf{M}$ of a Heisenberg ferromagnet is of fixed magnitude $|\mathbf{M}| = M_0$, but can point with equal energy in any direction. Thus the order parameter space for this magnet is a sphere (Fig. 9.33(b)). Mathematicians sometimes construct the sphere topologically by pasting together a square as in Fig. 9.33(a), with the corresponding sides matched according to the style of arrows (single or double), and with the sides pasted so that the arrows point in the same directions.

---

[44]One must admit that you cannot topologically lasso a cow, and you cannot practically lasso a doughnut.

(a) *To test that you understand this cutting and pasting, sketch directly onto Fig. 9.33(b) the seams where the edges are pasted together, turning the square (a) into the sphere (b). Include the arrows along the seams, with the correct styles and orientation. Assume that the seams are predominantly on the visible (front) side of the sphere, and that the north pole N and south pole S are as shown.* (Did you draw the arrows?) There are no topological line defects for Heisenberg ferromagnets. This is because $\Pi_1(\mathbb{S}^2) = 0$; any closed path on the sphere can be contracted to a point, so a loop in real space never surrounds something that cannot be smoothly filled in. We can give a tangible example of this—sometimes called "escape into the third dimension" (see Fig. 9.16).

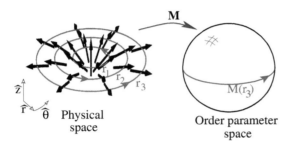

Fig. 9.34 Escape into the third dimension. The magnetic field for a planar two-dimensional material points upward (along $\hat{z}$) at the origin, and points almost radially outward (along $\hat{r}$) far away (radius r_3). As one circles $\vec{r}$ around the origin counter-clockwise at distance r_3, the magnetization $\mathbf{M}(\vec{r})$ circles the order parameter space around the equator, as shown.

A two-dimensional Heisenberg ferromagnet has an order parameter field

$$\mathbf{M} = M_0(\hat{z}\lambda(r) + \hat{r}\sqrt{1 - \lambda^2}), \qquad (9.30)$$

expressed in cylindrical coordinates (Fig. 9.34). Here $\lambda(r)$ decreases rapidly from one at $r = 0$ to zero at larger r, so that $\mathbf{M}$ points radially outward at long distances from the origin. One may thus measure a winding number $s = 1$ for this magnetization pattern at large r, but it nonetheless has a smooth magnetic field everywhere inside.

(b) *Just as the image $\mathbf{M}(\vec{r})$ of the magnetization along $r = r_3$ is shown in order parameter space,*

sketch on the right the approximate curves described by the magnetization at distances r_2 and r_1 from the origin, as denoted in real space on the left. Label these $M(r_1)$ and $M(r_2)$, in analogy with $M(r_3)$ shown.
(c) *Does the path in order parameter space contract to a point, as the loop in real space shrinks to $r = 0$? What is the homotopy group element of the sphere, $g \in \Pi_1(\mathbb{S}^2)$, corresponding to $M(r_3)$?*

(9.17) **Fingerprints.**[45] ③

Fig. 9.35 Whorl, from the Wikipedia article *Fingerprint.*

The framework of order parameters and broken symmetries can be used to explain the structure of many everyday phenomena, including some at your own fingertips (Fig. 9.35). In this problem we will use the tools we have developed to systematically describe the patterns they exhibit.
In many systems, there are multiple approaches to describing the same material system. Here we shall explore three approaches to fingerprints—studying the orientational symmetry breaking (as in a magnet or a nematic), studying the translational symmetry breaking (as in crystals), and studying them as families of equally spaced lines (as in smectic liquid crystals).
Let us first consider an idealization of fingerprint patterns on a flat plane, where the "ground state" looks like straight parallel ridges separated by a distance a.
The ridges break local rotational symmetry, $\mathbf{x} \mapsto R \cdot \mathbf{x}$ for rotations R.
(a) *Do any rotations give equivalent ground states? Is the space of rotations that yield*

[45]This exercise was developed in collaboration with Jaron Kent-Dobias. Copies of Fig. 9.36 and other fingerprints are available at the book website [181].

unique ground states[46] O(2), SO(2), *or* $\mathbb{R}P(1) =$ SO(2)/$\mathbb{Z}_2$? *Briefly defend your choice.*

(b) *Circle at least three orientational defects among Fig. 9.35, and label their topological charge (winding number). Make sure you find at least two different winding numbers.* (Copies available at [181]).

The ridges also break local translational symmetry, $\mathbf{x} \mapsto \mathbf{x} + \Delta\widehat{n}$, where $\widehat{n}$ points perpendicular to the local ridges.

(c) *Do any translations Δ yield an equivalent ground state? Ignoring the rotational symmetry breaking,*[47] *is the space of translations that yield unique ground states $\mathbb{R}^1/\mathbb{Z}$? $\mathbb{S}^1$? $\mathbb{T}^1$? Or all three? Briefly explain your reasoning.*

(d) *Circle at least three translational topological defects in Fig. 9.35, ignoring regions very close to the cores of any rotational defects. Do you see any with topological charge greater than the minimum (with Burgers vector $|\vec{b}| > a$)?*

Fig. 9.36 Transmuting defects. A translational defect (a dislocation) can change sign when pulled around an orientational defect.

Weird things happen when translational symmetry and orientational symmetry both break at the same time.

(e) *Consider two translational defects of the same sign, below an orientational defect as shown in Fig. 9.36. Consider smoothly pulling one defect around the orientational defect (lengthening the ridge). Can you get them to annihilate? Draw your path on a copy of Fig. 9.36 (available at [181]).*

(Similar charge changes can also arise in liquid crystals when a point defect circles a line defect—say, a hedgehog circles a nematic disclination line.)

Now let us view fingerprints as families of equally spaced lines. Smectic liquid crystals are composed of parallel 2D layers. The layers need not be flat, but they become quite precisely equally spaced. The condition of equal layer spacing is not topological! ("Equal spacing" is not preserved under continuous deformations.) Indeed, 3D smectic liquid crystals form exotic focal conic domains, whose defects are one-dimensional curves that form geometrically perfect ellipses and hyperbolas.

Fig. 9.37 Smectic fingerprint. One way of covering a model finger with equally spaced lines and only point defects.

The patterns on your fingers can be viewed as two-dimensional smectics on a curved surface, with the spacing between ridges a quite precisely equal. We shall model a finger as a cylinder with circumference $2na$ for some integer n, topped with a hemisphere of the same circumference.

There are many ways of covering a surface with equally spaced smectic ridges if one allows high-energy defects that form one-dimensional curves. (Defects are not always topological; they are regions where the local order is disrupted—here, where the ridges bend with curvature sharp compared to the ridge spacing a.) What happens when we restrict ourselves to equally spaced layers with only point defects? (See [3], which addresses this problem on a general curved surface.)

There are many ways to cover a cylinder with equally spaced smectic lines without defects. For example, the lines could form spirals around the cylinder. If you cut the cylinder vertically and flatten it, the lines would be straight and par-

---

[46]All three possibilities give orientational order parameter spaces that are topologically equivalent to a circle, S^1.
[47]We often ignore rotational symmetries when analyzing defects in crystals.

allel, separated by a at an angle that satisfies the periodic boundary conditions. On the other hand, to cover the sphere with equally spaced ridges and only point defects, one must have at least two defects. For example, ridges forming lines of latitude will have defects at the north and south pole.

(f) *Why are the lines of longitude not an allowed pattern of smectic lines for the sphere?*

Figure 9.37 shows one way of meshing equally spaced layers onto our model fingerprint, with only two point defects (one will be on the back side, under the fingernail).

(g) *Sketch a different covering with only point defects. Make sure your ridges match positions and slopes at the intersection, and that they are equally spaced everywhere.*

(9.18) Defects in crystals.[48] @

This exercise numerically explores the wide variety of defects formed in crystals. You will either need to be able to write on screenshots of the simulation (either electronically or after printing), or will need to make careful sketches of what you see.

Start up the simulator [32], and select Crystal (one of the **Initial Conditions** choices under the main simulator window). You should see a collection of atoms moving to lower their energy, *quenching* to a zero-temperature metastable state each time you hit **Restart**.[49] The system should settle into a mostly crystalline arrangement with a few vacancies, dislocations, walls, and more complicated defects. The grayscale is a measure of the stresses felt by the particles. The system has periodic boundary conditions, so some defects will stretch past the left or bottom and emerge through the right or top. Generate and admire the variety of defect patterns.

(a) *Generate a large crystalline region. Take a snapshot and crop it to show the hexagonal crystal. Draw the three axes along which the crystal has closely-packed rows of atoms. Regenerate other large crystal regions, and sketch the orientations of their axes (no need for screenshots). Are the axis orientations different under different quenches? Does the crystal spontaneously break orientational symmetry?*

(b) *Generate and crop a simulation showing a*

vacancy—a missing atom in an otherwise crystalline region.

These vacancies are often quite mobile in crystals (see Exercise 9.12). Their motion allows the atoms in the crystal to diffuse (a vacancy hop to the left is an atom hop to the right). Dopants and impurities can also use vacancies to diffuse through a crystal.

The regions of crystalline order are called *grains*, and two grains are separated by a *grain boundary*. A grain boundary is usually a few atoms thick, separating two regions of fairly undeformed hexagonal crystal with different orientations.

(c) *Generate a grain boundary stretching across the simulation, separating two crystals with different orientations. Draw their axes, and measure the misorientation angle—the smallest rotation needed to take one set of crystalline axes to the other. What is the largest misorientation angle possible in a hexagonal crystal?*

The grain boundary is an orientational defect. It is not a *topological* defect, because crystalline orientational order is a continuous symmetry—there are crystalline ground states at all angles. Crystals respond to bending with an abrupt wall, rather than with a gradual change in orientation; the translational crystalline order makes gradual changes in orientation energetically expensive.

(d) *Generate and crop an isolated dislocation (fairly deep inside a crystalline region). Note that the dislocations in this simulation are delocalized over a substantial distance along one of the crystal axes. Over roughly how many particle diameters does it extend (as measured by a noticeable skewing of the lattice hexagons, or by the stress measured as the grayscale coloring of the defects)?*

Something similar happens in an extreme way in materials like copper, where a dislocation can decompose into *partial dislocations* separated by a stacking fault.

The Burger's vector $\vec{b}$, by convention, is the displacement from the start to the end of a path that travels equal numbers of rows of atoms in a clockwise path around the defect.

(e) *Find and indicate the direction and magnitude of the Burger's vector of your dislocation in part (d), by tracking the atomic layers along a*

---

[48]This exercise uses the mosh pit simulator [32], developed to model humans dancing at heavy metal concerts: see [31, 188–190].
[49]You can speed up the relaxation by increasing Frameskip to draw fewer of the steps. You can remove the red "active" particle by changing **Fraction Red** to zero.

curve encircling the defect. On the square lattice (Fig. 9.11), the Burger's vector was perpendicular to an extra row of atoms that ended at the dislocation. *How is this different from the atomic configuration near your hexagonal-crystal dislocation?*

(9.19) **Defect entanglement.** (Mathematics) @

We argued in Section 9.4 that topological defect lines with homotopy elements α and β (Fig. 9.17) can cross one another only if their homotopy group elements commute, $\alpha\beta = \beta\alpha$. The argument is mainly the assertion in Fig. 9.18 that we can measure the topological charge of the trail connecting the two after they cross by looping the trail, wiggling the loop until it falls around the "legs" of the two defects, and then counting the paths around the two.

It is far from obvious that the wiggling does indeed leave the loop encircling both defects twice (Fig. 9.19). The most convincing demonstration is to construct two defect lines and a loop and try it.

If you are not provided with one, construct a model of the defects in Fig. 9.18, showing their configuration after one has attempted to cross the other. (Wire from metal hangers is stiff, but works well when pushed into holes in a base of some sort. Use pliers to bend over the sharp points to avoid puncture wounds.) *Form a loop long enough to encircle the two wires at the base twice with some slack.* (Tape stuck to itself is good, in that one can draw arrows around the loop.)

Arrange the loop so that it encircles the trail between the two defects. Wiggle and twist it so that it encircles the two defects below the trail, without pulling it over the top or bottom of the model (the defects are meant to extend to infinity). Does the resulting configuration follow the pattern in Fig. 9.19, passing around both defect counter-clockwise and then passing around both clockwise? Do you agree that the trail can be erased only if

$$\beta\alpha\beta^{-1}\alpha^{-1} = 1, \qquad (9.31)$$

a contractible loop? Finally, show that this implies $\beta\alpha = \alpha\beta$, that the two elements commute.

Nonabelian defects also behave in complex ways when they are dragged around one another. Analogous nonabelian *braiding* in quantum systems[50] could lead to a topologically protected method for implementing quantum computers [69].

(9.20) **Number and phase in superfluids.** (Quantum) ③

In Exercises 9.5, 9.6, and 9.13 we used symmetry to derive the emergent laws for a physical system. In this exercise, we explore a different, more microscopic approach to deriving emergent dynamical laws. It was developed in this context by Anderson [8, 9] (Exercise 9.8); similar ideas are used as part of Hohenberg and Halperin's [86] classification of dynamic critical phenomena.

In a superfluid, the order parameter field is a complex number $\psi(x, t)$ evolving in space and time. The local density of superfluid particles is given by $\psi^*(x)\psi(x)$. The free energy is well described by Ginzburg–Landau theory (as in Exercise 9.5)

$$\mathcal{F} = \tfrac{1}{2}K|\nabla\psi|^2 + V(|\psi|), \qquad (9.32)$$

where K is the superfluid stiffness per particle[51] and where $V(|\psi|) = \alpha|\psi^2| + \tfrac{1}{2}\beta|\psi|^4$ is independent of the phase of the complex order parameter (important later). The particle current of the superconducting fluid—our emergent law—is given by

$$\mathbf{J} = K/(2i\hbar)(\psi^*\nabla\psi - \psi\nabla\psi^*). \qquad (9.33)$$

Notice that this formula for the current is the same as that for a noninteracting quantum particle, if we substitute $\hbar/m$ for $K/\hbar$. This makes sense for the case of a Bose condensate,[52] where the particles are quenched into a single noninteracting quantum state. But superfluid Helium is strongly interacting! What deep principle tells us that the current is continued to be given by this formula, with just a different stiffness K?

The superfluid order parameter $\psi = |\psi| \exp(i\phi)$ has a phase ϕ that expresses the spontaneous breaking of gauge symmetry; a well-defined

---

[50]Figure 9.18 can represent the 2D space-time paths of two quantum defects (time being vertical). Two fermions will gain a minus sign as one orbits the other; two bosons will give a plus sign; anyons can have other changes in phase, sometimes nonabelian.

[51]Often $\tfrac{1}{2}K$ is written $\hbar^2/2m^*$ with m^* an effective mass of the particles. This choice makes apparent the analogy with quantum mechanics.

[52]We use this analogy in Exercise 9.7 to derive currents around vortices and in Exercise 9.15 to derive second sound.

phase $\phi(x)$ is associated with complete uncertainty in the local number density $\rho(x)$ (Exercise 9.8). More plainly, the particle density ρ and the phase ϕ are conjugate variables, like x and p; the uncertainty described above is analogous to the Heisenberg uncertainty relation. Also, just as $\dot{x} = \partial \mathcal{H}/\partial p$ and $\dot{p} = -\partial \mathcal{H}/\partial x$, we can write the emergent evolution equations for ρ and ϕ:

$$\frac{\partial \rho}{\partial t} = \frac{\partial \mathcal{F}}{\partial \phi}, \tag{9.34}$$

$$\hbar \frac{\partial \phi}{\partial t} = -\frac{\partial \mathcal{F}}{\partial \rho}. \tag{9.35}$$

Since $|\psi| = \sqrt{\rho}$, these two equations determine the evolution of both the magnitude and phase of ψ, and thus completely determine its evolution in time.

Do eqn 9.34 and $\partial \rho/\partial t = -\partial J/\partial x$ give us the supercurrent of eqn 9.33? Let us specialize to the one-dimensional case.

(a) Show that $\mathcal{F} = \frac{1}{2}K|\psi|^2(\phi')^2$ plus terms independent of ϕ.

(b) Using the calculus of variations, take $\phi(x) \to \phi(x) + \delta(x)$ to show that

$$\int \mathrm{d}x \, \frac{\partial \mathcal{F}}{\partial \phi} \delta(x) = \int \mathrm{d}x \, K|\psi|^2 \phi' \delta'(x) \tag{9.36}$$

(Hint: The integrand on the left is $\mathcal{F}[\phi+\delta] - \mathcal{F}[\phi]$ to lowest order in δ.)

(c) Show that J in eqn 9.33 is $K/\hbar|\psi|^2\phi'$. Integrate eqn 9.36 by parts to show that

$$\frac{\partial \mathcal{F}}{\partial \phi} = -\hbar \frac{\partial J}{\partial x} = \hbar \frac{\partial \rho}{\partial t}. \tag{9.37}$$

This confirms that eqn 9.34, derived from the fact that ρ and ϕ are conjugate variables, yields a supercurrent eqn 9.33 of the form given by single-particle quantum mechanics.

What about the other emergent equation of motion, eqn 9.35 for $\partial \phi/\partial t$? It can be used to derive the oscillation frequency in superconducting Josephson junctions (the Josephson frequency relation)...

Correlations, response, and dissipation

10

In this chapter we study how systems wiggle, and how they yield and dissipate energy when kicked.[1]

A material in thermal equilibrium may be macroscopically homogeneous and static, but it wiggles on the microscale from thermal fluctuations. We measure how systems wiggle and evolve in space and time using *correlation functions*. In Section 10.1 we introduce correlation functions, and in Section 10.2 we note that scattering experiments (of X-rays, neutrons, and electrons) directly measure correlation functions. In Section 10.3 we use statistical mechanics to calculate equal-time correlation functions using the ideal gas as an example. In Section 10.4 we use Onsager's *regression hypothesis* to derive the time-dependent correlation function.

We often want to know how a system behaves when kicked in various fashions. *Linear response theory* is a broad, systematic method developed for equilibrium systems in the limit of gentle kicks. We can use statistical mechanics to calculate the response when a system is kicked by elastic stress, electric fields, magnetic fields, acoustic waves, or light. The space-time dependent linear response to the space-time dependent influence is described in each case by a *susceptibility* (Section 10.5).

There are powerful relationships between the wiggling, yielding,[2] and dissipation in an equilibrium system. In Section 10.6 we show that yielding and dissipation are precisely the real and imaginary parts of the susceptibility. In Section 10.7 we show that the static susceptibility is proportional to the equal-time correlation function. In Section 10.8 we derive the *fluctuation-dissipation theorem*, giving the dynamic susceptibility in terms of the time-dependent correlation function. Finally, in Section 10.9 we use causality (the fact that the response cannot precede the perturbation) to derive the *Kramers–Kronig* relation, relating the real and imaginary parts of the susceptibility (yielding and dissipation).

[1]More information on these topics can be found in the classical context in [41, chapter 8], and for quantum systems in [64, 127].

[2]We use "yielding" informally for the in-phase, reactive response (Section 10.6), for which there appears not to be a standard term.

10.1 Correlation functions: motivation

We have learned how to derive the laws giving the equilibrium states of a system (Fig. 10.3) and the evolution laws of systems as they approach equilibrium (Figs. 10.1 and 10.2). How, though, do we characterize the resulting behavior? How do we extract from our ensemble of systems

Statistical Mechanics: Entropy, Order Parameters, and Complexity. James P. Sethna, Oxford University Press (2021). © Oxford University Press. DOI:10.1093/oso/9780198865247.003.0010

Fig. 10.1 Phase separation in an Ising model, quenched (abruptly cooled) from high temperatures to zero temperature [173]. The model quickly separates into local blobs of up- and down-spins, which grow and merge, coarsening to larger blob sizes (Section 11.4.1).

Fig. 10.2 Surface annealing. An STM image of a surface, created by bombarding a close-packed gold surface with noble-gas atoms, and then allowing the irregular surface to thermally relax (Tatjana Curcic and Barbara H. Cooper [49]). The figure shows individual atomic-height steps; the arrows each show a single step pit inside another pit. The characteristic sizes of the pits and islands grow as the surface evolves and flattens.

[3]We will discuss coarsening in more detail in Section 11.4.1.

some testable numbers or functions (measuring the patterns in space and time) that we can use to compare experiment and theory?

Figure 10.1 is the Ising model evolving at low temperature, showing the spin $S(\mathbf{x}, t)$ at position $\mathbf{x}$ and time t; the up-spin and down-spin regions are competing [173] to determine which will take over as the broken-symmetry ground state. Figure 10.2 is a gold surface that is thermally flattening from an irregular initial shape [49], showing the height $h(\mathbf{x}, t)$. These visual images incorporate a full, rich description of individual members of the ensemble of models—but it is hard to quantify whether experiments and theory agree by comparing snapshots of a random environment. In these two evolving systems, we might quantify the evolution with a measure of the typical feature size as a function of time. Figure 10.3 shows the Ising model at T_c, where fluctuations occur on all length and time scales. In this equilibrium system we might want a function that describes how likely a black region will extend a distance $\mathbf{r}$, or survive for a time τ.

We typically measure the space and time coherence in a system using *correlation functions*. Consider the alignment of two Ising spins $S(\mathbf{x}, t)$ and $S(\mathbf{x} + \mathbf{r}, t)$ in the coarsening figure (Fig. 10.1); spins measured at the same time t, but separated by a distance $\mathbf{r}$. If $|\mathbf{r}|$ is much larger than a typical blob size $L(t)$, the spins will have a 50/50 chance of being aligned or misaligned, so their average product will be near zero. If $|\mathbf{r}|$ is much smaller than a typical blob size L, the spins will usually be aligned parallel to one another (both $+1$ or both -1), so their average product will be near one. The equal-time spin–spin correlation function

$$C_t^{\mathrm{coar}}(\mathbf{r}) = \langle S(\mathbf{x}, t) S(\mathbf{x} + \mathbf{r}, t) \rangle \tag{10.1}$$

will go from one at $\mathbf{r} = \mathbf{0}$ to zero at $|\mathbf{r}| \gg L(t)$, and will cross $1/2$ at a characteristic blob size $L(t)$. In nonequilibrium problems like this one, the system is evolving in time, so the equal-time correlation function also evolves.[3]

The correlation function in general contains more information than just the typical blob size. Consider the equilibrium correlation function, say, for the Ising model

$$C(\mathbf{r}, \tau) = \langle S(\mathbf{x}, t) S(\mathbf{x} + \mathbf{r}, t + \tau) \rangle. \tag{10.2}$$

The *equal-time* correlation function $C(\mathbf{r}, 0)$ contains information about how much a spin influences its distant neighbors. Even at high temperatures, if a spin is up its immediate neighbors are more likely to point up than down. As the temperature approaches the ferromagnetic transition temperature T_c, this preference extends to further neighbors (Fig. 10.4). Below T_c we have *long-range order*; even very distant neighbors will tend to align with our spin, since the two broken-symmetry equilibrium states each have net magnetization per spin m. Above T_c the equal-time correlation function goes to zero at long distances r; below T_c it goes to m^2, since the fluctuations of two distant spins about the mean magnetization become uncorrelated:

$$C(\infty, 0) = \lim_{r \to \infty} \langle S(\mathbf{x}, t) S(\mathbf{x}+\mathbf{r}, t) \rangle = \langle S(\mathbf{x}, t) \rangle \langle S(\mathbf{x}+\mathbf{r}, t) \rangle = m^2. \tag{10.3}$$

What happens at the critical temperature? At T_c the equal-time correlation function decays as a power law $C(\mathbf{r}, 0) \sim \mathbf{r}^{-(d-2+\eta)}$ at long distances (Fig. 10.5), representing the fact that there are correlations at all length scales (a compromise between short- and infinite-range order). Similarly, the (equal-position) spin–spin correlation function $C(\mathbf{0}, \tau) = \langle s(t)s(t + \tau)\rangle$ at long times τ goes to zero for $T > T_c$, to m^2 for $T < T_c$, and at T_c decays as a power law with a different exponent $C(\mathbf{0}, \tau) \sim \tau^{-(d-2+\eta)/z}$. We will see how to explain these power laws in Chapter 12, when we study continuous phase transitions.

In other systems, one might study the atomic density–density correlation functions $C(\mathbf{r}, \tau) = \langle \rho(\mathbf{x} + \mathbf{r}, t + \tau)\rho(\mathbf{x}, t)\rangle$,[4] or the height–height correlation function for a surface (Fig. 10.2), or the phase–phase correlations of the superfluid order parameter, ...

10.2 Experimental probes of correlations

Many scattering experiments directly measure correlation functions. X-rays measure the electron density–density correlation function, neutrons can measure spin–spin correlation functions, and so on. Elastic scattering gives the equal-time correlation functions, while inelastic scattering can give the time-dependent correlation functions.

Let us briefly summarize how this works for X-ray elastic scattering. In X-ray diffraction[5] (Fig. 10.6) a plane-wave beam of wavevector $\mathbf{k}_0$ scatters off of the sample, with the emitted radiation along wavevector $\mathbf{k}_0 + \mathbf{k}$ proportional to $\widetilde{\rho}_e(\mathbf{k})$, the Fourier transform of the electron density $\rho_e(\mathbf{x})$ in the sample. The intensity of the scattered beam $|\widetilde{\rho}_e(\mathbf{k})|^2$ can be measured, for example, by exposing photographic film. But this intensity is given by the Fourier transform of the equal-time electron[6] density–density correlation function $C_{ee}(\mathbf{r}) = \langle \rho_e(\mathbf{x} + \mathbf{r}, t)\rho_e(\mathbf{x}, t)\rangle$ (eqn A.21):

$$
\begin{aligned}
|\widetilde{\rho}(\mathbf{k})|^2 = \widetilde{\rho}(\mathbf{k})^*\widetilde{\rho}(\mathbf{k}) &= \int \mathrm{d}\mathbf{x}'\, e^{i\mathbf{k}\cdot\mathbf{x}'}\rho(\mathbf{x}') \int \mathrm{d}\mathbf{x}\, e^{-i\mathbf{k}\cdot\mathbf{x}}\rho(\mathbf{x}) \\
&= \int \mathrm{d}\mathbf{x}\,\mathrm{d}\mathbf{x}'\, e^{-i\mathbf{k}\cdot(\mathbf{x}-\mathbf{x}')}\rho(\mathbf{x}')\rho(\mathbf{x}) \\
&= \int \mathrm{d}\mathbf{r}\, e^{-i\mathbf{k}\cdot\mathbf{r}} \int \mathrm{d}\mathbf{x}'\, \rho(\mathbf{x}')\rho(\mathbf{x}' + \mathbf{r}) \\
&= \int \mathrm{d}\mathbf{r}\, e^{-i\mathbf{k}\cdot\mathbf{r}} V \langle \rho(\mathbf{x})\rho(\mathbf{x} + \mathbf{r})\rangle = V \int \mathrm{d}\mathbf{r}\, e^{-i\mathbf{k}\cdot\mathbf{r}} C(\mathbf{r}) \\
&= V\widetilde{C}(\mathbf{k}).
\end{aligned}
\tag{10.4}
$$

In the same way, other scattering experiments also measure two-point correlation functions, averaged over the entire illuminated sample.

Real-space microscopy experiments and k-space diffraction experiments provide complementary information about a system. The real-space images are direct and easily appreciated and comprehended by the human mind. They are invaluable for studying unusual events (which

Fig. 10.3 Critical fluctuations. The two-dimensional square-lattice Ising model near the critical temperature T_c. Here the "islands" come in all sizes, and the equilibrium fluctuations happen on all time scales; see Chapter 12.

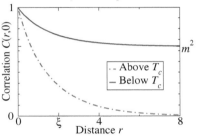

Fig. 10.4 Equal-time correlation function. A schematic equal-time correlation function $C(r, \tau = 0)$ at a temperature just above and just below the critical temperature T_c. At $r = 0$ the correlation function is $\langle S^2\rangle = 1$. (The distance $\xi(T)$ after which the correlation function decays exponentially to its long-distance value (zero or m^2) is the *correlation length*. At T_c the correlation length diverges, leading to fluctuations on all scales, Chapter 12).

[4]Here $\rho(\mathbf{x}, t) = \sum_j \boldsymbol{\delta}(\mathbf{x} - \mathbf{x}_j)$ measures the positions of the atoms.

[5]Medical X-rays and CAT scans measure the penetration of X-rays, not their diffraction.

[6]Since the electrons are mostly tied to atomic nuclei, $C_{ee}(\mathbf{r})$ is writable in terms of atom–atom correlation functions (Exercise 10.2). This is done using *form factors* [13, chapter 6].

Fig. 10.5 Power-law correlations. The schematic correlation function of Fig. 10.4 on a log–log plot, both at T_c (straight line, representing the power law $C \sim r^{-(d-2+\eta)}$) and above T_c (where the dependence shifts to $C \sim e^{-r/\xi(T)}$ at distances beyond the correlation length $\xi(T)$). See Chapter 12.

Fig. 10.6 X-ray scattering. A beam of wavevector $\mathbf{k}_0$ scatters off a density variation $\rho(\mathbf{x})$ with wavevector $\mathbf{k}$ to a final wavevector $\mathbf{k}_0 + \mathbf{k}$; the intensity of the scattered beam is proportional to $|\tilde{\rho}(\mathbf{k})|^2$ [13, chapter 6].

[7] $\delta\mathcal{F}/\delta\rho$ is a *variational derivative*. $\mathcal{F}$ is a mapping taking functions to other functions; under a small change $\delta\rho$ in its argument, $\mathcal{F}\{\rho + \delta\rho\} - \mathcal{F}\{\rho\} = \int (\delta\mathcal{F}/\delta\rho)\,\delta\rho\,\mathrm{d}\mathbf{x}$. The integral can be viewed as a kind of dot product of $\delta\rho$ with $\delta\mathcal{F}/\delta\rho$ (see Section A.3 for inner products in function space), so the variational derivative is just like a gradient of a function, where $f(\mathbf{x} + \boldsymbol{\delta}) = \nabla f \cdot \boldsymbol{\delta}$.

would be swamped in a bulk average), distributions of local geometries (individual ensemble elements rather than averages over the ensemble), and physical dynamical mechanisms. The k-space methods, on the other hand, by averaging over the entire sample, can provide great precision, and have close ties with calculational and analytical methods (as presented in this chapter). Indeed, often one will computationally Fourier transform measured real-space data in order to generate correlation functions (Exercise 10.1).

10.3 Equal-time correlations in the ideal gas

For the rest of this chapter we will consider systems which are in equilibrium and close to equilibrium. In these cases, we shall be able to find surprisingly tight relations between the correlations, response, and dissipation. We focus on the ideal gas, which is both the simplest and the most difficult case. In the exercises, you can calculate correlation functions that are algebraically more challenging (Exercise 10.8) but the ideal gas case is both conceptually subtle and fundamental. Let us start by calculating the equal-time correlation function $C^{\text{ideal}}(\mathbf{r}, 0)$.

The Helmholtz free energy density of the ideal gas is

$$\mathcal{F}^{\text{ideal}}(\rho(\mathbf{x}), T) = \rho(\mathbf{x})k_B T \left[\log(\rho(\mathbf{x})\lambda^3) - 1\right] \quad (10.5)$$

(eqn 6.62). The probability $P\{\rho(\mathbf{x})\}$ of finding a particular density profile $\rho(\mathbf{x})$ as a fluctuation is proportional to

$$P\{\rho(\mathbf{x})\} \propto e^{-\beta \int \mathcal{F}(\rho(\mathbf{x}))\,\mathrm{d}\mathbf{x}}. \quad (10.6)$$

Let us assume the fluctuations are small, and expand about $\langle\rho\rangle = \rho_0$:

$$\mathcal{F}(\rho(\mathbf{x})) = \mathcal{F}_0 + \left.\frac{\delta\mathcal{F}}{\delta\rho}\right|_{\rho_0} (\rho - \rho_0) + \frac{1}{2}\left.\frac{\delta^2\mathcal{F}}{\delta\rho^2}\right|_{\rho_0} (\rho - \rho_0)^2$$

$$= \mathcal{F}_0 + \mu_0(\rho - \rho_0) + \frac{1}{2}\alpha(\rho - \rho_0)^2, \quad (10.7)$$

where[7] $\mu_0 = (\delta\mathcal{F}/\delta\rho)|_{\rho_0}$ is the chemical potential and the coefficient of the quadratic term is

$$\alpha = \left.\frac{\partial^2\mathcal{F}}{\partial\rho^2}\right|_{\rho_0} = k_B T/\rho_0 = P_0/\rho_0^2 \quad (10.8)$$

(since the pressure $P_0 = Nk_B T/V = \rho_0 k_B T$). Only the integral of the free energy matters, so

$$\int \mathcal{F}(\rho(\mathbf{x}))\,\mathrm{d}\mathbf{x} = V\mathcal{F}_0 + \mu_0 \int (\rho - \rho_0)\,\mathrm{d}\mathbf{x} + \int \frac{1}{2}\alpha(\rho - \rho_0)^2\,\mathrm{d}\mathbf{x}, \quad (10.9)$$

where μ_0 drops out because the average of ρ equals ρ_0. We can also drop $\mathcal{F}_0$ because it changes the free energy of all configurations by a constant, and does not change their relative probabilities.[8] So the effective free energy of the ideal gas, for small density fluctuations, is

$$\mathcal{F}(\rho) = \frac{1}{2}\alpha(\rho - \rho_0)^2, \tag{10.10}$$

and the probability of finding a density fluctuation is

$$P\{\rho(\mathbf{x})\} \propto e^{-\beta \int \frac{1}{2}\alpha(\rho - \rho_0)^2 \, d\mathbf{x}}. \tag{10.11}$$

We can now calculate the expectation value of the equal-time density–density correlation function:

$$
\begin{aligned}
C^{\text{ideal}}(\mathbf{r}, 0) &= \langle \rho(\mathbf{x}, t)\rho(\mathbf{x} + \mathbf{r}, t)\rangle - \rho_0^2 \\
&= \langle (\rho(\mathbf{x}, t) - \rho_0)(\rho(\mathbf{x} + \mathbf{r}, t) - \rho_0)\rangle.
\end{aligned} \tag{10.12}
$$

Here we subtract off the square of the average density, so that we measure the correlations between the fluctuations of the order parameter about its mean value. (Subtracting the means gives us the *connected* correlation function.) If we break up the ideal gas into tiny boxes of size ΔV, the probability of having density $\rho(\mathbf{x}_j)$ in volume j is

$$P_j(\rho) \propto \exp\left(-\frac{\frac{1}{2}\alpha(\rho - \rho_0)^2 \Delta V}{k_B T}\right) = \exp\left(-\frac{(\rho - \rho_0)^2}{2/(\beta\alpha\,\Delta V)}\right). \tag{10.13}$$

This is a Gaussian with RMS $\sigma = \sqrt{1/(\beta\alpha\,\Delta V)}$, so the mean square fluctuations inside a single box is

$$\langle (\rho - \rho_0)^2 \rangle = \frac{1}{\beta\alpha\,\Delta V}. \tag{10.14}$$

The density fluctuations in different boxes are uncorrelated. This means $C^{\text{ideal}}(\mathbf{r}, 0) = 0$ for $\mathbf{r}$ reaching between two boxes,[9] and $C^{\text{ideal}}(\mathbf{0}, 0) = 1/(\beta\alpha\,\Delta V)$ within one box.

What does it mean for C^{ideal} to depend on the box size ΔV? The fluctuations become stronger as the box gets smaller. We are familiar with this; we saw earlier using the microcanonical ensemble (Section 3.2.1 and eqn 3.66) that the square of the number fluctuations in a small subvolume of ideal gas was equal to the expected number of particles $\langle (N - \langle N \rangle)^2 \rangle = N$,[10] so the fractional fluctuations $1/\sqrt{N}$ get larger as the volume gets smaller.

How do we write the correlation function, though, in the limit $\Delta V \to 0$? It must be infinite at $\mathbf{r} = \mathbf{0}$, and zero for all nonzero $\mathbf{r}$. More precisely, it is proportional the Dirac δ-function, this time in three dimensions $\boldsymbol{\delta}(\mathbf{r}) = \delta(r_x)\delta(r_y)\delta(r_z)$. The equal-time connected correlation function for the ideal gas is

$$C^{\text{ideal}}(\mathbf{r}, 0) = \frac{1}{\beta\alpha}\boldsymbol{\delta}(\mathbf{r}). \tag{10.16}$$

The correlation function of an uncorrelated system is white noise.[11]

[8]Each Boltzmann factor shifts by $e^{-\beta V \mathcal{F}_0}$, so Z shifts by the same factor, and so the ratio $e^{-\beta V \mathcal{F}_0}/Z$ giving the probability is independent of $\mathcal{F}_0$.

[9]Thus the correlation length ξ for the ideal gas is zero (Fig. 10.4).

[10]Do the two calculations agree? Using eqn 10.8 and the ideal gas law $P_0 V = N k_B T = N/\beta$, the density fluctuations

$$
\begin{aligned}
\langle (\rho - \rho_0)^2 \rangle &= \frac{(N - \langle N \rangle)^2}{(\Delta V)^2} \\
&= \frac{N}{(\Delta V)^2} = \frac{\rho_0}{\Delta V} \\
&= \frac{1}{(\rho_0/P_0)(P_0/\rho_0^2)\Delta V} \\
&= \frac{1}{(N/P_0 V)\,\alpha\Delta V} \\
&= \frac{1}{\beta\alpha\Delta V} \tag{10.15}
\end{aligned}
$$

are just as we computed from $\mathcal{F}^{\text{ideal}}$.

[11]White light is a mixture of all frequencies of light with equal amplitude and random phases (Exercise A.8). Our noise has the same property. The Fourier transform of $C(r, 0)$, $\widetilde{C}(\mathbf{k}, t = 0) = (1/V)|\widetilde{\rho}(k)|^2$ (as in eqn 10.4), is constant, independent of the wavevector k. Hence all modes have equal weight. To show that the phases are random, we can express the free energy (eqn 10.9) in Fourier space, where it is a sum over uncoupled harmonic modes; hence in equilibrium they have random relative phases (see Exercise 10.8).

We can see that the constant outside is indeed $1/\beta\alpha$, by using C^{ideal} to calculate the mean square fluctuations of the integral of ρ inside the box of volume ΔV:

$$
\begin{aligned}
\langle(\rho-\rho_0)^2\rangle_{\text{box}} &= \left\langle\left(\frac{1}{\Delta V}\int_{\Delta V}(\rho(\mathbf{x})-\rho_0)\,\mathrm{d}\mathbf{x}\right)^2\right\rangle \\
&= \frac{1}{(\Delta V)^2}\int_{\Delta V}\mathrm{d}\mathbf{x}\int_{\Delta V}\mathrm{d}\mathbf{x}'\,\langle(\rho(\mathbf{x})-\rho_0)(\rho(\mathbf{x}')-\rho_0)\rangle \\
&= \frac{1}{(\Delta V)^2}\int_{\Delta V}\mathrm{d}\mathbf{x}\int_{\Delta V}\mathrm{d}\mathbf{x}'\,\frac{1}{\beta\alpha}\boldsymbol{\delta}(\mathbf{x}-\mathbf{x}') \\
&= \frac{1}{\beta\alpha(\Delta V)^2}\int_{\Delta V}\mathrm{d}\mathbf{x} = \frac{1}{\beta\alpha\,\Delta V},
\end{aligned}
\tag{10.17}
$$

in agreement with our earlier calculation (eqn 10.14).

10.4 Onsager's regression hypothesis and time correlations

Equilibrium statistical mechanics does not determine the dynamics. Air and perfume are both good ideal gases, but density fluctuations in air lead to sound waves, while they lead to diffusion in perfume (which scatters off the air molecules). We need to supplement the free energy with more information in order to calculate time correlations. There are two basic choices. We could work with the microscopic laws; indeed, most treatments of this topic start from quantum mechanics [64]. Instead, here we will rely on the *macroscopic* evolution laws to specify our dynamics.

How are the density fluctuations in an ideal gas of perfume correlated in time? In particular, suppose at $t = 0$ there is a rare fluctuation, increasing the density of perfume at one point (Fig. 10.7). How will it decay to a more typical density profile as time passes?

Macroscopically our perfume obeys the diffusion equation of Chapter 2. There we derived the evolution laws for imposed initial nonequilibrium density profiles, and ignored the spontaneous thermal fluctuations. A macro-scale initial condition (Fig. 10.8) will evolve according to the diffusion equation $\partial\rho/\partial t = D\nabla^2\rho$. The confusing point about the microscopic density (Fig. 10.7) is that it introduces new spontaneous thermal fluctuations while it flattens old ones.

In this text, we have been rather casual in denoting averages, using the same symbol $\langle\cdot\rangle$ for time averages, spatial averages, and averages over microcanonical, canonical, and grand canonical ensembles. In this section we will be doing several different kinds of averages, and we need to distinguish between them. Our microcanonical, canonical, and grand canonical ensemble averages cannot be used to calculate quantities depending on more than one time (because the equilibrium ensembles are independent of dynamics). Let us write $\langle\cdot\rangle_{\text{eq}}$ for these equal-time equilibrium averages. The time–time correlation functions are defined by an

Fig. 10.7 Noisy decay of a fluctuation. An unusual fluctuation at $t = 0$ will slowly decay to a more typical thermal configuration at a later time τ.

Fig. 10.8 Deterministic decay of an initial state. An initial condition with the same density will slowly decay to zero.

equilibrium ensemble of time evolutions, which may include noise from the environment. Let us denote these averages by $\langle \cdot \rangle_{\mathrm{ev}}$. Thus $\langle \cdot \rangle_{\mathrm{ev}}$ generalizes $\langle \cdot \rangle_{\mathrm{eq}}$ to work on quantities that depend on more than one time. Finally, let us write $[\cdot]_{\rho_i}$ for the noisy evolution average fixing the initial condition $\rho(\mathbf{x}, 0) = \rho_i(\mathbf{x})$ at time zero. This averages over all the new spontaneous density fluctuations, allowing us to examine the decay of an initial spontaneous density fluctuation, or perhaps an initial imposed density profile.

We will assume that this last average fixing the initial condition obeys the same diffusion equation that governs the macroscopic time evolution (Figs. 10.7 and 10.8). For our diffusion of perfume, this means

$$\frac{\partial}{\partial t} [\rho(\mathbf{x}, t)]_{\rho_i} = D \nabla^2 [\rho(\mathbf{x}, t)]_{\rho_i} . \tag{10.18}$$

This assumption is precisely Onsager's *regression hypothesis* [147]:

> ... we may assume that a spontaneous deviation from the equilibrium decays according to the same laws as one that has been produced artificially.

Let us calculate the correlation function $\langle \rho(\mathbf{r} + \mathbf{r}', t + t')\rho(\mathbf{r}', t') \rangle_{\mathrm{ev}}$ by taking our evolution ensemble for fixed initial condition $[\rho(\mathbf{x}, t)]_{\rho_i}$ and then taking a thermal average over initial conditions $\rho(\mathbf{x}, 0) = \rho_i(\mathbf{x})$. We may use the fact that our system is homogeneous and time independent to measure our correlation function starting at the origin:

$$\begin{aligned}
C(\mathbf{r}, \tau) &= \langle (\rho(\mathbf{x} + \mathbf{r}, t + \tau) - \rho_0)(\rho(\mathbf{x}, t) - \rho_0) \rangle_{\mathrm{ev}} \\
&= \langle (\rho(\mathbf{r}, \tau) - \rho_0)(\rho(\mathbf{0}, 0) - \rho_0) \rangle_{\mathrm{ev}} \\
&= \left\langle ([\rho(\mathbf{r}, \tau)]_{\rho_i} - \rho_0)(\rho_i(\mathbf{0}) - \rho_0) \right\rangle_{\mathrm{eq}} .
\end{aligned} \tag{10.19}$$

In words, averaging over both initial conditions and noise $\langle \cdot \rangle_{\mathrm{ev}}$ is the same as first averaging over noise $[\cdot]_i$ and then over initial conditions $\langle \cdot \rangle_{\mathrm{eq}}$. We know from Onsager's regression hypothesis that

$$\begin{aligned}
\frac{\partial C^{\mathrm{ideal}}}{\partial t} &= \left\langle \frac{\partial}{\partial t} [\rho(\mathbf{r}, t)]_{\rho_i} (\rho_i(\mathbf{0}) - \rho_0) \right\rangle_{\mathrm{eq}} \\
&= \left\langle D \nabla^2 [\rho(\mathbf{r}, t)]_{\rho_i} (\rho_i(\mathbf{0}) - \rho_0) \right\rangle_{\mathrm{eq}} \\
&= D \nabla^2 \left\langle [\rho(\mathbf{r}, t)]_{\rho_i} (\rho_i(\mathbf{0}) - \rho_0) \right\rangle_{\mathrm{eq}} \\
&= D \nabla^2 \langle (\rho(\mathbf{r}, t) - \rho_0)(\rho(\mathbf{0}, 0) - \rho_0) \rangle_{\mathrm{ev}} \\
&= D \nabla^2 C^{\mathrm{ideal}}(\mathbf{r}, t).
\end{aligned} \tag{10.20}$$

The correlation function C obeys the same equation as the decays of imposed initial conditions. This is true in general.[12]

Thus to solve in a general system for the correlation function C, we must calculate as the initial condition the instantaneous correlations $C(\mathbf{x}, 0)$ using equilibrium statistical mechanics, and evolve it according

[12] We can use Onsager's regression hypothesis to calculate the correlation function C for a general order parameter $s(\mathbf{x}, t)$. Suppose that the macroscopic time evolution of $s(\mathbf{x}, t)$, to linear order in deviations away from its average value $\bar{s}$, is given by some Green's function (Section 2.4.2):

$$\begin{aligned}
s_{\mathrm{macro}}(\mathbf{x}, t) \\
= \bar{s} - \int d\mathbf{x}' \, G(\mathbf{x} - \mathbf{x}', t) \\
\times (s_{\mathrm{macro}}(\mathbf{x}', 0) - \bar{s}).
\end{aligned} \tag{10.21}$$

For convenience, let us set $\bar{s} = 0$. This convolution simplifies if we Fourier transform in position $\mathbf{x}$ but not in time t, using the convolution theorem for Fourier transforms (eqn A.23):

$$\widehat{s}_{\mathrm{macro}}(\mathbf{k}, t) = \widehat{G}(\mathbf{k}, t)\widehat{s}(\mathbf{k}, 0), \tag{10.22}$$

where we use a hat to denote the Fourier transform confined to position space. Onsager's regression hypothesis says that a spontaneous initial thermal fluctuation s_i will evolve according to the same law,

$$\begin{aligned}
[\widehat{s}(\mathbf{k}, t)]_{\widehat{s}_i} &= \widehat{G}(\mathbf{k}, t)\widehat{s}_i(\mathbf{k}) \\
&= \widehat{G}(\mathbf{k}, t)\widehat{s}(\mathbf{k}, t = 0),
\end{aligned} \tag{10.23}$$

so the correlation function

$$\begin{aligned}
C(\mathbf{r}, t) &= \langle s(\mathbf{r}, t)s(\mathbf{0}, 0) \rangle_{\mathrm{ev}} \\
&= \left\langle [s(\mathbf{r}, t)]_{s_i} s_i(\mathbf{0}) \right\rangle_{\mathrm{eq}}
\end{aligned} \tag{10.24}$$

evolves by

$$\begin{aligned}
\widehat{C}(\mathbf{k}, t) &= \left\langle [\widehat{s}(\mathbf{k}, t)]_{\widehat{s}_i} s_i(\mathbf{0}) \right\rangle_{\mathrm{eq}} \\
&= \left\langle \widehat{G}(\mathbf{k}, t)\widehat{s}(\mathbf{k}, 0)s_i(\mathbf{0}) \right\rangle_{\mathrm{eq}} \\
&= \widehat{G}(\mathbf{k}, t)\widehat{C}(\mathbf{k}, 0). \tag{10.25}
\end{aligned}$$

Again, the correlation function obeys the same evolution law as the decay of an imposed initial condition.

to the macroscopic evolution law. In the case of the ideal perfume gas, the equal-time correlations (eqn 10.16) are $C^{\text{ideal}}(\mathbf{r}, 0) = 1/(\beta\alpha)\,\boldsymbol{\delta}(\mathbf{r})$, and the evolution law is given by the diffusion equation. We know how an initial δ-function distribution evolves under the diffusion equation:[13] it is given by the Green's function (Section 2.4.2). The Green's function for the diffusion equation in one dimension (eqn 2.32) is $G(x, t) = (1/\sqrt{4\pi Dt})\,\mathrm{e}^{-x^2/4Dt}$. In three dimensions we take the product along x, y, and z to get G, and then divide by $\beta\alpha$, to get the correlation function

$$C^{\text{ideal}}(\mathbf{r}, \tau) = \frac{1}{\beta\alpha}G(\mathbf{r}, \tau) = \frac{1}{\beta\alpha}\left(\frac{1}{\sqrt{4\pi D\tau}}\right)^3 \mathrm{e}^{-\mathbf{r}^2/4D|\tau|}. \tag{10.26}$$

> [13] Notice that Onsager's regression hypothesis allows us to evolve $C(\tau)$ forward in time using the Green's function. For negative times, we note that for a time-translation-invariant system $C(\tau) \sim \langle\rho(t+\tau)\rho(t)\rangle = \langle\rho(t')\rho(t'-\tau)\rangle = C(-\tau)$, hence the absolute value of τ in eqn 10.26. In linear response theory, this can be handled formally by introducing *retarded and advanced Green's functions*.

This is the correlation function for an ideal gas satisfying the diffusion equation.

10.5 Susceptibility and linear response

How will our system yield when we kick it? The *susceptibility* $\chi(\mathbf{r}, \tau)$ gives the response at a distance $\mathbf{r}$ and time τ from a (gentle) kick. Let us formulate susceptibility for a general order parameter $s(\mathbf{x}, t)$, kicked by an external field $f(\mathbf{x}, t)$. That is, we assume that f appears in the free energy functional

$$F = F_0 + F_f \tag{10.27}$$

as a term

$$F_f(t) = -\int \mathrm{d}\mathbf{x}\, f(\mathbf{x}, t)s(\mathbf{x}, t). \tag{10.28}$$

You can think of f as a force density pulling s upward. If s is the particle density ρ, then f is minus an external potential $-V(\mathbf{x})$ for the particles; if s is the magnetization M of an Ising model, then f is the external field H; if s is the polarization $\mathbf{P}$ of a dielectric material, then f is an externally applied vector electric field $\mathbf{E}(\mathbf{x}, t)$. For convenience, we will assume in this section that $\bar{s} = 0$.[14]

> [14] Or rather, we can define s to be the deviation from the average value of the order parameter in the absence of a field.

How will the order parameter field s respond to the force f? If the force is a weak perturbation, we can presume a linear response, but perhaps one which is nonlocal in space and time. So, $s(\mathbf{x}, t)$ will depend upon $f(\mathbf{x}', t')$ at all earlier times $t' < t$:

$$s(\mathbf{x}, t) = \int \mathrm{d}\mathbf{x}' \int_{-\infty}^{t} \mathrm{d}t'\, \chi(\mathbf{x} - \mathbf{x}', t - t')f(\mathbf{x}', t'), \tag{10.29}$$

where $\chi(\mathbf{r}, \tau) = 0$ for $\tau < 0$ (causality: see note 21 on p. 295).

This nonlocal relation becomes much simpler if we Fourier transform[15] s, f, and χ in space and time. The AC susceptibility[16] $\widetilde{\chi}(\mathbf{k}, \omega)$ satisfies

> [15] We will use a tilde $\widetilde{A}(\mathbf{k}, \omega)$ to represent the Fourier transform of the function $A(\mathbf{x}, t)$ with respect to both space and time. We will also use a tilde $\widetilde{B}(\mathbf{k})$ to represent the Fourier transform of the static function $B(\mathbf{x})$ with respect to space. But we will use a hat $\widehat{A}(\mathbf{k}, t)$ to represent the Fourier transform of $A(\mathbf{x}, t)$ in space $\mathbf{x}$ alone.

$$\widetilde{s}(\mathbf{k}, \omega) = \widetilde{\chi}(\mathbf{k}, \omega)\widetilde{f}(\mathbf{k}, \omega), \tag{10.30}$$

> [16] AC stands for "alternating current", the kind of electricity that is used in most buildings; the voltage fluctuates periodically in time. The current from batteries is DC or "direct current", which does not vary in time. Somehow we have started using AC for all frequency-dependent systems.

since as usual the Fourier transform of the convolution is the product of the Fourier transforms (eqn A.23). The function χ is the *susceptibility* of the order parameter s to the external field f. For example, the polarization versus field is defined in terms of the polarizability α:[17] $\mathbf{P}(\mathbf{k},\omega) = \widetilde{\alpha}(\mathbf{k},\omega)\widetilde{\mathbf{E}}(\mathbf{k},\omega)$, the magnetization from an external field is $\widetilde{\mathbf{M}}(\mathbf{k},\omega) = \widetilde{\chi}(\mathbf{k},\omega)\widetilde{\mathbf{H}}(\mathbf{k},\omega)$, and so on.

10.6 Dissipation and the imaginary part

The time–time susceptibility $\chi(\mathbf{x},t)$ is real, but the AC susceptibility

$$\widetilde{\chi}(\mathbf{k},\omega) = \int d\mathbf{x}\,dt\;e^{i\omega t}e^{-i\mathbf{k}\cdot\mathbf{x}}\chi(\mathbf{x},t) = \chi'(\mathbf{k},\omega) + i\chi''(\mathbf{k},\omega) \qquad (10.31)$$

has a real part $\chi' = \mathrm{Re}[\widetilde{\chi}]$ and an imaginary part $\chi'' = \mathrm{Im}[\widetilde{\chi}]$.[18] It is clear from the definition that $\widetilde{\chi}(-\mathbf{k},-\omega) = \widetilde{\chi}^*(\mathbf{k},\omega)$; for a system with inversion symmetry $\mathbf{x} \leftrightarrow -\mathbf{x}$ we see further that $\chi(\mathbf{x},t) = \chi(-\mathbf{x},t)$ and hence $\widetilde{\chi}(\mathbf{k},-\omega) = \widetilde{\chi}^*(\mathbf{k},\omega)$, so χ' is even in ω and χ'' is odd. χ' gives the in-phase response to a sinusoidal force, and χ'' gives the response that lags in phase by $\pi/2$.[19]

The imaginary part χ'' in general gives the amount of *dissipation* induced by the external field.[20] The dissipation can be measured directly (for example, by measuring the resistance as a function of frequency of a wire) or by looking at the decay of waves in the medium (optical and ultrasonic attenuation and such). We know that "energy" is the integral of "force" f times "distance" ∂s, or force times velocity $\partial s/\partial t$ integrated over time. Ignoring the spatial dependence for simplicity, the time average of the power p dissipated per unit volume is

$$p = \lim_{T\to\infty}\frac{1}{T}\int_0^T f(t)\frac{\partial s}{\partial t}\,dt = \lim_{T\to\infty}\frac{1}{T}\int_0^T -s(t)\frac{\partial f}{\partial t}\,dt, \qquad (10.33)$$

where we have averaged over a time T and integrated by parts, assuming the boundary terms are negligible for $T\to\infty$. Assuming an AC force $f(t) = \mathrm{Re}[f_\omega e^{-i\omega t}] = \frac{1}{2}(f_\omega e^{-i\omega t} + f_\omega^* e^{i\omega t})$, we have

$$p(\omega) = \lim_{T\to\infty}\frac{1}{T}\int_0^T s(t)\frac{i\omega}{2}(f_\omega e^{-i\omega t} - f_\omega^* e^{i\omega t})\,dt, \qquad (10.34)$$

where the motion $s(t)$ is in turn due to the forcing at earlier times:[21]

$$s(t) = \int_{-\infty}^{\infty} dt'\;\chi(t-t')f(t')$$

$$= \int_{-\infty}^{\infty} d\tau\;\chi(\tau)f(t-\tau) \qquad (10.35)$$

$$= \int_{-\infty}^{\infty} d\tau\;\frac{\chi(\tau)}{2}(f_\omega e^{-i\omega(t-\tau)} + f_\omega^* e^{i\omega(t-\tau)}).$$

[17]In electromagnetism, one usually uses the dielectric permittivity ϵ rather than the polarizability α. In SI/MKS units, $\alpha = \epsilon - \epsilon_0$, subtracting off the "permittivity of the vacuum" ϵ_0; the dielectric constant is ϵ/ϵ_0. In Gaussian CGS units, $\alpha = (\epsilon - 1)/4\pi$, (and the dielectric constant is also ϵ). Note also α and ϵ are tensors (matrices) in anisotropic materials, and $\mathbf{P}$ need not be parallel to $\mathbf{E}$.

[18]Some use the complex conjugate of our formulæ for the Fourier transform (see Section A.1), substituting $-i$ for i in the time Fourier transforms. Their χ'' is the same as ours, because they define it to be *minus* the imaginary part of their Fourier-transformed susceptibility.

[19]If we apply $f(t) = \cos(\omega_0 t) = 1/2\pi \int \widetilde{f}(\omega)\exp(-i\omega t)dt$ (eqn A.7), so $\widetilde{f}(\omega) = \pi(\delta(\omega-\omega_0) + \delta(\omega+\omega_0))$, then the response is $\widetilde{s}(\omega) = \widetilde{\chi}(\omega)\widetilde{f}(\omega)$, so

$$s(t) = \frac{1}{2\pi}\int e^{-i\omega t}\widetilde{s}(\omega)\,d\omega$$
$$= \frac{1}{2}\left(e^{-i\omega_0 t}\chi(\omega_0) + e^{i\omega_0 t}\chi(-\omega_0)\right)$$
$$= \frac{1}{2}\big(e^{-i\omega_0 t}(\chi'(\omega_0) + i\chi''(\omega_0))$$
$$\qquad + e^{i\omega_0 t}(\chi'(\omega_0) - i\chi''(\omega_0)))$$
$$= (\chi'(\omega_0)\cos(\omega_0 t) \qquad (10.32)$$
$$\qquad + \chi''(\omega_0)\sin(\omega_0 t)).$$

Hence χ' gives the immediate in-phase response, and χ'' gives the response delayed by $\pi/2$.

[20]The real part is sometimes called the *reactive* response, whereas the imaginary part is the *dissipative* response.

[21]We define $\chi(\tau) = 0$ for $\tau < 0$. Negative times τ would correspond to responses χ to kicks in the future, which are forbidden by causality.

Substituting eqn 10.35 into eqn 10.34, we get

$$
\begin{aligned}
p(\omega) &= \lim_{T \to \infty} \frac{1}{T} \int_0^T \mathrm{d}t \int_{-\infty}^{\infty} \mathrm{d}\tau \, \frac{i\omega\chi(\tau)}{4} \\
&\quad \times (f_\omega e^{-i\omega(t-\tau)} + f_\omega^* e^{i\omega(t-\tau)})(f_\omega e^{-i\omega t} - f_\omega^* e^{i\omega t}) \quad (10.36) \\
&= \int_{-\infty}^{\infty} \mathrm{d}\tau \, \frac{i\omega\chi(\tau)}{4} \lim_{T \to \infty} \frac{1}{T} \int_0^T \mathrm{d}t \\
&\quad \times \left[f_\omega^2 e^{-i\omega(2t-\tau)} - f_\omega^{*\,2} e^{i\omega(2t-\tau)} + |f_\omega|^2 (e^{-i\omega\tau} - e^{i\omega\tau}) \right].
\end{aligned}
$$

The first and second terms are zero, and the third gives a sine,[22] so

$$
\begin{aligned}
p(\omega) &= \frac{\omega|f_\omega|^2}{2} \int_{-\infty}^{\infty} \mathrm{d}\tau \, \chi(\tau) \sin(\omega\tau) = \frac{\omega|f_\omega|^2}{2} \mathrm{Im}[\widetilde{\chi}(\omega)] \\
&= \frac{\omega|f_\omega|^2}{2} \chi''(\omega).
\end{aligned} \quad (10.37)
$$

Since the power dissipated must be positive, we find $\omega\chi''(\omega)$ is positive.[23]

Let us interpret this formula in the familiar case of electrical power dissipation in a wire. Under a (reasonably low-frequency) AC voltage $V(t) = V_\omega \cos(\omega t)$, a wire of resistance R dissipates average power $\langle P \rangle = \langle V^2/R \rangle = V_\omega^2 \langle \cos^2(\omega t) \rangle / R = \frac{1}{2} V_\omega^2 / R$ by Ohm's law. A wire of length L and cross-section A has electric field $E_\omega \cos \omega t$ with $E_\omega = V_\omega / L$, and it has resistance $R = L/(\sigma A)$, where σ is the conductivity of the metal. So the average dissipated power per unit volume $p(\omega) = \langle P \rangle / (LA) = \frac{1}{2}((E_\omega L)^2/(L/\sigma A))(1/LA) = \frac{1}{2}\sigma E_\omega^2$. Remembering that E_ω is the force f_ω and the polarizability $\alpha(\omega)$ is the susceptibility $\chi(\omega)$, eqn 10.37 tells us that the DC conductivity is related to the limit of the AC polarizability at zero frequency: $\sigma = \lim_{\omega \to 0} \omega \, \alpha''(\omega)$.

10.7 Static susceptibility

In many cases, we are interested in how a system responds to a static external force—rather than kicking a system, we lean on it. Under a point-like force, the dimple formed in the order parameter field is described by the *static susceptibility* $\chi_0(\mathbf{r})$.

If the external force is time independent (so $f(\mathbf{x}', t') = f(\mathbf{x}')$) the system will reach a perturbed equilibrium. The nonlocal relation between s and a small field f is given by the *static susceptibility*, χ_0:

$$
s(\mathbf{x}) = \int \mathrm{d}\mathbf{x}' \, \chi_0(\mathbf{x} - \mathbf{x}') f(\mathbf{x}'). \quad (10.38)
$$

If we take the Fourier series of s and f, we may represent this relation in terms of the Fourier transform of χ_0 (eqn A.23):

$$
\widetilde{s}_\mathbf{k} = \widetilde{\chi}_0(\mathbf{k}) \widetilde{f}_\mathbf{k}. \quad (10.39)
$$

[22] In particular,

$$i(e^{-i\omega\tau} - e^{i\omega\tau}) = 2\sin(\omega\tau),$$

$$\lim_{T \to \infty} (1/T) \int_0^T \mathrm{d}t \, e^{\pm 2i\omega t} = 0, \text{ and}$$

$$\lim_{T \to \infty} (1/T) \int_0^T \mathrm{d}t \, e^{0i\omega t} = 1.$$

[23] We knew already (beginning of this section) that χ'' was odd; now we know also that it is positive for $\omega > 0$.

We may use equilibrium statistical mechanics to find the resulting static change in the average order parameter field $s(\mathbf{x})$. As an example,[24] the free energy density for the ideal gas, in the linearized approximation of Section 10.3, is $\mathcal{F} = \frac{1}{2}\alpha(\rho - \rho_0)^2$. For a spatially varying static external potential $f(\mathbf{x}) = -V(\mathbf{x})$, this is minimized by $\rho(\mathbf{x}) = \rho_0 + f(\mathbf{x})/\alpha$, so (comparing with eqn 10.38) we find the static susceptibility is

$$\chi_0^{\text{ideal}}(\mathbf{r}) = \delta(\mathbf{r})/\alpha, \tag{10.43}$$

and in Fourier space it is $\widetilde{\chi}_0(\mathbf{k}) = 1/\alpha$. Here χ_0 is the "spring constant" giving the response to a constant external force.

Notice for the ideal gas that the static susceptibility $\chi_0^{\text{ideal}}(\mathbf{r})$ and the equal-time correlation function $C^{\text{ideal}}(\mathbf{r}, 0) = \delta(\mathbf{r})/(\beta\alpha)$ are proportional to one another: $\chi_0^{\text{ideal}}(\mathbf{r}) = \beta C^{\text{ideal}}(\mathbf{r}, 0)$. This can be shown to be true in general for equilibrium systems[24]

$$\chi_0(\mathbf{r}) = \beta C(\mathbf{r}, 0). \tag{10.44}$$

That is, the $\omega = 0$ static susceptibility χ_0 is given by dividing the instantaneous correlation function C by $k_B T$—both in real space and also in Fourier space:

$$\widetilde{\chi}_0(\mathbf{k}) = \beta\widehat{C}(\mathbf{k}, 0). \tag{10.45}$$

This *fluctuation-response* relation should be intuitively reasonable; a system or mode which is easy to perturb will also have big fluctuations.

---

[24]We can derive eqn 10.45 in general, by a (somewhat abstract) calculation. We find the expectation value $\langle \widetilde{s}_{\mathbf{k}} \rangle_{\text{eq}}$ for a given $\widetilde{f}_{\mathbf{k}}$, and then take the derivative with respect to $\widetilde{f}_{\mathbf{k}}$ to get $\widetilde{\chi}_0(\mathbf{k})$. The interaction term in the free energy eqn 10.28 reduces in the case of a static force to

$$F_f = -\int d\mathbf{x}\, f(\mathbf{x})s(\mathbf{x}) = -V\sum_{\mathbf{k}'} \widetilde{f}_{\mathbf{k}'}\widetilde{s}_{-\mathbf{k}'}, \tag{10.40}$$

where V is the volume (periodic boundary conditions) and the sum over $\mathbf{k}'$ is the sum over allowed wavevectors in the box (Appendix A). (We use Fourier series here instead of Fourier transforms because it makes the calculations more intuitive; we get factors of the volume rather than δ-functions and infinities.) The expectation value of the order parameter in the field is

$$\langle \widetilde{s}_{\mathbf{k}} \rangle_{\text{eq}} = \text{Tr}[\widetilde{s}_{\mathbf{k}} e^{-\beta(F_0 - V\sum_{\mathbf{k}'} \widetilde{f}_{\mathbf{k}'}\widetilde{s}_{-\mathbf{k}'})}] \Big/ \text{Tr}[e^{-\beta(F_0 - V\sum_{\mathbf{k}'} \widetilde{f}_{\mathbf{k}'}\widetilde{s}_{-\mathbf{k}'})}] = \frac{1}{\beta V}\frac{\partial \log Z}{\partial f_{-\mathbf{k}'}}, \tag{10.41}$$

where Tr integrates over all order parameter configurations s (formally a *path integral* over function space). The susceptibility is given by differentiating eqn 10.41:

$$\widetilde{\chi}_0(\mathbf{k}) = \frac{\partial \langle \widetilde{s}_{\mathbf{k}} \rangle_{\text{eq}}}{\partial \widetilde{f}_{\mathbf{k}}}\Bigg|_{f=0}$$

$$= \frac{\text{Tr}[\widetilde{s}_{\mathbf{k}}(\beta V \widetilde{s}_{-\mathbf{k}}) e^{-\beta(F_0 - V\sum_{\mathbf{k}'} \widetilde{f}_{\mathbf{k}'}\widetilde{s}_{-\mathbf{k}'})}]}{\text{Tr}[e^{-\beta(F_0 - V\sum_{\mathbf{k}'} \widetilde{f}_{\mathbf{k}'}\widetilde{s}_{-\mathbf{k}'})}]}\Bigg|_{f=0} - \frac{\text{Tr}[\widetilde{s}_{\mathbf{k}} e^{-\beta(F_0 - V\sum_{\mathbf{k}'} \widetilde{f}_{\mathbf{k}'}\widetilde{s}_{-\mathbf{k}'})}]\text{Tr}[(\beta V \widetilde{s}_{-\mathbf{k}}) e^{-\beta(F_0 - V\sum_{\mathbf{k}'} \widetilde{f}_{\mathbf{k}'}\widetilde{s}_{-\mathbf{k}'})}]}{\text{Tr}[e^{-\beta(F_0 - V\sum_{\mathbf{k}'} \widetilde{f}_{\mathbf{k}'}\widetilde{s}_{-\mathbf{k}'})}]^2}\Bigg|_{f=0}$$

$$= \beta V\left(\langle \widetilde{s}_{\mathbf{k}}\widetilde{s}_{-\mathbf{k}} \rangle - \langle \widetilde{s}_{\mathbf{k}} \rangle\langle \widetilde{s}_{-\mathbf{k}} \rangle\right) = \beta V\left\langle (\widetilde{s}_{\mathbf{k}} - \langle \widetilde{s}_{\mathbf{k}} \rangle)^2 \right\rangle = \beta\widehat{C}(\mathbf{k}, 0), \tag{10.42}$$

where the last equation is the Fourier equality of the correlation function to the absolute square of the fluctuation (eqn A.21, except (i) because we are using Fourier series instead of Fourier transforms there are two extra factors of V, and (ii) the $\langle s_{\mathbf{k}} \rangle$ subtraction gives us the connected correlation function, with $\bar{s}$ subtracted off).

Note that everything again is calculated by taking derivatives of the partition function; in eqn 10.41 $\langle \widetilde{s}_{\mathbf{k}} \rangle = (1/\beta)\,\partial \log Z/\partial f_{-\mathbf{k}}$ and in eqn 10.42 $\widehat{C}(\mathbf{k}, 0) = (1/\beta^2)\,\partial^2 \log Z/\partial f_{\mathbf{k}}\partial f_{-\mathbf{k}}$. The higher connected correlation functions can be obtained in turn by taking higher derivatives of $\log Z$. This is a common theoretical technique; to calculate correlations in an ensemble, add a force coupled to the corresponding field and take derivatives.

How is the static susceptibility $\chi_0(\mathbf{r})$ related to our earlier dynamic susceptibility $\chi(\mathbf{r}, t)$? We can use the dynamic susceptibility (eqn 10.29) in the special case of a time-independent force

$$s(\mathbf{x}, t) = \int d\mathbf{x}' \int_{-\infty}^{t} dt' \, \chi(\mathbf{x} - \mathbf{x}', t - t') f(\mathbf{x}')$$
$$= \int d\mathbf{x}' \int_{0}^{\infty} d\tau \, \chi(\mathbf{x} - \mathbf{x}', \tau) f(\mathbf{x}'), \qquad (10.46)$$

to derive a formula for χ_0:

$$\chi_0(\mathbf{r}) = \int_{0}^{\infty} dt \, \chi(\mathbf{r}, t) = \int_{-\infty}^{\infty} dt \, \chi(\mathbf{r}, t). \qquad (10.47)$$

Here we use the fact that the physical world obeys *causality* (effects cannot precede their cause) to set $\chi(\mathbf{r}, t) = 0$ for $t < 0$ (see Fig. 10.9). The integral over time in eqn 10.47 extracts the $\omega = 0$ Fourier component, so the $\mathbf{k}$-dependent static susceptibility is the zero-frequency limit of the AC susceptibility:

$$\widetilde{\chi}_0(\mathbf{k}) = \widetilde{\chi}(\mathbf{k}, \omega = 0). \qquad (10.48)$$

Often, one discusses the uniform susceptibility of a system—the response to an external field uniform not only in time but also in space. The specific heat of Section 6.1 is the uniform $\mathbf{k} = \mathbf{0}$ value of the static susceptibility to a change in temperature, as the magnetic susceptibility of Exercise 8.1 is the susceptibility to a uniform change in field. For the uniform static susceptibility, $s = \int d\mathbf{x}' \, \chi_0(\mathbf{x} - \mathbf{x}') f = \widetilde{\chi}_0(\mathbf{k} = \mathbf{0}) f$, so the uniform susceptibility is given by $\widetilde{\chi}_0(\mathbf{k})$ at $\mathbf{k} = \mathbf{0}$. Knowing $\widetilde{\chi}_0(\mathbf{k}) = \beta \widehat{C}(\mathbf{k}, t = 0)$ (eqn 10.45), we can relate the uniform susceptibility to the $\mathbf{k} = \mathbf{0}$ component of the equal-time correlation function. But at $\mathbf{k} = \mathbf{0}$, the correlation function is given by the mean square of the spatially averaged order parameter $\langle s \rangle_{\text{space}} = (1/V) \int s(\mathbf{x}) \, d\mathbf{x}$:[25]

$$k_B T \, \widetilde{\chi}_0(\mathbf{k} = \mathbf{0}) = \widehat{C}(\mathbf{k} = \mathbf{0}, t = 0) = \int d\mathbf{r} \, \langle s(\mathbf{r} + \mathbf{x}) s(\mathbf{x}) \rangle$$
$$= \int d\mathbf{r} \, \frac{1}{V} \left\langle \int d\mathbf{x} \, s(\mathbf{r} + \mathbf{x}) s(\mathbf{x}) \right\rangle$$
$$= V \left\langle \frac{1}{V} \int d\mathbf{r}' \, s(\mathbf{r}') \frac{1}{V} \int d\mathbf{x} \, s(\mathbf{x}) \right\rangle$$
$$= V \left\langle \langle s \rangle_{\text{space}}^2 \right\rangle. \qquad (10.49)$$

We have thus connected a uniform linear response to the fluctuations of the whole system. We have done this in special cases twice before, in Exercise 8.2(b) where the fluctuations in magnetization gave the susceptibility in the Ising model, and eqn 6.13 where the energy fluctuations were related to the specific heat. Equation 10.49 shows in general that fluctuations in spatially averaged quantities vanish in the thermodynamic limit $V \to \infty$:

$$\left\langle \langle s \rangle_{\text{space}}^2 \right\rangle = \frac{k_B T \widetilde{\chi}_0(\mathbf{0})}{V} \qquad (10.50)$$

so long as the uniform susceptibility stays finite.

Fig. 10.9 The susceptibility $\chi(\mathbf{r} = \mathbf{0}, t)$ for a hypothetical system with two characteristic relaxation times. Here $\chi(\mathbf{r}, t)$ gives the response to an impulse at a time t in the past; causality requires that there be no response *preceding* the impulse, so $\chi(\mathbf{r}, t) = 0$ for $t < 0$.

[25] Remember that the susceptibility and the correlated fluctuations are with respect to the average properties of the phase. So in eqn 10.49, if the average magnetization $\langle s \rangle$ is not zero (in a ferromagnet or in an external field), one must use the correlation function for the fluctuations about the mean $\widehat{C}(\mathbf{k}, t = 0) = \int d\mathbf{r} \langle (s(\mathbf{r} + \mathbf{x}) - \langle s \rangle)(s(\mathbf{x}) - \langle s \rangle) \rangle$. This is called the *connected* correlation function. This subtraction does not usually affect the susceptibility except at $\mathbf{k} = \mathbf{0}$.

10.8   The fluctuation-dissipation theorem

Now we turn to computing the dynamic susceptibility. It too is related to the correlation function, via the *fluctuation-dissipation* theorem.

How can we compute $\chi(\mathbf{r}, t)$, the space-time evolution after we kick the system at $\mathbf{r} = t = 0$? We know the time evolution starting from an imposed initial condition is given by the Green's function $G(\mathbf{r}, t)$. We can impose an initial condition using a static force $f(\mathbf{x}, t) = f(\mathbf{x})$ for $t < 0$, and release it at $t = 0$ so $f(\mathbf{x}, t) = 0$ for $t > 0$. We can then match the Green's function time evolution $s(\mathbf{x}, t) = \int \mathrm{d}x'\, G(\mathbf{x} - \mathbf{x}', t)s(\mathbf{x}', 0)$ with that given by the susceptibility $s(\mathbf{x}, t) = \int_{-\infty}^{0} \mathrm{d}t' \int \mathrm{d}x'\, f(\mathbf{x}')\chi(\mathbf{x} - \mathbf{x}', t - t')$.

Let us work it out for the ideal gas,[26] where $\chi_0(\mathbf{r}) = \delta(\mathbf{r})/\alpha$ (eqn 10.43), so $\rho(\mathbf{x}, 0) = f(\mathbf{x})/\alpha$. The subsequent time evolution is given by the Green's function $G(\mathbf{x}, t)$, which we have seen for the ideal gas gives the correlation function $C^{\text{ideal}}(\mathbf{x}, t) = G(\mathbf{x}, t)/(\beta\alpha)$ by Onsager's regression hypothesis (eqn 10.26):

$$\rho(\mathbf{x}, t) = \int \mathrm{d}\mathbf{x}'\, \rho(\mathbf{x}', 0)G(\mathbf{x} - \mathbf{x}', t) = \int \mathrm{d}\mathbf{x}'\, \frac{f(\mathbf{x}')}{\alpha} G(\mathbf{x} - \mathbf{x}', t)$$
$$= \int \mathrm{d}\mathbf{x}'\, f(\mathbf{x}')\beta C^{\text{ideal}}(\mathbf{x} - \mathbf{x}', t). \tag{10.56}$$

We match against $\rho(\mathbf{x}, t)$ written using the dynamical susceptibility. Since $f(\mathbf{x}, t) = 0$ for $t > 0$ the formula involves integrals up to time

---

[26] We can do this for a general order parameter field $s(\mathbf{x}, t)$. We start with an initial condition defined by a static external field $f(\mathbf{x})$, which is given by $\widehat{s}(\mathbf{k}, t = 0) = \widetilde{\chi}_0(\mathbf{k})\widetilde{f}(\mathbf{k})$. The subsequent time evolution is given by convolving with the Green's function $G(\mathbf{x}, t)$ (eqn 2.34), which is the same as multiplying by $\widehat{G}(\mathbf{k}, t)$:

$$\widehat{s}(\mathbf{k}, t) = \widetilde{\chi}_0(\mathbf{k})\widetilde{f}(\mathbf{k})\widehat{G}(\mathbf{k}, t). \tag{10.51}$$

We can also find an equation for $\widehat{s}(\mathbf{k}, t)$ by using the dynamic susceptibility, eqn 10.29, and the fact that $f(t') = 0$ for $t' > 0$:

$$s(\mathbf{x}, t) = \int \mathrm{d}\mathbf{x}' \int_{-\infty}^{t} \mathrm{d}t'\, \chi(\mathbf{x} - \mathbf{x}', t - t')f(\mathbf{x}', t') = \int \mathrm{d}\mathbf{x}' \int_{-\infty}^{0} \mathrm{d}t'\, \chi(\mathbf{x} - \mathbf{x}', t - t')f(\mathbf{x}') = \int \mathrm{d}\mathbf{x}' \int_{t}^{\infty} \mathrm{d}\tau\, \chi(\mathbf{x} - \mathbf{x}', \tau)f(\mathbf{x}'), \tag{10.52}$$

so

$$\widehat{s}(\mathbf{k}, t) = \int_{t}^{\infty} \mathrm{d}\tau\, \widehat{\chi}(\mathbf{k}, \tau)\widetilde{f}(\mathbf{k}). \tag{10.53}$$

This is true for any $\widetilde{f}(\mathbf{k})$, so with eqn 10.51, we find $\int_{t}^{\infty} \mathrm{d}\tau\, \widehat{\chi}(\mathbf{k}, \tau) = \widetilde{\chi}_0(\mathbf{k})\widehat{G}(\mathbf{k}, t)$. Now from the last section, eqn 10.45, we know $\widetilde{\chi}_0(\mathbf{k}) = \beta\widehat{C}(\mathbf{k}, 0)$. From the Onsager regression hypothesis, the Green's function $\widehat{G}(\mathbf{k}, t)$ for s has the same evolution law as is obeyed by the correlation function C (eqn 10.25), so $\widehat{C}(\mathbf{k}, 0)\widehat{G}(\mathbf{k}, t) = \widehat{C}(\mathbf{k}, t)$. Hence

$$\int_{t}^{\infty} \mathrm{d}\tau\, \widehat{\chi}(\mathbf{k}, \tau) = \beta\widehat{C}(\mathbf{k}, 0)\widehat{G}(\mathbf{k}, t) = \beta\widehat{C}(\mathbf{k}, t). \tag{10.54}$$

Differentiating both sides with respect to time yields the fluctuation-dissipation theorem in $\mathbf{k}$-space:

$$\widehat{\chi}(\mathbf{k}, t) = -\beta\frac{\partial\widehat{C}(\mathbf{k}, t)}{\partial t}. \tag{10.55}$$

zero; we change variables to $\tau = t - t'$:

$$\rho(\mathbf{x}, t) = \int d\mathbf{x}' \int_{-\infty}^{0} dt'\; f(\mathbf{x}')\chi(\mathbf{x} - \mathbf{x}', t - t')$$

$$= \int d\mathbf{x}' f(\mathbf{x}') \int_{t}^{\infty} d\tau\; \chi(\mathbf{x} - \mathbf{x}', \tau). \tag{10.57}$$

Comparing these two formulæ, we see that

$$\beta C^{\text{ideal}}(\mathbf{r}, t) = \int_{t}^{\infty} d\tau\; \chi(\mathbf{r}, \tau). \tag{10.58}$$

Taking the derivative of both sides, we derive one form of the *fluctuation-dissipation theorem*:

$$\chi^{\text{ideal}}(\mathbf{r}, t) = -\beta \frac{\partial C^{\text{ideal}}}{\partial t} \qquad (t > 0). \tag{10.59}$$

The fluctuation-dissipation theorem in this form is true in general for the linear response of classical equilibrium systems (see note 26). The linear dynamic susceptibility χ of a general order parameter field $s(\mathbf{x}, t)$ with correlation function $C(\mathbf{x}, t)$ is given by

$$\chi(\mathbf{x}, t) = -\beta \frac{\partial C(\mathbf{x}, t)}{\partial t} \qquad (t > 0). \tag{10.60}$$

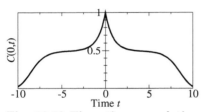

Fig. 10.10 Time–time correlation function. The time–time correlation function $C(\mathbf{r} = \mathbf{0}, \tau)$ for the same hypothetical system whose susceptibility was shown in Fig. 10.9.

What happens for $t < 0$? The correlation function must be symmetric in time (Fig. 10.10) since the equilibrium state is invariant under time-reversal symmetry:

$$C(\mathbf{r}, \tau) = \langle s(\mathbf{x}, t)s(\mathbf{x} + \mathbf{r}, t + \tau)\rangle$$

$$= \langle s(\mathbf{x}, t)s(\mathbf{x} + \mathbf{r}, t - \tau)\rangle = C(\mathbf{r}, -\tau). \tag{10.61}$$

But χ must be zero for $t < 0$ (Fig. 10.9) by causality:

$$\chi(\mathbf{r}, t) = 0 \qquad (t < 0). \tag{10.62}$$

We can see why it is called the fluctuation-*dissipation* theorem by looking at the AC version of the law. Again, for convenience, we ignore the spatial degrees of freedom. Using eqns 10.60 and 10.62, and integrating by parts, we find

$$\tilde{\chi}(\omega) = \int_{-\infty}^{\infty} dt\; \chi(t)e^{i\omega t} = -\beta \int_{0}^{\infty} dt\; \frac{\partial C}{\partial t} e^{i\omega t}$$

$$= -\beta C(t)e^{i\omega t}\Big|_{0}^{\infty} + i\omega\beta \int_{0}^{\infty} dt\; C(t)e^{i\omega t}. \tag{10.63}$$

Now, the first term is real and $C(t) = C(-t)$, so we may write the imaginary part of the susceptibility as

$$\chi''(\omega) = \text{Im}[\tilde{\chi}(\omega)] = \beta\omega \int_{0}^{\infty} dt\; C(t)\cos(\omega t)$$

$$= \frac{\beta\omega}{2} \int_{-\infty}^{\infty} dt\; C(t)e^{i\omega t} = \frac{\beta\omega}{2}\tilde{C}(\omega). \tag{10.64}$$

This is the AC version of the (classical) fluctuation-dissipation theorem, which we state again:

$$\chi''(\omega) = \frac{\beta\omega}{2}\widetilde{C}(\omega). \tag{10.65}$$

Using this result and eqn 10.37 relating the power dissipated $p(\omega)$ to χ'', we find

$$p(\omega) = \frac{\omega|f_\omega|^2}{2}\chi''(\omega) = \frac{\omega|f_\omega|^2}{2}\frac{\beta\omega}{2}\widetilde{C}(\omega)$$
$$= \frac{\beta\omega^2|f_\omega|^2}{4}\widetilde{C}(\omega). \tag{10.66}$$

This tells us that the power dissipated $p(\omega)$ under an external forcing f_ω is given in terms of the correlation function of the spontaneous fluctuations $\widetilde{C}(\omega)$; hence the name fluctuation-dissipation theorem.

Notice that the fluctuation-dissipation theorem applies only to equilibrium systems. (There are several interesting but much more speculative attempts to generalize it to nonequilibrium systems.) Also notice that we have ignored quantum mechanics in our derivation.[27] Indeed there are quantum-mechanical corrections; the fully quantum version of the fluctuation-dissipation theorem is

$$\chi''(\mathbf{k},\omega) = \text{Im}[\widetilde{\chi}(\mathbf{k},\omega)] = \frac{1}{2\hbar}(1 - e^{-\beta\hbar\omega})\widetilde{C}(\mathbf{k},\omega). \tag{10.67}$$

At high temperatures, $1 - e^{-\beta\hbar\omega} \sim \beta\hbar\omega$ and we regain our classical result, eqn 10.65.

[27]One can also derive the fluctuation-dissipation theorem quantum mechanically, and then use it to derive the Onsager regression hypothesis [64].

10.9   Causality and Kramers–Kronig

The susceptibility $\chi(t)$ (again, dropping the positions for simplicity) is a real-valued function on the half-line $t > 0$. The frequency-dependent susceptibility is composed of two real-valued functions $\chi'(\omega)$ and $\chi''(\omega)$ on the entire line. We can use the symmetries $\widetilde{\chi}(-\omega) = \widetilde{\chi}^*(\omega)$ to reduce this to two real-valued functions on the half-line $\omega > 0$, but it still seems like $\widetilde{\chi}(\omega)$ contains twice the information of $\chi(t)$. It makes it plausible that χ' and χ'' might be related somehow. Suppose we measure the frequency-dependent absorption of the material, and deduce $\chi''(\mathbf{k},\omega)$. Can we find the real part of the susceptibility $\chi'(\mathbf{k},\omega)$?

It is a remarkable fact that we can find a formula for $\chi'(\omega)$ in terms of $\chi''(\omega)$. This relation is called the *Kramers–Kronig relation*, and it follows from *causality*. For this argument, you will need to know some complex analysis.[28]

We know that $\chi(t) = 0$ for $t < 0$, because the laws of nature are causal; the response cannot precede the perturbation. What does this imply about $\chi(\omega) = \int dt\,\chi(t)e^{i\omega t}$? Consider the function χ as a function of a complex frequency $\omega = u + iv$:

$$\chi(u + iv) = \int_0^\infty dt\,\chi(t)e^{iut}e^{-vt}. \tag{10.68}$$

[28]If you have not heard of Cauchy's theorem, read on—you will be getting a preview of the key result in complex analysis.

It converges nicely for $v > 0$, but looks dubious for $v < 0$. In the complex ω plane, the fast convergence for $v > 0$ implies that $\chi(\omega)$ is analytic in the upper half-plane.[29] Also, it seems likely that there will be singularities (e.g. poles) in $\widetilde{\chi}(\omega)$ in the lower half-plane ($v < 0$).

We now apply a deep theorem of complex analysis. If C is a closed curve (or *contour*) in the complex z plane, and $f(z)$ is analytic everywhere in a region that encloses the contour, *Cauchy's theorem* states that the line integral of $f(z)$ around C is zero:

$$\oint_C f(z')\, dz' = 0. \tag{10.69}$$

Cauchy's theorem (which we shall not prove) is amazing, but it has a simple interpretation: it is the condition for the integral of $f(z)$ to exist as a complex function (Fig. 10.11).

Now consider the integral

$$\oint_{C_\omega} \frac{\widetilde{\chi}(\omega')}{\omega' - \omega}\, d\omega' = 0 \tag{10.70}$$

along the contour C_ω of Fig. 10.12. The integral is zero because $\widetilde{\chi}(\omega')$ is analytic in the upper half-plane, and thus so also is $\widetilde{\chi}(\omega')/(\omega' - \omega)$, except at the point $\omega' = \omega$ which is dodged by the small semicircle (of radius ϵ). The contribution of the large semicircle to this contour integral can be shown to vanish as its radius $R \to \infty$. The contribution of the small clockwise semicircle $\omega' = \omega + \epsilon \exp(i\theta)$, $\pi > \theta > 0$, is

$$\int_{\substack{\text{small} \\ \text{semicircle}}} \frac{\widetilde{\chi}(\omega')}{\omega' - \omega}\, d\omega' \approx \widetilde{\chi}(\omega) \int_{\substack{\text{small} \\ \text{semicircle}}} \frac{1}{\omega' - \omega}\, d\omega'$$
$$= \widetilde{\chi}(\omega) \log(\omega' - \omega)\Big|_{\omega' = \omega + \epsilon \exp(i\pi)}^{\omega' = \omega + \epsilon \exp(i0)}$$
$$= \widetilde{\chi}(\omega)[\log \epsilon - (\log \epsilon + i\pi)] = -i\pi\widetilde{\chi}(\omega). \tag{10.71}$$

The two horizontal segments, as $R \to \infty$ and $\epsilon \to 0$, converge to the integral over the whole real axis:

$$\lim_{\epsilon \to 0}\left[\int_{-\infty}^{\omega' - \epsilon} + \int_{\omega' + \epsilon}^{\infty}\right] \frac{\widetilde{\chi}(\omega')}{\omega' - \omega}\, d\omega' = \mathrm{PV}\int_{-\infty}^{\infty} \frac{\widetilde{\chi}(\omega')}{\omega' - \omega}\, d\omega', \tag{10.72}$$

where the left-hand side is the definition of the *principal value* (PV), the limit as the pole is approached symmetrically from either side.

Since the contribution of the semicircles and the horizontal segments sum to zero by Cauchy's theorem, eqn 10.71 must equal minus eqn 10.72, so[30]

$$\widetilde{\chi}(\omega) = \frac{1}{\pi i}\int_{-\infty}^{\infty} \frac{\widetilde{\chi}(\omega')}{\omega' - \omega}\, d\omega', \tag{10.74}$$

where the right-hand side is understood as the principal value.

Notice the i in the denominator. This implies that the real part of the integral gives the imaginary part of $\widetilde{\chi}$, and vice versa. In particular,

$$\chi'(\omega) = \mathrm{Re}[\widetilde{\chi}(\omega)] = \frac{1}{\pi}\int_{-\infty}^{\infty} \frac{\mathrm{Im}[\widetilde{\chi}(\omega')]}{\omega' - \omega}\, d\omega' = \frac{1}{\pi}\int_{-\infty}^{\infty} \frac{\chi''(\omega')}{\omega' - \omega}\, d\omega'. \tag{10.75}$$

[29]Analytic means that Taylor series expansions converge. It is amazing how many functions in physics are analytic; it seems we almost always can assume power series make sense. We have discussed in Section 8.3 that material properties are analytic functions of the parameters inside phases; we have discussed in Exercise 9.5 that the free energy for finite systems (and for finite coarse-grainings) is an analytic function of the order parameter and its gradients. Here we find yet another excuse for finding analytic functions: causality!

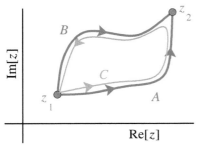

Fig. 10.11 Cauchy's theorem. For $f(z)$ to have a well-defined complex integral from z_1 to z_2, the contour integral over any two paths A and B connecting the points must agree. Hence the integral along a closed curve C formed by traversing A forward and B backward must be zero. Cauchy's theorem is thus the condition for the complex integral to be well defined, independent of the path from z_1 to z_2.

[30]Many will recognize this as being related to Cauchy's integral formula, which states that

$$\oint_C \frac{f(z')}{z' - z}\, dz' = 2\pi i f(z) N, \tag{10.73}$$

where N is the *winding number* of the path (the number of times it encircles z counter-clockwise). A counterclockwise loop plowing straight through z is the special case with $N = \frac{1}{2}$.

It is traditional to simplify it a bit more, by noticing that $\chi''(\omega) = -\chi''(-\omega)$, so

$$\chi'(\omega) = \frac{1}{\pi} \int_0^\infty \chi''(\omega') \left(\frac{1}{\omega' - \omega} - \frac{1}{-\omega' - \omega} \right) d\omega'$$

$$= \frac{2}{\pi} \int_0^\infty \chi''(\omega') \frac{\omega'}{\omega'^2 - \omega^2} \, d\omega'. \qquad (10.76)$$

Hence in principle one can measure the imaginary, dissipative part of a frequency-dependent susceptibility and do a simple integral to get the real, reactive part. Conversely,

$$\chi''(\omega) = -\frac{2\omega}{\pi} \int_0^\infty \chi'(\omega') \frac{1}{\omega'^2 - \omega^2} \, d\omega'. \qquad (10.77)$$

These are the *Kramers–Kronig relations*. In experiments they are a challenge to use; to deduce the real part, one must measure (or approximate) the dissipative imaginary part at all frequencies, from deep infrared to X-ray.

Fig. 10.12 Kramers–Kronig contour. A contour C_ω in the complex ω' plane. The horizontal axis is $\mathrm{Re}[\omega']$ and the vertical axis is $\mathrm{Im}[\omega']$. The integration contour runs along the real axis from $-\infty$ to ∞ with a tiny semicircular detour near a pole at ω. The contour is closed with a semicircle back at infinity, where $\chi(\omega')$ vanishes rapidly. The contour encloses no singularities, so Cauchy's theorem tells us the integral around it is zero.

Exercises

Five exercises introduce us to correlation functions. *Human correlations* and *Subway bench Monte Carlo* explore the statistics of seating on subway cars, experimentally and computationally. *Pair distributions and molecular dynamics* explores the spatial atom–atom pair distribution function in molecular dynamics simulations of liquids and gases. *Liquid free energy* and *Telegraph noise in nanojunctions* calculates correlation functions in space and time from microscopic models. *Cosmic microwave background radiation* studies the first of all correlation functions, left to us from the Big Bang.

Liquid dynamics and *Onsager regression hypothesis* apply and critique Onsager's relation between the response to kicks and dynamical fluctuations. *Harmonic susceptibility and dissipation* calculates the static and dynamic susceptibility for an oscillator, and uses the latter to compute the power dissipated under external forcing. *Harmonic fluctuation dissipation*, *Fluctuation-dissipation: Ising*, *Noise and Langevin equations*, and *Susceptibilities and correlations* apply the fluctuation-dissipation theorem in a variety of physical contexts.

Harmonic Kramers–Kronig, *Damped oscillator*, *Spin*, *Magnetic dynamics*, *Critical point response*, and *Quasi-particle poles and Goldstone's theorem* explore the various relations between correlation, response, and dissipation in space and time for the oscillator, a single spin, a continuum magnet, a system at its critical point, and damped phonons, illustrating the power and richness of the tools we introduce in this chapter.

(10.1) **Cosmic microwave background radiation.**[31] (Astrophysics) ③
In Exercise 7.15, we studied the the microwave background radiation. This radiation was emitted about 380,000 years after the Big Bang, when the electrically charged electrons and protons combined into neutral atoms and hence almost completely stopped scattering the photons [207], making the Universe transparent; the background radiation we see are these original photons. We saw there that the spectrum is a nearly perfectly a black-body spectrum, indicating that the photons at the decoupling time were in thermal equilibrium.
The Universe was not only in local thermal equilibrium at that time, but seemed to have reached precisely the same temperature in all directions (a global equilibrium). Apart from

---

[31]This exercise was developed with assistance from Ira Wasserman, Eanna Flanagan, Rachel Bean, and Dale Fixsen.

our galaxy (which emits microwaves) and a few other non cosmological point sources, the early experiments could not detect any angular fluctuations in the microwave temperature. This uniformity lends support to a model called *inflation*. In the inflationary scenario, space-time goes through a period of rapid expansion, leaving the observable Universe with a uniform density.

More discriminating experiments next detected an overall Doppler shift of the temperature field, which they used to measure the absolute velocity of the Sun.[32] The Sun has a net velocity of about a million miles per hour in the direction of the constellation Leo.

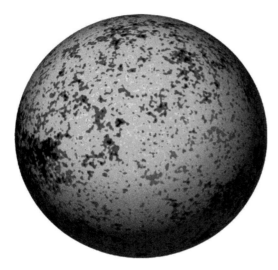

Fig. 10.13 Map of the microwave background radiation, from the NASA/WMAP Science Team [139]. Variation in temperature of the microwave background radiation, after the constant term, dipole term, and the radiation from our galaxy are subtracted out, from WMAP, the Wilkinson Microwave Anisotropy Probe. The satellite is at the center of the sphere, looking outward. The fluctuations are about one part in 100,000.

Subsequent experiments finally saw more interesting variations in the microwave background

radiation, shown in Fig. 10.13. We interpret these fluctuations as the ripples in the temperature, evolving according to a wave equation[33]

$$(1 + R)\frac{\partial^2 \Theta}{\partial t^2} = \frac{c^2}{3}\nabla^2\Theta, \qquad (10.78)$$

where c is the speed of light in a vacuum, Θ is the temperature fluctuation $\Delta T/T$, t is the time, and R is due to the contribution of matter to the total density (see Exercise 7.15 and [89]).

Fig. 10.14 Correlation function of microwave radiation in Fourier space, from the NASA/WMAP Science Team [139]. Temperature variations of the microwave background radiation, written in spherical harmonics (roughly an angular Fourier transform). You can think of l for the multipole moment as roughly corresponding to wavevector $k_l = 2\pi l/L$.[34] The curve through the data is a more complete theoretical fit, similar in spirit to this exercise.

What kind of initial correlations in the fluctuations does the inflationary scenario predict? The inflation theory predicts that at very early times the Universe was left in a state which we can think of as having tiny, random variations in temperature and density, and with zero initial velocities. These initial variations were

---

[32]Einstein's theory states that all motion is relative: the laws of physics do not depend on how fast the Sun is moving with respect to the distant galaxies. The Big Bang, through the microwave background radiation, establishes a preferred reference frame. We noted in Section 5.1 that the Big Bang is also responsible for the arrow of time (see also Exercise 5.24); the Big Bang dramatically broke time-reversal invariance and subtly broke Lorentz invariance.

[33]This equation is for *co-moving time-orthogonal gauge*, I am told.

[34]The first peak is at $l \sim 220$; they estimate the second peak to be at $l = 546 \pm 10$. Note that the horizontal axis in this plot is neither linear nor logarithmic.

not uncorrelated (white noise). Instead, inflation leaves the Universe in a state with *scale-invariant* initial velocity fields, starting with a power law

$$\widehat{C}(\mathbf{k}, t=0) = |\widehat{\Theta}(\mathbf{k}, t=0)|^2/V = Ak^{n_s-3}, \quad (10.79)$$

with $n_s \approx 1$ for most inflation theories. (Remember that the absolute square of the transform is the transform of the correlation function, $|\widehat{\Theta}(\mathbf{k}, t)|^2 \propto \widehat{C}(\mathbf{k}, t)$, eqn A.21.)

Let us model the Universe as a one-dimensional box with length L and periodic boundary conditions. Let us solve for the time evolution of the equal-time correlation function[35] $C(x-x', t) = \langle \Theta(x, t)\Theta(x', t) \rangle$ at $t = 380,000$ years given initial conditions $(\partial\Theta/\partial t)|_{t=0} = 0$ and $\Theta(x, t=0)$ random but with scale-invariant correlations (eqn 10.79). We can then evolve each Fourier component, and calculate the correlation function in $\mathbf{k}$-space at the decoupling time.

(a) *Given an initial $\widehat{\Theta}(\mathbf{k}, t=0)$ and assuming $(\partial\widehat{\Theta}/\partial t)|_{t=0} = 0$, calculate $\widehat{\Theta}(\mathbf{k}, t)$ from eqn 10.78. Calculate $\widehat{C}(\mathbf{k}, t) = \left\langle |\widehat{\Theta}(\mathbf{k}, t)|^2 \right\rangle$ in terms of A, c, and R given $n_s = 1$. For what value of L (in light years and in centimeters) will k_{220} be the first peak[36] of $k^2\widehat{C}(\mathbf{k}, t)$ at the decoupling time, if $R = 0.7$?* (Hint: $c = 3 \times 10^{10}$ cm/s, $t = 380,000$ years, and there happen to be about $\pi \times 10^7$ seconds in a year.) L can be interpreted as roughly the circumference of that part of the Universe visible in the cosmic background radiation at the time of decoupling.

(b) *Plot $l(l+1)\widehat{C}(k_l, t)$ at the decoupling time, for $R = 0.7$ and $A = 1$, from $l = 40$ to $l = 1,500$. Compare with Fig. 10.14.*

Your correlation function should look much less structured than the actual one. We have ignored many important phenomena in this exercise. At wavelengths long compared to the speed of sound times the age of the Universe ($l < 160$, roughly up to the first peak) the wave equation has not had time to start working.

At shorter wavelengths ($l > 1,100$) the waves become damped because the photon mean-free path among the baryons gets large. A calculation much like the one you solved, but including these effects and others (like gravity), shifts the higher peak positions and changes their amplitudes resulting in the curve in Fig. 10.14. For further details, see [89, 139].

(10.2) **Pair distributions and molecular dynamics.**[37] (Computation) ③

Many scattering experiments measure the correlations between atomic positions. Let our system have N particles in a volume V.[38] The mean density in a system with atoms at positions $\mathbf{x}_i$

$$\rho(\mathbf{x}) = \left\langle \sum_i \delta(\mathbf{x} - \mathbf{x}_i) \right\rangle, \quad (10.80)$$

and the density–density correlation function[39]

$$C(\mathbf{x}, \mathbf{x}') = \left\langle \sum_{i, j \neq i} \delta(\mathbf{x} - \mathbf{x}_i)\delta(\mathbf{x}' - \mathbf{x}_j) \right\rangle \quad (10.81)$$

are used to define the *pair correlation function* $g(\mathbf{x}, \mathbf{x}')$:

$$g(\mathbf{x}, \mathbf{x}') = \frac{C(\mathbf{x}, \mathbf{x}')}{\rho(\mathbf{x})\rho(\mathbf{x}')} = \frac{C(\mathbf{x} - \mathbf{x}')}{(N/V)^2}. \quad (10.82)$$

Here the last equality is valid for homogeneous systems like liquids and gases.[40]

(a) *Show analytically from eqn 10.80 that $\rho(\mathbf{x})$ for a homogeneous system is the indeed the average density N/V. Show that $g(\mathbf{x}, \mathbf{x}') = g(\mathbf{x} - \mathbf{x}') = g(\mathbf{r})$, where*

$$g(\mathbf{r}) = \left\langle \frac{V}{N^2} \sum_{i, j \neq i} \delta(\mathbf{r} - \mathbf{r}_{ij}) \right\rangle \quad (10.83)$$

and $\mathbf{r}_{ij} = \mathbf{x}_i - \mathbf{x}_j$ is the vector separating the positions of atoms i and j. If the system is homogeneous and the atoms are uncorrelated (i.e., an ideal gas), show that $g(\mathbf{r}) \equiv 1$. If the potential energy is the sum of pair interactions with potential $E(\mathbf{r}_{ij})$, write the potential energy as

---

[35]Time here is not the time lag between the two measurements, but rather the time evolved since the Universe stopped inflating (similar to the coarsening correlation function $C_{\text{coar}}(\mathbf{x}, t)$ in Section 10.1).

[36]We multiply by k^2 to correspond roughly to the vertical axis $l(l+1)C_l$ of Fig. 10.14.

[37]This exercise was developed in collaboration with Neil W. Ashcroft and Christopher R. Myers. Computational hints can be found at the book website [181].

[38]We will use periodic boundary conditions, so the edges of the container do not break the translational symmetry.

[39]Warning: In the theory of liquids, C is used for another function, the Ornstein–Zernike direct correlation function.

[40]The pair correlation function represents the degree to which atomic positions fluctuate together, beyond the clumping implied by the possibly inhomogeneous average density. For example, three-dimensional crystals have a long-range broken translational symmetry; $\rho(\mathbf{x})$ will have peaks at the lattice sites even after averaging over thermal vibrations.

an integral over three-dimensional space **r** *involving* N, V, $g(\mathbf{r})$, *and* $E(\mathbf{r})$.

For liquids and gases, $g(\mathbf{r}) \to 1$ as $r \to \infty$—the correlations die away as the separation grows. Liquids and gases are also isotropic; the pair correlation function g must be rotation invariant, and hence can only depend on the distance $r = |\mathbf{r}|$. A typical molecular dynamics code will have a fast `NeighborLocator` routine, which will return the `HalfNeighbor` pairs of atoms $j < i$ with $|r_{ij}|$ less than a given cutoff. A histogram of the distances between these nearby points, suitably rescaled, is a convenient way of numerically estimating the pair correlation function. Let the histogram $h(r_n)$ give the number of such pairs in our system with $r_n < r < r_n + \Delta r$.

(b) *For an isotropic, homogeneous system in three dimensions, show*

$$g(\mathbf{r}) = \frac{2V}{N^2}\frac{h(r)}{4\pi r^2\,\Delta r}. \qquad (10.84)$$

What is the corresponding formula in two dimensions?

Download our molecular dynamics software [14] from the book website [181]. In our simulation, we use the Lennard–Jones potential, in which all pairs of atoms interact with an energy with a short-range hard-core repulsion $\sim 1/r^{12}$ and a long-range (van der Waals) attraction $\sim 1/r^6$:

$$E_{\text{pair}}(r) = 4\epsilon\left(\left(\frac{\sigma}{r}\right)^{12} - \left(\frac{\sigma}{r}\right)^6\right). \qquad (10.85)$$

Lennard–Jones is a reasonable approximation for the interatomic forces between noble gas atoms like argon. In our simulation, we choose the length scale σ and the energy scale ϵ both equal to one.

(c) *Plot the Lennard–Jones pair potential as a function of* r, *choosing a vertical scale so that the attractive well is visible. Where is the minimum-energy spacing between two atoms? Can you see why the repulsion is called "hard core"?*

Gas. We start with a simulated gas, fairly near the vapor pressure.

(d) *Simulate a two-dimensional gas of Lennard–Jones atoms, at a temperature* $T = 0.5\epsilon$ *and a density* $\rho = 0.05/\sigma^2$. *Measure the pair distribution function for the gas for* $0 < r < 4\sigma$. *Note the absence of pairs at close distances. Can you observe the effects of the*

attractive well in the potential, both visually and in the pair correlation function?

In a low-density, high-temperature gas the correlations between atoms primarily involve only two atoms at once. The interaction energy of a pair of atoms in our system is $E_{\text{pair}}(r)$, so in this limit

$$g_{\text{theory}}(r) \propto \exp(-E(r)/k_B T). \qquad (10.86)$$

Since $E(r) \to 0$ and $g(r) \to 1$ as $r \to \infty$, the constant of proportionality should be one.

(e) *Compare* $g(r)$ *from part (d) with* g_{theory} *for a system with density* $\rho = 0.05/\sigma^2$. *Do they agree well? Do they agree even better at higher temperatures or lower densities, where multiparticle interactions are less important?*

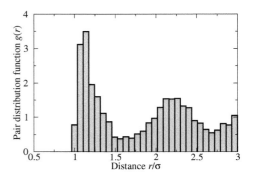

Fig. 10.15 **Pair distribution function** $g(r)$ for a two-dimensional Lennard–Jones liquid.

Liquid. We now turn to a liquid, at the same temperature as our gas but at a higher density.

(f) *Simulate a liquid, at* $\rho = 0.75/\sigma^2$ *and* $T = 0.5$. *Note from the animation that each atom in the two-dimensional simulation has around six nearest-neighbors, at nearly the minimum-energy distance. Calculate the pair distribution function (as in Fig. 10.15). Can you explain the features you see in terms of nearest-neighbors and second-nearest neighbors?*

(g) *If we define the coordination number of a liquid atom as all those with distances less than the position of the dip between nearest and next-nearest neighbors in* $g(r)$, *what is the mean number of near neighbors for your two-dimensional liquid?*

In most (three-dimensional) simple elemental liquids, the coordination number defined by this criterion is between 11 and 11.5. (The close-packed crystals have twelve nearest neighbors). The main exceptions are the group IV elements (carbon, silicon, germanium, ...) where the bonding is strongly angle dependent and the number of liquid near neighbors is smaller, around 5.5; their crystalline phases have three or four covalently bonded neighbors.

Crystal. In a three-dimensional crystal, the atoms vibrate around their equilibrium lattice positions (with only rare hops between lattice sites). If these vibrations become large compared to the lattice constant, then surely the crystal will melt. The *Lindemann criterion* says that for simple crystals, melting usually occurs when the thermal vibrations away from the lattice positions are about 10% of the interatomic spacing. The Lindemann criterion of course also implies that the typical variation in the separations for distant atoms in three-dimensional crystals (measured by the correlation function) stays much smaller than the lattice constant. The equilibrium correlation function will thus have oscillations extending forever. (Note how weird this is. Two atoms, tens of millions of atoms apart at ends of a one centimeter crystal that is just about to melt, are held at a separation precise to within a fraction of an Ångstrom.)

This is not true in two dimensions, where the lattice is not as stiff and thermal fluctuations are more severe. (The theory of two-dimensional crystals and how they melt has spawned many beautiful theoretical and experimental studies; look for works on the *Kosterlitz–Thouless–Halperin–Nelson–Young* transition.) Will the correlation function have long-range oscillations in our two-dimensional crystal?

(h) *Simulate a 2D crystal at $T = 0.1$, starting from a hexagonal crystal with interatomic spacing approximating the minimum of the pair potential. Calculate the isotropic spatial average of the pair correlation function.*[41] *By what percentage does the nearest-neighbor separation fluctuate? Are they small compared to the lattice constant? Also, can you identify which neighbors on the hexagonal lattice correspond to the second-nearest-neighbor and the third-nearest-neighbor peaks in $g(r)$?*

(10.3) **Damped oscillator.** ③
Let us explore further the fluctuating mass-on-a-spring (Section 6.5; see also Exercises 10.13, 10.15, 10.16, and 10.18). The coupling of the macroscopic motion to the internal degrees of freedom eventually damps any initial macroscopic oscillation; the remaining motions are microscopic thermal fluctuations. These fluctuations can be important, however, for nanomechanical and biological systems. In addition, the damped harmonic oscillator is a classic model for many atomic-scale physical processes, such as dielectric loss in insulators. (See [127] for a treatment by an originator of this subject.) Consider a damped, simple harmonic oscillator, forced with an external force f, obeying the equation of motion

$$\frac{d^2\theta}{dt^2} = -\omega_0^2\theta - \gamma\frac{d\theta}{dt} + \frac{f(t)}{m}. \qquad (10.87)$$

(a) Susceptibility. *Find the AC susceptibility $\tilde{\chi}(\omega)$ for the oscillator. Plot χ' and χ'' for $\omega_0 = m = 1$ and $\gamma = 0.2$, 2, and 5.* (Hint: Fourier transform the equation of motion, and solve for $\tilde{\theta}$ in terms of $\tilde{f}$.)

(b) Causality and critical damping. *Check, for positive damping γ, that your $\chi(\omega)$ is causal ($\chi(t) = 0$ for $t < 0$), by examining the singularities in the complex ω plane (Section 10.9). At what value of γ do the poles begin to sit on the imaginary axis?* The system is overdamped, and the oscillations disappear, when the poles are on the imaginary axis.

It would be natural now to ask you to verify the Kramers–Kronig relation (eqn 10.76), by writing χ' in terms of χ''. That turns out to be tricky both analytically and numerically. If you are ambitious, try it.

(c) Dissipation and susceptibility. *Given a forcing $f(t) = A\cos(\omega t)$, solve the equation and calculate $\theta(t)$. Calculate the average power dissipated by integrating your resulting formula for $f\, d\theta/dt$. Do your answers for the power and χ'' agree with the general formula for power dissipation, eqn 10.37?*

(d) Correlations and thermal equilibrium. *Use the fluctuation-dissipation theorem to calculate the correlation function $\tilde{C}(\omega)$ from $\chi''(\omega)$ (see eqn 10.65), where*

$$C(t - t') = \langle\theta(t)\theta(t')\rangle. \qquad (10.88)$$

---

[41]That is, use the routines you've developed for liquids and gases, ignoring the spatially dependent $\rho(x)$ in eqn 10.82 and discussed in note 40. This average still gives the correct potential energy.

Find the equal-time correlation function $C(0) = \langle\theta^2\rangle$, and show that it satisfies the equipartition theorem. (Hints: Our oscillator is in a potential well $V(\theta) = \frac{1}{2}m\omega_0^2\theta^2$. Also, $(\omega_0^2 - \omega^2)^2 + C^2\omega^2 = (\omega_0^2 - \omega^2 + iC\omega)(\omega_0^2 - \omega^2 - iC\omega)$; do contour integration.)

(10.4) **Spin.**[42] (Condensed matter) @
A spin in a solid has two states, $s_z = \pm 1/2$ with magnetic moments $M = g\mu_B s_z$, where $g\mu_B$ is a constant.[43] Due to fluctuations in its environment, the spin can flip back and forth thermally between these two orientations. In an external magnetic field H, the spin has energy $-M \cdot H$. Let $M(t)$ be the magnetization of the spin at time t. Given a time-dependent small external field $H(t)$ along z, the expectation value of M satisfies

$$\mathrm{d}\,[M(t)]_{M_i}/\mathrm{d}t = -\Gamma\,[M(t)]_{M_i} + \Gamma\chi_0 H(t),$$
(10.89)

where Γ is the spin equilibration rate, χ_0 is the static susceptibility of the magnetic moment, and the averaging $[\cdot]_{M_i}$ is over the noise provided by the environment, fixing the initial condition $M_i = M(0)$.

(a) *In the case that the field H is time independent, use equilibrium statistical mechanics to determine $\langle M(H)\rangle$. Using this formula for small H, determine χ_0 (which should be independent of H but dependent on temperature).*

(b) *Use the Onsager regression hypothesis to compute $C(t) = \langle M(t)M(0)\rangle_{\mathrm{ev}}$ at zero external field $H = 0$.* (Hint: What is $C(0)$?) *What should $C(t)$ be for times $t < 0$? What is $\widetilde{C}(\omega)$, the Fourier transform of $C(t)$?*

(c) *Assuming the classical fluctuation-dissipation theorem, derive the frequency-dependent susceptibility $\chi(t)$ and $\widetilde{\chi}(\omega)$. What are $\chi'(\omega)$ and $\chi''(\omega)$?*

(d) *Compute the power dissipated by the oscillator for an external magnetic field $H(t) = H_0 \cos(\omega t)$.*

(10.5) **Telegraph noise in nanojunctions.** (Condensed matter) ③
Many systems in physics exhibit *telegraph noise*, hopping between two states at random intervals (like a telegraph key going on and then off at different intervals for dots and dashes).

The nanojunction in Fig. 10.16 has two states, α and β. It makes transitions at random, with rate $\Gamma_{\beta\leftarrow\alpha} = \Gamma_{\beta\alpha}$ from α to β and rate $\Gamma_{\alpha\beta}$ from β to α.

Fig. 10.16 Telegraph noise in a metallic nanojunction. Resistance versus time $R(t)$ for a copper constriction, from [159]. We label α the state with low resistance R_α, and β the state with high resistance R_β. The two states probably represent a local shift of an atom or a small group of atoms in the constriction from one metastable state to another.

Master equation. Consider an ensemble of many identical copies of this system. Let the state of this ensemble at time t be given by $\boldsymbol{\rho}(t) = (\rho_\alpha, \rho_\beta)$, a vector of probabilities that the system is in the two states. This vector thus evolves according to the *master equation*

$$\mathrm{d}\boldsymbol{\rho}/\mathrm{d}t = M \cdot \boldsymbol{\rho}.$$
(10.90)

(a) *What is the 2×2 matrix M for our system, in terms of $\Gamma_{\alpha\beta}$ and $\Gamma_{\beta\alpha}$? At long times, when the system is in a stationary ensemble $\boldsymbol{\rho}^*$, what fraction of the time ρ_α^* will our system be in the α state?* (Notice that, unlike the Markov chains in Section 8.2, we now evolve continuously in time—making this a Markov *process*. Remember also that $\Gamma_{\alpha\beta}$ increases ρ_α and decreases ρ_β.)

(b) *Find the eigenvalue–eigenvector pairs for M.[44] Which eigenvector corresponds to the stationary state $\boldsymbol{\rho}^*$ from part (a)? Suppose that at $t = 0$ the system is known to be in the $\alpha = 1$ state, $\boldsymbol{\rho}(0) = \left(\begin{smallmatrix}1\\0\end{smallmatrix}\right)$. Write this initial condition in the basis of eigenvectors, and hence give a formula for the subsequent time evolution $\rho(t)$. What is the rate of decay to the stationary state?*

Let us define the Green's function $G_{\mu\nu}(\tau)$ to be the probability of being in the μ state at time $t + \tau$, given that it is in the ν state at time t. Thus in part (b), $\rho(t) = \left(\begin{smallmatrix}G_{\alpha\alpha}(t)\\G_{\beta\alpha}(t)\end{smallmatrix}\right)$.

---

[42]Adapted from exam question by Bert Halperin, Harvard University, 1976.
[43]Here g is the gyromagnetic ratio for the spin, and $\mu_B = e\hbar/2m_e$ is the Bohr magneton.
[44]More specifically, the right eigenvectors $M \cdot \boldsymbol{\rho}_\lambda = \lambda\boldsymbol{\rho}_\lambda$.

(c) *Solve for the 2×2 matrix $G(\tau)$. Check that* (i) *probability is conserved,* $\sum_\mu G_{\mu\nu}(\tau) = 1$; (ii) *evolution for zero time does not change the state,* $G_{\mu\nu}(0) = \delta_{\mu\nu}$; (iii) *the stationary state is stationary* $\sum_\nu G_{\mu\nu}(\tau)\rho_\nu^* = \rho_\mu^*$; *and* (iv) *the evolution goes at late times to the stationary state, independent of where it starts:* $G_{\mu\nu}(t) \to \langle \rho_\mu^* \rangle$ *as* $t \to \infty$.

Let $R(t)$ be the resistance as a function of time, hopping between R_α and R_β, as shown in Fig. 10.16, and let $\overline{R}$ be the time average of the resistance. In analogy to eqn 10.19 for equilibrium systems, the connected correlation function for the resistance fluctuations can be written as

$$
\begin{aligned}
C(\tau) &= \left\langle (R(t+\tau) - \overline{R})(R(t) - \overline{R}) \right\rangle_{\text{ev}} \\
&= \left\langle (R(\tau) - \overline{R})(R(0) - \overline{R}) \right\rangle_{\text{ev}} \quad (10.91) \\
&= \left\langle ([R(\tau)]_\nu - \overline{R})(R_\nu - \overline{R}) \right\rangle_{\text{eq}},
\end{aligned}
$$

in the (daunting) notation for different ensembles used in Section 10.4. Here $[R(\tau)]_\nu$ is the noise average of $R(\tau)$ given initial state ν.

(d) *Using the definition in the first line of eqn 10.91, argue directly that*

$$
C(\tau) = \sum_{\nu,\mu} G_{\mu\nu}(\tau)(R_\mu - \overline{R})(R_\nu - \overline{R})\rho_\nu^*,
$$

$$(10.92)$$

by considering the two possible initial states and the two possible final states a time τ later. Then calculate $[R(\tau)]_\nu$ in terms of G; use it and conservation of probability (part (c)) to give a second derivation of eqn 10.92 from the last line of eqn 10.91.

Fig. 10.17 Telegraph noise with three metastable states, from [158].

Nanojunctions, especially at higher temperatures, often show more than two metastable states in the experimental bandwidth.[45] Usually these form independent two-level fluctua-

tors (atomic rearrangements too far apart to interact substantially), but sometimes more complex behavior is seen. Figure 10.17 shows three resistance states, which we label α, β, and γ, from lowest resistance to highest. We notice from Fig. 10.17 that the rates $\Gamma_{\gamma\beta}$ and $\Gamma_{\beta\gamma}$ are the highest, followed by the rates $\Gamma_{\alpha\gamma}$ and $\Gamma_{\gamma\alpha}$. There are no transitions seen going between states α and β.

There is a large current flowing through the nanojunction, allowing the resistance to be measured. Whether these transitions are equilibrium fluctuations, perhaps with a field-dependent effective temperature, or whether they are nonequilibrium transitions mostly induced by the external current, could be tested if these last two rates could be measured. If detailed balance is violated, the system is out of equilibrium.

(e) Detailed balance. *Approximate the master equation rates $\Gamma_{\mu\nu}$ as a discrete-time Markov chain with a tiny step Δt and a 3×3 transition matrix P: $P_{\mu\nu} \approx \exp(\Gamma_{\mu\nu}\Delta t)$. What does the cyclic detailed balance condition (eqn 8.22) imply about the rates Γ? Assuming that the system satisfies detailed balance, what is the difference between the two unmeasured rates $\Gamma_{\alpha\beta}$ and $\Gamma_{\beta\alpha}$ in terms of the other four rates?*

(10.6) **Fluctuation-dissipation: Ising.**[46] (Condensed matter, Computation) ③
Let us consider the Ising Hamiltonian in a time-dependent external field $H(t)$,

$$
\mathcal{H} = -J \sum_{\langle ij \rangle} S_i S_j - H(t) \sum_i S_i, \quad (10.93)
$$

and look at the fluctuations and response of the time-dependent magnetization $M(t) = \sum_i S_i(t)$. The Ising model simulation should output both the time-dependent magnetization per spin $m(t) = (1/N) \sum_i S_i$ and the time–time correlation function[47] of the magnetization per spin,

$$
c(t) = \left\langle (m(0) - \langle m \rangle_{\text{eq}})(m(t) - \langle m \rangle_{\text{eq}}) \right\rangle_{\text{ev}}. \quad (10.94)
$$

We will be working above T_c, so $\langle m \rangle_{\text{eq}} = 0$. The time–time correlation function will start nonzero, and should die to zero over time. Suppose we start with a nonzero small external

---

[45]A junction is outside the bandwidth if it fluctuates either too fast or too slowly to measure with the experimental set-up.
[46]A link to the software can be found at the book website [181].
[47]Note that the formulæ in the text are in terms of the total magnetization $M = Nm$ and its correlation function $C = N^2 c$.

field, and turn it off at $t = 0$, so $H(t) = H_0\Theta(-t)$.[48] The magnetization $m(t)$ will be nonzero at $t = 0$, but will decrease to zero over time. By the Onsager regression hypothesis, $m(t)$ and $c(t)$ should decay with the same law. Run the Ising model, changing the size to 200×200. Equilibrate at $T = 3$ and $H = 0$, then do a good measurement of the time–time autocorrelation function and store the resulting graph. (Rescale it to focus on the short times before it equilibrates.) Now equilibrate at $T = 3$, $H = 0.05$, set $H = 0$, and run for a short time, measuring $m(t)$.

(a) *Does the shape and the time scale of the magnetization decay look the same as that of the autocorrelation function? Measure $c(0)$ and $m(0)$ and deduce the system-scale $C(0)$ and $M(0)$.*

Response functions and the fluctuation-dissipation theorem. The response function $\chi(t)$ gives the change in magnetization due to an infinitesimal impulse in the external field H. By superposition, we can use $\chi(t)$ to generate the linear response to any external perturbation. If we impose a small time-dependent external field $H(t)$, the average magnetization is

$$M(t) - \langle M \rangle_{\text{eq}} = \int_{-\infty}^{t} \mathrm{d}t'\, \chi(t-t')H(t'),$$

$$(10.95)$$

where $\langle M \rangle_{\text{eq}}$ is the equilibrium magnetization without the extra field $H(t)$ (zero for us, above T_c).

(b) *Using eqn 10.95, write $M(t)$ for the step down $H(t) = H_0\Theta(-t)$, in terms of $\chi(t)$.*

The fluctuation-dissipation theorem states

$$\chi(t) = -\beta\, \mathrm{d}C(t)/\mathrm{d}t, \qquad (10.96)$$

where $C(t) = \langle (M(0) - \langle M \rangle_{\text{eq}})(M(t) - \langle M \rangle_{\text{eq}}) \rangle_{\text{ev}}$.

(c) *Use eqn 10.96 and your answer to part (b) to predict the relationship between the demagnetization $M(t)$ and the correlation $C(t)$ you measured in part (a). How does your analytical ratio compare with the $t = 0$ ratio you noted down in part (a)?*

(10.7) Noise and Langevin equations. ③

We have never explicitly discussed how the energy removed from a system by damping is returned to the system to maintain thermal equilibrium. This energy input is through the thermal fluctuation noise introduced through the coupling to the heat bath. In this exercise we will derive a *Langevin equation* incorporating both noise and dissipation (see also Exercise 6.18 and [41, section 8.8]).

We start with a system with phase-space coordinates $\mathbb{P}, \mathbb{Q}$ and an internal potential energy $V(\mathbb{Q})$, coupled linearly to a heat bath through some coupling term $\mathbb{Q} \cdot \mathbb{F}$:

$$\mathcal{H} = \frac{\mathbb{P}^2}{2m} + V(\mathbb{Q}) + \mathcal{H}_{\text{bath}}(y_1, y_2, y_3, \dots)$$
$$- \mathbb{Q} \cdot \mathbb{F}(y_1, \dots). \qquad (10.97)$$

To our system, $\mathbb{F}$ acts as an external noise due to fluctuations in the bath. Here, however, we shall also view $\mathbb{F}$ as some collective coordinate of the bath, and $\mathbb{Q}(t)$ as the force from our system on the bath. (More traditionally, there is an interaction energy $-\Lambda\, \mathbb{Q} \cdot \mathbb{F}$; here we define the force to include the coupling strength Λ.) In the absence of the coupling to our system, assume that the bath would contribute an external noise $\mathbb{F}_b(t)$ with mean zero. In the presence of the coupling to the system, the mean value of the force will develop a nonzero expectation value

$$\langle \mathbb{F}(t) \rangle = \int_{-\infty}^{t} \mathrm{d}t'\, \chi_b(t-t')\mathbb{Q}(t'), \qquad (10.98)$$

where $\chi_b(t-t')$ is the susceptibility of the bath to the motion of the system $\mathbb{Q}(t)$. Our system then has an equation of motion with a random noise $\mathbb{F}$ and a time-retarded interaction due to χ_b:

$$m\ddot{\mathbb{Q}} = -\frac{\partial V}{\partial \mathbb{Q}} + \mathbb{F}_b + \int_{-\infty}^{t} \mathrm{d}t'\, \chi_b(t-t')\mathbb{Q}(t').$$
$$(10.99)$$

We can write this susceptibility in terms of the correlation function of the noise in the absence of the system:

$$C_b(t-t') = \langle \mathbb{F}_b(t)\mathbb{F}_b(t') \rangle \qquad (10.100)$$

using the fluctuation-dissipation theorem

$$\chi_b(t-t') = -\beta\frac{\partial C_b}{\partial t}, \qquad t > t'. \qquad (10.101)$$

---

[48]Here Θ is the Heaviside function: $\Theta(t) = 0$ for $t < 0$, and $\Theta(t) = 1$ for $t > 0$.

(a) *Integrating by parts and keeping the boundary terms, show that the equation of motion has the form*

$$m\ddot{Q} = -\frac{\partial \bar{V}}{\partial Q} + \mathbb{F}_b - \beta \int_{-\infty}^{t} dt'\, C_b(t - t')\dot{Q}(t').$$
(10.102)

What is the "potential of mean force" $\bar{V}$, in terms of V and C_b?

(b) *If the correlations in the bath are short-lived compared to the time scales of the system, we can approximate $\dot{Q}(t') \approx \dot{Q}(t)$ in eqn 10.102, leading to a viscous friction force $-\gamma\dot{Q}$. What is the formula for γ? Conversely, for a model system with a perfect viscous friction law $-\gamma\dot{Q}$ at temperature T, argue that the noise must be white $C_b(t - t') \propto \delta(t - t')$. What is the coefficient of the δ-function?* Notice that viscous friction implies a memoryless, Markovian heat bath, and vice versa.

Langevin equations are useful both in analytic calculations, and as one method for maintaining a constant temperature in molecular dynamics simulations.

(10.8) **Magnet dynamics.** (Condensed matter) ③
A one-dimensional classical magnet above its critical point is described by a free energy density

$$\mathcal{F}[M] = (C/2)(\nabla M)^2 + (B/2)M^2,$$
(10.103)

where $M(x)$ is the variation of the magnetization with position along the single coordinate x. The average magnetization is zero, and the total free energy of the configuration $M(x)$ is $F[M] = \int \mathcal{F}[M]\,dx$.

The methods we used to find the correlation functions and susceptibilities for the diffusion equation can be applied with small modifications to this (mathematically more challenging) magnetic system.

Assume for simplicity that the magnet is of length L, and that it has periodic boundary conditions. We can then write $M(x)$ in a Fourier series (eqn A.4)

$$M(x) = \sum_{n=-\infty}^{\infty} \widetilde{M}_n \exp(ik_n x),$$
(10.104)

with $k_n = 2\pi n/L$ and (eqn A.3)

$$\widetilde{M}_n = (1/L) \int_0^L M(x) \exp(-ik_n x).$$
(10.105)

As always, for linear systems with translation invariance (Section A.4) the free energy $F[M]$ decomposes into independent terms, one for each k_n.[49]

(a) *Calculate this decomposition of F: show that each term is quadratic.* (Hint: The only subtle case is $\widetilde{M}_n^2$; break it into real and imaginary parts.) *What is $\left\langle |\widetilde{M}_n|^2 \right\rangle_{eq}$, by equipartition? Argue that*

$$\left\langle \widetilde{M}_{-m} \widetilde{M}_n \right\rangle_{eq} = \frac{k_B T}{L(Ck_n^2 + B)} \delta_{mn}.$$
(10.106)

(b) *Calculate the equilibrium equal-time correlation function for the magnetization, $C(x, 0) = \langle M(x, 0)M(0, 0)\rangle_{eq}$.* (First, find the formula for the magnet of length L, in terms of a sum over n. Then convert the sum to an integral: $\int dk \leftrightarrow \sum_k \delta k = 2\pi/L \sum_k$.) *You will want to know the integral*

$$\int_{-\infty}^{\infty} e^{iuv}/(1 + a^2 u^2)\, du = (\pi/a) \exp(-|v|/a).$$
(10.107)

Assume the magnetic order parameter is not conserved, and is overdamped, so the time derivative of $[M]_{M_i}$ is given by a constant η times the variational derivative of the free energy:

$$\frac{\partial [M]_{M_i}}{\partial t} = -\eta \frac{\delta \mathcal{F}}{\delta M}.$$
(10.108)

M evolves in the direction of the total force on it.[50] The term $\delta\mathcal{F}/\delta M$ is the *variational derivative*:[51]

$$\delta F = F[M + \delta M] - F[M]$$
$$= \int (\mathcal{F}[M + \delta M] - \mathcal{F}[M])\, dx$$
$$= \int (\delta\mathcal{F}/\delta M)\delta M\, dx.$$
(10.109)

(c) *Calculate $\delta\mathcal{F}/\delta M$.* As in the derivation of the Euler–Lagrange equations [128, section 12.1] you will need to integrate one term by parts to factor out the δM.

---

[49]Notice that this independence implies that in equilibrium the phases of the different modes are uncorrelated.
[50]Remember that the average $[M]_{M_i}$ is over all future evolutions given the initial condition $M_i(x) = M(x, 0)$.
[51]This formula is analogous to taking the gradient of a scalar function of a vector, $f(\vec{y} + \vec{\delta}) - f(\vec{y}) \approx \nabla f \cdot \vec{\delta}$, with the dot product for vector gradient replaced by the integral over x for derivative in function space.

(d) *From your answer to part (c), calculate the Green's function for $G(x,t)$ for $[M]_{M_i}$, giving the time evolution of a δ-function initial condition $M_i(x) = M(x,0) = G(x,0) = \delta(x)$.* (Hint: You can solve this with Fourier transforms.)

The Onsager regression hypothesis tells us that the time evolution of a spontaneous fluctuation (like those giving $C(x,0)$ in part (b)) is given by the same formula as the evolution of an imposed initial condition (given by the Green's function of part (d)):

$$
\begin{aligned}
C(x,t) &= \langle M(x,t)M(0,0)\rangle_{\text{ev}} \\
&= \left\langle [M(x,t)]_{M_i} M(0,0)\right\rangle_{\text{eq}} \\
&= \left\langle \int_{-\infty}^{\infty} M(x',0)G(x-x',t)dx' M(0,0)\right\rangle_{\text{eq}} \\
&= \int_{-\infty}^{\infty} C(x',0)G(x-x',t)\,dx'.
\end{aligned}
$$
$$(10.110)$$

(e) *Using the Onsager regression hypothesis calculate the space-time correlation function $C(x,t) = \langle M(x,t)M(0,0)\rangle_{\text{ev}}$.* (This part is a challenge; your answer will involve the error function.) *If it is convenient, plot it for short times and for long times; does it look like $\exp(-|y|)$ in one limit and $\exp(-y^2)$ in another?*

The fluctuation-dissipation theorem can be used to relate the susceptibility $\chi(x,t)$ to the time-dependent impulse to the correlation function $C(x,t)$ (eqn 10.60). Let $\chi(x,t)$ represent the usual response of M to an external field $H(x',t')$ (eqn 10.29) with the interaction energy being given by $\int M(x)H(x)\,dx$.

(f) *Calculate the susceptibility $\chi(x,t)$ from $C(x,t)$. Start by giving the abstract formula, and then substitute in your answer from part (e).*

(10.9) **Quasiparticle poles and Goldstone's theorem.** (Condensed matter) ③
We saw in Exercise 9.14 that sound waves obey Goldstone's theorem. As the fundamental excitations associated with a continuous (translational) symmetry, they have low frequency excitations at long wavelengths, whose damping disappears as the frequency goes to zero. Here we want to explore the complex susceptibility of

damped sound, and look for analogues of quasiparticle in the behavior of sound.
We add forcing $f(x,t)/\rho$ to eqn 9.28 for the damping of sound waves:

$$
\frac{\partial^2 u}{\partial t^2} = c^2 \frac{\partial^2 u}{\partial x^2} + d^2 \frac{\partial}{\partial t}\frac{\partial^2 u}{\partial x^2} + \frac{f(x,t)}{\rho}. \quad (10.111)
$$

Here again c is the speed of sound c, ρ the density, and d^2 is the Kelvin damping term consistent with Galilean invariance (Exercise 9.6).
If we force the undamped system with a forcing $f(x,t) = \cos(kx)\cos(ckt)$ that matches the wavevector and frequency $\omega = ck$ of one of the phonons, the amplitude of that mode of vibration will grow without limit; the susceptibility diverges. Let us try it with damping.

(a) *What is the susceptibility $\widetilde{\chi}(k,\omega)$ for our damped system (eqn 10.111)?* (Hint: Change to Fourier variables k and ω as in Section A.1.) *Check that your answer without dissipation ($d = 0$) has poles[52] at $\omega_k = \pm ck$ for each wavevector k; the susceptibility diverges when excited on resonance. Show that $\Omega_k = \omega_k - i\Gamma_k$ when dissipation is included.* (If you have done Exercise 9.14, you may want to check your formula for Γ_k with the dispersion relation you derived there.)

(b) *Check that the poles in the susceptibility are all in the lower half-plane, as required by causality (Section 10.9).*

Neutron scattering can be used to measure the Fourier transform of the correlation function, $\widetilde{C}(k,\omega)$ in a material. Suppose in our material $c = 1{,}000\,\text{m/s}$, $d = 10^{-3}\,\text{m/s}^{\frac{1}{2}}$, $\rho = 1\,\text{g/cm}^3$, and $T = 300\,\text{K}$; for your convenience, $k_B = 1.3807 \times 10^{-23}\,\text{J/K}$.

(c) *Use the fluctuation-dissipation theorem*

$$
\chi''(\mathbf{k},\omega) = \frac{\beta\omega}{2}\widetilde{C}(\mathbf{k},\omega) \quad (10.112)
$$

(see eqn 10.65) to calculate $\widetilde{C}(k,\omega)$, the correlation function for our material. Plot $\widetilde{C}(k,\omega)$ at $\omega = 10^8\,\text{radians/sec}$ (about $16\,\text{MHz}$), for k from zero to 2×10^5. Does it appear difficult to estimate the dispersion relation ω_k from the correlation function?

In strongly interacting systems, the elementary excitations are *quasiparticles*: dressed electrons, photons, or vibrational modes which are not many-body eigenstates because they eventually decay (note 23 on p. 188). These

---

[52]A function $\psi(z)$ has a *pole* at z_0 if it diverges at z_0 like a constant times $z - z_0$. The poles are easy to find if ψ is a ratio of two functions which themselves do not go to infinity; the poles are at zeros of the denominator.

quasiparticles are *defined* in terms of poles in the propagator or quantum Green's function, closely related to our classical correlation function $\tilde{C}(k, \omega)$.

(10.10) **Human correlations.**[53] @

Consider a person-to-person correlation function for the center-of-mass positions $\mathbf{h}_\alpha = (x_\alpha, y_\alpha)$ of people labeled by α, in various public places. In particular, let $C_{\text{crowd}}(\mathbf{r}) = \langle \rho(\mathbf{h})\rho(\mathbf{h} + \mathbf{r}) \rangle$, where $\rho(\mathbf{h}) = \sum_\alpha \delta(\mathbf{h} - \mathbf{h}_\alpha)$, and the ensemble average $\langle \ldots \rangle$ is over a large number of similar events.

We shall consider the correlations between ten people seated in a short subway car (Fig. 10.18) of length $L = 2.5\,\text{m}$, on two benches along the edges $y = 0$ and $y = 2\,\text{m}$. (See also Exercise 10.11.)

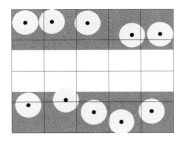

Fig. 10.18 Humans seated on a subway car.

(a) *Plot the displacement $(\Delta x, \Delta y)$ for each pair of people in a subway car, and plot them.* (You could print out Fig. 10.18 from [181], or outline the subway car with masking tape on the floor in class and place ten "people" with yellow stickies. Floor tiles form a nice grid.) *Assume five people on each side, randomly seated except with no two seated less than center-of-mass distance $\delta x = 0.3\,\text{m}$ between neighbors. Assume they have some randomness δy due to slouching (as in Fig. 10.18). What features do you see in your plot?*
(b) *Sketch a rough, two-dimensional contour plot from your measurements, as an estimate for $C_{\text{crowd}}(\mathbf{r})$.* (You may remember contour functions from topographic maps: they have

curves at constant height. Here we want curves at constant density, with contours separating regions with closely packed points from regions with sparser populations.) *Make sure your contour plot reflects the features you found in part (a).*
(c) *Sketch a rough contour plot for $C_{\text{crowd}}(\mathbf{r}, t)$ in the same subway car, where $t = 20$ minutes is long enough that most people will have gotten off at their stops.*
(d) *Sketch a rough contour plot for the Fourier transform $\tilde{C}_{\text{crowd}}(\mathbf{k})$ for the correlation function in part (b).*

(10.11) **Subway bench Monte Carlo.** ③

We shall study a model for the correlation function between the centers of passengers sitting on a bench alongside the windows in a subway car (explored in Exercise 10.10). We model the system as a one-dimensional equilibrium hard-disk gas of N disks with length a, contained in a car $0 < y < L$.[54]

If $Na = L$, there is no free space on the bench, and the passengers have centers of mass at packed positions

$$y_{\text{packed}} = [a/2, 3a/2, \ldots, (2N-1)a/2].$$
$$(10.113)$$

Call the free space $L_{\text{free}} = L - Na$. We can relate the ensemble of hard disks in one dimension in a box of length L and an ideal gas in a box of length L_{free}.
(a) *Show that the configuration $y_1 < y_2 < \cdots < y_N$ of passengers sitting on the bench of length L corresponds precisely to legal configurations $z_1 < \cdots < z_N$ of ideal gas particles in a box of length L_{free}, with z_n measuring the free space to the left[55] of passenger n, by giving an invertible linear mapping from $\mathbf{z} \to \mathbf{y}$. Is the position-space entropy the same for the ideal gas and the subway car? Why?* (Feel free to look up the entropy of the Tonks gas to check your answer. But do explain your reasoning. Would the entropy be the same if the mapping were nonlinear, but invertible?)

The correlation functions for the ideal gas and the subway car are definitely not the same. But the mapping to the ideal gas provides a very ef-

---
[53]Copies of Fig. 10.18 and a corresponding graph can be printed from the book website [181]
[54]In one dimension, a disk of diameter a is really a rod of length a, so this is often called a hard-rod gas. It is also known as the Tonks gas.
[55]That is, in the direction of decreasing y.

ficient way of sampling the equilibrium ensemble of subway passengers:

(b) *For $L = 10$, $a = 0.4$, and $N = 20$, generate a sample from the equilibrium ideal gas ensemble (that is, N uniformly distributed random numbers z_n in the range $[0, L_{\text{free}}]$). Generate a sample y_n from the subway car ensemble. (Hint: Use z_n and y_{packed}.) Plot a function that is one if a person is occupying the space, and zero if the space is empty, by generating a curve with four points $(0, y_n - a/2), (1, y_n - a/2), (1, y_n + a/2), (0, y_n + a/2)$ for each passenger.*

We can now collect the distances between each pair of passengers for this snapshot.

(c) *Take your sample y_n, and make a list of all the (signed) separations $\Delta y = y_n - y_m$ between pairs of passengers. Plot a histogram of this distribution, using the standard normalization (the number of counts in each bin) and a bin-width of one. Explain why it has an overall triangular shape. Zoom in to $-2 < \Delta y < 2$, using a bin width of 0.1. Explain briefly the two striking features you see near $\Delta y = 0$—the spike, and the gap.*

To explore the correlation function more quantitatively, we now want to take an average of an ensemble of many passenger packings.

(d) *Generate $n = 100$ ensembles of passengers as in part (b), and collect all the pair separations. Make a histogram of the pair separations, in the range $-2 < \Delta y < 2$ with bin width of 0.02. Adjust the vertical range of the plot to show the entire first bump (but clipping the central spike). Explain how the bumps in the histogram are related to the distances to neighbors seated m passengers to the right.*

At short distances, you can show that your histogram is related[56] to the density–density correlation function $C(r) = \langle \rho(y)\rho(y + r) \rangle$ evaluated at $\Delta y = r$, as follows.

(e) *In an infinite system with ρ particles per unit length, given there is a particle at y, write in terms of $C(r)$ the probability per unit length $P(y+r|y)$ that there is another particle at $y+r$. If we take n samples from the ensemble with N passengers each, and compute all the pair separations, what is the number per unit length of separations we expect at r, in terms of $C(r)$? Show that the expected number of separations*

in a narrow bin of width B centered at r is $(nNB/\rho)C(r)$. (Hint: The probability density $P(A|B)$ of A (person at r) given B (person at 0) is the probability density $P(A\&B)$ of both happening divided by the probability density $P(B)$ of B happening. Multiply by the number of people and the bin size to get the expected number in the bin.)

Next, we will decompose $C(r)$ between all pairs into a sum

$$C(r) = \sum_m C_m(r) \tag{10.114}$$

of terms collecting distances to the mth neighbor to the right (with negative m corresponding to the left). We calculate the shapes of these bumps for an infinite system at density ρ.

Let us start with $C_1(r)$. The separations between nearest neighbors $m = 1$ in the subway is just a larger than the separations between nearest neighbors in the corresponding one-dimensional ideal gas, $\Delta y_{\text{nn}} = a + \Delta z_{\text{nn}}$. The corresponding ideal gas, remember, has density $\rho_{\text{ideal}} = N/L_{\text{free}}$.

(f) *Argue that the $m = 1$ nearest-neighbor contribution to the correlation function is an exponential decay*

$$C_1(r) = \rho\,\rho_{\text{ideal}} \exp\left[-\rho_{\text{ideal}}(r - a)\right]\Theta(r - a), \tag{10.115}$$

where Θ is the Heaviside function (one for positive argument, zero otherwise). (Hint: How is it related to $C_1^{\text{ideal}}(r)$ for the ideal gas?) Using the relation between $C(r)$ and your histogram derived in part (d), plot your prediction against the first bump at positive Δy in your histogram.

(g) *Calculate $\widetilde{C}_1(k)$ from eqn 10.115, using the conventions of eqn A.9 in the Fourier appendix. (Hint: It is the integral of an exponential over a half-space.)*

The gaps between passengers in the infinite subway are just the same as the gaps between particle in the ideal gas—and hence are independent. The second-neighbor distance is thus the sum of two first-neighbor distances independently drawn from the same distribution, so the probability distribution $P_2(y + r|y)$ is the convolution of two first-neighbor probability distributions. Fourier transforms become useful here (see Exercise 12.11 and eqn A.23);

---

[56] $C(r)$ and our histogram are also closely related to the *pair distribution function* $g(r)$ we introduced in Exercise 10.2.

(h) *Using this and your answer to part (e), write the Fourier spectrum of the second neighbor peak $\widetilde{C}_2(k)$, first in terms of the Fourier spectrum of the first neighbor peak $\widetilde{C}_1(k)$, and then using your answer from part (g). What is $\widetilde{C}_n(k)$?* (Hint: Start with calculating $\widetilde{P}_2(k)$.)
By either doing inverse Fourier transforms back to r or convolutions, one may derive the correlation function for the subway problem ignoring the finite size of the car.[57] We present the answer here:

$$C_m(r) = \rho\, \rho_{\text{ideal}}^m \frac{(r - ma)^{m-1}}{(m-1)!}$$
$$e^{-\rho_{\text{ideal}}(r-ma)}\Theta(r - ma) \quad m > 0,$$
$$C_0(r) = \rho\, \delta(0)$$
$$C_m(r) = C_{(-m)}(-r) \quad m < 0. \tag{10.116}$$

(i) *Plot[58] the theoretical prediction of eqn 10.116 against your measured histogram, for $-2 < r < 2$. Explain any discrepancy you find.* (Hint: Remember the triangle plot in part (c).)

(10.12) **Liquid free energy.** ⓟ
The free energy F of a liquid[59] (unlike an ideal gas), resists rapid changes in its density fluctuations $\rho(x)$ around its average density. For small deviations about the average,

$$F(\rho) = \int \tfrac{1}{2}K(\nabla\rho)^2 + \tfrac{1}{2}\alpha\rho^2 \mathrm{d}x. \tag{10.117}$$

(See Exercise 9.4, where a similar term led to a surface tension for domain walls in magnets.)
Let us consider a one-dimensional fluid in a box of length L with periodic boundary conditions.
(a) *Show in detail that the free energy can be expressed in terms of the Fourier series of ρ:*

$$F(\widetilde{\rho}) = \sum_m \tfrac{1}{2}(Kk_m^2 + \alpha)|\widetilde{\rho}_m|^2 L, \tag{10.118}$$

with $k_m = 2\pi m/L$. (Hint: Use $\rho(x) = \sum_m \widetilde{\rho}_m \exp(\mathrm{i}kx)$ from our definition eqn A.4, substituting into eqn 10.117.)
(b) *Show that equal-time correlation function is $\widetilde{C}(k_m) = \langle|\widetilde{\rho}(k_m)|^2\rangle = k_BT/((Kk_m^2 + \alpha)L)$.*

(Hint: F is a sum of harmonic oscillators; use equipartition.) *Taking the limit $L \to \infty$ where the k-vectors become a continuum, show that $\widetilde{C}(k) = k_BT/(Kk^2 + \alpha)$.* (Hint: $\sum_m \equiv L/(2\pi)\int \mathrm{d}k$, since $\Delta k = k_{m+1} - k_m = 2\pi/L$. Use this and eqns A.4 and A.10 to relate the continuum $\widetilde{C}(k)$ to the discrete $\widetilde{C}(k_m)$ as $L \to \infty$.)
(c) *Transform $\widetilde{C}(k) = k_BT/(Kk^2 + \alpha)$ to real space. Does the term involving the stiffness K introduce correlations (as opposed to those in the ideal gas)? Do the correlations fall off to zero at long distances?* (Remember the $1/2\pi$ in the continuum inverse Fourier transform of eqn A.7. Hint: $\int_{-\infty}^{\infty} \exp(\mathrm{i}qy)/(1 + q^2)\mathrm{d}q = \pi \exp(-|y|)$.)[60]

(10.13) **Onsager regression hypothesis.** ⓐ
A harmonic oscillator with mass M, position Q and momentum P has frequency ω. The Hamiltonian for the oscillator is thus $\mathcal{H} = P^2/2M + \tfrac{1}{2}M\omega^2Q^2$. The correlation function $C_Q(\tau) = \langle Q(t + \tau)Q(t)\rangle$.
(a) *What is $C_Q(0)$?* (Hint: Use equipartition.)
The harmonic oscillator is coupled to a heat bath, which provides a dissipative force $-\dot{Q}/\eta$, so under an external force $F(t)$ we have the macroscopic equation of motion $\ddot{Q} = -\omega^2Q - \dot{Q}/\eta + F/M$ (ignoring thermal fluctuations). We assume the harmonic oscillator is overdamped, so

$$\mathrm{d}Q/\mathrm{d}t = -\lambda Q + (\lambda/M\omega^2)F, \tag{10.119}$$

where $\eta = \lambda/M\omega^2$. We first consider the system without an external force.
(b) *What is $C_Q(\tau)$, in terms of λ and your answer from (a)?* (Hint: Use the Onsager regression hypothesis to derive an equation for $\mathrm{d}C_Q/\mathrm{d}\tau$. Remember that C is symmetric in time.)
Now consider a system of two uncoupled harmonic oscillators as in Fig. 10.19, one as described above and another with mass m, position q, momentum p, and frequency Ω, coupled to a heat bath with overdamped decay

---

[57]These are just the mth neighbor correlations in the ideal gas, shifted to the right by ma, the sum of the widths of m passengers.
[58]In Mathematica, you may need to increase PlotPoints and MaxRecursion to avoid missing features in the curve.
[59]Real liquids are almost incompressible. Instead of the density variations we consider here, one usually describes inhomogeneous liquids as having a spatially varying pressure.
[60]The function $1/(1+q^2)$ is often called a Lorentzian or a Cauchy distribution. Its Fourier transform can be done using Cauchy's residue theorem, using the fact that $1/(1+q^2) = 1/((q+\mathrm{i})(q-\mathrm{i}))$ has a pole at $q = \pm\mathrm{i}$. So $\int_{-\infty}^{\infty} \exp(\mathrm{i}qy)/(1+q^2)\mathrm{d}q = \pi \exp(-|y|)$. The absolute value happens because one must close the contour differently for $y > 0$ and $y < 0$.

rate Λ. (Capital letters are meant to indicate size: this is a small mass with a high frequency and a quick decay time.) Here we only observe $X = Q + q$; Q and q represent microscopic degrees of freedom in a material, and X represents a macroscopic property whose correlation function Onsager is trying to estimate. We shall compute properties of this combination X, both here and in Exercise 10.16.

Fig. 10.19 Two springs. Two harmonic oscillators, with an external force $2F(t)$ coupled to the average displacement $(Q(t)+q(t))/2 = X(t)/2$ (equivalent to $F(t)$ coupled to the sum $Q + q = X$).

(c) *What is $C_X(0)$? Show that $C_X(\tau)$ can be written in terms of C_Q and C_q. (Hint: Q and q are completely uncoupled, and so are uncorrelated random variables.) Roughly sketch a graph of C_X as a function of τ, assuming $\lambda \approx \Lambda/10$ and $M\omega^2 \approx m\Omega^2/2$.*
A consequence of the Onsager regression hypothesis is that the correlation function decays "according to the same laws as [a deviation] that has been produced artificially". But clearly there are different ways in which $X(t)$ could evolve from an artificially imposed initial condition, depending on the way we start Q and q.
(d) *How would $X(t)$ decay from an initial state $Q(0) = X(0)$, $q(0) = 0$? From a state $q(0) = X(0)$, $Q(0) = 0$? Does either decay the way $C_X(t)$ decays from part (c)? How would $X(t)$ decay from an initial state prepared with a sudden impulsive force[61] $F(t) = J_{\mathrm{imp}}\delta(t)$? From an initial state prepared with a constant force $F(t) = F_0\Theta(-t)$? Does either decay in the same fashion as $C_X(t)$?*
Onsager's macroscopically perturbed system must be produced slowly compared to the internal dynamical relaxation times, for its fu-

ture evolution to be determined by its current state, and for his hypothesis to be well posed. Stated another way, Onsager hypothesizes that we have included all of the slow variables into the macroscopic observables whose dynamics and correlations are under consideration.

(10.14) **Liquid dynamics.** @
As a continuation of Exercise 10.12, let us consider the dynamics of our one-dimensional model liquid with free energy given by eqn 10.117.
Liquids in a porous environment (think water in the soil, or percolating through coffee grounds) are well described by overdamped macroscopic equations of motion, where the current is proportional to the gradient of the chemical potential $J = -\eta\nabla\mu$. Here the chemical potential is $\mu = (\delta\mathcal{F}/\delta\rho)$, the variational derivative of the free energy density with respect to liquid density.
(a) *Calculate μ for our liquid free energy (eqn 10.117). What is the macroscopic equation of motion $\partial\rho/\partial t = -\nabla \cdot J$? What is the equation giving the dynamics for wave vector k, $\partial\widehat{\rho}_k/\partial t$? Given an initial periodic density variation $\rho(t = 0) = \rho_0 + a\cos(kx)$, solve for $\rho(t)$.*
We found in Exercise 10.12 that the equal-time correlation function is $\widehat{C}(k_m, 0) = k_B T/((Kk_m^2 + \alpha)L)$. Onsager's regression hypothesis says that the decay of this correlation function will be the same as the decay of a macroscopic imposed initial state with the same wavevector (as in part a).
(b) *According to Onsager's regression hypothesis, what will the correlation $\widehat{C}(k, t)$ be?* (See note 12 on p. 293 and note 13 on p. 294.)
It is nontrivial to Fourier transform this back into real space, but the short-distance fluctuations will get smoothed out even faster than they do in the diffusion equation and the ideal gas.

(10.15) **Harmonic susceptibility, dissipation.** @
This is a continuation of Exercise 10.13, where we considered damped harmonic oscillators. Consider one harmonic oscillator (P, Q) with $\mathcal{H} = P^2/2M + \frac{1}{2}M\omega_0^2 Q^2 - F(t)Q(t)$, overdamped with decay rate λ.
The static susceptibility χ_0 is defined as the ratio of response Q to a small force F.

---

[61]The response to an impulsive force you calculated is the *susceptibility* (Section 10.5 and Exercise 10.15), which can be used to study the linear response to any perturbation.

(a) *Under a constant external force F_0, what is the equilibrium position Q_0? What is the static susceptibility χ_0?* (Hint: What is the spring constant for our harmonic oscillator?) Generally speaking, a susceptibility or *compliance* will be the inverse of the elastic modulus or stiffness—sometimes a matrix inverse.

Consider our harmonic oscillator held with a constant force F for all $t < 0$, and then released. For $t > 0$, the dynamics is clearly an exponential decay (releasing the overdamped harmonic oscillator), and can also be written as an integral of the time-dependent susceptibility $\chi(\tau)$ for the oscillator:

$$Q(t) = Q_0 \exp(-\lambda t)$$
$$= \int_{-\infty}^{0} dt' \chi(t-t')F \qquad (10.120)$$

(Note that the integral ends at $t' = 0$, where the force is released.)

(b) *Solve eqn 10.120 for $\chi(\tau)$.* (Hint: You can take derivatives of both sides and integrate the right-hand side by parts. Or you can plug in $\chi(\tau)$ as an exponential decay $A\exp(-Bt)$ and solve.)

We can now Fourier transform χ to find $\widetilde{\chi}(\omega) = \int dt \exp(i\omega t)\chi(t) = \chi'(\omega) + i\chi''(\omega)$.

(c) *What is $\widetilde{\chi}(\omega)$? $\chi'(\omega)$? $\chi''(\omega)$? Is the power dissipated for an arbitrary frequency dependent force f_ω (given by eqn 10.37) always positive?*

(10.16) **Harmonic fluctuation-dissipation.** @
This is a continuation of Exercise 10.13, where you derived the correlation function $C_Q(\tau) = A\exp(-\lambda|\tau|)$ for the overdamped harmonic oscillator with decay rate λ, and the correlation function $C_X(\tau) = A\exp(-\lambda|\tau|) + B\exp(-\Lambda|\tau|)$ for the sum $X = Q + q$ of two uncoupled oscillators with overdamped decay rates λ and Λ. In Exercise 10.15 we calculated the susceptibility directly. Now we shall calculate it using the fluctuation-dissipation theorem.

(a) *Using the classical fluctuation-dissipation theorem, eqn 10.60, derive the time-dependent susceptibilities $\chi_Q(\tau)$ and $\chi_X(\tau)$. You may leave your answer in terms of A and B.*

(b) *Sketch a graph of $\chi_X(\tau)$ for $-1/\lambda < \tau < 1/\lambda$, assuming that $\lambda \approx \Lambda/10$ and $A \approx 2B$.*

(10.17) **Susceptibilities and correlations.** @
An order parameter y is pulled upward slightly by a time-independent point force F at the origin; the resulting height $Y(x)$ is shown in Fig. 10.20. Without the force, the order parameter exhibits small fluctuations about height zero. These height fluctuations are measured at a temperature T, in the absence of an external force.

Fig. 10.20 Pulling.

(a) *Calculate the thermal ensemble average $\langle y(0,t)y(\Delta x,t)\rangle$ of these height fluctuations without the force, in terms of $Y(x)$, F, and $k_B T$. Plot it over $x \in (-10, 10)$, labeling the limits of your vertical axis in terms of these same variables.* (Hints: The fluctuations should be larger for higher temperature, and the response should be larger for higher forces. Section 10.7 may be useful.)

Fig. 10.21 Hitting.

Figure 10.21 shows the height $y(x,t)$ at the origin $x = 0$ after the same system is hit upward with an impulsive force $g\delta(x)\delta(t)$. The time decay of the height at the origin is measured to be exponential, $y(0,t) = y(0,0)\exp(-\gamma t)$ for $t > 0$. The equilibrium height fluctuations are measured in the system at $x = 0$ as a function of time in the absence of an external hit.

(b) *Calculate the thermal ensemble average $\langle y(0,t)y(0, t+\Delta t)\rangle$ in equilibrium (without the hit), in terms of $y(0,0)$, g, $k_B T$, Δt, and γ. Plot it for $t \in (-4/\gamma, 4/\gamma)$, labeling the lim-*

its of your vertical axis in terms of these same variables.

(c) *Give a formula for Y_0 in Fig. 10.20 (the response to a time-independent pull), in terms of $y(0,0)$ (the response to a δ-function hit), g, F, γ, and $k_B T$.*

(10.18) **Harmonic Kramers–Kronig.** @

In general, susceptibilities are causal: $\chi(\tau) = 0$ for $\tau < 0$ because the response cannot precede the force. We saw in general that this implied that $\widetilde{\chi}(\omega)$ had to be analytic in the upper half-plane, $\mathrm{Im}(\omega) > 0$.

Let us check this for the overdamped harmonic oscillator, where $\chi(\tau) = A\exp(-\lambda\tau)$ for $\tau > 0$. What is $\widetilde{\chi}(\omega)$? Where are its poles? Are they[62] in the upper half-plane? Are there any singularities in the lower half-plane?

(10.19) **Critical point response.** ③

An experimentalist is exploring the properties of a new two-dimensional magnet near its critical point. The magnetization vanishes continuously at a temperature T_c. Her theoretical colleague predicts that the magnetic fluctuations at a temperature $T > T_c$ will obey

$$C(\mathbf{r}, \tau) = \langle m(\mathbf{x}, t)m(\mathbf{x}+\mathbf{r}, t+\tau)\rangle \quad (10.121)$$
$$\approx (r^2 + (c\tau)^2)^{-\eta/2}e^{-\sqrt{r^2+(c\tau)^2}/\xi(T)}$$

where m is the local magnetization per unit area, c is a scale factor with units of velocity, η is a universal critical exponent, and $\xi(T)$ is the temperature-dependent correlation length. (See also Exercise 12.25.)

(a) *What is the equal-time correlation function? Is it isotropic (depending only on $r = |\mathbf{r}|$)? What is the time–time correlation function for a single site? If we define a three-component vector $\mathbf{R} = (r_1, r_2, c\tau)$, show that the correlation function is invariant under a 3D rotation of the vector $\mathbf{R}$.*

A space-time rotational symmetry like this sometimes happens at quantum critical points. We assume it here to make the calculation of $\widetilde{C}(\mathbf{k}, \omega)$ in part (h) feasible by hand.

The experimentalist expects that the fluctuations will become much larger at the critical point. The ensemble average of the total magnetization $\langle M(t)\rangle = \langle \int \mathrm{d}x \int \mathrm{d}y\, m(x, t)\rangle$ is zero

by symmetry above T_c (the symmetry is unbroken). But the average of the square $\langle M^2(t)\rangle$ must be nonzero. Above T_c the fluctuations are small and local, so $\langle M^2(t)\rangle$ will scale with the area of the sample.

(b) *Calculate the theoretical prediction for the mean-squared magnetization per unit area A, $\langle M^2\rangle/A$, as a function of ξ? Is the experimentalist correct, that $\langle M^2\rangle/A$ diverges at the critical point, when $\xi \to \infty$?* (Hint: Write M^2 as a double integral, and express $\langle M^2\rangle$ in terms of $C(\mathbf{r}, 0)$. You may use $\int_0^\infty x^{-a}\exp(-x)\mathrm{d}x = \Gamma(1-a)$, where the gamma function generalizes the factorial: $\Gamma(n) = (n-1)!$)

Now we couple to a uniform external field H (so the Hamiltonian includes a term MH, as usual).

(c) *Use your answer to part (b) to calculate the uniform susceptibility $\chi = \mathrm{d}M/\mathrm{d}H$ of their material. If $\xi \sim (T - T_c)^{-\nu}$, with what exponent $(T - T_c)^{-\gamma}$ does χ diverge?*

The experimentalist could use elastic neutron scattering to probe the material. Elastic neutron scattering probes the spatial Fourier transform $\widehat{C}(k_1, k_2, t = 0)$ of the equal-time magnetic correlation function $\langle m(\mathbf{x}, t)m(\mathbf{x}+\mathbf{r}, t)\rangle$, where $\mathbf{r} = (r_1, r_2)$. Note that the theoretical prediction for the equal-time magnetic correlation function is isotropic under rotations in space, so its Fourier transform $\widetilde{C}$ must be isotropic too, depending only on $|\mathbf{k}|$.

(d) *Show that $\widehat{C}(k_1, k_2, \tau = 0)$ at the critical point $(\xi \to \infty)$ is $\widehat{B}|k|^{\widehat{v}}$, for some constants $\widehat{B}$ and $\widehat{v}$. (Choose $\mathbf{k} = (k, 0)$ and change variables in the Fourier transform from (r_1, r_2) to $u = kr_1$, $v = kr_2$.) What is $\widehat{v}$? Write the constant $\widehat{B}$ as a double integral over u and v.*

One can evaluate this integral with some work: $\widehat{B} = 2\sqrt{\pi}\cos(\pi\eta/2)\Gamma(2-\eta)\Gamma((\eta+1)/2)/(\frac{1}{2}(1-\eta)\Gamma(\eta/2))$.

Neutrons often lose energy when scattering (*inelastic neutron scattering*); this energy loss probes the dynamic fluctuations in the material, measuring $\widetilde{C}(\mathbf{k}, \omega)$ (often called the *dynamic structure factor* $S(\mathbf{k}, \omega)$). Our experimentalist is also considering using inelastic neutron scattering to test her colleague's prediction.

(e) *Use the rotational symmetry of part (a)*

---

[62]Beware: The conventions on Fourier transforms are fussy (Appendix 13). On going from t to ω one multiplies by $\exp(i\omega t)$, not by $\exp(-i\omega t)$ (eqn A.6). Note that the opposite convention holds for spatial Fourier transforms (eqn A.9). Getting this wrong will swap the upper and lower half-plane.

and the strategy of part (d) to show that $\widetilde{C}(k_1, k_2, \omega) = \widetilde{B}|\mathbf{K}|^{\widetilde{\upsilon}}$ *for* $\mathbf{K} = (k_1, k_2, \omega/c)$ *at the critical point where* $\xi \to \infty$*. What is* $\widetilde{\upsilon}$ *for the inelastic scattering? What integral gives* $\widetilde{B}$*?*

Again, one can evaluate $\widetilde{B} = (4\pi/c)\Gamma(2 - \eta)\sin(\pi\eta/2)$.

The team is hoping to use this material as a damping mechanism, absorbing stray microwaves being emitted from a superconducting resonance cavity. They want to estimate the losses in the material due to a standing wave $H(x, y, t) = H_0 \cos(kx)\cos(\omega t)$.

(f) *Give a formula for* $\chi''(k, 0, \omega)$ *at the critical point using your answer for* $\widetilde{C}$ *and/or* $\widehat{C}$*, leaving your answer in terms of the constants* $\widetilde{\upsilon}$*,* $\widetilde{B}$*,* $\widehat{\upsilon}$*, and* $\widehat{B}$ *from parts (d) and (e). Calculate the power dissipated per unit area for the applied field* H*.* (The text largely ignores the spatial dependence of the driving force and the susceptibility; you will likely want to re-derive eqn 10.37 for the forcing $f(x, t) = H(x, t)$.)

Abrupt phase transitions

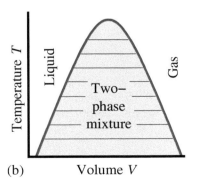

Most phase transitions are abrupt. At the transition, the system has discontinuities in most physical properties; the density, compressibility, viscosity, specific heat, dielectric constant, and thermal conductivity all jump to new values. Furthermore, in most cases these transitions happen with no precursors, no hint that a change is about to occur; pure water does not first turn to slush and then to ice,[1] and water vapor at $101\,°C$ has no little droplets of water inside.[2]

In Section 11.1 we explore the phase diagrams and free energies near abrupt phase transitions. In Section 11.2, we learn about the *Maxwell construction*, which is used to determine the coexistence point between two phases. In Section 11.3, we calculate how quickly the new phase nucleates at an abrupt transition. Finally, in Section 11.4, we introduce three kinds of microstructures associated with abrupt transitions—*coarsening*, *martensites*, and *dendrites*.

[1] Water with impurities like salt or sugar will form slush. The slush is ice that is mostly pure H_2O surrounded by rather salty water. The salt lowers the freezing point; as the ice freezes it expels the salt, lowering the freezing point of the remaining liquid.

[2] More precisely, there are only exponentially rare microscopic droplet fluctuations, which will be important for nucleation; see Section 11.3.

11.1 Stable and metastable phases

Water remains liquid as it is cooled, until at $0\,°C$ it abruptly freezes into ice. Water remains liquid as it is heated, until at $100\,°C$ it abruptly turns to vapor.

There is nothing abrupt, however, in boiling away a pan of water.[3] This is because one is not controlling the temperature directly, but rather is adding energy at a constant rate. Consider an insulated, flexible container of H_2O at fixed pressure, as we slowly add energy to it. When

[3] The *latent heat* is the change in enthalpy $E + PV$ at a phase transition. For boiling water it is $2,500\,J/g = 600\,calories/g$. Hence, at a constant energy input from your stove, it takes six times as long to boil away the water as it takes to raise it from freezing to boiling $((1\,cal/g\,°C) \times 100\,°C)$.

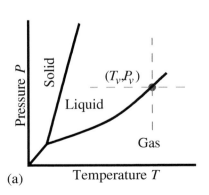

(a)

(b)

Fig. 11.1 $P–T$ **and** $T–V$ **phase diagrams** for a typical material. (a) The liquid–gas line gives the vapor pressure $P_v(T)$ or the boiling point $T_v(P)$. For H_2O the solid–liquid boundary would slope up and to the left, since ice is (atypically) less dense than water. (b) $T–V$ liquid–gas phase diagram at fixed N, showing the two-phase region. (See also Figs. 8.8 and 12.6(a)). The horizontal tie-lines represent phase separation; a system in this region will separate into domains of liquid and gas at the two endpoints of the corresponding tie-line. A $P–V$ phase diagram would look similar.

Statistical Mechanics: Entropy, Order Parameters, and Complexity. James P. Sethna, Oxford University Press (2021). © Oxford University Press. DOI:10.1093/oso/9780198865247.003.0011

Fig. 11.2 Stable and metastable states. (a) Helmholtz free energy as a function of volume. The dashed line represents metastable and unstable states (see Fig. 11.3(a)). In this range of volumes, the equilibrium state is nonuniform, a mixture of liquid and gas as shown by the dark straight line. (A mixture λN of stable gas and $(1-\lambda)N$ stable liquid will interpolate linearly on this plot: $V = \lambda V_{\text{gas}}^{\text{uniform}} + (1-\lambda)V_{\text{liq}}^{\text{uniform}}$, and $A = \lambda A_{\text{gas}} + (1-\lambda)A_{\text{liq}}$ plus an interfacial free energy which is negligible for large systems.) (b) Gibbs free energy for the liquid and gas phases along an isobar (constant pressure, horizontal dashed line in Fig. 11.1(a)). The phase transition occurs when the two curves cross.

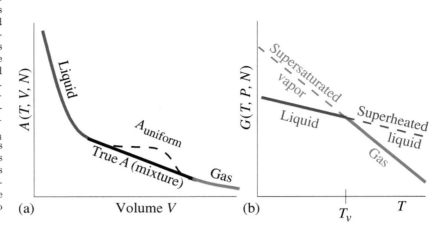

(a) Volume V (b) T_v T

[4]The straightness of the free energy in the two phase region says that the pressure $P = -(\partial A/\partial V)|_{T,N}$ is constant as the volume increases and the liquid turns to gas, just as the temperature stays constant as the liquid boils at fixed pressure. The free energy of a stable uniform state must always lie below any mixed state with the same total volume, number, and energy: equilibrium free energies are *convex functions*.

[5]We could also avoid the two-phase mixtures by using the grand free energy, $\Phi(T, V, \mu) = E - TS - \mu N$. The grand partition function allows the total number of particles to vary, so when the liquid turns to gas the molecules in the extra volume are simply removed.

[6]Minimizing G for the system maximizes the entropy for the Universe as a whole (the system plus the bath with which it exchanges energy and volume).

the system first reaches the liquid–gas transition, a small bubble of gas will form at the top; this bubble will gradually grow, inflating and filling the container over a range of energies. The transition from liquid to gas at fixed pressure passes through an intermediate *two-phase region*; the temperature of the system stays constant until the last liquid is gone. Alternatively, one can do the experiment fixing the temperature and varying the volume V of the container. The resulting phase diagram is shown in Fig. 11.1(b); the two-phase region results here from minimizing the Helmholtz free energy $A(T, V, N)$ as discussed in Fig. 11.2(a).[4]

To avoid these two-phase mixtures, we choose to work in the variables P and T, so we use the Gibbs free energy[5]

$$G(T, P, N) = E - TS + PV. \tag{11.1}$$

As usual (Section 6.5), whichever state minimizes G wins.[6] The Euler relation $E = TS - PV + \mu N$ tells us that $G = \mu N$ (Exercise 6.9). So, the state with the lower chemical potential will be favored, and the phase transition will occur when $\mu_{\text{liq}} = \mu_{\text{gas}}$. That makes sense; at a liquid–vapor boundary, one can exchange energy, volume, or particles. Energy will exchange until the temperatures are equal, volume will exchange until the pressures are equal, and then particles will move from liquid to gas until the chemical potentials are equal.

Remembering the shorthand thermodynamic relation $dE = T\,dS - P\,dV + \mu\,dN$, and applying it to eqn 11.1, we find $dG = -S\,dT + V\,dP + \mu\,dN$. Varying the temperature at fixed pressure and number of particles, we thus learn that

$$\left.\frac{\partial G}{\partial T}\right|_{P,N} = -S. \tag{11.2}$$

Figure 11.2(b) shows the Gibbs free energy versus temperature for the liquid and gas. At the phase boundary, the two free energies agree.

The difference in slopes of the two lines is given by the difference in entropies between the liquid and the gas (eqn 11.2). The thermodynamic definition of the entropy $S = \mathrm{d}Q/T$ (Section 5.1) tells us that the entropy difference is given by the latent heat per particle L times the number of particles N over the transition temperature T_v,

$$\Delta S = LN/T_v. \tag{11.3}$$

The fact that the Gibbs free energy has a kink at the phase transition reflects the jump in the entropy between liquid and gas; abrupt phase transitions will have jumps in the first derivatives of their free energies. This led early workers in the field to term these transitions *first-order transitions*.[7]

Notice that we continue to draw the free energy curves for the liquid and vapor on the "wrong" sides of the phase boundary. It is a common experimental fact that one can supercool vapors significantly beyond their condensation point.[8] With careful experiments on clean systems, one can also significantly superheat the liquid phase. Theoretically the issue is subtle. Some theories of these transitions have well-defined metastable phases (dashed lines in Fig. 11.2(a)). However, there certainly is no *equilibrium* vapor state below T_v. More sophisticated approaches (not discussed here) give an imaginary part to the free energy density of the metastable phase.[9]

11.2 Maxwell construction

Figure 11.3(a) shows the pressure versus volume as we expand our material at constant temperature. The liquid turns metastable as the volume increases, when the pressure reaches the vapor pressure for that temperature. The gas becomes metastable at that same pressure when the volume decreases. The metastable states are well defined only near the vapor pressure, where nucleation is slow (Section 11.3) and lifetimes are reasonably large. The dashed line shows a region which is completely unstable; a mixture of molecules prepared uniformly in space in this region will spontaneously separate into finely intertangled networks of the two phases[10] (Section 11.4.1).

How did we know to draw the coexistence line at the pressure we chose in Fig. 11.3(a)? How can we find the vapor pressure at which the liquid and gas coexist at this temperature? We know from the last section that the coexistence line occurs when their Gibbs free energies agree $G_{\mathrm{liq}} = G_{\mathrm{gas}}$. Again, $\mathrm{d}G = -S\,\mathrm{d}T + V\,\mathrm{d}P + \mu\,\mathrm{d}N$, so at constant temperature and number $(\partial G/\partial P)|_{T,N} = V$. Hence, we know that

$$\Delta G = \int_{P_{\mathrm{liq}}}^{P_{\mathrm{gas}}} V(P)\,\mathrm{d}P = 0. \tag{11.4}$$

Now this integral may seem zero by definition, because the limits of integration are both equal to the vapor pressure, $P_{\mathrm{liq}} = P_{\mathrm{gas}} = P_v(T)$.

[7]We avoid using this term, and the analogous term *second order* for continuous phase transitions. This is not only because their origin is obscure, but also because in the latter case it is misleading: the thermodynamic functions at a continuous phase transition have power-law singularities or essential singularities, not plain discontinuities in the second derivative (Chapter 12).

[8]That is precisely what occurs when the relative humidity goes beyond 100%.

[9]Just as the lifetime of a resonance in quantum mechanics is related to the imaginary part of its energy $E + i\hbar\Gamma$, so similarly is the rate per unit volume of nucleation of the new phase (Section 11.3) related to the imaginary part of the free energy density. We will not explain this here, nor will we discuss the corresponding essential singularity in the free energy (but see [35, 106, 107] and Exercise 11.15).

[10]This spontaneous separation is termed *spinodal decomposition*. In the past, the endpoints of the dashed curve were called *spinodal points*, but there is reason to doubt that there is any clear transition between nucleation and spontaneous separation, except in mean-field theories.

Fig. 11.3 Maxwell equal-area construction. (a) Pressure versus volume curve along an isotherm (dashed vertical line at constant temperature in Fig. 11.1(a)). At low volumes the material is liquid; as the volume crosses into the two-phase region in Fig. 11.1(b) the liquid becomes metastable. At high volumes the gas phase is stable, and again the metastable gas phase extends into the two-phase region. The dots represent the coexistence point where the pressures and chemical potentials of the two phases are equal. (b) The Gibbs free energy difference between two points at equal pressures is given by the difference in areas scooped out in the *P*–*V* curves shown (upward diagonal area minus downward diagonal area).

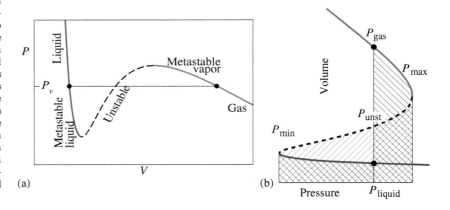

(a)

(b)

This formula instead represents the sum of four pieces (Fig. 11.3(b)):

$$\Delta G = \int_{P_{\mathrm{liq}}}^{P_{\mathrm{min}}} V(P)\,\mathrm{d}P + \int_{P_{\mathrm{min}}}^{P_{\mathrm{unst}}} V(P)\,\mathrm{d}P$$
$$+ \int_{P_{\mathrm{unst}}}^{P_{\mathrm{max}}} V(P)\,\mathrm{d}P + \int_{P_{\mathrm{max}}}^{P_{\mathrm{gas}}} V(P)\,\mathrm{d}P, \tag{11.5}$$

where the unstable point corresponds to the barrier top in Fig. 11.2(a) and $P_{\mathrm{liq}} = P_{\mathrm{unst}} = P_{\mathrm{gas}}$ at coexistence. Notice that the first and last terms are negative, since $P_{\mathrm{liq}} > P_{\mathrm{min}}$ and $P_{\mathrm{max}} > P_{\mathrm{gas}}$. These four integrals have a nice graphical interpretation, shown in Fig. 11.3(b): the first two subtract to give the area with stripes solely up and to the right and the last two subtract to give minus the area with stripes solely down and to the right. These two areas must be equal at the vapor pressure. This is the *Maxwell equal-area construction.*[11]

[11]The Maxwell construction only makes sense, one must remember, for theories like mean-field theories where one has an unstable branch for the $P(V)$ curve (Fig. 11.2(a)).

11.3 Nucleation: critical droplet theory

On a humid night, as the temperature drops, the air may become supersaturated with water vapor. How does this metastable vapor turn into drops of dew, or into the tiny water droplets that make up clouds or fog?

We have seen (Fig. 11.2(b)) that the Gibbs free energy difference between the gas and the liquid grows as the temperature decreases below T_v. We can estimate the chemical potential difference driving the formation of drops; we know $\partial G/\partial T = -S$ (eqn 11.2), and $\Delta S = LN/T_v$

Fig. 11.4 Vapor bubble. The nucleation of a new phase happens through a rare thermal fluctuation, where a droplet of the new phase forms of sufficient size to get it over the free energy barrier of Fig. 11.5.

(eqn 11.3), so if we supercool by $\Delta T = T_v - T$,

$$\Delta\mu = (G_{\text{gas}} - G_{\text{liq}})/N = \left(\left.\frac{\partial(G_{\text{gas}} - G_{\text{liq}})}{\partial T}\right|_{P,N}(T - T_v)\right)\Big/N$$

$$= \Delta S \Delta T/N = (LN/T_v)(\Delta T/N) = L\Delta T/T_v. \quad (11.6)$$

What is the obstacle impeding the formation of the droplets? It is the *surface tension σ* between the liquid and the gas phase. The surface tension is the Gibbs free energy per unit area of the interface between liquid and gas. Like tension in a rope, the interface between two phases can exert a force, pulling inward to minimize its area.[12]

To make a large droplet you must grow it from zero radius (Fig. 11.4). Since the surface tension cost grows as the area A and the bulk free energy gain grows as the volume V, tiny droplets will cost the system more than they gain. Consider the energy of a spherical droplet of radius R. The surface Gibbs free energy is σA. If the liquid has ρ_{liq} particles per unit volume, and each particle provides a Gibbs free energy gain of $\Delta\mu = L\Delta T/T_v$, the bulk free energy gain is $V\rho_{\text{liq}}\Delta\mu$. Hence

$$G_{\text{droplet}}(R) = \sigma A - V\rho_{\text{liq}}\Delta\mu = 4\pi R^2\sigma - (\tfrac{4}{3}\pi R^3)\rho_{\text{liq}}(L\Delta T/T_v). \quad (11.7)$$

This free energy is plotted in Fig. 11.5. Notice that at small R where surface tension dominates it rises quadratically, and at large R where the bulk chemical potential difference dominates it drops as the cube of R. The gas will stay a gas until a rare thermal fluctuation pays the free energy cost to reach the top of the barrier, making a *critical droplet*. The critical droplet radius R_c and the free energy barrier B are found by finding the maximum of $G(R)$:

$$\left.\frac{\partial G_{\text{droplet}}}{\partial R}\right|_{R_c} = 8\pi\sigma R_c - 4\pi\rho_{\text{liq}}(L\Delta T/T_v)R_c^2 = 0, \quad (11.8)$$

$$R_c = \frac{2\sigma T_v}{\rho_{\text{liq}}L\Delta T}, \quad (11.9)$$

$$B = \frac{16\pi\sigma^3 T_v^2}{3\rho_{\text{liq}}^2 L^2}\frac{1}{(\Delta T)^2}. \quad (11.10)$$

The probability of finding a critical droplet per unit volume is given by $\exp(-B/k_B T)$ times a prefactor. The nucleation rate per unit volume is the net flux of droplets passing by R_c, which is the velocity over the barrier times a correction for droplets re-crossing, times this probability of being on top of the barrier.[13] The prefactors are important for detailed theories, but the main experimental dependence is the exponentially small probability for having a critical droplet. Our net droplet nucleation rate per unit volume Γ thus has the form

$$\Gamma = (\text{prefactors})e^{-B/k_B T}. \quad (11.11)$$

[12]How do we define the surface tension? Theoretically, we measure liquid and gas Gibbs free energies $G_{\ell\ell}$ and G_{gg} at the coexistence point by placing them into cylindrical constant-pressure containers with walls that prefer one phase over the other (e.g. in the real world wax prefers gas). We then cut each container in half and glue them together at their midpoint; this forces an interface to form of area A equal to the cross-section of the containers. We then measure the Gibbs free energy with the interface $G_{\ell g}$, and define the surface tension $\sigma = [G_{\ell g} - \tfrac{1}{2}(G_{\ell\ell} + G_{gg})]/A$. We need to do something this complicated in order to be sure to cancel out the interaction free energies with the side walls.

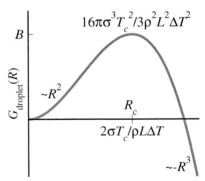

Fig. 11.5 Free energy barrier for droplet formation.

[13]See Exercise 6.11 for the similar problem of calculating chemical reaction rates.

[14]In previous chapters, we used statistical mechanics to compute average properties of a phase, and typical (Gaussian) fluctuations near the average. Critical droplet theory (and instantons, the quantum version) allows us to calculate rare fluctuations—far in the tail of the distribution—by asking for the typical fluctuations near the transition between two phases.

[15]The droplet nucleation rate has an essential singularity at T_v; it is zero to all orders in perturbation theory in ΔT. In some ways, this is why one can study the metastable states— perturbation theory naturally ignores the fact that they are unstable.

[16]Actually, in many clouds the temperature is low enough that ice crystals nucleate, rather than water droplets. Certain plant pathogens (*Pseudomonas syringae*) make proteins that are designed to efficiently nucleate ice crystals; the bacteria use the frost damage on the plants to invade. Humans use these proteins in snow-making machines at ski resorts.

Notice the following.[14]

- The critical droplet radius $R_c \propto 1/\Delta T$. If you undercool the gas only a tiny amount, you need a big droplet to overcome the surface tension.
- The barrier height $B \propto 1/(\Delta T)^2$. The energy barrier for nucleation diverges at T_v.
- The droplet nucleation rate $\Gamma \propto \exp(-C/(\Delta T)^2)$. It can be tiny for small undercoolings.[15]

The rates we have calculated are for *homogeneous nucleation*: the rate of forming the new phase in a perfectly clean system without boundaries. In practice, homogeneous nucleation rarely dominates. Because nucleation is so strongly suppressed by the surface tension, the system goes to great lengths to bypass at least part of the energy barrier. That is why dewdrops form on grass (or your windshield), rather than always forming in the air and dropping to the ground; the surface tension between water and grass is much lower than that between water and air, so a roughly hemispherical droplet can form—dividing the free energy barrier B in two. In cloud formation, nucleation occurs on small dust particles—again, lowering the interfacial area needed to get a droplet of a given curvature.[16]

Finally, we should mention that the nucleation of crystalline phases will not proceed with precisely spherical droplets. Because crystals have anisotropic surface tension, the maximum number of particles for a given surface energy is given not by a sphere, but by the *equilibrium crystal shape* (the same shape that a crystal will form in equilibrium at a constant number of particles; see Fig. 11.6).

11.4 Morphology of abrupt transitions

What happens after the phase transition is nucleated (or when the undercooling is so large that the transition occurs immediately)? This question leads us into a gigantic, rich subject that mostly belongs to geology, engineering, and materials science, rather than to statistical mechanics. We will give a brief introduction, with emphasis on topics where statistical mechanics is useful.

11.4.1 Coarsening

What do salad dressing, cast iron, and rocks have in common? *Coarsening* is crucial to all three. When you shake up oil and vinegar, they get jumbled together in small droplets. When you stop shaking, the tiny droplets merge into bigger ones, gradually making for a coarser and coarser mixture until all the oil is on the top.

Molten iron, before it is cast, has a fair percentage of carbon dissolved in it. As it cools, this dissolved carbon precipitates out (with many nuclei forming as in Section 11.3 and then growing until the carbon runs out), staying dispersed through the iron in particles whose size and number depend on the cooling schedule. The hardness and brittleness of cast iron depends on the properties of these carbon particles, and thus depends on how the iron is cooled.

Rocks often have lots of tiny grains of different materials: quartz, alkali feldspar, and plagioclase in granite; plagioclase feldspar and calcium-rich pyroxene in basalt, ... Different rocks have different sizes of these grains.

Rocks formed from the lava of erupting volcanoes have very fine grains; rocks deep underground cooled from magma over eons form large grains. For a particular grain to grow, the constituent atoms must diffuse through neighboring grains of other materials—a process that gets very slow as the grains get larger. Polycrystals also form from cooling single materials; different liquid regions will nucleate crystalline grains in different orientations, which then will grow and mush together. Here the grains can grow by stealing one another's molecules, rather than waiting for their brand of molecules to come from afar.

One can also see coarsening on the computer. Figure 11.7 shows two snapshots of the Ising model, quenched to zero temperature, at times differing by roughly a factor of ten. Notice that the patterns of up- and down-spins look statistically similar in the two figures, except that the overall length scale $L(t)$ has grown larger at later times. This is the characteristic feature that underlies all theories of coarsening: the system is statistically similar to itself at a later time, except for a time-dependent length scale $L(t)$.

The basic results about coarsening can be derived by arguments that are almost simplistic. Consider a snapshot (Fig. 11.8) of a coarsening system. In this snapshot, most features observed have a characteristic length scale $R \sim L(t)$. The coarsening process involves the smaller features shrinking to zero, so as to leave behind only features on larger scales. Thus we need to understand how long it takes for spheres and protrusions on a scale R to vanish, to derive how the size $L(t)$ of the remaining features grows with time. $L(t)$ is the size R_0 of the smallest original feature that has not yet shrunk to zero.

The driving force behind coarsening is surface tension: the system can lower its free energy by lowering the interfacial area between the different domains. We will focus on the evolution of a sphere as a solvable case.

Fig. 11.6 Equilibrium crystal shape of lead at about 300 °C [161, 169]. Notice the flat facets, which correspond to low-energy high-symmetry surfaces. There are three interesting statistical mechanics problems associated with these facets that we will not discuss in detail. (1) The crystalline orientations with flat facets are in a different phase than the rounded regions; they are below their *roughening transition*. (2) The equilibrium crystal shape, which minimizes the free energy, can be viewed as a Legendre transform of that free energy (the *Wulff construction*). (3) At lower temperatures, for some interactions, the entire equilibrium crystal shape can be faceted (below the *edge* and *corner rounding* transitions); we predict that the coarsening length will grow only logarithmically in time in this phase (Fig. 11.10).

(a)   (b)

Fig. 11.7 **Coarsening**. The spin configuration of the Ising model at $T = 0$ with nonconserved dynamics, after a time of (a) roughly twenty sweeps through the lattice, and (b) roughly 200 sweeps. Notice that the characteristic morphologies look similar, except that the later picture has a length scale roughly three times larger ($\sqrt{10} \approx 3$).

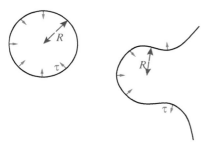

Fig. 11.8 Curvature-driven interface motion. The surface tension σ at the interface produces a traction (force per unit area) $\tau = 2\sigma\kappa$ that is proportional to the local mean curvature of the surface κ at that point. The coarsening morphology has a characteristic length R, so it has a characteristic mean curvature $\kappa \sim 1/R$. For nonconserved order parameters, these forces will lead to a length scale $L(t) \sim t^{1/2}$.

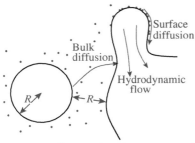

Fig. 11.9 Coarsening for conserved order parameter. Differences in local mean curvature drives the growth in the case of a conserved order parameter. Atoms will diffuse from regions of high positive curvature to regions of low or negative curvature. Bulk diffusion dominates on long length scales ($L(t) \sim t^{1/3}$); surface diffusion can be important when the scales are small ($L(t) \sim t^{1/4}$). For liquids, hydrodynamic flow makes things more complicated [187].

[17]Remember Einstein's relation, $D = \gamma k_B T$ (eqns 2.22 and 6.67) relating the mobility to the diffusion constant D.

The surface tension energy for a sphere of radius R is $F_{\text{surface}} = 4\pi R^2 \sigma$, so there is an inward force per unit area, (or *traction*) τ:

$$\tau = \frac{\partial F_{\text{surface}}}{\partial R}\bigg/(4\pi R^2) = 2\sigma/R. \tag{11.12}$$

A general surface has two radii of curvature R_1 and R_2, which can be positive or negative; the traction τ is perpendicular to the surface and given by the same formula 11.12 with $1/R$ replaced with the mean curvature $\frac{1}{2}\left[1/R_1 + 1/R_2\right]$.

There are two broad classes of coarsening problems: ones with *conserved* and *nonconserved* order parameters. Oil and vinegar, cast iron, and granite have conserved order parameters; to grow a domain one must pull molecules through the other materials. The single-component polycrystals and the Ising model shown in Fig. 11.7 are nonconserved; spins are free to flip from one orientation to the other, and molecules are free to shift from one grain orientation to another.

For a nonconserved order parameter, the interface will generally move with a velocity proportional to the traction and an interface mobility η:

$$\frac{dR}{dt} = -\eta\tau = -\eta\frac{2\sigma}{R}. \tag{11.13}$$

We can solve for the time t_f it takes for the sphere to disappear, and hence find out how $L(t)$ grows for the nonconserved case:

$$\int_{R_0}^0 R\,dR = \int_0^{t_f} -2\sigma\eta\,dt,$$
$$R_0^2/2 = 2\sigma\eta t_f, \tag{11.14}$$
$$L(t) \sim R_0 = \sqrt{4\sigma\eta t} \propto \sqrt{t}.$$

More complex geometries with protrusions and necks and such are not possible to solve explicitly, but in general features with a length scale R evolve on a time scale $t \propto R^2$, so the typical length scales grow as $L(t) \sim t^\beta$ with $\beta = 1/2$.

The argument for the case of a conserved order parameter is quite similar in spirit (Fig. 11.9). Here the curvature sets up a gradient in the chemical potential $\partial\mu/\partial x$, which causes molecules to diffuse from regions of high positive curvature to regions of low or negative curvature. The velocity of a particle will be given by the particle mobility[17] $\gamma = D/k_B T$ times the "particle force" gradient of the chemical potential,

$$v = \gamma\nabla\mu \Rightarrow J = \rho v = \rho\gamma\nabla\mu \tag{11.15}$$

where J is the current per unit area and ρ is the particle density. The chemical potential change for moving a molecule from our sphere of radius R to some flat interface is just the free energy change for removing one particle; since the number of particles in our sphere is $N = \frac{4}{3}\pi R^3\rho$,

$$\Delta\mu = \frac{dF_{\text{surface}}}{dR}\bigg/\frac{dN}{dR} = (8\pi\sigma R)/(4\pi R^2\rho) = \frac{2\sigma}{R\rho}. \tag{11.16}$$

The distance ΔR from the surface of our sphere to another flatter surface of the same phase is (by our assumption of only one characteristic length scale) also of order R, so

$$J \sim \rho\gamma\frac{\Delta\mu}{\Delta R} \sim \frac{2\gamma\sigma}{R^2}. \tag{11.17}$$

The rate of change of volume of the droplet is number flux per unit area J times the surface area, divided by the number per unit volume ρ:

$$\frac{\mathrm{d}V_{\mathrm{droplet}}}{\mathrm{d}t} = \frac{4}{3}\pi\left(3R^2\frac{\mathrm{d}R}{\mathrm{d}t}\right)$$
$$= -\frac{A_{\mathrm{droplet}}J}{\rho} = -(4\pi R^2)\frac{2\gamma\sigma}{\rho R^2} = -\frac{8\pi\gamma\sigma}{\rho},$$
$$\frac{\mathrm{d}R}{\mathrm{d}t} = -\frac{2\gamma\sigma}{\rho}\frac{1}{R^2}, \tag{11.18}$$
$$\int_{R_0}^{0} R^2\,\mathrm{d}R = \int_{0}^{t_f} -\frac{2\gamma\sigma}{\rho}\,\mathrm{d}t,$$
$$\frac{R_0^3}{3} = \frac{2\gamma\sigma}{\rho}t_f,$$

and so

$$L(t) \sim R_0 = \left(\frac{6\gamma\sigma}{\rho}t\right)^{1/3} \propto t^{1/3}. \tag{11.19}$$

This crude calculation—almost dimensional analysis—leads us to the correct conclusion that conserved order parameters should coarsen with $L(t) \sim t^\beta$ with $\beta = \frac{1}{3}$, if bulk diffusion dominates the transport.

The subject of coarsening has many further wrinkles.

- **Surface diffusion.** Often the surface diffusion rate is much higher than the bulk diffusion rate: the activation energy to hop across a surface is much lower than to remove a molecule completely. The current in surface diffusion goes as J times a perimeter (single power of R) instead of JA; repeating the analysis above gives $L(t) \sim t^{1/4}$. In principle, surface diffusion will always be less important than bulk diffusion as $t \to \infty$, but often it will dominate in the experimental range of interest.

- **Hydrodynamics.** In fluids, there are other important mechanisms for coarsening. For example, in binary liquid mixtures (oil and water) near 50/50, the two phases form continuous interpenetrating networks. Different regions of the network can have different curvatures and pressures, leading to coarsening via hydrodynamic flow [187].

- **Glassy logarithmic coarsening.** A hidden assumption in much of the coarsening theory is that the barriers to rearrangements involve only a few degrees of freedom. If instead you need to remove a whole layer at a time to reach a lower free energy, the dynamics may slow down dramatically as the sizes of the layers grow. This is precisely what happens in the three-dimensional Ising model with antiferromagnetic next-neighbor bonds mentioned earlier (Fig. 11.10, [185]). The

Fig. 11.10 Logarithmic growth of an interface. A simplified model of an interface perpendicular to a body diagonal in a frustrated next-neighbor Ising model with barriers to coarsening that diverge with length [185]. Notice that the interface is lowering its energy by poking out into facets along the cubic directions (a kind of facet coarsening). This process gets much slower as the faces get longer, because the energy barrier needed to flip a face grows linearly with its length. This slowdown happens below the *corner rounding transition* described in the caption to Fig. 11.6.

energy barrier needed to flip a layer of spins grows proportionally to L, leading to a logarithmic growth law $L(t) \sim \log(t)$. Some speculate that similar growing barriers are responsible for the slow relaxation in glass-forming liquids (Section 12.3.4).

- **Nonuniversality.** Much of the motivation for studying coarsening by physicists has been the close analogies with the scaling and power laws seen in continuous phase transitions (Chapter 12 and Exercise 12.3). However, there are important differences. The power laws in coarsening are simpler and in a way more universal—the ½ and ⅓ power laws we derived above are independent of the dimension of space, for example. On the other hand, for coarsening in crystals the scaling behavior and morphology is not universal; it will depend upon the anisotropic free energies and mobilities of the interfaces [162]. Basically each combination of materials and temperatures will have different scaling functions at late times. In retrospect this is reassuring; there is such a bewildering variety of microstructures in materials science and mineralogy that it made no sense for one scaling function to rule them all.

Fig. 11.11 Martensite. A particularly nice example of a martensitic structure, from Chu [44]. The light and dark stripes represent two different martensitic variants—that is, the crystal going through the phase transition can stretch in different directions, and the two shades indicate that the local lattice is stretching along two different axes. The tilted square region occupying most of the photograph could not change its overall shape without putting incompatible strains on the neighboring domains. By making this striped pattern, or *laminate*, the martensite can form an average of the different stretching directions that gives zero net strain.

[18]Iron and steel thus have both the complications of carbon particle coarsening and martensitic domain structure, both of which are important for structural properties and both of which depend in detail on the heating and beating they undergo during its manufacture.

[19]The pathological functions you find in real analysis—continuous but nowhere differentiable functions—are practical tools for studying martensites.

11.4.2 Martensites

Many crystals will undergo abrupt structural rearrangements as they are cooled—phase transitions between different crystal structures. A good example might be a cubic crystal stretching along one axis and shrinking along the other two. These transitions are often problematic; when part of the sample has transformed and the rest has not, the tearing stress at the interface often shatters the crystal.

In many materials (such as iron[18]) the crystalline shape transition bypasses large-scale stress build-up in the crystal by developing intricate layered structures forming a *martensite*. Figure 11.11 shows how martensites form a patterned microstructure in order to stretch locally without an overall net strain.

The tool used to study martensites is not statistical mechanics, but mathematics.[19] The basic goal, however, is the same: to minimize the (nonconvex) free energy for fixed boundary conditions (see Exercises 11.7 and 11.8).

11.4.3 Dendritic growth

Why do snowflakes form [115]? New ice crystals up in the atmosphere initially nucleate as roughly spherical crystals of ice. As the ice crystals continue to grow, however, an instability develops. The tips of the ice crystals that extend furthest into the surrounding supersaturated vapor will grow fastest, both because they see the highest concentration of water vapor and because the heat released by freezing diffuses away fastest at the tips (Exercise 11.9). The characteristic six-fold patterns arise because each snowflake is a single crystal with six-fold symmetry, and different crystal surface orientations grow at different rates. The

Fig. 11.12 Dendrites. Growing crystals will typically form branched structures (dendrites), because the tips grow faster than the grooves. Here are shown some dendrites growing into a melt, in a thin film being pulled through a temperature gradient (Bodenschatz, Utter, Ragnarson [201]).

immense variety of snowflake shapes reflects the different thermal and humidity variations experienced by each snowflake as it grows.

The same kind of branched patterns, called *dendrites*,[20] also form in other growing crystals for precisely the same reasons. Frost on your window is one obvious example; Fig. 11.12 shows another example: a solvent crystal growing into a mixture of solvent and polymer. Here, instead of heat being trapped in the grooves, the slowly diffusing polymer is being trapped and slowing down the solidification process. Many practical metals and alloys used in manufacturing are composed microscopically of tiny dendritic structures packed together.

[20]Dendron is Greek for tree.

Exercises

Unstable to what?, *Maxwell and van der Waals*, *The van der Waals critical point*, *Interfaces and van der Waals*, and *Gibbs free energy barrier* use an early, classic model for the liquid–gas phase transition to illustrate general features of abrupt phase transitions. Four exercises discuss nucleation. *Nucleation in 2D* and *Nucleation in the Ising model* discuss critical droplets in two-dimensional systems from analytical and numerical perspectives. *Nucleation of dislocation pairs* uses nucleation to explore how crystals are fundamentally more rigid than liquids. *Nucleation of cracks* explores the metastability of bridges, and *Elastic theory does not converge* explores the radius of convergence of elastic theory and the imaginary part of the free energy of metastable states.

We then illustrate the morphological structure and evolution developed during first-order phase transitions in six exercises. *Coarsening in the Ising model* numerically explores the standard model used to study the growing correlations after rapid cooling, and *Mosh pits* explores coarsening in an active matter system found at heavy metal concerts. *Origami microstructure* and *Minimizing sequences and microstructure* introduce us to the remarkable methods used by mathematicians and engineers to study martensites and other boundary-condition-induced microstructure. Finally, *Linear stability of a growing interface* and *Snowflakes and linear stability* introduce linear stability analysis in the context of dendrite formation.

(11.1) **Maxwell and van der Waals.** (Chemistry) @

The van der Waals (vdW) equation

$$(P + N^2 a/V^2)(V - Nb) = Nk_B T \quad (11.20)$$

is often applied as an approximate equation of state for real liquids and gases. The term $V - Nb$ arises from short-range repulsion between molecules (Exercise 3.5); the term $N^2 a/V^2$ incorporates the leading effects[21] of long-range attraction between molecules.

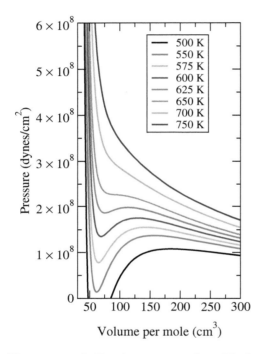

Fig. 11.13 *P–V* **plot: van der Waals**. Van der Waals (vdW) approximation (eqn 11.20) to H$_2$O, with $a = 0.55 \, \mathrm{J \, m^3/mol^2}$ ($a = 1.52 \times 10^{-35} \, \mathrm{erg \, cm^3/molecule}$), and $b = 3.04 \times 10^{-5} \, \mathrm{m^3/mol}$ ($b = 5.05 \times 10^{-23} \, \mathrm{cm^3/molecule}$), fit to the critical temperature and pressure for water.

Figure 11.13 shows the pressure versus volume curves for one mole of H$_2$O, within the vdW model. A real piston of water held at constant temperature would in equilibrium pass through three regimes as it expanded—first a decompressing liquid, then coexisting liquid and vapor at constant pressure (as the liquid evaporates or boils to fill the piston), and then a

decompressing gas. The Maxwell construction tells us what the vapor pressure and the two densities for the coexisting liquid and gas is at each temperature.

Get a copy of Fig. 11.13. By hand, roughly implement the Maxwell construction for each curve, and sketch the region in the P–V plane where liquid and gas can coexist.

You have not only found the coexistence region in the $P - -V$ diagram as temperature is varied, but also sketched the *tie-lines* connecting the two coexisting liquid and gas states at each temperature: a system at V, P in the coexistence region will decompose into the liquid and gas at the endpoints of its tie-line, with fractions determined by the position along the tie-line. In other phase diagrams with coexistence regions, these tie-lines may not be parallel (Fig. 8.8).

(11.2) **The van der Waals critical point.** (Chemistry) @

The top of the coexistence curve in Fig. 11.13 is the pressure, density, and temperature at which the distinction between liquid and gas disappears. It is the focus of much study, as the prime example of a critical point, with self-similar fluctuations and scaling behavior.

(a) *Identify this point on a sketch of Fig. 11.13. The vdW constants are fit to the critical temperature $T_c = 647.3 \, \mathrm{K}$ and pressure $P_c = 22.09 \, \mathrm{MPa} = 220.9 \times 10^6 \, \mathrm{dyne/cm^2}$; check that your estimate for the critical point roughly agrees with the values quoted. I have found few references that quote the critical volume per mole, and the two I have found disagree; one says around $50 \, \mathrm{cm^3/mol}$ and one says around 55. Plot the true critical point on your sketch. Is the location of the critical density of water predicted well by the vdW equation of state?*

Your sketch from Exercise 11.1 may not be precise enough to tell this, but the vdW phase boundaries meet at the critical point with a quadratic maximum: $1/\rho_\ell - 1/\rho_g \sim (P - P_c)^{1/2}$, where ρ_ℓ and ρ_g are the densities on the coexistence boundary (moles per volume) at the pressure P. Indeed, the peak of any mean-field prediction like van der Waals, gotten from analytic formulas, will have a Taylor series whose quadratic term gives an exponent $1/2$. Similarly, one can show that the vdW equation of

[21]These corrections are to leading orders in the density; they are small for dilute gases.

state implies that

$$\rho_\ell - \rho_g \sim (T_c - T)^{1/2} \sim (-t)^{1/2}. \quad (11.21)$$

(b) *Compare this latter prediction with Fig. 12.6(a). What critical exponent β does the van der Waals equation predict, assuming eqn 11.21? How does it compare to the actual value, $\beta = 0.326419\ldots$?*

This technical-seeming error in the shape of the maximum of the liquid–gas line was an early puzzle that led eventually to the discovery of self-similarity and universality, and the invention of the renormalization group (Chapter 12).

(11.3) **Interfaces and van der Waals.** (Chemistry) @

The chemical potential per particle for the vdW equation of state is

$$\mu[\rho] = -k_B T + P/\rho - a\rho + k_B T \log(\lambda^3 \rho) - k_B T \log(1 - b\rho), \quad (11.22)$$

where $\rho = N/V$ is the density.

Fig. 11.14 Chemical potential: van der Waals. Chemical potential $\mu[\rho]$ of water fit with the van der Waals equation, at the boiling temperature of water $T = 373\,\mathrm{K}$ and the corresponding van der Waals coexistence pressure $P = 1.5 \times 10^7$ dynes/cm^2.

(a) *Show that μ is minimized when ρ satisfies the vdW equation of state, eqn 11.20.*

(b) *According to the caption to Fig. 11.14, what is the vdW approximation to the vapor pressure at $373\,\mathrm{K} = 100\,°C$? How close is the vdW approximation to the true vapor pressure of water?* (Hint: Atmospheric pressure is around

one bar $= 0.1\,\mathrm{MPa} = 10^6$ dynes/cm^2. What happens when the vapor pressure hits atmospheric pressure?)

We can view Fig. 11.14 as a kind of free energy barrier for the formation of a liquid–gas interface. If μ_0 is the common chemical potential shared by the water and the vapor at this temperature, the extra Gibbs free energy for a density fluctuation $\rho(x)$ is

$$\Delta G = \int \rho(x) \left(\mu[\rho(x)] - \mu_0\right) \mathrm{d}^3 x \quad (11.23)$$

since $\rho(x)\,\mathrm{d}^3 x$ is the number of particles that suffer the chemical potential rise $\mu[\rho(x)]$ in the volume $\mathrm{d}^3 x$.

(c) *At room temperature, the interface between water and water vapor is very sharp: perhaps a molecule thick. This makes the whole idea of using a coarse-grained free energy problematical. Nonetheless, assuming an interfacial width of two or three Ångstroms, use the vdW model for the chemical potential (Fig. 11.14) and eqn 11.23 to roughly estimate the surface tension of water (the extra Gibbs free energy per unit area, roughly the barrier height times thickness). How does your answer compare with the measured value at the boiling point, 59 dynes/cm?* (One mole $= 6.023 \times 10^{23}$ molecules.)

(11.4) **Nucleation in the Ising model.**[22] (Computation) ③

The Ising model (Section 8.1) is not only our archetype for a continuous phase transition; it is also an excellent model for abrupt transitions. In particular, it is an excellent model for nucleation (this exercise) and for the dynamics of phase separation (Exercise 11.6). Supercooling water and waiting for an ice crystal nucleus, or waiting for a raindrop when the humidity is over 100%, can be shown to be quite analogous to changing a magnet from external field $H_{ext} > 0$ to $H_{ext} < 0$ at a temperature $T < T_c$. Start up the Ising model simulation. Run at $T = 1.5$ (below T_c) at size 40×40, initialized with all spins up. Set $H_{ext} = -0.3$ and watch the spins. They should eventually flop over to point down, with the new phase starting in a small, roughly circular cluster of spins, which then grows to fill the system.[23]

---

[22]A link to the software can be found at the book website [181].

[23]The system has periodic boundary conditions, so a cluster which starts near a boundary or corner may falsely look like more than one simultaneous nucleation event.

(a) *Using the graph of magnetization versus time, measure the average time it takes to cross zero (which we will call the time to nucleate the down phase), averaging over ten measurements. (You may want to reduce the graphics refresh rate to speed up the simulation.) Similarly measure the average time to nucleate the down phase for $H_{ext} = -0.2$.* Since the nucleation center can be located at any site on the lattice, the nucleation rate scales with the number of spins in the system. *Calculate, for both fields, the nucleation rate per spin $\Gamma_{exp}(H)$.*

We can use critical droplet theory (Section 11.3) to estimate the nucleation rate. Small droplets of the stable phase will shrink due to surface tension σ; large ones grow due to the free energy difference per unit area $H_{ext}\Delta M(T)$, where ΔM is the magnetization difference between the two states. Presuming that the temperature is high and the droplet large and the times long (so that continuum theories are applicable), one can estimate the critical radius R_c for nucleation.

(b) *Give the formula for the free energy of a flipped 2D Ising cluster of radius R as a function of σ, H, and ΔM. Give formulæ for R_c (the critical droplet size where the free energy is a local maximum), the resulting barrier B to nucleation, and the predicted rate $\Gamma_{theory} = \exp(-B/T)$ (assuming a prefactor of roughly one attempt per sweep per spin). At low temperatures, $\sigma \sim 2J \equiv 2$ and $\Delta M \approx 2$, since the system is almost fully magnetized and σ is the number of broken bonds (2J each) per unit length of interface. Make a table with rows for the two fields you simulated and with columns for H, R_c, B, Γ_{theory}, and Γ_{exp} from (a).*

This should work pretty badly. Is the predicted droplet size large enough (several lattice constants) so that the continuum theory should be valid?

If your software allows, test these ideas by starting with pre-flipped droplets of down-spins (white) in an up background. Use a small system (down to 40×40).

(c) *Start with $H = -0.2$, $T = 1.5$ and a down-spin droplet of radius five (diameter of ten), and run ten times. Does it grow more often than it shrinks, or vice versa? (Testing this should be fast.) On the magnetization curve, count the shrinking fraction f. Make a table of the values of H and f you measure. Vary the field H until the probabilities roughly match; find the field for $R_c = 5$ to within 0.1. For what field is the theoretical critical droplet radius $R_c = 5$ at $T = 1.5$?*

In part (b) we found that critical droplet theory worked badly for predicting the nucleation rate. In part (c) we found that it worked rather well (within a factor of two) at predicting the relationship between the critical droplet size and the external field. This is mostly because the nucleation rate depends exponentially on the barrier, so a small error in the barrier (or critical droplet radius) makes a big error in the nucleation rate. You will notice that theory papers rarely try to predict rates of reactions. They will almost always instead compare theoretical and experimental barrier heights (or here, critical droplet radii). This avoids embarrassment.

This free energy barrier to nucleation is what allows supercooled liquids and supersaturated vapor to be stable for long periods.

(11.5) **Nucleation of dislocation pairs.** (Engineering) @

Consider a two-dimensional crystal under shear shown in Fig. 11.15.[24] The external force is being relieved by the motion of the upper half of the crystal to the left with respect to the bottom half of the crystal by one atomic spacing a. If the crystal is of length L, the energy released by this shuffle when it is complete will be $|F|a = \sigma_{xy}La$. This shuffling has only partially been completed; only the span R between the two edge dislocations has been shifted (the dislocations are denoted by the conventional "tee" representing the end of the extra column of atoms). Thus the strain energy released by the dislocations so far is

$$|F|aR/L = \sigma_{xy}Ra. \qquad (11.24)$$

This energy is analogous to the bulk free energy gained for a liquid droplet in a supercooled gas. The dislocations, however, cost energy (analogous to the surface tension of the vapor droplet). They have a fixed core energy C that

---

[24]Exercise 9.12 discusses the general question of why crystals resist shear; Exercise 11.15 considers the alternative failure mechanism of fracture. An analysis similar to that of this exercise, but applied to superfluids, has been analyzed in detail [5, 6, 195]. The complete solution is made more complex by the effects of other dislocation pairs renormalizing the elastic constants at high temperatures (the Kosterlitz–Thouless–Halperin–Young transition).

depends on the details of the interatomic interaction, and a long-range interaction energy which, for the geometry shown in Fig. 11.15, is

$$\frac{\mu}{2\pi(1-\nu)} a^2 \log(R/a). \qquad (11.25)$$

Here μ is the 2D shear elastic constant[25] and ν is Poisson's ratio. Assume the temperature is low (so that the energies given by eqns 11.24 and 11.25 are good approximations for the appropriate free energies). By subtracting the energy gained from the dislocation from the energy cost, one finds in analogy to other critical droplet problems a critical distance R_c and a barrier height for thermally nucleated dislocation formation B.

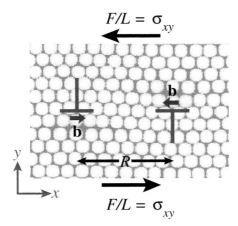

Fig. 11.15 Dislocation pair in a 2D hexagonal crystal. A loop around the defect on the right shows an extra row of atoms coming in from the bottom. By the conventions used in materials physics (assuming here that the dislocation points up out of the paper, see [83, Fig. 1.20, p. 23]) this *edge dislocation* has Burgers vector $\mathbf{b} = a\hat{x}$, where a is the distance between neighboring atoms. Similarly, the defect on the left has an extra row of atoms coming in from the bottom, and has $\mathbf{b} = -a\hat{x}$. The defects are centered at the same height, separated by a distance R. The crystal is under a shear stress $\sigma_{xy} = F/L$, where the force $F = \pm\sigma_{xy}L\hat{y}$ is applied to the top and bottom as shown (and the crystal is kept from rotating). (Figure by Nicholas Bailey.)

Of the following statements, which are true?
(T) (F) The critical distance R_c is proportional to $1/\sigma_{xy}$.

(T) (F) The energy barrier to thermal nucleation is proportional to $1/\sigma_{xy}^2$.
(T) (F) The rate Γ of thermal nucleation of dislocations predicted by our critical droplet calculation is of the form $\Gamma = \Gamma_0(T) (\sigma_{xy}/\mu)^{D/k_BT}$, for a suitable material-dependent function $\Gamma_0(T)$ and constant D.

Dislocations mediate plastic shear. For a small sample, each pair of dislocations nucleated will travel to opposite boundaries of the system and lead to a net shear of one lattice constant. Thus, at any nonzero temperature and external stress, a (two-dimensional) crystal will shear at a nonzero rate. How is the crystal, then, different in its response from a liquid?

(T) (F) According to our calculation, the response of a two-dimensional crystal under stress is indistinguishable from that of a liquid; even at low temperatures, the strain rate due to an external shear force is proportional to the stress.

(11.6) **Coarsening in the Ising model.**[26] (Computation) @

Coarsening is the process by which phases separate from one another; the surface tension drives tiny fingers and droplets to shrink, leading to a characteristic length scale that grows with time.

Start up the Ising model. Run with a fairly large system, demagnetize the system to a random initial state ($T = \infty$), set $T = 1.5$ (below T_c), and observe the behavior. Notice the coarsening into up- and down-spin regions, each of which has a "gas" of single spins of the opposite sign sprinkled throughout.

(a) *If we think of the up-spins as vinegar and the down-spins as oil, does it look like the separation of salad dressing after a vigorous shake? Would one expect a few molecules of oil to dissolve in the vinegar, and vice versa? Consider the overall domain shapes as they coarsen. Do they look statistically the same at different times, except for the growing overall coarsening length?*
(b) *By eye, measure the typical length scale of the domains after 2^n sweeps, for $n = 0, 1, 2, \ldots$ until the system begins to have only a few domains. (For early times, take single steps on a small system to measure the tiny growing do-*

---

[25]The 2D elastic constants μ and ν can be related to their 3D values; in our notation μ has units of energy per unit area.
[26]Ising simulation software can be found at [28].

mains. *For late times, set the graphics refresh rate to a large power of two, and pause when you hit* 2^n*.) Plot your estimate for* $L(t)$ *versus* t *on a log–log plot. With what power does it grow? What power did we expect?*

For a more quantitative analysis, we can measure the typical length scale $L(t)$ of this pattern using the total energy, if we can ignore the "vapor".

(c) *Argue that at zero temperature the total energy above the ground-state energy is proportional to the perimeter separating up-spin and down-spin regions.*[27] *(At higher temperatures, the single flipped spins corresponding to oil molecules in the vinegar add to the energy, but should not count as part of the perimeter.) Argue that the inverse of the perimeter per unit area is a reasonable definition for the length scale of the pattern.*

(d) *With a random initial state, set temperature and external field to zero. Measure the mean energy* $E(t)$ *per spin as a function of time. Plot your estimate for* $L(t)$ *in part (c) versus* t *on a log–log plot. (You might try plotting a few such curves, to estimate errors.) With what power law does it grow?*

(11.7) **Origami microstructure.**[28] (Mathematics, Engineering) @

Figure 11.11 shows the domain structure in a thin sheet of material that has undergone a *martensitic* phase transition. These phase transitions change the shape of the crystalline unit cell; for example, the high-temperature phase might be cubic, and the low-temperature phase might be stretched along one of the three axes and contracted along the other two. These three possibilities are called *variants*. A large single crystal at high temperatures thus can transform locally into any one of the three variants at low temperatures.

The order parameter for the martensitic transition is a *deformation field* $\mathbf{y}(\mathbf{x})$, representing the final position $\mathbf{y}$ in the martensite of an original position $\mathbf{x}$ in the undeformed, unrotated austenite. The variants differ by their defor-

mation gradients $\nabla \mathbf{y}$ representing the stretch, shear, and rotation of the unit cells during the crystalline shape transition.

In this exercise, we develop an analogy between martensites and paper folding. Consider a piece of graph paper, white on one side and gray on the other, lying flat on a table. We will explore the ways in which this paper can be folded. We will consider only folding in two dimensions—with the final state lying flat on the table. This piece of paper has two distinct low-energy states, one variant with white side up and one variant with gray side up.

The (free) energy density for the paper is independent of rotations, but grows quickly when the paper is stretched or sheared. The paper, like martensites, can be represented as a deformation field $\mathbf{y}(\mathbf{x})$, representing the final position $\mathbf{y}$ of a point $\mathbf{x}$ of the paper placed horizontally on the table with the gray side up. Here, because our folded paper lies on the plane, both $\mathbf{x}$ and $\mathbf{y}$ are two-component vectors. Naturally $\mathbf{y}(\mathbf{x})$ must be a continuous function to avoid ripping the paper. Since the energy is independent of an overall translation of the paper on the table, it can depend only on gradients of the deformation field. To lowest order,[29] the energy density can be written in terms of the deformation gradient $\nabla \mathbf{y} = \partial_j y_i$:

$$\mathcal{F} = \alpha |(\nabla \mathbf{y})^\top \nabla \mathbf{y} - \mathbb{I}|^2 = \alpha (\partial_i y_j \partial_i y_k - \delta_{jk})^2. \tag{11.26}$$

The constant α is large, since paper is hard to stretch. In this problem, we will be interested in the zero-energy ground states for the free energy.

(a) *Show that the zero-energy ground states of the paper free energy density (eqn 11.26) include the two variants and rotations thereof, as shown in Fig. 11.16. Specifically, show* (1) *that any rotation* $y_i(x_j) = R_{ij} x_j$ *of the gray side up position is a ground state, where* $R_{ij} = \begin{pmatrix} \cos\theta_\ell & -\sin\theta_\ell \\ \sin\theta_\ell & \cos\theta_\ell \end{pmatrix}$*, and* (2) *that flipping the paper to the white side up and then rotating,* $y_i(x_j) = R_{ik} P_{kj} x_j = \begin{pmatrix} \cos\theta_w & -\sin\theta_w \\ \sin\theta_w & \cos\theta_w \end{pmatrix} \begin{pmatrix} 1 & 0 \\ 0 & -1 \end{pmatrix} \begin{pmatrix} x \\ y \end{pmatrix}$ *also gives a ground state. (Hint: If* $\mathbf{y}(\mathbf{x}) =$

---

[27]The energy on a square lattice measures the *Manhattan length* of the perimeter, reflecting the rectangular grid of the streets in Manhattan (ignoring Broadway).

[28]This exercise was developed in collaboration with Richard D. James. A one-sided patterned sheet suitable for creasing and a printable version of Fig. 11.18 (both the full version and the foldable version with two levels) can be found at the book website [181].

[29]Including higher derivatives of the deformation field into the energy density would lead to an energy per unit length for the creases.

$M\mathbf{x} = M_{ij}x_j$ for a linear transformation M, what is $\nabla\mathbf{y} = \partial_j y_i$?)

Hence our two variants are the rotations $\mathbb{I} = \begin{pmatrix} 1 & 0 \\ 0 & 1 \end{pmatrix}$ and $P = \begin{pmatrix} 1 & 0 \\ 0 & -1 \end{pmatrix}$.

In the real martensite, there are definite rules (or *compatibility conditions*) for boundaries between variants: given one variant, only certain special orientations are allowed for the boundary and the other variant. A boundary in our piece of paper between a gray-up and white-up variant lying flat on the table is simply a crease (Fig. 11.17).

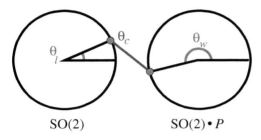

Fig. 11.16 Paper order parameter space. The allowed zero-energy deformation gradients for a piece of paper lying flat on a table. Let θ be the angle between the x axis of the graph paper and the near edge of the table. The paper can be rotated by any angle θ_ℓ (so the deformation gradient is a pure rotation in the group[30] SO(2)). Or, it can be flipped over horizontally ($(x,y) \to (x,-y)$, multiplying by $P = \begin{pmatrix} 1 & 0 \\ 0 & -1 \end{pmatrix}$) and then rotated by θ_w (deformation gradient in the set SO(2)·P). Our two variants are hence given by the identity rotation $\mathbb{I}$ and the reflection P; the ground states rotate the two variants. An interface between two of these ground states is a straight crease at angle θ_c (Fig. 11.17).

(b) *Place a piece of paper long edge downward on the table. Holding the left end fixed $\theta_\ell = 0$, try folding it along crease lines at different angles θ_c. Find the relation between the crease angle θ_c and the angle θ_w of the right-hand portion of the paper. The right-hand portion has been flipped and rotated. Find $\nabla\mathbf{y}^w = R_{\theta_w} P$ in terms of θ_c.*

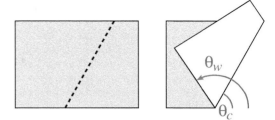

Fig. 11.17 Paper crease. An interface between two ground states $\theta_\ell = 0$ and θ_w for our paper on the table is a straight crease with angle θ_c.

Finding the displacement $\mathbf{y}(\mathbf{x})$ for the right-hand portion of the paper demands that we know where the crease is. Suppose the crease is along an axis $\mathbf{c}$. We can derive the compatibility condition governing a crease by noting that $\mathbf{y}$ along the crease must agree for the white and the gray faces, so the directional derivative $(\mathbf{c} \cdot \nabla)\mathbf{y}$ must agree.

(c) *Given the relation you deduced for the geometry in part (b), show that the difference in the directional derivatives $(\nabla\mathbf{y}^\ell - \nabla\mathbf{y}^w)$ is zero along $\mathbf{c}$, $(\mathbf{c}\cdot\nabla)\mathbf{y}^\ell - (\mathbf{c}\cdot\nabla)\mathbf{y}^w = (\partial_j y_i^\ell - \partial_j y_i^w)c_j = \mathbf{0}$. (Hints: $\nabla\mathbf{y}^\ell$ is the identity. $\cos(2\theta) = \cos^2\theta - \sin^2\theta$, $\sin(2\theta) = 2\sin\theta\cos\theta$.)*

In general, two variants with deformation gradients A and B of a martensite can be connected together along a flat boundary perpendicular to $\mathbf{n}$ if there are rotation matrices $R^{(1)}$ and $R^{(2)}$ such that[31]

$$R^{(1)}B - R^{(2)}A = \mathbf{a} \otimes \mathbf{n},$$
$$\sum_k R_{ik}^{(1)}B_{kj} - \sum_k R_{ik}^{(2)}A_{kj} = a_i n_j, \quad (11.27)$$

where $\mathbf{a} \otimes \mathbf{n}$ is the outer product of a and n. This compatibility condition ensures that the directional derivatives of y along a boundary direction $\mathbf{c}$ (perpendicular to $\mathbf{n}$) will be the same for the two variants, $(\nabla\mathbf{y}_1 - \nabla\mathbf{y}_2)\mathbf{c} = (R^{(1)}B - R^{(2)}A)\mathbf{c} = \mathbf{a}(\mathbf{n} \cdot \mathbf{c}) = 0$ and hence that the deformation field is continuous at the boundary. For our folded paper, $\nabla\mathbf{y}$ is either $R\mathbb{I}$

---

[30]A matrix M is a rotation matrix if $M^\top M = \mathbb{I}$, which means that the columns of M (the images of the coordinate axes) are orthonormal. Such a matrix is called *orthogonal*. The set of all $n \times n$ orthogonal matrices forms a group, O(n). Since $\det(AB) = \det(A)\det(B)$, and $\det(M^\top) = \det(M)$, orthogonal matrices M either have $\det(M) = 1$ (so-called *special orthogonal* matrices, in the group SO(n)) or $\det(M) = -1$ (in which case they are the product of a special orthogonal matrix times the reflection P). Thus O(n) as a set or manifold always comes in two distinct components. Hence in part (a) you are showing that all elements of O(2) are ground states for the paper.

[31]That is, the difference is a *rank one* matrix, with zero eigenvalues along all directions perpendicular to $\mathbf{n}$.

or RP for some proper rotation R, and hence eqn 11.27 is just what you proved in part (c). As can be seen in Fig. 11.11, the real martensite did not transform by stretching uniformly along one axis. Instead, it formed multiple thin layers of two of the variants. It can do so for a modest energy cost because the surface energy of the boundary between two variants is low.

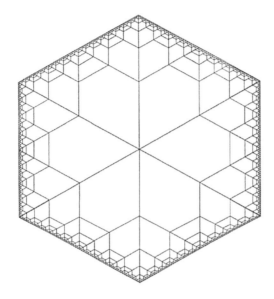

Fig. 11.18 Origami microstructure. Two-dimensional origami example of microstructure formation, by Richard D. James.

The martensite is driven to this laminated structure to satisfy *boundary conditions*. Steels go through a martensitic transition; as the blacksmith cools the horseshoe, local crystalline regions of the iron stretch along one of several possible axes. The red-hot horseshoe does not change shape overall as it is plunged into the water, though. This is for two reasons. First, if part of the horseshoe started stretching before the rest, there would be big stresses at the boundary between the transformed and untransformed regions. Second, a horseshoe is made up of many different crystalline grains, and the stretching is along different axes in different grains. Instead, the horseshoe, to a good approximation, picks a local mixture between the different variants that overall produces no net average stretch.

This is done by creating finely divided structures, like the laminated structure seen in Fig. 11.11.[32] At the boundaries of the square region, the martensite must not stretch, so it produces a fine laminated structure where the stretching in one domain cancels the contraction for its neighbors.

Our paper folding example forms a similar microstructure when we insist that the boundary lie along a curve other than the natural one.

(d) *Print out a full-sized simplified version of Fig. 11.18. Cut out the hexagon, and fold along the edges. Where does the boundary go?*[33]

The mathematicians and engineers who study these problems take the convenient limit where the energy of the boundaries between the variants (the crease energy in our exercise) goes to zero. In that limit, the microstructures can become infinitely fine, and only quantities like the relative mixtures between variants are well defined. It is a wonderful example where the pathological functions of real analysis describe important physical phenomena.

(11.8) **Minimizing sequences and microstructure.**[34] (Mathematics, Engineering) ⓟ
The martensitic morphology seen in Fig. 11.11 is a finely divided mixture between two different crystal variants. This layered structure (or *laminate*) is produced by the material to minimize the strain energy needed to glue the different domains together. If the interfacial energy needed to produce the boundaries between the domains were zero, the layering could become infinitely fine, leading to a mathematically strange function. The displacement field $\mathbf{y}(\mathbf{x})$ in this limit would be continuous, and at

---

[32]The laminated microstructure of the real martensite is mathematically even more strange than that of the paper. The martensite, in the limit where the boundary energy is ignored, has a deformation gradient which is discontinuous everywhere in the region; our folded paper has a deformation gradient which is discontinuous only everywhere along the boundary. See Exercise 11.8.

[33] Deducing the final shape of the boundary can be done by considering how the triangles along the edge overlap after being folded. Note that the full structure of Fig. 11.18 cannot be folded in three dimensions, because the paper must pass through itself (Chen Wang, private communication, see [181]). The analogous problem does not arise in martensites.

[34]This exercise was developed in collaboration with Richard D. James.

each point **x** it would have a gradient which agrees with one of the ground states.[35] However, the gradient would be discontinuous everywhere, jumping from one variant to the next each time a boundary between domains is crossed.

It is in this weird limit that the theory of martensites becomes elegant and comprehensible. If you are thinking that no such function **y**(**x**) exists, you are correct; one can approach zero strain energy with finer and finer laminates, but no function **y**(**x**) can actually have zero energy. Just as for the function in Fig. 11.19, the greatest lower bound of the martensitic energy *exists*, but is not *attained*.

A *minimizing sequence* for a function $g(x)$ with lower bound g_0 is a sequence of arguments $x_1, x_2, \ldots$ for which $g(x_n) > g(x_{n+1})$ and $\lim g(x_n) = g_0$.

Fig. 11.19 Function with no minimum. The function $g(x) = \begin{cases} x^2, & x \neq 0, \\ 1, & x = 0 \end{cases}$ can get arbitrarily close to zero, but never attains that minimum.

(a) *Find a minimizing sequence for the somewhat silly function g in Fig. 11.19.*

This kind of microstructure often arises in systems with *nonconvex* free energy densities. Consider a problem where the energy of a function $y(x)$ is given by

$$\mathcal{F}[y] = \int_0^1 [(y'^2 - 1)^2 + y^2] \, dx, \qquad (11.28)$$

with boundary conditions $y(0) = y(1) = 0$. This energy is low if $y(x)$ stays near zero and the slope $dy/dx = y'(x)$ stays near ± 1. The latter is why it is nonconvex: there are two values of the slope which have low energy density,

but intermediate values of the slope have higher energy density.[36] This free energy is similar to that for two-dimensional paper folding (Exercise 11.7); you could think of it as the folding of a one-dimensional sheet of paper ($y' = \pm 1$ representing face-up and face-down states) in a potential y^2 pulling all edge points to the origin, crinkling the boundary to a point.

Microstructure Theorem 1. $\mathcal{F}[y]$ *of eqn 11.28 does not attain its minimum.*

(b) *Prove Microstructure Theorem 1. i) Show that zero is a lower bound for the energy $\mathcal{F}$. ii) Construct a minimizing sequence of functions $y_n(x)$ for which $\lim \mathcal{F}[y_n] = 0$. iii) Show that the second term of $\mathcal{F}[y]$ is zero only for $y(x) = 0$, which does not minimize $\mathcal{F}$.*

(Advanced) *Young measures.* It is intuitively clear that any minimizing sequence for the free energy of eqn 11.28 must have slopes that approach $y' \approx \pm 1$, and yet have values that approach $y \approx 0$. Mathematically, we introduce a probability distribution (the *Young measure*) $\nu_x(S)$ giving the probability of having slope $S = y'(x + \epsilon)$ for points $x + \epsilon$ near x.

(c) *Argue that the Young measure which describes minimizing sequences for the free energy in eqn 11.28 is $\nu_x(S) = \frac{1}{2}\delta(S-1) + \frac{1}{2}\delta(S+1)$. Hint: The free energy is the sum of two squares. Use the first term to argue that the Young measure is of the form $\nu_x(S) = a(x)\delta(S-1) + (1 - a(x))\delta(S+1)$. Then write $\langle y(x) \rangle$ as an integral involving $a(x)$, and use the second term in the free energy to show $a(x) = \frac{1}{2}$.*

(11.9) **Snowflakes and linear stability.** (Condensed matter) ③

Consider a rather clunky two-dimensional model for the nucleation and growth of an ice crystal, or more generally a crystal growing in a supercooled liquid. As in coarsening with conserved order parameters, the driving force is given by the supercooling, and the bottleneck to motion is diffusion. For ice crystal formation in the atmosphere, the growing ice crystal consumes all of the water vapor near the interface; new atoms must diffuse in from afar. In other systems the bottleneck might be diffusion of latent heat away from the interface, or diffusion of impurity atoms (like salt) that prefer the liquid phase.

---

[35]Except on the boundaries between domains, which, although dense, still technically have measure zero.

[36]A function $f[x]$ is convex if $f[\lambda a + (1-\lambda)b] \leq \lambda f[a] + (1-\lambda)f[b]$; graphically, the straight line segment between the two points $(a, f[a])$ and $(b, f[b])$ lies above f if f is convex. The free energy $\mathcal{F}$ in eqn 11.28 is nonconvex as a function of the slope y'.

The current shape of the crystal is given by a curve $\mathbf{x}(\lambda, t)$ giving the current solid–liquid interface, parameterized by λ. (We are free to choose any parameterization λ we wish—the the angle θ in polar coordinates, the arc length around the interface ...; the equations of motion 11.29 must be *gauge invariant*—carefully written to be independent of how we measure (gauge) the position λ around the curve.) If $\widehat{n}$ is the local unit normal pointing outward from crystal into liquid, and $\mathcal{S}(\kappa)$ is the local growth speed of the crystal as a function of the local curvature κ of the interface, then[37]

$$\widehat{n} \cdot \frac{\partial \mathbf{x}}{\partial t} = \mathcal{S}(\kappa) = A + B\kappa - C\kappa|\kappa|. \quad (11.29)$$

Equation 11.29 has been chosen to reproduce the physics of nucleation and coarsening of circular droplets of radius $R(t)$ and curvature $\kappa = 1/R$.

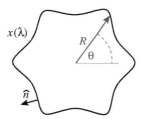

Fig. 11.20 Crystal shape coordinates A crystalline nucleus with six incipient fingers (see Fig. 11.12 for a fully formed dendrite). We describe the crystal–liquid interface in two dimensions either with a parameterized curve $\mathbf{x}(\lambda)$ or, more specifically, with the radius $R(\theta)$.

(a) *Generalize the coarsening law for conserved-order parameters (eqn 11.18) and the calculation for the critical nucleus size (eqn 11.9) to two-dimensional droplets. Setting $A = B = 0$ in eqn 11.29, what value of C reproduces the coarsening law in 2D? What value for A then yields the correct critical nucleus radius?*
Hence A represents the effects of undercooling (favoring crystal over vapor) and C represents the effects of surface tension.

(b) *Consider an interface with regions of both positive and negative curvature (as in Fig. 11.21). What direction should the surface tension push the fingertip regions of positive curvature κ? What direction should the surface tension push the interface in the channels (negative κ)? Would an analytic term $C\kappa^2$ in eqn 11.29 have given the correct behavior?*
The term $B\kappa$ speeds up regions of positive curvature (crystalline fingers) and slows down regions of negative curvature (channels left behind). It crudely mimics the diffusion of heat or impurities away from the interface (Fig. 11.21): the growing ice tips probe colder, more humid regions.
Our model is clunky because it tries to model the nonlocal effects of the diffusion [108] into a local theory [34]. More microscopic models include an explicit thermal boundary layer [16, 17].[38] The basic physical picture, however, nicely mimics the more realistic treatments.

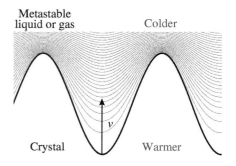

Fig. 11.21 Diffusion field ahead of a growing interface. Contours of temperature, water vapor concentration, or salt concentration in the liquid ahead of a growing crystal (dark curve). Notice that the gradients are low in the troughs and high at the tips of the fingers; this allows the tips to grow faster. We do not plot the contours in the crystal.

We will take the growing circular droplet solution, and look to see if a small oscillation of the interface will grow or shrink with time. We can parameterize a nearly circular droplet with

---

[37]The component of $\partial \mathbf{x}/\partial t$ parallel to the interface does not affect the growth; it only affects the time-dependent parameterization of the curve.
[38]In particular, this is why we need a nonanalytic term $\kappa|\kappa|$. Also, our growth rate A for a flat interface is not realistic. For $\Delta T < L/c$, for example, the latent heat is too large to be absorbed by the undercooling; the final state is a mixture of ice and water, so only fingered interfaces can grow (and never close).

a curve $R(\theta, t)$. In these coordinates the curvature is given by

$$\kappa = \frac{R^2 + 2 \left(\partial R / \partial \theta\right)^2 - R(\partial^2 R / \partial \theta^2)}{\left(R^2 + (\partial R / \partial \theta)^2\right)^{3/2}}. \quad (11.30)$$

Check that $\kappa = 1/R$ for a perfect circle.

(c) *Write $\widehat{n} \cdot \partial \mathbf{x}/\partial t$, the left-hand side of eqn 11.29, in terms of $\partial R/\partial t$, R, and $\partial R/\partial \theta$.* (Hints: $\widehat{n}$ is the unit vector perpendicular to the local tangent to the interface $[\partial(R \cos\theta)/\partial\theta, \ \partial(R \sin\theta)/\partial\theta]$, pointing outward. Here $\partial \mathbf{x}/\partial t$ points along $\widehat{r}$, because we chose θ as our parameterization.[39] Your answer should not have an explicit dependence on θ.)

Small ice crystals shrink, larger ones grow. Even larger ones grow fingers, or *dendrites* (Fig. 11.12). We can use *linear stability analysis* to find the size at which the circular droplet starts growing fingers. Linear stability analysis takes an exact solution and sees whether it is stable to small perturbations.

Expand $R(\theta, t) = \sum_{m=-\infty}^{\infty} r_m(t) \exp(im\theta)$. We want to know, for each number of fingers m, at what average radius[40] r_0 the fingers will start to grow.

Fig. 11.22 Snowflake. (© Ken Libbrecht, from snowcrystals.com [114, 115].)

(d) *Assume $r_m(t)$ is small for $m \neq 0$, and expand eqn 11.29 (written in polar coordinates as in eqn 11.30 and part (c)) to linear order in r_m and its derivatives. Write the evolution law for $\partial r_m/\partial t$ in terms of r_m and r_0 (but not r_n for other integers n). In terms of A, B, and C, give the radius r_0 at which the fingers will start to grow.* (Hint: After linearizing, multiply by $e^{-im\theta}$ and integrate over θ; then use the orthogonality properties of the Fourier transform, eqn A.27.)

Whenever you find a simple exact solution to a problem, you can test the stability using linear stability analysis, as we did here. Add a small perturbation, linearize, and see whether all of the different Fourier modes decay.

The six fold structure of snowflakes (Fig. 11.22) is due to the six fold molecular crystal structure of ice; the growth rate of the surface depends on angle, another effect that we have ignored in our model.

(11.10) **Gibbs free energy barrier.** (Chemistry) ③

Figure 11.14 shows the chemical potential for a theory (due to van der Waals) for the liquid–gas transition, as a function of density $\rho = N/V$, at a pressure and temperature where gas and liquid coexist.

(a) *Which well is the gas? Which is the liquid? How can we tell that the two coexist?*

This figure is analogous to Fig. 11.2, except that instead of the Helmholtz potential A we plot the Gibbs free energy per particle $\mu = G/N$. In Exercise 11.3, we use the chemical potential barrier between the two wells to estimate the surface tension of water.

But what does this plot mean? The caption says the temperature and pressure are fixed. That explains why we use the Gibbs free energy $G(T, P, N)$. We know $G(T, P, N) = N\mu(T, P)$. (Remember, G is extensive, and T and P are intensive, so it makes sense that it must be linear in the only other extensive variable N.) But the caption tells us that the graph is for fixed $T = 373\,\text{K}$ and $P = 1.5 \times 10^7\,\text{dynes/cm}^2$. How can μ also depend on $\rho = N/V$? Because most of the points on Fig. 11.14 represent states that are not in equilibrium with the external world.

(b) *In equilibrium, show that $(\mathrm{d}/\mathrm{d}V (A + PV))|_{T,N} = 0$, and hence $\mathrm{d}\mu/\mathrm{d}\rho = 0$. Which*

---

[39]The coordinate θ does not change as we evolve t, because the parameterization θ varies to keep it fixed, if that helps.
[40]That is, r_m for $m = 0$.

points on this graph are in local equilibrium? Which local equilibrium is unstable to uniform changes in density?

We often want to study the free energy of systems that are not in a global equilibrium. In Section 6.7, for example, we studied the free energy of a spatially varying ideal gas by constraining the local density $\rho(x)$ and minimizing the free energy with respect to other degrees of freedom. Here we want to study the free energy inside a domain wall between the liquid and the gas, interpolating at densities between the two equilibrium states. Usually one would use the Helmholtz free energy $A(T, V, N)$ to study systems with fixed $\rho = N/V$. But here we also have an external pressure, which biases the free energy toward the liquid.

Knowing the equilibrium Helmholtz free energy $A(T, V, N)$ for a local volume $\rho = N/V$, can we calculate the nonequilibrium, constrained Gibbs free energy density $G(T, P, N, V)$ in terms of the four variables? We can think of this as two subsystems (as in Fig. 6.4), (the gas/liquid mixture and the environment) each in internal equilibrium, but where we have not yet allowed the subsystems to exchange volume. (c) *Argue that $G(T, P, N, V) = A(T, V, N) + PV$, where we do not vary V to minimize the sum, is the free energy for the local subsystem incorporating the cost of stealing volume from the environment at pressure P.*

The Helmholtz free energy for the van der Waals gas is

$$A(T, V, N) = -aN^2/V \qquad (11.31)$$
$$+ Nk_BT \left(\log \left(\frac{N\lambda^3}{V - bN} \right) - 1 \right).$$

(d) *Verify eqn 11.22 from Exercise 11.3 using eqn 11.31 and the discussion in part (c).*

(11.11) **Unstable to what?** ⓟ
In Fig. 11.3(a), suppose a system held at constant pressure $P < P_v$ and constant temperature starts in the metastable liquid state. Draw this initial state on a copy of the figure, and draw an arrow to the final state.

(11.12) **Nucleation in 2D.** ⓟ
Abrupt phase transitions can also arise in two-dimensional systems. Lipid bilayer membranes, such as the ones surrounding your cells, often undergo a miscibility phase transition. At high temperatures, two types of lipids may dissolve in one another, but at low temperatures they may be immiscible (see Fig. 8.8). If the transition is abrupt, it will happen via the same kind of critical droplet nucleation that Section 11.3 analyzed in three dimensions.

Calculate the critical radius R_c and the free energy barrier B for a circular critical droplet in a two-dimensional abrupt phase transition. Let the transition be at temperature T_m, with latent heat per unit area ℓ, and line tension λ per unit length between the two phases.

The lipid membranes in your cells appear to be nearly at the transition where the two types of lipids phase separate. When cooled, experiments show that they do not nucleate droplets. They appear instead to be going through a continuous phase transition, leading to self-similar fluctuations (Fig. 12.27, [202]) described by the Ising universality class (Chapter 12).

(11.13) **Linear stability of a growing interface.** (Surface science) ⓟ
A crystal has atoms slowly added to its surface, forming what amounts to a two-dimensional gas. A single step on the surface grows from one edge of the crystal to the other. That is, the surface is one atom taller for the region $y < u(x)$, where $u(x)$ starts nearly constant (a nearly flat front). As the gas of surface atoms on the lower level diffuse and hit the step edge, they can stick and increase the local height u. (Assume the gas of atoms on the top surface cannot jump over the edge, and does not affect the growth.) The edge thus grows at different speeds, depending on the local curvature of $u(x)$.

(a) *Compare regions of the step edge where $u''(x) > 0$ to regions where $u''(x) < 0$. Which have more directions from which a diffusing atom on the bottom surface can stick? Which will grow faster?*

We model the initial time evolution of the front with a simple, linear growth law:

$$\partial u/\partial t = v_0 + w\partial^2 u/\partial x^2. \qquad (11.32)$$

(b) *According to your answer in part (a), is w positive or negative?*

To solve this evolution law, it is convenient to work in Fourier space, writing

$$\tilde{u}(k) = (1/L) \int_0^L u(x) \exp(-ikx)dx, \quad (11.33)$$

where the field extends from 0 to L. You may assume that all derivatives of $u(x)$ are zero at the edges of the field (that is, ignore the boundary terms).

(c) *Find the differential equation for $\partial \tilde{u}(k_m)/\partial t$, giving the evolution of the wiggles in the step edge as it grows, where $k_m = 2\pi m/L$.* (Hint: You may want to consider $k_m = 0$ separately.)

(d) *Solve the differential equation for $\tilde{u}_k(t)$ in terms of the initial $\tilde{u}_k$ at $t = 0$. For what values of w does an initial irregularity in the front smoothen out with time? For what values does it grow?* (Hint: With the assumption that the regions where the step edge pokes out grow fastest, the tips of a sinusoidal $u(x)$ should advance faster than the valleys, leading to the growth of small initial waves.)

(11.14) **Nucleation of cracks.**[41] (Engineering, Condensed matter) ③

Fig. 11.23 Stretched block of elastic material, length L and width W, elongated vertically by a force F per unit area A, with free side boundaries. The block will stretch a distance $\Delta L/L = F/YA$ vertically and shrink by $\Delta W/W = \sigma \Delta L/L$ in both horizontal directions, where Y is Young's modulus and σ is Poisson's ratio, linear elastic constants characteristic of the material. For an isotropic material, the other elastic constants can be written in terms of Y and σ; for example, the (linear) bulk modulus $\kappa_{\mathrm{lin}} = Y/3(1 - 2\sigma)$.

Crystals are rigid (Exercise 9.12); under external stress they bend elastically. If we pull them too hard, how do they fail? Some crystals, like most metals, are ductile; they bend at larger stresses through the motion of topological defects (Exercise 11.5). Others are brittle; they fracture in two under shear. Here we study how

thermal fluctuations can, in principle, nucleate fracture no matter how gently we stretch it. Consider a large steel cube, stretched by a moderate strain $\epsilon = \Delta L/L$ (Fig. 11.23). You may assume $\epsilon \ll 0.1\%$, where we can ignore plastic deformation.

Fig. 11.24 Fractured block of elastic material, as in Fig. 11.23 but broken in two. The free energy here is $2\gamma A$, where γ is the free energy per unit area A of (undeformed) fracture surface.

(a) *At nonzero temperature, what is the equilibrium ground state for the cube as $L \to \infty$ for fixed ϵ?* (Hints: Remember, or show, that the free energy per unit (undeformed) volume of the cube is $\frac{1}{2}Y\epsilon^2$. Notice Fig. 11.24 as an alternative candidate for the ground state.) *For steel, with $Y = 2 \times 10^{11}\,\mathrm{N/m^2}$, $\gamma \approx 2.5\,\mathrm{J/m^2}$,[42] and density $\rho = 8{,}000\,\mathrm{kg/m^3}$, how much can we stretch a beam of length $L = 10\,\mathrm{m}$ before the equilibrium length is broken in two? How does this compare with the amount the beam stretches under a load equal to its own weight?*

Why don't bridges fall down? The beams in the bridge are in a *metastable state*. What is the barrier separating the stretched and fractured beam states? Consider a crack in the beam, of length ℓ (Fig. 11.25). Your intuition may tell you that tiny cracks will be harmless, but a long crack will tend to grow at small external stress. For convenient calculations, we will now switch problems from a stretched steel beam to a taut two-dimensional membrane under an isotropic tension, a negative pressure $P < 0$. That is, we are calculating the rate at which a balloon will spontaneously pop due to thermal fluctuations. The crack costs a surface free energy $2\alpha\ell$, where α is the free energy per unit length of membrane perimeter. A detailed elastic theory calculation

---

[41] This exercise draws heavily on Alex Buchel's work [35, 36].

[42] This is the energy for a clean, flat (100) surface, a bit more than 1eV/surface atom [204]. The surface left by a real fracture in (ductile) steel will be rugged and severely distorted, with a much higher energy per unit area. This is why steel is much harder to break than glass, which breaks in a brittle fashion with much less energy left in the fracture surfaces.

shows that a straight crack of length ℓ will release a (Gibbs free) energy $\pi P^2(1-\sigma^2)\ell^2/4Y$.

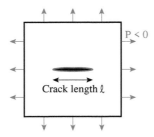

P < 0

Crack length ℓ

Fig. 11.25 Critical crack of length ℓ, in a two-dimensional material under isotropic tension (negative hydrostatic pressure $P < 0$).

(b) *What is the critical length ℓ_c of the crack, at which it will spontaneously grow rather than heal? What is the barrier $B(P)$ to crack nucleation? Write the net free energy change in terms of ℓ, ℓ_c, and α. Graph the net free energy change ΔG due to the the crack, versus its length ℓ.*

The point at which the crack is energetically favored to grow is called the *Griffiths threshold*, of considerable importance in the study of brittle fracture.

The predicted fracture nucleation rate $R(P)$ per unit volume from homogeneous thermal nucleation of cracks is thus

$$R(P) = (\text{prefactors})\exp(-B(P)/k_B T).$$
(11.34)

One should note that thermal nucleation of fracture in an otherwise undamaged, undisordered material will rarely be the dominant failure mode. The surface tension is of order an eV per bond ($> 10^{3}\,^{\circ}\text{K}/\text{Å}$), so thermal cracks of area larger than tens of bond lengths will have insurmountable barriers even at the melting point. Corrosion, flaws, and fatigue will ordinarily lead to structural failures long before thermal nucleation will arise.

(11.15) **Elastic theory does not converge.**[43] (Engineering, Condensed matter) ④
In this exercise, we shall use methods from quantum field theory to tie together two topics

that American science and engineering students study in their first year of college: Hooke's law and the convergence of infinite series. We do so by studying how things break, continuing the analysis started in Exercise 11.14.

Many perturbative expansions in physics have zero radius of convergence. The most precisely calculated quantity in physics is the gyromagnetic ratio of the electron [172]

$$\begin{aligned}(g-2)_{\text{theory}} = {}& \alpha/(2\pi) - 0.328478\ldots(\alpha/\pi)^2 \\ & + 1.181241\ldots(\alpha/\pi)^3 \\ & - 1.4092\ldots(\alpha/\pi)^4 \qquad (11.35) \\ & + 4.396\ldots \times 10^{-12}\end{aligned}$$

a power series in the fine structure constant $\alpha = e^2/\hbar c = 1/137.035999\ldots$. (The last term is an α-independent correction due to other kinds of interactions.) Freeman Dyson gave a wonderful argument that this power-series expansion, and quantum electrodynamics as a whole, has zero radius of convergence. He noticed that the theory is sick (unstable) for any negative α (corresponding to a pure imaginary electron charge e). The series must have zero radius of convergence since any circle in the complex plane about $\alpha = 0$ includes part of the sick region.

How does Dyson's argument connect to fracture nucleation? Fracture at $P < 0$ is the kind of instability that Dyson was worried about for quantum electrodynamics for $\alpha < 0$. It has implications for the convergence of nonlinear elastic theory.

Hooke's law tells us that a spring stretches a distance proportional to the force applied: $x - x_0 = F/K$, defining the spring constant $1/K = dx/dF$. Under larger forces, Hooke's law will have corrections with higher powers of F. We could define a "nonlinear spring constant" $K(F)$ by

$$\frac{1}{K(F)} = \frac{x(F) - x(0)}{F} = k_0 + k_1 F + \ldots .$$
(11.36)

Instead of a spring constant, we'll calculate a nonlinear version of the bulk modulus $\kappa_{\text{nl}}(P)$ giving the pressure needed for a given fractional change in volume, $\Delta P = -\kappa \Delta V/V$. The linear isothermal bulk modulus[44] is given by

[43] This exercise draws heavily on Alex Buchel's work [35, 36].

[44] Warning: For many purposes (e.g. sound waves) one must use the *adiabatic* elastic constant $1/\kappa = -(1/V)(\partial V/\partial P)|_S$. For most solids and liquids these are nearly the same.

$1/\kappa_{\mathrm{lin}} = -(1/V)(\partial V/\partial P)|_T$; we can define a nonlinear generalization by

$$\frac{1}{\kappa_{\mathrm{nl}}(P)} = -\frac{1}{V(0)}\frac{V(P) - V(0)}{P}$$
$$= c_0 + c_1 P + c_2 P^2 + \cdots + c_N P^N + \cdots \tag{11.37}$$

This series can be viewed as higher and higher-order terms in a nonlinear elastic theory.

(a) *Given your argument in Exercise 11.14 about the stability of materials under tension, would Dyson argue that the series in eqn 11.37 has a zero or a nonzero radius of convergence?* In Exercise 1.5 we saw the same argument holds for Stirling's formula for $N!$, when extended to a series in $z = 1/N$; any circle in the complex z plane around the origin contains poles from large negative integers $-N$, so Stirling too has zero radius of convergence. These series are *asymptotic expansions*. Convergent expansions $\sum c_n x^n$ converge for fixed x as $n \to \infty$; asymptotic expansions need only converge to order $O(x^{n+1})$ as $x \to 0$ for fixed n. Hooke's law, Stirling's formula, and quantum electrodynamics are examples of how important, powerful, and useful asymptotic expansions can be.

Buchel [35, 36], using a clever trick from field theory [215, chapter 40], was able to calculate the large-order terms in elastic theory, essentially by doing a Kramers–Kronig transformation on your formula for the decay rate (eqn 11.34) in part (b) of Exercise 11.14. His logic works as follows:

• The Gibbs free energy density $\mathcal{G}$ of the metastable state is complex for negative P. The real and imaginary parts of the free energy for complex P form an analytic function (at least in our calculation) except along the negative P axis, where there is a branch cut [106, 107].

• Our isothermal bulk modulus for $P > 0$ can be computed in terms of $\mathcal{G} = G/V(0)$. Since $dG = -S\,dT + V\,dP + \mu\,dN$, $V(P) = (\partial G/\partial P)|_T$ and hence[45]

$$\frac{1}{\kappa_{\mathrm{nl}}(P)} = -\frac{1}{V(0)}\frac{(\partial G/\partial P)|_T - V(0)}{P}$$
$$= -\frac{1}{P}\left(\left.\frac{\partial \mathcal{G}}{\partial P}\right|_T - 1\right). \tag{11.38}$$

(b) *Write the coefficients c_n of eqn 11.37 in terms of the coefficients g_m in the nonlinear expansion*

$$\mathcal{G}(P) = \sum g_m P^m. \tag{11.39}$$

• The decay rate $R(P)$ per unit volume is proportional to the imaginary part of the free energy $\mathrm{Im}[\mathcal{G}(P)]$, just as the decay rate Γ for a quantum state is related to the imaginary part $i\hbar\Gamma$ of the energy of the resonance. More specifically, for $P < 0$ the imaginary part of the free energy jumps as one crosses the real axis:

$$\mathrm{Im}[\mathcal{G}(P \pm i\epsilon)] = \pm(\text{prefactors})R(P). \tag{11.40}$$

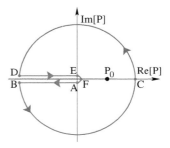

Fig. 11.26 Contour integral in complex pressure. The free energy density $\mathcal{G}$ of the elastic membrane is analytic in the complex P plane except along the negative P axis. This allows one to evaluate $\mathcal{G}$ at positive pressure P_0 (where the membrane is stable and $\mathcal{G}$ is real) with a contour integral as shown.

• Buchel then used Cauchy's formula to evaluate the real part of $\mathcal{G}$ in terms of the imaginary part (Fig. 11.26), which is given by the decay rate R per unit volume:

$$\mathcal{G}(P_0) = \frac{1}{2\pi i}\oint_{ABCDEF}\frac{\mathcal{G}(P)}{P - P_0}\,dP$$
$$= \frac{1}{2\pi i}\int_B^0 \frac{\mathcal{G}(P + i\epsilon) - \mathcal{G}(P - i\epsilon)}{P - P_0}\,dP$$
$$+ \int_{EFA} + \int_{BCD}$$
$$= \frac{1}{\pi}\int_B^0 \frac{\mathrm{Im}[\mathcal{G}(P + i\epsilon)]}{P - P_0}\,dP$$
$$+ (\text{unimportant}) \tag{11.41}$$

---

[45]Notice that this is not the (more standard) pressure-dependent linear bulk modulus, $\kappa_{\mathrm{lin}}(P)$, which is given by $1/\kappa_{\mathrm{lin}}(P) = -(1/V)(\partial V/\partial P)|_T = -(1/V)(\partial^2 \mathcal{G}/\partial P^2)|_T$. This would also have a Taylor series in P with zero radius of convergence at $P = 0$, but it has a different interpretation; $\kappa_{\mathrm{nl}}(P)$ is the nonlinear response at $P = 0$, while $\kappa_{\mathrm{lin}}(P)$ is the pressure-dependent linear response.

where the integral over the small semicircle vanishes as its radius $\epsilon \to 0$ and the integral over the large circle is convergent and hence unimportant to high-order terms in perturbation theory.

The decay rate (eqn 11.34) for $P < 0$ should be of the form

$$R(P) \propto \text{(prefactors)} \exp(-D/P^2), \quad (11.42)$$

where D is some constant characteristic of the material.

(c) *Using eqns 11.40, 11.41, and 11.42, and assuming the prefactors combine into a constant A, write the free energy for $P_0 > 0$ as an integral involving the decay rate over $-\infty < P < 0$. Expanding $1/(P - P_0)$ in a Taylor series in powers of P_0, and assuming one may exchange sums and integration, find and evaluate the integral for g_m in terms of D and m. Calculate from g_m the coefficients c_n, and then use the ratio test to calculate the radius of convergence of the expansion for $1/\kappa_{\text{nl}}(P)$, eqn 11.37.* (Hints: Use a table of integrals, a computer algebra package, or change variable $P = -\sqrt{D/t}$ to make your integral into the Γ function,

$$\Gamma(z) = (z-1)! = \int_0^\infty t^{z-1} \exp(-t)\mathrm{d}t. \quad (11.43)$$

If you wish, you may use the ratio test on every second term, so the radius of convergence is the value $\lim_{n\to\infty} \sqrt{|c_n/c_{n+2}|}$.)
(Why is this approximate calculation trustworthy? Your formula for the decay rate is valid only up to prefactors that may depend on the pressure; this dependence (some power of P) won't change the asymptotic ratio of terms c_n. Your formula for the decay rate is an approximation, but one which becomes better and better for smaller values of P; the integral for the high-order terms g_m (and hence c_n) is concentrated at small P, so your approximation is asymptotically correct for the high-order terms.)
Thus the decay rate of the metastable state can be used to calculate the high-order terms in perturbation theory in the stable phase! This is a general phenomena in theories of metastable states, both in statistical mechanics and in quantum physics.

(11.16) **Mosh pits.**[46] (Active matter) ③
Crowds of humans enjoying heavy metal concerts are exposed to extreme conditions of crowding, loud music, bright flashing lights, and frequent intoxication. They often engage in a collective behavior known as *moshing*, where ten to a few hundred people move and collide randomly in a region called a *mosh pit*. In this exercise, we shall use the mosh pit simulator [32] developed to study this emergent behavior [188–190]. We shall use it to illustrate some common features of *active matter*.
Active matter is the study of the collective motion of particles (or *agents*) that are self-propelled. Many of the tools we rely upon in equilibrium statistical mechanics (maximizing entropy, equilibration, free energies) are challenging or impossible to extend to active matter systems. These systems nonetheless exhibit the same bewildering variety of emergent behaviors that we find in equilibrium systems. There is much experimental activity in biological systems, from birds and wildebeests (Exercise 2.20), to bacterial motility, to microtubule dynamics within cells. Artificially engineered systems are also attracting widespread attention, from bots and gyroscopes to Janus particles (Exercise 1.12). Theoretical progress in this field relies in part upon directly developing the emergent dynamical laws. The apparent success of simplistic theoretical models and primitive engineering systems to capture the emergent collective behavior of complex biological systems (bacteria, birds, and humans) is inspiring.
Launch the mosh pit simulator [32]. It will exhibit a central region filled with red, active agents (representing moshers dancing) surrounded by black, passive agents (representing spectators). Depending upon the parameters, the red region may be a mosh pit, where the velocities are close to a "thermal" equilibrium as seen in actual concerts [190], or a circle pit, with a collective rotational motion also seen in heavy metal concerts [31].
Select the Moshpit initial condition. The graph at bottom right is the speed distribution of red, active agents.
(a) *Is the speed distribution roughly as expected from an equilibrium two-dimensional system?*

---

[46]This exercise makes use of the mosh pit simulator [32] developed by Matt Bierbaum for [190].

Visually, are the black particles also in thermal equilibrium, at the same temperature (as required if the system were in equilibrium)?
Select the Circlepit initial condition. The two left graphs show the net angular momentum in the circle pit.

(b) *Notice the large fluctuations in the motion of the active moshers.* Find Bierbaum's video of a large circle pit [27, 31], which runs the simulation with ~200 times the number of agents. *Are the fluctuations in the larger system noticeably smaller? Do they scale as $1/\sqrt{N}$, where N is the number of agents?*

Giant number fluctuations are often studied in active matter systems. Note also the spontaneous emergence of a broken symmetry in the angular momentum, reminiscent of that seen in continuous phase transitions. Nothing in the simulation is biased between clockwise and counter-clockwise. (One should note that there is a bias for counter-clockwise rotation in actual concerts [190], in both the northern and southern hemispheres.)

Return to the simulator, and select Circlepit again. Remove the selection of Circular I[nitial]C[onditions], and hit 'Restart'. Now the active agents are initially spread uniformly through the system. (You can accelerate the simulation by increasing Frameskip.) Observe how the active agents self-segregate—not always into a circle pit, but sometimes into a river flowing across the periodic boundary conditions. Note that the interaction energies between pairs of agents (black or red) are all the same. In equilibrium systems, phase separation is the triumph of energy over entropy; here the segregation is driven by the nonequilibrium dynamics.

Find and view Bierbaum's larger self-segregating simulation [26, 31]. The segregation is reminiscent of coarsening (Section 11.4.1), except that the individual clusters here appear almost alive as they ooze and spin around. Coarsening in traditional near-local-equilibrium systems depends on whether the total fractions of the separating components are conserved or not. Nonconserved particles phase separate with a length scale $L \sim t^{1/2}$, and conserved particles coarsen more slowly, as $L \sim t^{1/3}$.

(c) *Measure by eye the typical diameter of the clusters in the large animation, at times separated by factors of two (0.10, 0.20, 0.40, 1.20, 2.40).* (One strategy is to estimate the number of clusters, and presume all active particles are contained in a cluster. Remember the square root.) *Plot diameter versus time on a log–log plot: estimate the power law with which the clusters grow.*

(d) *Are the particles locally conserved in these simulation? Is particle diffusion between clusters the dominant mechanism for coarsening, as it is for many systems in local equilibrium? If not, observe the large simulation and describe the dominant mechanism you see.*

Segregation of particle types is studied both in active matter systems and also in nonequilibrium granular systems. Reference is often made to the *Brazil nut effect*, where in cans of mixed nuts the largest are observed to segregate to the top.

Continuous phase transitions

12

Continuous phase transitions are fascinating. As we raise the temperature of a magnet, the magnetization will vanish continuously at a critical temperature T_c. At T_c we observe large fluctuations in the magnetization (Fig. 12.1); instead of picking one of the up-spin, down-spin, or zero-magnetization states, this model magnet at T_c is a kind of fractal[1] blend of all three. This fascinating behavior is not confined to equilibrium thermal phase transitions. Figure 12.2 shows the *percolation transition*. An early paper which started the widespread study of this topic [112] described punching holes at random places in a conducting sheet of paper and measuring the conductance. Their measurement fell to a very small value as the number of holes approached the critical concentration, because the conducting paths were few and tortuous just before the sheet fell apart. Thus this model too shows a continuous transition: a qualitative change in behavior at a point where the properties are singular but continuous.

Many physical systems involve events of a wide range of sizes, the largest of which are often catastrophic. Figure 12.3(a) shows the energy released in earthquakes versus time during 1995. The Earth's crust responds to the slow motion of the tectonic plates in continental drift through a series of sharp, impulsive earthquakes. The same kind of crackling noise arises in many other systems, from crumpled paper [88] to Rice Krispies™ [102], to magnets [180]. The number of these impulsive *avalanches* for a given size often forms a power law $D(s) \sim s^{-\tau}$ over many decades of sizes (Fig. 12.3(b)). In the last few decades, it has been recognized that many of these systems also can be studied as critical points—continuous transitions between qualitatively different states. We can understand the properties of large avalanches in these systems using the same tools developed for studying equilibrium phase transitions.

The renormalization-group and scaling methods we use to study these critical points are deep and powerful. Much of the history and practice in the field revolves around complex schemes to implement these methods for various specific systems. In this chapter we will focus on the key ideas most useful in exploring experimental systems and new theoretical models, and will not cover the methods for calculating critical exponents.

In Section 12.1 we will examine the striking phenomenon of *universality*: two systems, microscopically completely different, can exhibit pre-

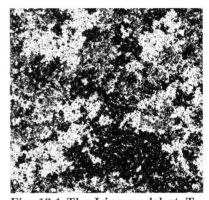

Fig. 12.1 The Ising model at T_c, the critical temperature separating the magnetized phase $T < T_c$ from the zero-magnetization phase $T > T_c$. The white and black regions represent positive and negative magnetizations $s = \pm 1$. Unlike the abrupt transitions studied in Chapter 11, here the magnetization goes to zero continuously as $T \to T_c$ from below.

[1]The term *fractal* was coined to describe sets which have characteristic dimensions that are not integers; it roughly corresponds to noninteger Hausdorff dimensions in mathematics. The term has entered the popular culture, and is associated with strange, rugged sets like those depicted in the figures here.

Statistical Mechanics: Entropy, Order Parameters, and Complexity. James P. Sethna, Oxford University Press (2021). © Oxford University Press. DOI:10.1093/oso/9780198865247.003.0012

Fig. 12.2 Percolation transition.
A percolation model on the computer, where bonds between grid points are removed rather than circular holes. Let the probability of removing a bond be $1 - p$; then for p near one (no holes) the conductivity is large, but decreases as p decreases. After enough holes are punched (at $p_c = 1/2$ for this model), the biggest cluster just barely hangs together, with holes on all length scales. At larger probabilities of retaining bonds $p = 0.51$, the largest cluster is intact with only small holes (bottom left); at smaller $p = 0.49$ the sheet falls into small fragments (bottom right; shadings denote clusters). Percolation has a phase transition at p_c, separating a connected phase from a fragmented phase (Exercises 2.13 and 12.12).

Fig. 12.3 Earthquake sizes. (a) Earthquake energy release in 1995 versus time. This time series, when sped up, sounds like crackling noise [102]. (b) Histogram of the number of earthquakes in 1995 as a function of their size S. Notice the logarithmic scales; the smallest earthquakes shown are a million times smaller and a thousand times more probable than the largest earthquakes. The fact that this distribution is well described by a power law is the Gutenberg–Richter law $\sim S^{-2/3}$.

cisely the same critical behavior near their phase transitions. We will provide a theoretical rationale for universality in terms of a *renormalization-group* flow in a space of all possible systems.

In Section 12.2 we will explore the characteristic *self-similar* structures found at continuous transitions. Self-similarity is the explanation for the fractal-like structures seen at critical points: a system at its critical point looks the same when rescaled in length (and time). We will show that *power laws* and *scaling functions* are simply explained from the assumption of self-similarity.

Finally, in Section 12.3 we will give an overview of the many types of systems that are being understood using renormalization-group and scaling methods.

12.1   Universality

Quantitative theories of physics are possible because macroscale phenomena are often independent of microscopic details. We saw in Chapter 2 that the diffusion equation is largely independent of the underlying random collision processes. Fluid mechanics relies upon the emergence of simple laws—the Navier-Stokes equations—from complex underlying microscopic interactions; if the macroscopic fluid motions depended in great detail on the shapes and interactions of the constituent molecules, we could not write simple continuum laws. Ordinary quantum mechanics relies on the fact that the behavior of electrons, nuclei, and photons are largely independent of the details of how the nucleus is assembled—nonrelativistic quantum mechanics is an effective theory which emerges out of more complicated unified theories at low energies. High-energy particle theorists developed the original notions of *renormalization* in order to understand how these effective theories emerge in relativistic quantum systems. Lattice quantum chromodynamics (simulating the strong interaction which assembles the nucleus) is useful only because a lattice simulation which breaks translational, rotational, and Lorentz symmetries can lead on long length scales to a behavior that nonetheless exhibits these symmetries. In each of these fields of physics, many dif-

Fig. 12.4 The Burridge–Knopoff model of earthquakes, with the earthquake fault modeled by blocks pulled from above and sliding with friction on a surface below. It was later realized by Carlson and Langer [40] that this model evolves into a state with a large range of earthquake sizes even for regular arrays of identical blocks.

Fig. 12.5 A medium-sized avalanche (flipping 282,785 domains) in a model of avalanches and hysteresis in magnets [180] (see Exercises 8.13 and 12.13, and Fig. 12.11). The shading depicts the time evolution: the avalanche started in the dark region in the back, and the last spins to flip are in the upper, front region. The sharp changes in shading are real, and represent sub-avalanches separated by times where the avalanche almost stops (see Fig. 8.18).

[2]Here $B = T_c^M/T_c^{\ell g}$ is as usual the rescaling of temperature and $A(M,T) = a_1 M + a_2 + a_3 T = (\rho_c\rho_0/M_0)M + \rho_c(1+s) - (\rho_c s/T_c^{\ell g})T$ is a simple shear coordinate transformation from $(\rho, T^{\ell g})$ to (M, T^M). As it happens, there is another correction proportional to $(T_c - T)^{1-\alpha}$, where $\alpha \sim 0.1$ is the specific heat exponent. It can also be seen as a kind of tilt, from a pressure-dependent effective Ising-model coupling strength. It is small for the simple molecules in Fig. 12.6(a), but significant for liquid metals [70]. Both the tilt and this $1-\alpha$ correction are *subdominant*, meaning that they vanish faster as we approach T_c than the order parameter $(T_c - T)^\beta$.

[3]The term *generic* is a mathematical term which roughly translates as "except for accidents of zero probability", like finding a function with zero second derivative at the maximum.

ferent microscopic models lead to the same low-energy, long-wavelength theory.

The behavior near continuous transitions is unusually independent of the microscopic details of the system—so much so that we give a new name to it: *universality*. Figure 12.6(a) shows that the liquid and gas densities $\rho_\ell(T)$ and $\rho_g(T)$ for a variety of atoms and small molecules appear quite similar when rescaled to the same critical density and temperature. This similarity is partly for mundane reasons: the interaction between the molecules is roughly the same in the different systems up to overall scales of energy and distance. Hence argon and carbon monoxide satisfy

$$\rho^{CO}(T) = A\rho^{Ar}(BT) \tag{12.1}$$

for some overall changes of scale A, B. However, Fig. 12.6(b) shows a completely different physical system—interacting electronic spins in manganese fluoride, going through a ferromagnetic transition. The fit curves through the magnetic and liquid–gas data are the same if we allow ourselves to not only fit T and the scale of the order parameter (ρ and M, respectively), but also allow ourselves to use a more general coordinate change

$$\rho^{Ar}(T) = A(M(BT), T), \tag{12.2}$$

which tilts the vertical axis.[2] Nature does not anticipate our choice of ρ and T for variables. At the liquid–gas critical point the natural measure of density is temperature dependent, and $A(M,T)$ is the coordinate change to the natural coordinates. Apart from this choice of variables, this magnet and these liquid–gas transitions all behave the same at their critical points.

This would perhaps not be a surprise if these two phase diagrams had parabolic tops; the local maximum of an analytic curve generically[3] looks parabolic. But the jumps in magnetization and density near T_c both vary as $(T_c-T)^\beta$ with the same exponent $\beta \approx 0.3264\ldots$, distinctly different from the square root singularity $\beta = 1/2$ of a generic analytic function.

Also, there are many other properties (susceptibility, specific heat, correlation lengths) which have power-law singularities at the critical point, and all of the exponents of these power laws for the liquid–gas systems agree with the corresponding exponents for the magnets. This is universality. When two different systems have the same singular properties at their critical points, we say they are in the same *universality class*. Importantly, the theoretical Ising model (despite its drastic simplification of the interactions and morphology) is also in the same universality class as these experimental uniaxial ferromagnets and liquid–gas systems—allowing theoretical models (e.g. Figs. 12.4 and 12.5) to be directly testable in real experiments.

To get a clear feeling about how universality arises, consider site and bond percolation in Fig. 12.7. Here we see two microscopically different systems (left) from which basically the same behavior emerges (right) on long length scales. Just as the systems approach the threshold of falling

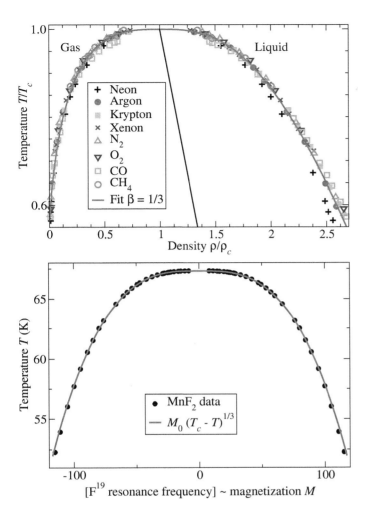

Fig. 12.6 Universality. (a) Universality at the liquid–gas critical point. The liquid–gas coexistence lines $(\rho(T)/\rho_c$ versus $T/T_c)$ for a variety of atoms and small molecules, near their critical points (T_c, ρ_c) [78]. The curve is a fit to the argon data, $\rho/\rho_c = 1 + s(1 - T/T_c) \pm \rho_0(1 - T/T_c)^\beta$ with $s = 0.75$, $\rho_0 = 1.75$, and $\beta = 1/3$ [78]. (b) Universality: ferromagnetic–paramagnetic critical point. Magnetization versus temperature for a uniaxial antiferromagnet MnF_2 [82]. We have shown both branches $\pm M(T)$ and swapped the axes so as to make the analogy with the liquid–gas critical point (above) apparent. Notice that both the magnet and the liquid–gas critical point have order parameters that vary as $(1 - T/T_c)^\beta$ with $\beta \approx 1/3$. The liquid–gas coexistence curves are tilted; the two theory curves would align if we defined an effective magnetization for the liquid–gas critical point $\rho_{\text{eff}} = \rho - 0.75\rho_c(1 - T/T_c)$ (thin midline, above). This is not an accident; both are in the same universality class, along with the three-dimensional Ising model, with the current estimate for $\beta = 0.3264\ldots$.

apart, they become similar to one another! In particular, all signs of the original lattice structure and microscopic rules have disappeared.[4]

Thus we observe in these cases that different microscopic systems look the same near critical points, if we ignore the microscopic details and confine our attention to long length scales. To study this systematically, we need a method to take a kind of continuum limit, but in systems which remain inhomogeneous and fluctuating even on the largest scales. This systematic method is called the *renormalization group*.[5]

The renormalization group starts with a remarkable abstraction: it

[4]Notice in particular the *emergent symmetries* in the problem. The large percolation clusters at p_c are statistically both translation invariant and rotation invariant, independent of the grids that underly them. In addition, we will see that there is an emergent *scale invariance*—a kind of symmetry connecting different length scales (as we also saw for random walks, Fig. 2.2).

[5]Often abbreviated "RG". The word renormalization grew out of quantum electrodynamics, where the effective charge on the electron changes size (norm) as a function of length scale. The word group is usually thought to refer to the family of coarse-graining operations that underly the method (with the group product being repeated coarse-graining). However, there is no inverse operation to coarse-graining, so the renormalization group does not satisfy the definition of a mathematical group.

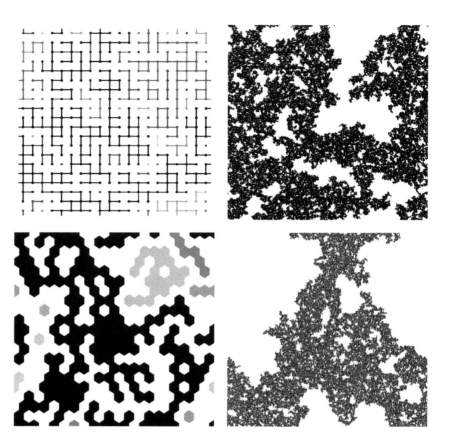

Fig. 12.7 Universality in percolation. Universality suggests that the entire morphology of the percolation cluster at p_c should be independent of microscopic details. On the top, we have bond percolation, where the bonds connecting nodes on a square lattice are occupied at random with probability p; the top right shows the infinite cluster on a 1,024 × 1,024 lattice at $p_c = 0.5$. On the bottom, we have site percolation on a triangular lattice, where it is the hexagonal sites that are occupied with probability $p = p_c = 0.5$. Even though the microscopic lattices and occupation rules are completely different, the resulting clusters look statistically identical. (One should note that the site percolation cluster is slightly less dark. Universality holds up to overall scale changes, here up to a change in the density.)

works in an enormous *system space*. Different points in system space represent different materials under different experimental conditions, and different physical models of these materials with different interactions and evolution rules. So, for example, in Fig. 12.8 we can consider the space of all possible models for hysteresis and avalanches in three-dimensional systems. There is a different dimension in this system space for each possible parameter in a theoretical model (disorder, coupling, next-neighbor coupling, dipole fields, ...) and also for each parameter in an experiment (chemical composition, temperature, annealing time, ...). A given experiment or theoretical model will traverse a line in system space as a parameter is varied; the line at the top of the figure might represent an avalanche model (Exercise 8.13) as the strength of the disorder R is varied.

The renormalization group studies the way in which system space maps into itself under *coarse-graining*. The coarse-graining operation shrinks the system and removes microscopic degrees of freedom. Ignoring the microscopic degrees of freedom yields a new physical system with identical long-wavelength physics, but with different (renormalized) values of the parameters. As an example, Fig. 12.9 shows a

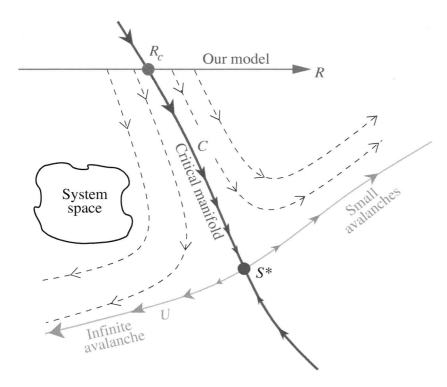

Fig. 12.8 The renormalization group defines a mapping from the space of physical systems into itself using a coarse-graining procedure. Consider the system space of all possible models of avalanches in hysteresis [180]. Each model can be coarse-grained into a new model, removing some fraction of the microscopic degrees of freedom and introducing new rules so that the remaining domains still flip at the same external fields. A fixed point S^* under this coarse-graining mapping will be self-similar (Fig. 12.11) because it maps into itself under a change in length scale. Points like R_c that flow into S^* will also show the same self-similar behavior (except on short length scales that are coarse-grained away during the flow to S^*). Models at R_c and S^* share the same *universality class*. Systems near to their critical point coarse-grain away from S^* along the unstable curve U; hence they too share universal properties (Fig. 12.12).

real-space renormalization-group "majority rule" coarse-graining procedure applied to the Ising model.[6] Several detailed mathematical techniques have been developed to implement this coarse-graining operation: not only real-space renormalization groups, but momentum-space ϵ-expansions, Monte Carlo renormalization groups, etc. These implementations are both approximate and technically challenging; we will not pursue them in this chapter (but see Exercises 12.9 and 12.11).

Under coarse graining, we often find a fixed point S^* for this mapping in system space. All the systems that flow into this fixed point under coarse-graining will share the same long-wavelength properties, and will hence be in the same universality class.

Figure 12.8 depicts the flows in system space. It is a two-dimensional picture of an infinite-dimensional space. You can think of it as a planar cross-section in system space, which we have chosen to include the line for our model and the fixed point S^*; in this interpretation the arrows and flows denote projections, since the real flows will point somewhat out of the plane. Alternatively, you can think of it as the curved surface swept out by our model in system space as it coarse-grains, in which case you should ignore the parts of the figure below the curve U.[7]

Figure 12.8 shows the case of a fixed point S^* that has one unstable direction, leading outward along U. Points deviating from S^* in that direction will not flow to it under coarse-graining, but rather will flow

[6]We will not discuss the methods used to generate effective interactions between the coarse-grained spins.

[7]The unstable manifold of the fixed point.

Fig. 12.9 Ising model at T_c: coarse-graining. Coarse-graining of a snapshot of the two-dimensional Ising model at its critical point. Each coarse-graining operation changes the length scale by a factor $B = 3$. Each coarse-grained spin points in the direction given by the majority of the nine fine-grained spins it replaces. This type of coarse-graining is the basic operation of the real-space renormalization group.

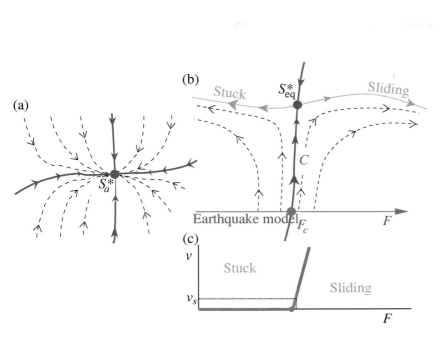

Fig. 12.10 Generic and self-organized criticality. (a) Often there will be fixed points that attract in all directions. These fixed points describe phases rather than phase transitions. Most phases are rather simple, with fluctuations that die away on long length scales. When fluctuations remain important, they will exhibit self-similarity and power laws called *generic scale invariance*. (b) The critical manifold C in this earthquake model separates a phase of stuck faults from a phase of sliding faults, with the transition due to the external stress F across the fault. Only along C does one find self-similar behavior and a broad spectrum of earthquakes. (c) The velocity of the fault will vary as a power law $v \sim (F - F_c)^\beta$ near the critical force F_c. The motion of the continental plates, however, drives the fault at a constant, very slow velocity v_s, automatically setting F to F_c and yielding earthquakes of all sizes; the model exhibits *self-organized criticality*.

away from it. Fixed-points with unstable directions correspond to continuous transitions between qualitatively different states. In the case of hysteresis and avalanches, there is a phase consisting of models where all the avalanches remain small, and another phase consisting of models where one large avalanche sweeps through the system, flipping most of the domains. The surface C which flows into S^* represents systems at their critical points; hence our model exhibits avalanches of all scales at R_c where it crosses C.[8]

Cases like the liquid–gas transition with two tuning parameters (T_c, P_c) determining the critical point will have fixed points with two unstable directions in system space. What happens when we have no unstable directions? The fixed point S_a^* in Fig. 12.10 represents an entire region of system space that shares long-wavelength properties; it represents a *phase* of the system. Usually phases do not show fluctuations on all scales. Fluctuations arise near transitions because the system does not know which of the available neighboring phases to prefer. However, there are cases where the fluctuations persist even inside phases, leading to *generic scale invariance*. A good example is the case of the random walk[9] where a broad range of microscopic rules lead to the same long-wavelength random walks, and fluctuations remain important on all scales without tuning any parameters (Fig. 2.2).

Sometimes the external conditions acting on a system naturally drive it to stay near or at a critical point, allowing one to spontaneously observe fluctuations on all scales. A good example is provided by certain models of earthquake fault dynamics. Figure 12.10(b) shows the

[8]Because S^* has only one unstable direction, C has one less dimension than system space (mathematically we say C has *co-dimension* one) and hence can divide system space into two phases. Here C is the *stable manifold* for S^*.

[9]See Section 2.1 and Exercises 12.10 and 12.11.

renormalization-group flows for these earthquake models. The horizontal axis represents the external stress on the earthquake fault. For small external stresses, the faults remain stuck, and there are no earthquakes. For strong external stresses, the faults slide with an average velocity v, with some irregularities but no large events. The earthquake fixed point S^*_{eq} describes the transition between the stuck and sliding phases, and shows earthquakes of all scales. The Earth, however, does not apply a constant stress to the fault; rather, continental drift applies a constant, extremely small velocity v_s (of the order of centimeters per year). Figure 12.10(c) shows the velocity versus external force for this transition, and illustrates how forcing at a small external velocity naturally sets the earthquake model at its critical point—allowing spontaneous generation of critical fluctuations, called *self-organized criticality*.

12.2 Scale invariance

The other striking feature of continuous phase transitions is self-similarity, or scale invariance. We can see this vividly at the critical points of the Ising model (Fig. 12.1), percolation (Fig. 12.2), and the avalanche model (Figs. 12.5 and 12.11). Each shows roughness, irregularities, and holes on all scales at the critical point. This rough, fractal structure stems at root from a hidden symmetry in the problem: these systems are (statistically) invariant under a change in *length scale*.

Consider Figs. 2.2 and 12.11, depicting the self-similarity in a random walk and a cross-section of the avalanches in the hysteresis model. In each set, the upper-left figure shows a large system, and each succeeding picture zooms in by another factor of two. In the hysteresis model, all the figures show a large avalanche spanning the system (black), with a variety of smaller avalanches of various sizes, each with the same kind of irregular boundary (Fig. 12.5). If you blur your eyes a bit, the figures should look roughly alike. This rescaling and eye-blurring process is the renormalization-group coarse-graining transformation. Figure 12.9 shows one tangible rule sometimes used to implement this coarse-graining operation, applied repeatedly to a snapshot of the Ising model at T_c. Again, the correlations and fluctuations look the same after coarse-graining; the Ising model at T_c is statistically self-similar.

How does the renormalization group explain self-similarity? The fixed point S^* under the renormalization group is the same after coarse-graining (that's what it means to be a fixed point). Any other system that flows to S^* under coarse-graining will also look self-similar (except on the microscopic scales that are removed in the first few steps of coarse-graining, during the flow to S^*). Hence systems at their critical points naturally exhibit self-similarity.

This scale invariance can be thought of as an emergent symmetry: invariance under changes of length scale. In a system invariant under translations, the expectation of any function of two positions x_1,

Fig. 12.11 Avalanches: scale invariance. Magnifications of a cross-section of all the avalanches in a run of our hysteresis model (Exercises 8.13 and 12.13) each one the lower right-hand quarter of the previous. The system started with a billion domains $(1,000^3)$. Each avalanche is shown in a different shade. Again, the larger scales look statistically the same.

[10]In some cases, such as the 2D Ising model, the standard power-law scaling will become more complex, including logarithms or exponentials.

[11] Usually, the important *relevant* directions will grow under coarse-graining, with positive exponents. Here the correlation length ξ shrinks, as would the total system length L for a finite-size sample, or the avalanche size S in eqn 12.5 and Fig. 12.13. Thus the inverses $1/L$ and $1/S$ are relevant variables, becoming more important to the behavior as we coarse-grain.

[12]If the size of the avalanche were the cube of its length, then c would equal 3 since $(1 + \mathrm{d}\ell)^3 = 1 + 3\mathrm{d}\ell + O(\mathrm{d}\ell^2)$. Instead, c is the fractal dimension of the avalanche.

[13]This factor A is partly due to coarse-graining; the same avalanches occur independent of the units of length with which we measure, but the probability density $D(S)$ per unit size per unit volume will change. It is also partly due to the rescaling factor allowed by the renormalization group—here D' must omit the smallest avalanches $S < S_{\mathrm{smallest}}$ (now invisible after blurring our eyes), so the overall normalization factor dividing D changes as well under rescaling.

[14]Note that the flow equation 12.5 for D specifies the total derivative $\mathrm{d}D/\mathrm{d}\ell = -aD$ (see Section 4.1), while being at a fixed point (eqn 12.4) specifies the partial derivative $\partial D/\partial \ell = 0$. The total derivative is $\mathrm{d}D_\ell(S_\ell)/\mathrm{d}\ell$; it gives the change of $D'(S') - D(S)$, where the partial derivative $\partial D/\partial \ell$ gives the change $D'(S) - D(S)$.

x_2 can be written in terms of the separation between the two points $\langle g(x_1, x_2)\rangle = \mathcal{G}(x_2 - x_1)$. In just the same way,[10] scale invariance will allow us to write functions of N variables in terms of *scaling functions* of $N - 1$ variables—except that these scaling functions are typically multiplied by power laws in one of the variables.

Let us begin with the case of functions of one variable. Consider the avalanche size distribution $D(S)$ for a model, say the real earthquakes in Fig. 12.3(a), or our model for hysteresis, at the critical point. Imagine taking the same system, but increasing the units of length with which we measure the system—stepping back, blurring our eyes, and looking at the system on a coarse-grained level. Imagine that we multiply the spacing between markings on our rulers by a small amount $B = 1 + \mathrm{d}\ell$. After coarsening, any length scales in the problem (like the correlation length ξ) will be divided[11] by B.

$$\xi' = \xi/B = \xi/(1 + \mathrm{d}\ell) = \xi - \xi\mathrm{d}\ell + O(\mathrm{d}\ell^2),$$
$$\mathrm{d}\xi/\mathrm{d}\ell = -\xi, \tag{12.3}$$
$$\xi[\ell] = \xi_\ell = \xi_0 \exp(-\ell).$$

Thus the factor by which length has been coarse-grained at stage ℓ is $\exp(\ell)$. We shall denote X_ℓ the value of a quantity X after coarse-graining by $\exp(\ell)$. This means X_0 is the initial condition to our renormalization-group flow equations, and hence is the "true" value in the simulation or experiment.

We assume that the system at its critical point flows to a fixed point under our renormalization group, so for large avalanches we expect

$$D'(S) = D(S); \qquad \partial D/\partial \ell = 0. \tag{12.4}$$

How do D and S change under the renormalization group? The avalanche sizes S after coarse-graining will be smaller by some factor[12] $C = 1 + c\mathrm{d}\ell$. The overall scale of $D(S)$ will change by some factor $A = 1 + a\mathrm{d}\ell$.[13] Thus our renormalization-group equations are

$$S' = S/C = S/(1 + c\mathrm{d}\ell), \qquad \mathrm{d}S/\mathrm{d}\ell = -cS,$$
$$D'(S') = AD(S) = D(S)(1 + a\mathrm{d}\ell), \quad \mathrm{d}D/\mathrm{d}\ell = aD. \tag{12.5}$$

Here $D'(S')$ is the distribution measured with the new ruler: a smaller avalanche with a larger probability density.[14]

Solving eqn 12.5 for the flow of S, we find $S_\ell = S_0 \exp(-c\ell)$. Here S_0 is both the initial condition for our differential equation at $\ell = 0$, and is the physical avalanche size for which we want to know the probability $D_0(S_0)$. Similarly, $D_\ell(S_\ell) = \exp(a\ell)D_0(S_0)$. We want to solve for the probability $D_0(S_0)$ in terms of S_0. We can get rid of the exponential factor by noticing $\exp(a\ell) = \exp(-c\ell)^{-a/c} = (S_\ell/S_0)^{-a/c}$, so $D_\ell(S_\ell) = (S_\ell/S_0)^{-a/c}D_0(S_0)$. We can then choose to flow until ℓ^* such that $S_{\ell^*} = 1$ (along the $r = 0$ axis in Fig. 12.13). Thus we find $D_0(S_0) = D_{\ell^*}(1)S_0^{-a/c}$, or

$$D(S) = NS^{-a/c}, \tag{12.6}$$

where N is the constant $D_{\ell*}(1)$. Thus the probability density of large avalanches is a power law $\sim S^{-a/c}$. This last result is quite general. Not only avalanches and earthquakes (Fig. 12.3), but usually anything which rescales by a constant factor near criticality under the renormalization group exhibits a power law.

Because the properties shared in a universality class only hold up to overall scales, the constant N is system dependent. (In this case, $\int_{S_{\text{smallest}}}^{\infty} D(S)\mathrm{d}S = 1$ because D is a probability distribution, so the normalization factor $N = (\tau-1)S_{\text{smallest}}^{(1-\tau)}$.) However, the exponents a, c, and a/c are *universal*—independent of experiment (with the universality class). Some of these exponents have standard names: the exponent c gives the fractal dimension of the avalanche, and is usually called d_f or $1/\sigma\nu$. The exponent a/c giving the size distribution law is called τ in percolation and in most models of avalanches in magnets[15] and is related to the Gutenberg–Richter exponent for earthquakes[16] (Fig. 12.3(b) and Exercise 12.17).

Most measured quantities depending on one variable will have similar power-law singularities in systems with emergent scale invariance. Thus the correlation function of the Ising model at T_c (Fig. 10.4) decays with distance x in dimension d as $C(x) \propto x^{-(d-2+\eta)}$ and the distance versus time for random walks (Section 2.1) grows as $t^{1/2}$, both because these systems are self-similar.[17]

Self-similarity is also expected near to the critical point. Here as one coarsens the length scale a system will be statistically similar to itself *at a different set of parameters* (Fig. 12.12). Thus a system undergoing phase separation (Section 11.4.1, Exercise 12.3), when coarsened, is similar to itself at an earlier time (when the domains were smaller), and a percolation cluster just above p_c (Fig. 12.2 (bottom left)) when coarsened is similar to one generated further from p_c (hence with smaller holes).

For a magnet slightly below T_c, a system coarsened by a factor $B = 1 + \mathrm{d}\ell$ will be similar to one farther from T_c by a factor $E = 1 + e\mathrm{d}\ell$. Here the standard Greek letter for the length rescaling exponent is $\nu = 1/e$. Similar to the case of the avalanche size distribution, the coarsened system must have its magnetization rescaled upward by $F = (1 + f\mathrm{d}\ell)$ (with $f = \beta/\nu$) to match that of the lower-temperature original magnet (Fig. 12.12):

$$M'(T_c - t) = FM(T_c - t) = M(T_c - Et),$$
$$(1 + f\mathrm{d}\ell)\,M(T_c - t) = M\Big(T_c - t(1 + e\mathrm{d}\ell)\Big), \tag{12.7}$$

so

$$\mathrm{d}t/\mathrm{d}\ell = et,$$
$$\mathrm{d}M/\mathrm{d}\ell = fM. \tag{12.8}$$

Again, $M \propto t^{f/e} = t^\beta$, providing a rationale for the power laws we saw in magnetism and the liquid–gas transition (Fig. 12.6). Similarly, the specific heat, correlation length, correlation time, susceptibility, and surface tension of an equilibrium system will have power-law divergences

Fig. 12.12 Scaling near criticality. If two points in system space flow towards one another under coarse-graining, their behavior must be similar on long length scales. Here we measure the magnetization $M(T)$ for our system (top line) at two different temperatures, $T_c - t$ and $T_c - Et$. The dots represent successive coarse-grainings by a factor B; under this renormalization group $M \to M' \to M'' \to M^{[3]} \ldots$. Here $M(T_c - t)$ after four coarse-grainings maps to nearly the same system as $M(T_c - Et)$ after three coarse-grainings. We thus know, on long length scales, that $M'(T_c - t)$ must agree with $M(T_c - Et)$; the system is similar to itself *at a different set of external parameters* (eqns 12.7 and 12.8). All systems near to criticality first are attracted near to the fixed point, and then flow away along a common trajectory (here the horizontal axis, in Fig. 12.8 the unstable curve U). Their properties are universal because they all escape along the same path.

[15]Except ours (Figs. 12.11 and 12.14), where we used τ to denote the avalanche size law at the critical field and disorder; integrated over the hysteresis loop $D_{\text{int}} \propto S^{-\bar{\tau}}$ with $\bar{\tau} = \tau + \sigma\beta\delta$.

[16]We must not pretend that we have found the final explanation for the Gutenberg–Richter law. There are many different models that give exponents $\approx 2/3$, but it remains controversial which of these, if any, are correct for real-world earthquakes.

[17]This is because power laws are the only self-similar function. If $f(x) = x^{-\alpha}$, then on a new scale multiplying x by B, $f(Bx) = B^{-\alpha}x^{-\alpha} \propto f(x)$. (See [142] for more on power laws.)

$(T - T_c)^{-X}$, where by definition X is α, ν, $z\nu$, γ, and $-\mu = -(d-1)\nu$, respectively. One can also vary the field H away from the critical point and measure the resulting magnetization, which varies as $H^{1/\delta}$.

To specialists in critical phenomena, these exponents are central; whole conversations often rotate around various combinations of Greek letters. We know how to calculate critical exponents from various analytical approaches,[18] and they are simple to measure (although hard to measure well [123]).

Critical exponents are not everything, however. Universality should extend even to those properties that we have not been able to write formulæ for. In particular, there are an abundance of functions of two and more variables that one can measure. Figure 12.14 shows the distribution of avalanche sizes $D_{\text{int}}(S, R)$ in our model of hysteresis, integrated over the hysteresis loop (Fig. 8.15), at various disorders R above R_c (Exercise 8.13). Notice that only at $R_c \approx 2.16$ do we get a power-law distribution of avalanche sizes; at larger disorders there are extra small avalanches, and a strong decrease in the number of avalanches beyond a certain size $S_{\text{max}}(R)$.

Let us derive the scaling form for $D_{\text{int}}(S, R)$. By using scale invariance, we will be able to write this function of two variables as a power of one of the variables times a universal, one-variable function of a combined *invariant scaling combination*. As in Fig. 12.8 we expect that systems will flow away from criticality: a system at $R = R_c + r$ after coarse-graining will be similar to a system further from the critical disorder, say at $R' = R_c + Er = R_c + (1 + ed\ell)r$. Together with our treatment of the avalanche sizes at R_c (eqn 12.5) we know that

$$\begin{aligned} \mathrm{d}r/\mathrm{d}\ell &= er, \\ \mathrm{d}S/\mathrm{d}\ell &= -cS, \\ \mathrm{d}D/\mathrm{d}\ell &= aD. \end{aligned} \tag{12.9}$$

First, how does the deviation of the disorder r vary away from criticality? Solving $\mathrm{d}r/\mathrm{d}\ell = er$, we find $r_\ell = r_0 \exp(e\ell)$. Similarly $S_\ell = S_0 \exp(-c\ell)$, so we may write r_ℓ in terms of S_0 as $r_\ell = r_0(S_\ell/S_0)^{-e/c}$. Hence

$$r_\ell S_\ell^\sigma = r_0 S_0^\sigma \tag{12.10}$$

where $1/\sigma = c/e$ will be the exponent governing how the cutoff in the avalanche size distribution varies with disorder (inset, Fig. 12.14). The combination $X = rS^\sigma$ in eqn 12.10 is *invariant* under the renormalization-group flow. That is, each renormalization-group trajectory (dashed curve in Fig. 12.13) corresponds to a different value of this invariant scaling combination. It makes sense that such invariants would be important. The avalanches of size S_0 in a system with disorder shifted by r_0 away from criticality will have properties (duration, or their average shape, or the probability $D_0(S_0, r_0)$ we calculate here), which are similar to other points on the same trajectory.

Now, how does D renormalize? Solving $\mathrm{d}D/\mathrm{d}\ell = aD$, $D_\ell(S_\ell, r_\ell) = \exp(a\ell)D_0(S_0, r_0) = (S_\ell/S_0)^{-a/c}D_0(S_0, r_0)$, so $D_0 = S_0^{-a/c}D_{\ell*}(1, r_{\ell*})$,

[18]They can be derived from the eigenvalues of the linearization of the renormalization-group flow around the fixed point S^* in Fig. 12.8 (see Exercises 12.7 and 12.11).

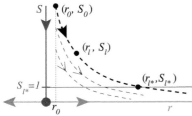

Fig. 12.13 Disorder and avalanche size: renormalization-group flows. Here S is the avalanche size and $r = R - R_c$ is the change in the disorder from the critical point. Under coarse-graining, the disorder grows and the avalanche sizes shrink (see note 11 on p. 360). The renormalization group allows us to describe properties of large avalanches S near the critical disorder $r \ll 1$ as rescaled versions of the avalanches at the corresponding crossing point along, say, $S = 1$. Equation 12.6 shows that, at criticality, the size distribution $D(S, R_c)$ can be written in terms of the constant $N = D(1, R_c)$ rescaled by a power law (eqn 12.6) (since all points on the vertical axis flow to $S = 1$, $R = R_c$). Later, we show that $D(S_0, R_c + r_0)$ can be written as the same power law times $D(1, R_c + r_{\ell*})$, where $r_{\ell*}(S_0)$ is the intersection of the trajectory starting at (r_0, S_0) with the line $S_{\ell*} = 1$.

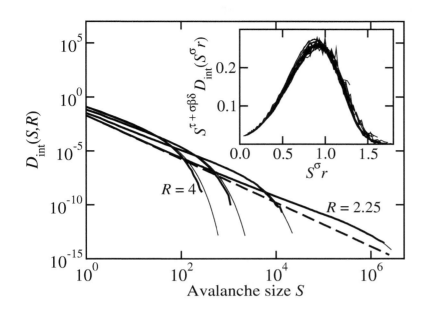

Fig. 12.14 Avalanche size distribution for our model of hysteresis. Notice the large range of probability densities. (We can measure a $D(S)$ value of 10^{-14} by running billions of spins and binning over ranges $\Delta S \sim 10^5$.) (i) Although only at $R_c \approx 2.16$ do we get a pure power law (dashed line, $D(S) \propto S^{-\bar{\tau}}$), we have large avalanches with hundreds of spins even a factor of two away from the critical point. (ii) The curves have the wrong slope except very close to the critical point; be warned that a power law over two decades (although often publishable [123]) may not yield a reliable exponent. (iii) The scaling curves (thin lines) work well even far from R_c. Inset: We plot $D(S)/S^{-\bar{\tau}}$ (here with $\bar{\tau} = \tau + \sigma\beta\delta$), versus $S^\sigma(R - R_c)/R$ to extract the universal scaling curve $\mathcal{D}(X)$ (eqn 12.11). Varying the critical exponents and R_c to get a good collapse allows us to measure the exponents far from R_c, where power-law fits are still unreliable (Exercise 12.12(g)).

where we again renormalize until the value ℓ^* such that $S_{\ell^*} = 1$ (see Fig. 12.13). Using $a/c = \bar{\tau}$ and eqn 12.10, we find for the "physical" variables $D = D_0$, $S = S_0$, and $r = r_0$ that

$$D(S, R) = D_0(S_0, R_0) = S^{-\bar{\tau}}\mathcal{D}(rS^\sigma) = S^{-\bar{\tau}}\mathcal{D}((R - R_c)S^\sigma) \quad (12.11)$$

where we derive *scaling function* $\mathcal{D}(X) = D_{\ell^*}(1, X)$. This scaling function is another universal prediction of the theory (up to an overall choice of units for S, r, and D). In Fig. 12.13, we have expressed the avalanche size distribution $D(S, R_c + r)$ for any point in the (r, S) plane in terms of the values $D(1, R_c + rS^\sigma)$ along $S = 1$. The emergent scale invariance allows us to write a function of two variables in terms of a universal function of one variable.

We can use a *scaling collapse* of the experimental or numerical data to extract this universal function, by plotting $D/S^{-\bar{\tau}}$ against $X = S^\sigma(R - R_c)$; the inset of Fig. 12.14 shows this scaling collapse.

Similar universal scaling functions appear in many contexts. Considering just the equilibrium Ising model, there are scaling functions for the magnetization $M(H, T) = (T_c - T)^\beta \mathcal{M}\left(H/(T_c - T)^{\beta\delta}\right)$, for the correlation function $C(x, t, T) = x^{-(d-2+\eta)}\mathcal{C}\left(x/|T - T_c|^{-\nu}, t/|T - T_c|^{-z\nu}\right)$, and for finite-size effects $M(T, L) = (T_c - T)^\beta \mathcal{M}\left(L/(T_c - T)^{-\nu}\right)$ in a system confined to a box of size L^d.

12.3 Examples of critical points

Ideas from statistical mechanics have found broad applicability in sciences and intellectual endeavors far from their roots in equilibrium thermal systems. The scaling and renormalization-group methods introduced in this chapter have seen a particularly broad range of applications; we will touch upon a few in this conclusion to our text.

12.3.1 Equilibrium criticality: energy versus entropy

Scaling and renormalization-group methods originated in the study of continuous phase transitions in equilibrium systems. Ising models, Potts models,[19] Heisenberg models, phase transitions in liquid crystals, wetting transitions, equilibrium crystal shapes (Fig. 11.6), two-dimensional melting—these are the grindstones on which our renormalization-group tools were sharpened.

The transition in all of these systems represents the competition between energy and entropy, with energy favoring order at low temperatures and entropy destroying it at high temperatures. Figure 12.15 shows the results of a classic, amazing experiment—the analysis of the superfluid transition in helium (the same order parameter, and also the same universality class, as the XY model). The superfluid density is expected to have the form

$$\rho_s \propto (T_c - T)^\beta (1 + d(T_c - T)^x), \qquad (12.12)$$

where x is a universal, subdominant *singular correction to scaling* (Exercise 12.31). Since here $\beta \approx \frac{2}{3}$, they plot $\rho_s / (T - T_c)^{2/3}$ so that deviations from the simple expectation are highlighted. The slope in the top, roughly straight curve reflects the difference between their measured value of $\beta = 0.6749 \pm 0.0007$ and their multiplier $\frac{2}{3}$. The other curves show the effects of the subdominant correction, whose magnitude d increases with increasing pressure. Recent experiments improving on these results were done on the International Space Station, in order to reduce the effects of gravity.

Fig. 12.15 Superfluid density in helium: scaling plot. This classic experiment [1, 77] in 1980 measured the superfluid density $\rho_s(T)$ in helium to great precision. Notice the logarithmic scale on the horizontal axis; the lowest pressure data (saturated vapor pressure ≈ 0.0504 bar) spans three decades of temperature shift from T_c. This plot emphasizes the deviations from the expected power law.

[19]Potts models are Ising-like models with N states per site, rather than two.

12.3.2 Quantum criticality: zero-point fluctuations versus energy

Thermal fluctuations do not exist at zero temperature, but there are many well-studied *quantum phase transitions* which arise from the competition of potential energy and quantum fluctuations. Many of the earliest studies focused on the metal–insulator transition and the phenomenon of *localization*, where disorder can lead to insulators even when there are states at the Fermi surface. Scaling and renormalization-group methods played a central role in this early work; for example, the states near the *mobility edge* (separating localized from extended states) are self-similar and fractal. Other milestones include the Kondo effect, macroscopic quantum coherence (testing the fundamentals of quantum measurement theory), transitions between quantum Hall plateaux, and superconductor–normal metal transitions. Figure 12.16 shows an experiment studying a transition directly from a superconductor to an

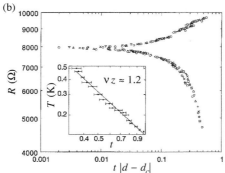

Fig. 12.16 The superconductor–insulator transition. (a) Thin films of amorphous bismuth are insulators (resistance grows to infinity at zero temperature), while films above about $12\,\text{Å}$ are superconducting (resistance goes to zero at a temperature above zero). (b) Scaling collapse. Resistance plotted against the scaled thickness for the superconductor–insulator transition, with each thickness rescaled by an independent factor t to get a good collapse. The top scaling curve $\mathcal{F}_-$ is for the insulators $d < d_c$, and the bottom one $\mathcal{F}_+$ is for the superconductors $d > d_c$. The inset shows $t \equiv T^{-1/\nu z}$, with $\nu z \sim 1.2$. (From [126].)

insulator, as the thickness of a film is varied. The resistance is expected to have the scaling form

$$R(d, T) = R_c \mathcal{F}_\pm\Big((d - d_c)T^{-1/\nu z}\Big); \tag{12.13}$$

the authors plot $R(d, T)$ versus $t(d - d_c)$, vary t until the curves collapse (main part of Fig. 12.16(b)), and read off $1/\nu z$ from the plot of t versus T (inset). In the following decades, the phase diagrams of resistance varying with thickness, temperature, and magnetic field in different materials have been studied, and several renormalization-group universality classes have been proposed to explain the various experiments.

12.3.3 Dynamical systems and the onset of chaos

Much of statistical mechanics focuses on systems with large numbers of particles, or systems connected to a large external environment. Continuous transitions also arise in isolated or simply driven systems with only a few important degrees of freedom, where they are called *bifurcations*. A bifurcation is a qualitative change in behavior which arises when a parameter in a set of differential equations passes through a critical value. The study of these bifurcations is the theory of *normal forms* (Exercise 12.4). Bifurcation theory contains analogies to universality classes, critical exponents, and analytic corrections to scaling.

Dynamical systems, even when they contain only a few degrees of freedom, can exhibit immensely complex, chaotic behavior. The mathematical trajectories formed by chaotic systems at late times—the *attractors*—are often fractal in structure, and many concepts and methods from statistical mechanics are useful in studying these sets.[20]

It is in the study of the onset of chaos where renormalization-group methods have had a spectacular impact. Figure 12.17 shows a simple dynamical system undergoing a series of bifurcations leading to a chaotic state. Feigenbaum (Exercises 12.15, 12.16, 12.9, 12.29, and 12.30) analyzed the series using a renormalization group, coarse-graining not in

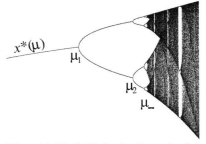

Fig. 12.17 Self-similarity at the onset of chaos. The attractor as a function of μ for the Feigenbaum logistic map $f(x) = 4\mu x(1 - x)$. For small $\mu < \mu_1$, repeatedly iterating f converges to a fixed point $x^*(\mu)$. As μ is raised past μ_1, the map converges into a two-cycle; then a four-cycle at μ_2, an eight-cycle at $\mu_3 \ldots$ These period-doubling bifurcations converge geometrically: $\mu_\infty - \mu_n \propto \delta^{-n}$ where $\delta = 4.669201609102990\ldots$ is a universal constant. At μ_∞ the system goes chaotic. (Exercise 12.9).

[20]For example, statistical mechanical ensembles become *invariant measures* (Exercise 4.3), and the attractors are characterized using concepts related to entropy (Exercise 5.16).

space but in *time*. Again, this behavior is universal—exactly the same series of bifurcations (up to smooth coordinate changes) arise in other maps and in real physical systems. Other renormalization-group calculations have been important for the study of the transition to chaos from quasiperiodic motion, and for the breakdown of the last non chaotic region in Hamiltonian systems (see Exercise 4.4).

12.3.4 Glassy systems: random but frozen

Let us conclude with a common continuous transition for which our understanding remains incomplete: glass transitions.

Glasses are out of equilibrium; their relaxation times diverge as they are cooled, and they stop rearranging at a typical temperature known as the glass transition temperature. Many other disordered systems also appear to be glassy, in that their relaxation times get very slow as they are cooled, and they freeze into disordered configurations.[21] This freezing process is sometimes described as developing *long-range order in time*, or as a *broken ergodicity* (see Section 4.2).

The basic reason that many of the glassy systems freeze into random states is *frustration*. Frustration was defined first for spin glasses, which are formed by randomly substituting magnetic atoms into a nonmagnetic host. The magnetic spins are coupled to one another at random; some pairs prefer to be parallel (ferromagnetic couplings) and some antiparallel (antiferromagnetic). Whenever strongly interacting spins form a loop with an odd number of antiferromagnetic bonds (Fig. 12.18) they are frustrated; one of the bonds will have to be left in an unhappy state, since there must be an even number of spin inversions around the loop (Fig. 12.18). It is believed in many cases that frustration is also important for configurational glasses (Fig. 12.19).

The study of disordered magnetic systems is mathematically and computationally sophisticated. The equilibrium ground state for the three-dimensional random-field Ising model,[22] for example, has been rigorously proven to be ferromagnetic (boring); nonetheless, when cooled in zero external field we understand why it freezes into a disordered state, because the coarsening process develops diverging free energy barriers to relaxation. Methods developed to study the spin glass transition have seen important applications in neural networks (which show a forgetting transition as the memory becomes overloaded) and more recently in guiding algorithms for solving computationally hard (**NP**–complete) problems (see Exercises 1.8 and 8.15). Some basic conceptual questions, however, remain unanswered. For example, we still do not know whether spin glasses have a finite or infinite number of equilibrium states—whether, upon infinitely slow cooling, one still has many glassy configurations.[23]

In real configurational glasses the viscosity and relaxation times grow by ten to fifteen orders of magnitude in a relatively small temperature range, until the cooling rate outpaces the equilibration. We fundamentally do not know why the viscosity diverges so rapidly in so many materials. There are at least three competing pictures for the glass

[21]Glasses are different from disordered systems. The randomness in disordered systems is fixed, and occurs in both the high- and low-temperature phases; the disorder in the traditional *configurational* glasses freezes in as it cools. See also Section 5.2.2.

Fig. 12.18 Frustration. A spin glass has a collection of magnetic ions with interactions of random sign. Here we see a triangle of Ising ±1 spins with one antiferromagnetic bond—one of the three bonds must be unsatisfied in any spin configuration. Hence the system is said to be *frustrated*.

[22]Our model for hysteresis and avalanches (Figs. 8.18, 12.5, 12.11, and 12.14; Exercises 8.13, 8.14, and 12.13) is this same random-field Ising model, but in a growing external field and out of equilibrium.

[23]There are "cluster" theories which assume two (spin-flipped) ground states, competing with "replica" and "cavity" methods applied to infinite-range models which suggest many competing ground states. Some rigorous results are known.

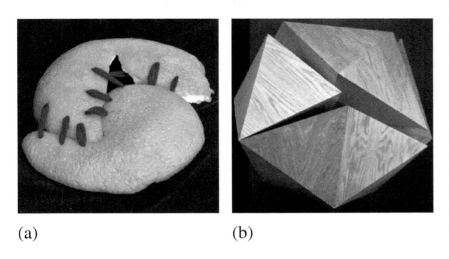

(a)   (b)

Fig. 12.19 Frustration and curvature. One kind of frustration arises when the energetically favorable local packing of atoms or molecules is incompatible with the demands of building large structures. Here we show two artistic renditions (courtesy of Pamela Davis Kivelson [99]). (a) The classic problem faced by map-makers: the peel of an orange (or the crust of the Earth) cannot be mapped smoothly onto a flat space without stretching and tearing it. (b) The analogous problem faced in many metallic glasses, whose atoms locally prefer to form nice compact tetrahedra: twenty tetrahedra cannot be glued together to form an icosahedron. Just as the orange peel segments can be nicely fit together on the sphere, the metallic glasses are unfrustrated in curved space [167].

transition: (1) It reflects an underlying equilibrium transition to an ideal, zero-entropy glass state, which would be formed under infinitely slow cooling; (2) It is a purely dynamical transition (where the atoms or molecules jam together); (3) It is not a transition at all, but just a crossover where the liquid viscosity jumps rapidly (say, because of the formation of semipermanent covalent bonds).

12.3.5 Perspectives

Many of the physicists who read this text will spend their careers outside of traditional physics. Physicists continue to play significant roles in the financial world (econophysics, computational finance, derivative trading), in biology (bioinformatics, models of ecology and evolution), computer science (traffic models, algorithms for solving hard problems), and to some extent in social science modeling (models of voting behavior and consensus building). The tools and methods of statistical mechanics (particularly the scaling methods used to study continuous transitions) are perhaps the most useful tools that we bring to these disparate subjects. Conversely, I hope that this text will prove useful as an introduction of these tools and methods to computer scientists, biologists, engineers, and finance professionals, as they continue to broaden and fertilize the field of statistical mechanics.

Exercises

Ising self-similarity, *Crackling noises*, and *Hearing chaos* will expose your eyes to the hierarchy of structure in the Ising model, and then your ears to crackling noise and the onset of chaos. *Scaling and corrections to scaling* and *The Gutenberg–Richter law* confirm that you can identify critical exponents and scaling functions. *Renormalization-group trajectories* challenges you to derive universal exponents and scaling functions directly from a renormalization-group (RG) flow. *Random walks and universal exponents*, *Diffusion equation and universal scaling functions*, and *RG and the central limit theorem* apply RG methods to familiar physics from Chapter 2; *Critical correlations* explores how properties introduced in Chapter 10 behave at critical points. Higher-order corrections to power laws at critical points (as in Fig. 12.15) are explored in *Scaling and corrections to scaling* and *Singular corrections to scaling*; some are due to *irrelevant* directions in system space that shrink under coarsegraining, and others are due to the terms we dropped as we linearized.

Mean-field theory is an effective approximation when fluctuations on all scales is *not* important—in high dimensions, in systems with long-range interactions and many neighbors, in systems with only a few degrees of freedom (as in *Bifurcation theory*), or in systems far from critical points. Mean-field theory can be derived and motivated in several ways: (1) presuming that each neighboring degree of freedom exhibits the mean behavior of the system (*Mean-field theory*); (2) generating a rigorous variational upper bound to the free energy by the use of a trial Hamiltonian (*Ising mean-field derivation* and *Mean-field bound for free energy*); (3) ignoring fluctuations (Exercise 9.5, *Landau theory for the Ising model*); (4) coupling the sites together in a branching tree (the Bethe lattice), removing the loops (used also in computer science to analyze machine learning algorithms); or (5) using an infinite-range model, coupling each site to all of the other sites (*The onset of lasing* and *Avalanche size distribution*).

Several exercises provide applications of the RG in unusual contexts: *Scaling and coarsening* and *The onset of lasing* to nonequilibrium behavior; *Superconductivity and the renormalization group* to the foundations of Fermi liquid theory; *Activated rates and the saddle-node transition* to chemical reaction rates; *Earthquakes and wires* to plastic yielding in metals; *Pandemic!* to the epidemiology of disease outbreaks; and *Biggest of bunch: Gumbel* and *Extreme values: Gumbel, Weibull, and Fréchet* to extreme value statistics (vital to the insurance industry). Several exercises explore in depth examples of critical behavior introduced in the text. You will explore avalanche power laws in *Hysteresis and Barkhausen noise*, derive a universal size distribution in *Avalanche size distribution*, and simulate them numerically in *Hysteresis and avalanches: scaling*; *Percolation and universality* will prompt another simulation. Power laws are explored in *Period doubling and the onset of chaos*; *Period doubling and the RG* probes scaling and universality and motivates Feigenbaum's exact RG transformation, *The onset of chaos: lowest order RG* allows you to estimate the universal exponents almost by hand, and *The onset of chaos: full RG* prompts you to re-implement Feigenbaum's entire calculation. Finally, some amazing mathematical and physical consequences of self-similarity can be visually explored in *Conformal invariance*.

(12.1) **Ising self-similarity.**[24] (Computation) @
Start up the Ising model. Run a large system at zero external field and $T = T_c = 2/\log(1 + \sqrt{2}) \approx 2.26919$. Set the refresh rate low enough that graphics is not the bottleneck, and run for at least a few hundred sweeps to equilibrate. You should see a fairly self-similar structure, with fractal-looking up-spin clusters inside larger down-spin structures inside ...
Can you find a nested chain of three clusters? Four?

(12.2) **Scaling and corrections to scaling.** ⓟ
Near critical points, the self-similarity under rescaling leads to characteristic power-law singularities. These dependences may be disguised, however, by less-singular corrections to scaling (see Exercise 12.31).
An experiment measures the susceptibility $\chi(T)$ in a magnet for temperatures T slightly above the ferromagnetic transition temperature T_c. They find their data is fit well by the form

$$\chi(T) = A(T - T_c)^{-1.25} + B + C(T - T_c) + D(T - T_c)^{1.77}. \quad (12.14)$$

(a) *Assuming this is the correct dependence near T_c, what is the critical exponent γ?*

[24]Ising simulation software can be found at [28].

When measuring functions of two variables near critical points, one finds universal scaling functions. The whole function is a prediction of the theory.

The pair correlation function $C(r, T) = \langle S(x)S(x + r)\rangle$ is measured in a three-dimensional system just above T_c. It is found to be spherically symmetric, and of the form

$$C(r, T) = r^{-1.026}\mathcal{C}(r(T - T_c)^{0.65}), \quad (12.15)$$

where the function $\mathcal{C}(x)$ is found to be roughly $\exp(-x)$.

(b) *What is the critical exponent ν? The exponent η?*

(12.3) **Scaling and coarsening.** (Condensed matter) ⓟ

During coarsening, we found that the system changed with time, with a length scale that grows as a power of time: $L(t) \sim t^{1/2}$ for a nonconserved order parameter, and $L(t) \sim t^{1/3}$ for a conserved order parameter. These exponents, unlike critical exponents, are simple rational numbers that can be derived from arguments akin to dimensional analysis (Section 11.4.1). Associated with these diverging length scales there are scaling functions. Coarsening does not lead to a system which is self-similar to itself at equal times, but it does lead to a system which at two different times looks the same—apart from a shift of length scales.

An Ising model with nonconserved magnetization is quenched to a temperature T well below T_c. After a long time t_0, the equal-time correlation function is $C_{t_0}^{\text{coar}}(\mathbf{r}, T)$.

Assume that the correlation function at short distances $C_t^{\text{coar}}(\mathbf{0}, T)$ will be time independent, and that the correlation function at later times will have the same functional form apart from a rescaling of the length. Write the correlation function at time twice t_0, $C_{2t_0}^{\text{coar}}(\mathbf{r}, T)$, in terms of $C_{t_0}^{\text{coar}}(\mathbf{r}, T)$. Write a scaling form

$$C_t^{\text{coar}}(\mathbf{r}, T) = t^{-\omega}\mathcal{C}(\mathbf{r}/t^{\rho}, T). \quad (12.16)$$

Use the time independence of $C_t^{\text{coar}}(\mathbf{0}, T)$ and the fact that the order parameter is not conserved (Section 11.4.1) to predict the numerical values of the exponents ω and ρ. The scaling function $\mathcal{C}$ for coarsening *does* depend on temperature (and is, in particular, anisotropic for low temperature, with domain walls lining up with lattice planes).

Low-temperature coarsening is not as universal as continuous phase transitions are (Section 11.4.1); even in one model, different temperatures have different scaling functions.

(12.4) **Bifurcation theory.** (Dynamical systems) @
Dynamical systems theory is the study of the time evolution given by systems of differential equations. Let $\mathbf{x}(t)$ be a vector of variables evolving in time t, let $\boldsymbol{\lambda}$ be a vector of parameters governing the differential equation, and let $\mathbf{F}_\lambda(\mathbf{x})$ be the differential equations

$$\dot{\mathbf{x}} \equiv \frac{\partial \mathbf{x}}{\partial t} = \mathbf{F}_\lambda(\mathbf{x}). \quad (12.17)$$

The typical focus of the theory is not to solve the differential equations for general initial conditions, but to study the qualitative behavior. In general, they focus on *bifurcations*—special values of the parameters $\boldsymbol{\lambda}$ where the behavior of the system changes qualitatively.

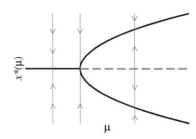

Fig. 12.20 Pitchfork bifurcation diagram. The flow diagram for the pitchfork bifurcation (eqn 12.18). The dashed line represents unstable fixed points, and the solid thick lines represent stable fixed points. The thin lines and arrows represent the dynamical evolution directions. It is called a pitchfork because of the three tines on the right emerging from the handle on the left.

(a) *Consider the differential equation in one variable $x(t)$ with one parameter μ:*

$$\dot{x} = \mu x - x^3. \quad (12.18)$$

Show that there is a bifurcation at $\mu_c = 0$, by showing that an initial condition with small, nonzero $x(0)$ will evolve qualitatively differently at late times for $\mu > 0$ versus for $\mu < 0$. Hint: Although you can solve this differential equation explicitly, we recommend instead that you argue this qualitatively from the bifurcation diagram in Fig. 12.20; a few words should suffice. Dynamical systems theory has much in common with equilibrium statistical mechanics of phases and phase transitions. The liquid–gas transition is characterized by external parameters $\boldsymbol{\lambda} = (P, T, N)$, and has a current state described by $\mathbf{x} = (V, E, \mu)$. Equilibrium phases correspond to fixed points ($x^*(\lambda)$ with $\dot{x}^* = 0$) in the dynamics, and phase transitions correspond to bifurcations.[25]

For example, the power laws we find near continuous phase transitions have simpler analogues the dynamical systems.

(b) *Find the critical exponent β for the pitchfork bifurcation, defined by $x^*(\mu) \propto (\mu - \mu_c)^\beta$ as $\mu \to \mu_c$.*

Bifurcation theory also predicts universal behavior; all pitchfork bifurcations have the same scaling behavior near the transition.

(c) *At what value λ_c does the differential equation*

$$\dot{m} = \tanh(\lambda m) - m \qquad (12.19)$$

have a bifurcation? Does the fixed point value $m^*(\lambda)$ behave as a power law $m^* \sim |\lambda - \lambda_c|^\beta$ near λ_c (up to corrections with higher powers of $\lambda - \lambda_c$)? Does the value of β agree with that of the pitchfork bifurcation in eqn 12.18? (Hint: the Taylor series for $\tanh(x) = x - x^3/3 + 2x^5/15 + O(x^7)$.) Make sure to explain why a Taylor series is justified.

Just as there are different universality classes for continuous phase transitions with different renormalization-group fixed points, there are different classes of bifurcations each with its own *normal form*. Some of the other important normal forms include the saddle-node bifurcation (Exercise 12.22),

$$\dot{x} = \mu - x^2, \qquad (12.20)$$

the transcritical exchange of stability,

$$\dot{x} = \mu x - x^2, \qquad (12.21)$$

and the Hopf bifurcation,

$$\begin{aligned} \dot{x} &= (\mu - (x^2 + y^2))x - y, \\ \dot{y} &= (\mu - (x^2 + y^2))y + x. \end{aligned} \qquad (12.22)$$

Thus normal form theory describes universality classes for continuous transitions in low-dimensional dynamical systems. It also describes continuous transitions in mean-field theory (Exercise 12.5). Finally, the renormaliza-

tion group transforms coarse-graining of statistical mechanical systems into an ordinary differential equation in system space—the domain of dynamical systems theory. Normal form theory explains why one can linearize this flow (by changing variables) if the fixed point is suitably hyperbolic, and the normal forms for those cases that cannot be linearized can be used to group the exceptions into universality *families* [157].

(12.5) **Mean-field theory.** (Condensed matter) @

In Chapter 11 and Exercise 9.5, we make reference to mean-field theories, a term which is often loosely used for any theory which approximately averages over local fluctuations of the order parameter field to form in an effective free energy.

In the Ising model on a square lattice, this amounts to assuming each spin $s_j = \pm 1$ has four neighbors which are magnetized with the average magnetization $m = \langle s_j \rangle$, leading to a one-spin mean-field Hamiltonian

$$\mathcal{H} = -4Jms_j. \qquad (12.23)$$

(a) *At temperature k_BT, what is the value for $\langle s_j \rangle$ in eqn 12.23, given m? At what temperature T_c is the phase transition, in mean field theory? (Hint: At what temperature is a nonzero $m = \langle s \rangle$ self-consistent?) Argue as in Exercise 12.4 part (c) that $m \propto (T_c - T)^\beta$ near T_c. Is this value for the critical exponent β correct for the Ising model in either two dimensions ($\beta = 1/8$) or three dimensions ($\beta \approx 0.32641\ldots$)?*

(b) *Show that the mean-field solution you found in part (a) is the minimum in an effective temperature-dependent free energy*

$$V(m) = 2Jm^2 \qquad (12.24)$$
$$- k_BT \log(\cosh(4Jm/k_BT)).$$

(The first term comes from the mean-field approximation of the energy per bond $J\langle s_is_j \rangle \approx J\langle s_i \rangle \langle s_j \rangle = m^2$, times the number of bonds per spin. The second term is the free energy of a spin in the mean field $4Jm$ due to the four neighbors. See Exercise 12.27 for a derivation.) *On a single graph, plot $V(m)$ for $J = 1$ and $1/(k_BT) = 0.1$, 0.25, and 0.5, for $-2 < m < 2$, showing the continuous phase transition. Compare with Fig. 9.23.*

---

[25]In Section 8.3, we noted that inside a phase all properties are analytic in the parameters. Here, properties are analytic except at bifurcations, which are values of λ where nonanalyticities in the long-time dynamics are observed.

(c) *What would the mean-field Hamiltonian be for the square-lattice Ising model including an external field H? At $T = T_c$ from part (a), what is the magnetization $m(H) \propto H^\delta$ at $T = T_c$, near $H = 0$? How does the mean-field critical exponent δ compare with the known values $\delta_{2D} = 15$ and $\delta_{3D} \approx 4.7898$ for the Ising model in two and three dimensions? Is the agreement improving for higher dimensions?*

(d) *Show that the mean-field magnetization is given by the minima in*

$$V(m, H) = 2Jm^2 \qquad (12.25)$$
$$- k_B T \log \left(\cosh\left(\frac{4Jm + H}{k_B T}\right) \right)$$

(derived in Exercise 12.26). On a single graph, plot $V(m, H)$ for $\beta = 0.5$ and $H = 0$, 0.5, 1.0, and 1.5, showing metastability and an abrupt transition. At what value of H does the metastable state become completely unstable? Compare with Fig. 11.2(a).

(e) *Changing variables to $t = T - T_c$, take a Taylor series of $V(m, H)$ about $m = t = H = 0$, up to fourth order in m and linear order in t and H. Compare with the Landau theory free energy for the Ising model, eqn 9.18. What is $\mu(T)$, the coefficient of m^2? What is f_0 and g at the critical point $t = H = 0$?*

(12.6) **The onset of lasing.**[26] (Quantum, Optics, Mathematics) ③

Lasers represent a stationary, condensed state. It is different from a phase of matter not only because it is made up out of energy, but also because it is intrinsically a nonequilibrium state. In a laser entropy is not maximized, free energies are not minimized—and yet the state has a robustness and integrity reminiscent of phases in equilibrium systems.

In this exercise, we will study a system of excited atoms coupled to a photon mode just before it begins to lase. We will see that it exhibits the diverging fluctuations and scaling that we have studied near critical points.

Let us consider a system of atoms weakly coupled to a photon mode. We assume that N_1 atoms are in a state with energy E_1, N_2 atoms are in a higher energy E_2, and that these atoms

are strongly coupled to some environment that keeps these populations fixed.[27] Below the onset of lasing, the probability $\rho_n(t)$ that the photon mode is occupied by n photons obeys

$$\frac{d\rho_n}{dt} = a\big(n\rho_{n-1}N_2 - n\rho_n N_1 - (n+1)\rho_n N_2$$
$$+ (n+1)\rho_{n+1}N_1\big). \qquad (12.26)$$

The first term on the right-hand side represents the rate at which one of the N_2 excited atoms experiencing $n - 1$ photons will emit a photon; the second term represents the rate at which one of the N_1 lower-energy atoms will absorb one of n photons; the third term represents emission in an environment with n photons, and the last represents absorption with $n + 1$ photons. The fact that absorption in the presence of m photons is proportional to m and emission is proportional to $m + 1$ is a property of bosons (Exercises 7.8(c) and 7.9). The constant $a > 0$ depends on the lifetime of the transition, and is related to the Einstein A coefficient (Exercise 7.8).

(a) *Find a simple expression for $d\langle n \rangle / dt$, where $\langle n \rangle = \sum_{m=0}^{\infty} m\rho_m$ is the mean number of photons in the mode. (Hint: Collect all terms involving ρ_m.) Show for $N_2 > N_1$ that this mean number grows indefinitely with time, leading to a macroscopic occupation of photons into this single state—a laser.*[28]

Now, let us consider our system just before it begins to lase. Let $\epsilon = (N_2 - N_1)/N_1$ be our measure of how close we are to the lasing instability. We might expect the value of $\langle n \rangle$ to diverge as $\epsilon \to 0$ like $\epsilon^{-\nu}$ for small ϵ. Near a phase transition, one also normally observes *critical slowing-down*: to equilibrate, the phase must communicate information over large distances of the order of the correlation length, which takes a time which diverges as the correlation length diverges. Let us define a critical-slowing-down exponent ζ for our lasing system, where the typical relaxation time is proportional to $|\epsilon|^{-\zeta}$ as $\epsilon \to 0$.

(b) *For $\epsilon < 0$, below the instability, solve your equation from part (a) for the long-time stationary value of $\langle n \rangle$. What is ν for our system? For a general initial condition for the*

---

[26]This exercise was developed with the help of Alex Gaeta and Al Sievers.
[27]That is, we assume that the atoms are being pumped into state N_2 to compensate for both decays into our photon mode and decays into other channels. Pumping means exciting atoms into state E_2 by other means.
[28]The number of photons will eventually stop growing when they begin to pull energy out of the N_2 excited atoms faster than the pumping can replace them—invalidating our equations.

mean number of photons, solve for the time evolution. It should decay to the long-time value exponentially. Does the relaxation time diverge as $\epsilon \to 0$? What is ζ?

(c) *Solve for the stationary state $\boldsymbol{\rho}^*$ for $N_2 < N_1$. (Your formula for ρ_n^* should not involve $\boldsymbol{\rho}^*$.) If N_2/N_1 is given by a Boltzmann probability at temperature T, is $\boldsymbol{\rho}^*$ the thermal equilibrium distribution for the quantum harmonic oscillator at that temperature?* Warning: The number of bosons in a phonon mode is given by the Bose–Einstein distribution, but the probability of different occupations in a quantum harmonic oscillator is given by the Boltzmann distribution (see Section 7.2 and Exercise 7.2). We might expect that near the instability the probability of getting n photons might have a scaling form

$$\rho_n^*(\epsilon) \sim n^{-\tau} \mathcal{D}(n|\epsilon|^\nu). \qquad (12.27)$$

(d) *Show for small ϵ that there is a scaling form for $\boldsymbol{\rho}^*$, with corrections that go to zero as $\epsilon \to 0$, using your answer to part (c). What is τ? What is the function $\mathcal{D}(x)$?* (Hint: In deriving the form of $\mathcal{D}$, ϵ is small, but $n\epsilon^\nu$ is of order one. If you were an experimentalist doing scaling collapses, you would plot $n^\tau \rho_n$ versus $x = n|\epsilon|^\nu$; try changing variables in $n^\tau \rho_n$ to replace ϵ by x, and choose τ to eliminate n for small ϵ.)

(12.7) **Renormalization-group trajectories.** @
This exercise provides an early introduction to how we will derive power laws and universal scaling functions in Section 12.2 from universality and coarse-graining.

An Ising model near its critical temperature T_c is described by two variables: the distance to the critical temperature $t = (T - T_c)/T_c$, and the external field $h = H/J$. Under coarse-graining, changing lengths to $x' = (1 - \epsilon)\, x$, the system is observed to be similar to itself at a shifted temperature $t' = (1 + a\epsilon)\, t$ and a shifted external field $h' = (1 + b\epsilon)\, h$, with ϵ infinitesimal and $a > b > 0$ (so there are two relevant eigendirections, with the temperature more strongly relevant than the external field). Assume $a > b > 0$.

The curves shown here connect points that are similar up to some rescaling factor.

(a) *Which diagram has curves consistent with our renormalization-group flow, for $a > b > 0$?*

Is the flow under coarse-graining inward or outward from the origin? (No math should be required. Hint: After coarse-graining, how does h/t change?)

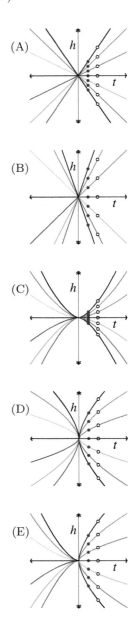

The solid dots are at temperature t_0; the open circles are at temperature $t = 2t_0$.

(b) *In terms of ϵ and a, by what factor must x be rescaled by to relate the systems at t_0 and*

$2t_0$? (Algebraic tricks: Use $(1 + \delta) \approx \exp(\delta)$ everywhere. If you rescale multiple times until $\exp(na\epsilon) = 2$, you can solve for $(1 - \epsilon)^n \approx \exp(-n\epsilon)$ without solving for n.) *If one of the solid dots in the appropriate figure from part (a) is at (t_0, h_0), what is the field $\hat{h}$ for the corresponding open circle, in terms of a, b, ϵ, and the original coordinates?* (You may use the relation between $\hat{h}$ and h_0 to check your answer for part (a).)

The magnetization $M(t, h)$ is observed to rescale under this same coarse-graining operation to $M' = (1 + c\epsilon) M$, so $M((1 + a\epsilon) t, (1 + b\epsilon) h) = (1 + c\epsilon) M(t, h)$.

(c) *Suppose $M(t, h)$ is known at (t_0, h_0), one of the solid dots. Give a formula for $M(2t_0, \hat{h})$ at the corresponding open circle, in terms of $M(t_0, h_0)$, the original coordinates, a, b, c, and ϵ. (Hint: Again, rescale n times.) Substitute your formula for $\hat{h}$ into the formula, and solve for $M(t_0, h_0)$.*

You have now basically derived the key result of the renormalization group; the magnetization curve at t_0 can be found from the magnetization curve at $2t_0$. In Section 12.2, we shall coarse-grain not to $t = 2t_0$, but to $t = 1$. We shall see that the magnetization everywhere can be predicted from the magnetization where the invariant curve crosses $t = 1$.

(d) *There was nothing about the factor of two in our shift in temperature that was special. Substitute $1/t_0$ for 2 in your formula from part (c). Show that $M(t, h) = t^\beta \mathcal{M}(h/t^{\beta\delta})$ (the standard scaling form for the magnetization in the Ising model). What are β and δ in terms of a, b, and c? How is $\mathcal{M}$ related to $M(t, h)$ where the renormalization-group trajectory crosses $t = 1$?*

Note that we have succeeded in writing $M(t, h)$ in the two-dimensional plane in terms of its value $M(1, h)$ along the line $t = 1$. A property depending on n variables near a critical point has a singular part that can be written as a power law in one variable times a scaling function $\mathcal{M}$ of $n - 1$ variables. What is more, the power law and the scaling function is universal—shared between all systems that can flow to the same renormalization-group fixed point.

(12.8) **Superconductivity and the renormalization group.** (Condensed matter) @

Ordinary superconductivity happens at a rather low temperature. In contrast to phonon energy scales (hundreds of degrees Kelvin times k_B) or electronic energy scales (tens of thousands of degrees Kelvin), phonon-mediated superconductivity in most materials happens below a few Kelvin. This is largely explained by the Bardeen–Cooper–Schrieffer (BCS) theory of superconductivity, which predicts that the transition temperature for weakly coupled superconductors is

$$T_c = 1.764 \, \hbar \omega_D \exp\left(-1/V g(\varepsilon_F)\right), \quad (12.28)$$

where ω_D is a characteristic phonon frequency, V is an attraction between electron pairs mediated by the phonons, and $g(\varepsilon_F)$ is the density of states (DOS) of the electron gas (eqn 7.73) at the Fermi energy. If V is small, $\exp(-1/V g(\varepsilon_F))$ is exponentially small, explaining why materials often have to be so cold to go superconducting.

Superconductivity was discovered decades before it was explained. Many looked for explanations which would involve interactions with phonons, but there was a serious obstacle. People had studied the interactions of phonons with electrons, and had shown that the system stays metallic (no superconductivity) *to all orders in perturbation theory.*

(a) *Taylor expand T_c (eqn 12.28) about $V = 0^+$ (about infinitesimal positive V). Guess the value of all the terms in the Taylor series. Can we expect to explain superconductivity at positive temperatures by perturbing in powers of V?*

There are two messages here:

- Proving something to all orders in perturbation theory does not make it true.

- Since phases are regions in which perturbation theory converges (see Section 8.3), the theorem is not a surprise. It is true inside a metallic phase, but stops being true at the phase transition.

In recent times, people have developed a renormalization-group description of the Fermi liquid state and its instabilities[29] (see note 23 on p. 188). Discussing Fermi liquid theory,

---

[29]There are also other instabilities of Fermi liquids. Charge-density waves, for example, also have the characteristic $\exp(-1/aV)$ dependence on the coupling V.

the BCS theory of superconductivity, or this renormalization-group description would take us far into rather technical subjects. However, we can illustrate all three by analyzing a rather unusual renormalization-group flow.

Roughly speaking, the renormalization-group treatment of Fermi liquids says that the Fermi surface is a fixed point of a coarse-graining in *energy*. That is, they start with a system space consisting of a partially filled band of electrons with an energy width W, including all kinds of possible electron–electron repulsions and attractions. They coarse-grain by perturbatively eliminating (integrating out) the electronic states near the edges of the band,

$$W' = (1 - \delta)W, \qquad (12.29)$$

incorporating their interactions and effects into altered interaction strengths among the remaining electrons. These altered interactions give the renormalization-group flow in the system space. The equation for W gives the change under one iteration ($n = 1$); we can pretend n is a continuous variable and take $\delta n \to 0$, so $(W' - W)/\delta \to \mathrm{d}W/\mathrm{d}n$, and hence

$$\mathrm{d}W/\mathrm{d}n = -W. \qquad (12.30)$$

When they do this calculation, they find the following.

- The noninteracting Fermi gas we studied in Section 7.7 is a *fixed point of the renormalization group*. All interactions are zero at this fixed point. Let V represent one of these interactions.[30]

- The fixed point is unstable to an attractive interaction $V > 0$, but is stable to a repulsive interaction $V < 0$.

- Attractive forces between electrons grow under coarse-graining and lead to new phases, but repulsive forces shrink under coarse-graining, leading back to the metallic free Fermi gas.

This is quite different from our renormalization-group treatment of phase transitions, where *relevant* directions like the temperature and field were unstable under coarse-graining, whether shifted up or down from the fixed point, and other directions were *irrelevant* and stable

(Fig. 12.8). For example, the temperature of our Fermi gas is a relevant variable, which rescales under coarse-graining like

$$\begin{aligned} T' &= (1 + a\delta)T, \\ \mathrm{d}T/\mathrm{d}n &= aT. \end{aligned} \qquad (12.31)$$

Here $a > 0$, so the effective temperature becomes larger as the system is coarse-grained. How can they get a variable V which grows for $V > 0$ and shrinks for $V < 0$?

- When they do the coarse-graining, they find that the interaction V is *marginal*: to linear order it neither increases nor decreases.

Fig. 12.21 Fermi liquid theory renormalization-group flows. The renormalization flows defined by eqns 12.31 and 12.32. The temperature T is relevant at the free Fermi gas fixed point; the coupling V is marginal. The distinguished curve represents a phase transition boundary $T_c(V)$. Below T_c, for example, the system is superconducting; above T_c it is a (finite-temperature) metal.

The next allowed term in the Taylor series near the fixed point gives us the coarse-grained equation for the interaction:

$$\begin{aligned} V' &= (1 + b\delta V)V, \\ \mathrm{d}V/\mathrm{d}n &= bV^2. \end{aligned} \qquad (12.32)$$

- They find $b > 0$.

(b) *True or false? (See Fig. 12.21.)*
(T) (F) For $V > 0$ (attractive interactions), the interactions get stronger with coarse-graining.
(T) (F) For $V < 0$ (repulsive interactions), coarse-graining leads us back to the free Fermi gas, explaining why the Fermi gas describes metals (Section 7.7).

---

[30]V will be the pairing between opposite-spin electrons near the Fermi surface for superconductors.

(T) (F) Temperature is an irrelevant variable, but dangerous.

(T) (F) The invariant scaling variable

$$x = TV^{1/\beta\delta} \qquad (12.33)$$

is unchanged by the coarse-graining (second equations in 12.31 and 12.32), where β and δ are universal critical exponents;[31] hence x labels the progress along the curves in Fig. 12.21 (increasing in the direction of the arrows).

(T) (F) The invariant scaling variable

$$y = T \exp\left(a/(bV)\right) \qquad (12.34)$$

is unchanged by the coarse-graining, so each curve in Fig. 12.21 has a fixed value for y.

Now, without knowing anything about superconductivity, let us presume that our system goes superconducting at some temperature $T_c(V)$ when the interactions are attractive. When we coarse-grain a system that is at the superconducting transition temperature, we must get another system that is at its superconducting transition temperature.

(c) What value for a/b must they calculate in order to get the BCS transition temperature (eqn 12.28) from this renormalization group? What is the value of the scaling variable (whichever you found in part (b)) along $T_c(V)$?

Thus the form of the BCS transition temperature at small V, eqn 12.28, can be explained by studying the Fermi gas *without reference to the superconducting phase*!

(12.9) **Period doubling and the RG.**[32] (Mathematics, Complexity, Computation, Dynamical systems) ④

In this exercise, we use renormalization-group and scaling methods to study the *onset of chaos* (see Exercise 12.16). There are several routes by which a dynamical system can start exhibiting chaotic motion; this exercise studies the *period-doubling cascade*, and introduces Feigenbaum's renormalization group (which is implemented in Exercises 12.29 and 12.30).

Chaos is often associated with dynamics which stretch and fold; when a batch of taffy is being pulled, the motion of a speck in the taffy

depends sensitively on the initial conditions. A simple representation of this physics is provided by the map[33]

$$f(x) = 4\mu x(1 - x) \qquad (12.35)$$

restricted to the domain $(0, 1)$. It takes $f(0) = f(1) = 0$, and $f(\frac{1}{2}) = \mu$. Thus, for $\mu = 1$ it precisely folds the unit interval in half, and stretches it to cover the original domain.

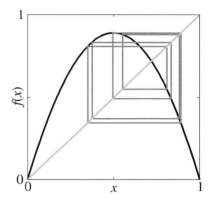

Fig. 12.22 Period-eight cycle. Iterating around the attractor of the Feigenbaum map at $\mu = 0.89$.

The study of dynamical systems (e.g., differential equations and maps like eqn 12.35) often focuses on the behavior after long times, where the trajectory moves along the *attractor*. We can study the onset and behavior of chaos in our system by observing the evolution of the attractor as we change μ. For small enough μ, all points shrink to the origin; the origin is a stable fixed point which attracts the entire interval $x \in (0, 1)$. For larger μ, we first get a stable fixed point inside the interval, and then *period doubling*.

(a) Iteration: Set $\mu = 0.2$; iterate f for some initial points x_0 of your choosing, and convince yourself that they all are attracted to zero. Plot f and the diagonal $y = x$ on the same plot. Are there any fixed points other than $x = 0$? Repeat for $\mu = 0.3$, $\mu = 0.7$, and 0.8. What happens? On the same graph, plot f, the diagonal $y = x$, and the segments $\{x_0, x_0\}$,

---

[31] Note that here δ is not the infinitesimal change in parameter.

[32] This exercise and the associated software were developed in collaboration with Christopher Myers. Hints for the computations can be found at the book website [181]. There are many exercises exploring this chaotic logistic map (see Index).

[33] Parts (a) and (b) overlap somewhat with Exercise 4.3.

$\{x_0, f(x_0)\}$, $\{f(x_0), f(x_0)\}$, $\{f(x_0), f(f(x_0))\}$, ... *(representing the convergence of the trajectory to the attractor; see Fig. 12.22). See how $\mu = 0.7$ and 0.8 differ. Try other values of μ. By iterating the map many times, find a point a_0 on the attractor. As above, then plot the successive iterates of a_0 for $\mu = 0.7$, 0.75, 0.8, 0.88, 0.89, 0.9, and 1.0.*

You can see at higher μ that the system no longer settles into a stationary state at long times. The fixed point where $f(x) = x$ exists for all $\mu > \frac{1}{4}$, but for larger μ it is no longer *stable*. If x^* is a fixed point (so $f(x^*) = x^*$) we can add a small perturbation $f(x^* + \epsilon) \approx f(x^*) + f'(x^*)\epsilon = x^* + f'(x^*)\epsilon$; the fixed point is stable (perturbations die away) if $|f'(x^*)| < 1$.[34]

In this particular case, once the fixed point goes unstable the motion after many iterations becomes periodic, repeating itself after *two* iterations of the map—so $f(f(x))$ has two new fixed points. This is called *period doubling*. Notice that by the chain rule $\mathrm{d}f(f(x))/\mathrm{d}x = f'(x)f'(f(x))$, and indeed for a period N orbit x_0, x_1, ..., x_{N-1} we find the slope of the iteration around the cycle

$$\frac{\mathrm{d}f^{[N]}}{\mathrm{d}x} = \frac{\mathrm{d}f(f(\dots f(x)\dots))}{\mathrm{d}x} \qquad (12.36)$$
$$= f'(x)f'(f(x))\dots f'(f(\dots f(x)\dots)),$$
$$= f'(x_0)f'(x_1)f'(x_2)\dots f'(x_{N-1}),$$

so the stability of a period-N orbit is determined by the product of the derivatives of f at each point along the orbit.

(b) Analytics: *Find the fixed point $x^*(\mu)$ of the map 12.35, and show that it exists and is stable for $1/4 < \mu < 3/4$. If you are ambitious or have a computer algebra program, show that the period-two cycle is stable for $3/4 < \mu < (1 + \sqrt{6})/4$.*

(c) Bifurcation diagram: *Plot the attractor as a function of μ, for $0 < \mu < 1$; compare with Fig. 12.17. (Pick regularly spaced $\delta\mu$, run $N_{\text{transient}}$ steps, record N_{cycles} steps, and plot. After the routine is working, you should be able to push $N_{\text{transient}}$ and N_{cycles} both larger than 100, and $\delta\mu < 0.01$.) Also on the bifurcation diagram, plot the line $x = 1/2$ where $f(x)$ reaches its maximum. Finally, plot the bifurcation dia-*

gram for another one-humped map

$$f_{\sin}(x) = B\sin(\pi x), \qquad (12.37)$$

for $0 < B < 1$. Do the bifurcation diagrams appear similar to one another?

Fig. 12.23 Self-similarity in period-doubling bifurcations. The period doublings occur at geometrically spaced values of the control parameter $\mu_\infty - \mu_n \propto \delta^{-n}$, and the attractor during the period-2^n cycle is similar to one-half of the attractor during the 2^{n+1}-cycle, except inverted and larger, rescaling x by a factor of α and μ by a factor of δ. The boxes shown in the diagram illustrate this self-similarity; each box looks like the next, except expanded by δ along the horizontal μ axis and flipped and expanded by α along the vertical axis.

Notice the complex, structured, chaotic region for large μ (which we study in Exercise 4.3). How do we get from a stable fixed point $\mu < \frac{3}{4}$ to chaos? The onset of chaos in this system occurs through a cascade of *period doublings*. There is the sequence of bifurcations as μ increases—the period-two cycle starting at $\mu_1 = \frac{3}{4}$, followed by a period-four cycle starting at μ_2, period-eight at μ_3—a whole period-doubling cascade. The convergence appears geometrical, to a fixed point μ_∞:

$$\mu_n \approx \mu_\infty - A\delta^{-n}, \qquad (12.38)$$

so

$$\delta = \lim_{n \to \infty} (\mu_{n-1} - \mu_{n-2})/(\mu_n - \mu_{n-1}) \qquad (12.39)$$

and there is a similar geometrical self-similarity along the x axis, with a (negative) scale factor α relating each generation of the tree (Fig. 12.23).

---

[34]In a continuous evolution, perturbations die away if the Jacobian of the derivative at the fixed point has all negative eigenvalues. For mappings, perturbations die away if all eigenvalues of the Jacobian have magnitude less than one.

In Exercise 4.3, we explained the boundaries in the chaotic region as images of $x = \frac{1}{2}$. These special points are also convenient for studying period-doubling. Since $x = \frac{1}{2}$ is the maximum in the curve, $f'(\frac{1}{2}) = 0$. If it were a fixed point (as it is for $\mu = \frac{1}{2}$), it would not only be stable, but unusually so: a shift by ϵ away from the fixed point converges after one step of the map to a distance $\epsilon f'(\frac{1}{2}) + \epsilon^2/2 f''(\frac{1}{2}) = O(\epsilon^2)$. We say that such a fixed point is *superstable*. If we have a period-N orbit that passes through $x = \frac{1}{2}$, so that the Nth iterate $f^N(\frac{1}{2}) \equiv f(\dots f(\frac{1}{2})\dots) = \frac{1}{2}$, then the orbit is also superstable, since (by eqn 12.36) the derivative of the iterated map is the product of the derivatives along the orbit, and hence is also zero.

These superstable points happen roughly halfway between the period-doubling bifurcations, and are easier to locate, since we know that $x = \frac{1}{2}$ is on the orbit. Let us use them to investigate the geometrical convergence and self-similarity of the period-doubling bifurcation diagram from part (d). We will measure both the superstable values of μ and the size of the centermost "leaf" in the bifurcation diagram (crossed by the line $x = \frac{1}{2}$ where $f(x)$ takes its maximum). For this part and part (h), you will need a routine that finds the roots $G(y) = 0$ for functions G of one variable y.

(d) *The Feigenbaum numbers and universality: Numerically, find the values of μ_n^s at which the 2^n-cycle is superstable, for the first few values of n.* (Hint: Define a function $G(\mu) = f_\mu^{[2^n]}(\frac{1}{2}) - \frac{1}{2}$, and find the root as a function of μ. In searching for μ_{n+1}^s, you will want to search in a range $(\mu_n^s + \epsilon, \mu_n^s + (\mu_n^s - \mu_{n-1}^s)/A)$ where $A \sim 3$ works pretty well. Calculate μ_0^s and μ_1^s by hand.) *Calculate the ratios $(\mu_{n-1}^s - \mu_{n-2}^s)/(\mu_n^s - \mu_{n-1}^s)$; do they appear to converge to the Feigenbaum number $\delta = 4.6692016091029909\dots$? Estimate μ_∞ by using your last two values of μ_n^s, your last ratio estimate of δ, and eqns 12.38 and 12.39. In the superstable orbit with 2^n points, the nearest point to $x = \frac{1}{2}$ is $f^{[2^{n-1}]}(\frac{1}{2})$.[35] Calculate the ratios of the amplitudes $f^{[2^{n-1}]}(\frac{1}{2}) - \frac{1}{2}$ at successive values of n; do they appear to converge to the universal value $\alpha = -2.50290787509589284\dots$? Cal-* culate *the same ratios for the map $f_2(x) = B\sin(\pi x)$; do α and δ appear to be universal (independent of the mapping)?*

The limits α and δ are independent of the map, so long as it folds (one hump) with a quadratic maximum. They are the same, also, for experimental systems with many degrees of freedom which undergo the period-doubling cascade. This self-similarity and universality suggests that we should look for a renormalization-group explanation.

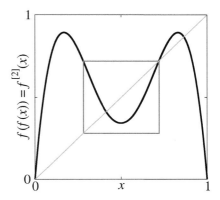

Fig. 12.24 Renormalization-group transformation. The renormalization-group transformation takes $g(g(x))$ in the small window with upper corner x^* and inverts and stretches it to fill the whole initial domain and range $(0,1) \times (0,1)$.

(e) *Coarse-graining in time. Plot $f(f(x))$ vs. x for $\mu = 0.8$, together with the line $y = x$ (or see Fig. 12.24). Notice that the period-two cycle of f becomes a pair of stable fixed points for $f^{[2]}$.* (We are coarse-graining in time—removing every other point in the time series, by studying $f(f(x))$ rather than f.) *Compare the plot with that for $f(x)$ vs. x for $\mu = 0.5$. Notice that the region zoomed in around $x = \frac{1}{2}$ for $f^{[2]} = f(f(x))$ looks quite a bit like the entire map f at the smaller value $\mu = 0.5$. Plot $f^{[4]}(x)$ at $\mu = 0.875$; notice again the small one-humped map near $x = \frac{1}{2}$.*

The fact that the one-humped map reappears in smaller form just after the period-doubling bifurcation is the basic reason that succeeding bifurcations so often follow one another. The fact that many things are universal is due to

[35]This is true because, at the previous superstable orbit, 2^{n-1} iterates returned us to the original point $x = \frac{1}{2}$.

the fact that the little one-humped maps have a shape which becomes *independent of the original map* after several period-doublings.

Let us define this renormalization-group transformation T, taking function space into itself. Roughly speaking, T will take the small upside-down hump in $f(f(x))$ (Fig. 12.24), invert it, and stretch it to cover the interval from $(0,1)$. Notice in your graphs for part (g) that the line $y = x$ crosses the plot $f(f(x))$ not only at the two points on the period-two attractor, but also (naturally) at the old fixed point $x^*[f]$ for $f(x)$. This unstable fixed point plays the role for $f^{[2]}$ that the origin played for f; our renormalization-group rescaling must map $(x^*[f], f(x^*)) = (x^*, x^*)$ to the origin. The corner of the window that maps to $(1,0)$ is conveniently located at $1 - x^*$, since our map happens to be symmetric[36] about $x = \frac{1}{2}$. For a general one-humped map $g(x)$ with fixed point $x^*[g]$ the side of the window is thus of length $2(x^*[g] - \frac{1}{2})$. To invert and stretch, we must thus rescale by a factor $\alpha[g] = -1/(2(x^*[g] - \frac{1}{2}))$. Our renormalization-group transformation is thus a mapping $T[g]$ taking function space into itself, where

$$T[g](x) = \alpha[g]\left(g\left(g(x/\alpha[g] + x^*[g])\right) - x^*[g]\right).$$
$$(12.40)$$

(This is just rescaling x to squeeze into the window, applying g twice, shifting the corner of the window to the origin, and then rescaling by α to fill the original range $(0,1) \times (0,1)$.)

(f) Scaling and the renormalization group: *Write routines that calculate $x^*[g]$ and $\alpha[g]$, and define the renormalization-group transformation $T[g]$. Plot $T[f]$, $T[T[f]]$,... and compare them. Are we approaching a fixed point f^* in function space?*

This explains the self-similarity; in particular, the value of $\alpha[g]$ as g iterates to f^* becomes the Feigenbaum number $\alpha = -2.5029\ldots$

(g) Universality and the renormalization group: *Using the sine function of eqn 12.37, compare $T[T[f_{\sin}]]$ to $T[T[f]]$ at their onsets of chaos. Are they approaching the same fixed point?*

By using this rapid convergence in function space, one can prove both that there will (often) be an infinite geometrical series of period-doubling bifurcations leading to chaos, and that

this series will share universal features (exponents α and δ and features) that are independent of the original dynamics.

(12.10) **RG and the central limit theorem: short.** (Mathematics) ④

If you are familiar with the renormalization group and Fourier transforms, this problem can be stated very quickly. If not, you are probably better off doing the long version (Exercise 12.11).

Write a renormalization-group transformation T taking the space of probability distributions into itself, that takes two random variables, adds them, and rescales the width by the square root of two [42]. Show that the Gaussian of width σ is a fixed point. Find the eigenfunctions f_n and eigenvectors λ_n of the linearization of T at the fixed point. (Hint: It is easier in Fourier space.) Describe physically what the relevant and marginal eigenfunctions represent. By subtracting the fixed point distribution from a binomial distribution, find the leading correction to scaling, as a function of x. Which eigenfunction does it represent? Why is the leading irrelevant eigenvalue not dominant here?

(12.11) **RG and the central limit theorem: long.** (Mathematics) ④

In this exercise, we will develop a renormalization group in *function space* to derive the central limit theorem [42]. We will be using maps (like our renormalization transformation T) that take a function ρ of x into another function of x; we will write $T[\rho]$ as the new function, and $T[\rho](x)$ as the function evaluated at x. We will also make use of the Fourier transform (eqn A.6)

$$\mathcal{F}[\rho](k) = \int_{-\infty}^{\infty} e^{-ikx} \rho(x)\, dx; \qquad (12.41)$$

$\mathcal{F}$ maps functions of x into functions of k. When convenient, we will also use the tilde notation: $\tilde{\rho} = \mathcal{F}[\rho]$, so for example (eqn A.7)

$$\rho(x) = \frac{1}{2\pi} \int_{-\infty}^{\infty} e^{ikx} \tilde{\rho}(k)\, dk. \qquad (12.42)$$

The central limit theorem states that the sum of many independent random variables tends to a Gaussian whatever the original distribution might have looked like (just as they do for random walks, Chapter 2). That is, the Gaussian

---

[36] For asymmetric maps, we would need to locate this other corner $f(f(x_c)) = x^*$ numerically. As it happens, breaking this symmetry is irrelevant at the fixed point.

distribution is the *fixed point function* for large random sums. When summing many random numbers, the details of the distributions of the individual random variables becomes unimportant; simple behavior emerges. We will study this using the renormalization group, giving an example where we can explicitly implement the coarse-graining transformation. Here our system space is the space of probability distributions $\rho(x)$. There are four steps in the procedure:

(1) **Coarse-grain.** Reduce the number of the degrees of freedom by some fraction. Here, we will add pairs of random variables (reducing the number by half); the probability distribution for sums of N independent random variables of distribution f is the same as the distribution for sums of $N/2$ random variables of distribution $f * f$, where $*$ denotes convolution.

(a) *Argue that if $\rho(x)$ is the probability that a random variable has value x, then the probability distribution of the sum of two random variables drawn from this distribution is the convolution*

$$C[\rho](x) = (\rho * \rho)(x) = \int_{-\infty}^{\infty} \rho(x-y)\rho(y) \, \mathrm{d}y. \tag{12.43}$$

Remember (eqn A.23) the Fourier transform of the convolution is the product of the Fourier transforms, so

$$\mathcal{F}[C[\rho]](k) = (\widetilde{\rho}(k))^2. \tag{12.44}$$

(2) **Rescale.** The behavior at larger lengths will typically be similar to that of smaller lengths, but some of the constants will shift (or *renormalize*). Here the mean and width of the distributions will increase as we coarse-grain. We confine our main attention to distributions of zero mean. Remember that the width (standard deviation) of the sum of two random variables drawn from ρ will be $\sqrt{2}$ times the width of one variable drawn from ρ, and that the overall height will have to shrink by $\sqrt{2}$ to stay normalized. We define a rescaling operator $S_{\sqrt{2}}$ which reverses this spreading of the probability distribution:

$$S_{\sqrt{2}}[\rho](x) = \sqrt{2}\rho(\sqrt{2}x). \tag{12.45}$$

(b) *Show that if ρ is normalized (integrates to one), so is $S_{\sqrt{2}}[\rho]$. Show that the Fourier transform is*

$$\mathcal{F}[S_{\sqrt{2}}[\rho]](k) = \widetilde{\rho}(k/\sqrt{2}). \tag{12.46}$$

Our renormalization-group transformation is the composition of these two operations,

$$T[\rho](x) = S_{\sqrt{2}}[C[\rho]](x) \tag{12.47}$$
$$= \sqrt{2} \int_{-\infty}^{\infty} \rho(\sqrt{2}x - y)\rho(y) \, \mathrm{d}y.$$

Adding two Gaussian random variables (convolving their distributions) and rescaling the width back should give the original Gaussian distribution; the Gaussian should be a *fixed point*.

(c) *Show that the Gaussian distribution*

$$\rho^*(x) = (1/\sqrt{2\pi}\sigma) \exp(-x^2/2\sigma^2) \tag{12.48}$$

is indeed a fixed point in function space under the operation T. You can do this either by direct integration, or by using the known properties of the Gaussian under convolution.

(d) *Use eqns 12.44 and 12.46 to show that*

$$\mathcal{F}[T[\rho]](k) = \widetilde{T}[\widetilde{\rho}](k) = \widetilde{\rho}(k/\sqrt{2})^2. \tag{12.49}$$

Calculate the Fourier transform of the fixed point $\widetilde{\rho}^*(k)$ (or see Exercise A.4). Using eqn 12.49, show that $\widetilde{\rho}^*(k)$ is a fixed point in Fourier space under our coarse-graining operator $\widetilde{T}$.[37]

These properties of T and ρ^* should allow you to do most of the rest of the exercise without any messy integrals.

The central limit theorem tells us that sums of random variables have probability distributions that approach Gaussians. In our renormalization-group framework, to prove this we might try to show that our Gaussian fixed point is *attracting*: that all nearby probability distributions flow under iterations of T to ρ^*.

(3) **Linearize about the fixed point.** Consider a function near the fixed point: $\rho(x) = \rho^*(x) + \epsilon f(x)$. In Fourier space, $\widetilde{\rho}(k) = \widetilde{\rho}^*(k) + \epsilon \widetilde{f}(k)$. We want to find the eigenvalues λ_n and eigenfunctions f_n of the derivative of the mapping T. That is, they must satisfy

$$T[\rho^* + \epsilon f_n] = \rho^* + \lambda_n \epsilon f_n + O(\epsilon^2),$$
$$\widetilde{T}[\widetilde{\rho}^* + \epsilon \widetilde{f}_n] = \widetilde{\rho}^* + \lambda_n \epsilon \widetilde{f}_n + O(\epsilon^2). \tag{12.50}$$

---

[37]To be explicit, the operator $\widetilde{T} = \mathcal{F} \circ T \circ \mathcal{F}^{-1}$ is a renormalization-group transformation that maps Fourier space into itself.

(e) *Show using eqns 12.49 and 12.50 that the transforms of the eigenfunctions satisfy*

$$\widetilde{f_n}(k) = (2/\lambda_n)\widetilde{\rho^*}(k/\sqrt{2})\,\widetilde{f_n}(k/\sqrt{2}). \quad (12.51)$$

(4) **Find the eigenvalues and calculate the universal critical exponents.**
(f) *Show that*

$$\widetilde{f_n}(k) = (ik)^n\,\widetilde{\rho^*}(k) \quad (12.52)$$

is the Fourier transform of an eigenfunction (i.e. that it satisfies eqn 12.51.) What is the eigenvalue λ_n?
Our fixed point actually does not attract all distributions near it. The directions with eigenvalues greater than one are called *relevant*; they are dangerous, corresponding to deviations from our fixed point that grow under coarse-graining. The directions with eigenvalues equal to one are called *marginal*; they do not get smaller (to linear order) and are thus also potentially dangerous. When you find relevant and marginal operators, you always need to understand each of them on physical grounds.
(g) *The eigenfunction $f_0(x)$ with the biggest eigenvalue corresponds to an unphysical perturbation; why?* (Hint: Probability distributions must be normalized to one.) *The next two eigenfunctions f_1 and f_2 have important physical interpretations. Show that $\rho^* + \epsilon f_1$ to is equivalent to a shift in the mean of ρ, and $\rho^* + \epsilon f_2$ is a shift in the standard deviation σ of ρ^*, to linear order in the shifts.*
In this case, the relevant perturbations do not take us to qualitatively new phases—just to other Gaussians with different means and variances. All other eigenfunctions should have eigenvalues λ_n less than one. This means that a perturbation in that direction will shrink under the renormalization-group transformation:

$$T^N(\rho^* + \epsilon f_n) - \rho^* \sim \lambda_n^N \epsilon f_n. \quad (12.53)$$

Corrections to scaling and coin flips. Does anything really new come from all this analysis? One nice thing that comes out is the *leading corrections to scaling.* The fixed point of the renormalization group explains the Gaussian shape of the distribution of N coin flips in the limit $N \to \infty$, but the linearization about the fixed point gives a systematic understanding of the corrections to the Gaussian distribution for large but not infinite N.

Usually, the largest eigenvalues are the ones which dominate. In our problem, consider adding a small perturbation to the fixed point f^* along the two leading irrelevant directions f_3 and f_4:

$$\rho(x) = \rho^*(x) + \epsilon_3 f_3(x) + \epsilon_4 f_4(x). \quad (12.54)$$

These two eigenfunctions can be inverse-transformed from their k-space form (eqn 12.52):

$$\begin{aligned} f_3(x) &\propto \rho^*(x)(3x/\sigma - x^3/\sigma^3), \\ f_4(x) &\propto \rho^*(x)(3 - 6x^2/\sigma^2 + x^4/\sigma^4). \end{aligned} \quad (12.55)$$

What happens to these perturbations under multiple applications of our renormalization-group transformation T? After ℓ applications (corresponding to adding together 2^ℓ of our random variables), the new distribution should be given by

$$T^\ell(\rho)(x) \sim \rho^*(x) + \lambda_3^\ell \epsilon_3 f_3(x) + \lambda_4^\ell \epsilon_4 f_4(x). \quad (12.56)$$

Since $1 > \lambda_3 > \lambda_4 \ldots$, the leading correction should be dominated by the perturbation with the largest eigenvalue.
(h) *Plot the difference between the binomial distribution giving the probability of m heads in N coin flips, and a Gaussian of the same mean and width, for $N = 10$ and $N = 20$.* (The Gaussian has mean of $N/2$ and standard deviation $\sqrt{N}/2$, as you can extrapolate from the case $N = 1$.) *Does it approach one of the eigenfunctions f_3 or f_4 (eqns 12.55)?*
(i) *Why did a perturbation along $f_3(x)$ not dominate the asymptotics? What symmetry forced $\epsilon_3 = 0$? Should flips of a biased coin break this symmetry?*
Using the renormalization group to demonstrate the central limit theorem might not be the most efficient route to the theorem, but it provides quantitative insights into how and why the probability distributions approach the asymptotic Gaussian form.

(12.12) **Percolation and universality.**[38] (Complexity, Computation) ④

Figure 12.2 vividly illustrates how a sheet of paper develops big gaping holes (lower left), becomes fractal (top), and then falls into pieces (lower right) as sites are randomly deleted with probability $(1-p)$. In this exercise, we first explore the the pieces that fell out by $p = p_c$ (the cluster size distribution), and then the fraction of paper left as a function of p for $p > p_c$ (the size of the infinite cluster). The former naturally gives a pretty good power law, but we need to do *finite-size scaling* to convince ourselves that the latter will give the expected power law for an infinite system. See Exercise 2.13 to simulate percolation and measure these quantities.

Cluster size distribution: power laws at $\mathbf{p_c}$. A system at its percolation threshold p_c is self-similar. When looked at on a longer length scale (say, with a ruler with notches spaced $1+\epsilon$ farther apart, for infinitesimal ϵ), the statistical behavior of the large percolation clusters should be unchanged, if we simultaneously rescale various measured properties according to certain rules. Let x be the length and S be the size (number of nodes) in a percolation cluster, and let $n(S)$ be the probability that a given cluster will be of size S at p_c.[39] The cluster measured with the new ruler will have a length $x' = x/(1-\epsilon)$, a size $S' = S/(1+c\epsilon)$, and will occur with probability $n' = (1+a\epsilon)n$.

(a) *In precise analogy to our analysis of the avalanche size distribution (eqns 12.5–12.6), show that the probability is a power law, $n(S) \propto S^{-\tau}$. What is τ, in terms of a and c?*

In two dimensions, there are exact results known for many properties of percolation. In particular, it is known that[40] $\tau = 187/91$. You can test this numerically, either with the code you developed for Exercise 2.13, or by using the software provided.

(b) *Calculate the cluster size distribution $n(S)$, both for bond percolation on the square lattice*

and for site percolation on the triangular lattice, for a large system size (perhaps $L \times L$ with $L = 400$) at $p = p_c$.[41] Plot $\log(n(S))$ versus $\log(S)$ for both bond and site percolation, together with the power law $n(S) \propto S^{-187/91}$ predicted by the exact result. You will rapidly find sizes S that have no avalanches; we need to bin the avalanches into larger groups, especially for larger sizes where the data is sparse. Do the plots again, now with all the data included, using bins that start at size ranges $1 \le S < 2$ and grow by a factor of 1.2 for each bin. You should see clear evidence that the distribution of clusters does look like a power law (a straight line on your log–log plot), and fairly convincing evidence that the power law is converging to the exact result at large S and large system sizes.

The size of the infinite cluster: power laws near p_c. Much of the physics of percolation above p_c revolves around the connected piece left after the small clusters fall out, often called the *percolation cluster*. For $p > p_c$ this largest cluster occupies a fraction of the whole system, often called $P(p)$.[42] The fraction of nodes in this largest cluster for $p > p_c$ is closely analogous to the $T < T_c$ magnetization $M(T)$ in magnets (Fig. 12.6(b)) and the density difference $\rho_l(T) - \rho_g(T)$ near the liquid–gas critical point (Fig. 12.6(a)). In particular, the value $P(p)$ goes to zero continuously as $p \to p_c$.

Systems that are not at p_c are not self-similar. However, there is a scaling relation between systems at differing values of $p-p_c$ (see Fig. 12.12): a system coarsened by a small factor $1+\epsilon$ will be similar to one farther from p_c by a factor $1+\epsilon/\nu$, except that the percolation cluster fraction P must be rescaled upward by $1 + \beta\epsilon/\nu$.[43] This last rescaling reflects the fact that the percolation cluster becomes more dense as you coarsegrain, filling in or blurring away the smaller holes. You may check, just as for the magnetization (eqn 12.7), that

$$P(p) \sim (p_c - p)^\beta. \qquad (12.57)$$

---

[38] This exercise and the associated software were developed in collaboration with Christopher Myers. Computational hints can be found at the book website [181].

[39] Hence the probability that a given node is in a cluster of size S is proportional to $Sn(S)$.

[40] A non obvious result!

[41] Conveniently, the critical probability $p_c = \frac{1}{2}$ for both these systems; see Exercise 2.13, part(c). This enormously simplifies the scaling analysis, since we do not need to estimate p_c as well as the critical exponents.

[42] For $p < p_c$, there will still be a largest cluster, but it will not grow much bigger as the system size grows and the fraction $P(p) \to 0$ for $p < p_c$ as the system length $L \to \infty$.

[43] These particular combinations of Greek letters are chosen to give the conventional names for the critical exponents.

For percolation in two dimensions, $\beta = 5/36$ and $\nu = 4/3$.

(c) *Calculate the fraction of nodes $P(p)$ in the largest cluster, for both bond and site percolation, at a series of points $p = p_c + 2^{-n}$ for as large a percolation lattice as is convenient, and a good range of n. (Once you get your method debugged, $n = 10$ on an $L \times L$ lattice with $L = 200$ should be not be a challenge.) Do a log–log plot of $P(p)$ versus $p - p_c$, and compare along with the theory prediction, eqn 12.57 with $\beta = 5/36$.*

You should find that the numerics in part (c) are not compelling, even for rather large system sizes. The two curves look a bit like power laws, but the slopes β_{eff} on the log–log plot do not agree with one another or with the theory. Worse, as you get close to p_c the curves, although noisy, definitely are not going to zero. This is natural; there will always be a largest cluster, and it is only as the system size $L \to \infty$ that the largest cluster can vanish as a fraction of the system size.

Finite-size scaling (advanced). We can extract better values for β from small simulations by explicitly including the length L into our analysis. Let $P(p, L)$ be the mean fraction of nodes[44] in the largest cluster for a system of size L.

(d) *On a single graph, plot $P(p, L)$ versus p for bond percolation $L = 5$, 10, 20, 50, and 100, focusing on the region around $p = p_c$ where they differ from one another. (At $L = 10$ you will want p to range from 0.25 to 0.75; for $L = 50$ the range should be from 0.45 to 0.55 or so.) Five or ten points will be fine. You will discover that the sample-to-sample variations are large (another finite-size effect), so average each curve over perhaps ten or twenty realizations.*

Each curve $P(p, L)$ is rounded near p_c, as the characteristic cluster lengths reach the system box length L. Thus this rounding is itself a symptom of the universal long-distance behavior, and we can study the dependence of the rounding on L to extract better values of the critical exponent β. We will do this using a *scaling collapse*, rescaling the horizontal and vertical axes so as to make all the curves fall onto a single scaling function.

First, we must derive the scaling function for $P(p, L)$. We know that

$$L' = L/(1+\epsilon),$$
$$(p_c - p)' = (1 + \epsilon/\nu)(p_c - p),$$ (12.58)

since the system box length L rescales like any other length. It is convenient to change variables from p to $X = (p_c - p)L^{1/\nu}$; let $P(p, L) = \bar{P}(L, (p_c - p)L^{1/\nu})$.

(e) *Show that X is unchanged under coarse-graining (eqn 12.58). (You can either show $X' = X$ up to terms of order ϵ^2, or you can show $\mathrm{d}X/\mathrm{d}\epsilon = 0$.)*

The combination $X = (p_c - p)L^{1/\nu}$ is another *scaling variable*. The combination $\xi = |p - p_c|^{-\nu}$ is the way in which lengths diverge at the critical point, and is called the *correlation length*. Two systems of different lengths and different values of p should be similar if the lengths are the same when measured in units of ξ. L in units of ξ is $L/\xi = X^\nu$, so different systems with the same value of the scaling variable X are statistically similar. We can turn this verbal assertion into a mathematical scaling form by studying how $\bar{P}(L, X)$ coarse-grains.

(f) *Using eqns 12.58 and the fact that P rescales upward by $(1 + \beta\epsilon/\nu)$ under coarse-graining, write the similarity relationship for $\bar{P}$. Following our derivation of the scaling form for the avalanche size distribution (through eqn 12.11), show that $\bar{P}(L, X) = L^{-\beta/\nu}\mathcal{P}(X)$ for some function $\mathcal{P}(X)$, and hence*

$$P(p, L) \propto L^{-\beta/\nu}\mathcal{P}((p - p_c)L^{1/\nu}).   (12.59)$$

Presuming that $\mathcal{P}(X)$ goes to a finite value as $X \to 0$, derive the power law giving the percolation cluster size $L^2 P(p_c, L)$ as a function of L. Derive the power-law variation of $\mathcal{P}(X)$ as $X \to \infty$ using the fact that $P(p, \infty) \propto (p - p_c)^\beta$.

Now, we can use eqn 12.59 to deduce how to rescale our data. We can find the finite-sized scaling function $\mathcal{P}$ by plotting $L^{\beta/\nu}P(p, L)$ versus $X = (p - p_c)L^{1/\nu}$, again with $\nu = 4/3$ and $\beta = 5/36$.

(g) *Plot $L^{\beta/\nu}P(p, L)$ versus X for $X \in [-0.8, +0.8]$, plotting perhaps five points for each curve, for both site percolation and bond percolation. Use system sizes $L = 5$, 10, 20,*

---

[44]You can take a microcanonical-style ensemble over all systems with exactly L^2p sites or $2L^2p$ bonds, but it is simpler just to do an ensemble average over random number seeds.

and 50. *Average over many clusters for the smaller sizes (perhaps 400 for $L = 5$), and over at least ten even for the largest.*

Your curves should collapse onto two scaling curves, one for bond percolation and one for site percolation.[45] Notice here that the finite-sized scaling curves collapse well for small L, while we would need to go to much larger L to see good power laws in $P(p)$ directly (part (c)). Notice also that both site percolation and bond percolation collapse for the same value of β, even though the rough power laws from part (c) seemed to differ. In an experiment (or a theory for which exact results were not available), one can use these scaling collapses to estimate p_c, β, and ν.

(12.13) **Hysteresis and avalanches: scaling.**[46] (Complexity, Computation) ③
In this chapter we have studied the behavior of the nonequilibrium random-field Ising model extensively as a system with avalanches, power laws, and universal scaling functions (see also Exercises 12.14 and 12.20 for more introductions to crackling noise and scaling, and Exercise 12.28 to calculate the size distribution analytically in a simplified model). In Exercises 8.13 and 8.14, we explored how to simulate this model on the computer. Here we use the simulation to explore the scaling behavior. Run the simulation in two dimensions on a lattice at least $1,000 \times 1,000$ with disorder $R = 0.9$, or a three-dimensional simulation on a lattice at least 100^3 at $R = 2.16$. The simulation is a simplified model of magnetic hysteresis, described in [180]; see also [179]. The spins s_i begin all pointing down, and flip upward as the external field H grows from minus infinity, depending on the spins of their neighbors and a local random field h_i. The flipped spins are colored as they flip, with spins in the same *avalanche* sharing the same color. An avalanche is a collection of spins which flip together, all triggered from the same original spin. The *disorder* is the ratio R of the RMS width $\sqrt{\langle h_i^2 \rangle}$ to the ferromagnetic coupling J between spins:

$$R = \sqrt{\langle h^2 \rangle}/J. \qquad (12.60)$$

Examine the $M(H)$ curve for our model and the dM/dH curve. The individual avalanches should be visible on the first graph as jumps, and on the second graph as spikes. This kind of time series (a set of spikes or pulses with a broad range of sizes) we hear as *crackling noise*. To hear the crackling noises from our model, as well as crackling noise we have assembled from crumpling paper, fires, Rice Krispies™, and the Earth, check out either our educational website [102] or Exercise 12.14.

Examine the avalanche size distribution. The (unlabeled) vertical axis on the log–log plot gives the number of avalanches $D(S, R)$; the horizontal axis gives the size S (with $S = 1$ on the left-hand side). Equivalently, $D(S, R)$ is the probability distribution that a given avalanche during the simulation will have size S. The graph is created as a histogram, and the curve changes color beginning with the first bin with zero entries (after which the data becomes much less useful, and should be ignored).

If available, examine the spin–spin correlation function $C(x, R)$. It shows a log–log plot of the probability (vertical axis) that an avalanche initiated at a point $\mathbf{x}_0$ will extend to include a spin $\mathbf{x}_1$ a distance $x = \sqrt{(\mathbf{x}_1 - \mathbf{x}_0)^2}$ away.

Two dimensions is fun to watch, but the scaling behavior is subtle and has only recently been understood [81]. In three dimensions we have good evidence for scaling and criticality at a phase transition in the dynamical evolution. There is a phase transition in the dynamics at $R_c \sim 2.16$ on the three-dimensional cubic lattice. Well below R_c one large avalanche flips most of the spins. Well above R_c all avalanches are fairly small; at very high disorder each spin flips individually. The critical disorder is the point, as $L \to \infty$, where one first finds *spanning avalanches*, which extend from one side of the simulation to the other.

Simulate a 3D system at $R = R_c = 2.16$ with $L = 100$ (one million spins, or larger, if you have a fast machine). It will be fastest if you use the *sorted list* algorithm (Exercise 8.14). The display will show an $L \times L$ cross-section of the 3D avalanches. Notice that there are many tiny avalanches, and a few large ones. Below R_c

---

[45] These two curves should also have collapsed onto one another, given a suitable rescaling of the horizontal and vertical axes, had we done the triangular lattice in a square box instead of a rectangular box (which we got from shearing an $L \times L$ lattice). The finite-size scaling function will in general depend on the boundary condition, and in particular on the shape of the box.
[46] Computational hints and software can be found at the book website [181].

you will find one large colored region forming the background for the others; this is the spanning, or *infinite* avalanche. Look at the $M(H)$ curve (the bottom half of the hysteresis loop). It has many small vertical jumps (avalanches), and one large one (corresponding to the spanning avalanche).

(a) *What fraction of the system is flipped by the one largest avalanche, in your simulation? Compare this with the hysteresis curve at $R = 2.4 > R_c$. Does it have a similar big jump, or is it continuous?*

Below R_c we get a big jump; above R_c all avalanches are small compared to the system size. If the system size were large enough, we believe the fraction of spins flipped by the spanning avalanche at R_c would go to zero. The largest avalanche would nonetheless span the system—just like the percolation cluster at p_c spans the system but occupies zero volume in the limit of large systems (Fig. 12.2 and Exercise 12.12).

The other avalanches form a nice power-law size distribution; let us measure it carefully. Do a set of 10 runs at $L = 100$ and $R = R_c = 2.16$. Watch the avalanches. Notice that sometimes the second-largest avalanche in the view (the largest being the background color) is sometimes pretty small; this is often because the cross-section we view missed it. Look at the avalanche size distribution. Notice that at R_c you find a pretty good power-law distribution (a straight line on the log–log plot). We denote this critical exponent $\bar{\tau} = \tau + \sigma\beta\delta$:

$$D(S, R_c) \sim S^{-\bar{\tau}} = S^{-(\tau+\sigma\beta\delta)}. \quad (12.61)$$

(b) *From your plot, measure this exponent combination from your simulation. It should be close to two. Is your estimate larger or smaller than two?*

This power-law distribution is to magnets roughly as the Gutenberg–Richter law (Fig. 12.3(b)) is to earthquakes (but see Exercise 12.17). The power law stems naturally from the self-similarity.

We want to explore how the avalanche size distribution changes as we move above R_c. We will do a series of three or four runs at different values of R, and then graph the avalanche size distributions after various transformations.

Do a run at $R = 6$ and $R = 4$ with $L = 100$, and make sure your data files are properly output. Do runs at $R = 3$, $R = 2.5$, and $R = 2.16$ at $L = 200$.

(c) *Copy and edit your avalanche size distribution files, removing the data after the first bin with zero avalanches in it. Start up a graphics program, and plot the curves on a log–log plot; they should look like power laws for small S, and cut off exponentially at larger S. Enclose a copy of your plot.*

We expect the avalanche size distribution to have the scaling form

$$D(S, R) = S^{-(\tau+\sigma\beta\delta)}\mathcal{D}(S(R-R_c)^{1/\sigma}) \quad (12.62)$$

sufficiently close to R_c. This reflects the similarity of the system to itself at a different set of parameters; a system at $2(R - R_c)$ has the same distribution as a system at $R - R_c$ except for an overall change A in probability and B in the size scale of the avalanches, so $D(S, R - R_c) \approx AD(BS, 2(R - R_c))$.

(d) *What are A and B in this equation for the scaling form given by eqn 12.62?*

At $R = 4$ and 6 we should expect substantial corrections! Let us see how well the collapse works anyhow.

(e) *Multiply the vertical axis of each curve by $S^{\tau+\sigma\beta\delta}$. This then should give four curves $\mathcal{D}(S(R-R_c)^{1/\sigma})$ which are (on a log–log plot) roughly the same shape, just shifted sideways horizontally (rescaled in S by the typical largest avalanche size, proportional to $1/(R-R_c)^{1/\sigma}$). Measure the peak of each curve. Make a table with columns R, S_{peak}, and $R - R_c$ (with $R_c \sim 2.16$). Do a log–log plot of $R - R_c$ versus S_{peak}, and estimate σ in the expected power law $S_{peak} \sim (R-R_c)^{-1/\sigma}$.*

(f) *Do a scaling collapse: plot $S^{\tau+\sigma\beta\delta}D(S, R)$ versus $(R-R_c)^{1/\sigma}S$ for the avalanche size distributions with $R > R_c$. How well do they collapse onto a single curve?*

The collapses become compelling only near R_c, where you need very large systems to get good curves.

(12.14) **Crackling noises.**[47] @

Listen to the sound file for the earthquakes in 1995 (shown graphically in Fig. 12.3). Each click represents an earthquake, with the sound energy proportional to the energy release and

---

[47]The computer exercises link on the book website [181] provides links to audio files of various types of crackling noise for this exercise.

with the year compressed into a couple of seconds.

(a) *Roughly how many clicks can you hear? Consider the histogram in Fig. 12.3(b), down to what magnitude do you estimate you can perceive?*

Now listen to paper crumpling, Rice Krispies[TM], fire, and to our model of magnetic noise. All of these sound files share common features—they are composed of brief pulses or avalanches with a broad range of sizes. Just as for earthquakes, the largest avalanches are rare, and the smaller avalanches are more common, with avalanches of size S happening with probability proportional to $S^{-\tau}$.

These sound files differ, however, in their values for the exponent τ. For earthquakes, τ can vary from one fault type to another; our data set (Fig. 12.3(b)) shows τ between 1.5 and 1.8 (see Exercise 12.17). Milk invading bubbles in puffed rice is likely a special case of the well studied problem of fluid invasion into porous media (also known as *imbibition*), where in three dimensions estimates of τ range between around 1.3 and 1.5. Paper crumpling and fire are not well understood, but we have an excellent understanding of our model of magnetic noise,[48] where $\tau \sim 2$.

Changing τ should change the sound of the crackling noise—changing the predominance of the loudest crackles over the rest.

(b) *Should larger τ correspond to fewer very large avalanches, or more? Does that correspond roughly to your perception of the difference between the sounds from earthquakes and magnets? What about Rice Krispies[TM]?*

(12.15) **Hearing chaos.**[49] (Dynamical systems) @
This exercise listens to a system as it transitions from regular to chaotic behavior. Chaos scrambles our knowledge of the initial state of a system (Exercise 5.9), allowing entropy to increase and mediating equilibration. Also, the onset of chaos (Exercises 12.16 and 12.9) is studied using the same renormalization-group methods we use to study continuous phase transitions (Chapter 12).

The English word chaos makes us think of billiard balls, bumble bees, bumper cars, or turbulence—the realm of statistical mechanics. The mathematical term chaos in dynamical systems often reflects a more subtle irregularity of motion. Think of an unbalanced dryer, or a faucet that drips at irregular intervals.

One of the characteristics of chaotic motion is a continuous spectrum of frequencies. (Or, perhaps more precisely, systems that are non chaotic but non stationary usually have a spectrum that consists of sums and differences of a few frequencies.) Our ears are spectacularly sensitive to frequencies. Listen to the audio files from the book website [181]. Figure 12.17 shows the states visited at long times in the logistic map $f_\mu(x)$. The audio moves the loudspeaker diaphragm back and forth in time as $x(t + \Delta) = f_\mu(x(t))$; thus starting at μ_1 one hears a tone with frequency $\omega = 2\pi/\Delta$ that grows louder as one approaches μ_2. The audio continues to raise μ, slowly crossing through a sequence of period-doubling bifurcations and then into the chaotic region.

(a) *If the period of a sound wave doubles, how much does the pitch change? (One note? A perfect fourth? A perfect fifth? An octave?)*

Listen to the two audio files. (One starts closer to the onset of chaos.)

(b) *Does the pitch start sounding like a pure tone? Can you hear the first period doubling? Does the new pitch agree with your answer to part (a)?*

Shortly after the first period doubling, you should begin to hear a chaotic signal.

(c) *Describe the new sound. Is it musical, or noisy?*

Figure 12.17 shows the states visited at long times by the logistic map, as the parameter μ is varied along the horizontal axis. It does not show the dynamics as the system hops between one point and another along the curves, but you can reconstruct that. For example, between μ_1 and μ_2 the system hops between the upper and lower branches of the curve, giving a tone with period equal to one iteration of the map. Between μ_2 and μ_3 you heard a second tone one

---

[48]Barkhausen noise in magnets is usually studied not using sound emission, but by measuring jumps in the magnetization as the external field is increased. Also, real magnets are not described by our model, but are in the same universality class as fluid invasion.

[49] [181] provides links to Erich Mueller's audio files for this exercise. There are many exercises exploring this chaotic logistic map (see Index).

octave lower. You probably did not hear the next period doubling, but after μ_∞ you began to hear noise.

(d) *Note the "windows" of non chaotic motion deep inside the chaotic region. In the audio file, do you hear periods of tonal sounds interspersed in the noise?*

(12.16) **Period doubling and the onset of chaos.**[50] (Dynamical systems) ⓟ

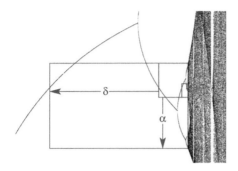

Fig. 12.25 Scaling in the period doubling bifurcation diagram. Shown are the points x on the attractor (vertical) as a function of the control parameter μ (horizontal), for the logistic map $f(x) = 4\mu x(1-x)$, near the transition to chaos.

Most of you will be familiar with the *period doubling route to chaos*, and the bifurcation diagram (Fig. 12.25). (See also Section 12.3.3). The self-similarity here is not in space, but in time. It is discrete instead of continuous; the behavior is similar if one rescales time by a factor of two, but not by a factor $1 + \epsilon$. Hence instead of power laws we find a discrete self-similarity as we approach the critical point μ_∞.

(a) *From Fig. 12.25, roughly estimate the values of the Feigenbaum numbers δ (governing the rescaling of $\mu - \mu_\infty$) and α (governing the rescaling of $x - x_p$, where $x_p = 1/2$ is the peak of the logistic map).* (Hint: be sure to check the signs.)

Remember that the relaxation time for the Ising model became long near the critical temperature; it diverges as $t^{-\zeta}$ where t measures the distance to the critical temperature. Remember that the correlation length diverges as $t^{-\nu}$. Can we define ζ and ν for period doubling?

(b) *If each rescaling shown doubles the period T of the map, and T grows as $T \sim (\mu_\infty - \mu)^{-\zeta}$ near the onset of chaos, write ζ in terms of α and δ. If ξ is the smallest typical length scale of the attractor, and we define $\xi \sim (\mu_\infty - \mu)^{-\nu}$ (as is traditional at thermodynamic phase transitions), what is ν in terms of α and δ?* (Hint: be sure to check the signs.)

(12.17) **The Gutenberg–Richter law.** (Engineering, Geophysics) ⓟ

Power laws often arise at continuous transitions in nonequilibrium extended systems, particularly when disorder is important. We don't yet have a complete understanding of earthquakes, but they seem clearly related to the transition between pinned and sliding faults as the tectonic plates slide over one another.

The size S of an earthquake (the energy radiated, shown in the upper axis of Figure 12.3(b)) is traditionally measured in terms of a *magnitude* $M \propto \log S$ (lower axis). The Gutenberg–Richter law tells us that the number of earthquakes of magnitude $M \propto \log S$ goes down as their size S increases. Figure 12.3(b) shows that the number of avalanches of magnitude between M and $M + 1$ is proportional to S^{-B} with $B \approx 2/3$. However, it is traditional in the physics community to consider the probability density $P(S)$ of having an avalanche of size S. If $P(S) \sim S^{-\tau}$, *give a formula for τ in terms of the Gutenberg–Richter exponent B.* (Hint: The bins in the histogram have different ranges of size S. Use $P(M)\,dM = P(S)\,dS$.)

(12.18) **Random walks and universal exponents.** ⓐ

Figure 2.2 shows that a random walk appears statistically self-similar. Random walks naturally have self-similarity and power laws: they are said to exhibit *generic scale invariance*.

(a) *Argue that the fractal dimension of the random walk is two, independent of dimension (see note 6, p. 25.)* (Hint: How does the mass of the ink T needed to draw the walk scale with the distance d between endpoints?)

DNA has a *persistence length* of 50 nm; one can model it roughly as a random walk with a step size of the persistence length. The diameter of a DNA molecule is 2 nm, and will not overlap with itself during the random walk. Will the thickness change the power law $d \sim T^\nu$ relating distance d versus length T, and hence the fractal dimension $1/\nu$? More generally, are *self-avoiding* random walks a different universality class? (See Exercise 2.10.)

(b) *Show that $d_f = \nu = 1$ for a one-dimensional self-avoiding random walk.* (Hint: There are only two allowed walks.)

We can think about self-avoidance in general dimensions by asking whether a typical random walk will intersect itself. In the spirit of the renormalization group, let us divide a long polymer of N persistence lengths into M segments, with $1 \ll M \ll N$, each of which forms a fuzzy blob of dimension d_f. Suppose two of the blobs from distant parts of the polymer overlap. Each may have lots of empty space between the polymer segments, so they may or may not intersect even if they overlap. If they typically intersect, intersections would happen on all scales, and the critical exponents would likely change. If overlapping blobs rarely intersect, one would reasonably expect that the polymer would act like an ordinary random walk on long length scales.

(c) *What is the likely dimension of the intersection of two D-dimensional smooth surfaces in the same vicinity of $\mathbb{R}^d$?* (Hint: We can define a smooth D dimensional set in d dimensions as the solution of $d - D$ constraint equations. Check your answer for two curves on the plane, which typically intersect in isolated points. Check it for two curves in three dimensions, which typically will miss one another.) *Argue by analogy that* self-avoiding *random walks in dimensions two and three will likely have a different fractal dimension than ordinary random walks (see note 10, p. 27 and Exercise 2.10), but in dimensions above four they could be in the same universality class.*

Indeed, it turns out that self-avoiding random walks have $\nu_{2D} = \frac{3}{4}$ and $d_f = \frac{4}{3}$ in two dimensions, $\nu_{3D} \approx 0.588$ and $d_f = 1/\nu_{3D}$ in three dimensions, and logarithmic corrections in four dimensions; above four dimensions (the *upper critical dimension*) they obey the mean-field behavior of ordinary random walks at long length scales.

This pattern (varying critical exponents shifting to mean-field above an upper critical dimension) is common to most statistical mechanical models. Also, we should note that the self-avoiding random walk can be viewed as the limit of an n-component spin model as $n \to 0$; the Ising model ($n = 1$) also obeys mean-field theory above four dimensions.

(12.19) **Diffusion equation and universal scaling functions.**[51] ②
The diffusion equation universally describes microscopic hopping systems at long length scales. Here we investigate how to write the evolution in a universal scaling form.

The solution to a diffusion problem with a nonzero drift velocity is given by $\rho(x,t) = 1/\sqrt{4\pi Dt} \exp\left(-(x - vt)^2/(4Dt)\right)$. We will coarse-grain by throwing away half the time points. We will then rescale the distribution so it looks like the original distribution. We can just write these two operations as $t' = t/2$, $x' = x/\sqrt{2}$, $\rho' = \sqrt{2}\rho$.[52] These three together constitute our renormalization-group operation.

(a) *Write an expression for $\rho'(x',t')$ in terms of D, v, x', and t' (not in terms of D' and v'). Use it to determine the new renormalized velocity v' and diffusion constant D'. Are v and D relevant, irrelevant, or marginal variables?*

Typically, whenever writing properties in a scaling function, there is some freedom in deciding which invariant combinations to use. Here let us use the invariant combination of variables, $\mathcal{X} = x/\sqrt{t}$ and $\mathcal{V} = \sqrt{t}v$. We can then write

$$\rho(x,t) = t^{-\alpha}\mathcal{P}(\mathcal{X}, \mathcal{V}, D), \qquad (12.63)$$

a power law times a universal scaling function of invariant combination of variables.

(b) *Show that $\mathcal{X}$ and $\mathcal{V}$ are invariant under our renormalization-group operation. What is α? Write an expression for $\mathcal{P}$, in terms of $\mathcal{X}$, $\mathcal{V}$, and D (and not x, v, or t).*

(Note that we need to solve the diffusion equation to find the universal scaling function $\mathcal{P}$, but we can learn a lot from just knowing that it is a fixed point of the renormalization group. So, the universal exponent α and the invariant scaling combinations $\mathcal{X}$, $\mathcal{V}$, and D are determined just by the coarsening and rescaling

---

[51]This exercise was developed in collaboration with Archishman Raju.
[52]Because ρ is a density, we need to rescale $\rho'\mathrm{d}x' = \rho\mathrm{d}x$.

steps in the renormalization group. In experiments and simulations, one often uses data to extract the universal critical exponents and universal scaling functions, relying on emergent scale invariance to tell us that a scaling form like eqn 12.63 is expected.)

(12.20) **Hysteresis and Barkhausen noise.** (Complexity) @

Hysteresis is associated with abrupt phase transitions. Supercooling and superheating are examples (as temperature crosses T_c). Magnetic recording, the classic place where hysteresis is studied, is also governed by an abrupt phase transition—here the hysteresis in the magnetization, as the external field H is increased (to magnetize the system) and then decreased again to zero. Magnetic hysteresis is characterized by crackling (Barkhausen) electromagnetic noise. This noise is due to avalanches of spins flipping as the magnetic interfaces jerkily are pushed past defects by the external field (much like earthquake faults jerkily responding to the stresses from the tectonic plates). It is interesting that when dirt is added to this abrupt magnetic transition, it exhibits the power-law scaling characteristic of continuous transitions. Our model of magnetic hysteresis (unlike the experiments) has avalanches and scaling only at a special critical value of the disorder $R_c \sim 2.16$ (Fig. 12.14). The integrated probability distribution $D(S, R)$ has a power law $D(S, R_c) \propto S^{-\bar{\tau}}$ at the critical point (where $\bar{\tau} = \tau + \sigma\beta\delta$ for our model) but away from the critical point takes the *scaling form*

$$D(S, R) \propto S^{-\bar{\tau}} \mathcal{D}(S^{\sigma} (R - R_c)). \quad (12.64)$$

Note from eqn 12.64 that at the critical disorder $R = R_c$ the distribution of avalanche sizes is a power law $D(S, R_c) = S^{-\bar{\tau}}$. The scaling form controls how this power law is altered as R moves away from the critical point. From Fig. 12.14 we see that the main effect of moving above R_c is to cut off the largest avalanches at a typical largest size $S_{\max}(R)$, and another important effect is to form a bulge of extra avalanches just below the cutoff.

Using the scaling form from eqn 12.64, with what exponent does $S_{\max}$ diverge as $r = (R_c - R) \to 0$? (Hint: At what size S is $D(S, R)$, say, one millionth of $S^{-\bar{\tau}}$?) Given $\bar{\tau} \approx 2.03$, how does the mean $\langle S \rangle$ and the mean-square $\langle S^2 \rangle$ avalanche size scale with $r = (R_c - R)$? (Hint:

Your integral for the moments should have a lower cutoff S_0, the smallest possible avalanche, but no upper cutoff, since that is provided by the scaling function $\mathcal{D}$. Assume $\mathcal{D}(0) > 0$. Change variables to $Y = S^{\sigma} r$. Which moments diverge?)

(12.21) **Earthquakes and wires.** ⓟ

A wire will spring back when it is bent slightly, but will undergo irreversible plastic deformation when bent too much. An inexpensive fork will bend irreversibly if you use it to cut off a tough piece of meat. This irreversible deformation involves the motion of topological dislocations (Fig. 9.11) through the crystals in the metal. These dislocations shift in avalanches, with one destabilizing the next, leading to a rearrangement of large numbers of dislocations and metal atoms.

Why do we not notice these avalanches when we bend wires? Let us make an analogy to earthquakes—irreversible *plastic deformation* of the Earth's crust, due to the stress from the motion of the tectonic plates.

First, is it true that most of the Earth's deformation happens in small events? Or do the biggest dominate? In 2011 the Tōhokua earthquake (magnitude 9.0) led to a disaster at the Fukushima Daiichi nuclear power plant in Japan. This earthquake is estimated to have a rupture length of 500 km, width of 200 km, and a slip distance of roughly 2.4 m; it released about 1.9×10^{17} Joules.

(a) *There are many more small earthquakes than large ones. Do you expect that the energy $S_{Tohokua}$ released by the Tōhokua earthquake was more or less than that released by the sum of all the smaller avalanches since the last big earthquake in this region (the Sanriku earthquake of 869, magnitude 8.4, smaller by a factor $S_{Tohokua}/S_{Sanriku} \approx 5$)? How would one use the critical exponent τ to answer this question?* (Presume that the energy release is measured by the size S with $P(S) \propto S^{-\tau}$. Most avalanche models have $1 < \tau < 2$; the earthquake value for τ (Exercise 12.17) is in that range.)

Crackling noise is not self-averaging, unlike most of the statistical mechanics systems we study. The largest events stand out; fluctuations dominate the behavior.

If the biggest dislocation avalanches in wires dominate the deformation, why would we not notice the jerky motion?

(b) *Would an earthquake that spanned the circumference of the Earth be noticeable from outer space?* Presume the width (limited by the slip angle and the depth of the Earth's crust) and the slip distance (perhaps limited by the width?) remain the same as for the Tōhokua earthquake.

It has been argued [48, 175] that plasticity avalanches are macroscopically unobservable in most materials because they have fractal dimension d_f equal to two, and so the biggest avalanche in a crystal of length L has $S \lesssim L^{d_f} = L^2$. The "thin direction" of the avalanche is set by the atomic scale, so one can view the biggest avalanche as roughly shifting one atomic layer (of area L^2) by one atom. This has been measured in nanopillars. Dislocation avalanches have been observed in human-scale ice crystals [209] and zinc; perhaps the fractal dimension is different for these hexagonal crystals.

Crackling noise also arises in the fracture of disordered brittle materials like bone and seashells, where microcracks of a large range of sizes happen [184]. There, the microcracking is cut off when a crack grows out of control, breaking the bone in two.

(12.22) **Activated rates and the saddle-node transition.**[53] (Computation) ③

The renormalization group and scaling are used to describe continuous transitions between different states. Can we apply it to organize our understanding of chemical reaction rates (Exercises 6.11 and 8.22)? How is the rate of transition over the barrier between a bound and unbound molecule (Fig. 12.26 left), related to the behavior where the barrier E disappears (right)?

Mathematicians call the point where E vanishes the *saddle-node* transition (Exercise 12.4). At finite temperature with overdamped dynamics, the average time to escape the barrier, for large barriers compared to the temperature, is given

by

$$\tau \approx 2\pi\eta(K\widetilde{K})^{-1/2}\exp(E/k_B T). \qquad (12.65)$$

Here the exponential $\tau \approx A\exp(E/k_B T)$ was proposed by Arrhenius in 1889, and follows generally from the probability density for being at the top of the barrier relative to the bottom of the bound state. The prefactor[54] $A = 2\pi\eta(K\widetilde{K})^{-1/2}$ was derived by Kramers, in terms of the damping η and the harmonic expansion coefficients at the bottom of the well $V(x) \approx V_0 + \frac{1}{2}K(x - x_{\min})^2$ and the top of the barrier $V(x) \approx V_0 + E - \frac{1}{2}\widetilde{K}(x - x_{\max})^2$.

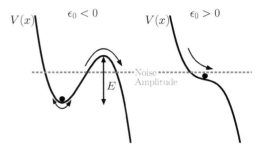

Fig. 12.26 Noisy saddle-node transition. The saddle-node transition describes overdamped motion in a potential depending on a parameter ϵ_0. For $\epsilon_0 < 0$ the potential has a minimum (node) and a local maximum (saddle). These two merge at $\epsilon_0 = 0$, and for $\epsilon_0 > 0$ the potential has everywhere a downhill slope. We get an important model of chemical reaction rates if we add noise.

In this problem we will examine a scaling analysis of this *noisy saddle-node transition*, describing the behavior near the transition in terms of power laws and universal scaling functions.[55] We shall examine both the mean barrier crossing time τ and also the slowest decaying mode $\rho(x, t) \sim \rho_0(x)\exp(-\lambda_0 t)$ for the particles at long times.

To compute the time τ, one can study the diffusive motion of the particle from $x = -\infty$ to $x = \infty$. For a noisy overdamped system, the

---

[53]This exercise was developed in collaboration with David Hathcock, and is closely related to his manuscript [80]. Hints for the computational portions are available at the book website [181].

[54] In Exercise 6.11 you derived this prefactor for a system which was not overdamped—the particles flew over the barrier, rather than diffusing. This changes the prefactor.

[55]Here the coarsening and rescaling just change the units of time and space; the universal eigenvalues come from dimensional analysis. The renormalization group gives us an excellent framework to discuss the behavior, though, with universal scaling functions like eqn 12.74.

motion of a particle in potential $V(x)$ satisfies the Langevin equation[56]

$$\dot{x} = f(x) + g\,\xi(t), \qquad (12.66)$$

where $f(x) = -\eta^{-1}dV/dx$, η is the damping constant, g is the noise amplitude, and $\xi(t)$ is white noise with zero mean $\langle \xi(t) \rangle = 0$ and no time correlations $\langle \xi(t)\xi(t') \rangle = \delta(t - t')$.

How big a noise g should we apply to mimic a temperature T? We can discover this by ignoring the potential well, and finding the diffusion constant for the equation $\dot{x} = g\,\xi(t)$. We then use the Einstein relation between D and T.

(a) *If $f(x) = 0$ in eqn 12.66, calculate how far x will diffuse in a time τ. Show that $D = g^2/2$. Use the Einstein relation to show that the temperature $k_BT = \eta g^2/2$.* (Hint: $\langle x^2 \rangle \sim 2D\tau$, and $x(\tau) = \int_0^\tau \dot{x}$. The double integral for x^2 should involve $\langle \xi(t)\xi(t') \rangle$.)

To study the slowest decaying mode, we shall study the evolution of the probability density $\rho(x,t)$, which obeys a diffusion equation[57] in the potential V:

$$\frac{\partial \rho}{\partial t} = \frac{\partial}{\partial x}\left(\frac{1}{\eta}\frac{\partial V}{\partial x}\rho\right) + \frac{g^2}{2}\frac{\partial^2 \rho}{\partial x^2}. \qquad (12.67)$$

(b) *Show that the Boltzmann distribution, $\rho^*(x) \propto \exp(-V(x)/k_BT)$, is a steady state solution to eqn 12.67. Is the Boltzmann distribution normalizable for the potential shown in the left side of Fig. 12.26?*

When the potential grows fast enough as $x \to \pm\infty$, the Boltzmann distribution is the equilibrium distribution for the system. In our case the potential is not bounded, so particles eventually escape to infinity.

The last term of eqn 12.67 is the diffusion you derived in part (a). Can we derive the other term from our Langevin equation 12.66?

(c) *If we evolve $\rho(x)$ with eqn 12.66 setting the noise $g = 0$, what is the current $J(x)$? What is the resulting equation for $\partial\rho/\partial t$?*

We parameterize the space of barrier-crossing problems by the noise g and the coefficients ϵ_n of the Taylor series of $f(x) = -\eta^{-1}dV/dx$:

$$f(x) = ax^2 + \epsilon_0 + \sum_{n=3}^{\infty} \epsilon_n x^n \qquad (12.68)$$

Here ϵ_2, coming from the cubic term in the potential, is set to a. Also, we choose $x = 0$ to be the point where the slope of the potential is smallest, (most negative in the case with a barrier); thus $d^2V/dx^2 = 0$ so $f'(0) = 0$ and hence $\epsilon_1 = 0$. So eqn 12.66 becomes

$$\dot{x} = ax^2 + \epsilon_0 + \sum_{n=3}^{\infty} \epsilon_n x^n + g\,\xi(t). \qquad (12.69)$$

When $\epsilon < 0$, the potential has a well, while $\epsilon > 0$ gives a potential that is downward sloping (Fig. 12.26). The higher order terms distort the shape of the barrier or sloping ramp.

It is useful to define the escape time as the time $\tau(g, \epsilon_0, \epsilon_1, \dots)$ for particles in a noisy potential to reach $x = \infty$ starting from $x = -\infty$. (This definition also makes sense when there is no barrier.) If we coarse-grain time $t \to t' = t/b$ by a factor of b, your answer to part (a) should tell you that the noise grows $g\xi(t) \to gb^{1/2}\xi(t')$. We then rescale $x \to x' = bx$ to keep the cubic term a in the potential fixed, giving [80]

$$g' = b^{3/2}g$$
$$\epsilon_0' = b\,\epsilon_0$$
$$\epsilon_3' = b^{-1}\epsilon_3 \qquad (12.70)$$
$$\dots$$
$$\epsilon_n' = b^{2-n}\epsilon_n.$$

One might expect, near the transition where the barrier vanishes, that the important terms in the Taylor series would be the lowest ones.

(d) *Which are the relevant variables? Which are irrelevant? Do the relevant variables correspond to those defining the cubic potential?*

In the usual way, this implies that τ takes the scaling form

$$\tau = \frac{1}{(g\,a)^{2/3}}\mathcal{T}\big(\epsilon_0/(g^{4/3}a^{1/3}), \epsilon_3/(g^{-2/3}a^{4/3}),$$
$$\dots, \epsilon_n/(g^{2(2-n)/3}a^{(n+1)/3})\big), \qquad (12.71)$$

where $\mathcal{T}$ is a universal scaling function.[58]

As the noise g becomes small, all but the first argument $\alpha = \epsilon_0/(g^{4/3}a^{1/3})$ go to zero. Let us set them to zero and focus on α, which leaves us with a cubic potential $V(x)/\eta = -ax^3/3 - \epsilon_0x$ in a noise of strength g, and scaling form

$$\tau = \frac{1}{(g\,a)^{2/3}}\mathcal{T}(\alpha). \qquad (12.72)$$

---

[56]Exercises 6.18 and 10.7 study Langevin dynamics for underdamped systems, where the *momentum* of the particle is noisy.
[57]This is called the Fokker–Planck equation.
[58]The arguments of $\mathcal{T}$ are a complete set of dimensionless parameter ratios, so this is true by dimensional analysis.

(e) *Show that the classic Kramer's formula eqn 12.65 follows from our universal scaling form 12.72 if, in the high barrier limit $\alpha \to -\infty$*

$$\mathcal{T}(\alpha) \sim \frac{\pi}{\sqrt{|\alpha|}} \exp\left(\frac{8|\alpha|^{3/2}}{3}\right). \quad (12.73)$$

Another limit in which calculations are manageable is $\alpha \to \infty$. In this limit, there is no barrier and noise can be ignored, so we can solve for the time τ by calculating the classical path $x(t)$.

(f) *Solve for τ using the deterministic equation of motion, and show that $\mathcal{T}(\alpha) \sim \pi/\sqrt{\alpha}$ for large α.*

One can derive [80] the universal scaling function

$$\mathcal{T}(\alpha) = 2^{1/3}\pi^2 \left[\text{Ai}^2(-2^{2/3}\alpha) + \text{Bi}^2(-2^{2/3}\alpha)\right], \quad (12.74)$$

Here Ai and Bi are the first and second Airy functions.

(g) *Plot your two asymptotic scaling functions together with the exact formula eqn 12.74. Do your calculations agree with this expression in the appropriate limits?* (Hint: Your plot should duplicate Fig. 2 in [80].)

(12.23) **Biggest of bunch: Gumbel.**[59] (Mathematics, Statistics, Computation, Engineering) ③
Much of statistical mechanics focuses on the average behavior in an ensemble, or the mean square fluctuations about that average. In many cases, however, we are far more interested in the extremes of a distribution.

Engineers planning dike systems are interested in the highest flood level likely in the next hundred years. Let the high water mark in year j be H_j. Ignoring long-term weather changes (like global warming) and year-to-year correlations, let us assume that each H_j is an independent and identically distributed (IID) random variable with probability density $\rho_1(H_j)$. The *cumulative distribution function* (cdf) is the probability that a random variable is less than a given threshold. Let the cdf for a single year be $F_1(H) = P(H' < H) = \int^H \rho_1(H')\,\mathrm{d}H'$.

(a) *Write the probability $F_N(H)$ that the highest flood level (largest of the high-water marks) in the next $N = 1{,}000$ years will be less than H, in terms of the probability $F_1(H)$ that the*

high-water mark in a single year is less than H.

The distribution of the largest or smallest of N random numbers is described by *extreme value statistics*. Extreme value statistics is a valuable tool in engineering (reliability, disaster preparation), in the insurance business, and recently in bioinformatics (where it is used to determine whether the best alignments of an unknown gene to known genes in other organisms are significantly better than that one would generate randomly).

(b) *Suppose that $\rho_1(H) = \exp(-H/H_0)/H_0$ decays as a simple exponential ($H > 0$). Using the formula*

$$(1 - A) \approx \exp(-A) \quad \text{small A} \quad (12.75)$$

show that the cumulative distribution function F_N for the highest flood after N years is

$$F_N(H) \approx \exp\left[-\exp\left(\frac{\mu - H}{\beta}\right)\right]. \quad (12.76)$$

for large H. (Why is the probability $F_N(H)$ small when H is not large, at large N?) What are μ and β for this case?

The constants β and μ just shift the scale and zero of the ruler used to measure the variable of interest. Thus, using a suitable ruler, the largest of many events is given by a Gumbel distribution

$$F(x) = \exp(-\exp(-x))$$
$$\rho(x) = \partial F/\partial x = \exp(-(x + \exp(-x))). \quad (12.77)$$

How much does the probability distribution for the largest of N IID random variables depend on the probability density of the individual random variables? Surprisingly little! It turns out that the largest of N Gaussian random variables also has the same Gumbel form that we found for exponentials. Indeed, any probability distribution that has unbounded possible values for the variable, but that decays faster than any power law, will have extreme value statistics governed by the Gumbel distribution [133, section 8.3]. In particular, suppose

$$F_1(H) \approx 1 - A\exp(-BH^\delta) \quad (12.78)$$

as $H \to \infty$ for some positive constants A, B, and δ. It is in the region near $H^*[N]$, defined by $F_1(H^*[N]) = 1 - 1/N$, that F_N varies in an interesting range (because of eqn 12.75).

---

[59]Computational hints can be found at the book website [181].

(c) *Show that the extreme value statistics $F_N(H)$ for this distribution is of the Gumbel form (eqn 12.76) with $\mu = H^*[N]$ and $\beta = 1/(B\delta\, H^*[N]^{\delta-1})$.* (Hint: Taylor expand $F_1(H)$ at H^* to first order.)

The Gumbel distribution is *universal*. It describes the extreme values for any unbounded distribution whose tails decay faster than a power law.[60] (This is quite analogous to the central limit theorem, which shows that the normal or Gaussian distribution is the universal form for sums of large numbers of IID random variables, so long as the individual random variables have non infinite variance.)

The Gaussian or standard normal distribution $\rho_1(H) = (1/\sqrt{2\pi})\exp(-H^2/2)$, for example, has a cumulative distribution $F_1(H) = (1/2)(1 + \mathrm{erf}(H/\sqrt{2}))$ which at large H has asymptotic form $F_1(H) \sim 1 - (1/\sqrt{2\pi}H)\exp(-H^2/2)$. This is of the general form of eqn 12.78 with $B = \frac{1}{2}$ and $\delta = 2$, except that A is a slowly varying function of H. This slow variation does not change the asymptotics.

(d) *Generate $M = 10{,}000$ lists of $N = 1{,}000$ random numbers distributed with this Gaussian probability distribution. Plot a normalized histogram of the largest entries in each list. Plot also the predicted form $\rho_N(H) = \mathrm{d}F_N/\mathrm{d}H$ from part (c).* (Hint: $H^*(N) \approx 3.09023$ for $N = 1{,}000$; check this if it is convenient.)

Other types of distributions can have extreme value statistics in different universality classes (see Exercise 12.24). Distributions with power-law tails (like the distributions of earthquakes and avalanches described in Chapter 12) have extreme value statistics described by *Fréchet distributions* (Exercise 12.24). Distributions that have a strict upper or lower bound[61] have extreme value distributions that are described by Weibull statistics (see Exercise 1.9).

(12.24) **Extreme values: Gumbel, Weibull, and Fréchet.** (Mathematics, Statistics, Engineering) ③

Extreme value statistics is the study of the maximum or minimum of a collection of random numbers. It has obvious applications in the insurance business (where one wants to know the biggest storm or flood in the next decades, see Exercise 12.23) and in the failure of large systems (where the weakest component or flaw leads to failure; see Exercise 1.9). Recently, extreme value statistics has become of significant importance in bioinformatics. (In guessing the function of a new gene, one often searches entire genomes for good matches (or *alignments*) to the gene, presuming that the two genes are evolutionary descendents of a common ancestor and hence will have similar functions. One must understand extreme value statistics to evaluate whether the best matches are likely to arise simply at random.)

The limiting distribution of the biggest or smallest of N random numbers as $N \to \infty$ takes one of three *universal forms*, depending on the probability distribution of the individual random numbers. In this exercise we understand these forms as fixed points in a renormalization group.

Given a probability distribution $\rho_1(x)$, we define the *cumulative distribution function* (cdf) as $F_1(x) = \int_{-\infty}^{x} \rho(x')\,\mathrm{d}x'$. Let us define $\rho_N(x)$ to be the probability density that, out of N random variables, the largest is equal to x. Let $F_N(x)$ to be the corresponding cdf.

(a) *Write a formula for $F_{2N}(x)$ in terms of $F_N(x)$. If $F_N(x) = \exp(-g_N(x))$, show that $g_{2N}(x) = 2g_N(x)$.*

Our renormalization-group coarse-graining operation will remove half of the variables, throwing away the smaller of every pair, and returning the resulting new probability distribution. In terms of the function $g(x) = -\log\int_{-\infty}^{x} \rho(x')dx'$, it therefore will return a rescaled version of $2g(x)$. This rescaling is necessary because, as the sample size N increases, the maximum will drift upward—only the form of the probability distribution stays the same, the mean and width can change. Our renormalization-group coarse-graining operation thus maps function space into itself, and is of the form

$$T[g](x) = 2g(ax + b). \qquad (12.79)$$

(This renormalization group is the same as that we use for sums of random variables in Exer-

---

[60]The Gumbel distribution can also describe extreme values for a bounded distribution, if the probability density at the boundary goes to zero faster than a power law.

[61]More specifically, bounded distributions that have power-law asymptotics have Weibull statistics; see note 60 and Exercise 1.9, part (d).

cise 12.11 where $g(k)$ is the logarithm of the Fourier transform of the probability density.) There are three distinct types of fixed point distributions for this renormalization-group transformation, which (with an appropriate linear rescaling of the variable x) describe most extreme value statistics. The Gumbel distribution (Exercise 12.23) is of the form

$$F_{\text{gumbel}}(x) = \exp(-\exp(-x)),$$
$$\rho_{\text{gumbel}}(x) = \exp(-x)\exp(-\exp(-x)),$$
$$g_{\text{gumbel}}(x) = \exp(-x).$$

The Weibull distribution (Exercise 1.9) is of the form

$$F_{\text{weibull}}(x) = \begin{cases} \exp(-(-x)^\alpha) & x < 0 \\ 1 & x \geq 0 \end{cases}$$
$$g_{\text{weibull}}(x) = \begin{cases} (-x)^\alpha & x < 0 \\ 0 & x \geq 0, \end{cases} \qquad (12.80)$$

and the Fréchet distribution is of the form

$$F_{\text{fréchet}}(x) = \begin{cases} 0 & x \leq 0 \\ \exp(-x^{-\alpha}) & x > 0 \end{cases}$$
$$g_{\text{fréchet}}(x) = \begin{cases} \infty & x < 0 \\ x^{-\alpha} & x \geq 0, \end{cases} \qquad (12.81)$$

where $\alpha > 0$ in each case.

(b) *Show that these distributions are fixed points for our renormalization-group transformation eqn 12.79. What are a and b for each distribution, in terms of α?*
In parts (c) and (d) you will show that there are only these three fixed points $g^*(x)$ for the renormalization transformation, $T[g^*](x) = 2g^*(ax + b)$, up to an overall linear rescaling of the variable x, with some caveats...
(c) *First, let us consider the case $a \neq 1$. Show that the rescaling $x \to ax + b$ has a fixed point $x = \mu$. Show that the most general form for the fixed point function is*

$$g^*(\mu \pm z) = z^{\alpha'} p_\pm(\gamma \log z) \qquad (12.82)$$

for $z > 0$, where $p_\pm$ is periodic and α' and γ are constants such that $p_\pm$ has period equal to one. (Hint: Assume $p(y) \equiv 1$, find α', and then show $g^*/z^{\alpha'}$ is periodic.) What are α' and γ? Which

choice for a, p_+, and p_- gives the Weibull distribution? The Fréchet distribution?
Normally the periodic function $p(\gamma \log(x - \mu))$ is assumed or found to be a constant (sometimes called $1/\beta$, or $1/\beta^{\alpha'}$). If it is not constant, then the probability density must have an infinite number of oscillations as $x \to \mu$, forming a weird essential singularity.
(d) *Now let us consider the case $a = 1$. Show again that the fixed point function is*

$$g^*(x) = e^{-x/\beta} p(x/\gamma) \qquad (12.83)$$

with p periodic of period one, and with suitable constants β and γ. What are the constants in terms of b? What choice for p and β yields the Gumbel distribution?
Again, the periodic function p is often assumed a constant (e^μ), for reasons which are not as obvious as in part (c).
What are the domains of attraction of the three fixed points? If we want to study the maximum of many samples, and the initial probability distribution has $F(x)$ as its cdf, to which universal form will the extreme value statistics converge? Mathematicians have sorted out these questions. If $\rho(x)$ has a power-law tail, so $1 - F(x) \propto x^{-\alpha}$, then the extreme value statistics will be of the Frechet type, with the same α. If the initial probability distribution is bounded by μ and if $1 - F(\mu - |y|) \propto y^\alpha$ (where $\alpha = 0$ is a jump to zero, Exercise 1.9), then the extreme value statistics will be of the Weibull type. If the probability distribution decays faster than any polynomial (say, exponentially), or if a bounded distribution vanishes to zero faster than a power law at the bound, then the extreme value statistics will be of the Gumbel form (Exercise 12.23).

(12.25) **Critical correlations.**[62] ③
In Exercise 10.19, we discussed the correlation function predicted by a theorist for a magnet near its critical point T_c, where the net magnetization vanishes and the correlation length $\xi(T)$ diverges. She predicted a correlation function of the form[63]

$$C(\mathbf{r}, \tau) = \langle m(\mathbf{x}, t) m(\mathbf{x} + \mathbf{r}, t + \tau) \rangle \qquad (12.84)$$
$$\approx (r^2 + (c\tau)^2)^{-\eta/2} e^{-\sqrt{r^2 + (c\tau)^2}/\xi(T)}.$$

---

[62]Ising simulation software can be found at [28].
[63]To be careful, $\langle m(\mathbf{x}, t) m(\mathbf{x} + \mathbf{r}, t + \tau) \rangle = m_s^2 ((r/r_s)^2 + (\tau/\tau_s)^2)^{-\eta} e^{-\sqrt{(r/r_s)^2 + (\tau/\tau_s)^2}/(\xi(T)/\xi_s)}$, where r_s, τ_s, ξ_s, and m_s are scale factors that depend on the material. $c = r_s/\tau_s$ is the only constant we keep, for simplicity.

In that exercise, we explored her implied predictions for various experiments measuring different susceptibilities and losses near the critical point. Here we shall explore the scaling properties of her proposed correlation function.

Is the prediction of the theoretical colleague reasonable for magnets at their critical point? Run a simulation of the Ising model. Equilibrate it at its critical point (perhaps using the Wolff algorithm), and observe the qualitative dynamical features it shows under Metropolis dynamics. Pay particular attention to features that can motivate your discussion in part (b).

(a) *Describe qualitatively the lifetimes of clusters as a function of their size. Observe a single spin deep inside a large cluster of up-spins. Does it on average remain pointing up, for roughly the lifetime of the cluster? Does it occasionally flip down (so its average value is less than one)? Qualitatively describe the net magnetization per spin of a cluster, in terms of its size. As the cluster size gets huge, what must happen to its magnetization?* (Hint: We are only at the border of the magnetized phase.)

(b) *Consider the relation between your observations and $C(r, \tau)$, at the critical point where $\xi \to \infty$. Do a log–log graph of C versus r for $1 < r < 1,000$ with $\eta = 1/4$, for three values of $c\tau = 0, 10, 100$. Explain how $C(\tau)$ changes as time elapses, using your observations of the lifetime of the clusters and the behavior of spins contained in the clusters.*

(c) *Show that the theorist's prediction for $\langle m(\mathbf{x}, t)m(\mathbf{x} + \mathbf{r}, t + \tau)\rangle$ is of the scaling form $C(\mathbf{r}, \tau) = r^{-\eta}\mathcal{C}(r/\xi, \tau/\xi^z)$ mentioned in Section 12.2. What is her prediction for z?[64] What is her prediction for the universal scaling function $\mathcal{C}(Y, Z)$, as a function of its arguments $Y = r/\xi$ and $Z = \tau/\xi^z$?*

(12.26) **Ising mean field derivation.**[65] (Mathematics) ③

In this exercise, we derive the mean-field free energy F^{MF} for the Ising model used in Exercise 12.5. Our mean-field approximation will be a rigorous upper bound on the true free energy

F, as shown by a theorem attributed to Gibbs, Bogoliubov, and Feynman (GBF)

$$F \leq F^{\mathrm{MF}} = F^{[0]} + \langle \mathcal{H} - \mathcal{H}^{[0]}\rangle_{[0]} \quad (12.85)$$

(see Exercise 12.27 for a derivation). Here $\mathcal{H}^{[0]}$ is a simple trial Hamiltonian with the same states α (say spin configurations on a square lattice) as the true Hamiltonian $\mathcal{H}$. $F^{[0]}$ is the free energy of that trial Hamiltonian, and $\langle X\rangle_{[0]}$ is the thermal expectation of the operator X in the thermal ensemble for the trial Hamiltonian

$$\langle X\rangle_{[0]} = (1/Z^{[0]})\sum_\alpha X_\alpha \exp(-\beta\mathcal{H}_\alpha^{[0]}). \quad (12.86)$$

If the original Hamiltonian depends on parameters (here $\mathcal{H}(J, H) = -J\sum_{\langle ij\rangle} s_i s_j - H\sum_i s_i$), and the new Hamiltonian has parameters (here $\mathcal{H}^{[0]}(h) = -h\sum_i s_i$), the new parameter(s) can be varied to minimize F^{MF}, leading to a temperature-dependent mean-field variational bound $F^{\mathrm{MF}}(\beta, J, H)$ that is remarkably useful for practical calculations away from critical points.

The key challenge is to find a trial Hamiltonian simple enough that one can calculate both the free energy $F^{[0]}$ and the expectation values $\langle\mathcal{H}\rangle_{[0]}$ and $\langle\mathcal{H}^{[0]}\rangle_{[0]}$. For the N-spin nearest-neighbor Ising model on a square lattice of this problem, let us use as a trial Hamiltonian a set of N uncoupled spins, each in an external field h.

Derive the mean field equations (eqn 12.25 used in Exercise 12.5) for the square lattice Ising ferromagnet $\mathcal{H} = -J\sum_{\langle ij\rangle} s_i s_j - H\sum s_i$ using the GBF inequality (eqn 12.85), where $\mathcal{H}^{[0]} = -\sum_i h S_i$ is a trial Hamiltonian, with h chosen to minimize the right-hand side.

(12.27) **Mean-field bound for free energy.**[66] ③

Exercise 12.26 uses a rigorous upper bound eqn 12.85, $F \leq F^{\mathrm{MF}} = F^{[0]} + \langle\mathcal{H} - \mathcal{H}^{[0]}\rangle_{[0]}$ to generate the standard mean-field theory for the Ising model. Again, $\mathcal{H}^{[0]}$ is a trial Hamiltonian with the same states α as the true Hamiltonian $\mathcal{H}$, $F^{[0]}$ is the trial Hamiltonian free energy, and $\langle X\rangle_{[0]} = (1/Z^{[0]})\sum_\alpha X_\alpha e^{-\beta\mathcal{H}_\alpha^{[0]}}$ is average of X

---

[64] The dynamic critical exponent z, like the coarsening exponent, depends on the dynamical rules for equilibrating the model. For example, the two-dimensional Ising model has $\eta = 1/4$. With Metropolis dynamics, $z \approx 2.2$; with conserved order parameter (Kawasaki) dynamics z is $15/4$. The Wolff algorithm for the 2D Ising model has $z \approx 0.3$ (correlation times which grow much more slowly as the critical point is approached). The quantum transverse 1D Ising model, which has the same critical behavior as the 2D Ising model, has $z = 1$.

[65] See Cardy [39], equation 2.7 and exercise 2.1.

[66] See Cardy [39], eqn 2.7 and Exercise 2.1.

in the trial thermal ensemble, since the probability in the trial ensemble of being in state α is $\rho_\alpha^{[0]} = \exp(-\beta \mathcal{H}_\alpha^{[0]})/Z^{[0]}$. In this exercise, we derive this bound, which is attributed to Gibbs, Bogoliubov, and Feynman (GBF). This will be done in two steps.

(a) *Show that*

$$
\begin{aligned}
Z &= \sum_\alpha e^{-\beta \mathcal{H}_\alpha} \\
&= Z^{[0]} \langle e^{-\beta(\mathcal{H} - \mathcal{H}^{[0]})} \rangle_{[0]}.
\end{aligned}
\tag{12.87}
$$

(Hint: Work backward from the answer)

Now $\exp(x)$ is a convex function: the line connecting two points on the graph lies above the curve: $\lambda \exp(x_1) + (1-\lambda) \exp(x_2) \geq \exp(\lambda x_1 + (1-\lambda)x_2)$. This is true for sums of many terms (see note 37 on p. 110):

$$
\begin{aligned}
\langle \exp(X) \rangle_{[0]} = \sum_\alpha \rho_\alpha^{[0]} \exp(X_\alpha) &\geq \exp(\rho_\alpha^{[0]} X_\alpha) \\
&= \exp(\langle x \rangle_{[0]}).
\end{aligned}
$$

(b) *Use this to derive the GBF inequality of eqn 12.85, $F \leq F^{\mathrm{MF}} = F^{[0]} + \langle \mathcal{H} - \mathcal{H}^{[0]} \rangle_{[0]}$.* Mean field theories are of great value away from critical points, where the fluctuations are small, or even at critical points when the fluctuations are irrelevant (in high dimensions or when long-range forces are important).

(12.28) **Avalanche size distribution.** ③

One can develop a mean-field theory for avalanches in nonequilibrium disordered systems by considering a system of N Ising spins coupled to one another by an infinite-range interaction of strength J/N, with an external field H and each spin also having a local random field h:

$$
\mathcal{H} = -J_0/N \sum_{i,j} S_i S_j - \sum_i (H + h_i) S_i. \tag{12.88}
$$

We assume that each spin flips over when it is pushed over; i.e., when its change in energy

$$
\begin{aligned}
H_i^{\mathrm{loc}} = \frac{\partial \mathcal{H}}{\partial S_i} &= J_0/N \sum_j S_j + H + h_i \\
&= J_0 m + H + h_i
\end{aligned}
$$

changes sign.[67] Here $m = (1/N) \sum_j S_j$ is the average magnetization of the system. All spins

start by pointing down. A new avalanche is launched when the least stable spin (the unflipped spin of largest local field) is flipped by increasing the external field H. Each spin flip changes the magnetization by $2/N$. If the magnetization change from the first spin flip is enough to trigger the next-least-stable spin, the avalanche will continue.

We assume that the probability density for the random field $\rho(h)$ during our avalanche is a constant

$$
\rho(h) = (1 + t)/(2J_0). \tag{12.89}
$$

(a) *Show that at $t = 0$ each spin flip will trigger on average one other spin to flip, for large N. Can you qualitatively explain the difference between the two phases with $t < 0$ and $t > 0$?* Thus the constant t measures how close the density is to the critical density $1/(2J_0)$.

We can solve exactly for the probability $D(S, t)$ of having an avalanche of size S. To have an avalanche of size S triggered by a spin with random field h, you must have precisely $S - 1$ spins with random fields in the range $\{h, h + 2J_0 S/N\}$ (triggered during the avalanche, which increases the magnetization by $2S/N$). The probability of this is given by the Poisson distribution. In addition, the random fields must be arranged so that the first spin triggers the rest. The probability of this turns out to be $1/S$.

(b) *(Optional) By imagining putting periodic boundary conditions on the interval $\{h, h + 2J_0 S/N\}$, argue that exactly one spin out of the group of S spins will trigger the rest as a single avalanche.* (Hint from Ben Machta: For simplicity, we may assume[68] the avalanche starts at $H = m = 0$. Try plotting the local field $H^{\mathrm{loc}}(h') = J_0 m(h') + h'$ that a spin with random field h' would feel if the spins between h' and h were flipped. How would this function change if we shuffle all the random fields around the periodic boundary conditions?)

(c) *Using or assuming the result (b), show that the distribution of avalanche sizes is*

$$
D(S, t) = \frac{S^{S-1}}{S!}(t + 1)^{S-1} e^{-S(t+1)}. \tag{12.90}
$$

---

[67] We ignore the self-interaction, which is unimportant at large N.

[68] Equivalently, measure the random fields with respect to h_0 of the triggering spin, and let m be the magnetization change since the avalanche started.

With t small (near to the critical density) and for large avalanche sizes S we expect this to have a scaling form:

$$D(S,t) = S^{-\tau} \mathcal{D}(S/t^{-x}) \qquad (12.91)$$

for some mean-field exponent x. That is, taking $t \to 0$ and $S \to \infty$ along a path with St^x fixed, we can expand $D(S,t)$ to find the scaling function.

(d) *Show that $\tau = 3/2$ and $x = 2$. What is the scaling function $\mathcal{D}$? Hint: You'll need to use Stirling's formula $S! \sim \sqrt{2\pi S}(S/\mathrm{e})^S$ for large S, and that $1 + t = \exp(\log(1 + t)) \approx \mathrm{e}^{t - t^2/2 + t^3/3\cdots}$.*

This is a bit tricky to get right. Let us check it by doing the plots.

(e) *Plot $S^\tau D(S,t)$ versus $Y = S/t^{-x}$ for $t = 0.2, 0.1,$ and 0.05 in the range $Y \in (0, 10)$. Does it converge to $\mathcal{D}(Y)$?*

See [178] for more information.

(12.29) **The onset of chaos: lowest order RG.**[69] (Dynamical systems) @

In this exercise, we set up a low-order approximate renormalization group to study the period-doubling route to chaos of Fig. 12.17. Our goal is to estimate the scaling factors $\alpha \sim -2.5029$ and $\delta \sim 4.6692$ governing the self-similarity in space x and control parameter μ near the onset of chaos μ_∞, as discussed in Exercise 12.16. In Exercise 12.30, we shall add higher orders to this rough approximation, implementing Feigenbaum's original calculation.

The period-doubling route to chaos is understood by a renormalization group that (as usual) has a coarse-graining step and a rescaling. The coarse-graining step "decimates" the time series $\{x, g(x), g(g(x)), \dots\}$ by dropping every other point, replacing the function $g(x)$ by $g(g(x))$. The rescaling expands x about its maximum by a factor α. If we choose the maximum value of $g(x)$ to be at zero, the renormalization group sends g to the function $T[g]$, where

$$T[g](x) = \alpha g\left(g(x/\alpha)\right). \qquad (12.92)$$

(We explore the scaling and universality implied by this renormalization group in Exercise 12.9; our eqn 12.92 is the same as eqn 12.40 in that exercise, except there our functions took their maximum at $x = \frac{1}{2}$.)

Our functions $g(x)$ are *one-humped maps*, with a parabolic maximum at zero. To lowest order, let us approximate $g(x)$ by a parabola centered at the origin

$$g(x) \approx 1 + G_1 x^2 \qquad (12.93)$$

where $G_1 < 0$.

We shall approximate the fixed point g^* of our renormalization group by demanding that $T[g^*](x) = g^*(x)$ at two points, $x = 0$ and $x = 1$. These two equations will determine our two unknowns, α and G_1^*.

(a) *Use the fixed point condition at $x = 0$ to show that $\alpha = 1/g^*(1) = 1/(1 + G_1^*)$, and hence $G_1^* = (1 - \alpha)/\alpha$.*

(b) *Use the fixed point condition at $x = 1$ to give an equation for α, substituting in your equation for G_1^* above.* (Hint: The equation simplifies to a sixth-order polynomial divided by α^6.) We expect $\alpha \approx -2.5$.

(c) *Plot the quantity from part (b) that must be zero: does it have a root near $\alpha \approx -2.5$? Numerically solve your equation from part (b) for the root closest to -2.5. What is α in our approximation?* (Your approximation for α should be within a few percent of the correct value $\alpha = -2.5029\dots$; your value for G_1^* should be within a few percent of -1.4 and not too far from the true value of the quadratic term at the fixed point, $-1.5276\dots$)

In statistical mechanics, we find the universal critical exponents by linearizing the renormalization-group flows about the fixed point and finding directions that grow. Here the exponent $\delta = 4.669\dots$ describes the fastest growing direction in function space: $T[g^* + \epsilon\psi](x) - g^*(x) = \delta\epsilon\psi(x)$. That is, we add a perturbation $g(x) = g^*(x) + \epsilon\psi(x)$ and study to linear order in ϵ how the perturbation grows under T. The lowest-order perturbation to our parabola adds an overall constant $G_0 \to 1 + \epsilon$. Our function $g^*(x)$ is fixed at both $x = 0$ and $x = 1$. Let us check the growth of our perturbation at $x = 0$, which should grow by a factor of approximately δ when our renormalization-group transformation is applied.

(d) *Using $g(x) = (1 + \epsilon) + G_1 x^2$, write the formula for the term in $T[g](0)$ linear in ϵ as a function of G_1 and α. Insert your fixed point values for G_1 and α from part (c). What is*

---

[69] There are many exercises exploring this chaotic logistic map (see Index).

your estimate for δ? (Your approximation for δ should be within a few percent of the correct value $\delta = 4.669\ldots$.)

(12.30) **The onset of chaos: full RG.**[70] (Computation, Dynamical systems) ③

In this exercise, we implement Feigenbaum's numerical scheme [56, pp. 693-4] for finding high-precision values of the universal constants

$$\alpha = -2.50290787509589282228390287322$$
$$\delta = 4.66920160910299067185320382158,$$
$$(12.94)$$

that quantify the scaling properties of the period-doubling route to chaos (Fig. 12.17, Exercise 12.16). This extends the lowest-order calculation of Exercise 12.29.

Our renormalization-group operation (Exercises 12.9 and 12.29) coarse-grains in time taking $g \to g \circ g$, and then rescales distance x by a factor of α. Centering our functions at $x = 0$, this leads to $T[g](x) = \alpha g\left(g(x/\alpha)\right)$ (eqn 12.92). We shall solve for the properties at the onset of chaos by analyzing our function-space renormalization group by expanding our functions in a power series

$$g(x) \approx 1 + \sum_{n=1}^{N} G_n x^{2n}. \quad (12.95)$$

Notice that we only keep even powers of x; the fixed point is known to be symmetric about the maximum, and the unstable mode responsible for the exponent δ will also be symmetric. Keeping odd perturbations would double the number of eigenvectors and eigenvalues.

First, we must approximate the fixed point $g^*(x)$ and the corresponding value of the universal constant α. At order N, we must solve for α and the N polynomial coefficients G_n^*. We can use the $N + 1$ equations fixing the function at equally spaced points in the positive unit interval:

$$T[g^*](x_m) = g^*(x_m),$$
$$x_m = m/N, \quad (12.96)$$
$$m = \{0, \ldots, N\}.$$

We can use the first of these equations to solve for α.

(a) *Show that the equation for $m = 0$ sets $\alpha = 1/g^*(1)$.*

We can use a root-finding routine to solve for G_n^*.

(b) *Implement the other N constraint equations of eqn 12.96 in a form appropriate for your method of finding roots of nonlinear equations, substituting your value for α from part (a). Check that your routine at $N = 1$ gives values for $\alpha \approx -2.5$ and $G_1^* \approx -1.5$.* (These should reproduce the values from Exercise 12.29(c).)

(c) *Use a root-finding routine to calculate α for $N = 1, \ldots, 9$. Start the search at $G_1^* = -1.5$, $G_n^* = 0$ ($n > 1$) to avoid landing at the wrong fixed point.* (If it is convenient for you to use high-precision arithmetic, continue to higher N.) *To how many decimal places can you reproduce the correct value for α in eqn 12.94?*

Now we need to solve for the renormalization-group flows $T[g]$, linearized about the fixed point $g(x) = g^*(x) + \epsilon\psi(x)$. Feigenbaum tells us that $T[g^* + \epsilon\psi] = T[g^*] + \epsilon\mathcal{L}[\psi]$, where $\mathcal{L}$ is the linear operator taking $\psi(x)$ to

$$\mathcal{L}[\psi](x) = \alpha\psi(g^*(x/\alpha)) + \alpha g^{*\prime}(g(x/\alpha))\psi(x/\alpha).$$
$$(12.97)$$

(d) *Derive eqn 12.97.*

We want to find eigenfunctions that satisfy $\mathcal{L}[\psi] = \lambda\psi$, confining ourselves to the even subspace. Again, we can expand $\psi(x)$ in an even polynomial

$$\psi(x) = \sum_{n=0}^{N-1} \psi_n x^{2n} \qquad (\psi_0 \equiv 1). \quad (12.98)$$

We then approximate the action of $\mathcal{L}$ on ψ by its action at N points $\widetilde{x}_i$, that need not be the same as the N points x_m we used to find g^*. We shall use $\widetilde{x}_i = (i-1)/(N-1)$, $i = 1, \ldots, N$. (For $N = 1$, we use $\widetilde{x}_1 = 0$.) This leads us to a linear system of N equations for the coefficients ψ_n, using eqns 12.98 and 12.99:

$$\sum_{n=0}^{N-1} \left[\alpha g(\widetilde{x}_i/\alpha)^{2n} + \alpha g'(g(\widetilde{x}_i/\alpha))(\widetilde{x}_i/\alpha)^{2n}\right]\psi_n$$
$$= \lambda \sum_{n=0}^{N-1} \widetilde{x}_i^{2n}\psi_n.$$
$$(12.99)$$

These equations for the coefficients ψ_n of the eigenfunctions of $\mathcal{L}$ is in the form of a *generalized eigenvalue problem*

$$\sum_n L_{in}\psi_n = \lambda \sum_n X_{in}\psi_n. \quad (12.100)$$

---

[70]Computer hints can be found at the book website [181]. There are many exercises exploring this chaotic logistic map (see Index).

The solution to the generalized eigenvalue problem can be found from the eigenvalues of $X^{-1}L$, but most eigenvalue routines provide a more efficient and accurate option for directly solving the generalized equation given L and X.

(e) *Write a routine that calculates the matrices L and X implicitly defined by eqns 12.100 and 12.99. For $N = 1$ you should generate 1×1 matrices. For $N = 1$, what is your prediction for δ?* (These should reproduce the values from Exercise 12.29(d).)

(f) *Solve the generalized eigenvalue problem for L and X for $N = 1, \ldots, 9$. To how many decimal places can you reproduce the correct value for δ in eqn 12.94? Are the other eigenvalues you calculate all irrelevant?*

(12.31) **Singular corrections to scaling.** ③
The renormalization group says that the number of relevant directions at the fixed point in system space is the number of parameters we need to tune to see a critical point, and that the critical exponents depend on the eigenvalues of these relevant directions. Do the irrelevant directions matter? Here we discuss the *singular corrections to scaling* we saw, e.g., in Fig. 12.15. Let the Ising model in zero field be described by flow equations

$$dt_\ell/d\ell = t_\ell/\nu, \qquad du_\ell/d\ell = -yu_\ell, \quad (12.101)$$

where t_ℓ describes the renormalization of the reduced temperature $t = (T_c - T)/T_c$ after a coarse-graining by a factor $b = \exp(\ell)$, and u and u_ℓ represent a slowly decaying irrelevant perturbation under the renormalization group. In Fig. 12.8, one may view t as the expanding eigendirection running roughly horizontally, and u as the contracting, irrelevant coordinate running roughly vertically. Thus our model starts with a value u_0 associated to the distance in system space between our critical point R_c and the RG fixed point S^* along the irrelevant coordinate.

(a) *What is the invariant combination $z = ut^\omega$ that stays constant under the renormalization group? What is ω in terms of the eigenvalues $-y$ and $1/\nu$?*

Properties near critical points have universal power-law singularities, but the corrections to these power laws also have universal properties predicted by the renormalization group. These come in two types—*analytic* corrections to scaling and *singular* corrections to scaling.

Let us consider corrections to the susceptibility. In analogy with other systems we have studied, we would expect that the susceptibility

$$\chi(t, u) = t^{-\gamma} X(z) \qquad (12.102)$$

with $X(z)$ a universal function of the invariant combination you found in part (a). As a function of t, $\chi(t, u)$ has singularities at small t. But we expect properties to be analytic as we vary u, since the irrelevant direction is not being tuned to a special value, so we expect that a Taylor series of $\chi(t, u)$ in powers of u should make sense. Since $z \propto u$, we thus expect that $X(z)$ will be an analytic function of z for small z.[71]

(b) *Show that for small t, your z from part (a) goes to zero. Taylor expand $X(z)$. What corrections do you predict for the susceptibility from the first and second-order terms in the series?* These are the singular corrections to scaling due to the irrelevant perturbation u.

An Ising magnet on a sample holder is loaded into a magnetometer, and the susceptibility is measured[72] at zero external field as a function of reduced temperature $t = (T - T_c)/T_c$. It is found to be well approximated by

$$\chi(T) = At^{-1.24} + Bt^{-0.83} + Ct^{-0.42} \\ + D + Et + \ldots. \qquad (12.103)$$

You may ignore any errors due to the magnetometer.

(c) *The exponent $\omega \approx 0.407$ for the 3D Ising universality class, and $\gamma \approx 1.237$. Which terms are explained as singular corrections to scaling?*

(d) *Can you provide a physical interpretation for the terms in eqn 12.103 that are not explained by singular corrections to scaling? For example, how do we expect the susceptibility of the sample holder to depend on temperature?* These are examples of *analytic* corrections to scaling.

One must note that it is normally completely infeasible in an experiment or simulation to

---

[71]Had we used a scaling variable $Z = tu^{1/\omega}$, for example, we would not have expected the corresponding scaling function to be analytic in small Z.

[72]The accuracy of the quoted exponents is not experimentally realistic.

measure quantities with sufficiently accuracy to identify so many simultaneous corrections to scaling.

(12.32) **Conformal invariance.**[73] (Mathematics, Biology, Computation) ③

Emergence in physics describes new laws that arise from complex underpinnings, and often the new laws exhibit a larger symmetry than the original model. The diffusion equation emerges as a continuum limit from complex random microscopic motion, and diffusion on a square lattice has circular symmetry. Critical phenomena emerges near continuous phase transitions, and the resultant symmetry under dilations is exploited by the renormalization group to predict universal power laws and scaling functions.

The Ising model on a square lattice at the critical point, like the diffusion equation, also has an emergent circular symmetry: the complex patterns of up- and down-spins look the same on long length scales also when rotated by an angle. Indeed, making use of the symmetries under changes of length scale, position, and angle (plus one spatially nonuniform transformation), systems at their critical points have a *conformal* symmetry group.

Fig. 12.27 Two proteins in a critical membrane. Membranes in living cells have many types of proteins floating in a two-dimensional fluid of different lipids. This two-dimensional fluid is close to a phase transition in the Ising universality class. Similar to a liquid–gas critical point where a uniform fluid separates into liquid and gas regions, the lipids and proteins separate into two different components—darker and lighter here, corresponding to up- and down-spins in the Ising model.

In two dimensions, the conformal symmetry group becomes huge. Roughly speaking, any complex analytic function $f(z) = u(x + iy) + iv(x + iy)$ takes a snapshot of an Ising model $M(x, y)$ and warps it into a new magnetization pattern at (u, v) that "looks the same". (Here u, v, x, and y are all real.)

You may remember that most ordinary functions (like z^2, $\sqrt{z}$, $\sin(z)$, $\log(z)$, and $\exp(iz)$) are analytic. All of them yield cool transformations of the Ising model—weird and warped when magnified until you see the pixels, but recognizably Ising-like on long scales. This exercise will generate an example.

(a) *What analytic function shrinks a region uniformly by a factor b, holding $z = 0$ fixed? What analytic function translates the lattice by a vector $\mathbf{r}_0 = (u_0, v_0)$? What analytic function rotates a region by an angle θ?*

(b) *Expanding $f(z + \delta) = f(z) + \delta f'(z)$, show that an analytic function f transforms a local region about z to linear order in δ by first rotating and dilating δ about z and then translating. What complex number gives the net translation?*

Figure 12.28 shows how one can use this conformal symmetry to study the interactions between circular "proteins" embedded in a two-dimensional membrane at an Ising critical point [121] (Fig. 12.27).

In the renormalization group, we first coarse-grain the system (shrinking by a factor b) and then rescale the magnetization (by some power b^{y_M}) in order to return to statistically the same critical state: $\widehat{M} = b^{y_M} M$. This rescaling turns the larger pixels in Fig. 12.28 more gray; a mostly up-spin white region with tiny pixels is mimicked by a large single pixel with the statistically averaged gray color.

We can discover the correct power for M by examining the rescaling of the correlation function $C(\mathbf{r}) = \langle M(\mathbf{x})M(\mathbf{x} + \mathbf{r})\rangle$. In the Ising model at its critical point the correlation function $C(r) \sim r^{-(2-d+\eta)}$. In dimension $d = 2$, $\eta = 1/4$. The correlation function for the conformally transformed magnetization should be the same as the original correlation function.

(c) *If we coarse-grain by a constant factor b, by what power of b must we multiply M to make*

---

[73]This exercise was developed in collaboration with Benjamin Machta. Ising snapshots and hints for the computations can be found at the book website [181].

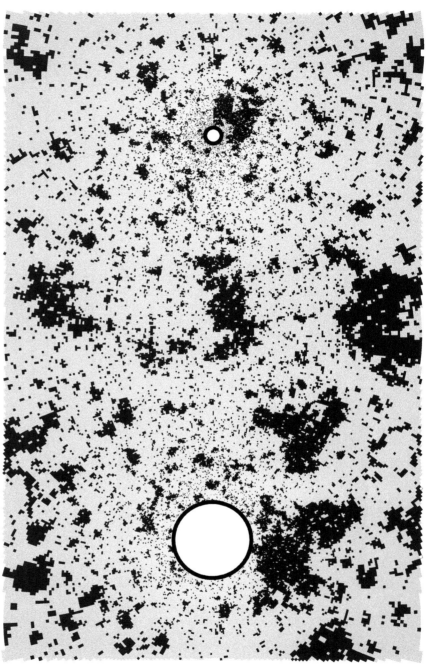

Fig. 12.28 Two proteins in a critical membrane. The figure shows the pixels of a critical Ising model simulated in a square, conformally warping the square onto the exterior of two circles (representing two proteins in a cell membrane). The warped pixels vary in size—largest along the left and right edges, smallest near the smaller circle. They also locally rotate and translate the square lattice, but notice the pixels remain looking square—the angles and aspect ratios remain unchanged. The pixels are gray rather than black and white, with only the smallest pixels pure black and white; we must not only warp the pixels conformally, but also rescale the magnetization. Ignoring the pixelation, the different regions look statistically similar. A gray large pixel mimics the average shade of similarly sized regions of tiny pixels.

$C(\mathbf{r}) = \langle \widehat{M}(f(z))\widehat{M}(f(z) + \mathbf{r}) \rangle$? *Explain your reasoning.*

This implies that if our conformal transformation takes a pixel at $M(z)$ to a warped pixel of area A at $f(z)$, it must rescale the magnetization by

$$\widehat{M}(f(z)) = |A|^{-1/16} M(z). \qquad (12.104)$$

You may use this to check your answer to part (c). (Note that the pixel area for a locally uniform compression by b changes by $|\mathrm{d}f/\mathrm{d}z|^2 = 1/b^2$.)

Figure 12.29 illustrates the self-similarity for the Ising model under rescaling, in analogy with the treatment of scale invariance in Fig. 12.11 for the random-field avalanche model. Here, unlike in that figure, as we zoom in we incorporate the renormalization of the magnetization by changing the grayscale.

Let us now explore the image of the Ising model under various analytic functions. Download [181] or generate [28] a snapshot of an Ising model, equilibrated at the critical temperature. Large systems will be slow. Debug everything with small systems (cropped from your main image). For larger systems, try saving your images without displaying them first. On current machines, 512×512 will be a stretch; anything over 128×128 is acceptable, but larger systems will make the physics a bit clearer.

The hints files will help you to import an Ising image, convert it into a two-dimensional array $S_{mn} = \pm 1$, and select an $L \times L$ subregion (especially while you debug your code). We imagine these spins spread over the unit square in the complex plane; our code generates a list of spins and square polygons associated to each, with the spin S_{mn} sitting at the center[74] $z_{mn} = ((m + \frac{1}{2}) + \mathrm{i}(n + \frac{1}{2}))/L$ of a $1/L \times 1/L$ square. The code will allow you to provide a function $f(z)$ of your choice, and will return the deformed quadrilaterals[75] centered at $f(z_{mn})$ with areas A_{mn}, and their associated rescaled spins $A_{mn}^{-1/16} S_{mn}$.[76] The software will also provide example routines showing how to output and shade in these quadrilaterals.

(d) *Generate a conformal transformation with*

a nonlinear complex analytic function of your choice, warping the Ising model in an interesting way. Zoom in to regions occupied by lots of small pixels, where the individual pixels are not obvious.[77] Include both your entire plot and a cropped figure showing the expanded zoom. Discuss your zoomed plot critically—does it appear that the Ising correlations and visual structures have been faithfully retained by your analytic transformation?

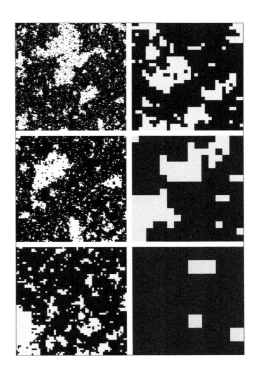

Fig. 12.29 Ising model under conformal rescaling. The "powers of two" rescaling of the avalanches in Fig. 12.11 ignored this rescaling of M. Here we show the Ising model, again with the lowest right-hand quarter of each square inflated to fill the next—but now properly faded according to eqn 12.104.

(e) *Load an Ising model equilibrated above the critical temperature at $T = 100$ (random noise), and one at $T = 3$ (short-range correlations).*

---

[74]If you work in Mathematica or Fortran, where the indices of arrays run from $(1 \ldots \mathrm{L})$, $z_{mn} = ((m - \frac{1}{2}) + \mathrm{i}(n - \frac{1}{2}))/L$.

[75]The routine will drop quadrilaterals that extend to infinity, and also will remove quadrilaterals with "negative area".

[76]The graphics routines will want one-dimensional lists of quadrilaterals and spins, even though they are on a grid.

[77]You can zoom either using the graphics software or (sometimes more efficient) by saving a vector-graphics figure (like pdf) and viewing it separately.

Distort and zoom each using your chosen conformal transformation. Analyze each and include your images. If you "blur your eyes" enough to ignore the individual pixels, can you tell how much the system has been dilated? Are your conformally transformed images faithfully retaining the correlations and visual structures away from the critical point? Away from T_c the original Ising model snapshot looks critical on scales small compared to the correlation length, ξ, and looks random on scales large compared to ξ. *For $T = 3$, which regions of your conformally transformed figure look qualitatively like T_c? Which regions look like $T = 100$? Why?*

(f) *Invent a* nonanalytic *function, and use it to distort your Ising model.* (Warning: most functions you write down, like $\log(\cosh^4(z + 1/z^2))$ will be analytic except at a few singularities. Two suggested methods: invent two unrelated real-valued functions $u(x, y)$ and $v(x, y)$ and form $f = u + iv$, or pick two different complex analytic functions $g(z)$ and $h(z)$ and use $f(z) = \text{Re}(g(z)) + \text{iIm}(h(z))$. Make sure your function is nonanalytic almost everywhere (e.g. violates the Cauchy–Riemann equations), not just at a point.)[78] *Find an example that makes for an interesting picture; include your images, including a zoom into a region with many pixels that range in size. As above, examine the images critically—does it appear that the Ising correlations and visual structures have been faithfully retained by your nonanalytic transformation? Describe the distortions you see.* (Are the pixels still approximately square?)

Conformal symmetry in two dimensions was studied in an outgrowth of string theory. The representations of the conformal group allowed them to deduce the exact critical exponents for all the usual two-dimensional statistical mechanical models—reproducing Onsager's result for the 2D Ising model, known solutions for the 2D tricritical Ising model, 2D percolation, etc. More recently, field theorists [101] have used conformal invariance in higher dimensions (with a strategy called *conformal bootstrap*) to produce bounds for critical exponents. They now hold the record on accuracy for the exponents of the 3D Ising model, giving $\beta = 0.326419(3)$, $\nu = 0.629971(4)$, and

$\delta = 4.78984(1)$.

(12.33) **Pandemic.**[79] (Epidemiology, Computation) ③ Perhaps the most substantive contribution to public health provided by physics is the application of statistical mechanics ideas to model disease propagation. In this exercise, we shall introduce a few categories of epidemiological models, discuss how they can inspire and inform public health strategies (once adapted to real-world data), and then study one model as a continuous phase transition. You should leave this exercise empowered to think about the public health responses and modeling of potential pandemics—Ebola, SARS, and now COVID-19. Perhaps a few of us will contribute to the field.

Pandemics can undergo a phase transition. For diseases like measles, a single contagious child in an environment where nobody is immune will infect between twelve and eighteen people before recovering, depending on details like population density. For influenza, this number is around two to three. We define the *basic reproduction number* R_0 to be this average number of people infected by a contagious person before recovering, in a fully susceptible community. R_0 thus is 12–18 for measles, 2–3 for influenza. For a new pathogen, where nobody is immune, $R_0 < 1$ will mean that an outbreak will eventually die out, and $R_0 > 1$ means that a large initial outbreak will spread globally until reaching a significant fraction of the entire population. Much effort is spent during a pandemic to lower R_0 into the safe range.

This transition is a continuous phase transition, with fluctuations on all scales near the critical threshold $R_0 = 1$. In this exercise, you will briefly consider three types of epidemic models (compartmental models, network models, and lattice models), compare different social interventions designed to lower R_0, and explore the fluctuations and critical behavior very close to threshold.

Compartmental models use coupled differential equations to model the disease spread between different *compartments* of the population. The classic SIR model (see Exercise 6.25) involves

---

[78]Warning: Quadrilaterals with negative area are not drawn. This will always happen for the conjugate $f^*(z)$ of an analytic function. Try $u - iv$ if $u + iv$ causes problems.

[79]This exercise was developed in collaboration with David Hathcock. Computer hints can be found at the book website [181].

three coupled compartments,

$$\frac{\mathrm{d}S}{\mathrm{d}t} = -\beta IS, \quad \frac{\mathrm{d}I}{\mathrm{d}t} = \beta IS - \gamma I, \quad \frac{\mathrm{d}R}{\mathrm{d}t} = \gamma I, \tag{12.105}$$

where $S(t)$, $I(t)$, and $R(t)$ are the proportions of the population that are susceptible, infected, and recovered. The parameter β measures the rate of infection spreading contact between people and γ is the rate at which people recover. Network models treat people as nodes, connected to their contacts with edges. They assume a *transmissibility* T, the average probability that a victim will infect each of their contacts. For low T the epidemics die out; there will be a critical T_c above which a large outbreak will continue to grow exponentially. There are a variety of networks studied: fully connected networks (where SIR models become valid), loopless branching tree networks where everyone has k neighbors, real-world networks gleaned from data on households and school attendance [132], and scale-free networks with a power-law distribution $p(k) \propto k^{-\alpha}$ for the probability that a person has k contacts (has *degree k*). (Scale-free networks have been found to approximate the pattern of interactions between proteins in cells and nodes on the Internet, and serve as our model for populations with wide variation in the number of social contacts with potential for disease transmission.) Lattice models—networks in two dimensions where only near neighbors are contacts—are sometimes used in agricultural settings, where the plant victims are immobile and the disease is spread only by direct proximity.

(a) *In the SIR model, if a person is first infected at t_0, what is the probability they are still infected, and contagious, at time t, given the recovery rate γ? If they infect people at a rate βS, how many will they infect before they recover, in a fully susceptible large population? Write R_0 for the SIR model in terms of β and γ.*

Network models usually ignore the long-range correlations between nodes: except for real-world networks, the contacts are usually picked at random so there are very few short loops. In that limit, Meyers et al. [132] express R_0 in terms of the moments $\langle k^n \rangle = \sum k^n p(k)$ of the degree distribution, which they solve for using generating functions (see Exercise 2.23):

$$R_0 = T \left(\frac{\langle k^2 \rangle}{\langle k \rangle} - 1 \right). \tag{12.106}$$

People like nurses and extroverts with a lot of contacts can act as *super-spreaders*, infecting large numbers of colleagues. Scale-free networks explore what happens with a range of contacts: the smaller the exponent α, the larger the range of variation.

(b) *What is the critical transmissibility T_c predicted by the network model in eqn 12.106? Show that, for a scale-free network with $\alpha \leq 3$ the critical transmissibility $T_c = 0$; no matter how unlikely a contact will cause disease spread, there are rare individuals with so many contacts that they (on average) will cause exponential growth of the pathogen. If our population had $\alpha = 3$, what percentage of the people would we need to vaccinate to immunize everyone with more than 100 contacts? What would the resulting T_c, the maximum safe transmissibility, be?* (If you find that the first percentage is small, use that fact to simplify your calculation of T_c. Hint: $\sum_1^\infty k^{-z} = \zeta(z)$, the Riemann zeta function, which diverges at $z = 1$.)

An important limitation of these network results is that they assume the population is structureless: apart from the degree distribution, the network is completely random. This is not the case in a 2D square lattice, for example. It has degree distribution $p_k = \delta_{k,4}$, but connections between nodes are defined by the lattice, and not randomly assigned. As you might expect, disease spread is closely related to percolation. In the mean-field theory, percolation predicts that the epidemic size distribution exponent is $\tau = 3/2 = 1.5$; you will explore this in parts (e) and (f). In 2D, the lattice structure changes the universality class, the epidemic sizes are given by the cluster-size distribution exponent $\tau = 187/91 \approx 2.055$.

Besides exhibiting different power-law scaling, the value of the critical transmissibility can be quite different in structured populations.

(c) *What is T_c for a branching tree with $k = 4$ branches at each node (so $p(k) = \delta_{k,4}$)?* (Hint: You can derive this directly, or use your answer to the first question in part (b).) *Compare that to the critical transmissibility for a 2D square lattice, $T_c = 0.5384$ [197]. Which is more resistant to disease spread?*

One might imagine that a lattice model would mimic the effect of travel restrictions to prevent disease spread. Travel restrictions reduce the contact numbers, but do not change the qualitative behavior. This is due to the *small world phenomenon*: a surprisingly small number of long-range contacts can change the qualitative behavior of a network (see Exercise 1.7). Only a few long-distance travelers are needed to make our world well connected.

Finally, let us numerically explore the fluctuations and scaling behavior exhibited by epidemics at their critical points. We shall assume (correctly) that our population is well connected. We shall also assume that our population does not have system-scale heterogeneities: we ignore cities, subpopulations of vulnerable and crowded people, and events like the Olympics. Given these assumptions, one can argue that the qualitative behavior near enough to the critical point $R_0 = 1$ is *universal*, and controlled not by the details of the network or SIR model but only by the distance $1 - R_0$ to the critical point.

Let us organize our victims in "generations" of infected people, with I_{n+1} the number of victims infected by the I_n people in generation n; we shall view the generation as roughly corresponding to the time evolution of the pandemic. The mean $\langle I_{n+1} \rangle = R_0 I_n$, but it will fluctuate about that value with a Poisson distribution, so I_{n+1} is a random integer chosen from a Poisson distribution with mean $R_0 I_n$.

(d) *Write a routine* `pandemicInstance`*, that returns the evolution vector* $[I_0, I_1 \ldots I_n \ldots]$ *and the total size* $S = \sum_n I_n$*. Iterate your routine with with* $R_0 = 0.9999$ *and* $I_0 = 1$ *in a loop until you find an epidemic with size* $S \geq 10^5$*. Plot the trajectory of this epidemic,* I_n *vs.* n*. Does your epidemic nearly halt during the time interval? Do the pieces of the epidemic before and after this near halt appear statistically similar to the entire epidemic?*

One might presume that these large fluctuations could pose a real challenge to guessing whether social policies designed to suppress a growing pandemic are working. We must note, however, that the fluctuations are important only near $R_0 = 1$, or when the infected population becomes small.

At $R_0 = 1$, the size of the epidemic S has a power-law probability density $P(S) \propto S^{-\tau}$ for large avalanches S.

(e) *Write a routine* `pandemicEnsemble` *that does not store the trajectory, but instead runs* N *epidemics at a given value of* R_0*, and returns a list of their sizes. Plot a histogram of the sizes of* 10^4 *epidemics with* $R_0 = 0.99$*, with, say, 100 bins.*

Regular histograms here are not useful; our distribution has a long but important tail of large events. Most epidemics subside quickly at this value of R_0, but a few last for hundreds of generations and infect tens of thousands of people. We need to convert to *logarithmic binning*.

(f) *Change the bins used in your histogram to increase logarithmically, and be sure to normalize so that the counts are divided by the "bin width" (the number of integers in that bin) and the number of epidemics being counted. Present the distribution of sizes for* 10^4 *epidemics at* $R_0 = 0.99$ *on log–log plots. On the same plot, show the power-law prediction* $\tau = 3/2$ *at the critical point.*

In Exercise 12.28 we derived the universal scaling form for the avalanche size distribution in the random-field Ising model. This calculation also applies to our pandemic model. It predicts that the probability $P(S)$ of an epidemic of size S for small distances $r = (1 - R_0)$ below the critical point is given by

$$P(S) = CS^{-3/2}e^{-Sr^2/2}, \qquad (12.107)$$

where the nonuniversal constant C is around 0.4 to 0.5 (depending on the small S cutoff). Note that this gives the predicted power law $\tau = 3/2$, and is cut off above a typical size that grows quadratically in $1/r$.

(g) *Multiply your data by* $S^{3/2}$ *to make it near constant for small sizes. Plot it and the scaling prediction (eqn 12.107) on a log–log plot. Does the universal scaling function describe your simulated epidemic ensemble?*

The tools we learn in statistical mechanics—generating functions, universality, power laws, and scaling functions—make tangible predictions for practical models of disease propagation. They work best in the region of greatest societal importance $R_0 \approx 1$, where costly efforts to contain the pandemic are minimized while avoiding uncontrolled growth.

Fourier methods

Why are Fourier methods important? Why is it so useful for us to transform functions of time and space $y(\mathbf{x}, t)$ into functions of frequency and wavevector $\widetilde{y}(\mathbf{k}, \omega)$?

- *Humans hear frequencies.* The human ear analyzes pressure variations in the air into different frequencies. Large frequencies ω are perceived as high pitches; small frequencies are low pitches. The ear, very roughly, does a Fourier transform of the pressure $P(t)$ and transmits $|\widetilde{P}(\omega)|^2$ to the brain.[1]

- *Diffraction experiments measure Fourier components.* Many experimental methods diffract waves (light, X-rays, electrons, or neutrons) off of materials (Section 10.2). These experiments typically probe the absolute square of the Fourier amplitude of whatever is scattering the incoming beam.

- *Common mathematical operations become simpler in Fourier space.* Derivatives, correlation functions, and convolutions can be written as simple products when the functions are Fourier transformed. This has been important to us when calculating correlation functions (eqn 10.4), summing random variables (Exercises 1.2 and 12.11), and calculating susceptibilities (eqns 10.30, 10.39, and 10.53, and Exercises 9.14 and 10.9). In each case, we turn a calculus calculation into algebra.

- *Linear differential equations in translationally invariant systems have solutions in Fourier space.*[2] We have used Fourier methods for solving the diffusion equation (Section 2.4.1), and more broadly to solve for correlation functions and susceptibilities (Chapter 10).

In Section A.1 we introduce the conventions typically used in physics for the Fourier series, Fourier transform, and fast Fourier transform. In Section A.2 we derive their integral and differential properties. In Section A.3, we interpret the Fourier transform as an orthonormal change-of-basis in function space. And finally, in Section A.4 we explain why Fourier methods are so useful for solving differential equations by exploring their connection to translational symmetry.

[1] Actually, this is how the ear *seems* to work, but not how it *does* work. First, the signal to the brain is time dependent, with the tonal information changing as a word or tune progresses; it is more like a wavelet transform, giving the frequency content in various time slices. Second, the phase information in $\widetilde{P}$ is not completely lost; power and pitch are the primary signal, but the relative phases of different pitches are also perceptible. Third, experiments have shown that the human ear is very nonlinear in its mechanical response.

[2] Translation invariance in Hamiltonian systems implies momentum conservation. This is why in quantum mechanics Fourier transforms convert position-space wavefunctions into momentum-space wavefunctions—even for systems which are not translation invariant.

A.1 Fourier conventions

Here we define the Fourier series, the Fourier transform, and the fast Fourier transform, as they are commonly defined in physics and as they are used in this text.

Statistical Mechanics: Entropy, Order Parameters, and Complexity. James P. Sethna, Oxford University Press (2021). © Oxford University Press. DOI:10.1093/oso/9780198865247.003.0013

The Fourier series for functions of time, periodic with period T, is

$$\widetilde{y}_m = \frac{1}{T} \int_0^T y(t) \exp(i\omega_m t)\, dt, \tag{A.1}$$

where $\omega_m = 2\pi m/T$, with integer m. The Fourier series can be resummed to retrieve the original function using the *inverse Fourier series*:

$$y(t) = \sum_{m=-\infty}^{\infty} \widetilde{y}_m \exp(-i\omega_m t). \tag{A.2}$$

Fourier series of functions in space are defined with the *opposite* sign convention[3] in the complex exponentials. Thus in a three-dimensional box of volume $V = L \times L \times L$ with periodic boundary conditions, these formulæ become

$$\widetilde{y}_{\mathbf{k}} = \frac{1}{V} \int y(\mathbf{x}) \exp(-i\mathbf{k} \cdot \mathbf{x})\, dV, \tag{A.3}$$

and

$$y(\mathbf{x}) = \sum_{\mathbf{k}} \widetilde{y}_{\mathbf{k}} \exp(i\mathbf{k} \cdot \mathbf{x}), \tag{A.4}$$

where the $\mathbf{k}$ run over a lattice of wavevectors

$$\mathbf{k}_{(m,n,o)} = [2\pi m/L, 2\pi n/L, 2\pi o/L] \tag{A.5}$$

in the box.

The Fourier transform is defined for functions on the entire infinite line:

$$\widetilde{y}(\omega) = \int_{-\infty}^{\infty} y(t) \exp(i\omega t)\, dt, \tag{A.6}$$

where now ω takes on all values.[4] We regain the original function by doing the inverse Fourier transform:

$$y(t) = \frac{1}{2\pi} \int_{-\infty}^{\infty} \widetilde{y}(\omega) \exp(-i\omega t)\, d\omega. \tag{A.7}$$

This is related to the inverse Fourier series by a continuum limit (Fig. A.1):

$$\frac{1}{2\pi} \int d\omega \approx \frac{1}{2\pi} \sum_{\omega} \Delta\omega = \frac{1}{2\pi} \sum_{\omega} \frac{2\pi}{T} = \frac{1}{T} \sum_{\omega}, \tag{A.8}$$

where the $1/T$ here compensates for the factor of T in the definitions of the forward Fourier series. In three dimensions the Fourier transform formula A.6 is largely unchanged,

$$\widetilde{y}(\mathbf{k}) = \int y(\mathbf{x}) \exp(-i\mathbf{k} \cdot \mathbf{x})\, dV, \tag{A.9}$$

while the inverse Fourier transform gets the cube of the prefactor:

$$y(\mathbf{x}) = \frac{1}{(2\pi)^3} \int_{-\infty}^{\infty} \widetilde{y}(\mathbf{k}) \exp(i\mathbf{k} \cdot \mathbf{x})\, d\mathbf{k}. \tag{A.10}$$

[3]This inconsistent convention allows waves of positive frequency to propagate forward rather than backward. A single component of the inverse transform, $e^{i\mathbf{k} \cdot \mathbf{x}} e^{-i\omega t} = e^{i(\mathbf{k} \cdot \mathbf{x} - \omega t)}$ propagates in the $+\mathbf{k}$ direction with speed $\omega/|\mathbf{k}|$; had we used a $+i$ for Fourier transforms in both space and time $e^{i(\mathbf{k} \cdot \mathbf{x} + \omega t)}$ would move *backward* (along $-\mathbf{k}$) for $\omega > 0$.

[4]Why do we divide by T or L for the series and not for the transform? Imagine a system in an extremely large box. Fourier series are used for functions which extend over the entire box; hence we divide by the box size to keep them finite as $L \to \infty$. Fourier transforms are usually used for functions which vanish quickly, so they remain finite as the box size gets large.

Fig. A.1 Approximating the integral as a sum. By approximating the integral $\int \widetilde{y}(\omega) \exp(i\omega t)\, d\omega$ as a sum over the equally spaced points ω_m, $\sum_m \widetilde{y}(\omega) \exp(i\omega_m t)\Delta\omega$, we can connect the formula for the Fourier transform to the formula for the Fourier series, explaining the factor $1/2\pi$ in eqn A.7.

The fast Fourier transform (FFT) starts with N equally spaced data points y_ℓ, and returns a new set of complex numbers $\widetilde{y}_m^{\mathrm{FFT}}$:

$$\widetilde{y}_m^{\mathrm{FFT}} = \sum_{\ell=0}^{N-1} y_\ell \exp(\mathrm{i}2\pi m\ell/N), \qquad (\mathrm{A.11})$$

with $m = 0, \ldots, N-1$. The inverse of the FFT is given by

$$y_\ell = \frac{1}{N} \sum_{m=0}^{N-1} \widetilde{y}_m^{\mathrm{FFT}} \exp(-\mathrm{i}2\pi m\ell/N). \qquad (\mathrm{A.12})$$

The FFT essentially samples the function $y(t)$ at equally spaced points $t_\ell = \ell T/N$ for $\ell = 0, \ldots, N-1$:

$$\widetilde{y}_m^{\mathrm{FFT}} = \sum_{\ell=0}^{N-1} y_\ell \exp(\mathrm{i}\omega_m t_\ell). \qquad (\mathrm{A.13})$$

It is clear from eqn A.11 that $\widetilde{y}_{m+N}^{\mathrm{FFT}} = \widetilde{y}_m^{\mathrm{FFT}}$, so the FFT is periodic with period $\omega_N = 2\pi N/T$. The inverse transform can also be written

$$y_\ell = \frac{1}{N} \sum_{m=-N/2+1}^{N/2} \widetilde{y}_m^{\mathrm{FFT}} \exp(-\mathrm{i}\omega_m t_\ell), \qquad (\mathrm{A.14})$$

where we have centered[5] the sum ω_m at $\omega = 0$ by using the periodicity.[6]

Often the values $y(t)$ (or the data points y_ℓ) are real. In this case, eqns A.1 and A.6 show that the negative Fourier amplitudes are the complex conjugates of the positive ones: $\widetilde{y}(\omega) = \widetilde{y}^*(-\omega)$. Hence for real functions the real part of the Fourier amplitude will be even and the imaginary part will be odd.[7]

The reader may wonder why there are so many versions of roughly the same Fourier operation.

(1) The function $y(t)$ can be defined on a finite interval with periodic boundary conditions on $(0, T)$ (series, FFT) or defined in all space (transform). In the periodic case, the Fourier coefficients are defined only at discrete wavevectors $\omega_m = 2\pi m/T$ consistent with the periodicity of the function; in the infinite system the coefficients are defined at all ω.

(2) The function $y(t)$ can be defined at a discrete set of N points $t_n = n\Delta t = nT/N$ (FFT), or at all points t in the range (series, transform). If the function is defined only at discrete points, the Fourier coefficients are periodic with period $\omega_N = 2\pi/\Delta t = 2\pi N/T$.[8]

There are several arbitrary choices made in defining these Fourier methods, that vary from one field to another:

• Some use the notation $\mathrm{j} = \sqrt{-1}$ instead of i.

[5] If N is odd, to center the FFT the sum should be taken over $-(N-1)/2 \leq m \leq (N-1)/2$.

[6] Notice that the FFT returns the negative ω Fourier coefficients as the last half of the array, $m = N/2 + 1, N/2 + 2, \ldots$. (This works because $-N/2 + j$ and $N/2 + j$ differ by N, the periodicity of the FFT.) One must be careful about this when using Fourier transforms to solve calculus problems numerically. For example, to evolve a density $\rho(x)$ under the diffusion equation (Section 2.4.1) one must multiply the first half of the array $\widetilde{\rho}_m$ by $\exp(-Dk_m^2 t) = \exp(-D[m(2\pi/L)]^2 t)$, but multiply the second half by $\exp(-D(K - k_m)^2 t) = \exp(-D[(N - m)(2\pi/L)]^2 t)$.

[7] This allows one to write slightly faster FFTs specialized for real functions. One pays for the higher speed by an extra programming step unpacking the resulting Fourier spectrum.

[8] There is one more logical possibility: a discrete set of points that fill all space; the atomic displacements in an infinite crystal is the classic example. In Fourier space, such a system has continuous k, but periodic boundary conditions at $\pm K/2 = \pm \pi/a$ (the edges of the *Brillouin zone*).

[9]The real world is invariant under the transformation i ↔ −i, but complex quantities will get conjugated. Swapping i for −i in the time series formulæ, for example, make χ analytic in the *lower* half-plane (Fig. 10.12). Some who do swap i ↔ −i also choose to define $\chi''(\omega) = -\text{Im}[\chi(\omega)]$ instead of $+\text{Im}[\chi(\omega)]$ (eqn 10.31) so it remains the same under both conventions.

- More substantively, some use the complex conjugate of our formulæ, substituting −i for i in the time or space transform formulæ. This alternative convention makes no change for any real quantity.[9]
- Some use a $1/\sqrt{T}$ and $1/\sqrt{2\pi}$ factor symmetrically on the Fourier and inverse Fourier operations.
- Some use frequency and wavelength ($f = 2\pi\omega$ and $\lambda = 2\pi/k$) instead of angular frequency ω and wavevector k. This makes the transform and inverse transform more symmetric, and avoids some of the prefactors.

Our Fourier conventions are those most commonly used in physics.

A.2   Derivatives, convolutions, and correlations

The important differential and integral operations become multiplications in Fourier space. A calculus problem in t or x thus becomes an algebra exercise in ω or k.

Integrals and derivatives. Because $(\text{d}/\text{d}t)\,e^{-i\omega t} = -i\omega e^{i\omega t}$, the Fourier coefficient of the derivative of a function $y(t)$ is $-i\omega$ times the Fourier coefficient of the function:

$$\text{d}y/\text{d}t = \sum \widetilde{y}_m \left(-i\omega_m \exp(-i\omega_m t)\right) = \sum (-i\omega_m \widetilde{y}_m) \exp(-i\omega_m t),$$

(A.15)

so

$$\left.\frac{\widetilde{\text{d}y}}{\text{d}t}\right|_\omega = -i\omega \widetilde{y}_\omega.$$

(A.16)

This holds also for the Fourier transform and the fast Fourier transform. Since the derivative of the integral gives back the original function, the Fourier series for the indefinite integral of a function y is thus given by dividing by $-i\omega$:

$$\widetilde{\int y(t)\,\text{d}t} = \frac{\widetilde{y}_\omega}{-i\omega} = i\frac{\widetilde{y}_\omega}{\omega}$$

(A.17)

[10]Either the mean $\widetilde{y}(\omega = 0)$ is zero or it is nonzero. If the mean of the function is zero, then $\widetilde{y}(\omega)/\omega = 0/0$ is undefined at $\omega = 0$. This makes sense; the indefinite integral has an arbitrary integration constant, which gives its Fourier series an arbitrary value at $\omega = 0$. If the mean of the function $\bar{y}$ is not zero, then the integral of the function will have a term $\bar{y}(t - t_0)$. Hence the integral is not periodic and has no Fourier series. (On the infinite interval the integral has no Fourier transform because it is not in $\mathbb{L}^2$.)

except at $\omega = 0$.[10]

These relations are invaluable in the solution of many linear partial differential equations. For example, we saw in Section 2.4.1 that the diffusion equation

$$\frac{\partial\rho}{\partial t} = D\frac{\partial^2\rho}{\partial x^2}$$

(A.18)

becomes manageable when we Fourier transform x to k:

$$\frac{\partial\widetilde{\rho}_k}{\partial t} = -Dk^2\widetilde{\rho},$$

(A.19)

$$\widetilde{\rho}_k(t) = \rho_k(0)\exp(-Dk^2 t).$$

(A.20)

[11]The absolute square of the Fourier transform of a time signal is called the *power spectrum*.

Correlation functions and convolutions. The absolute square of the Fourier transform[11] $|\widetilde{y}(\omega)|^2$ is given by the Fourier transform of the correlation function $C(\tau) = \langle y(t)y(t+\tau)\rangle$:

$$
\begin{aligned}
|\widetilde{y}(\omega)|^2 = \widetilde{y}(\omega)^*\widetilde{y}(\omega) &= \int \mathrm{d}t'\,\mathrm{e}^{-\mathrm{i}\omega t'}y(t')\int \mathrm{d}t\,\mathrm{e}^{\mathrm{i}\omega t}y(t) \\
&= \int \mathrm{d}t\,\mathrm{d}t'\,\mathrm{e}^{\mathrm{i}\omega(t-t')}y(t')y(t) = \int \mathrm{d}\tau\,\mathrm{e}^{\mathrm{i}\omega\tau}\int \mathrm{d}t'\,y(t')y(t'+\tau) \\
&= \int \mathrm{d}\tau\,\mathrm{e}^{\mathrm{i}\omega\tau}T\langle y(t)y(t+\tau)\rangle = T\int \mathrm{d}\tau\,\mathrm{e}^{\mathrm{i}\omega\tau}C(\tau) \\
&= T\widetilde{C}(\omega), && \text{(A.21)}
\end{aligned}
$$

where T is the total time t during which the Fourier spectrum is being measured. Thus diffraction experiments, by measuring the square of the **k**-space Fourier transform, give us the spatial correlation function for the system (Section 10.2).

The convolution[12] $h(z)$ of two functions $f(x)$ and $g(y)$ is defined as

$$h(z) = \int f(x)g(z-x)\,\mathrm{d}x. \qquad \text{(A.22)}$$

The Fourier transform of the convolution is the product of the Fourier transforms. In three dimensions,[13]

$$
\begin{aligned}
\widetilde{f}(\mathbf{k})\widetilde{g}(\mathbf{k}) &= \int \mathrm{e}^{-\mathrm{i}\mathbf{k}\cdot\mathbf{x}}f(\mathbf{x})\,\mathrm{d}\mathbf{x}\int \mathrm{e}^{-\mathrm{i}\mathbf{k}\cdot\mathbf{y}}g(\mathbf{y})\,\mathrm{d}\mathbf{y} \\
&= \int \mathrm{e}^{-\mathrm{i}\mathbf{k}\cdot(\mathbf{x}+\mathbf{y})}f(\mathbf{x})g(\mathbf{y})\,\mathrm{d}\mathbf{x}\,\mathrm{d}\mathbf{y} = \int \mathrm{e}^{-\mathrm{i}\mathbf{k}\cdot\mathbf{z}}\,\mathrm{d}\mathbf{z}\int f(\mathbf{x})g(\mathbf{z}-\mathbf{x})\,\mathrm{d}\mathbf{x} \\
&= \int \mathrm{e}^{-\mathrm{i}\mathbf{k}\cdot\mathbf{z}}h(\mathbf{z})\,\mathrm{d}\mathbf{z} = \widetilde{h}(\mathbf{k}). && \text{(A.23)}
\end{aligned}
$$

[12]Convolutions show up in sums and Green's functions. The sum $\mathbf{z} = \mathbf{x} + \mathbf{y}$ of two random vector quantities with probability distributions $f(\mathbf{x})$ and $g(\mathbf{y})$ has a probability distribution given by the convolution of f and g (Exercise 1.2). An initial condition $f(\mathbf{x}, t_0)$ propagated in time to $t_0 + \tau$ is given by convolving with a Green's function $g(\mathbf{y}, \tau)$ (Section 2.4.2).

[13]The convolution and correlation theorems are closely related; we do convolutions in time and correlations in space to illustrate both the one-dimensional and vector versions of the calculation.

A.3 Fourier methods and function space

There is a nice analogy between the space of vectors $\mathbf{r}$ in three dimensions and the space of functions $y(t)$ periodic with period T, which provides a simple way of thinking about Fourier series. It is natural to define our function space to including all complex functions $y(t)$. (After all, we want the complex Fourier plane-waves $\mathrm{e}^{-\mathrm{i}\omega_m t}$ to be in our space.) Let us list the following common features of these two spaces.

- **Vector space.** A vector $\mathbf{r} = (r_1, r_2, r_3)$ in $\mathbb{R}^3$ can be thought of as a real-valued function on the set $\{1, 2, 3\}$. Conversely, the function $y(t)$ can be thought of as a vector with one complex component for each $t \in [0, T)$.

  Mathematically, this is an evil analogy. Most functions which have independent random values for each point t are undefinable, unintegrable, and generally pathological. The space becomes well defined if we confine ourselves to functions $y(t)$ whose absolute squares $|y(t)|^2 = y(t)y^*(t)$ can be integrated. This vector space of functions is called $\mathbb{L}^2$.[14]

[14]More specifically, the Fourier transform is usually defined on $\mathbb{L}^2[\mathbb{R}]$, and the Fourier series is defined on $\mathbb{L}^2[0, T]$.

- **Inner product.** The analogy to the dot product of two three-dimensional vectors $\mathbf{r} \cdot \mathbf{s} = r_1 s_1 + r_2 s_2 + r_3 s_3$ is an inner product between two functions y and z:

$$y \cdot z = \frac{1}{T} \int_0^T y(t) z^*(t) \, \mathrm{d}t. \tag{A.24}$$

You can think of this inner product as adding up all the products $y_t z_t^*$ over all points t, except that we weight each point by $\mathrm{d}t/T$.

- **Norm.** The distance between two three-dimensional vectors $\mathbf{r}$ and $\mathbf{s}$ is given by the *norm* of the difference $|\mathbf{r} - \mathbf{s}|$. The norm of a vector is the square root of the dot product of the vector with itself, so $|\mathbf{r} - \mathbf{s}| = \sqrt{(\mathbf{r} - \mathbf{s}) \cdot (\mathbf{r} - \mathbf{s})}$. To make this inner product norm work in function space, we need to know that the inner product of a function with itself is never negative. This is why, in our definition A.24, we took the complex conjugate of $z(t)$. This norm on function space is called the L^2 norm:

$$\|y\|_2 = \sqrt{\frac{1}{T} \int_0^T |y(t)|^2 \, \mathrm{d}t}. \tag{A.25}$$

Thus our restriction to square-integrable functions makes the norm of all functions in our space finite.[15]

[15]Another important property is that the only vector whose norm is zero is the zero vector. There are many functions whose absolute squares have integral zero, like the function which is zero except at $T/2$, where it is one, and the function which is zero on irrationals and one on rationals. Mathematicians finesse this difficulty by defining the vectors in $\mathbb{L}^2$ not to be functions, but rather to be *equivalence classes* of functions whose relative distance is zero. Hence the zero vector in $\mathbb{L}^2$ includes all functions with norm zero.

- **Basis.** A natural basis for $\mathbb{R}^3$ is given by the three unit vectors $\hat{\mathbf{x}}_1$, $\hat{\mathbf{x}}_2$, $\hat{\mathbf{x}}_3$. A natural basis for our space of functions is given by the functions $\hat{f}_m = \mathrm{e}^{-\mathrm{i}\omega_m t}$, with $\omega_m = 2\pi m/T$ to keep them periodic with period T.

- **Orthonormality.** The basis in $\mathbb{R}^3$ is orthonormal, with $\hat{\mathbf{x}}_i \cdot \hat{\mathbf{x}}_j$ equaling one if $i = j$ and zero otherwise. Is this also true of the vectors in our basis of plane waves? They are normalized:

$$\|\hat{f}_m\|_2^2 = \frac{1}{T} \int_0^T |\mathrm{e}^{-\mathrm{i}\omega_m t}|^2 \, \mathrm{d}t = 1. \tag{A.26}$$

They are also orthogonal, with

$$\begin{aligned} \hat{f}_m \cdot \hat{f}_n &= \frac{1}{T} \int_0^T \mathrm{e}^{-\mathrm{i}\omega_m t} \mathrm{e}^{\mathrm{i}\omega_n t} \, \mathrm{d}t = \frac{1}{T} \int_0^T \mathrm{e}^{-\mathrm{i}(\omega_m - \omega_n)t} \, \mathrm{d}t \\ &= \frac{1}{-\mathrm{i}(\omega_m - \omega_n)T} \left. \mathrm{e}^{-\mathrm{i}(\omega_m - \omega_n)t} \right|_0^T = 0 \end{aligned} \tag{A.27}$$

(unless $m = n$) since $\mathrm{e}^{-\mathrm{i}(\omega_m - \omega_n)T} = \mathrm{e}^{-\mathrm{i}2\pi(m-n)} = 1 = \mathrm{e}^{-\mathrm{i}0}$.

- **Coefficients.** The coefficients of a three-dimensional vector are given by taking dot products with the basis vectors: $r_n = \mathbf{r} \cdot \hat{\mathbf{x}}_n$. The analogy in function space gives us the definition of the Fourier coefficients, eqn A.1:

$$\tilde{y}_m = y \cdot \hat{f}_m = \frac{1}{T} \int_0^T y(t) \exp(\mathrm{i}\omega_m t) \, \mathrm{d}t. \tag{A.28}$$

- **Completeness.** We can write an arbitrary three-dimensional vector $\mathbf{r}$ by summing the basis vectors weighted by the coefficients: $\mathbf{r} =$

$\sum r_n \hat{\mathbf{x}}_n$. The analogy in function space gives us the formula A.2 for the inverse Fourier series:

$$y = \sum_{m=-\infty}^{\infty} \widetilde{y}_m \widehat{f}_m,$$

$$y(t) = \sum_{m=-\infty}^{\infty} \widetilde{y}_m \exp(-i\omega_m t). \tag{A.29}$$

One can show that any function in $\mathbb{L}^2$ can be expanded in our basis $\widehat{f}_m$; eqn A.29 converges.

Our coefficient eqn A.28 follows from our completeness eqn A.29 and orthonormality:

$$\widetilde{y}_\ell \overset{?}{=} y \cdot \widehat{f}_\ell = \left(\sum_m \widetilde{y}_m \widehat{f}_m \right) \cdot \widehat{f}_\ell$$

$$= \sum_m \widetilde{y}_m \left(\widehat{f}_m \cdot \widehat{f}_\ell \right) = \widetilde{y}_\ell \tag{A.30}$$

or, writing things out,

$$\widetilde{y}_\ell \overset{?}{=} \frac{1}{T} \int_0^T y(t) e^{i\omega_\ell t} \, dt$$

$$= \frac{1}{T} \int_0^T \left(\sum_m \widetilde{y}_m e^{-i\omega_m t} \right) e^{i\omega_\ell t} \, dt$$

$$= \sum_m \widetilde{y}_m \left(\frac{1}{T} \int_0^T e^{-i\omega_m t} e^{i\omega_\ell t} \, dt \right) = \widetilde{y}_\ell. \tag{A.31}$$

Our function space, together with our inner product (eqn A.24), is a *Hilbert space* (a complete inner product space).

A.4   Fourier and translational symmetry

Why are Fourier methods so useful? Specifically, why are the solutions to linear differential equations so often given by plane waves: sines and cosines and e^{ikx}?[16] Most of our basic equations are derived for systems with a *translational symmetry*. Time-translational invariance holds for any system without an explicit external time-dependent force; invariance under spatial translations holds for all homogeneous systems.

Why are plane waves special for systems with translational invariance? *Plane waves are the eigenfunctions of the translation operator.* Define $\mathcal{T}_\Delta$ (Fig. A.2), an operator which takes function space into itself, and acts to shift the function a distance Δ to the right:[17]

$$\mathcal{T}_\Delta \{f\}(x) = f(x - \Delta). \tag{A.32}$$

Any solution $f(x,t)$ to a translation-invariant equation will be mapped by $\mathcal{T}_\Delta$ onto another solution. Moreover, $\mathcal{T}_\Delta$ is a linear operator (translating the sum is the sum of the translated functions). If we think of the

Fig. A.2 The mapping $\mathcal{T}_\Delta$ takes function space into function space, shifting the function to the right by a distance Δ. For a physical system that is translation invariant, a solution translated to the right is still a solution.

[16]It is true, we are making a big deal about what is usually called the separation of variables method. But why does separation of variables so often work, and why do the separated variables so often form sinusoids and exponentials?

[17]That is, if $g = \mathcal{T}_\Delta \{f\}$, then $g(x) = f(x - \Delta)$, so g is f shifted to the right by Δ.

[18]You are familiar with eigenvectors of 3×3 symmetric matrices M, which transform into multiples of themselves when multiplied by M, $M \cdot \mathbf{e}_n = \lambda_n \mathbf{e}_n$. The translation $\mathcal{T}_\Delta$ is a linear operator on function space just as M is a linear operator on $\mathbb{R}^3$.

[19]The real exponential e^{Ax} is also an eigenstate, with eigenvalue $\mathrm{e}^{-A\Delta}$. This is also allowed. Indeed, the diffusion equation is time-translation invariant, and it has solutions which decay exponentially in time ($\mathrm{e}^{-\omega_k t}\mathrm{e}^{ikx}$, with $\omega_k = Dk^2$). Exponentially decaying solutions in space also arise in some translation-invariant problems, such as quantum tunneling and the penetration of electromagnetic radiation into metals.

[20]Written out in equations, this simple idea is even more obscure. Let $\mathcal{U}_t$ be the time-evolution operator for a translationally invariant equation (like the diffusion equation of Section 2.2). That is, $\mathcal{U}_t\{\rho\}$ evolves the function $\rho(x,\tau)$ into $\rho(x, \tau + t)$. ($\mathcal{U}_t$ is not translation in time, but evolution in time.) Because our system is translation invariant, translated solutions are also solutions for translated initial conditions: $\mathcal{T}_\Delta\{\mathcal{U}_t\{\rho\}\} = \mathcal{U}_t\{\mathcal{T}_\Delta\{\rho\}\}$. Now, if $\rho_k(x,0)$ is an eigenstate of $\mathcal{T}_\Delta$ with eigenvalue λ_k, is $\rho_k(x,t) = \mathcal{U}_t\{\rho_k\}(x)$ an eigenstate with the same eigenvalue? Yes indeed:

$$\mathcal{T}_\Delta\{\rho_k(x,t)\} = \mathcal{T}_\Delta\{\mathcal{U}_t\{\rho_k(x,0)\}\}$$
$$= \mathcal{U}_t\{\mathcal{T}_\Delta\{\rho_k(x,0)\}\}$$
$$= \mathcal{U}_t\{\lambda_k\rho_k(x,0)\}$$
$$= \lambda_k\mathcal{U}_t\{\rho_k(x,0)\}$$
$$= \lambda_k\rho_k(x,t) \qquad \text{(A.34)}$$

because the evolution law $\mathcal{U}_t$ is linear.

translation operator as a big matrix acting on function space, we can ask for its eigenvalues[18] and eigenvectors (or *eigenfunctions*) f_k:

$$\mathcal{T}_\Delta\{f_k\}(x) = f_k(x - \Delta) = \lambda_k f_k(x). \qquad \text{(A.33)}$$

This equation is solved by our complex plane waves $f_k(x) = \mathrm{e}^{ikx}$, with $\lambda_k = \mathrm{e}^{-ik\Delta}$.[19]

Why are these eigenfunctions useful? The time evolution of an eigenfunction must have the same eigenvalue λ! The argument is something of a tongue-twister: translating the time-evolved eigenfunction gives the same answer as time evolving the translated eigenfunction, which is time evolving λ times the eigenfunction, which is λ times the time-evolved eigenfunction.[20]

The fact that the different eigenvalues do not mix under time evolution is precisely what made our calculation work; time evolving $A_0\mathrm{e}^{ikx}$ had to give a multiple $A(t)\mathrm{e}^{ikx}$ since there is only one eigenfunction of translations with the given eigenvalue. Once we have reduced the partial differential equation to an ordinary differential equation for a few eigenstate amplitudes, the calculation becomes feasible.

Quantum physicists will recognize the tongue-twister above as a statement about simultaneously diagonalizing commuting operators: since translations commute with time evolution, one can find a complete set of translation eigenstates which are also time-evolution solutions. Mathematicians will recognize it from group representation theory: the solutions to a translation-invariant linear differential equation form a representation of the translation group, and hence they can be decomposed into irreducible representations of that group. These approaches are basically equivalent, and very powerful. One can also use these approaches for systems with other symmetries. For example, just as the invariance of homogeneous systems under translations leads to plane-wave solutions with definite wavevector k, it is true that:

- the invariance of isotropic systems (like the hydrogen atom) under the rotation group leads naturally to solutions involving spherical harmonics with definite angular momenta ℓ and m;

- the invariance of the strong interaction under SU(3) leads naturally to the *8-fold way* families of mesons and baryons; and

- the invariance of the Universe under the Poincaré group of space-time symmetries (translations, rotations, and Lorentz boosts) leads naturally to particles with definite mass and spin!

Exercises

We begin with three Fourier series exercises, *Sound wave*, *Fourier cosines* (numerical), and *Double sinusoids*. We then explore Fourier transforms with *Fourier Gaussians* (numerical) and *Uncertainty*, focusing on the effects of translating and scaling the width of the function to be transformed. *Fourier relationships* analyzes the normalization needed to go from the FFT to the Fourier series, and *Aliasing and windowing* explores two common numerical inaccuracies associated with the FFT. *White noise* explores the behavior of Fourier methods on random functions. *Fourier matching* is a quick, visual test of one's understanding of Fourier methods. Finally, *Gibbs phenomenon* explores what happens when you torture a Fourier series by insisting that smooth sinusoids add up to a function with a jump discontinuity.

(A.1) **Sound wave.** ①

A musical instrument playing a note of frequency ω_1 generates a pressure wave $P(t)$ periodic with period $2\pi/\omega_1$: $P(t) = P(t + 2\pi/\omega_1)$. The complex Fourier series of this wave (eqn A.2) is zero except for $m = \pm 1$ and ± 2, corresponding to the fundamental ω_1 and the first overtone. At $m = 1$, the Fourier amplitude is $2 - i$, at $m = -1$ it is $2 + i$, and at $m = \pm 2$ it is 3. What is the pressure $P(t)$:
(a) $\exp\left((2 + i)\omega_1 t\right) + 2\exp\left(3\omega_1 t\right)$,
(b) $\exp\left(2\omega_1 t\right)\exp\left(i\omega_1 t\right) \times 2\exp\left(3\omega_1 t\right)$,
(c) $\cos 2\omega_1 t - \sin\omega_1 t + 2\cos 3\omega_1 t$,
(d) $4\cos\omega_1 t - 2\sin\omega_1 t + 6\cos 2\omega_1 t$,
(e) $4\cos\omega_1 t + 2\sin\omega_1 t + 6\cos 2\omega_1 t$?

(A.2) **Fourier cosines.** (Computation) ②

In this exercise, we will use the computer to illustrate features of Fourier series and discrete fast Fourier transforms using sinusoidal waves. First, we will take the Fourier series of periodic functions $y(x) = y(x + L)$ with $L = 20$. We will sample the function at $N = 32$ points, and use a FFT to approximate the Fourier series. The Fourier series will be plotted as functions of k, at $-k_{N/2}$, ..., $k_{N/2-2}$, $k_{N/2-1}$. (Remember that the negative m points are given by the last half of the FFT.)
(a) *Analytically (that is, with paper and pencil) derive the Fourier series $\tilde{y}_m$ in this interval for*
$\cos(k_1 x)$ *and* $\sin(k_1 x)$. *Hint: They are zero except at the two values* $m = \pm 1$. *Use the spatial transform (eqn A.3).*
(b) *What spacing δk between k-points k_m do you expect to find? What is $k_{N/2}$? Evaluate each analytically as a function of L and numerically for* $L = 20$.
Numerically (on the computer) choose a cosine wave $A\cos(k(x - x_0))$, evaluated at 32 points from $x = 0$ to 20 as described previously, with $k = k_1 = 2\pi/L$, $A = 1$, and $x_0 = 0$. Examine its Fourier series.
(c) *Check your predictions from part (a) for the Fourier series for* $\cos(k_1 x)$ *and* $\sin(k_1 x)$. *Check your predictions from part (b) for δk and for* $k_{N/2}$.
Decrease k to increase the number of wavelengths, keeping the number of data points fixed. Notice that the Fourier series looks fine, but that the real-space curves quickly begin to vary in amplitude, much like the patterns formed by beating (superimposing two waves of different frequencies). By increasing the number of data points, you can see that the beating effect is due to the small number of points we sample. Even for large numbers of sampled points N, though, beating will still happen at very small wavelengths (when we get close to $k_{N/2}$). Try various numbers of waves m up to and past $m = N/2$.

(A.3) **Double sinusoid.** ②

Which plot in A–E gives the spatial Fourier series (eqn A.4) associated with the function $f(x) = 3\sin(x) + \cos(2x)$? *(The solid line is the real part, the dashed line is the imaginary part.)*

(B)

(C)

(D)

(E)

(A.4) **Fourier Gaussians.** (Computation) ②

In this exercise, we will use the computer to illustrate features of Fourier transforms, focusing on the particular case of Gaussian functions, but illustrating general properties.

The Gaussian distribution (also known as the normal distribution) has the form

$$G(x) = \frac{1}{\sqrt{2\pi}\sigma} \exp\left(-(x-x_0)^2/2\sigma^2\right), \quad (A.35)$$

where σ is the standard deviation and x_0 is the

center. Let

$$G_0(x) = \frac{1}{\sqrt{2\pi}} \exp\left(-x^2/2\right) \quad (A.36)$$

be the Gaussian of mean zero and $\sigma = 1$. The Fourier transform of G_0 is another Gaussian, of standard deviation one, but no normalization factor:[21]

$$\widetilde{G}_0(k) = \exp(-k^2/2). \quad (A.37)$$

In this exercise, we study how the Fourier transform of $G(x)$ varies as we change σ and x_0.

Widths. As we make the Gaussian narrower (smaller σ), it becomes more pointy. Shorter lengths mean higher wavevectors, so we expect that its Fourier transform will get wider.

(a) *Starting with the Gaussian with $\sigma = 1$, numerically measure the width of its Fourier transform at some convenient height. The full width at half maximum (FWHM) is a sensible choice. Change σ to 2 and to 0.1, and measure the widths, to verify that the Fourier space width goes inversely with the real width.*

(b) *Analytically show that this rule is true in general. Change variables in eqn A.6 to show that if $z(x) = y(Ax)$ then $\widetilde{z}(k) = \widetilde{y}(k/A)/A$. Using eqn A.36 and this general rule, write a formula for the Fourier transform of a Gaussian centered at zero with arbitrary width σ.*

(c) *Analytically compute the product $\Delta x \, \Delta k$ of the FWHM of the Gaussians in real and Fourier space. (Your answer should be independent of the width σ.)* This is related to the Heisenberg uncertainty principle, $\Delta x \, \Delta p \sim \hbar$, which you learn about in quantum mechanics.

Translations. Notice that a narrow Gaussian centered at some large distance x_0 is a reasonable approximation to a δ-function. We thus expect that its Fourier transform will be similar to the plane wave $\widetilde{G}(k) \sim \exp(-ikx_0)$ we would get from $\delta(x - x_0)$.

(d) *Numerically change the center of the Gaussian. How does the Fourier transform change? Convince yourself that it is being multiplied by the factor $\exp(-ikx_0)$. How does the power spectrum $|\widetilde{G}(\omega)|^2$ change as we change x_0?*

(e) *Analytically show that this rule is also true in general. Change variables in eqn A.6 to show that if $z(x) = y(x - x_0)$ then $\widetilde{z}(k) = \exp(-ikx_0)\widetilde{y}(k)$. Using this general rule, extend*

---

[21]Here is an elementary-*looking* derivation. We complete the square inside the exponent, and change from x to $y = x + ik$: $\frac{1}{\sqrt{2\pi}}\int_{-\infty}^{\infty} e^{-ikx}e^{-x^2/2}\,dx = \frac{1}{\sqrt{2\pi}}\int_{-\infty}^{\infty} e^{-(x+ik)^2/2}\,dx\,e^{(ik)^2/2} = [\int_{-\infty+ik}^{\infty+ik}\frac{1}{\sqrt{2\pi}}e^{-y^2/2}\,dy]\,e^{-k^2/2} = (1)e^{-k^2/2}$. But the normalized Gaussian is being integrated over the contour $\mathrm{Im}[y] = k$; we need Cauchy's theorem (Fig. 10.11) to shift the contour to the real axis to show that it is equal to one.

your answer from part (b) to write the formula for the Fourier transform of a Gaussian of width σ and center x_0.

(A.5) **Uncertainty.** ②

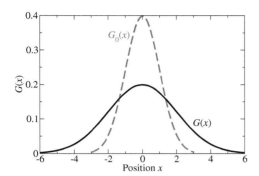

Fig. A.3 Real-space Gaussians.

The dashed line in Fig. A.3 shows

$$G_0(x) = 1/\sqrt{2\pi} \exp(-x^2/2). \qquad (A.38)$$

The dark line shows another function $G(x)$. The areas under the two curves $G(x)$ and $G_0(x)$ are the same. The dashed lines in the choices (A–E) represent the Fourier transform $\widetilde{G}_0(k) = \exp(-k^2/2)$. Which has a solid curve that represents the Fourier transform of G?

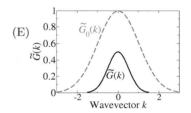

(A.6) **Fourier relationships.** ②

In this exercise, we explore the relationships between the Fourier series and the fast Fourier transform. The first is continuous and periodic in real space, and discrete and unbounded in Fourier space; the second is discrete and periodic both in real and in Fourier space. Thus, we must again convert integrals into sums (as in Fig. A.1).

As we take the number of points N in our FFT to ∞ the spacing between the points gets smaller and smaller, and the approximation of the integral as a sum gets better and better.

Let $y_\ell = y(t_\ell)$ where $t_\ell = \ell(T/N) = \ell(\Delta t)$. Approximate the Fourier series integral A.1 above as a sum over y_ℓ, $(1/T)\sum_{\ell=0}^{N-1} y(t_\ell) \exp(-\mathrm{i}\omega_m t_\ell)\Delta t$. For small positive m, give the constant relating $\tilde{y}_m^{\mathrm{FFT}}$ to the Fourier series coefficient $\tilde{y}_m$.

(A.7) **Aliasing and windowing.** (Computation) ③

In this exercise, we will use the computer to illustrate numerical challenges in using the FFT.

The Fourier series $\widetilde{y}_m$ runs over all integers m. The FFT runs only over $0 \leq m < N$. There are three ways to understand this difference: function-space dimension, wavelengths, and aliasing.

Function-space dimension. The space of periodic functions $y(t)$ on $0 \leq t < T$ is infinite, but we are sampling them only at N points. The space of possible fast Fourier series must also have N dimensions. Now, each coefficient of the FFT is complex (two dimensions), but the negative frequencies are complex conjugate to their positive partners (giving two net dimensions for the two frequencies ω_m and $\omega_{-m} \equiv \omega_{N-m}$). If you are fussy, $\widetilde{y}_0$ has no partner, but is real (only one dimension), and if N is even $\widetilde{y}_{-N/2}$ also is partnerless, but is real. So N k-points are generated by N real points.

Wavelength. The points at which we sample the function are spaced $\Delta = T/N$ apart. It makes sense that the FFT would stop when the wavelength becomes close to Δ; we cannot resolve wiggles shorter than our sample spacing.

(a) *If the number of points N is even, show that the FFT stops when the wavelength becomes twice Δ.* (Hint: Which m has the shortest wavelength? Show that this wave alternates between $\pm C$ in time.) *Show that this component of $\widetilde{y}$ in Fourier space is partnerless. If $y(t)$ is real, show that this component is real as well.*

So, the FFT returns Fourier components only until $\omega_{N/2}$ when there is one point per bump (half-period) in the cosine wave.

Aliasing. Suppose our function really does have wiggles with shorter distances than our sampling time Δ. Then its FFT will have contributions to the long-wavelength coefficients $\widetilde{y}_m^{\mathrm{FFT}}$ from these shorter-wavelength wiggles; specifically $\widetilde{y}_{m\pm N}$, $\widetilde{y}_{m\pm 2N}$, etc.

(b) *On our sampled points t_ℓ, analytically show that $\exp(\mathrm{i}\omega_{m\pm N} t_\ell) = \exp(\mathrm{i}\omega_m t_\ell)$. Show that the FFT $\widetilde{y}_m$ can be written in terms of the Fourier series $\widetilde{y}_m$, up to an overall constant factor, as*

$$\widetilde{y}_m^{\mathrm{FFT}} \propto \sum_{n=-\infty}^{\infty} \widetilde{y}_{m+nN}.$$

(Hint: Use the "comb" function $\sum_{j=-\infty}^{\infty} \delta(t - j\Delta) = (1/\Delta)\sum_{n=-\infty}^{\infty} e^{2\pi \mathrm{i} nt/\Delta}$.) *Which term is the desired answer? Thus show that the FFT*

has bogus contributions from all frequencies outside its cutoff, folded in (or aliased into) the low frequency window.

If you sample a function at N points with Fourier components beyond $\omega_{N/2}$, their contributions get added to Fourier components at smaller frequencies. This is called *aliasing*, and is an important source of error in Fourier methods. We always strive to sample enough points to avoid it.

(c) *Form a 32-point wave packet $y(t) = 1/(\sqrt{2\pi}\sigma)\exp(-t^2/2\sigma^2)$. Change the width σ of the packet to make it thinner, until the packet begins to look ratty (almost as thin as the spacing between the sample points t_ℓ). Plot the function and its FFT. Notice how the Fourier series aliases as it hits the edges and overlaps.*

(d) *Windowing. One often needs to take Fourier series of functions which are not periodic in the interval. Set the number of data points N to 256 (powers of two are faster) and plot $y(t) = \cos \omega_m t$ for $m = 20$ with an illegal non-integer value $m = 20.5$. Note that the plot of the real-space function $y(t)$ is not periodic in the interval $[0, T)$ for $m = 20.5$. Plot its Fourier series. Note that each of the two peaks has broadened into a whole staircase. Try looking at the power spectrum (which is proportional to $|\widetilde{y}|^2$), and again compare $m = 20$ with $m = 20.5$. This is a numerical problem known as windowing, and there are various schemes to minimize its effects as well.*

(A.8) **White noise.** (Computation) ②
White light is a mixture of light of all frequencies. White noise is a mixture of all sound frequencies, with constant average power per unit frequency. The hissing noise you hear on radio and TV between stations is approximately white noise; there are a lot more high frequencies than low ones, so it sounds high-pitched.

What kind of time signal would generate white noise? Generate independent random numbers $y_\ell = y(\ell L/N)$ chosen from a Gaussian[22] distribution $\rho(y) = (1/\sqrt{2\pi})\exp(-y^2/2\sigma)$. You should see a jagged, random function. Set the number of data points to, say, 1,024.

Examine the Fourier transform of the noise signal. The Fourier transform of the white noise looks amazingly similar to the original signal. It is different, however, in two important ways.

---

[22]We choose the numbers with probability given by the Gaussian distribution, but it would look about the same if we took numbers with a uniform probability in, say, the range $(-1, 1)$.

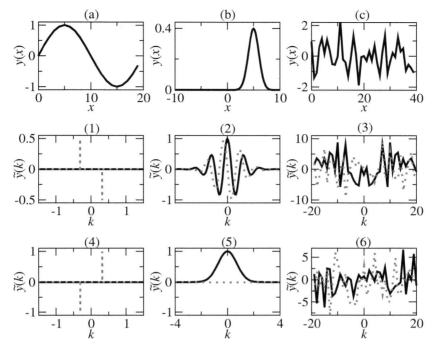

Fig. A.4 Fourier matching.

First, it is complex: there is a real part and an imaginary part. The second is for you to discover.

Examine the region near $k = 0$ on the Fourier plot, and describe how the Fourier transform of the noisy signal is different from a random function. In particular, what symmetry do the real and imaginary parts have? Can you show that this is true for any real function $y(x)$?

Now examine the power spectrum $|\widetilde{y}|^2$.[23] Check that the power is noisy, but on average is crudely independent of frequency. (You can check this best by varying the random number seed.) White noise is usually due to random, uncorrelated fluctuations in time.

(A.9) **Fourier matching.** ②

The top three plots (a)–(c) in Fig. A.4 are functions $y(x)$ of position. For each, pick out which of the six plots (1)–(6) are the corresponding

function $\widetilde{y}$ in Fourier space? (Dark line is real part, lighter dotted line is imaginary part.) (This exercise should be fairly straightforward after doing Exercises A.2, A.4, and A.8.)

(A.10) **Gibbs phenomenon.** (Mathematics) ③

In this exercise, we will look at the Fourier series for the step function and the triangle function. They are challenging because of the sharp corners, which are hard for sine waves to mimic. Consider a function $y(x)$ which is A in the range $0 < x < L/2$ and $-A$ in the range $L/2 < x < L$ (Fig. A.5). It is a kind of step function, since it takes a step downward at $L/2$ (Fig. A.5).[24]

(a) *As a crude approximation, the step function looks a bit like a chunky version of a sine wave, $A \sin(2\pi x/L)$. In this crude approximation, what would the complex Fourier series be (eqn A.4)?*

---

[23]For a time signal $f(t)$, the average power at a certain frequency is proportional to $|\tilde{f}(\omega)|^2$; ignoring the proportionality constant, the latter is often termed the power spectrum. This name is sometimes also used for the square of the amplitude of spatial Fourier transforms as well.

[24]It can be written in terms of the standard Heaviside step function $\Theta(x) = 0$ for $x < 0$ and $\Theta(x) = 1$ for $x > 0$, as $y(x) = A(1 - 2\Theta(x - L/2))$.

(b) *Show that the odd coefficients for the complex Fourier series of the step function are $\widetilde{y}_m = -2Ai/(m\pi)$ (m odd). What are the even ones? Check that the coefficients $\widetilde{y}_m$ with $m = \pm 1$ are close to those you guessed in part (a).*

(c) *Setting $A = 2$ and $L = 10$, plot the partial sum of the Fourier series (eqn A.1) for $m = -n, -n+1, \ldots, n$ with $n = 1$, 3, and 5. (You are likely to need to combine the coefficients $\widetilde{y}_m$ and $\widetilde{y}_{-m}$ into sines or cosines, unless your plotting package knows about complex exponentials.) Does it converge to the step function? If it is not too inconvenient, plot the partial sum up to $n = 100$, and concentrate especially on the overshoot near the jumps in the function at 0, $L/2$, and L. This overshoot is called the Gibbs phenomenon, and occurs when you try to approximate functions $y(x)$ that have discontinuities.*

One of the great features of Fourier series is that it makes taking derivatives and integrals easier. What does the integral of our step function look like? Let us sum the Fourier series for it!

(d) *Calculate the Fourier series of the integral of the step function, using your complex Fourier series from part (b) and the formula A.17 for the Fourier series of the integral. Plot your results, doing partial sums up to $\pm m = n$, with $n = 1$, 3, and 5, again with $A = 2$ and $L = 10$. Would the derivative of this function look like the step function? If it is convenient, do $n = 100$, and notice there are no overshoots.*

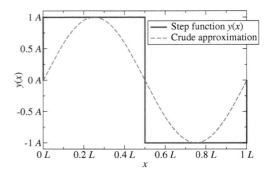

Fig. A.5 Step function.

References

[1] Ahlers, G. (1980). Critical phenomena at low temperatures. *Reviews of Modern Physics*, **52**, 489–503. 364

[2] Alemi, A. A., Bierbaum, M. K., Myers, C. R., and Sethna, J. P. (2015). You can run, you can hide: The epidemiology and statistical mechanics of zombies. *Physical Review E*, **92**, 052801. 175

[3] Alexander, G. P., Kamien, R. D., and Mosna, R. A. (2012). Conformal smectics and their many metrics. *Physical Review E*, **85**, 050701. 283

[4] Allen, M. P. and Tildesley, D. J. (2017). *Computer simulation of liquids* (2nd edition). Oxford University Press, New York. 217

[5] Ambegaokar, V., Halperin, B. I., Nelson, D., and Siggia, E. (1978). Dissipation in two-dimensional superfluids. *Physical Review Letters*, **40**, 783–6. 334

[6] Ambegaokar, V., Halperin, B. I., Nelson, D., and Siggia, E. (1980). Dynamics of superfluid films. *Physical Review B*, **21**, 1806–26. 334

[7] Anderson, M. H., Ensher, J. R., Matthews, M. R., Wieman, C. E., and Cornell, E. A. (1995). Observation of Bose–Einstein condensation in a dilute atomic vapor. *Science*, **269**, 198. http://jilawww.colorado.edu/bec/. 203

[8] Anderson, P. W. (1965). Coherent matter field phenomena in superfluids. In *Some recent definitions in the basic sciences*, Volume 2, pp. 21–40. New York. Reprinted in [11, p. 229]. 274, 285

[9] Anderson, P. W. (1966). Considerations on the flow of superfluid helium. *Reviews of Modern Physics*, **38**, 298–310. Reprinted in [11, p. 249]. 274, 285

[10] Anderson, P. W. (1972). More is different. *Science*, **177**(4047), 393–6. 15

[11] Anderson, P. W. (1984). *Basic notions of condensed matter physics*. Benjamin–Cummings, Menlo Park, CA. 227, 419

[12] Anderson, P. W. (1999). Solving problems in finite time. *Nature*, **400**, 115–6. 240

[13] Ashcroft, N. W. and Mermin, N. D. (1976). *Solid state physics*. Hold, Rinehart, and Wilson, Philadelphia. 188, 195, 202, 289, 290

[14] Bailey, N. P., Cretegny, T., Sethna, J. P., Coffman, V. R., Dolgert, A. J., Myers, C. R., Schiøtz, J., and Mortensen, J. J. (2006). Digital material: A flexible atomistic simulation code. http://arXiv.org/abs/cond-mat/0601236/. 306

[15] BekensteinD, J. D. (1973). Black holes and entropy. *Physical Review D*, **7**(8), 2333–46. 130

[16] Ben-Jacob, E., Goldenfeld, N., Langer, J. S., and Schön, G. (1983). Dynamics of interfacial pattern formation. *Physical Review Letters*, **51**, 1930–2. 340

[17] Ben-Jacob, E., Goldenfeld, N., Langer, J. S., and Schön, G. (1984). Boundary-layer model of pattern formation in solidification. *Physical Review A*, **29**, 330–40. 340

[18] Bender, C. M. and Orszag, S. A. (1978). *Advanced mathematical methods for scientists and engineers*. McGraw-Hill, New York. 8

[19] Bennett, C. H. (1982). The thermodynamics of computation—a review. *International Journal of Theoretical Physics*, **21**, 905–40. 114, 115

[20] Berman, G. J., Bialek, W., and Shaevitz, J. W. (2016). Predictability and hierarchy in drosophila behavior. *Proceedings of the National Academy of Sciences*, **113**(42), 11943–8. 245

[21] Berman, G. J., Choi, D. M., Bialek, W., and Shaevitz, J. W. (2014). Mapping the stereotyped behaviour of freely moving fruit flies. *Journal of The Royal Society Interface*, **11**(99), 20140672. 245

[22] Berman, G. J., Choi, D. M., Bialek, W., and Shaevitz, J. W. (2014). Mapping the structure of drosophilid behavior. *bioRxiv*. 245

[23] Berman, H. M., Westbrook, J., Feng, Z., Gilliland, G., Bhat, T. N., Weissig, H., Shindyalov, I. N., and Bourne, P. E. (2000). The Protein Data Bank. *Nucleic Acids Research*, **28**(1), 235–42. 105

[24] Bhanot, G., Creutz, M., Horvath, I., Lacki, J., and Weckel, J. (1994). Series expansions without diagrams. *Physical Review E*, **49**, 2445–53. 226

[25] Bialek, W. (2012). *Biophysics: Searching for principles*. Princeton University Press, Princeton, NJ. 248

[26] Bierbaum, M. K. (2013). Formation of circle pits. https://www.youtube.com/watch?v=h46wpNX0N1E, or see [181]. 347

[27] Bierbaum, M. K. (2013). Simulation of large circle pit. https://www.youtube.com/watch?v=YLggpXiFJWg, or see [181]. 347

[28] Bierbaum, M. K. (2015). Ising simulation. http://mattbierbaum.github.io/ising.js, or see [181]. 228, 231, 335, 368, 393, 401

[29] Bierbaum, M. K., Alemi, A. A., Myers, C. R., and Sethna, J. P. (2015). Zombietown USA. http://hey.runat.me/pages/programming/zombietown_usa.html, or see [181]. 175

[30] Bierbaum, M. K., Alemi, A. A., Myers, C. R., and Sethna, J. P. (2015). Zombietown USA simulator. http://mattbierbaum.github.io/zombies-usa/, or see [181]. 175

[31] Bierbaum, M. K., Silverberg, J. L., Sethna, J. P., and Cohen, I. (2013). Collective motion of mosh pits: Bierbaum web site. http://hey.runat.me/pages/science/moshpits_project.html, or see [181]. 16, 45, 170, 347

[32] Bierbaum, M. K., Silverberg, J. L., Sethna, J. P., and Cohen, I. (2013). Moshpits simulation. http://mattbierbaum.github.io/moshpits.js/, or see [181]. 16, 45, 170, 284, 346

[33] Bortz, A. B., Kalos, M. H., and Lebowitz, J. L. (1975). A new algorithm for Monte Carlo simulation of Ising spin systems. *Journal of Computational Physics*, **17**, 10–8. 233

[34] Brower, R. C., Kessler, D. A., Koplik, J., and Levine, H. (1983). Geometrical approach to moving interface dynamics. *Physical Review Letters*, **51**, 1111–4. 340

[35] Buchel, A. and Sethna, J. P. (1996). Elastic theory has zero radius of convergence. *Physical Review Letters*, **77**, 1520. 8, 323, 343, 345

[36] Buchel, A. and Sethna, J. P. (1997). Statistical mechanics of cracks: Thermodynamic limit, fluctuations, breakdown, and asymptotics of elastic theory. *Physical Review E*, **55**, 7669. 8, 343, 345

[37] Carathéodory, C. (1909). Untersuchungen über die Grundlagen der Thermodynamyk. *Mathematische Annalen*, **67**, 355–86. 149

[38] Carathéodory, C. (1976). Investigation into the foundations of thermodynamics. In *The second law of thermodynamics*, pp. 229–56. Dowden, Hutchinson and Ross, Stroudsberg, PA. 149

[39] Cardy, J. (1996). *Scaling and renormalization in statistical physics*. Cambridge University Press, Cambridge. 41, 394

[40] Carlson, J. M. and Langer, J. S. (1989). Properties of earthquakes generated by fault dynamics. *Physical Review Letters*, **62**, 2632. 351

[41] Chandler, D. (1987). *Introduction to modern statistical mechanics*. Oxford University Press, Oxford. 287, 310

[42] Chayes, J. T., Chayes, L., Sethna, J. P., and Thouless, D. J. (1986). A mean-field spin glass with short-range interactions. *Communications in Mathematical Physics*, **106**, 41. 378

[43] Choy, T.-S., Naset, J., Chen, J., Hershfield, S., and Stanton, C. (2000). The Fermi surface database. http://www.phys.ufl.edu /fermisurface/. 194, 195

[44] Chu, C. (1993). *Hysteresis and microstructures: A study of biaxial loading on compound twins of copper–aluminum–nickel single crystals*. Ph. D. thesis, Aerospace Engineering, University of Minnesota. 330

[45] Chui, S. T. and Weeks, J. D. (1978). Dynamics of the roughening transition. *Physical Review Letters*, **40**, 733–6. 278

[46] Coleman, S. (1985). Secret symmetry: An introduction to spontaneous symmetry breakdown and gauge fields. In *Aspects of symmetry, selected Erice lectures*, pp. 113–84. Cambridge University Press, Cambridge. 253

[47] Crooks, G. E. (1998). Nonequilibrium measurements of free energy differences for microscopically reversible Markovian systems. *Journal of Statistical Physics*, **90**(5), 1481–7. 91

[48] Csikor, F. F., Motz, C., Weygand, D., Zaiser, M., and Zapperi, S. (2007). Dislocation avalanches, strain bursts, and the problem of plastic forming at the micrometer scale. *Science*, **318**(5848), 251–4. 389

[49] Curcic, T. and Cooper, B. H. (1995). STM images of nanoscale pit decay. By permission. 288

[50] deGennes, P. G. and Prost, J. (1993). *The physics of liquid crystals,* 2nd edition. Clarendon Press, Oxford. 253

[51] Domb, C. (1974). Ising model. In *Phase transitions and critical phenomena,* Vol. 3, pp. 357–478. Academic Press, New York. 226

[52] Domb, C. and Guttman, A. J. (1970). Low-temperature series for the Ising model. *Journal of Physics C*, **8**, 1652–60. 226

[53] Dyson, F. (1960). Search for artificial stellar sources of infrared radiation. *Science*, **131**, 1667–8. 129

[54] Dyson, F. J. (1979). Time without end: Physics and biology in an open universe. *Reviews of Modern Physics*, **51**, 447. 113

[55] Elowitz, M. B. and Leibler, S. (2000). A synthetic oscillatory network of transcriptional regulators. *Nature*, **403**, 335–8. 235

[56] Feigenbaum, M. (1979). The universal metric properties of nonlinear transformations. *Journal of Statistical Physics*, **21**, 669–706. 397

[57] Feinberg, J. (1982). Self-pumped continuous wave phase conjugator using internal reflection. http://www.youtube.com /watch?v=gAy39ErqV34. 135

[58] Fermi, E., Pasta, J., and Ulam, S. (1965). Studies of nonlinear problems. I. In *E. Fermi, collected papers,* Vol. II, pp. 978. University of Chicago Press, Chicago. (Reprinted from Los Alamos report LA-1940, 1955). 84

[59] Feynman, R. P. (1972). *Statistical mechanics, a set of lectures.* Addison-Wesley, Menlo Park, CA. 198

[60] Feynman, R. P. (1996). *Feynman lectures on computation.* Westview Press, Boulder, CO. 114, 115

[61] Feynman, R. P., Leighton, R. B., and Sands, M. (1963). *The Feynman lectures on physics.* Addison-Wesley, Menlo Park, CA. 35, 47, 158, 173, 200, 212

[62] Fisher, I. R., Cheon, K. O., Panchula, A. F., Canfield, P. C., Chernikov, M., Ott, H. R., and Dennis, K. (1999). Magnetic and transport properties of single-grain $R-Mg-Zn$ icosahedral quasicrystals [$R = Y$, $(Y_{1-x}Gd_x)$, $(Y_{1-x}Tb_x)$, Tb, Dy, Ho, and Er]. *Physical Review B*, **59**, 308–321. 253

[63] Fixsen, D. J., Cheng, E. S., Gales, J. M., Mather, J. C., Shafer, R. A., and Wright, E. L. (1996). The cosmic microwave background spectrum from the full COBE FIRAS data set. *The Astrophysical Journal*, **473**, 576–87. 204

[64] Forster, D. (1975). *Hydrodynamic fluctuations, broken symmetry, and correlation functions.* Benjamin–Cummings, Reading, MA. 287, 292, 301

[65] Frenkel, D. and Louis, A. A. (1992). Phase separation in a binary hard-core mixture. An exact result. *Physical Review Letters*, **68**, 3363. 167

[66] Gauthier, G., Reeves, M. T., Yu, X., Bradley, A. S., Baker, M. A., Bell, T. A., Rubinsztein-Dunlop, H., Davis, M. J., and Neely,

T. W. (2019). Giant vortex clusters in a two-dimensional quantum fluid. *Science*, **364**(6447), 1264–7. 95

[67] Gillespie, D. T. (1976). A general method for numerically simulating the stochastic time evolution of coupled chemical reactions. *Journal of Computational Physics*, **22**, 403–34. 233

[68] Girvan, M. and Newman, M. E. J. (2002). Community structure in social and biological networks. *Proceedings of the National Academy of Sciences*, **12**, 7821–6. 13

[69] Goldbart, P. M. and Kamien, R. D. (2019). Tying it all together. *Physics Today*, **72**, 46. 285

[70] Goldstein, R. E. and Ashcroft, N. W. (1985). Origin of the singular diameter in coexistence curve of a metal. *Physical Review Letters*, **55**, 2164–7. 352

[71] Gomes, C. P. and Selman, B. (2002). Satisfied with physics. *Science*, **297**, 784–5. 243

[72] Gomes, C. P., Selman, B., Crato, N., and Kautz, H. (2000). Heavy-tailed phenomena in satisfiability and constraint satisfaction problems. *Journal of Automated Reasoning*, **24**, 67–100. 242

[73] Goss, P. J. E. and Peccoud, J. (1998). Quantitative modeling of stochastic systems in molecular biology by using stochastic Petri nets. *Proceedings of the National Academy of Sciences*, **95**, 6750–5. 232, 234

[74] Gottlieb, M. (1966). *Seven states of matter.* Walker & Co., New York. 253

[75] Gregersen, J. L., Mattle, D., Fedosova, N. U., Nissen, P., and Reinhard, L. (2016). Isolation, crystallization and crystal structure determination of bovine kidney Na^+,K^+-ATPase. *Acta Crystallographica Section F*, **72**(4), 282–7. 105

[76] Greiner, W., Neise, L., and Stöcker, H. (1995). *Thermodynamics and statistical mechanics.* Springer, New York. 71, 151

[77] Greywall, D. S. and Ahlers, G. (1973). Second-sound velocity and superfluid density in ^4He under pressure near T_λ. *Physical Review A*, **7**, 2145–62. 364

[78] Guggenheim, E. A. (1945). The principle of corresponding states. *Journal of Chemical Physics*, **13**, 253–61. 353

[79] Hänggi, P., Talkner, P., and Borkovec, M. (1990). Reaction-rate theory: Fifty years after Kramers. *Reviews of Modern Physics*, **62**, 251. 154, 165

[80] Hathcock, D. and Sethna, J. P. (2020). Renormalization group for barrier escape: Crossover to intermittency. https://arxiv.org/pdf/1902.07382.pdf. 225, 390

[81] Hayden, L. X., Raju, A., and Sethna, J. P. (2019). Unusual scaling for two-dimensional avalanches: Curing the faceting and scaling in the lower critical dimension. *Physical Review Research*, **1**, 033060. 383

[82] Heller, P. and Benedek, G. B. (1962). Nuclear magnetic resonance in MnF_2 near the critical point. *Physical Review Letters*, **8**, 428–32. 353

[83] Hirth, J. P. and Lothe, J. (1982). *Theory of dislocations*, 2nd edition. John Wiley & Sons, New York. 335

[84] Hodgdon, J. A. and Sethna, J. P. (1993). Derivation of a general three-dimensional crack-propagation law: A generalization of the principle of local symmetry. *Physical Review B*, **47**, 4831–40. 271

[85] Hohenberg, P. C. (1967). Existence of long-range order in one and two dimensions. *Physical Review*, **158**, 383–6. 46, 260

[86] Hohenberg, P. C. and Halperin, B. I. (1977). Theory of dynamic critical phenomena. *Reviews of Modern Physics*, **49**, 435–79. 285

[87] Hopfield, J. J. (1974). Kinetic proofreading: A new mechanism for reducing errors in biosynthetic processes requiring high specificity. *Proceedings of the National Academy of Sciences*, **71**(10), 4135–9. 248

[88] Houle, P. A. and Sethna, J. P. (1996). Acoustic emission from crumpling paper. *Physical Review E*, **54**, 278. 349

[89] Hu, W. (2001). Ringing in the new cosmology (introduction to the acoustic peaks and polarization). http://background.uchicago.edu/~whu/intermediate/intermediate.html. 205, 304

[90] Huberman, S., Duncan, R. A., Chen, K., Song, B., Chiloyan, V., Ding, Z., Maznev, A. A., Chen, G., and Nelson, K. A. (2019). Observation of second sound in graphite at temperatures above 100 k. *Science*, **364**(6438), 375–9. 281

[91] Hull, J. C. (2005). *Options, futures, and other derivatives*. Prentice Hall, Upper Saddle River, NJ. 40

[92] Jacobsen, J., Jacobsen, K. W., and Sethna, J. P. (1997). Rate theory for correlated processes: Double-jumps in adatom diffusion. *Physical Review Letters*, **79**, 2843. 165

[93] Jarzynski, C. (1997). Nonequilibrium equality for free energy differences. *Physical Review Letters*, **78**(14), 2690. 93

[94] Jensen, R. V. and Myers, C. R. (1985). Images of the critical points of nonlinear maps. *Physical Review A*, **32**, 1222–4. 88

[95] Johnstone, S. P., Groszek, A. J., Starkey, P. T., Billington, C. J., Simula, T. P., and Helmerson, K. (2019). Evolution of large-scale flow from turbulence in a two-dimensional superfluid. *Science*, **364**(6447), 1267–71. 95

[96] Kardar, M., Parisi, G., and Zhang, Y.-C. (1986). Dynamic scaling of growing interfaces. *Physical Review Letters*, **56**, 889–92. 271

[97] Kent-Dobias, J. and Sethna, J. P. (2018). Cluster representations and the Wolff algorithm in arbitrary external fields. *Physical Review E*, **98**, 063306. 232

[98] Khinchin, A. I. (1957). *Mathematical foundations of information theory*. Dover, New York. 126

[99] Kivelson, P. D. (2002). Neur-on. http://www.neur-on.com. 367

[100] Koehn, P. (2005). Europarl: A parallel corpus for statistical machine translation. In *MT summit*, Volume 5, pp. 79–86. 174

[101] Kos, F., Poland, D., Simmons-Duffin, D., and Vichi, A. (2016). Precision islands in the Ising and O(N) models. *Journal of High Energy Physics*, **2016**(8), 36. 402

[102] Kuntz, M. C., Houle, P., and Sethna, J. P. (1998). Crackling noise. http://www.lassp.cornell.edu/sethna/SimScience/crackling/. 237, 349, 351, 383

[103] Kuntz, M. C., Perković, O., Dahmen, K. A., and Sethna, J. P. (1999). Hysteresis, avalanches, and noise: Numerical methods. *Computing in Science and Engineering*, **1**, 73–81. 237, 240

[104] Landau, L. D. and Lifshitz, E. M. (1965). *Quantum mechanics, non-relativistic theory,* 2nd edition. Pergamon Press, Oxford. 198

[105] Landau, L. D. and Lifshitz, E. M. (1980). *Statistical physics.* Butterworth Heinemann, Oxford. 153

[106] Langer, J.S (1967). Theory of the condensation point. *Annals of Physics*, **41**(1), 108–57. 323, 345

[107] Langer, J. S. (1969). Statistical theory of the decay of metastable states. *Annals of Physics (NY)*, **54**, 258–75. 323, 345

[108] Langer, J. S. (1980). Instabilities and pattern formation in crystal growth. *Reviews of Modern Physics*, **52**, 1–28. 340

[109] Langer, S. A., Grannan, E. R., and Sethna, J. P. (1990). Nonequilibrium entropy and entropy distributions. *Physical Review B*, **41**, 2261. 121

[110] Langer, S. A. and Sethna, J. P. (1986). Textures in a chiral smectic liquid crystal film. *Physical Review A*, **34**, 5305. 270

[111] Langer, S. A. and Sethna, J. P. (1988). Entropy of glasses. *Physical Review Letters*, **61**, 570. (N.B.: M. Goldstein noticed the bounds earlier). 120

[112] Last, B. J. and Thouless, D. J. (1971). Percolation theory and electrical conductivity. *Physical Review Letters*, **27**, 1719–21. 349

[113] Lebowitz, J. L. and Penrose, O. (1973). Modern ergodic theory. *Physics Today*, **26**, 23–9. 79, 82

[114] Libbrecht, K. (2003). Snowflakes: Photographed by Kenneth Libbrecht. http://snowcrystals.com. 341

[115] Libbrecht, K. and Rasmussen, P. (2003). *The snowflake: Winter's secret beauty.* Voyageur Press, Stillwater, MN. 330, 341

[116] Lifshitz, R. (2011). Symmetry breaking and order in the age of quasicrystals. *Israel Journal of Chemistry*, **51**(1112), 1156–67. 253

[117] Liphardt, J., Onoa, B., Smith, S. B., Tinoco, I., and Bustamante, C. (2001). Reversible unfolding of single RNA molecules by mechanical force. *Science*, **292** (5517), 733–7. 161

[118] Lucas, A. Kinetic proofreading. Former website. 249

[119] Ma, Jie, Tan, Chuang, Gao, Xiang, Fulbright, Robert M., Roberts, Jeffrey W., and Wang, Michelle D. (2019). Transcription factor regulation of RNA polymerase's torque generation capacity. *Proceedings of the National Academy of Sciences*, **116**(7), 2583–2588. 45

[120] Machta, B. B. (2015). Dissipation bound for thermodynamic control. *Physical Review Letters*, **115**, 260603. 100, 114, 173

[121] Machta, B. B., Veatch, S. L., and Sethna, J. P. ((2012)). Critical Casimir forces in cellular membranes. *Physical Review Letters*, **109**, 138101. 399

[122] Maier, R. S. and Stein, D. L. (1993). Escape problem for irreversible systems. *Physical Review E*, **48**, 931–38. 165

[123] Malcai, O., Lidar, D. A., Biham, O., and Avnir, D. (1997). Scaling range and cutoffs in empirical fractals. *Physical Review E*, **56**, 2817–28. 362, 363

[124] Malkiel, B. G. (2003). *The random walk guide to investing: Ten rules for financial success*. Norton, New York. 27

[125] Malkiel, B. G. (2004). *A random walk down Wall Street*. Norton, New York. 27

[126] Marković, N., Christiansen, C., and Goldman, A. M. (1998). Thickness–magnetic field phase diagram at the superconductor–insulator transition in 2D. *Physical Review Letters*, **81**, 5217–20. 365

[127] Martin, P. C. (1968). Measurements and correlation functions. In *Problème à N Corps (Many-Body Physics), Proceedings, Ecole d'Eté de Physique Théorique, Les Houches, France, 1967*, pp. 37–136. Gordon and Breach, New York. 287, 307

[128] Mathews, J. and Walker, R. L. (1964). *Mathematical methods of physics*. Addison-Wesley, Redwood City, CA. 161, 162, 268, 311

[129] McGath, G. and Buldyrev, S. (1996). The self-avoiding random walk. http://polymer.bu.edu/java/java/saw/saw.html. BU Center for Polymer Studies. 27

[130] Mermin, N. D. (1979). The topological theory of defects in ordered media. *Reviews of Modern Physics*, **51**, 591–648. 253, 260

[131] Mermin, N. D. and Wagner, H. (1966). Absence of ferromagnetism or antiferromagnetism in one- or two-dimensional isotropic heisenberg models. *Physical Review Letters*, **17**, 1133–6. 46, 260

[132] Meyers, L. A., Pourbohloul, B., Newman, M. E. J., Skowronski, D. M., and Brunham, R. C. (2005). Network theory and SARS: Predicting outbreak diversity. *Journal of Theoretical Biology*, **232**(1), 71–81. 403

[133] Mézard, M. and Montanari, A. (2009). *Information, physics, and computation*. Oxford University Press, New York. 77, 391

[134] Mézard, M., Parisi, G., and Zeccina, R. (2002). Analytic and algorithmic solution of random satisfiability problems. *Science*, **297**, 812. 243

[135] Miller, J., Weichman, P. B., and Cross, M. C. (1992). Statistical mechanics, Euler's equation, and Jupiter's red spot. *Physical Review A*, **45**(4), 2328–59. 95

[136] Monasson, R., Zeccina, R., Kirkpatrick, S., Selman, B., and Troyansky, L. (1999). Determining computational complexity from characteristic "phase transitions". *Nature*, **400**, 133–7. 240

[137] NASA and ESA and M. Postman (STScI) and the CLASH Team (2011). Galaxy cluster MACS 1206. https://apod.nasa.gov/apod/ap111017.html. 130

[138] NASA/JPL (2014). Image from Voyager I, NASA's Goddard space flight center. http://www.nasa.gov/content/jupiters-great-red-spot-viewed-by-voyager-i. 94

[139] NASA/WMAP Science Team (2004). Wilkinson microwave anisotropy probe. http://map.gsfc.nasa.gov/. 205, 304

[140] Newman, M. E. J. (2000). Models of the small world. *Journal of Statistical Physics*, **101**, 819–41. 10

[141] Newman, M. E. J. (2001). Scientific collaboration networks. II. Shortest paths, weighted networks, and centrality. *Physical Review E*, **64**, 016132. 13

[142] Newman, M. E. J. (2005). Power laws, Pareto distributions and Zipf's law. *Contemporary Physics*, **46**, 323–51. http://arXiv.org/abs/cond-mat/0412004/. 173, 361

[143] Newman, M. E. J. and Barkema, G. T. (1999). *Monte Carlo methods in statistical physics*. Oxford University Press, Oxford. 232

[144] Newman, M. E. J. and Watts, D. J. (1999). Renormalization-group analysis of the small-world network model. *Physics Letters A*, **263**, 341–6. 12

[145] Nielsen, O. H., Sethna, J. P., Stoltze, P., Jacobsen, K. W., and Nørskov, J. K. (1994). Melting a copper cluster: Critical-droplet theory. *Europhysics Letters*, **26**, 51–6. 116

[146] Nova (2000). Trillion dollar bet. http://www.pbs.org/wgbh/nova/stockmarket/. 40

[147] Onsager, L. (1969). Motion of ions—principles and concepts. *Science*, **166** (3911), 1359. (1968 Nobel Prize lecture). 293

[148] Pananos, A. D., Bury, T. M., Wang, C., Schonfeld, J., Mohanty, S. P., Nyhan, B., Salathé, M., and Bauch, C. T. (2017). Critical dynamics in population vaccinating behavior. *Proceedings of the National Academy of Sciences*, **114**(52), 13762–7. 175

[149] Parisi, G. (1988). *Statistical field theory*. Perseus, Redding, MA. 226

[150] Piantadosi, S. T. (2014). Zipf's word frequency law in natural language: A critical review and future directions. *Psychonomic Bulletin & Review*, **21**, 1112–30. 173

[151] Polyakov, A. M. (1993). The theory of turbulence in two dimensions. *Nuclear Physics B*, **396**(2–3), 367–85. 94

[152] Poon, L. Cat map. University of Maryland Chaos Group, http://www-chaos.umd.edu/misc/catmap.html. 118

[153] Press, W. H., Teukolsky, S. A., Vetterling, W. T., and Flannery, B. P. (2002). *Numerical recipes in C++ [C, Fortran, ...], the art of scientific computing,* 2nd edition. Cambridge University Press, Cambridge. 71

[154] Purcell, E. M. (1977). Life at low Reynolds number. *American Journal of Physics*, **45**, 3–11. 44

[155] Quinn, K. N., Clement, C. B., De Bernardis, F., Niemack, M. D., and Sethna, J. P. (2019). Visualizing probabilistic models and data with intensive principal component analysis. *Proceedings of the National Academy of Sciences*, **116**(28), 13762–7. 21

[156] Rajagopal, K. and Wilczek, F. (2001). Enforced electrical neutrality of the color-flavor locked phase. *Physical Review Letters*, **86**, 3492–5. 206

[157] Raju, A., Clement, C. B., Hayden, L. X., Kent-Dobias, J. P., Liarte, D. B., Rocklin, D. Z., and Sethna, J. P. (2019). Normal form for renormalization groups. *Physical Review X*, **9**, 021014. 370

[158] Ralls, K. S. and Buhrman, R. A. (1991). Microscopic study of 1/f noise in metal nanobridges. *Physical Review B*, **44**, 5800–17. 309

[159] Ralls, K. S., Ralph, D. C., and Buhrman, R. A. (1989). Individual-defect electromigration in metal nanobridges. *Physical Review B*, **40**, 11561–70. 308

[160] Rasmussen, K. O., Cretegny, T., Kevrekidis, P. G., and Gronbech-Jensen, N. (2000). Statistical mechanics of a discrete nonlinear system. *Physical Review Letters*, **84**, 3740. 84

[161] Rottman, C., Wortis, M., Heyraud, J. C., and Métois, J. J. (1984). Equilibrium shapes of small lead crystals: Observation of Pokrovsky–Talapov critical behavior. *Physical Review Letters*, **52**, 1009–12. 327

[162] Rutenberg, A. D. and Vollmayr-Lee, B. P. (1999). Anisotropic coarsening: Grain shapes and nonuniversal persistence exponents. *Physical Review Letters*, **83**, 3772–5. 330

[163] Schroeder, D. V. (2000). *Thermal physics*. Addison-Wesley Longman, San Francisco, CA. 38

[164] Schwarz, U. T., English, L. Q., and Sievers, A. J. (1999). Experimental generation and observation of intrinsic localized spin wave modes in an antiferromagnet. *Physical Review Letters*, **83**, 223. 84

[165] Sehnal, D., Rose, A. S., Koča, J., Burley, S. K., and Velankar, S. (2018). Mol*: Towards a common library and tools for web molecular graphics. In *Proceedings of the workshop on molecular graphics and visual analysis of molecular data* (ed. J. Byska, M. Krone, and B. Sommer), MolVA '18, Goslar, DEU, pp. 29–33. The Eurographics Association, ACM, New York. 105

[166] Selman, B., Krautz, H., and Cohen, B. (1996). Local search strategies for satisfiability testing. *Dimacs Series in Discrete Mathematics and Theoretical Computer Science*, **26**, 521–32. 243

[167] Sethna, J. P. (1985). Frustration, curvature, and defect lines in metallic glasses and cholesteric blue phases. *Physical Review B*, **31**, 6278. 270, 367

[168] Sethna, J. P. (1992). Order parameters, broken symmetry, and topology. In *1991 lectures in complex systems: The proceedings of the 1991 complex systems summer school, Santa Fe, New Mexico*, Volume XV, pp. 243. Addison–Wesley. http://www.lassp.cornell.edu/sethna/OrderParameters. 253

[169] Sethna, J. P. (1995). Equilibrium crystal shapes. http://www.lassp.cornell.edu/sethna/CrystalShapes. 327

[170] Sethna, J. P. (1996). Jupiter! The 3-body problem. http://www.physics.cornell.edu/sethna/teaching/sss/jupiter/jupiter.htm. 90

[171] Sethna, J. P. (1997). Cracks and elastic theory. http://www.lassp.cornell.edu/sethna/Cracks/Zero_Radius_of_Convergence.html. 8

[172] Sethna, J. P. (1997). Quantum electrodynamics has zero radius of convergence. http://www.lassp.cornell.edu/sethna/Cracks/QED.html. 8, 344

[173] Sethna, J. P. (1997). What is coarsening? http://www.lassp.cornell.edu/sethna/Coarsening/What_Is_Coarsening.html. 288

[174] Sethna, J. P. (1997). What is the radius of convergence? http://www.lassp.cornell.edu/sethna/Cracks/What_Is_Radius_of_Convergence.html. 8

[175] Sethna, J. P. (2007). Crackling wires. *Science*, **318**(5848), 207–8. 389

[176] Sethna, J. P., Bierbaum, M. K., Dahmen, K. A., Goodrich, C. P., Greer, J. R., Hayden, L. X., Kent-Dobias, J. P., Lee, E. D., Liarte, D. B., Ni, X., Quinn, K. N., Raju, A., Rocklin, D. Z., Shekhawat, A., and Zapperi, S. (2017). Deformation of crystals: Connections with statistical physics. *Annual Review of Materials Research*, **47**, 217–46. 278

[177] Sethna, J. P., Chachra, R., Machta, B. B., and Transtrum, M. K. (2013). Why is science possible? http://www.lassp.cornell.edu/sethna/Sloppy/WhyIsSciencePossible.html. 15

[178] Sethna, J. P., Dahmen, K. A., Kartha, S., Krumhansl, J. A., Roberts, B. W., and Shore, J. D. (1993). Hysteresis and hierarchies: Dynamics of disorder-driven first-order phase transformations. *Physical Review Letters*, **70**, 3347–50. 396

[179] Sethna, J. P., Dahmen, K. A., and Kuntz, M. C. (1996). Hysteresis and avalanches. http://www.lassp.cornell.edu/sethna/hysteresis/. 383

[180] Sethna, J. P., Dahmen, K. A., and Myers, C. R. (2001). Crackling noise. *Nature*, **410**, 242. 349, 352, 355, 383

[181] Sethna, J. P. and Myers, C. R. (2020). *Entropy, order parameters, and complexity*, computer exercises and course materials. http://www.physics.cornell.edu/sethna/StatMech/EOPCHintsAndMaterials.html, backup site http://www.lassp.cornell.edu/sethna/StatMech/EOPCHintsAndMaterials.html. 12, 39, 89, 95, 213, 237, 246, 266, 276, 278, 283, 306, 313, 338, 385, 401, 420

[182] Sethna, J. P., Shore, J. D., and Huang, M. (1991). Scaling theory for the glass transition. *Physical Review B*, **44**, 4943. Acknowledgement of D. S. Fisher in *PRB* **47**, 14661 (1993). 258

[183] Shannon, C. E. (1948). A mathematical theory of communication. *The Bell System Technical Journal*, **27**, 379–423. 109

[184] Shekhawat, A., Zapperi, S., and Sethna, J. P. (2013). From damage percolation to crack nucleation through finite-size criticality. *Physical Review Letters*, **110**, 185505. 389

[185] Shore, J. D., Holzer, M., and Sethna, J. P. (1992). Logarithmically slow domain growth in nonrandom frustrated systems: Ising models with competing interactions. *Physical Review B*, **46**, 11376–404. 329

[186] Sievers, A. J. and Takeno, S. (1988). Intrinsic localized modes in anharmonic crystals. *Physical Review Letters*, **61**, 970–3. 84

[187] Siggia, E. D. (1979). Late stages of spinodal decomposition in binary mixtures. *Physical Review A*, **20**, 595. 328, 329

[188] Silverberg, J. L. (2015). Moshing with physics. TED talk at https://www.youtube.com/watch?v=rjvaiiXIySc. 16, 45, 170, 346

[189] Silverberg, J. L., Bierbaum, M. K., Sethna, J. P., and Cohen, I. (2013). Collective motion in mosh pits: Cohen group web site, links to press. http://cohengroup.lassp.cornell.edu/projects/collective-motion-mosh-pits/. 16, 45, 170, 346

[190] Silverberg, J. L., Bierbaum, M. K., Sethna, J. P., and Cohen, I. (2013). Collective motion of humans in mosh and circle pits at heavy metal concerts. *Physical Review Letters*, **110**, 228701. 16, 45, 170, 346

[191] Südkamp, T. and Bracht, H. (2016). Self-diffusion in crystalline silicon: A single diffusion activation enthalpy down to 755 °C. *Physical Review B*, **94**, 125208. 279

[192] Swendsen, R. H. and Wang, J.-S. (1987). Nonuniversal critical dynamics in Monte Carlo simulations. *Physical Review Letters*, **58**, 86. 231

[193] Szilard, L. (1929). Über die Entropieverminderung in einem thermodynamischen System bei Eingriffen intelligenter Wesen. *Zeitschrift für Physik*, **53**(11), 840–56. 113

[194] Szilard, L. (1990). On the decrease of entropy in a thermodynamic system by the intervention of intelligent beings. In *Maxwell's demon: Entropy, information, and computing* (ed. H. S. Leff and A. F. Rex), pp. 124–33. Princeton University Press, Princeton, NJ. 113

[195] Tang, L.-H. and Chen, Q.-H. (2003). Finite-size and boundary effects on the I–V characteristics of two-dimensional superconducting networks. *Physical Review B*, **67**, 024508. 334

[196] Thomas, S. B. and Parks, G. S. (1931). Studies on glass VI. Some specific heat data on boron trioxide. *Journal of Physical Chemistry*, **35**, 2091–102. 121

[197] Tomé, T. and Ziff, R. M. (2010). Critical behavior of the susceptible-infected-recovered model on a square lattice. *Physical Review E*, **82**, 051921. 403

[198] Toner, J. and Tu, Y. (1995). Long-range order in a two-dimensional dynamical XY model: How birds fly together. *Physical Review Letters*, **75**, 4326–9. 46, 271

[199] Toner, J., Tu, Y., and Ramaswamy, S. (2005). Hydrodynamics and phases of flocks. *Annals of Physics*, **318**(1), 170–244. 45

[200] Trefethen, L. N. and Trefethen, L. M. (2000). How many shuffles to randomize a deck of cards? *Proceedings of the Royal Society of London A*, **456**, 2561–8. 122

[201] Utter, B., Ragnarsson, R., and Bodenschatz, E. (2001, May). Alternating tip splitting in directional solidification. *Phys. Rev. Lett.*, **86**, 4604–4607. 331

[202] Veatch, S. L., Cicuta, P., Sengupta, P., Honerkamp-Smith, A., Holowka, D., and Baird, B. (2008). Critical fluctuations in plasma

membrane vesicles. *ACS Chemical Biology*, **3**(5), 287–93. PMID: 18484709. 342

[203] Vegge, T., Sethna, J. P., Cheong, S. A., Jacobsen, K. W., Myers, C. R., and Ralph, D. C. (2001). Calculation of quantum tunneling for a spatially extended defect: The dislocation kink in copper has a low effective mass. *Physical Review Letters*, **86**, 1546–9. 209

[204] Vitos, L., Ruban, A.V., Skriver, H.L., and Kollr, J. (1998). The surface energy of metals. *Surface Science*, **411**(1), 186 – 202. 343

[205] Wallace, D. (2010). Gravity, entropy, and cosmology: In search of clarity. *British Journal for the Philosophy of Science*, **61**, 513–40. 103, 131

[206] Watts, D. J. and Strogatz, S. H. (1998). Collective dynamics of 'small-world' networks. *Nature*, **393**, 440–42. 10

[207] Weinberg, S. (1977). *The first three minutes.* Basic Books (Perseus), New York. 103, 204, 303

[208] Weinberg, S. (2015). *Lectures on quantum mechanics,* 2nd edition. Cambridge University Press, Cambridge. 210

[209] Weiss, Jérôme and Marsan, David (2003). Three-dimensional mapping of dislocation avalanches: Clustering and space/time coupling. *Science*, **299**(5603), 89–92. 389

[210] Wells, H. G. (1895). *The time machine.* http://www.gutenberg. org/files/35/35-h/35-h.htm. 103

[211] Wigner, E. (1960). The unreasonable effectiveness of mathematics in the natural sciences. *Communications in Pure and Applied Mathematics*, **13**, 1–14. http://www.dartmouth.edu/~matc /MathDrama/reading/Wigner.html. 23

[212] Wilczek, F. and Shapere, A. (1989). *Geometric Phases in Physics.* World Scientific, Singapore. 44

[213] Wolff, U. (1989). Collective Monte Carlo updating for spin systems. *Physical Review Letters*, **62**, 361. 231

[214] Yeomans, J. M. (1992). *Statistical mechanics of phase transitions.* Oxford University Press, Oxford. 219

[215] Zinn-Justin, J. (1996). *Quantum field theory and critical phenomena* (3rd edition). Oxford University Press, Oxford. 345

[216] Zunger, A., Wang, L. G., Hart, G. L. W., and Sanatai, M. (2002). Obtaining Ising-like expansions for binary alloys from first principles. *Modelling and Simulation in Materials Science and Engineering*, **10**, 685–706. 219

Index

Index entry format examples:

- Figure captions: f5.6 84 would be in Fig. 5.6 on p. 84.
- Exercises: e5.15 102 would be in Exercise 5.15 on p. 102.
- Notes and footnotes: n31 88 would be in note 31 on p. 88.